T0189041

Lecture Notes in Computer Science 14227

The series Lecture Notes in Computer Science (LNCS), including its subseries Lecture Notes in Artificial Intelligence (LNAI) and Lecture Notes in Bioinformatics (LNBI), has established itself as a medium for the publication of new developments in computer science and information technology research, teaching, and education.

LNCS enjoys close cooperation with the computer science R & D community, the series counts many renowned academics among its volume editors and paper authors, and collaborates with prestigious societies. Its mission is to serve this international community by providing an invaluable service, mainly focused on the publication of conference and workshop proceedings and postproceedings. LNCS commenced publication in 1973.

Hayit Greenspan · Anant Madabhushi ·
Parvin Mousavi · Septimiu Salcudean ·
James Duncan · Tanveer Syeda-Mahmood ·
Russell Taylor
Editors

Medical Image Computing and Computer Assisted Intervention – MICCAI 2023

26th International Conference
Vancouver, BC, Canada, October 8–12, 2023
Proceedings, Part VIII

Editors
Hayit Greenspan
Icahn School of Medicine, Mount Sinai,
NYC, NY, USA

Tel Aviv University
Tel Aviv, Israel

Parvin Mousavi
Queen's University
Kingston, ON, Canada

James Duncan
Yale University
New Haven, CT, USA

Russell Taylor
Johns Hopkins University
Baltimore, MD, USA

Anant Madabhushi
Emory University
Atlanta, GA, USA

Septimiu Salcudean
The University of British Columbia
Vancouver, BC, Canada

Tanveer Syeda-Mahmood
IBM Research
San Jose, CA, USA

ISSN 0302-9743 ISSN 1611-3349 (electronic)
Lecture Notes in Computer Science
ISBN 978-3-031-43992-6 ISBN 978-3-031-43993-3 (eBook)
https://doi.org/10.1007/978-3-031-43993-3

This Springer imprint is published by the registered company Springer Nature Switzerland AG
The registered company address is: Gewerbestrasse 11, 6330 Cham, Switzerland

Paper in this product is recyclable.

Preface

We are pleased to present the proceedings for the 26th International Conference on Medical Image Computing and Computer-Assisted Intervention (MICCAI). After several difficult years of virtual conferences, this edition was held in a mainly in-person format with a hybrid component at the Vancouver Convention Centre, in Vancouver, BC, Canada October 8–12, 2023. The conference featured 33 physical workshops, 15 online workshops, 15 tutorials, and 29 challenges held on October 8 and October 12. Co-located with the conference was also the 3rd Conference on Clinical Translation on Medical Image Computing and Computer-Assisted Intervention (CLINICCAI) on October 10.

MICCAI 2023 received the largest number of submissions so far, with an approximately 30% increase compared to 2022. We received 2365 full submissions of which 2250 were subjected to full review. To keep the acceptance ratios around 32% as in previous years, there was a corresponding increase in accepted papers leading to 730 papers accepted, with 68 orals and the remaining presented in poster form. These papers comprise ten volumes of Lecture Notes in Computer Science (LNCS) proceedings as follows:

- Part I, LNCS Volume 14220: Machine Learning with Limited Supervision and Machine Learning – Transfer Learning
- Part II, LNCS Volume 14221: Machine Learning – Learning Strategies and Machine Learning – Explainability, Bias, and Uncertainty I
- Part III, LNCS Volume 14222: Machine Learning – Explainability, Bias, and Uncertainty II and Image Segmentation I
- Part IV, LNCS Volume 14223: Image Segmentation II
- Part V, LNCS Volume 14224: Computer-Aided Diagnosis I
- Part VI, LNCS Volume 14225: Computer-Aided Diagnosis II and Computational Pathology
- Part VII, LNCS Volume 14226: Clinical Applications – Abdomen, Clinical Applications – Breast, Clinical Applications – Cardiac, Clinical Applications – Dermatology, Clinical Applications – Fetal Imaging, Clinical Applications – Lung, Clinical Applications – Musculoskeletal, Clinical Applications – Oncology, Clinical Applications – Ophthalmology, and Clinical Applications – Vascular
- Part VIII, LNCS Volume 14227: Clinical Applications – Neuroimaging and Microscopy
- Part IX, LNCS Volume 14228: Image-Guided Intervention, Surgical Planning, and Data Science
- Part X, LNCS Volume 14229: Image Reconstruction and Image Registration

The papers for the proceedings were selected after a rigorous double-blind peer-review process. The MICCAI 2023 Program Committee consisted of 133 area chairs and over 1600 reviewers, with representation from several countries across all major continents. It also maintained a gender balance with 31% of scientists who self-identified as women. With an increase in the number of area chairs and reviewers, the reviewer load on the experts was reduced this year, keeping to 16–18 papers per area chair and about 4–6 papers per reviewer. Based on the double-blinded reviews, area chairs' recommendations, and program chairs' global adjustments, 308 papers (14%) were provisionally accepted, 1196 papers (53%) were provisionally rejected, and 746 papers (33%) proceeded to the rebuttal stage. As in previous years, Microsoft's Conference Management Toolkit (CMT) was used for paper management and organizing the overall review process. Similarly, the Toronto paper matching system (TPMS) was employed to ensure knowledgeable experts were assigned to review appropriate papers. Area chairs and reviewers were selected following public calls to the community, and were vetted by the program chairs.

Among the new features this year was the emphasis on clinical translation, moving Medical Image Computing (MIC) and Computer-Assisted Interventions (CAI) research from theory to practice by featuring two clinical translational sessions reflecting the real-world impact of the field in the clinical workflows and clinical evaluations. For the first time, clinicians were appointed as Clinical Chairs to select papers for the clinical translational sessions. The philosophy behind the dedicated clinical translational sessions was to maintain the high scientific and technical standard of MICCAI papers in terms of methodology development, while at the same time showcasing the strong focus on clinical applications. This was an opportunity to expose the MICCAI community to the clinical challenges and for ideation of novel solutions to address these unmet needs. Consequently, during paper submission, in addition to MIC and CAI a new category of "Clinical Applications" was introduced for authors to self-declare.

MICCAI 2023 for the first time in its history also featured dual parallel tracks that allowed the conference to keep the same proportion of oral presentations as in previous years, despite the 30% increase in submitted and accepted papers.

We also introduced two new sessions this year focusing on young and emerging scientists through their Ph.D. thesis presentations, and another with experienced researchers commenting on the state of the field through a fireside chat format.

The organization of the final program by grouping the papers into topics and sessions was aided by the latest advancements in generative AI models. Specifically, Open AI's GPT-4 large language model was used to group the papers into initial topics which were then manually curated and organized. This resulted in fresh titles for sessions that are more reflective of the technical advancements of our field.

Although not reflected in the proceedings, the conference also benefited from keynote talks from experts in their respective fields including Turing Award winner Yann LeCun and leading experts Jocelyne Troccaz and Mihaela van der Schaar.

We extend our sincere gratitude to everyone who contributed to the success of MICCAI 2023 and the quality of its proceedings. In particular, we would like to express our profound thanks to the MICCAI Submission System Manager Kitty Wong whose meticulous support throughout the paper submission, review, program planning, and proceeding preparation process was invaluable. We are especially appreciative of the effort and dedication of our Satellite Events Chair, Bennett Landman, who tirelessly coordinated the organization of over 90 satellite events consisting of workshops, challenges and tutorials. Our workshop chairs Hongzhi Wang, Alistair Young, tutorial chairs Islem

Rekik, Guoyan Zheng, and challenge chairs, Lena Maier-Hein, Jayashree Kalpathy-Kramer, Alexander Seitel, worked hard to assemble a strong program for the satellite events. Special mention this year also goes to our first-time Clinical Chairs, Drs. Curtis Langlotz, Charles Kahn, and Masaru Ishii who helped us select papers for the clinical sessions and organized the clinical sessions.

We acknowledge the contributions of our Keynote Chairs, William Wells and Alejandro Frangi, who secured our keynote speakers. Our publication chairs, Kevin Zhou and Ron Summers, helped in our efforts to get the MICCAI papers indexed in PubMed. It was a challenging year for fundraising for the conference due to the recovery of the economy after the COVID pandemic. Despite this situation, our industrial sponsorship chairs, Mohammad Yaqub, Le Lu and Yanwu Xu, along with Dekon's Mehmet Eldegez, worked tirelessly to secure sponsors in innovative ways, for which we are grateful.

An active body of the MICCAI Student Board led by Camila Gonzalez and our 2023 student representatives Nathaniel Braman and Vaishnavi Subramanian helped put together student-run networking and social events including a novel Ph.D. thesis 3-minute madness event to spotlight new graduates for their careers. Similarly, Women in MICCAI chairs Xiaoxiao Li and Jayanthi Sivaswamy and RISE chairs, Islem Rekik, Pingkun Yan, and Andrea Lara further strengthened the quality of our technical program through their organized events. Local arrangements logistics including the recruiting of University of British Columbia students and invitation letters to attendees, was ably looked after by our local arrangement chairs Purang Abolmaesumi and Mehdi Moradi. They also helped coordinate the visits to the local sites in Vancouver both during the selection of the site and organization of our local activities during the conference. Our Young Investigator chairs Marius Linguraru, Archana Venkataraman, Antonio Porras Perez put forward the startup village and helped secure funding from NIH for early career scientist participation in the conference. Our communications chair, Ehsan Adeli, and Diana Cunningham were active in making the conference visible on social media platforms and circulating the newsletters. Niharika D'Souza was our cross-committee liaison providing note-taking support for all our meetings. We are grateful to all these organization committee members for their active contributions that made the conference successful.

We would like to thank the MICCAI society chair, Caroline Essert, and the MICCAI board for their approvals, support and feedback, which provided clarity on various aspects of running the conference. Behind the scenes, we acknowledge the contributions of the MICCAI secretariat personnel, Janette Wallace, and Johanne Langford, who kept a close eye on logistics and budgets, and Diana Cunningham and Anna Van Vliet for including our conference announcements in a timely manner in the MICCAI society newsletters. This year, when the existing virtual platform provider indicated that they would discontinue their service, a new virtual platform provider Conference Catalysts was chosen after due diligence by John Baxter. John also handled the setup and coordination with CMT and consultation with program chairs on features, for which we are very grateful. The physical organization of the conference at the site, budget financials, fund-raising, and the smooth running of events would not have been possible without our Professional Conference Organization team from Dekon Congress & Tourism led by Mehmet Eldegez. The model of having a PCO run the conference, which we used at

MICCAI, significantly reduces the work of general chairs for which we are particularly grateful.

Finally, we are especially grateful to all members of the Program Committee for their diligent work in the reviewer assignments and final paper selection, as well as the reviewers for their support during the entire process. Lastly, and most importantly, we thank all authors, co-authors, students/postdocs, and supervisors for submitting and presenting their high-quality work, which played a pivotal role in making MICCAI 2023 a resounding success.

With a successful MICCAI 2023, we now look forward to seeing you next year in Marrakesh, Morocco when MICCAI 2024 goes to the African continent for the first time.

October 2023

Tanveer Syeda-Mahmood
James Duncan
Russ Taylor
General Chairs

Hayit Greenspan
Anant Madabhushi
Parvin Mousavi
Septimiu Salcudean
Program Chairs

Organization

General Chairs

Tanveer Syeda-Mahmood	IBM Research, USA
James Duncan	Yale University, USA
Russ Taylor	Johns Hopkins University, USA

Program Committee Chairs

Hayit Greenspan	Tel-Aviv University, Israel and Icahn School of Medicine at Mount Sinai, USA
Anant Madabhushi	Emory University, USA
Parvin Mousavi	Queen's University, Canada
Septimiu Salcudean	University of British Columbia, Canada

Satellite Events Chair

Bennett Landman	Vanderbilt University, USA

Workshop Chairs

Hongzhi Wang	IBM Research, USA
Alistair Young	King's College, London, UK

Challenges Chairs

Jayashree Kalpathy-Kramer	Harvard University, USA
Alexander Seitel	German Cancer Research Center, Germany
Lena Maier-Hein	German Cancer Research Center, Germany

Tutorial Chairs

Islem Rekik	Imperial College London, UK
Guoyan Zheng	Shanghai Jiao Tong University, China

Clinical Chairs

Curtis Langlotz	Stanford University, USA
Charles Kahn	University of Pennsylvania, USA
Masaru Ishii	Johns Hopkins University, USA

Local Arrangements Chairs

Purang Abolmaesumi	University of British Columbia, Canada
Mehdi Moradi	McMaster University, Canada

Keynote Chairs

William Wells	Harvard University, USA
Alejandro Frangi	University of Manchester, UK

Industrial Sponsorship Chairs

Mohammad Yaqub	MBZ University of Artificial Intelligence, Abu Dhabi
Le Lu	DAMO Academy, Alibaba Group, USA
Yanwu Xu	Baidu, China

Communication Chair

Ehsan Adeli	Stanford University, USA

Publication Chairs

Ron Summers	National Institutes of Health, USA
Kevin Zhou	University of Science and Technology of China, China

Young Investigator Chairs

Marius Linguraru	Children's National Institute, USA
Archana Venkataraman	Boston University, USA
Antonio Porras	University of Colorado Anschutz Medical Campus, USA

Student Activities Chairs

Nathaniel Braman	Picture Health, USA
Vaishnavi Subramanian	EPFL, France

Women in MICCAI Chairs

Jayanthi Sivaswamy	IIIT, Hyderabad, India
Xiaoxiao Li	University of British Columbia, Canada

RISE Committee Chairs

Islem Rekik	Imperial College London, UK
Pingkun Yan	Rensselaer Polytechnic Institute, USA
Andrea Lara	Universidad Galileo, Guatemala

Submission Platform Manager

Kitty Wong	The MICCAI Society, Canada

Virtual Platform Manager

John Baxter	INSERM, Université de Rennes 1, France

Cross-Committee Liaison

Niharika D'Souza	IBM Research, USA

Program Committee

Sahar Ahmad	University of North Carolina at Chapel Hill, USA
Shadi Albarqouni	University of Bonn and Helmholtz Munich, Germany
Angelica Aviles-Rivero	University of Cambridge, UK
Shekoofeh Azizi	Google, Google Brain, USA
Ulas Bagci	Northwestern University, USA
Wenjia Bai	Imperial College London, UK
Sophia Bano	University College London, UK
Kayhan Batmanghelich	University of Pittsburgh and Boston University, USA
Ismail Ben Ayed	ETS Montreal, Canada
Katharina Breininger	Friedrich-Alexander-Universität Erlangen-Nürnberg, Germany
Weidong Cai	University of Sydney, Australia
Geng Chen	Northwestern Polytechnical University, China
Hao Chen	Hong Kong University of Science and Technology, China
Jun Cheng	Institute for Infocomm Research, A*STAR, Singapore
Li Cheng	University of Alberta, Canada
Albert C. S. Chung	University of Exeter, UK
Toby Collins	Ircad, France
Adrian Dalca	Massachusetts Institute of Technology and Harvard Medical School, USA
Jose Dolz	ETS Montreal, Canada
Qi Dou	Chinese University of Hong Kong, China
Nicha Dvornek	Yale University, USA
Shireen Elhabian	University of Utah, USA
Sandy Engelhardt	Heidelberg University Hospital, Germany
Ruogu Fang	University of Florida, USA

Aasa Feragen	Technical University of Denmark, Denmark
Moti Freiman	Technion - Israel Institute of Technology, Israel
Huazhu Fu	IHPC, A*STAR, Singapore
Adrian Galdran	Universitat Pompeu Fabra, Barcelona, Spain
Zhifan Gao	Sun Yat-sen University, China
Zongyuan Ge	Monash University, Australia
Stamatia Giannarou	Imperial College London, UK
Yun Gu	Shanghai Jiao Tong University, China
Hu Han	Institute of Computing Technology, Chinese Academy of Sciences, China
Daniel Hashimoto	University of Pennsylvania, USA
Mattias Heinrich	University of Lübeck, Germany
Heng Huang	University of Pittsburgh, USA
Yuankai Huo	Vanderbilt University, USA
Mobarakol Islam	University College London, UK
Jayender Jagadeesan	Harvard Medical School, USA
Won-Ki Jeong	Korea University, South Korea
Xi Jiang	University of Electronic Science and Technology of China, China
Yueming Jin	National University of Singapore, Singapore
Anand Joshi	University of Southern California, USA
Shantanu Joshi	UCLA, USA
Leo Joskowicz	Hebrew University of Jerusalem, Israel
Samuel Kadoury	Polytechnique Montreal, Canada
Bernhard Kainz	Friedrich-Alexander-Universität Erlangen-Nürnberg, Germany and Imperial College London, UK
Davood Karimi	Harvard University, USA
Anees Kazi	Massachusetts General Hospital, USA
Marta Kersten-Oertel	Concordia University, Canada
Fahmi Khalifa	Mansoura University, Egypt
Minjeong Kim	University of North Carolina, Greensboro, USA
Seong Tae Kim	Kyung Hee University, South Korea
Pavitra Krishnaswamy	Institute for Infocomm Research, Agency for Science Technology and Research (A*STAR), Singapore
Jin Tae Kwak	Korea University, South Korea
Baiying Lei	Shenzhen University, China
Xiang Li	Massachusetts General Hospital, USA
Xiaoxiao Li	University of British Columbia, Canada
Yuexiang Li	Tencent Jarvis Lab, China
Chunfeng Lian	Xi'an Jiaotong University, China

Jianming Liang — Arizona State University, USA
Jianfei Liu — National Institutes of Health Clinical Center, USA
Mingxia Liu — University of North Carolina at Chapel Hill, USA
Xiaofeng Liu — Harvard Medical School and MGH, USA
Herve Lombaert — École de technologie supérieure, Canada
Ismini Lourentzou — Virginia Tech, USA
Le Lu — Damo Academy USA, Alibaba Group, USA
Dwarikanath Mahapatra — Inception Institute of Artificial Intelligence, United Arab Emirates
Saad Nadeem — Memorial Sloan Kettering Cancer Center, USA
Dong Nie — Alibaba (US), USA
Yoshito Otake — Nara Institute of Science and Technology, Japan
Sang Hyun Park — Daegu Gyeongbuk Institute of Science and Technology, South Korea
Magdalini Paschali — Stanford University, USA
Tingying Peng — Helmholtz Munich, Germany
Caroline Petitjean — LITIS Université de Rouen Normandie, France
Esther Puyol Anton — King's College London, UK
Chen Qin — Imperial College London, UK
Daniel Racoceanu — Sorbonne Université, France
Hedyeh Rafii-Tari — Auris Health, USA
Hongliang Ren — Chinese University of Hong Kong, China and National University of Singapore, Singapore
Tammy Riklin Raviv — Ben-Gurion University, Israel
Hassan Rivaz — Concordia University, Canada
Mirabela Rusu — Stanford University, USA
Thomas Schultz — University of Bonn, Germany
Feng Shi — Shanghai United Imaging Intelligence, China
Yang Song — University of New South Wales, Australia
Aristeidis Sotiras — Washington University in St. Louis, USA
Rachel Sparks — King's College London, UK
Yao Sui — Peking University, China
Kenji Suzuki — Tokyo Institute of Technology, Japan
Qian Tao — Delft University of Technology, Netherlands
Mathias Unberath — Johns Hopkins University, USA
Martin Urschler — Medical University Graz, Austria
Maria Vakalopoulou — CentraleSupelec, University Paris Saclay, France
Erdem Varol — New York University, USA
Francisco Vasconcelos — University College London, UK
Harini Veeraraghavan — Memorial Sloan Kettering Cancer Center, USA
Satish Viswanath — Case Western Reserve University, USA
Christian Wachinger — Technical University of Munich, Germany

Reviewers

Alaa Eldin Abdelaal
John Abel
Kumar Abhishek
Shahira Abousamra
Mazdak Abulnaga
Burak Acar
Abdoljalil Addeh
Ehsan Adeli
Sukesh Adiga Vasudeva
Seyed-Ahmad Ahmadi
Euijoon Ahn
Faranak Akbarifar
Alireza Akhondi-asl
Saad Ullah Akram
Daniel Alexander
Hanan Alghamdi
Hassan Alhajj
Omar Al-Kadi
Max Allan
Andre Altmann
Pablo Alvarez
Charlems Alvarez-Jimenez
Jennifer Alvén
Lidia Al-Zogbi
Kimberly Amador
Tamaz Amiranashvili
Amine Amyar
Wangpeng An
Vincent Andrearczyk
Manon Ansart
Sameer Antani
Jacob Antunes
Michel Antunes
Guilherme Aresta
Mohammad Ali Armin
Kasra Arnavaz
Corey Arnold
Janan Arslan
Marius Arvinte
Muhammad Asad
John Ashburner
Md Ashikuzzaman
Shahab Aslani
Mehdi Astaraki
Angélica Atehortúa
Benjamin Aubert
Marc Aubreville
Paolo Avesani
Sana Ayromlou
Reza Azad
Mohammad Farid Azampour
Qinle Ba
Meritxell Bach Cuadra
Hyeon-Min Bae
Matheus Baffa
Cagla Bahadir
Fan Bai
Jun Bai
Long Bai
Pradeep Bajracharya
Shafa Balaram
Yaël Balbastre
Yutong Ban
Abhirup Banerjee
Soumyanil Banerjee
Sreya Banerjee
Shunxing Bao
Omri Bar
Adrian Barbu
Joao Barreto
Adrian Basarab
Berke Basaran
Michael Baumgartner
Siming Bayer
Roza Bayrak
Aicha BenTaieb
Guy Ben-Yosef
Sutanu Bera
Cosmin Bercea
Jorge Bernal
Jose Bernal
Gabriel Bernardino
Riddhish Bhalodia
Jignesh Bhatt
Indrani Bhattacharya
Binod Bhattarai
Lei Bi
Qi Bi
Cheng Bian
Gui-Bin Bian
Carlo Biffi
Alexander Bigalke
Benjamin Billot
Manuel Birlo
Ryoma Bise
Daniel Blezek
Stefano Blumberg
Sebastian Bodenstedt
Federico Bolelli
Bhushan Borotikar
Ilaria Boscolo Galazzo
Alexandre Bousse
Nicolas Boutry
Joseph Boyd
Behzad Bozorgtabar
Nadia Brancati
Clara Brémond Martin
Stéphanie Bricq
Christopher Bridge
Coleman Broaddus
Rupert Brooks
Tom Brosch
Mikael Brudfors
Ninon Burgos
Nikolay Burlutskiy
Michal Byra
Ryan Cabeen
Mariano Cabezas
Hongmin Cai
Tongan Cai
Zongyou Cai
Liane Canas
Bing Cao
Guogang Cao
Weiguo Cao
Xu Cao
Yankun Cao
Zhenjie Cao

Jaime Cardoso
M. Jorge Cardoso
Owen Carmichael
Jacob Carse
Adrià Casamitjana
Alessandro Casella
Angela Castillo
Kate Cevora
Krishna Chaitanya
Satrajit Chakrabarty
Yi Hao Chan
Shekhar Chandra
Ming-Ching Chang
Peng Chang
Qi Chang
Yuchou Chang
Hanqing Chao
Simon Chatelin
Soumick Chatterjee
Sudhanya Chatterjee
Muhammad Faizyab Ali Chaudhary
Antong Chen
Bingzhi Chen
Chen Chen
Cheng Chen
Chengkuan Chen
Eric Chen
Fang Chen
Haomin Chen
Jianan Chen
Jianxu Chen
Jiazhou Chen
Jie Chen
Jintai Chen
Jun Chen
Junxiang Chen
Junyu Chen
Li Chen
Liyun Chen
Nenglun Chen
Pingjun Chen
Pingyi Chen
Qi Chen
Qiang Chen
Runnan Chen
Shengcong Chen
Sihao Chen
Tingting Chen
Wenting Chen
Xi Chen
Xiang Chen
Xiaoran Chen
Xin Chen
Xiongchao Chen
Yanxi Chen
Yixiong Chen
Yixuan Chen
Yuanyuan Chen
Yuqian Chen
Zhaolin Chen
Zhen Chen
Zhenghao Chen
Zhennong Chen
Zhihao Chen
Zhineng Chen
Zhixiang Chen
Chang-Chieh Cheng
Jiale Cheng
Jianhong Cheng
Jun Cheng
Xuelian Cheng
Yupeng Cheng
Mark Chiew
Philip Chikontwe
Eleni Chiou
Jungchan Cho
Jang-Hwan Choi
Min-Kook Choi
Wookjin Choi
Jaegul Choo
Yu-Cheng Chou
Daan Christiaens
Argyrios Christodoulidis
Stergios Christodoulidis
Kai-Cheng Chuang
Hyungjin Chung
Matthew Clarkson
Michaël Clément
Dana Cobzas
Jaume Coll-Font
Olivier Colliot
Runmin Cong
Yulai Cong
Laura Connolly
William Consagra
Pierre-Henri Conze
Tim Cootes
Teresa Correia
Baris Coskunuzer
Alex Crimi
Can Cui
Hejie Cui
Hui Cui
Lei Cui
Wenhui Cui
Tolga Cukur
Tobias Czempiel
Javid Dadashkarimi
Haixing Dai
Tingting Dan
Kang Dang
Salman Ul Hassan Dar
Eleonora D'Arnese
Dhritiman Das
Neda Davoudi
Tareen Dawood
Sandro De Zanet
Farah Deeba
Charles Delahunt
Herve Delingette
Ugur Demir
Liang-Jian Deng
Ruining Deng
Wenlong Deng
Felix Denzinger
Adrien Depeursinge
Mohammad Mahdi Derakhshani
Hrishikesh Deshpande
Adrien Desjardins
Christian Desrosiers
Blake Dewey
Neel Dey
Rohan Dhamdhere

Maxime Di Folco
Songhui Diao
Alina Dima
Hao Ding
Li Ding
Ying Ding
Zhipeng Ding
Nicola Dinsdale
Konstantin Dmitriev
Ines Domingues
Bo Dong
Liang Dong
Nanqing Dong
Siyuan Dong
Reuben Dorent
Gianfranco Doretto
Sven Dorkenwald
Haoran Dou
Mitchell Doughty
Jason Dowling
Niharika D'Souza
Guodong Du
Jie Du
Shiyi Du
Hongyi Duanmu
Benoit Dufumier
James Duncan
Joshua Durso-Finley
Dmitry V. Dylov
Oleh Dzyubachyk
Mahdi (Elias) Ebnali
Philip Edwards
Jan Egger
Gudmundur Einarsson
Mostafa El Habib Daho
Ahmed Elazab
Idris El-Feghi
David Ellis
Mohammed Elmogy
Amr Elsawy
Okyaz Eminaga
Ertunc Erdil
Lauren Erdman
Marius Erdt
Maria Escobar
Hooman Esfandiari
Nazila Esmaeili
Ivan Ezhov
Alessio Fagioli
Deng-Ping Fan
Lei Fan
Xin Fan
Yubo Fan
Huihui Fang
Jiansheng Fang
Xi Fang
Zhenghan Fang
Mohammad Farazi
Azade Farshad
Mohsen Farzi
Hamid Fehri
Lina Felsner
Chaolu Feng
Chun-Mei Feng
Jianjiang Feng
Mengling Feng
Ruibin Feng
Zishun Feng
Alvaro Fernandez-Quilez
Ricardo Ferrari
Lucas Fidon
Lukas Fischer
Madalina Fiterau
Antonio Foncubierta-Rodríguez
Fahimeh Fooladgar
Germain Forestier
Nils Daniel Forkert
Jean-Rassaire Fouefack
Kevin François-Bouaou
Wolfgang Freysinger
Bianca Freytag
Guanghui Fu
Kexue Fu
Lan Fu
Yunguan Fu
Pedro Furtado
Ryo Furukawa
Jin Kyu Gahm
Mélanie Gaillochet
Francesca Galassi
Jiangzhang Gan
Yu Gan
Yulu Gan
Alireza Ganjdanesh
Chang Gao
Cong Gao
Linlin Gao
Zeyu Gao
Zhongpai Gao
Sara Garbarino
Alain Garcia
Beatriz Garcia Santa Cruz
Rongjun Ge
Shiv Gehlot
Manuela Geiss
Salah Ghamizi
Negin Ghamsarian
Ramtin Gharleghi
Ghazal Ghazaei
Florin Ghesu
Sayan Ghosal
Syed Zulqarnain Gilani
Mahdi Gilany
Yannik Glaser
Ben Glocker
Bharti Goel
Jacob Goldberger
Polina Golland
Alberto Gomez
Catalina Gomez
Estibaliz Gómez-de-Mariscal
Haifan Gong
Kuang Gong
Xun Gong
Ricardo Gonzales
Camila Gonzalez
German Gonzalez
Vanessa Gonzalez Duque
Sharath Gopal
Karthik Gopinath
Pietro Gori
Michael Götz
Shuiping Gou

Maged Goubran
Sobhan Goudarzi
Mark Graham
Alejandro Granados
Mara Graziani
Thomas Grenier
Radu Grosu
Michal Grzeszczyk
Feng Gu
Pengfei Gu
Qiangqiang Gu
Ran Gu
Shi Gu
Wenhao Gu
Xianfeng Gu
Yiwen Gu
Zaiwang Gu
Hao Guan
Jayavardhana Gubbi
Houssem-Eddine Gueziri
Dazhou Guo
Hengtao Guo
Jixiang Guo
Jun Guo
Pengfei Guo
Wenzhangzhi Guo
Xiaoqing Guo
Xueqi Guo
Yi Guo
Vikash Gupta
Praveen Gurunath Bharathi
Prashnna Gyawali
Sung Min Ha
Mohamad Habes
Ilker Hacihaliloglu
Stathis Hadjidemetriou
Fatemeh Haghighi
Justin Haldar
Noura Hamze
Liang Han
Luyi Han
Seungjae Han
Tianyu Han
Zhongyi Han
Jonny Hancox
Lasse Hansen
Degan Hao
Huaying Hao
Jinkui Hao
Nazim Haouchine
Michael Hardisty
Stefan Harrer
Jeffry Hartanto
Charles Hatt
Huiguang He
Kelei He
Qi He
Shenghua He
Xinwei He
Stefan Heldmann
Nicholas Heller
Edward Henderson
Alessa Hering
Monica Hernandez
Kilian Hett
Amogh Hiremath
David Ho
Malte Hoffmann
Matthew Holden
Qingqi Hong
Yoonmi Hong
Mohammad Reza Hosseinzadeh Taher
William Hsu
Chuanfei Hu
Dan Hu
Kai Hu
Rongyao Hu
Shishuai Hu
Xiaoling Hu
Xinrong Hu
Yan Hu
Yang Hu
Chaoqin Huang
Junzhou Huang
Ling Huang
Luojie Huang
Qinwen Huang
Sharon Xiaolei Huang
Weijian Huang
Xiaoyang Huang
Yi-Jie Huang
Yongsong Huang
Yongxiang Huang
Yuhao Huang
Zhe Huang
Zhi-An Huang
Ziyi Huang
Arnaud Huaulmé
Henkjan Huisman
Alex Hung
Jiayu Huo
Andreas Husch
Mohammad Arafat Hussain
Sarfaraz Hussein
Jana Hutter
Khoi Huynh
Ilknur Icke
Kay Igwe
Abdullah Al Zubaer Imran
Muhammad Imran
Samra Irshad
Nahid Ul Islam
Koichi Ito
Hayato Itoh
Yuji Iwahori
Krithika Iyer
Mohammad Jafari
Srikrishna Jaganathan
Hassan Jahanandish
Andras Jakab
Amir Jamaludin
Amoon Jamzad
Ananya Jana
Se-In Jang
Pierre Jannin
Vincent Jaouen
Uditha Jarayathne
Ronnachai Jaroensri
Guillaume Jaume
Syed Ashar Javed
Rachid Jennane
Debesh Jha
Ge-Peng Ji

Luping Ji
Zexuan Ji
Zhanghexuan Ji
Haozhe Jia
Hongchao Jiang
Jue Jiang
Meirui Jiang
Tingting Jiang
Xiajun Jiang
Zekun Jiang
Zhifan Jiang
Ziyu Jiang
Jianbo Jiao
Zhicheng Jiao
Chen Jin
Dakai Jin
Qiangguo Jin
Qiuye Jin
Weina Jin
Baoyu Jing
Bin Jing
Yaqub Jonmohamadi
Lie Ju
Yohan Jun
Dinkar Juyal
Manjunath K N
Ali Kafaei Zad Tehrani
John Kalafut
Niveditha Kalavakonda
Megha Kalia
Anil Kamat
Qingbo Kang
Po-Yu Kao
Anuradha Kar
Neerav Karani
Turkay Kart
Satyananda Kashyap
Alexander Katzmann
Lisa Kausch
Maxime Kayser
Salome Kazeminia
Wenchi Ke
Youngwook Kee
Matthias Keicher
Erwan Kerrien
Afifa Khaled
Nadieh Khalili
Farzad Khalvati
Bidur Khanal
Bishesh Khanal
Pulkit Khandelwal
Maksim Kholiavchenko
Ron Kikinis
Benjamin Killeen
Daeseung Kim
Heejong Kim
Jaeil Kim
Jinhee Kim
Jinman Kim
Junsik Kim
Minkyung Kim
Namkug Kim
Sangwook Kim
Tae Soo Kim
Younghoon Kim
Young-Min Kim
Andrew King
Miranda Kirby
Gabriel Kiss
Andreas Kist
Yoshiro Kitamura
Stefan Klein
Tobias Klinder
Kazuma Kobayashi
Lisa Koch
Satoshi Kondo
Fanwei Kong
Tomasz Konopczynski
Ender Konukoglu
Aishik Konwer
Thijs Kooi
Ivica Kopriva
Avinash Kori
Kivanc Kose
Suraj Kothawade
Anna Kreshuk
AnithaPriya Krishnan
Florian Kromp
Frithjof Kruggel
Thomas Kuestner
Levin Kuhlmann
Abhay Kumar
Kuldeep Kumar
Sayantan Kumar
Manuela Kunz
Holger Kunze
Tahsin Kurc
Anvar Kurmukov
Yoshihiro Kuroda
Yusuke Kurose
Hyuksool Kwon
Aymen Laadhari
Jorma Laaksonen
Dmitrii Lachinov
Alain Lalande
Rodney LaLonde
Bennett Landman
Daniel Lang
Carole Lartizien
Shlomi Laufer
Max-Heinrich Laves
William Le
Loic Le Folgoc
Christian Ledig
Eung-Joo Lee
Ho Hin Lee
Hyekyoung Lee
John Lee
Kisuk Lee
Kyungsu Lee
Soochahn Lee
Woonghee Lee
Étienne Léger
Wen Hui Lei
Yiming Lei
George Leifman
Rogers Jeffrey Leo John
Juan Leon
Bo Li
Caizi Li
Chao Li
Chen Li
Cheng Li
Chenxin Li
Chnegyin Li

Dawei Li
Fuhai Li
Gang Li
Guang Li
Hao Li
Haofeng Li
Haojia Li
Heng Li
Hongming Li
Hongwei Li
Huiqi Li
Jian Li
Jieyu Li
Kang Li
Lin Li
Mengzhang Li
Ming Li
Qing Li
Quanzheng Li
Shaohua Li
Shulong Li
Tengfei Li
Weijian Li
Wen Li
Xiaomeng Li
Xingyu Li
Xinhui Li
Xuelu Li
Xueshen Li
Yamin Li
Yang Li
Yi Li
Yuemeng Li
Yunxiang Li
Zeju Li
Zhaoshuo Li
Zhe Li
Zhen Li
Zhenqiang Li
Zhiyuan Li
Zhjin Li
Zi Li
Hao Liang
Libin Liang
Peixian Liang
Yuan Liang
Yudong Liang
Haofu Liao
Hongen Liao
Wei Liao
Zehui Liao
Gilbert Lim
Hongxiang Lin
Li Lin
Manxi Lin
Mingquan Lin
Tiancheng Lin
Yi Lin
Zudi Lin
Claudia Lindner
Simone Lionetti
Chi Liu
Chuanbin Liu
Daochang Liu
Dongnan Liu
Feihong Liu
Fenglin Liu
Han Liu
Huiye Liu
Jiang Liu
Jie Liu
Jinduo Liu
Jing Liu
Jingya Liu
Jundong Liu
Lihao Liu
Mengting Liu
Mingyuan Liu
Peirong Liu
Peng Liu
Qin Liu
Quan Liu
Rui Liu
Shengfeng Liu
Shuangjun Liu
Sidong Liu
Siyuan Liu
Weide Liu
Xiao Liu
Xiaoyu Liu
Xingtong Liu
Xinwen Liu
Xinyang Liu
Xinyu Liu
Yan Liu
Yi Liu
Yihao Liu
Yikang Liu
Yilin Liu
Yilong Liu
Yiqiao Liu
Yong Liu
Yuhang Liu
Zelong Liu
Zhe Liu
Zhiyuan Liu
Zuozhu Liu
Lisette Lockhart
Andrea Loddo
Nicolas Loménie
Yonghao Long
Daniel Lopes
Ange Lou
Brian Lovell
Nicolas Loy Rodas
Charles Lu
Chun-Shien Lu
Donghuan Lu
Guangming Lu
Huanxiang Lu
Jingpei Lu
Yao Lu
Oeslle Lucena
Jie Luo
Luyang Luo
Ma Luo
Mingyuan Luo
Wenhan Luo
Xiangde Luo
Xinzhe Luo
Jinxin Lv
Tianxu Lv
Fei Lyu
Ilwoo Lyu
Mengye Lyu

Qing Lyu
Yanjun Lyu
Yuanyuan Lyu
Benteng Ma
Chunwei Ma
Hehuan Ma
Jun Ma
Junbo Ma
Wenao Ma
Yuhui Ma
Pedro Macias Gordaliza
Anant Madabhushi
Derek Magee
S. Sara Mahdavi
Andreas Maier
Klaus H. Maier-Hein
Sokratis Makrogiannis
Danial Maleki
Michail Mamalakis
Zhehua Mao
Jan Margeta
Brett Marinelli
Zdravko Marinov
Viktoria Markova
Carsten Marr
Yassine Marrakchi
Anne Martel
Martin Maška
Tejas Sudharshan Mathai
Petr Matula
Dimitrios Mavroeidis
Evangelos Mazomenos
Amarachi Mbakwe
Adam McCarthy
Stephen McKenna
Raghav Mehta
Xueyan Mei
Felix Meissen
Felix Meister
Afaque Memon
Mingyuan Meng
Qingjie Meng
Xiangzhu Meng
Yanda Meng
Zhu Meng
Martin Menten
Odyssée Merveille
Mikhail Milchenko
Leo Milecki
Fausto Milletari
Hyun-Seok Min
Zhe Min
Song Ming
Duy Minh Ho Nguyen
Deepak Mishra
Suraj Mishra
Virendra Mishra
Tadashi Miyamoto
Sara Moccia
Marc Modat
Omid Mohareri
Tony C. W. Mok
Javier Montoya
Rodrigo Moreno
Stefano Moriconi
Lia Morra
Ana Mota
Lei Mou
Dana Moukheiber
Lama Moukheiber
Daniel Moyer
Pritam Mukherjee
Anirban Mukhopadhyay
Henning Müller
Ana Murillo
Gowtham Krishnan Murugesan
Ahmed Naglah
Karthik Nandakumar
Venkatesh Narasimhamurthy
Raja Narayan
Dominik Narnhofer
Vishwesh Nath
Rodrigo Nava
Abdullah Nazib
Ahmed Nebli
Peter Neher
Amin Nejatbakhsh
Trong-Thuan Nguyen
Truong Nguyen
Dong Ni
Haomiao Ni
Xiuyan Ni
Hannes Nickisch
Weizhi Nie
Aditya Nigam
Lipeng Ning
Xia Ning
Kazuya Nishimura
Chuang Niu
Sijie Niu
Vincent Noblet
Narges Norouzi
Alexey Novikov
Jorge Novo
Gilberto Ochoa-Ruiz
Masahiro Oda
Benjamin Odry
Hugo Oliveira
Sara Oliveira
Arnau Oliver
Jimena Olveres
John Onofrey
Marcos Ortega
Mauricio Alberto Ortega-Ruíz
Yusuf Osmanlioglu
Chubin Ou
Cheng Ouyang
Jiahong Ouyang
Xi Ouyang
Cristina Oyarzun Laura
Utku Ozbulak
Ece Ozkan
Ege Özsoy
Batu Ozturkler
Harshith Padigela
Johannes Paetzold
José Blas Pagador Carrasco
Daniel Pak
Sourabh Palande
Chengwei Pan
Jiazhen Pan

Jin Pan
Yongsheng Pan
Egor Panfilov
Jiaxuan Pang
Joao Papa
Constantin Pape
Bartlomiej Papiez
Nripesh Parajuli
Hyunjin Park
Akash Parvatikar
Tiziano Passerini
Diego Patiño Cortés
Mayank Patwari
Angshuman Paul
Rasmus Paulsen
Yuchen Pei
Yuru Pei
Tao Peng
Wei Peng
Yige Peng
Yunsong Peng
Matteo Pennisi
Antonio Pepe
Oscar Perdomo
Sérgio Pereira
Jose-Antonio Pérez-Carrasco
Mehran Pesteie
Terry Peters
Eike Petersen
Jens Petersen
Micha Pfeiffer
Dzung Pham
Hieu Pham
Ashish Phophalia
Tomasz Pieciak
Antonio Pinheiro
Pramod Pisharady
Theodoros Pissas
Szymon Płotka
Kilian Pohl
Sebastian Pölsterl
Alison Pouch
Tim Prangemeier
Prateek Prasanna
Raphael Prevost
Juan Prieto
Federica Proietto Salanitri
Sergi Pujades
Elodie Puybareau
Talha Qaiser
Buyue Qian
Mengyun Qiao
Yuchuan Qiao
Zhi Qiao
Chenchen Qin
Fangbo Qin
Wenjian Qin
Yulei Qin
Jie Qiu
Jielin Qiu
Peijie Qiu
Shi Qiu
Wu Qiu
Liangqiong Qu
Linhao Qu
Quan Quan
Tran Minh Quan
Sandro Queirós
Prashanth R
Febrian Rachmadi
Daniel Racoceanu
Mehdi Rahim
Jagath Rajapakse
Kashif Rajpoot
Keerthi Ram
Dhanesh Ramachandram
João Ramalhinho
Xuming Ran
Aneesh Rangnekar
Hatem Rashwan
Keerthi Sravan Ravi
Daniele Ravì
Sadhana Ravikumar
Harish Raviprakash
Surreerat Reaungamornrat
Samuel Remedios
Mengwei Ren
Sucheng Ren
Elton Rexhepaj
Mauricio Reyes
Constantino Reyes-Aldasoro
Abel Reyes-Angulo
Hadrien Reynaud
Razieh Rezaei
Anne-Marie Rickmann
Laurent Risser
Dominik Rivoir
Emma Robinson
Robert Robinson
Jessica Rodgers
Ranga Rodrigo
Rafael Rodrigues
Robert Rohling
Margherita Rosnati
Łukasz Roszkowiak
Holger Roth
José Rouco
Dan Ruan
Jiacheng Ruan
Daniel Rueckert
Danny Ruijters
Kanghyun Ryu
Ario Sadafi
Numan Saeed
Monjoy Saha
Pramit Saha
Farhang Sahba
Pranjal Sahu
Simone Saitta
Md Sirajus Salekin
Abbas Samani
Pedro Sanchez
Luis Sanchez Giraldo
Yudi Sang
Gerard Sanroma-Guell
Rodrigo Santa Cruz
Alice Santilli
Rachana Sathish
Olivier Saut
Mattia Savardi
Nico Scherf
Alexander Schlaefer
Jerome Schmid

Adam Schmidt
Julia Schnabel
Lawrence Schobs
Julian Schön
Peter Schueffler
Andreas Schuh
Christina Schwarz-Gsaxner
Michaël Sdika
Suman Sedai
Lalithkumar Seenivasan
Matthias Seibold
Sourya Sengupta
Lama Seoud
Ana Sequeira
Sharmishtaa Seshamani
Ahmed Shaffie
Jay Shah
Keyur Shah
Ahmed Shahin
Mohammad Abuzar Shaikh
S. Shailja
Hongming Shan
Wei Shao
Mostafa Sharifzadeh
Anuja Sharma
Gregory Sharp
Hailan Shen
Li Shen
Linlin Shen
Mali Shen
Mingren Shen
Yiqing Shen
Zhengyang Shen
Jun Shi
Xiaoshuang Shi
Yiyu Shi
Yonggang Shi
Hoo-Chang Shin
Jitae Shin
Keewon Shin
Boris Shirokikh
Suzanne Shontz
Yucheng Shu
Hanna Siebert
Alberto Signoroni
Wilson Silva
Julio Silva-Rodríguez
Margarida Silveira
Walter Simson
Praveer Singh
Vivek Singh
Nitin Singhal
Elena Sizikova
Gregory Slabaugh
Dane Smith
Kevin Smith
Tiffany So
Rajath Soans
Roger Soberanis-Mukul
Hessam Sokooti
Jingwei Song
Weinan Song
Xinhang Song
Xinrui Song
Mazen Soufi
Georgia Sovatzidi
Bella Specktor Fadida
William Speier
Ziga Spiclin
Dominik Spinczyk
Jon Sporring
Pradeeba Sridar
Chetan L. Srinidhi
Abhishek Srivastava
Lawrence Staib
Marc Stamminger
Justin Strait
Hai Su
Ruisheng Su
Zhe Su
Vaishnavi Subramanian
Gérard Subsol
Carole Sudre
Dong Sui
Heung-Il Suk
Shipra Suman
He Sun
Hongfu Sun
Jian Sun
Li Sun
Liyan Sun
Shanlin Sun
Kyung Sung
Yannick Suter
Swapna T. R.
Amir Tahmasebi
Pablo Tahoces
Sirine Taleb
Bingyao Tan
Chaowei Tan
Wenjun Tan
Hao Tang
Siyi Tang
Xiaoying Tang
Yucheng Tang
Zihao Tang
Michael Tanzer
Austin Tapp
Elias Tappeiner
Mickael Tardy
Giacomo Tarroni
Athena Taymourtash
Kaveri Thakoor
Elina Thibeau-Sutre
Paul Thienphrapa
Sarina Thomas
Stephen Thompson
Karl Thurnhofer-Hemsi
Cristiana Tiago
Lin Tian
Lixia Tian
Yapeng Tian
Yu Tian
Yun Tian
Aleksei Tiulpin
Hamid Tizhoosh
Minh Nguyen Nhat To
Matthew Toews
Maryam Toloubidokhti
Minh Tran
Quoc-Huy Trinh
Jocelyne Troccaz
Roger Trullo

Chialing Tsai
Apostolia Tsirikoglou
Puxun Tu
Samyakh Tukra
Sudhakar Tummala
Georgios Tziritas
Vladimír Ulman
Tamas Ungi
Régis Vaillant
Jeya Maria Jose Valanarasu
Vanya Valindria
Juan Miguel Valverde
Fons van der Sommen
Maureen van Eijnatten
Tom van Sonsbeek
Gijs van Tulder
Yogatheesan Varatharajah
Madhurima Vardhan
Thomas Varsavsky
Hooman Vaseli
Serge Vasylechko
S. Swaroop Vedula
Sanketh Vedula
Gonzalo Vegas Sanchez-Ferrero
Matthew Velazquez
Archana Venkataraman
Sulaiman Vesal
Mitko Veta
Barbara Villarini
Athanasios Vlontzos
Wolf-Dieter Vogl
Ingmar Voigt
Sandrine Voros
Vibashan VS
Trinh Thi Le Vuong
An Wang
Bo Wang
Ce Wang
Changmiao Wang
Ching-Wei Wang
Dadong Wang
Dong Wang
Fakai Wang
Guotai Wang
Haifeng Wang
Haoran Wang
Hong Wang
Hongxiao Wang
Hongyu Wang
Jiacheng Wang
Jing Wang
Jue Wang
Kang Wang
Ke Wang
Lei Wang
Li Wang
Liansheng Wang
Lin Wang
Ling Wang
Linwei Wang
Manning Wang
Mingliang Wang
Puyang Wang
Qiuli Wang
Renzhen Wang
Ruixuan Wang
Shaoyu Wang
Sheng Wang
Shujun Wang
Shuo Wang
Shuqiang Wang
Tao Wang
Tianchen Wang
Tianyu Wang
Wenzhe Wang
Xi Wang
Xiangdong Wang
Xiaoqing Wang
Xiaosong Wang
Yan Wang
Yangang Wang
Yaping Wang
Yi Wang
Yirui Wang
Yixin Wang
Zeyi Wang
Zhao Wang
Zichen Wang
Ziqin Wang
Ziyi Wang
Zuhui Wang
Dong Wei
Donglai Wei
Hao Wei
Jia Wei
Leihao Wei
Ruofeng Wei
Shuwen Wei
Martin Weigert
Wolfgang Wein
Michael Wels
Cédric Wemmert
Thomas Wendler
Markus Wenzel
Rhydian Windsor
Adam Wittek
Marek Wodzinski
Ivo Wolf
Julia Wolleb
Ka-Chun Wong
Jonghye Woo
Chongruo Wu
Chunpeng Wu
Fuping Wu
Huaqian Wu
Ji Wu
Jiangjie Wu
Jiong Wu
Junde Wu
Linshan Wu
Qing Wu
Weiwen Wu
Wenjun Wu
Xiyin Wu
Yawen Wu
Ye Wu
Yicheng Wu
Yongfei Wu
Zhengwang Wu
Pengcheng Xi
Chao Xia
Siyu Xia
Wenjun Xia
Lei Xiang

Tiange Xiang
Deqiang Xiao
Li Xiao
Xiaojiao Xiao
Yiming Xiao
Zeyu Xiao
Hongtao Xie
Huidong Xie
Jianyang Xie
Long Xie
Weidi Xie
Fangxu Xing
Shuwei Xing
Xiaodan Xing
Xiaohan Xing
Haoyi Xiong
Yujian Xiong
Di Xu
Feng Xu
Haozheng Xu
Hongming Xu
Jiangchang Xu
Jiaqi Xu
Junshen Xu
Kele Xu
Lijian Xu
Min Xu
Moucheng Xu
Rui Xu
Xiaowei Xu
Xuanang Xu
Yanwu Xu
Yanyu Xu
Yongchao Xu
Yunqiu Xu
Zhe Xu
Zhoubing Xu
Ziyue Xu
Kai Xuan
Cheng Xue
Jie Xue
Tengfei Xue
Wufeng Xue
Yuan Xue
Zhong Xue
Ts Faridah Yahya
Chaochao Yan
Jiangpeng Yan
Ming Yan
Qingsen Yan
Xiangyi Yan
Yuguang Yan
Zengqiang Yan
Baoyao Yang
Carl Yang
Changchun Yang
Chen Yang
Feng Yang
Fengting Yang
Ge Yang
Guanyu Yang
Heran Yang
Huijuan Yang
Jiancheng Yang
Jiewen Yang
Peng Yang
Qi Yang
Qiushi Yang
Wei Yang
Xin Yang
Xuan Yang
Yan Yang
Yanwu Yang
Yifan Yang
Yingyu Yang
Zhicheng Yang
Zhijian Yang
Jiangchao Yao
Jiawen Yao
Lanhong Yao
Linlin Yao
Qingsong Yao
Tianyuan Yao
Xiaohui Yao
Zhao Yao
Dong Hye Ye
Menglong Ye
Yousef Yeganeh
Jirong Yi
Xin Yi
Chong Yin
Pengshuai Yin
Yi Yin
Zhaozheng Yin
Chunwei Ying
Youngjin Yoo
Jihun Yoon
Chenyu You
Hanchao Yu
Heng Yu
Jinhua Yu
Jinze Yu
Ke Yu
Qi Yu
Qian Yu
Thomas Yu
Weimin Yu
Yang Yu
Chenxi Yuan
Kun Yuan
Wu Yuan
Yixuan Yuan
Paul Yushkevich
Fatemeh Zabihollahy
Samira Zare
Ramy Zeineldin
Dong Zeng
Qi Zeng
Tianyi Zeng
Wei Zeng
Kilian Zepf
Kun Zhan
Bokai Zhang
Daoqiang Zhang
Dong Zhang
Fa Zhang
Hang Zhang
Hanxiao Zhang
Hao Zhang
Haopeng Zhang
Haoyue Zhang
Hongrun Zhang
Jiadong Zhang
Jiajin Zhang
Jianpeng Zhang

Jiawei Zhang
Jingqing Zhang
Jingyang Zhang
Jinwei Zhang
Jiong Zhang
Jiping Zhang
Ke Zhang
Lefei Zhang
Lei Zhang
Li Zhang
Lichi Zhang
Lu Zhang
Minghui Zhang
Molin Zhang
Ning Zhang
Rongzhao Zhang
Ruipeng Zhang
Ruisi Zhang
Shichuan Zhang
Shihao Zhang
Shuai Zhang
Tuo Zhang
Wei Zhang
Weihang Zhang
Wen Zhang
Wenhua Zhang
Wenqiang Zhang
Xiaodan Zhang
Xiaoran Zhang
Xin Zhang
Xukun Zhang
Xuzhe Zhang
Ya Zhang
Yanbo Zhang
Yanfu Zhang
Yao Zhang
Yi Zhang
Yifan Zhang
Yixiao Zhang
Yongqin Zhang
You Zhang
Youshan Zhang
Yu Zhang
Yubo Zhang
Yue Zhang
Yuhan Zhang
Yulun Zhang
Yundong Zhang
Yunlong Zhang
Yuyao Zhang
Zheng Zhang
Zhenxi Zhang
Ziqi Zhang
Can Zhao
Chongyue Zhao
Fenqiang Zhao
Gangming Zhao
He Zhao
Jianfeng Zhao
Jun Zhao
Li Zhao
Liang Zhao
Lin Zhao
Mengliu Zhao
Mingbo Zhao
Qingyu Zhao
Shang Zhao
Shijie Zhao
Tengda Zhao
Tianyi Zhao
Wei Zhao
Yidong Zhao
Yiyuan Zhao
Yu Zhao
Zhihe Zhao
Ziyuan Zhao
Haiyong Zheng
Hao Zheng
Jiannan Zheng
Kang Zheng
Meng Zheng
Sisi Zheng
Tianshu Zheng
Yalin Zheng
Yefeng Zheng
Yinqiang Zheng
Yushan Zheng
Aoxiao Zhong
Jia-Xing Zhong
Tao Zhong
Zichun Zhong
Hong-Yu Zhou
Houliang Zhou
Huiyu Zhou
Kang Zhou
Qin Zhou
Ran Zhou
S. Kevin Zhou
Tianfei Zhou
Wei Zhou
Xiao-Hu Zhou
Xiao-Yun Zhou
Yi Zhou
Youjia Zhou
Yukun Zhou
Zongwei Zhou
Chenglu Zhu
Dongxiao Zhu
Heqin Zhu
Jiayi Zhu
Meilu Zhu
Wei Zhu
Wenhui Zhu
Xiaofeng Zhu
Xin Zhu
Yonghua Zhu
Yongpei Zhu
Yuemin Zhu
Yan Zhuang
David Zimmerer
Yongshuo Zong
Ke Zou
Yukai Zou
Lianrui Zuo
Gerald Zwettler

Outstanding Area Chairs

Mingxia Liu	University of North Carolina at Chapel Hill, USA
Matthias Wilms	University of Calgary, Canada
Veronika Zimmer	Technical University Munich, Germany

Outstanding Reviewers

Kimberly Amador	University of Calgary, Canada
Angela Castillo	Universidad de los Andes, Colombia
Chen Chen	Imperial College London, UK
Laura Connolly	Queen's University, Canada
Pierre-Henri Conze	IMT Atlantique, France
Niharika D'Souza	IBM Research, USA
Michael Götz	University Hospital Ulm, Germany
Meirui Jiang	Chinese University of Hong Kong, China
Manuela Kunz	National Research Council Canada, Canada
Zdravko Marinov	Karlsruhe Institute of Technology, Germany
Sérgio Pereira	Lunit, South Korea
Lalithkumar Seenivasan	National University of Singapore, Singapore

Honorable Mentions (Reviewers)

Kumar Abhishek	Simon Fraser University, Canada
Guilherme Aresta	Medical University of Vienna, Austria
Shahab Aslani	University College London, UK
Marc Aubreville	Technische Hochschule Ingolstadt, Germany
Yaël Balbastre	Massachusetts General Hospital, USA
Omri Bar	Theator, Israel
Aicha Ben Taieb	Simon Fraser University, Canada
Cosmin Bercea	Technical University Munich and Helmholtz AI and Helmholtz Center Munich, Germany
Benjamin Billot	Massachusetts Institute of Technology, USA
Michal Byra	RIKEN Center for Brain Science, Japan
Mariano Cabezas	University of Sydney, Australia
Alessandro Casella	Italian Institute of Technology and Politecnico di Milano, Italy
Junyu Chen	Johns Hopkins University, USA
Argyrios Christodoulidis	Pfizer, Greece
Olivier Colliot	CNRS, France

Lei Cui	Northwest University, China
Neel Dey	Massachusetts Institute of Technology, USA
Alessio Fagioli	Sapienza University, Italy
Yannik Glaser	University of Hawaii at Manoa, USA
Haifan Gong	Chinese University of Hong Kong, Shenzhen, China
Ricardo Gonzales	University of Oxford, UK
Sobhan Goudarzi	Sunnybrook Research Institute, Canada
Michal Grzeszczyk	Sano Centre for Computational Medicine, Poland
Fatemeh Haghighi	Arizona State University, USA
Edward Henderson	University of Manchester, UK
Qingqi Hong	Xiamen University, China
Mohammad R. H. Taher	Arizona State University, USA
Henkjan Huisman	Radboud University Medical Center, the Netherlands
Ronnachai Jaroensri	Google, USA
Qiangguo Jin	Northwestern Polytechnical University, China
Neerav Karani	Massachusetts Institute of Technology, USA
Benjamin Killeen	Johns Hopkins University, USA
Daniel Lang	Helmholtz Center Munich, Germany
Max-Heinrich Laves	Philips Research and ImFusion GmbH, Germany
Gilbert Lim	SingHealth, Singapore
Mingquan Lin	Weill Cornell Medicine, USA
Charles Lu	Massachusetts Institute of Technology, USA
Yuhui Ma	Chinese Academy of Sciences, China
Tejas Sudharshan Mathai	National Institutes of Health, USA
Felix Meissen	Technische Universität München, Germany
Mingyuan Meng	University of Sydney, Australia
Leo Milecki	CentraleSupelec, France
Marc Modat	King's College London, UK
Tiziano Passerini	Siemens Healthineers, USA
Tomasz Pieciak	Universidad de Valladolid, Spain
Daniel Rueckert	Imperial College London, UK
Julio Silva-Rodríguez	ETS Montreal, Canada
Bingyao Tan	Nanyang Technological University, Singapore
Elias Tappeiner	UMIT - Private University for Health Sciences, Medical Informatics and Technology, Austria
Jocelyne Troccaz	TIMC Lab, Grenoble Alpes University-CNRS, France
Chialing Tsai	Queens College, City University New York, USA
Juan Miguel Valverde	University of Eastern Finland, Finland
Sulaiman Vesal	Stanford University, USA

Wolf-Dieter Vogl	RetInSight GmbH, Austria
Vibashan VS	Johns Hopkins University, USA
Lin Wang	Harbin Engineering University, China
Yan Wang	Sichuan University, China
Rhydian Windsor	University of Oxford, UK
Ivo Wolf	University of Applied Sciences Mannheim, Germany
Linshan Wu	Hunan University, China
Xin Yang	Chinese University of Hong Kong, China

Contents – Part VIII

Clinical Applications – Neuroimaging

Microscopy

Clinical Applications – Neuroimaging

CoactSeg: Learning from Heterogeneous Data for New Multiple Sclerosis Lesion Segmentation

Yicheng Wu[1(✉)], Zhonghua Wu[2], Hengcan Shi[1], Bjoern Picker[3,4], Winston Chong[3,4], and Jianfei Cai[1]

[1] Department of Data Science & AI, Faculty of Information Technology, Monash University, Melbourne, VIC 3168, Australia
yicheng.wu@monash.edu
[2] SenseTime Research, Singapore 069547, Singapore
[3] Alfred Health Radiology, Alfred Health, Melbourne, VIC 3004, Australia
[4] Central Clinical School, Faculty of Medicine, Nursing and Health Sciences, Monash University, Melbourne, VIC 3800, Australia

Abstract. New lesion segmentation is essential to estimate the disease progression and therapeutic effects during multiple sclerosis (MS) clinical treatments. However, the expensive data acquisition and expert annotation restrict the feasibility of applying large-scale deep learning models. Since single-time-point samples with all-lesion labels are relatively easy to collect, exploiting them to train deep models is highly desirable to improve new lesion segmentation. Therefore, we proposed a **coact**ion **seg**mentation (CoactSeg) framework to exploit the heterogeneous data (*i.e.,* new-lesion annotated two-time-point data and all-lesion annotated single-time-point data) for new MS lesion segmentation. The CoactSeg model is designed as a unified model, with the same three inputs (the baseline, follow-up, and their longitudinal brain differences) and the same three outputs (the corresponding all-lesion and new-lesion predictions), no matter which type of heterogeneous data is being used. Moreover, a simple and effective relation regularization is proposed to ensure the longitudinal relations among the three outputs to improve the model learning. Extensive experiments demonstrate that utilizing the heterogeneous data and the proposed longitudinal relation constraint can significantly improve the performance for both new-lesion and all-lesion segmentation tasks. Meanwhile, we also introduce an in-house MS-23v1 dataset, including 38 Oceania single-time-point samples with all-lesion labels. Codes and the dataset are released at https://github.com/ycwu1997/CoactSeg.

Keywords: Multiple Sclerosis Lesion · Longitudinal Relation · Heterogeneous Data

H. Greenspan et al. (Eds.): MICCAI 2023, LNCS 14227, pp. 3–13, 2023.
https://doi.org/10.1007/978-3-031-43993-3_1

1 Introduction

Multiple sclerosis (MS) is a common inflammatory disease in the central nervous system (CNS), affecting millions of people worldwide [7] and even leading to the disability of young population [19]. During the clinical treatment of MS, lesion changes, especially the emergence of new lesions, are crucial criteria for estimating the effects of given anti-inflammatory disease-modifying drugs [2]. However, MS lesions are usually small, numerous, and appear similar to Gliosis or other types of brain lesions, *e.g.,* ischemic vasculopathy [8]. Identifying MS lesion changes from multi-time-point data is still a heavy burden for clinicians. Therefore, automatically quantifying MS lesion changes is essential in constructing a computer-aided diagnosis (CAD) system for clinical applications.

Deep learning has been widely used for MS lesion segmentation from brain MRI sequences [20,25]. For example, the icobrain 5.1 framework [16] combined supervised and unsupervised approaches and designed manual rules to fuse the final segmentation results. Some works [10,28] further studied the complementary features from other MRI modalities for MS lesion segmentation. Meanwhile, to train a better deep model, class-imbalance issues [26] and prior brain structures [27] have been respectively investigated to improve the performance. With the impressive performance achieved by existing pure MS lesion segmentation methods [11], recent attention has been shifted to analyze the longitudinal MS changes [5,6], such as stable, new, shrinking, and enlarging lesions, with the focus on new MS lesion segmentation [9].

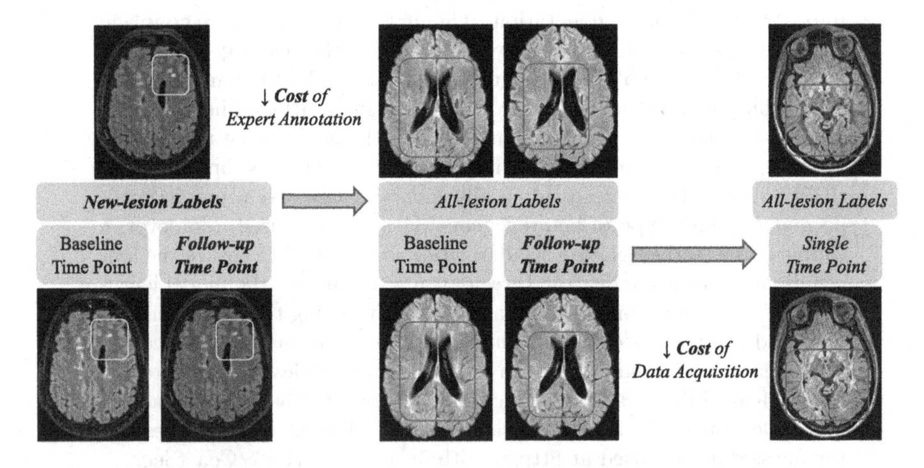

Fig. 1. Heterogeneous data and annotations: *new-lesion annotated two-time-point data (Left), all-lesion two-time-point data (Middle), and all-lesion single-time-point data (Right),* with different expert annotation and data acquisition costs. Here, we exploit additional single-time-point data (Right) to help new MS lesion segmentation.

However, collecting adequate well-labeled longitudinal MS lesion data for model learning is highly challenging since it needs multi-time-point data from the same set of patients, and requires costly and time-consuming expert annotations. Figure 1 shows the three types of heterogeneous MS lesion data: new-lesion annotated two-time-point data, all-lesion annotated two-time-point data, and all-lesion annotated single-time-point data, each of which is associated with different costs. New-lesion annotated two-time-point data is the ideal one for learning new lesion segmentation, but with the highest data acquisition and annotation costs. Annotating all lesions in two-time-point data can reduce the annotation cost, but it requires accurate brain registration and rule-based post-processing to identify lesion changes, which cannot avoid noise accumulation and often leads to sub-optimal performance. All-lesion annotated single-time-point data is with the cheapest data acquisition and annotation costs. This motivates us to raise the question: *"Can we leverage all-lesion annotated single-time-point data to promote the new MS lesion segmentation?"*

Therefore, in this paper, we proposed a deep **Coact**ion **Seg**mentation (CoactSeg) model that can unify heterogeneous data and annotations for the new MS lesion segmentation task. Specifically, CoactSeg takes three-channel inputs, including the baseline, follow-up, and corresponding differential brains, and produces all-lesion and new-lesion segmentation results at the same time. Moreover, a longitudinal relation constraint (*e.g.,* new lesions should only appear at the follow-up scans) is proposed to regularize the model learning in order to integrate the two tasks (new and all lesion segmentation) and boost each other. Extensive experiments on two MS datasets demonstrate that our proposed CoactSeg model is able to achieve superior performance for both new and all MS lesion segmentation, *e.g.,* obtaining 63.82% Dice on the public MICCAI-21 dataset [4] and 72.32% Dice on our in-house MS-23v1 dataset, respectively. It even outperforms two neuro-radiologists on MICCAI-21.

Overall, the contributions of this work are three-fold:

- We propose a simple unified model CoactSeg that can be trained on both new-lesion annotated two-time-point data and all-lesion annotated single-time-point data in the same way, with the same input and output format;
- We design a relation regularizer to ensure the longitudinal relations among all and new lesion predictions of the baseline, follow-up, and corresponding differential brains;
- We construct an in-house MS-23v1 dataset, which includes 38 Oceania single-time-point 3D FLAIR scans with manual all-lesion annotations by experienced human experts. We will release this dataset publicly.

Table 1. Details of the experimental datasets in this work. Note that the data split is fixed and shown in the 4th column (Order: Training, Validation).

Dataset	Region	Modality	# of subjects	# of time points	Annotation Type
MICCAI-21 [4]	France	FLAIR	40 (32, 8)	2	New lesions
MS-23v1 (Ours)	Oceania	FLAIR	38 (30, 8)	1	All lesions

2 Datasets

We trained and evaluated our CoactSeg model on two MS segmentation datasets, as shown in Table 1. On the public MICCAI-21 dataset[1], we only use its training set since it does not provide official labels of testing samples. Specifically, 40 two-time-point 3D FLAIR scans are captured by 15 MRI scanners at different locations. Among them, 11 scans do not contain any new MS lesions. The follow-up data were obtained around 1–3 years after the first examination. Four neuro-radiologists from different centers manually annotated new MS lesions, and a majority voting strategy was used to obtain the final ground truth. For pre-processing, the organizers only performed a rigid brain registration, and we further normalized all MRI scans to a fixed resolution of [0.5, 0.75, 0.75] mm.

Since the public MS lesion data is not adequate [1,3,4], we further collected 38 single-time-point 3D FLAIR sequences as a new MS dataset (MS-23v1). Specifically, all samples were anonymized and captured by a 3T Siemens scanner in Alfred Health, Australia. To the best of our knowledge, this will be *the first open-source dataset from Oceania* for MS lesion segmentation, contributing to the diversity of existing public MS data. Two neuro-radiologists and one senior neuro-scientist segmented all MS lesions individually and in consensus using the MRIcron segmentation tool[2]. The voxel spacing of all samples is then normalized to an isotropic resolution of [0.8, 0.8, 0.8] mm.

Finally, when conducting the mixed training, we used a fixed data split in this paper (*i.e.*, 62 samples for training and 16 for validation in total). Note that we followed the setting of the public challenge [4], which selects the new validation set from MICCAI-21 that does not include samples without any new MS lesions.

3 Method

3.1 Overview

Figure 2 illustrates the overall pipeline of our proposed CoactSeg model F_θ. We construct a quadruple set (X_b, X_{fu}, X_d, Y) for the model training. Here, the longitudinal difference map $x_d \in X_d$ is obtained by a subtraction operation between the baseline brain $x_b \in X_b$ and its follow-up $x_{fu} \in X_{fu}$ (*i.e.*, $x_d = x_{fu} - x_b$). Therefore, given heterogeneous annotations, *i.e.*, all-lesion labels $y^s_{al} \in Y^s_{al}$ in single-time-point data and new-lesion labels $y^t_{nl} \in Y^t_{nl}$ in two-time-point data, the CoactSeg model F_θ is designed to exploit both for the model training.

3.2 Multi-head Architecture

Figure 2 shows that new-lesion regions are highlighted in the brain difference map x_d. Hence, besides x_b and x_{fu}, CoactSeg also receives x_d as inputs. It generates

[1] https://portal.fli-iam.irisa.fr/msseg-2/.

[2] https://www.nitrc.org/projects/mricron/.

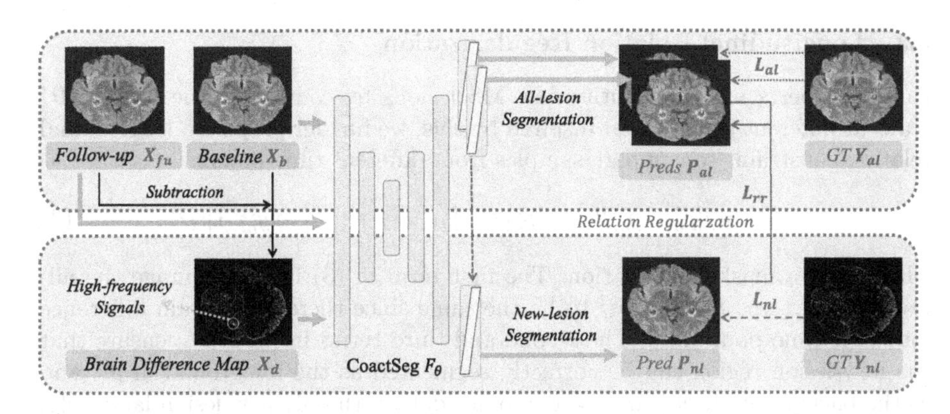

Fig. 2. Pipeline of our proposed CoactSeg model F_θ, which receives the baseline X_b, follow-up X_{fu}, and corresponding longitudinal brain differences X_d as inputs. F_θ segments all lesions P_{al} and predicts new lesions P_{nl} in the condition of X_b, X_{fu} and X_d. Note that new-lesion regions would have higher intensities in X_d.

all-lesion and new-lesion predictions as

$$\begin{aligned} p_{al}^{s1},\ p_{al}^{s2},\ p_{nl}^{s} &= F_\theta(x_b^s,\ x_{fu}^s,\ x_d^0) \\ p_{al}^{t1},\ p_{al}^{t2},\ p_{nl}^{t} &= F_\theta(x_b^t,\ x_{fu}^t,\ x_d^t). \end{aligned} \quad (1)$$

For single-time-point samples $x^s \in X^s$, x_b^s and x_{fu}^s are identical as x^s, and the difference map becomes an all-zero matrix x_d^0, with p_{al}^{s1}, p_{al}^{s2} and p_{nl}^s being the corresponding all-lesion and new-lesion predictions of x^s. For two-time-point data $x^t \in X^t$, x_b^t and x_{fu}^t respectively denote the first and second time-point data samples, with p_{al}^{t1}, p_{al}^{t2} and p_{nl}^t being the all-lesion segmentation results at the first and second time-point and the new-lesion results of x^t, respectively.

In this way, we unify the learning of both single and two-time-point data with heterogeneous annotations by using the same model F_θ, with the same input and output formats. Note that, inspired by semi-supervised learning [22–24], we mix x^s and x^t samples into each batch for training. Given the heterogeneous annotations, *i.e.*, all-lesion labels for single-time-point data and new-lesion labels for two-time-point data, we apply the following corresponding supervisions:

$$\begin{aligned} L_{al} &= Dice(p_{al}^{s1},\ y_{al}^s) + Dice(p_{al}^{s2},\ y_{al}^s) \\ L_{nl} &= Dice(p_{nl}^t,\ y_{nl}^t) \end{aligned} \quad (2)$$

where *Dice* refers to the common Dice loss for medical segmentation tasks. We use a 3D VNet [15] as the backbone of F_θ and three prediction heads are designed as individual convolutional blocks. Note that, the last prediction head also receives the features from the first two in order to capture the all-lesion information. Compared to the recent work [30] for exploiting heterogeneous data, our architecture avoids the complicated design of dynamic prediction heads.

3.3 Longitudinal Relation Regularization

Human experts usually identify new MS lesions by comparing the brain MRI scans at different time points. Inspired by this, we further propose a longitudinal relation constraint to compare samples from different time points:

$$L_{rr} = ||p_{al}^{s1},\ p_{al}^{s2}||_2 + ||p_{al}^{t1} \otimes y_{nl}^{t},\ 0||_2 + ||p_{al}^{t2} \otimes y_{nl}^{t}, 1||_2 \tag{3}$$

where $\otimes$ is a masking operation. The first term in (3) is to encourage the all-lesion predictions p_{al}^{s1} and p_{al}^{s2} to be the same since there is no brain difference for single-time-point data. The second and third terms in (3) are to ensure that the new-lesion region can be correctly segmented as the foreground in p_{al}^{t2} and as the background in p_{al}^{t1} in two-time-point data with only new lesion labels y_{nl}^{t}.

Finally, the overall loss function to train our CoactSeg model becomes a weighted sum of L_{al}, L_{nl}, and the regularization L_{rr}:

$$L = L_{al} + \lambda_1 \times L_{nl} + \lambda_2 \times L_{rr} \tag{4}$$

where λ_1 and λ_2 are constants to balance different tasks.

4 Results

Implementation Details. For training, we normalize all inputs as zero mean and unit variance. Then, among common augmentation operations, we use the random flip or rotation to perturb inputs. Since MS lesions are always small, we apply a weighted cropping strategy to extract 3D patches of size $80 \times 80 \times 80$ to relieve the class imbalance problem [26]. Specifically, if the input sample contains the foreground, we randomly select one of the foreground voxels as the patch center and shift the patch via a maximum margin of [-10, 10] voxels. Otherwise, we randomly crop 3D patches. The batch size is set as eight (*i.e.*, four new-lesion two-time-point samples and four all-lesion single-time-point samples). We apply Adam optimizer with a learning rate of 1e-2. The overall training iterations are 20k. In the first 10k iterations, λ_1 and λ_2 are set to 1 and 0, respectively, in order to train the model for segmenting MS lesions at the early training stage. After that, we set λ_2 as 1 to apply the relation regularization. During testing, we extract the overlapped patches by a stride of $20 \times 20 \times 20$ and then re-compose them into the entire results.

Note that we follow [18] to mask the non-brain regions and all experiments are only conducted in the brain regions with the same environment (Hardware: Single NVIDIA Tesla V100 GPU; Software: PyTorch 1.8.0, Python 3.8.10; Random Seed: 1337). The computational complexity of our model is 42.34 GMACs, and the number of parameters is 9.48 M.

Performance for MS Lesion Segmentation. Two MS tasks (*i.e.*, new-lesion segmentation on MICCAI-21 and all-lesion segmentation on our MS-23v1

Table 2. Comparisons of new-lesion segmentation on MICCAI-21. Note that the human experts' performance is shown based on their individually annotated results.

Method	Performance on MICCAI-21 (New MS Lesions)				
	Dice (%) ↑	Jaccard (%) ↑	95HD (voxel) ↓	ASD (voxel) ↓	F1 (%) ↑
SNAC [14]	53.07	39.19	66.57	26.39	30.71
SNAC (VNet) [14]	56.81	42.85	26.58	12.49	57.59
Neuropoly [12]	56.33	43.23	54.95	24.16	17.47
CoactSeg (Ours)	63.82	51.68	30.35	12.14	61.96
Human Expert #1	77.52	65.76	27.83	5.47	82.34
Human Expert #2	66.89	58.11	N/A	N/A	68.19
Human Expert #3	58.56	46.51	60.99	12.41	62.88
Human Expert #4	60.68	49.95	N/A	N/A	66.58

dataset) are used to evaluate the proposed CoactSeg. Besides common segmentation metrics [13] including Dice, Jaccard, 95% Hausdorff Distance (95HD), and Average Surface Distance (ASD), we further follow [3] to use the instance-level F1 score (F1) to denote the lesion-wise segmentation performance. Here, tiny lesions (*i.e.*, fewer than 11 voxels) are not included in the F1 calculation as [3].

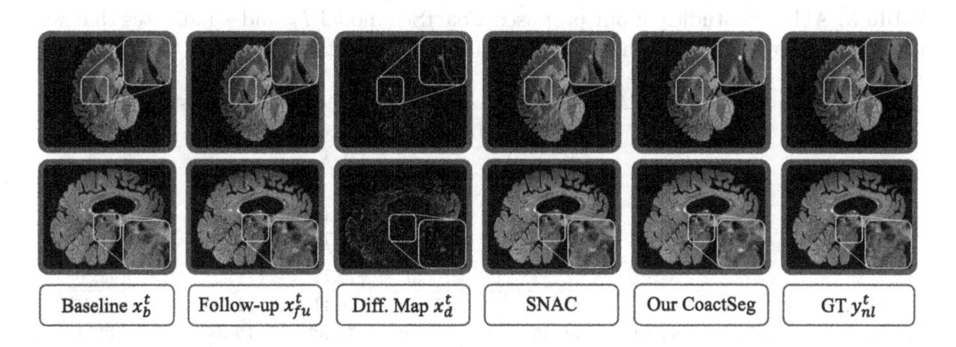

Fig. 3. Exemplar results for new MS lesion segmentation on the MICCAI-21 dataset.

Figure 3 illustrates that our proposed CoactSeg accurately segments the tiny new lesions on MICCAI-21. Compared to the recent work [14], our model can even predict new lesions with low contrast (indicated by the enlarged yellow rectangles in Fig. 3). Table 2 gives the quantitative results on MICCAI-21. We can see that: 1) Our model achieves good segmentation performance for new MS lesion segmentation and outperforms the second-best method [14] by 7.01% in Dice; 2) Compared with human experts, our proposed model also outperforms two of them (*i.e.*, #3 and #4) in terms of the segmentation and the shape-related metrics; 3) For the lesion-wise F1 score, our method significantly reduces the performance gap between deep models and human experts, achieving a comparable F1 with expert #3 (*i.e.*, 61.96% vs. 62.88%).

Figure 4 shows the all-lesion segmentation results of our CoactSeg model on our in-house MS-23v1 dataset. It can be seen that CoactSeg is able to segment

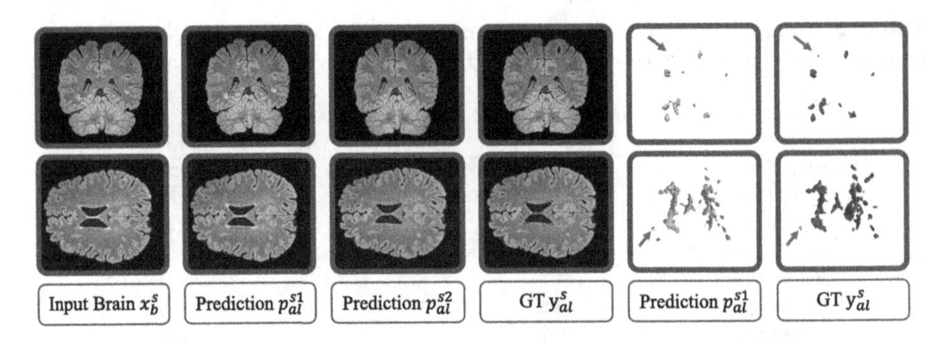

Fig. 4. Exemplar results for all MS lesion segmentation obtained by our CoactSeg model on our in-house MS-23v1 dataset (2D View: Left; 3D View: Right).

most MS lesions, even for very tiny ones (highlighted by red arrows). Moreover, we can see that the segmentation results of the first two prediction heads are relatively consistent (*i.e.*, the 2nd and 3rd columns of Fig. 4), demonstrating the effectiveness of our proposed relation regularization.

Table 3. Ablation studies of our proposed CoactSeg model F_θ and $*$ indicates that we apply a stage-by-stage training strategy in the experiments.

L_{rr}	Training Data		MICCAI-21 (New MS Lesions)			MS-23v1 (All MS Lesions)		
	MICCAI-21	MS-23v1	Dice (%) ↑	95HD (voxel) ↓	F1 (%) ↑	Dice (%) ↑	95HD (voxel) ↓	F1 (%) ↑
w/o	✓		59.91	35.73	45.61	N/A		
w/o		✓	N/A			70.94	14.46	**44.82**
w/o	✓	✓	61.53	43.05	51.54	69.41	18.93	34.43
w/	✓		58.49	50.17	51.35	N/A		
w/		✓	N/A			71.28	12.45	42.96
w/	✓	✓	62.15	43.26	56.97	70.44	12.88	44.04
w/	✓*	✓*	**63.82**	**30.35**	**61.96**	**72.32**	**12.38**	42.51

Ablation Study. Table 3 further shows the ablation study for both new and all MS lesion segmentation tasks. It reveals that: 1) Introducing the heterogeneous data significantly improves the performance of new-lesion segmentation on MICCAI-21 with an average Dice gain of 2.64%; 2) Exploiting the relation regularization for mixed training can further improve the performance on the two datasets; 3) The simple stage-by-stage training strategy (See the *Implementation Details* 4) can better balance two tasks and achieve the overall best segmentation performance for both tasks.

5 Conclusion

In this paper, we have presented a unified model CoactSeg for new MS lesion segmentation, which can predict new MS lesions according to the two-time-point inputs and their differences while at the same time segmenting all MS lesions. Our model effectively exploits heterogeneous data for training via a multi-head architecture and a relation regularization. Experimental results demonstrated that introducing all-lesion single-time-point data can significantly improve the new-lesion segmentation performance. Moreover, the relation constraint also facilitates the model to capture the longitudinal MS changes, leading to a further performance gain. Our in-house MS-23v1 dataset will be made public to help the MS lesion research. Future works will explore more longitudinal relations to study the fine-grained MS changes as well as consider more powerful constraints to address the domain gap [21] and fairness [29] problems. Moreover, we plan to collect and annotate more MS lesion data to improve the possibility of training large-scale deep models for clinical applications [17].

Acknowledgement. This work was supported in part by the Monash FIT Start-up Grant, in part by the Novartis (ID: 76765455), and in part by the Monash Institute of Medical Engineering (MIME) Project: 2022-13. We here appreciate the public repositories of SNAC [14] and Neuropoly [12], and also thanks for the efforts to collect and share the MS dataset [2] and the MS-23v1 dataset from Alfred Health, Australia.

References

1. Carass, A., et al.: Longitudinal multiple sclerosis lesion segmentation: resource and challenge. NeuroImage **148**, 77–102 (2017)
2. Commowick, O., Cervenansky, F., Cotton, F., Dojat, M.: Msseg-2 challenge proceedings: multiple sclerosis new lesions segmentation challenge using a data management and processing infrastructure. In: MICCAI 2021, p. 126 (2021)
3. Commowick, O., et al.: Objective evaluation of multiple sclerosis lesion segmentation using a data management and processing infrastructure. Sci. Rep. **8**(1), 13650 (2018)
4. Commowick, O., et al.: Multiple sclerosis lesions segmentation from multiple experts: the MICCAI 2016 challenge dataset. Neuroimage **244**, 118589 (2021)
5. Gessert, N., et al.: 4d deep learning for multiple-sclerosis lesion activity segmentation. In: MIDL 2020 (2020)
6. Gessert, N., et al.: Multiple sclerosis lesion activity segmentation with attention-guided two-path CNNs. Computer. Med. Imaging Graph. **84**, 101772 (2020)
7. Gold, R., et al.: Placebo-controlled phase 3 study of oral bg-12 for relapsing multiple sclerosis. N. Engl. J. Med. **367**(12), 1098–1107 (2012)
8. He, T., et al.: MS or not MS: T2-weighted imaging (t2wi)-based radiomic findings distinguish MS from its mimics. Multip. Sclerosis Relat. Disord. **61**, 103756 (2022)
9. Krüger, J., et al.: Fully automated longitudinal segmentation of new or enlarged multiple sclerosis lesions using 3d convolutional neural networks. NeuroImage: Clin. **28**, 102445 (2020)
10. La Rosa, F., et al.: Multiple sclerosis cortical and WM lesion segmentation at 3t MRI: a deep learning method based on flair and mp2rage. NeuroImage: Clin. **27**, 102335 (2020)
11. Ma, Y., et al.: Multiple sclerosis lesion analysis in brain magnetic resonance images: techniques and clinical applications. IEEE J. Biomed. Health Inf. **26**(6), 2680–2692 (2022)

12. Macar, U., Karthik, E.N., Gros, C., Lemay, A., Cohen-Adad, J.: Team neuropoly: description of the pipelines for the MICCAI 2021 MS new lesions segmentation challenge. arXiv preprint arXiv:2109.05409 (2021)
13. Maier-Hein, L., et al.: Metrics reloaded: pitfalls and recommendations for image analysis validation. arXiv preprint arXiv:2206.01653 (2022)
14. Mariano, C., Yuling, L., Kain, K., Linda, L., Chenyu, W., Michael, B.: Estimating lesion activity through feature similarity: a dual path UNET approach for the msseg2 MICCAI challenge. https://github.com/marianocabezas/msseg2
15. Milletari, F., Navab, N., Ahmadi, S.A.: V-net: fully convolutional neural networks for volumetric medical image segmentation. In: 3DV 2016, pp. 565–571 (2016)
16. Rakić, M., et al.: icobrain MS 5.1: combining unsupervised and supervised approaches for improving the detection of multiple sclerosis lesions. NeuroImage: Clin. **31**, 102707 (2021)
17. Reuter, M., Schmansky, N.J., Rosas, H.D., Fischl, B.: Within-subject template estimation for unbiased longitudinal image analysis. Neuroimage **61**(4), 1402–1418 (2012)
18. Schell, M., et al.: Automated brain extraction of multi-sequence MRI using artificial neural networks. In: Human Brain Mapping, pp. 1–13 (2019)
19. Sharmin, S., et al.: Confirmed disability progression as a marker of permanent disability in multiple sclerosis. Eur. J. Neurol. **29**(8), 2321–2334 (2022)
20. Tang, Z., Cabezas, M., Liu, D., Barnett, M., Cai, W., Wang, C.: LG-net: lesion gate network for multiple sclerosis lesion inpainting. In: de Bruijne, M. et al. (eds.) MICCAI 2021, pp. 660–669. Springer, Cham (2021). https://doi.org/10.1007/978-3-030-87234-2_62
21. Wolleb, J., et al.: Learn to ignore: domain adaptation for multi-site MRI analysis. In: Wang, L., Dou, Q., Fletcher, P.T., Speidel, S., Li, S. (eds.) MICCAI 2022, pp. 725–735. Springer, Cham (2022). https://doi.org/10.1007/978-3-031-16449-1_69
22. Wu, Y., Ge, Z., Zhang, D., Xu, M., Zhang, L., Xia, Y., Cai, J.: Mutual consistency learning for semi-supervised medical image segmentation. Med. Image Anal. **81**, 102530 (2022)
23. Wu, Y., Wu, Z., Wu, Q., Ge, Z., Cai, J.: Exploring smoothness and class-separation for semi-supervised medical image segmentation. In: Wang, L., Dou, Q., Fletcher, P.T., Speidel, S., Li, S. (eds.) MICCAI 2022, vol. 13435, pp. 34–43. Springer, Cham (2022). https://doi.org/10.1007/978-3-031-16443-9_4
24. Wu, Y., Xu, M., Ge, Z., Cai, J., Zhang, L.: Semi-supervised left atrium segmentation with mutual consistency training. In: de Bruijne, M., et al. (eds.) MICCAI 2021, pp. 297–306. Springer, Cham (2021). https://doi.org/10.1007/978-3-030-87196-3_28
25. Zeng, C., Gu, L., Liu, Z., Zhao, S.: Review of deep learning approaches for the segmentation of multiple sclerosis lesions on brain MRI. Front. Neuroinform. **14**, 610967 (2020)
26. Zhang, H., et al.: Qsmrim-net: imbalance-aware learning for identification of chronic active multiple sclerosis lesions on quantitative susceptibility maps. NeuroImage: Clin. **34**, 102979 (2022)
27. Zhang, H., Wang, R., Zhang, J., Liu, D., Li, C., Li, J.: Spatially covariant lesion segmentation. arXiv preprint arXiv:2301.07895 (2023)
28. Zhang, H., et al.: All-net: anatomical information lesion-wise loss function integrated into neural network for multiple sclerosis lesion segmentation. NeuroImage: Clin. **32**, 102854 (2021)

29. Zhang, H., Yuan, X., Nguyen, Q.V.H., Pan, S.: On the interaction between node fairness and edge privacy in graph neural networks. arXiv preprint arXiv:2301.12951 (2023)
30. Zhang, J., Xie, Y., Xia, Y., Shen, C.: Dodnet: learning to segment multi-organ and tumors from multiple partially labeled datasets. In: CVPR 2021, pp. 1195–1204 (2021)

Generating Realistic Brain MRIs via a Conditional Diffusion Probabilistic Model

Wei Peng[1], Ehsan Adeli[1], Tomas Bosschieter[1], Sang Hyun Park[2], Qingyu Zhao[1], and Kilian M. Pohl[1](✉)

[1] Stanford University, Stanford, CA 94305, USA
kpohl@stanford.edu
[2] Daegu Gyeongbuk Institute of Science and Technology, Daegu, South Korea

Abstract. As acquiring MRIs is expensive, neuroscience studies struggle to attain a sufficient number of them for properly training deep learning models. This challenge could be reduced by MRI synthesis, for which Generative Adversarial Networks (GANs) are popular. GANs, however, are commonly unstable and struggle with creating diverse and high-quality data. A more stable alternative is Diffusion Probabilistic Models (DPMs) with a fine-grained training strategy. To overcome their need for extensive computational resources, we propose a conditional DPM (cDPM) with a memory-efficient process that generates realistic-looking brain MRIs. To this end, we train a 2D cDPM to generate an MRI subvolume conditioned on another subset of slices from the same MRI. By generating slices using arbitrary combinations between condition and target slices, the model only requires limited computational resources to learn interdependencies between slices even if they are spatially far apart. After having learned these dependencies via an attention network, a new anatomy-consistent 3D brain MRI is generated by repeatedly applying the cDPM. Our experiments demonstrate that our method can generate high-quality 3D MRIs that share a similar distribution to real MRIs while still diversifying the training set. The code is available at https://github.com/xiaoiker/mask3DMRI_diffusion and also will be released as part of MONAI, at https://github.com/Project-MONAI/GenerativeModels.

1 Introduction

The synthesis of medical images has great potential in aiding tasks like improving image quality, imputing missing modalities [30], performing counterfactual analysis [17], and modeling disease progression [9,10,29]. However, synthesizing brain MRIs is non-trivial as they are of high dimension, yet the training data are relatively small in size (compared to 2D natural images). High-quality synthetic MRIs have been produced by conditional models based on real MRI of the same subject acquired with different MRI sequences [2,16,21,27]. However, such models require large data sets (which are difficult to get) and fail to significantly improve data diversity [11,26], i.e., producing MRIs substantially

H. Greenspan et al. (Eds.): MICCAI 2023, LNCS 14227, pp. 14–24, 2023.
https://doi.org/10.1007/978-3-031-43993-3_2

deviating from those in the training data; data diversity is essential to the generalizability of large-scale models [26]. Unconditional models based on Generative Adversarial Networks (GANs) bypass this drawback by generating new, independent MRIs from random noise [1,7]. However, these models often produce lower quality MRIs as they currently can only be trained on lower resolution MRIs or 2D slices due to their computational needs [6,12]. Furthermore, GAN-based models are known to be unstable during training and even can suffer from mode collapse [4]. An alternative is diffusion probabilistic models (DPMs) [8,22], which formulate the fine-grained mapping between data distribution and Gaussian noise as a gradual process modeled within a Markov chain. Due to their multi-step, fine-grained training strategy, DPMs tend to be more stable during training than GANs and therefore are more accurate for certain medical imaging applications, such as segmentation and anomaly detection [13,25]. However, DPMs tend to be computationally too expensive to synthesize brain MRI at full image resolution [3,18]. We address this issue by proposing a memory-efficient 2D conditional DPM (cDPM) that relies on learning the interdependencies between 2D slices to produce high-quality 3D MRI volumes.

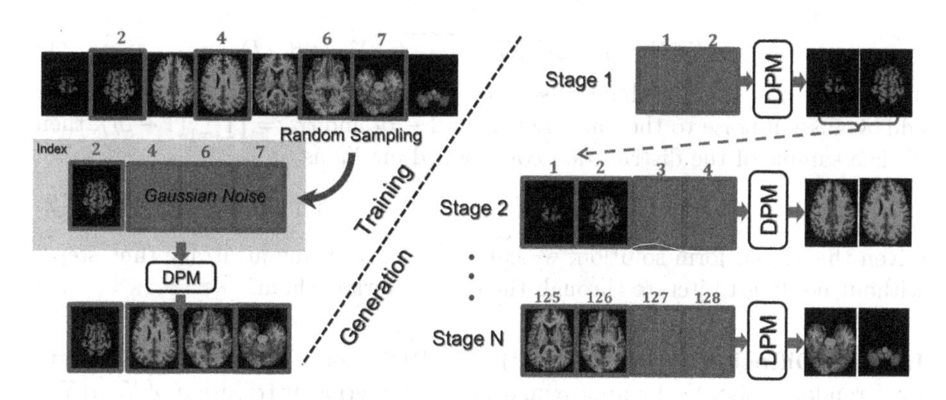

Fig. 1. A memory efficient DPM. **Left:** Based on 'condition' slices, cDPM learns to generate 'target' slices. **Right:** A new 3D MRI is created by repeatedly running the trained model to synthesize target slices conditioned on those it created in prior stages.

Unlike the sequence of 2D images defining a video, all 2D slices of an MRI are interconnected with each other as they define a 3D volume capturing brain anatomy. Our cDPM learns these interdependencies (even between distant slices) by training an attention network [24] on arbitrary combinations of condition and target slices. Once learned, the cDPM creates new samples while capturing brain anatomy in 3D. It does so by producing the first few slices from random noise and then using those slices to synthesize subsequent ones (see Fig. 1). We show that this computationally efficient conditional DPM can produce MRIs that are more realistic than those produced by GAN-based architectures. Furthermore, our experiments reveal that cDPM is able to generate synthetic MRIs, whose distribution matches that of the training data.

2 Methodology

We first review the basic DPM framework for data generation (Sect. 2.1). Then, we introduce our efficient strategy for generating 3D MRI slices (Sect. 2.2) and finally describe the neural architecture of cDPMs (Sect. 2.3).

2.1 Diffusion Probabilistic Model

The Diffusion Probabilistic Model (DPM) [8,22] generates MRIs from random noise by iterating between mapping 1) data gradually to noise (a.k.a., Forward Diffusion Process) and 2) noise back to data (a.k.a., Reverse Diffusion Process).

Forward Diffusion Process (FDP). Let real data $\mathbf{X_0} \sim \mathbf{q}$ sampled from the (real data) distribution q be the input to the FDP. FDP then simulates the diffusion process that turns X_0 after T perturbations into Gaussian noise $X_T \sim \mathcal{N}(0, I)$, where $\mathcal{N}$ is the Gaussian distribution with zero mean and the variance being the identity matrix I. This process is formulated as a Markov chain, whose transition kernel $q(X_t|X_{t-1})$ at time step $t \in \{0, \ldots, T\}$ is defined as

$$q(X_t|X_{t-1}) := \mathcal{N}(X_t; \sqrt{1-\beta_t} \cdot X_{t-1}, \beta_t \cdot I). \quad (1)$$

The weight $\beta_t \in (0, 1)$ is changed so that the chain gradually enforces drift, i.e., adds Gaussian noise to the data. Let $\alpha_t := 1 - \beta_t$ and $\bar{\alpha}_t := \prod_{s=1}^{t}(1 - \beta_t)$, then X_t is a sample of the distribution conditioned on X_0 as

$$q(X_t|X_0) := \mathcal{N}(X_t; \sqrt{\bar{\alpha}_t} \cdot X_0, (1 - \bar{\alpha}_t) \cdot I). \quad (2)$$

Given this closed-form solution, we can sample X_t at any arbitrary time step t without needing to iterate through the entire Markov chain.

Reverse Diffusion Process (RDP). The RDP aims to generate realistic data from random noise X_T by approximating the posterior distribution $p(X_{t-1}|X_t)$. It does so by going through the entire Markov chain from time step T to 0, i.e.,

$$p(X_{0:T}) := p(X_T) \prod_{t=1}^{T} p_\theta(X_{t-1}|X_t). \quad (3)$$

Defining the conditional distribution $p_\theta(X_{t-1}|X_t) := \mathcal{N}(X_{t-1}; \mu_\theta(X_t, t), \Sigma)$ with fixed variance Σ, then (according to [8]) the mean can be rewritten as

$$\mu_\theta(X_t, t) = \frac{1}{\sqrt{\alpha_t}} \left(X_t - \frac{\beta_t}{\sqrt{(1 - \bar{\alpha}_t)}} \epsilon_\theta(X_t, t) \right), \quad (4)$$

with $\epsilon_\theta(\cdot)$ being the estimate of a neural network defined by parameters θ. θ minimizes the reconstructing loss defined by the following expected value

$$\mathbb{E}_{\mathbf{X_0} \sim \mathbf{q}, \mathbf{t} \in [\mathbf{0}, \ldots, \mathbf{T}], \epsilon \sim \mathcal{N}(\mathbf{0}, \mathbf{I})} \left[||\epsilon - \epsilon_\theta(X_t, t)||_2^2 \right],$$

where $|| \cdot ||_2$ is the L2 norm, and X_t is inferred from Eq. (2) based on X_0.

2.2 Conditional Generation with DPM (cDPM)

To synthetically create high-resolution 3D MRI, we propose an efficient cDPM model that learns the interdependencies between 2D slices of an MRI so that it can generate slices based on another set of already synthesized ones (see Fig. 1).

Specifically, given an MRI $X \in \mathbb{R}^{D\times H\times W}$, we randomly sample two sets of slice indexes: the condition set $\mathcal{C}$ and the target set $\mathcal{P}$. Let len($\cdot$) be the number of slices in a set, then the 'condition' slices are defined as $X^{\mathcal{C}} \in \mathbb{R}^{\text{len}(\mathcal{C})\times H\times W}$ and the 'target' slices as $X^{\mathcal{P}} \in \mathbb{R}^{\text{len}(\mathcal{P})\times H\times W}$ with $\text{len}(\mathcal{P}) \geq 1$. Confining the FDP of Sect. 2.1 just to the target $X^{\mathcal{P}}$, the RDP now aims to reconstruct $X_t^{\mathcal{P}}$ for each time $t = T, T-1, \ldots, 0$ starting from random noise at $t = T$ and conditioned on $X^{\mathcal{C}}$. Let $\widetilde{X}_t$ be the subvolume consisting of $X_t^{\mathcal{P}}$ and $X^{\mathcal{C}}$, then the joint distribution of the Markov chain defined by Eq. (3) now reads

$$p(X_{0:T}^{\mathcal{P}}) := p(X_T^{\mathcal{P}}) \prod_{t=1}^{T} p_\theta(X_{t-1}^{\mathcal{P}}|\widetilde{X}_t). \tag{5}$$

Observe that Eq. (5) is equal to Eq. (3) in case $\text{len}(\mathcal{C}) = 0$.

To estimate $\mu_\theta(\widetilde{X}_t, t)$ as described in Eq. (4), we sample arbitrary index sets $\mathcal{C}$ and $\mathcal{P}$ so that $\text{len}(\mathcal{C}) + \text{len}(\mathcal{P}) \leq \tau_{\text{max}}$, where τ_{max} is the maximum number of slices based on the available resources. We then capture the dependencies across slices by feeding the index sets $\mathcal{C}$ and $\mathcal{P}$ and the corresponding slices (i.e., $X^{\mathcal{C}}$ and $X_t^{\mathcal{P}}$ built from $X_0 \sim q$) into an attention network [20]. The neural network aims to minimize the canonical loss function

$$\text{Loss}(\theta) := \mathbb{E}_{\mathbf{X_0}\sim\mathbf{q},\epsilon\sim\mathcal{N}(\mathbf{0},\mathbf{I}),\mathcal{C}+\mathcal{P}\leq\tau_{\text{max}},\mathbf{t}} \left[||\epsilon - \epsilon_\theta(X_t^{\mathcal{P}}, X^{\mathcal{C}}, \mathcal{C}, \mathcal{P}, t)||_2^2\right]. \tag{6}$$

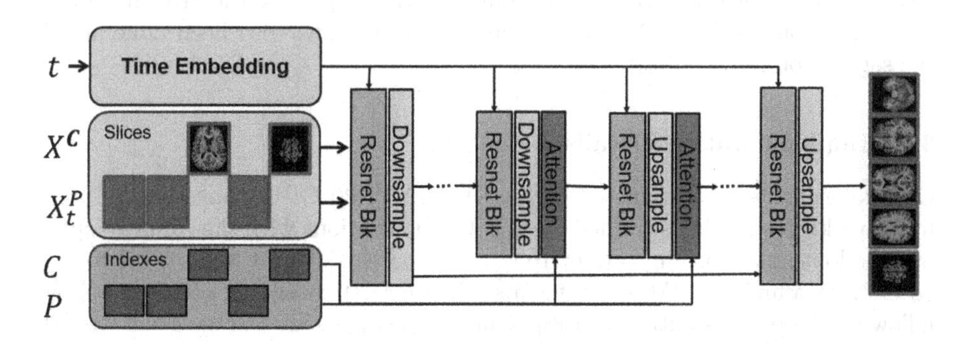

Fig. 2. The architecture of cDPM is a U-shape neural network with skip connections and the input at step 't' are slice indexes $\{\mathcal{C}, \mathcal{P}\}$, condition sub-volume $X^{\mathcal{C}}$, and current target sub-volume $X_t^{\mathcal{P}}$.

As the neural network can now be trained on many different (arbitrary) slice combinations (defined by $\mathcal{C}$ and $\mathcal{P}$), the cDPM only requires a relatively small number of MRIs for training. Furthermore, it will learn short- and long-range dependencies across slices as the spatial distance between slices from $\mathcal{C}$ and $\mathcal{P}$ varies. Learning these dependencies (after being trained for a sufficiently large number of iterations) enables cDPMs to produce 2D slices that, when put together, result in realistic looking, high-resolution 3D MRIs.

2.3 Network Architecture

As done by [8], cDPMs are implemented as a U-Net [19] with a time embedding module (see Fig. 2). We add a multi-head self-attention mechanism [24] to model the relationship between slices. After training the cDPM as in Fig. 1, a 3D MRI is generated in N stages. Specifically, the cDPM produces the initial set of slices of that MRI volume from random noise (i.e., unconditioned). Conditioned on those synthetic slices, the cDPM then runs again to produce a new set of slices. The process of synthetically creating slices based on ones generated during prior stages is repeated until an entire 3D MRI is produced.

3 Experiments

3.1 Data

We use 1262 t1-weighted brain MRIs of subjects from three different datasets: the Alzheimer's Disease Neuroimaging Initiative (ADNI-1), UCSF (PI: V. Valcour), and SRI International (PI: E.V. Sullivan and A. Pfefferbaum) [28]. Processing includes denoising, bias field correction, skull stripping, and affine registration to a template, and normalizing intensity values between 0 and 1. In addition, we padded and resized the MRIs to have dimensions $128 \times 128 \times 128$ resulting in a voxel resolution of $1.375\,mm \times 1.375\,mm \times 1.0\,mm$. Splitting the MRI along the axial direction results in 2D slices. Note, this could have also been done along the sagittal or coronal direction.

3.2 Implementation Details

Our experiments are conducted on an NVIDIA A100 GPU using the PyTorch framework. The model is trained using 200,000 iterations with the AdamW optimizer adopting a learning rate of 10^{-4} and a batch size of 3. $\tau_{\max}$ is set to 20. After the training, cDPM generates a synthetic MRI consisting of 128 slices by following the process outlined in Fig. 1 in N=13 stages. Each stage generates 10 slices starting with pure noise ($X^{\mathcal{C}} = \emptyset$) and (after the first stage) being conditioned on the 10 slices produced by the prior stage. After training on all real MRIs, we use the resulting conditional DPM to generate 500 synthetic MRIs.

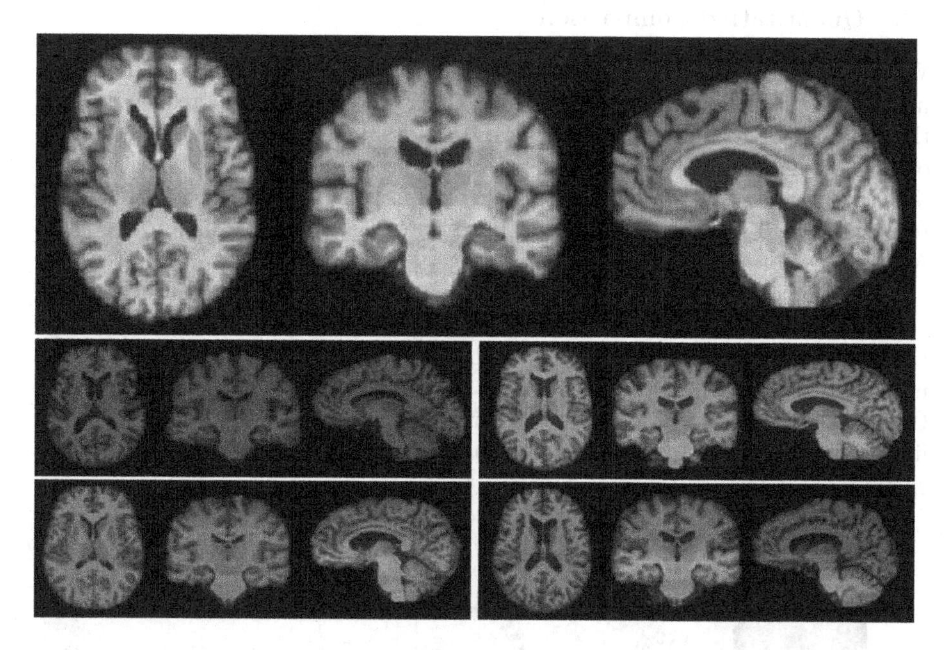

Fig. 3. 5 MRIs generated by our conditional DPM visualized in the axial, coronal, and sagittal plane. The example in the first row is enlarged to highlight the high quality of synthetic MRIs generated by our approach.

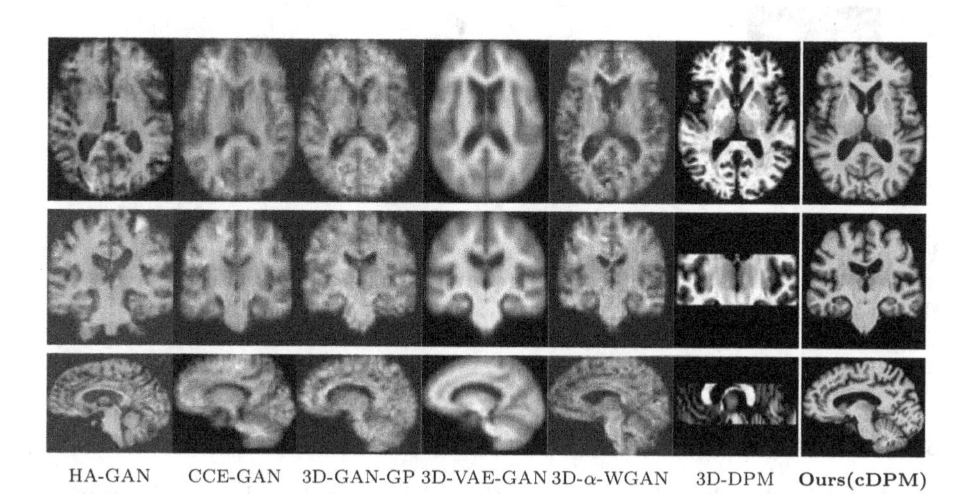

Fig. 4. 3 views of MRIs generated by 7 models. Compared to the MRIs produced by the other approaches, our cDPM model generates the most realistic MRI scans that provide more distinct gray matter boundaries and greater anatomical details. (Color figure online)

3.3 Quantitative Comparison

We evaluate the quality of synthetic MRIs based on 3 metrics: (i) computing the distance between synthetic and 500 randomly selected real MRIs via the Maximum-Mean Discrepancy (MMD) score [5], (ii) measuring the diversity of the synthetic MRIs via the pair-wise multi-scale Structure Similarity (MS-SSIM) [12], and (iii) comparing the distributions of synthetic to real MRIs with respect to the 3 views via the Frèchet Inception Distance (FID) [26] (a.k.a, FID-Axial, FID-Coronal, FID-Sagittal).

We compare those scores to ones produced by six recently published methods: (i) 3D-DPM [3], (ii) 3D-VAE-GAN [14], (iii) 3D-GAN-GP [6], (iv) 3D-α-WGAN [12], (v) CCE-GAN [26], and (vi) HA-GAN [23]. We needed to re-implement the first 5 methods and used the open-source code available for HA-GAN. 3D-DPM was only able to generate 32 slices at a time (due to GPU limitations) so that we computed its quality metrics by also cropping the corresponding real MRI to those 32 slices.

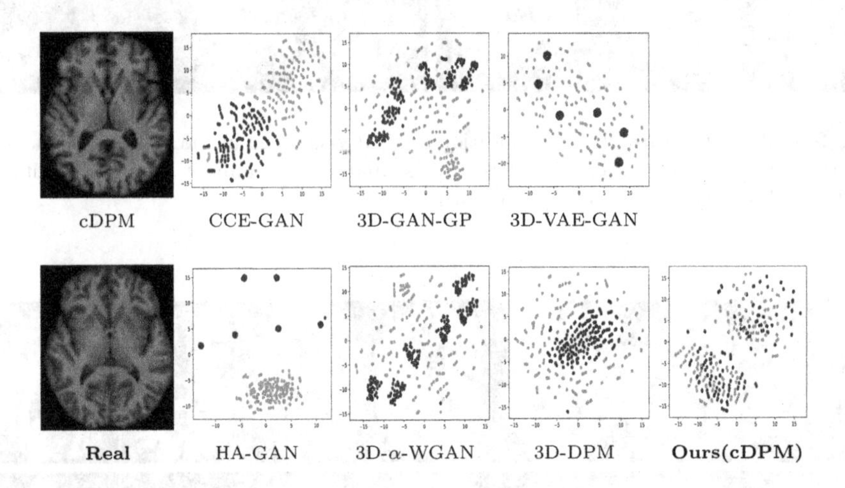

Fig. 5. Left: One MRI generated by our model and its closest real MRI based on MS-SSIM. **Right:** tSNE embedding of 200 generated samples (blue) of each model and their closest real MRIs (orange). Only our model generated independent and diverse samples as the data points overlay but are not identical to the training data. (Color figure online)

3.4 Results

Qualitative Results. The center of the axial, coronal, and sagittal views of five MRIs generated by cDPM shown in Fig. 3 look realistic. Compared to the MRIs produced by the other approaches other than 3D-DPM (see Fig. 4), the MRIs of cDPM are sharper; specifically, the gray matter boundaries are more distinct and the scan provides greater anatomical details. As expected, 3D-DPM produced synthetic slices of similar quality as cDPM but failed to do so for the entire MRI.

The synthetic MRIs of cDPM shown in Fig. 3 are also substantially different from each other, suggesting that our method could be used to create an augmented data set that is anatomically diverse. Figure 5 further substantiates the claim, which plots the t-SNE embedding [15] of 200 synthetic MRIs (blue) and their closest real counterpart (orange) according to MS-SSIM for each method. Note, matching of all 500 synthetic MRIs was computationally too expensive to perform (takes days to complete per method). Based on those plots, cDPM is the only approach able to generate MRIs, whose distribution resembled that of the real MRIs. This finding is somewhat surprising given that the MRI subvolumes generated by 3D-DPM looked real. Unlike the real data, however, their distributions are clustered around the average. Thus, 3D-DPM fails to diversify the data set even if (in the future) more computational resources would allow the method to generate a complete 3D MRI.

Table 1. Measuring the quality of 500 synthetic MRIs. '( )' contains the absolute difference to the MS-SSIM score of the real MRIs, which was 0.792. In bold are the optimal scores among methods that generate the entire volume. Scores denoted with an asterisk '*' are only computed on 32 slices.

	MS-SSIM (%)	MMD↓ (10^3)	FID-A ↓	FID-C ↓	FID-S ↓
3D-VAE-GAN [14]	88.3 (9.1)	5.15	320	247	398
3D-GAN-GP [6]	81.0 (1.8)	15.7	141	127	281
3D-α-WGAN [12]	82.6 (3.4)	13.2	121	116	193
CCE-GAN [26]	81.5 (2.3)	3.54	69.4	869	191
HA-GAN [23]	36.8 (42.4)	226	477	1090	554
3D-DPM [3]	79.7 (0.5)*	15.2*	188	–	–
Ours (cDPM)	**78.6 (0.6)**	**3.14**	**32.4**	**45.8**	**91.1**

Quantitative Results. Table 1 lists the average scores of MS-SSIM, MMD, and FID for each method. Among all models that generated complete MRI volumes, cDPM performed best. Only the absolute difference between the MS-SSIM score of 3D-DPM and the real MRIs was slightly lower (i.e., 0.005) than the absolute difference for cDPM (i.e., 0.006). This comparison, however, is not fair as the MS-SSIM score for 3D-DPM was only computed on 32 slices. Further supporting this argument is that FID-A (the only score computed for the same slice across all methods) was almost 5 times worse for 3D-DPM than cDPM.

4 Conclusion

We propose a novel conditional DPM (cDPM) for efficiently generating 3D brain MRIs. Starting with random noise, our model can progressively generate MRI slices based on previously generated slices. This conditional scheme enables training the cDPM with limited computational resources and training data. Qualitative and quantitative results demonstrate that the model is able to produce high-fidelity 3D MRIs and outperform popular and recent generative models such as the CCE-GAN and 3D-DPM. Our framework can easily be extended to other imaging modalities and can potentially assist in training deep learning models on a small number of samples.

Acknowledgement. This work was partly supported by funding from the National Institute of Health (MH113406, DA057567, AA021697, AA017347, AA010723, AA005965, and AA028840), the DGIST R&D program of the Ministry of Science and ICT of KOREA (22-KUJoint-02), Stanford School of Medicine Department of Psychiatry and Behavioral Sciences Faculty Development and Leadership Award, and by the Stanford HAI Google Cloud Credit.

References

1. Bermudez, C., Plassard, A.J., Davis, L.T., Newton, A.T., Resnick, S.M., Landman, B.A.: Learning implicit brain MRI manifolds with deep learning. In: Medical Imaging 2018: Image Processing, vol. 10574, pp. 408–414. SPIE (2018)
2. Dar, S.U., Yurt, M., Karacan, L., Erdem, A., Erdem, E., Çukur, T.: Image synthesis in multi-contrast MRI with conditional generative adversarial networks. IEEE Trans. Med. Imaging **38**(10), 2375–2388 (2019)
3. Dorjsembe, Z., Odonchimed, S., Xiao, F.: Three-dimensional medical image synthesis with denoising diffusion probabilistic models. In: Medical Imaging with Deep Learning (2022). https://openreview.net/forum?id=Oz7lKWVh45H
4. Goodfellow, I., et al.: Generative adversarial networks. Commun. ACM **63**(11), 139–144 (2020)
5. Gretton, A., Borgwardt, K.M., Rasch, M.J., Schölkopf, B., Smola, A.: A kernel two-sample test. J. Mach. Learn. Res. **13**(1), 723–773 (2012)
6. Gulrajani, I., Ahmed, F., Arjovsky, M., Dumoulin, V., Courville, A.C.: Improved training of wasserstein GANs. In: NIPS, vol. 30, pp. 5769–5779 (2017)
7. Han, C., et al.: GAN-based synthetic brain MR image generation. In: IEEE International Symposium on Biomedical Imaging, pp. 734–738 (2018)
8. Ho, J., Jain, A., Abbeel, P.: Denoising diffusion probabilistic models. Adv. Neural Inf. Process. Syst. **33**, 6840–6851 (2020)
9. Jung, E., Luna, M., Park, S.H.: Conditional GAN with an attention-based generator and a 3D discriminator for 3D medical image generation. In: de Bruijne, M., et al. (eds.) MICCAI 2021. LNCS, vol. 12906, pp. 318–328. Springer, Cham (2021). https://doi.org/10.1007/978-3-030-87231-1_31
10. Jung, E., Luna, M., Park, S.H.: Conditional GAN with 3D discriminator for MRI generation of Alzheimer's disease progression. Pattern Recogn. **133**, 109061 (2023)
11. Karras, T., Laine, S., Aila, T.: A style-based generator architecture for generative adversarial networks. In: Proceedings of the IEEE/CVF Conference on Computer Vision and Pattern Recognition, pp. 4401–4410 (2019)

12. Kwon, G., Han, C., Kim, D.: Generation of 3D brain MRI using auto-encoding generative adversarial networks. In: Shen, D., et al. (eds.) MICCAI 2019. LNCS, vol. 11766, pp. 118–126. Springer, Cham (2019). https://doi.org/10.1007/978-3-030-32248-9_14
13. La Barbera, G., et al.: Anatomically constrained CT image translation for heterogeneous blood vessel segmentation. In: British Machine Vision Virtual Conference, p. 776 (2022)
14. Larsen, A.B.L., Sønderby, S.K., Larochelle, H., Winther, O.: Autoencoding beyond pixels using a learned similarity metric. In: International Conference on Machine Learning, pp. 1558–1566. Proceedings of Machine Learning Research (2016)
15. Van der Maaten, L., Hinton, G.: Visualizing data using t-SNE. J. Mach. Learn. Res. **9**(11) (2008)
16. Ouyang, J., Adeli, E., Pohl, K.M., Zhao, Q., Zaharchuk, G.: Representation disentanglement for multi-modal brain MRI Analysis. In: Feragen, A., Sommer, S., Schnabel, J., Nielsen, M. (eds.) IPMI 2021. LNCS, vol. 12729, pp. 321–333. Springer, Cham (2021). https://doi.org/10.1007/978-3-030-78191-0_25
17. Pawlowski, N., Coelho de Castro, D., Glocker, B.: Deep structural causal models for tractable counterfactual inference. Adv. Neural Inf. Process. Syst. **33**, 857–869 (2020)
18. Pinaya, W.H., et al.: Brain imaging generation with latent diffusion models. In: Deep Generative Models: DGM4MICCAI 2022, pp. 117–126 (2022)
19. Ronneberger, O., Fischer, P., Brox, T.: U-Net: convolutional networks for biomedical image segmentation. In: International Conference on Medical Image Computing and Computer-Assisted Intervention, vol. 9351, pp. 234–241 (2015)
20. Shaw, P., Uszkoreit, J., Vaswani, A.: Self-attention with relative position representations. In: Proceedings of the 2018 Conference of the North American Chapter of the Association for Computational Linguistics: Human Language Technologies, vol. 2, pp. 464–468. Association for Computational Linguistics, New Orleans (2018). https://doi.org/10.18653/v1/N18-2074, https://aclanthology.org/N18-2074
21. Shin, H.C., et al.: Medical image synthesis for data augmentation and anonymization using generative adversarial networks. In: International Workshop on Simulation and Synthesis in Medical Imaging, vol. 11037 (2018)
22. Sohl-Dickstein, J., Weiss, E., Maheswaranathan, N., Ganguli, S.: Deep unsupervised learning using nonequilibrium thermodynamics. In: International Conference on Machine Learning, pp. 2256–2265. PMLR (2015)
23. Sun, L., Chen, J., Xu, Y., Gong, M., Yu, K., Batmanghelich, K.: Hierarchical amortized GAN for 3D high resolution medical image synthesis. IEEE J. Biomed. Health Inf. **26**(8), 3966–3975 (2022)
24. Vaswani, A., et al.: Attention is all you need. In: Advances in Neural Information Processing Systems, vol. 30 (2017)
25. Wolleb, J., Bieder, F., Sandkühler, R., Cattin, P.C.: Diffusion models for medical anomaly detection. In: International Conference on Medical Image Computing and Computer-Assisted Intervention, vol. 13438, pp. 35–45 (2022)
26. Xing, S., Sinha, H., Hwang, S.J.: Cycle consistent embedding of 3D brains with auto-encoding generative adversarial networks. In: Medical Imaging with Deep Learning (2021)
27. Yu, B., Zhou, L., Wang, L., Fripp, J., Bourgeat, P.: 3D cGAN based cross-modality MR image synthesis for brain tumor segmentation. In: IEEE 15th International Symposium on Biomedical Imaging, pp. 626–630 (2018)
28. Zhang, J., et al.: Multi-label, multi-domain learning identifies compounding effects of HIV and cognitive impairment. Med. Image Anal. **75**, 102246 (2022)

29. Zhao, Q., Liu, Z., Adeli, E., Pohl, K.M.: Longitudinal self-supervised learning. Med. Image Anal. **71**, 102051 (2021)
30. Zheng, S., Charoenphakdee, N.: Diffusion models for missing value imputation in tabular data. In: NeurIPS Table Representation Learning (TRL) Workshop (2022)

Towards Accurate Microstructure Estimation via 3D Hybrid Graph Transformer

Junqing Yang[1(✉)], Haotian Jiang[2(✉)], Tewodros Tassew[1], Peng Sun[1], Jiquan Ma[2(✉)], Yong Xia[1], Pew-Thian Yap[3,4], and Geng Chen[1(✉)]

[1] National Engineering Laboratory for Integrated Aero-Space-Ground-Ocean Big Data Application Technology, School of Computer Science and Engineering, Northwestern Polytechnical University, Xi'an, China
geng.chen@nwpu.edu.cn

[2] School of Computer Science and Technology, Heilongjiang University, Harbin, China
majiquan@hlju.edu.cn

[3] Department of Radiology, University of North Carolina, Chapel Hill, NC, USA

[4] Biomedical Research Imaging Center, University of North Carolina, Chapel Hill, NC, USA

Abstract. Deep learning has drawn increasing attention in microstructure estimation with undersampled diffusion MRI (dMRI) data. A representative method is the hybrid graph transformer (HGT), which achieves promising performance by integrating q-space graph learning and x-space transformer learning into a unified framework. However, this method overlooks the 3D spatial information as it relies on training with 2D slices. To address this limitation, we propose 3D hybrid graph transformer (3D-HGT), an advanced microstructure estimation model capable of making full use of 3D spatial information and angular information. To tackle the large computation burden associated with 3D x-space learning, we propose an efficient q-space learning model based on simplified graph neural networks. Furthermore, we propose a 3D x-space learning module based on the transformer. Extensive experiments on data from the human connectome project show that our 3D-HGT outperforms state-of-the-art methods, including HGT, in both quantitative and qualitative evaluations.

Keywords: Microstructure Imaging · Graph Neural Network · Transformer · 3D Spatial Domain

---

J. Yang and H. Jiang–Contributed equally to this work. This work was supported in part by the National Natural Science Foundation of China through Grants 62201465 and 62171377, and the Natural Science Foundation of Heilongjiang Province through Grant LH2021F046. P.-T. Yap was supported in part by the United States National Institutes of Health (NIH) through Grants MH125479 and EB008374.

H. Greenspan et al. (Eds.): MICCAI 2023, LNCS 14227, pp. 25–34, 2023.
https://doi.org/10.1007/978-3-031-43993-3_3

1 Introduction

Diffusion microstructure imaging has drawn increasing research attention in recent years. A number of powerful microstructure models have been proposed and shown great success in both clinical and research sides. Typical examples include diffusion kurtosis imaging (DKI) [15], neurite orientation dispersion and density imaging (NODDI) [23], and spherical mean technique (SMT) [16]. However, these models usually rely on diffusion MRI (dMRI) data densely sampled in q-space with a sufficient angular resolution and multiple shells, which are impractical in clinical settings.

To resolve this problem, deep learning has been introduced to learn the mapping between high-quality microstructure indices and q-space undersampled dMRI data. For instance, Golkov et al. [12], for the first time, introduced deep learning to microstructure estimation from undersampled dMRI data with a multilayer perceptron (MLP). Gibbons et al. [11] trained a 2D CNN to generate NODDI and fractional anisotropy parameter maps from undersampled dMRI data. Tian et al. [18] proposed DeepDTI to predict high-fidelity diffusion tensor imaging metrics using a 10-layer 3D CNN. Ye et al. designed MEDN [21] and MEDN+ [22] based on a dictionary-inspired framework to learn the mapping between undersampled dMRI data and high-quality microstructure indices. Zheng et al. [24] proposed a three-stage microstructure estimation model that combines sparse encoding with a transformer. Chen et al. [4] introduced the graph convolutional neural network (GCNN) to learn q-space information for microstructure estimation, which considers the angular information in q-space and achieves promising performance.

Recently, inspired by the fact that dMRI data live in a joint x-q space [3,14], a powerful hybrid model, called HGT [5], was proposed to jointly learn spatial-angular information for accurate microstructure estimation. It achieves superior performance in comparison with various types of existing methods. However, a major limitation of HGT lies in the ignorance of the fact that the spatial domain of dMRI is a 3D space rather than a 2D one. A large number of studies have demonstrated that careful consideration of 3D space is the key to advancing the performance of different medical image analysis tasks [7,9].

To this end, we propose 3D-HGT, an advanced microstructure estimation model capable of making full use of 3D x-space information and q-space information jointly. Specifically, we design an efficient q-space learning module based on simple graph convolution (SGC) [20], which runs with less memory at high speed and is able to alleviate the large computational burden associated with 3D spatial information learning. We further propose a 3D x-space learning module based on a U-shape transformer, which is able to capture the long-range relationships in 3D spatial space for improved performance. Finally, we train two modules end-to-end for predicting microstructure index maps. Our 3D-HGT exploits 3D spatial information and angular information in an efficient manner for more accurate microstructure estimation. We perform extensive experiments on data from the human connectome project (HCP) [19]. The experimental results demonstrate that our 3D-HGT outperforms cutting-edge methods, including HGT, both quantitatively and qualitatively.

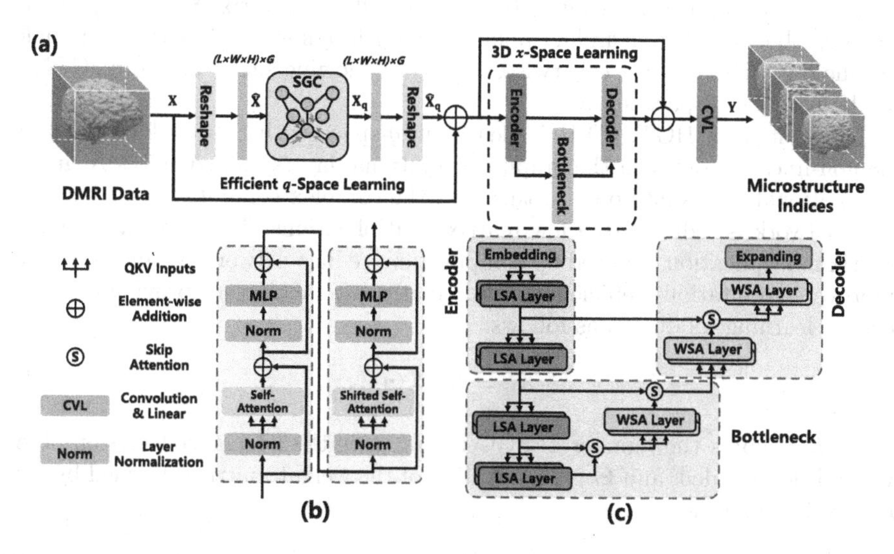

Fig. 1. Overview of 3D-HGT. (a) shows the overall architecture of the model, where the q-space learning module first learns the feature of dMRI data in q-space, and then the x-space learning module learns the 3D spatial information; (b) shows the architecture of the LSA/WSA component, which is a cascaded transformer group with two self-attention blocks; (c) shows the structure of the x-space learning module.

2 3D Hybrid Graph Transformer

2.1 Network Overview

As shown in Fig. 1, our 3D-HGT consists of two key modules: an efficient q-space learning module and a 3D x-space learning module. The q-space learning module is built with SGC, which can extract q-space features from voxel-wise dMRI data more efficiently. Motivated by the nnFormer [25], the x-space learning module mainly consists of a U-shaped network composed of local volume-based multi-head self-attention (LSA) and wide volume-based multi-head self-attention (WSA) components.

The input dMRI data $\mathbf{X} \in \mathbb{R}^{L \times W \times H \times G}$ has four dimensions, where L, W, H, and G denote the length, width, height, and number of gradient directions, respectively. In our model, the data is first reshaped into a two-dimensional tensor $\hat{\mathbf{X}} \in \mathbb{R}^{LWH \times G}$ and fed into the SGC network to learn the q-space features efficiently. The output $\mathbf{X}_\mathrm{q}$ is then reshaped to four dimensions $\hat{\mathbf{X}}_\mathrm{q} \in \mathbb{R}^{L \times W \times H \times G}$ and enters into the x-space learning module followed by convolutional layers. Finally, our 3D-HGT generates microstructure predictions $\mathbf{Y} \in \mathbb{R}^{L \times W \times H \times M}$, where M denotes the number of microstructure indices.

2.2 Efficient q-Space Learning Module

According to [2,6], the geometric structure of the q-space of dMRI data can be encoded as a graph $\mathcal{G}$ determined by a binary affinity matrix $\mathbf{A} = \{a_{i,j}\}$. If the angle $\theta_{i,j}$ between two sampling points i and j in q-space is less than an angle threshold θ, then $a_{i,j} = 1$; otherwise, $a_{i,j} = 0$. Following such processing, GCNN can be used for q-space learning.

Different from HGT, 3D-HGT learns in q-space with SGC, which removes the non-linearity between the graph convolutional layers and collapses K graph convolutional layers into one. In this way, the complex computation of a multi-layer network is reduced to a few matrix multiplications. Thus, we can use fast matrix multiplication to speed up computation, reduce network redundancy, and improve computational efficiency. Mathematically, the feature provided by our q-space learning module is as follows:

$$\mathbf{X}_{\mathrm{q}} = \hat{\mathbf{A}}^{K}\hat{\mathbf{X}}\boldsymbol{\Theta}, \tag{1}$$

where K denotes the number of SGC layers, $\hat{\mathbf{A}}$ is the normalized version of $\mathbf{A}$ with self-loop added, and $\boldsymbol{\Theta}$ is the product of the weight matrices learned by the q-space learning module.

2.3 3D x-Space Learning Module

The overall design of the x-space learning module is a U-shaped network composed of three parts: an encoder, a bottleneck, and a decoder. The three components are made up of cascaded self-attention transformer blocks.

The encoder and decoder are similar in structure, with two self-attention layers. However, there are two key differences between them. Firstly, the former is composed of two local volume-based layers, while the latter has two wide volume-based layers. Secondly, before entering the network, data from the encoder passes through an embedding layer, while data from decoders pass through an extension layer for feature integration before outputting results. The bottleneck layer comprises two LSAs and one WSA, which allows skip attention to integrate shallow and deep attention between the encoder and decoder layers. Notably, we omit the pooling or un-pooling operations in the encoder-decoder architecture since pooling can lead to the loss of features, especially when dealing with small-sized 3D patches.

The Embedding Layer. The embedding layer divides the input $\hat{\mathbf{X}}_{\mathrm{q}}$ into high-dimensional patches, which are then sent into the transformer block. Unlike the nnFormer, our embedding layer consists of two convolutional layers, a GELU [13] layer and a normalization layer [1], where both the convolutional kernel and step size are set to one.

Local Volume-Based Multi-head Self-attention (LSA). Based on Swin Transformer [17], LSA is an offset window self-retaining module that defines the window as a local 3D volume block. The multi-head self-attention is computed in this local volume, which reduces the computational cost significantly.

Let $\mathbf{Q}, \mathbf{K}, \mathbf{V} \in \mathbb{R}^{S_\mathrm{L} \times S_\mathrm{W} \times S_\mathrm{H} \times D}$ for LSA be the query, key, and value matrices, and $\mathbf{B} \in \mathbb{R}^{S_\mathrm{L} \times S_\mathrm{W} \times S_\mathrm{H}}$ be the relative position matrix, where $\{S_\mathrm{L}, S_\mathrm{W}, S_\mathrm{H}\}$ denotes the size of the local volume and D is the dimension of query/key. The self-attention is then computed in each 3D local volume as follows:

$$\mathrm{Attention}(\mathbf{Q}, \mathbf{K}, \mathbf{V}) = \mathrm{Softmax}(\frac{\mathbf{Q}\mathbf{K}^T}{\sqrt{D}} + \mathbf{B})\mathbf{V}. \quad (2)$$

The LSA computational complexity can be expressed as follows:

$$\Omega(\text{ LSA }) = 4N_\mathrm{P}C^2 + 2S_\mathrm{L}S_\mathrm{W}S_\mathrm{H}N_\mathrm{P}C, \quad (3)$$

where N_P is the size of a patch, and C is the length of the embedding sequence.

Wide Volume-Based Multi-head Self-attention (WSA). Although LSA is efficient, its receptive field is limited. Thus, by increasing the size of the window to $\{m \times S_\mathrm{L}, m \times S_\mathrm{W}, m \times S_\mathrm{H}\}$ with m denoting a constant (two by default), we have WSA, which improves the global context awareness ability of LSA. The computational complexity of WSA is as follows:

$$\Omega(\text{ WSA }) = 4N_\mathrm{P}C^2 + 2m^3S_\mathrm{L}S_\mathrm{W}S_\mathrm{H}N_\mathrm{P}C. \quad (4)$$

We extend the self-attention field of view in the bottleneck using three wide-field transformer blocks, where six WSA layers are used.

Skip Attention. This component effectively combines shallow and deep attention. A single-layer neural network decomposes the output $\mathbf{X}_l$ at layer l of the encoder into a key matrix $\mathbf{K}_{l'}$ and a value matrix $\mathbf{V}_{l'}$, while the output of $\mathbf{X}_{l'}$ at layer l' of the decoder is used as a query matrix $\mathbf{Q}_{l'}$. Mathematically, the self-attention is defined as:

$$\mathrm{Attention}(\mathbf{Q}_{l'}, \mathbf{K}_{l'}, \mathbf{V}_{l'}) = \mathrm{Softmax}(\frac{\mathbf{Q}_{l'}\mathbf{K}_{l'}^\mathrm{T}}{\sqrt{D_{l'}}} + \mathbf{B}_{l'})\mathbf{V}_{l'}, \quad (5)$$

where $\mathbf{B}_{l'}$ is the relative position encoding matrix, and $D_{l'}$ is the dimension of query/key.

3 Experiments

3.1 Implementation Details

Experimental Settings. We use the mean square error as the loss function, Adam as the optimizer, and an initial learning rate of 9e-4. We set the number of epochs to 100, batch size to 2, and iterations to 10. The angular threshold θ for constructing graph is set at 45°, and the number of graph convolutional layers (i.e., K) is set to 2. In the x-space learning module, the embedding dimension is set to 192, and the window sizes of the LSA and WSA layers are set to 4 and 8, respectively. The model was implemented using PyTorch 1.11 and PyTorch-Geometric 2.1.0, and trained on a server equipped with an RTX 3090 GPU.

Comparison Methods. We compare our 3D-HGT with various methods, including AMICO [8], MLP [12], GCNN [4], MEDN [21], MEDN+ [22], U-Net [10], U-Net++ [26], and HGT [5].

Table 1. Performance evaluation of microstructre estimation using single-shell undersampled data with b=1000 s/mm^2, 30 gradient directions in total. All: A combination of ICVF, ISOVF, and ODI.

Method	PSNR ↑				SSIM ↑			
	ICVF	ISOVF	ODI	All	ICVF	ISOVF	ODI	All
AMICO [8]	8.09	14.90	12.03	10.80	0.020	0.273	0.380	0.210
MLP [12]	19.15	24.07	20.31	20.72	0.670	0.706	0.667	0.681
MEDN [21]	19.18	24.43	20.54	20.87	0.674	0.737	0.684	0.698
GCNN [4]	19.27	24.48	20.55	20.92	0.676	0.742	0.681	0.700
MEDN+ [22]	19.14	24.67	21.30	21.14	0.684	0.754	0.717	0.719
U-Net [10]	19.95	25.11	20.41	21.28	0.727	0.787	0.651	0.722
U-Net++ [26]	19.94	25.49	20.87	21.51	0.750	0.798	0.688	0.745
HGT [5]	21.96	27.13	22.12	23.16	0.814	**0.850**	0.757	0.807
3D-HGT	**23.60**	**27.98**	**22.43**	**24.10**	**0.837**	0.845	**0.765**	**0.816**

3.2 Dataset and Evaluation Metrics

Dataset. Following [5], we randomly select 21 subjects from the HCP to construct our dataset. For each dMRI scan, q-space undersampling is performed on a single shell with b=1000 s/mm^2 to extract 30 gradients uniformly. The undersampled data is then normalized by dividing by the average of b_0 images. Finally, we extract patches from the normalized data with a dimension of $32\times32\times32\times30$. The validation and test sets are also processed in the same way. In our experiments, the ratio between training, validation, and test sets is 10:1:10. We train our model to predict NODDI-derived indices, including intracellular volume fraction (ICVF), isotropic volume fraction (ISOVF), and orientation dispersion index (ODI). The gold standard microstructural indices are computed using the complete HCP data with 270 gradients using AMICO [8].

Evaluation Metrics. We evaluate the quality of predicted NODDI-derived indices with the peak signal-to-noise ratio (PSNR) and structural similarity index measure (SSIM).

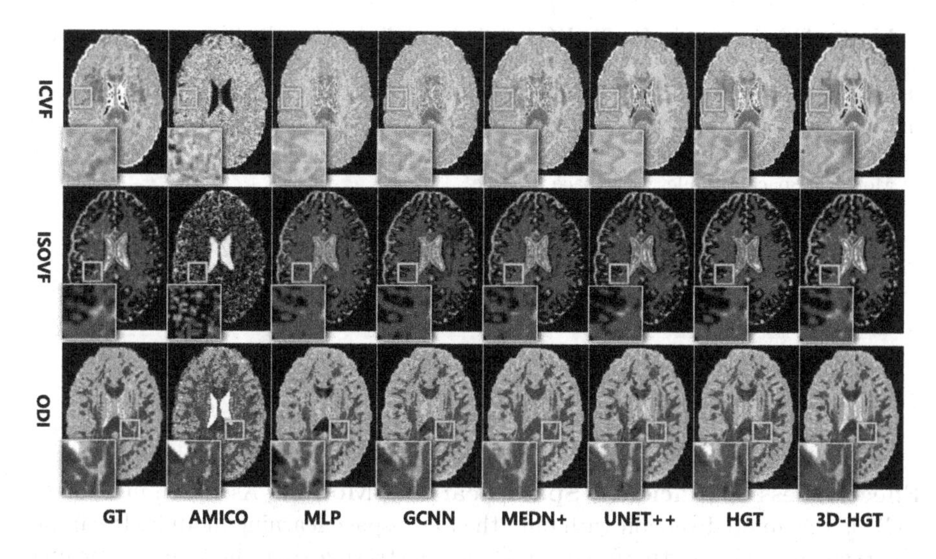

Fig. 2. Visual comparison of prediction results.

3.3 Experimental Results

Table 1 compares our 3D-HGT model with cutting-edge models in control experiments under the same conditions using various evaluation metrics. By taking into account the 3D x-space data from the dMRI data, our model is able to learn rich features across a broader range of spatial domains. 3D-HGT outperforms HGT in terms of PSNR and SSIM. The most notable improvements among all indices can be observed in the PSNR of the ICVF, which has increased from 21.96 to 23.60 dB, and the SSIM of the ICVF, which has increased from 0.814 to 0.837. Furthermore, compared with the deep learning models (i.e., MLP to HGT) in Table 1, 3D-HGT improves the PSNR of ALL by 16.3%, 15.5%, 15.2%, 14.0%, 13.3%, 12.0%, and 4.1%, respectively. Although HGT takes into account x-space and q-space, it only makes use of the 2D spatial features of the data. In contrast, 3D-HGT explicitly considers the 3D spatial domain and improves the PSNR and SSIM of ALL by 4.1% and 1.1%, respectively.

Figure 2 depicts the visual comparison of 3D-HGT and other models' microstructure predictions. In particular, the close-up views, shown in the bottom row of Fig. 2, indicate that our model provides the best result, which is much closer to the ground truth.

3.4 Ablation Study

To investigate the effectiveness of the proposed modules, we perform ablation experiments with different ablated versions. The ablation results are shown in Table 2. It should be noted that the ablated versions "(A)" and "(E)" correspond to HGT and the full version of 3D-HGT, respectively.

Table 2. Ablation study for 3D-HGT. "2D Trans." denotes the 2D x-space learning module of HGT. "3D Trans." denotes the 3D x-space learning module of 3D-HGT, as shown in Fig. 1(c). The metric "Time" indicates the time cost (in seconds) for training the corresponding ablation version in one epoch.

Model	q-Space Module		x-Space Module		PSNR ↑				Time ↓
	TAGCN	SGC	2D Trans.	3D Trans.	ICVF	ISOVF	ODI	All	
(A)	✓		✓		21.96	27.13	22.12	23.16	121
(B)		✓	✓		21.97	27.21	22.33	23.28	**109**
(C)				✓	23.38	27.78	22.40	23.97	122
(D)	✓			✓	23.44	27.80	22.35	23.98	226
(E)		✓		✓	**23.60**	**27.98**	**22.43**	**24.10**	197

Effectiveness of Efficient q-Space Learning Module. As shown in Table 2, "(C)" is the ablated version with only the 3D x-space learning module. It can be observed that "(E)" outperforms "(C)", demonstrating that our q-space learning module can effectively improve performance. Moreover, we perform additional experiments using the undersampled dMRI data with a higher angular resolution, i.e., 45 gradients. The results show that our q-space learning module provides a larger improvement when the dMRI data is with a higher angular resolution.

The q-space learning with SGC also shows advantages over TAGCN. Both "(B)" and "(E)" improve the performance and computational efficiency in comparison with their corresponding ablated versions equipped with TAGCN, i.e., "(A)" and "(D)". In particular, compared with "(D)", the training time cost of "(E)" is reduced by nearly 30 s for one epoch, verifying the high efficiency of our q-space learning module.

Effectiveness of 3D x-Space Learning Module. As shown in Table 2, two sets of comparisons, "(A)" vs. "(D)" and "(B)" vs. "(E)", indicate that modifying the x-space learning module to a 3D transformer improves performance. More specifically, compared with "(A)", "(D)" improves the PSNR of ICVF by 1.48 dB, ISOVF by 0.67 dB, ODI by 0.23 dB and ALL by 0.82 dB. Compared with "(B)", "(E)" improves the PSNR of ICVF by 1.63 dB, ISOVF by 0.77 dB, ODI by 0.10 dB and ALL by 0.82 dB. The improvement owes to the 3D x-space learning module, which is equipped with LSAs and WSAs, allowing the model to capture long-term dependencies with a large 3D receptive field.

4 Conclusion

In this paper, we proposed 3D-HGT, an improved microstructure estimation model that makes full use of 3D x-space information and q-space information. Our x-space learning is achieved with a 3D transformer architecture, allowing the model to thoroughly learn the long-term dependencies of features in the 3D spatial domain. To alleviate the large computational burden associated with 3D x-space learning, we further propose an efficient q-space learning module, which is built with a simplified graph learning architecture. Extensive experiments on the HCP demonstrate that, compared with HGT, 3D-HGT effectively improves the quality of microstructure estimation.

References

1. Ba, J.L., Kiros, J.R., Hinton, G.E.: Layer normalization. arXiv preprint arXiv:1607.06450 (2016)
2. Chen, G., Dong, B., Zhang, Y., Lin, W., Shen, D., Yap, P.T.: Denoising of diffusion MRI data via graph framelet matching in x-q space. IEEE Trans. Med. Imaging **38**(12), 2838–2848 (2019)
3. Chen, G., Dong, B., Zhang, Y., Lin, W., Shen, D., Yap, P.T.: XQ-SR: joint x-q space super-resolution with application to infant diffusion MRI. Med. Image Anal. **57**, 44–55 (2019)
4. Chen, G., et al.: Estimating tissue microstructure with undersampled diffusion data via graph convolutional neural networks. In: Martel, A.L., et al. (eds.) MICCAI 2020. LNCS, vol. 12267, pp. 280–290. Springer, Cham (2020). https://doi.org/10.1007/978-3-030-59728-3_28
5. Chen, G., et al.: Hybrid graph transformer for tissue microstructure estimation with undersampled diffusion MRI data. In: International Conference on Medical Image Computing and Computer-Assisted Intervention, LNCS, vol. 13431, pp. 113–122. Springer, Cham (2022). https://doi.org/10.1007/978-3-031-16431-6_11
6. Chen, G., Wu, Y., Shen, D., Yap, P.T.: Noise reduction in diffusion MRI using non-local self-similar information in joint x-q space. Med. Image Anal. **53**, 79–94 (2019)
7. Chen, H., Dou, Q., Yu, L., Qin, J., Heng, P.A.: VoxResNet: deep voxelwise residual networks for brain segmentation from 3D MR images. NeuroImage **170**, 446–455 (2018)
8. Daducci, A., et al.: Accelerated microstructure imaging via convex optimization (AMICO) from diffusion MRI data. NeuroImage **105**, 32–44 (2015)
9. Dou, Q., et al.: Automatic detection of cerebral microbleeds from MR images via 3D convolutional neural networks. IEEE Trans. Med. Imaging **35**(5), 1182–1195 (2016)
10. Falk, T., et al.: U-net: deep learning for cell counting, detection, and morphometry. Nat. Methods **16**(1), 67–70 (2019)
11. Gibbons, E.K., et al.: Simultaneous NODDI and GFA parameter map generation from subsampled q-space imaging using deep learning. Magnet. Resonan. Med. **81**(4), 2399–2411 (2019)
12. Golkov, V., et al.: q-Space deep learning: twelve-fold shorter and model-free diffusion MRI scans. IEEE Trans. Med. Imaging **35**(5), 1344–1351 (2016)
13. Hendrycks, D., Gimpel, K.: Gaussian error linear units (GELUs). arXiv preprint arXiv:1606.08415 (2016)
14. Hong, Y., Chen, G., Yap, P.-T., Shen, D.: Multifold acceleration of diffusion MRI via deep learning reconstruction from slice-undersampled data. In: Chung, A.C.S., Gee, J.C., Yushkevich, P.A., Bao, S. (eds.) IPMI 2019. LNCS, vol. 11492, pp. 530–541. Springer, Cham (2019). https://doi.org/10.1007/978-3-030-20351-1_41

15. Jensen, J.H., Helpern, J.A., Ramani, A., Lu, H., Kaczynski, K.: Diffusional kurtosis imaging: the quantification of non-gaussian water diffusion by means of magnetic resonance imaging. Magnet. Resonan. Med. **53**(6), 1432–1440 (2005)
16. Kaden, E., Kruggel, F., Alexander, D.C.: Quantitative mapping of the per-axon diffusion coefficients in brain white matter. Magnet. Resonan. Med. **75**(4), 1752–1763 (2016)
17. Liu, Z., et al.: Swin transformer: hierarchical vision transformer using shifted windows. In: Proceedings of the IEEE/CVF International Conference on Computer Vision, pp. 10012–10022 (2021)
18. Tian, Q., et al.: DeepDTI: high-fidelity six-direction diffusion tensor imaging using deep learning. NeuroImage **219**, 117017 (2020)
19. Van Essen, D.C., et al.: The WU-Minn human connectome project: an overview. NeuroImage **80**, 62–79 (2013)
20. Wu, F., Souza, A., Zhang, T., Fifty, C., Yu, T., Weinberger, K.: Simplifying graph convolutional networks. In: International Conference on Machine Learning, pp. 6861–6871. PMLR (2019)
21. Ye, C.: Estimation of tissue microstructure using a deep network inspired by a sparse reconstruction framework. In: Niethammer, M., et al. (eds.) IPMI 2017. LNCS, vol. 10265, pp. 466–477. Springer, Cham (2017). https://doi.org/10.1007/978-3-319-59050-9_37
22. Ye, C.: Tissue microstructure estimation using a deep network inspired by a dictionary-based framework. Med. Image Anal. **42**, 288–299 (2017)
23. Zhang, H., Schneider, T., Wheeler-Kingshott, C.A., Alexander, D.C.: NODDI: practical in vivo neurite orientation dispersion and density imaging of the human brain. NeuroImage **61**(4), 1000–1016 (2012)
24. Zheng, T., et al.: A microstructure estimation transformer inspired by sparse representation for diffusion MRI. Med. Image Anal. **86**, 102788 (2023)
25. Zhou, H.Y., et al.: nnFormer: volumetric medical image segmentation via a 3D transformer. IEEE Trans. Image Process. (2023)
26. Zhou, Z., Rahman Siddiquee, M.M., Tajbakhsh, N., Liang, J.: UNet++: a nested U-Net architecture for medical image segmentation. In: Stoyanov, D., et al. (eds.) DLMIA/ML-CDS -2018. LNCS, vol. 11045, pp. 3–11. Springer, Cham (2018). https://doi.org/10.1007/978-3-030-00889-5_1

Cortical Analysis of Heterogeneous Clinical Brain MRI Scans for Large-Scale Neuroimaging Studies

Karthik Gopinath(✉), Douglas N. Greve, Sudeshna Das, Steve Arnold, Colin Magdamo, and Juan Eugenio Iglesias

Athinoula A. Martinos Center for Biomedical Imaging, Massachusetts General Hospital and Harvard Medical School, Boston, USA
kgopinath@mgh.harvard.edu

Abstract. Surface analysis of the cortex is ubiquitous in human neuroimaging with MRI, e.g., for cortical registration, parcellation, or thickness estimation. The convoluted cortical geometry requires isotropic scans (e.g., 1 mm MPRAGEs) and good gray-white matter contrast for 3D reconstruction. This precludes the analysis of most brain MRI scans acquired for clinical purposes. Analyzing such scans would enable neuroimaging studies with sample sizes that cannot be achieved with current research datasets, particularly for underrepresented populations and rare diseases. Here we present the first method for cortical reconstruction, registration, parcellation, and thickness estimation for clinical brain MRI scans of any resolution and pulse sequence. The methods has a learning component and a classical optimization module. The former uses domain randomization to train a CNN that predicts an implicit representation of the white matter and pial surfaces (a signed distance function) at 1 mm isotropic resolution, independently of the pulse sequence and resolution of the input. The latter uses geometry processing to place the surfaces while accurately satisfying topological and geometric constraints, thus enabling subsequent parcellation and thickness estimation with existing methods. We present results on 5 mm axial FLAIR scans from ADNI and on a highly heterogeneous clinical dataset with 5,000 scans. Code and data are publicly available at https://surfer.nmr.mgh.harvard.edu/fswiki/recon-all-clinical.

1 Introduction

Clinical MRI exams account for the overwhelming majority of brain MRI scans acquired worldwide every year [19]. These exams comprise several scans acquired during a session with different orientations (axial, coronal, sagittal), resolutions, and MRI contrasts. The acquisition hardware and pulse sequence parameters differ significantly across (and even within) centers, leading to highly heterogeneous data. Since cortical thickness is a robust biomarker in the study of normal aging [25] and many brain disorders and diseases [21,22,24], methods that can

H. Greenspan et al. (Eds.): MICCAI 2023, LNCS 14227, pp. 35–45, 2023.
https://doi.org/10.1007/978-3-031-43993-3_4

extract parcellations and thickness measurements from clinical scans (while registering to a reference spherical coordinate frame) are highly desirable. However, cortical analysis of clinical scans is complex due to large slice spacing (resulting in incomplete cortex geometry description) and heterogeneous acquisitions (hindering supervised approaches leveraging image intensity distributions).

Existing neuroimaging research studies [12] rely on isotropic scans with good gray-white matter contrast (typically a 1 mm MPRAGE) and utilize prior information on tissue intensities. Classical cortical analysis pipelines like FreeSurfer [5,9] generate two triangle meshes per hemisphere, one for the white matter (WM) surface and one for the pial surface, while preventing self-intersections. The spherical topology of the surfaces enables mapping coordinates to a sphere, thus enabling computation of vertex-wise statistics across subjects in a common space.

Over the last two years, machine learning approaches for cortical reconstruction on 1 mm MPRAGEs have emerged. Methods based on signed distance functions (SDF) like DeepCSR [4] or SegRecon [11] predict voxel-wise SDFs for the WM and pial surfaces. The final meshes are computed as the SDF isosurfaces and do not guarantee topological correctness. PialNN [18] uses an explicit representation to project the pial surface from the WM surface, which is assumed to be topologically correct. Approaches based on surface deformation like TopoFit [14] or Vox2Cortex [3] use image and graph convolutions to predict a deformation that maps a topologically correct template mesh to an input MRI, thus generating WM and pial surfaces. However, these approaches neither prevent self intersections nor guarantee topological correctness.

Contribution: Our proposed method allows cortical analysis of brain MRI scans of any orientation, resolution, and MRI contrast without retraining, making it possible to use it out of the box for straightforward analysis of large datasets "in the wild". The proposed method combines two modules: a convolutional neural network (CNN) that estimates SDFs of the WM and pial surfaces, and a classical geometry processing module that places the surfaces while satisfying geometric constraints (no self-intersections, spherical topology, regularity). The CNN capitalizes on recent advances in domain randomization to provide robustness against changes in acquisition – in contrast with existing learning approaches that can only process images acquired with the same resolution and MRI contrast as the scans they were trained on. Finally, our method's classical geometry processing gives us geometric guarantees and grants instant access to an array of existing methods for cortical thickness estimation, registration, and parcellation [9].

Further Related Work: The parameterization of surfaces as SDFs has been combined with deep neural networks in several domains [20], including cortical reconstruction [4]. Our robustness to MRI contrast and resolution changes is achieved using ideas from the domain randomization literature [26], which involves training supervised CNNs with synthetic images generated from segmentations on the fly at every iteration. These techniques have been successfully applied to MRI analysis [1,13] and use random sampling of simulation param-

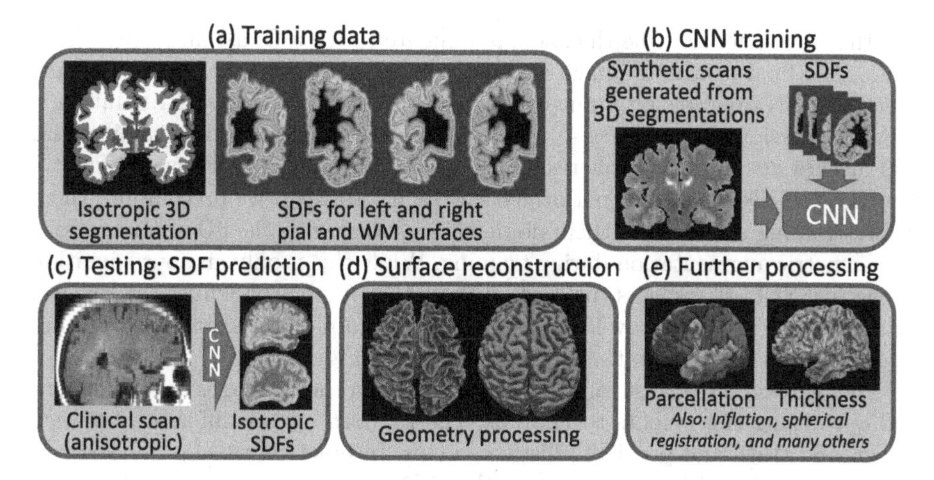

Fig. 1. Overview of our proposed approach for cortical analysis of clinical brain MRI scans of any resolution and MRI contrast, without retraining. The images shown in (c–e) correspond to a real axial FLAIR scan with 5 mm slice spacing and 5 mm thickness.

eters such as orientation, contrast, and resolution from uniform distributions at every mini-batch, which results in unrealistic appearance, making the CNN agnostic to these features.

2 Methods

Our proposed method (Fig. 1) has two distinct components: a learning module to estimate isotropic SDFs from anisotropic scans and a geometry processing module to place the WM and pial surfaces with topological constraints.

2.1 Learning of SDFs

This module estimates isotropic SDFs of the WM and pial surfaces of both hemispheres in a contrast- and resolution-independent fashion. It utilizes a domain randomization approach based on training a voxel-wise SDF regression CNN with synthetic data, which comprises volumetric segmentations and corresponding surfaces (real images are *not* used). Such training data can be obtained "for free" by running FreeSurfer on isotropic T1 scans (we used the HCP dataset [10]).

Given a 3D segmentation and four surface meshes (WM and pial surfaces for each hemisphere; see Fig. 1a), we compute the following input/target pairs at every training iteration. As input, we simulate a synthetic MRI scan of random orientation, resolution, and contrast from the 3D segmentation. For this purpose, we use a Gaussian mixture model conditioned on the (spatially augmented) labels, combined with models of bias field, resolution, and noise similar to [1]. We use random sampling to determine the orientation (coronal, axial, sagittal, or isotropic), slice spacing (between 1 and 9 mm), and slice thickness (between 1 mm and the slice spacing). The thickness is simulated with a Gaussian kernel across slices. The final synthetic image is upscaled to 1 mm isotropic resolution, such that the CNN operates on input-output pairs of the same size and resolution.

As regression targets, we use voxel-wise SDFs computed from the WM and pial meshes for both hemispheres. The computation of the SDFs would greatly slow down CNN training if performed on the fly. Instead, we precompute them before training and deform them nonlinearly (along with the 3D segmentation) for geometric augmentation during training. While this is only an approximation to the real SDF, it respects the zero-level-set that implicitly defines the surface, and we found it to work well in practice. An example of a synthetic scan and target SDFs used to train the CNN are shown in Fig. 1b.

The regression CNN is trained by feeding the synthetic images to the network and optimizing the weights to minimize the L1 norm of the difference between the ground truth and predicted distance maps. In practice, we clip the SDFs at an absolute value of 5 mm to prevent the CNN from wasting capacity trying to model relatively small variations far away from the surfaces of interest. At test time, the input scan is upscaled to the isotropic resolution of the training data and pushed through the CNN to obtain the predicted SDFs.

2.2 Geometry Processing for Surface Placement

To process real clinical scans, we first feed them to the trained CNN to predict the SDFs for the pial and WM surfaces for both hemispheres (Fig. 1c). To avoid generating topologically incorrect surfaces from these SDFs, we capitalize on the extensive literature on the geometry processing of cortical meshes with classical techniques. For reconstructing WM surfaces, we run SynthSeg [1] on the input scan to obtain two binary masks corresponding to the left and right WM labels. From this point on, processing happens independently for each hemisphere. First, we fill in the holes in the hemisphere's mask and tessellate it to obtain the initial WM mesh. Then, we smooth the mesh and use automated manifold surgery [8] to guarantee spherical topology. Next, we iteratively deform the WM mesh by minimizing an objective function consisting of a fidelity term and a regularizer.

Specifically: let $\mathcal{M} = (\boldsymbol{X}, \mathcal{K})$ denote a triangle mesh, where $\boldsymbol{X} = [\boldsymbol{x}_1, \ldots, \boldsymbol{x}_V]$ represents the coordinates of its V vertices ($\boldsymbol{x}_v \in \mathbb{R}^3$), and $\mathcal{K}$ represents the connectivity. Let $D_w(\boldsymbol{r})$ be the SDF for the WM surface estimated by our CNN, where $\boldsymbol{r}$ is the spatial location. The objective function ("energy") is the following:

$$E[\boldsymbol{X}; D_w(\boldsymbol{r}), \mathcal{K}] = \sum_{v=1}^{V} [\tanh D_w(\boldsymbol{x}_v)]^2 + \lambda_1 \sum_{v=1}^{V} \sum_{u \in \mathcal{N}_v} [\boldsymbol{n}_v^t(\boldsymbol{x}_v - \boldsymbol{x}_u)]^2$$
$$+ \lambda_2 \sum_{v=1}^{V} \sum_{u \in \mathcal{N}_v} \left\{ [\boldsymbol{e}_{1v}^t(\boldsymbol{x}_v - \boldsymbol{x}_u)]^2 + [\boldsymbol{e}_{2v}^t(\boldsymbol{x}_v - \boldsymbol{x}_u)]^2 \right\}. \quad (1)$$

The first term in Eq. 1 is the fidelity term, which encourages the SDF to be zero on the mesh vertices; we squash the SDF through a *tanh* function to prevent huge gradients far away from zero. The second and third terms are regularizers that endow the mesh with a spring-like behavior [5]: $\boldsymbol{n}_v$ is the surface normal at vertex v; $\boldsymbol{e}_{1v}$ and $\boldsymbol{e}_{2v}$ define an orthonormal basis for the tangent plane at vertex v; $\mathcal{N}_v$ is the neighborhood of v according to $\mathcal{K}$; and λ_1 and λ_2 are relative weights, which we define according to [5] ($\lambda_1 = 0.0006$, $\lambda_2 = 0.0002$). Optimization is performed with gradient descent. At every iteration, self-intersections are monitored and eliminated by reducing the step size as needed [5].

The pial surface is fitted with a very similar procedure, but using the predicted SDF of the pial surface. Figure 1d shows examples of reconstructed surfaces for the axial FLAIR scan from Fig. 1c. Given the fitted WM and pial surfaces, we use FreeSurfer to compute cortical thickness, parcellation, and registration to a common coordinate frame in spherical coordinates (Fig. 1e).

2.3 Implementation Details

Our voxel-wise regression CNN is a 3D U-net [23] trained with synthetic pairs generated on the fly as explained in Sect. 2.1 above. The U-net has 5 levels with 2 layers each, uses $3 \times 3 \times 3$ convolutions and exponential linear activations. The layers have 24^l features, where l is the level number. The last layer uses linear activation functions to model the SDFs. The CNN weights were optimized with stochastic gradient descent using a fixed step size of 0.0001 and 300,000 iterations (enough to converge). At test time, the run time is dominated by the geometry processing (2–3 h, depending on the complexity of the manifold surgery).

3 Experiments and Results

3.1 Datasets

- **HCP**: we used 150 randomly selected subjects (71 males, ages: 29.9±3.4 years) from HCP [10] to train the U-net. We ran FreeSurfer to obtain the segmentations and SDFs (images are discarded as they are not used in training).
- **ADNI**: in our first experiment, we used 1 mm MPRAGE and corresponding 5 mm axial FLAIR scans of 200 randomly selected subjects from the ADNI dataset [17] (95 males, ages 74.5 ± 7.4 years). This setup enables us to directly compare the results from research- and clinical-grade scans.
- **Clinical**: this dataset comprises 9,735 scans with a plethora of pulse sequence combinations and resolutions from 1,367 MRI sessions of distinct subjects with memory complaints (749 males, ages 18–90) from hospital [2]. Surfaces were successfully generated for 5,064 scans; the rest failed due to insufficient field of view. This dataset includes a wide range of MR contrasts and resolutions. We note that this dataset also includes 581 1 mm MPRAGE scans.

The availability of 1 mm MPRAGEs for some of the subjects enables us to process them with FreeSurfer and use the result as ground truth [16] (Fig. 2a).

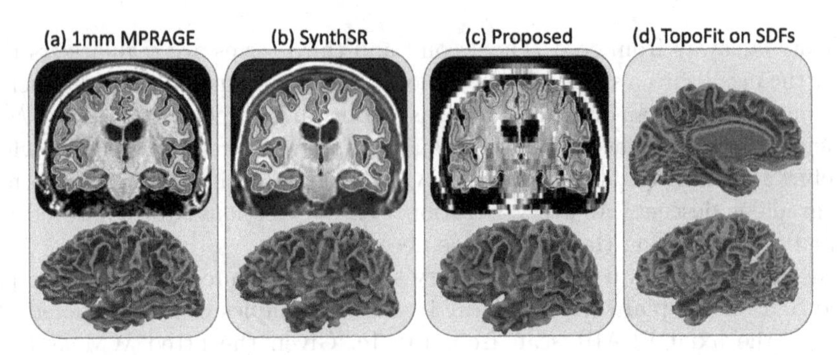

Fig. 2. Qualitative comparison on a 5 mm axial FLAIR scan from ADNI. (a) Ground truth T1 with WM (red) and pial (yellow) surfaces estimated by FreeSurfer (top); and 3D reconstruction of the left WM surface (bottom); (b) Synthetic T1 produced by SynthSR with FreeSurfer surfaces. (c) Our proposed method. (d) Examples of WM surfaces produced by TopoFit, trained to predict on the output of our SDF predictor, show small blobs in different areas (e.g., indicated by yellow arrows) (Color figure online)

3.2 Competing Methods

To the best of our knowledge, the only existing competing method for our proposed algorithm is SynthSR [15], which utilizes a synthetic data generator like ours to turn scans of any resolution and contrast into synthetic 1 mm MPRAGES – which can be subsequently processed with FreeSurfer to obtain surfaces (Fig. 2b). Compared with our proposed approach (Fig. 2c), this pipeline inherits the smoothness of the synthetic MPRAGE, leading to smoother surfaces that may miss larger folds. We also tried training TopoFit [14] on the synthetic images and predicted SDFs, but failed to produce neural networks with good generalization ability, as they led to small blobs on the surfaces at test time (Fig. 2d).

3.3 Results on the ADNI Dataset

Figure 3 summarizes the results on the ADNI dataset. While previous machine learning approaches focus evaluation on distance errors, these can be difficult to interpret. Instead, we evaluate our method using the performance on the downstream tasks that one is ultimately interested in. First, we computed the accuracy of the Desikan-Killiany parcellation [6] produced by SynthSR and our proposed method. Figure 3a shows the results on the inflated surface of the *fsaverage* template. Since the parcellation is computed from the curvature of the WM surface, it is a relatively easy problem. The overlap between the ground truth parcellation and the two competing methods is very high. Dice scores over 0.90 are obtained for almost every region in both methods, and the average across regions is almost identical for both methods (0.95).

We then used the obtained parcellations to study the effect of Alzheimer's disease (AD) on cortical thickness, using a group study between AD subjects

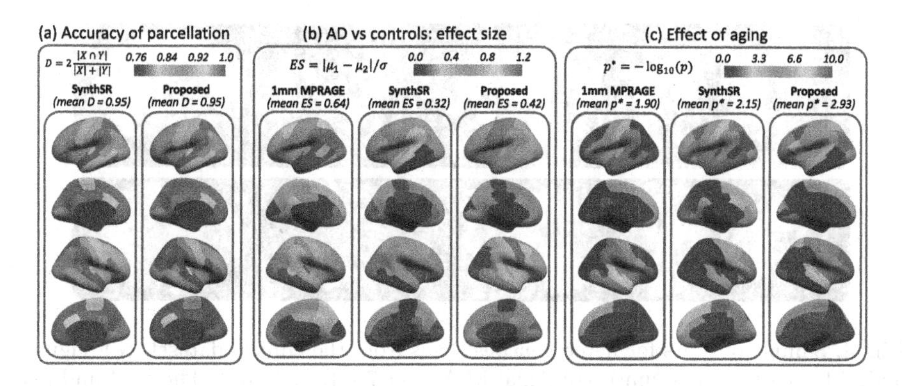

Fig. 3. Summary of results on the ADNI dataset, displayed on the inflated surface of FreeSurfer's average subject (*fsaverage*). (a) Accuracy of parcellation for SynthSR and our proposed method using Dice scores. (b) Ability to discriminate AD vs controls, measured with effect sizes. (c) Effect of aging, measured as the strength of the (negative) correlation between age and thickness. The strength of the correlation is represented by p-values of a Student's t test assessing whether the correlation significantly differs from zero; note that we log-transform the p-values for easier visualization.

and elderly controls. For this purpose, we first fit a general linear model (GLM) to the cortical thickness at every parcel, using age, gender, and AD status as covariates. We then used the model coefficients to correct the thickness estimates for age and gender, and compared the thicknesses of the two groups.

Figure 3b shows the effect sizes (ES) for the reference 1 mm MPRAGEs and the competing methods. The 1 mm scans yield the expected AD cortical thinning pattern [7], with strong atrophy in the temporal lobe (ES >1.0) and, to a lesser extent, in parietal and middle frontal areas (ES$\sim$1.0). The average ES across all regions is 0.64. As expected, the thickness estimates based on the FLAIR scans are less able to detect the differences between the two groups. SynthSR loses, on average, half of the ES (0.32 vs 0.64). Most worrying, it cannot detect the effect on the temporal areas (particularly middle temporal). Our method can detect these differences with ES>0.6 in all temporal regions. On average, our method recovers one third of the ES lost by SynthSR (0.42 vs 0.32).

Finally, we studied the effect of aging on cortical thickness using the same GLM as above. Figure 3c shows maps of p-values computed with Student's t distribution, where we have transformed $p^* = \log_{10}(p)$ for easier visualization. Once more, the 1 mm MPRAGEs display the expected pattern [25], with strongest atrophy in superior-temporal and, to less extent, the central and medial frontal gyri. SynthSR fails to detect the superior-temporal effect in the left hemisphere and barely discerns it in the right hemisphere. Our approach, on the other hand, successfully detects these effects. We also note that SynthSR and our method display false positives in frontal areas of the right hemisphere; further analysis (possibly with manual quality control) will be needed to elucidate this result.

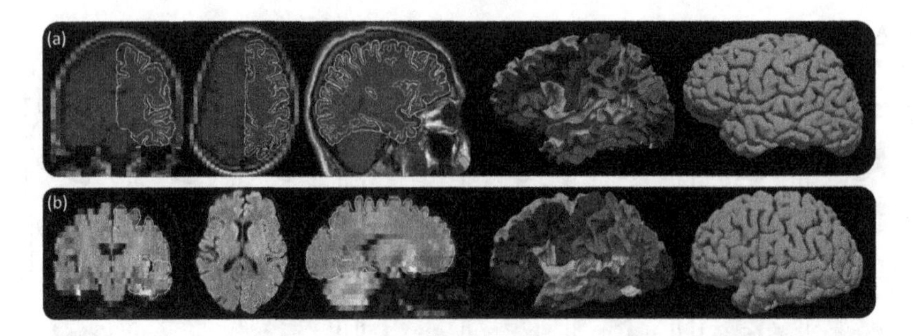

Fig. 4. Sample outputs for heterogeneous scans from the clinical dataset: (a) Sagittal TSE-T1 scan (.4×.4×6 mm). (b) Axial FLAIR (1.7 × 1.7× 6 mm). The WM and pial surfaces are shown on the right. The cortical parcellation is overlaid on the WM surface.

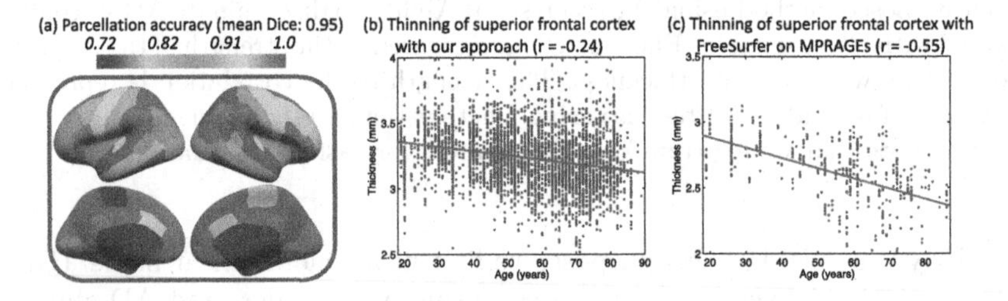

Fig. 5. Results on clinical dataset. (a) Dice scores for parcellation, using FreeSurfer on MPRAGEs as ground truth. (b–c) Thinning of superior frontal cortex in aging, as measured with our method on anisotropic scans (b) and FreeSurfer on MPRAGEs (c).

3.4 Results on the Clinical Dataset

The clinical dataset, despite not being clustered into well defined groups as ADNI, enables us to evaluate our method with the type of data that it is conceived for: a heterogeneous set of brain MRI scans acquired "in the wild". Samples of such scans and outputs produced by our method are shown in Fig. 4. In this experiment, we first used the 581 1 mm MPRAGEs to compute the Dice scores of the Desikan-Killiany parcellation on clinical acquisitions. The results are displayed in Fig. 5a, and show that our proposed method is able to sustain high accuracy in this task (the mean Dice is the same as for ADNI), despite the huge variability in the acquisition protocol of the input scans. As in the previous experiment, we also computed aging curves using all non-1 mm-MPRAGE scans (4,483 in total), while correcting for gender and slice spacing. The fitted curve for a representative region (the superior frontal area, which shows consistent effects in Fig. 3c) is shown in Fig. 5b. While the thinning trend exists, the data are rather noisy and the linear fit ($\rho = -0.24$) underestimates the effect of aging, i.e., the magnitude of the slope. This is apparent when comparing with the fit produced by the 581 MPRAGEs (Fig. 5c, $\rho = -0.55$).

3.5 Discussion and Conclusion

We have presented a novel method for cortical analysis of clinical brain scans of any MRI contrast and resolution that does not require retraining. To the best of our knowledge, this is the first method seeking to solve this difficult problem. The method runs in 2–3 h but could be sped up by replacing some modules (e.g., the spherical registration) with faster learning methods.

Our method provides accurate parcellation across the board, which is helpful in applications like diffusion MRI (e.g., for seeding or constraining tractography with surfaces and parcellations when a T1 scan is unavailable or is difficult to register due to geometric distortion of the diffusion-weighted images). However, we observed increased variability in cortical thickness when processing the highly heterogeneous clinical dataset. Future work will focus on improving the reliability of thickness measurements and provide a confidence for the quality of reconstruction and cortical thickness prediction for lower resolution scans. In such scenarios assessing and modeling geometric covariates (e.g., vertex-wise distance to the nearest slice or angle between surface and acquisition orientation) may help reduce such variability.

Our method and the clinical dataset are publicly available, which enables researchers worldwide to capitalize on millions of retrospective clinical scans to perform cortical analysis currently unattainable in research studies, particularly for rare diseases and underrepresented populations.

Acknowledgment. This work is primarily funded by the National Institute of Aging (1R01AG070988). Further support is provided by, BRAIN Initiative (1RF1MH123195, 1UM1MH130981), National Institute of Biomedical Imaging and Bioengineering (1R01EB031114), Alzheimer's Research UK (ARUK-IRG2019A-003), National Institute of Aging (P30AG062421)

References

1. Billot, B., et al.: SynthSeg: segmentation of brain MRI scans of any contrast and resolution without retraining. Med. Image Anal. **86**, 102789 (2023)
2. Billot, B., Magdamo, C., Cheng, Y., Arnold, S.E., Das, S., Iglesias, J.E.: Robust machine learning segmentation for large-scale analysis of heterogeneous clinical brain MRI datasets. Proc. Natl. Acad. Sci. **120**(9), e2216399120 (2023)
3. Bongratz, F., Rickmann, A.M., Pölsterl, S., Wachinger, C.: Vox2Cortex: fast explicit reconstruction of cortical surfaces from 3D MRI scans with geometric deep neural networks. In: CVPR, pp. 20773–20783 (2022)
4. Cruz, R.S., Lebrat, L., Bourgeat, P., Fookes, C., Fripp, J., Salvado, O.: DeepCSR: a 3D deep learning approach for cortical surface reconstruction. In: WACV. pp, 806–815 (2021)
5. Dale, A.M., Fischl, B., Sereno, M.I.: Cortical surface-based analysis: I. segmentation and surface reconstruction. Neuroimage **9**(2), 179–194 (1999)
6. Desikan, R.S., et al.: An automated labeling system for subdividing the human cerebral cortex on MRI scans into gyral based regions of interest. Neuroimage **31**(3), 968–980 (2006)

7. Dickerson, B.C., Bakkour, A., Salat, D.H., Feczko, E., et al.: The cortical signature of Alzheimer's disease: regionally specific cortical thinning relates to symptom severity in very mild to mild AD dementia and is detectable in asymptomatic amyloid-positive individuals. Cereb. Cortex **19**(3), 497–510 (2009)
8. Fischl, B., Liu, A., Dale, A.M.: Automated manifold surgery: constructing geometrically accurate and topologically correct models of the human cerebral cortex. IEEE Trans. Med. Imaging **20**(1), 70–80 (2001)
9. Fischl, B., Sereno, M., Dale, A.M.: Cortical surface-based analysis: II: inflation, flattening, and a surface-based coordinate system. Neuroimage **9**(2), 195–207 (1999)
10. Glasser, M., Sotiropoulos, S., Wilson, J.A., Coalson, T.S.: The minimal preprocessing pipelines for the human connectome project. Neuroimage **80**, 105–24 (2013)
11. Gopinath, K., Desrosiers, C., Lombaert, H.: SegRecon: learning joint brain surface reconstruction and segmentation from images. In: de Bruijne, M., et al. (eds.) MICCAI 2021. LNCS, vol. 12907, pp. 650–659. Springer, Cham (2021). https://doi.org/10.1007/978-3-030-87234-2_61
12. Hibar, D., Westlye, L.T., Doan, N.T., Jahanshad, N., et al.: Cortical abnormalities in bipolar disorder: an MRI analysis of 6503 individuals from the ENIGMA bipolar disorder working group. Mol. Psychiatry **23**(4), 932–942 (2018)
13. Hoffmann, M., Billot, B., Greve, D.N., Iglesias, J.E., Fischl, B., Dalca, A.V.: SynthMorph: learning contrast-invariant registration without acquired images. IEEE Trans. Med. Imaging **41**(3), 543–558 (2021)
14. Hoopes, A., Iglesias, J.E., Fischl, B., Greve, D., Dalca, A.V.: TopoFit: rapid reconstruction of topologically-correct cortical surfaces. In: MIDL (2021)
15. Iglesias, J., Billot, B., Balbastre, Y., Tabari, A., et al.: Joint super-resolution and synthesis of 1 mm isotropic MP-RAGE volumes from clinical MRI exams with scans of different orientation, resolution & contrast. Neuroimage **237**, 118206 (2021)
16. Iscan, Z., Jin, T.B., Kendrick, A., Szeglin, B., Lu, H., Trivedi, M., et al.: Test-retest reliability of Freesurfer measurements within and between sites: Effects of visual approval process. Hum. Brain Mapp. **36**(9), 3472–3485 (2015)
17. Jack, C.R., Jr., Bernstein, M.A., Fox, N.C., Thompson, P., Alexander, G., et al.: The Alzheimer's disease neuroimaging initiative (ADNI): MRI methods. J. Magn. Reson. Imaging **27**(4), 685–691 (2008)
18. Ma, Q., Robinson, E.C., Kainz, B., Rueckert, D., Alansary, A.: PialNN: A fast deep learning framework for cortical pial surface reconstruction. In: International Workshop on Machine Learning in Clinical Neuroimaging. pp. 73–81 (2021)
19. Oren, O., Kebebew, E., Ioannidis, J.P.: Curbing unnecessary and wasted diagnostic imaging. JAMA **321**(3), 245–246 (2019)
20. Park, J.J., Florence, P., Straub, J., Newcombe, R., Lovegrove, S.: DeepSDF: learning continuous signed distance functions for shape representation. In: CVPR, pp. 165–174 (2019)
21. Pereira, J.B., Ibarretxe, N., Marti, M.J., Compta, Y., et al.: Assessment of cortical degeneration in patients with Parkinson's disease by voxel-based morphometry, cortical folding, and cortical thickness. Hum. Brain Mapp. **33**, 2521–34 (2012)
22. Querbes, O., Aubry, F., Pariente, J., Lotterie, J.A., Démonet, J.F., et al.: Early diagnosis of Alzheimer's disease using cortical thickness: impact of cognitive reserve. Brain **132**(8), 2036–2047 (2009)
23. Ronneberger, O., Fischer, P., Brox, T.: U-net: convolutional networks for biomedical image segmentation. In: Navab, N., Hornegger, J., Wells, W.M., Frangi, A.F. (eds.) MICCAI 2015. LNCS, vol. 9351, pp. 234–241. Springer, Cham (2015). https://doi.org/10.1007/978-3-319-24574-4_28

24. Rosas, H., et al.: Regional and progressive thinning of the cortical ribbon in Huntington's disease. Neurology **58**(5), 695–701 (2002)
25. Salat, D.H., et al.: Thinning of the cerebral cortex in aging. Cereb. Cortex **14**(7), 721–730 (2004)
26. Tobin, J., Fong, R., Ray, A., Schneider, J., Zaremba, W., Abbeel, P.: Domain randomization for transferring deep neural networks from simulation to the real world. In: IROS, pp. 23–30 (2017)

Flow-Based Geometric Interpolation of Fiber Orientation Distribution Functions

Xinyu Nie[1,2] and Yonggang Shi[1,2](✉)

[1] USC Stevens Neuroimaging and Informatics Institute, University of Southern California, Los Angeles, CA 90033, USA
yonggans@usc.edu

[2] Department of Electrical and Computer Engineering, University of Southern California, Los Angeles, CA 90089, USA

Abstract. The fiber orientation distribution function (FOD) is an advanced model for high angular resolution diffusion MRI representing complex fiber geometry. However, the complicated mathematical structures of the FOD function pose challenges for FOD image processing tasks such as interpolation, which plays a critical role in the propagation of fiber tracts in tractography. In FOD-based tractography, linear interpolation is commonly used for numerical efficiency, but it is prone to generate false artificial information, leading to anatomically incorrect fiber tracts. To overcome this difficulty, we propose a flow-based and geometrically consistent interpolation framework that considers peak-wise rotations of FODs within the neighborhood of each location. Our method decomposes a FOD function into multiple components and uses a smooth vector field to model the flows of each peak in its neighborhood. To generate the interpolated result along the flow of each vector field, we develop a closed-form and efficient method to rotate FOD peaks in neighboring voxels and realize geometrically consistent interpolation of FOD components. By combining the interpolation results from each peak, we obtain the final interpolation of FODs. Experimental results on Human Connectome Project (HCP) data demonstrate that our method produces anatomically more meaningful FOD interpolations and significantly enhances tractography performance.

Keywords: Fiber Orientation Distribution · Interpolation · Tractography

1 Introduction

Diffusion MRI (dMRI) is the most widely used technique for studying human brain structural connectivity *in vivo* [1]. Significant improvements in imaging techniques dramatically increased the spatial and angular resolution of dMRI [2] and provided opportunities for advanced models such as fiber orientation distribution (FOD) [3], which facilitates the development of FOD-based fiber tracking for brain connectivity research. However, well-known challenges in current tractography methods generate large amounts of false positives and negatives [4]. While there have been considerable efforts in developing novel fiber tracking methods [5], a critical step in tractography, FOD interpolation, has received rare attention.

H. Greenspan et al. (Eds.): MICCAI 2023, LNCS 14227, pp. 46–55, 2023.
https://doi.org/10.1007/978-3-031-43993-3_5

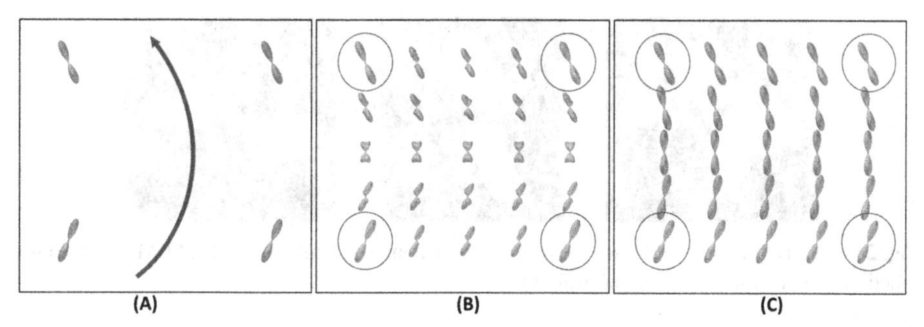

Fig. 1. An illustrative example of FOD interpolations. (A) FODs of four neighboring voxels from a bending fiber bundle were highlighted by red circles in (B) and (C). The blue arrow shows the direction of the fiber bundle. (B) Interpolated FODs by linear interpolation, where artificial peaks have been generated. (C) The interpolated result by our proposed method correctly accounts for rotation and follows the bending geometry of the fiber bundle.

In popular FOD-based tractography, linear interpolation is commonly adopted for numerical efficiency. Still, it often generates artificial directions and ignores rotations between neighboring FODs, as shown in Fig. 1. (B), which can lead to false positive streamlines. To enhance FOD interpolation, a Riemannian framework was proposed in [6]; under the square root reparameterization, the space of FOD functions can form the positive orthant of the unit Hilbert sphere. However, this framework is computationally expensive and sometimes fails to provide anatomically meaningful interpolations [7]. A rotation group action-based framework [7] was proposed that simultaneously averages the shape and rotation of FODs. A later work [8] proposed a rotation-induced Riemannian metric for FODs and introduced a weighted mean for FOD interpolation. However, since only one rotation is used for the whole FOD, these methods cannot handle more general situations where individual FOD peaks experience different rotations. More importantly, these methods have not been adopted in a tractography framework to advance fiber tracking performance due to their numerical complexity.

In this work, we develop a novel framework to perform geometrically consistent interpolation of FODs and demonstrate its effectiveness in enhancing the performance of fiber tracking. We decompose each FOD function with multiple peak lobes into components, each with only one peak lobe. Then, we locally model neighboring voxels' single-peak components, consistent in direction, as a vector field flow fitted by polynomials. Each vector field locally represents the geometry of an underlying fiber bundle and continuously determines the direction of single-peak components within the support. Then, a closed-form solution is developed to account for rotations of FODs represented as spherical harmonics and realize the geometrically consistent interpolation of each FOD component, as shown in Fig. 1. (C). The interpolation of a complete FOD function with multiple peak lobes is obtained by merging the single-peak interpolations from all the covering vector fields. In our experiments, we use HCP data to quantify the accuracy of the proposed FOD interpolation algorithm and show that it achieves superior performance than the commonly used linear interpolation approach. Furthermore, we apply our interpolation method to perform upsampling of FOD fields and significantly improve the performance of FOD-based tractography both qualitatively and quantitatively.

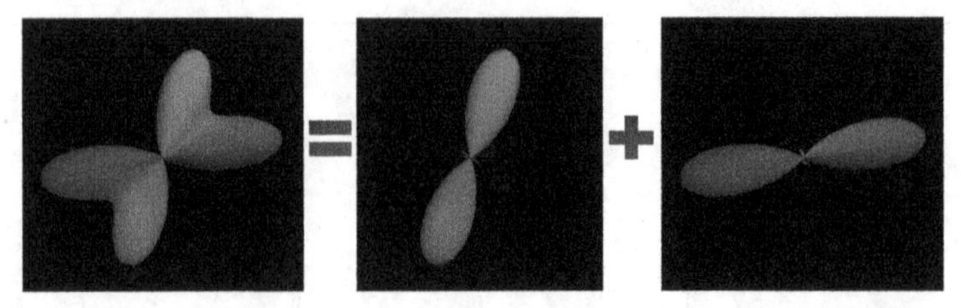

Fig. 2. A FOD function with two peak lobes is decomposed into two SPHARM-based FOD functions, each with only a salient peak lobe.

2 Method

2.1 FOD Decomposition

The fiber orientation distribution (FOD) is an advanced model representing the complicated crossing fiber's geometry [9]. However, the multiple peak lobes of the FOD function pose a challenge for image processing. Our solution is to decompose the FOD function into several independent components, each containing only one peak lobe. A FOD function is conventionally represented under the real spherical harmonics (SPHARMs) basis up to the order N:

$$FOD(\theta, \varphi) = \sum_{n,m} u_n^m Y_n^m(\theta, \varphi) = U^T Y(\theta, \varphi) \quad (1)$$

where Y_n^m is the mth $(-n \leq m \leq n)$ real SPHARM basis at the order n $(0 \leq n \leq N)$ and u_n^m is the coefficient for the corresponding basis, U is the vector that represents all the coefficients u_n^m, and θ and φ are the polar and azimuth angles of the spherical coordinates in R^3. For any FOD function, we expand it using (1) on a unit sphere represented by a triangular mesh, search the peaks on the mesh, and accept the peaks whose value is higher than a threshold THD (e.g., 0.1). For a FOD function with K peak lobes, we solve the following optimization problem for its decomposition:

$$\underset{U_1,\ldots,U_K}{\arg\min} \, || \sum_{k=1}^{K} U_k - U||_2^2 + \lambda_1 \sum_{k=1}^{K} ||A_k U_k - A_k U||_2^2 + \lambda_2 \sum_{k=1}^{K} || \sum_{j \neq k}^{K} A_j U_k||_2^2 \quad (2)$$

where U_k are the coefficients for the decomposed single-peak FOD components, and A_k is the matrix that represents the values of SPHARMs at neighboring directions around the kth peak (vertices within two rings of each peak). The first term enforces the sum of the decomposed single-peak components to equal the original FOD; the second term enforces each component to equal the original FOD near the corresponding peak; the third term suppresses each component around other peaks. We show an example of a FOD function decomposition in Fig. 2, where a FOD function is decomposed into two single-peak components.

2.2 Modeling Single Peak FOD Components as Flow of Vector Fields

For each single-peak FOD component, we model it with the flow of a smooth vector field, which supports geometrically consistent interpolations of FOD components.

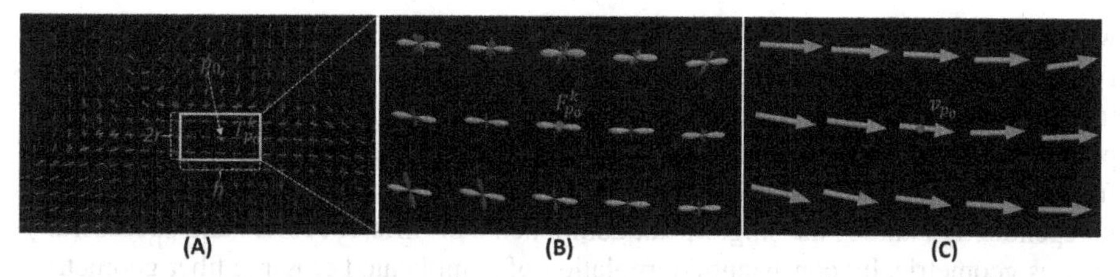

Fig. 3. A vector field within a tube centering at a voxel p_0. (A) Tube $T^k_{p_0}$ (yellow box) centering at p_0 along the peak direction v_{p0} of the kth FOD component $F^k_{p_0}$ (green). (B) The FODs within the tube and the $\{F^k_{p_t}\}$ are in green. (C) The vectors $\{v_{pt}\}$ are picked at voxels within the tube, and these vectors are used to solve (3) and compute the supporting vector field $V^k_{p_0}$. (color figure online)

We represent the kth single-peak component of the FOD function at a voxel p_0 as $F^k_{p_0}$. We choose the peak direction of $F^k_{p_0}$ as the seeding vector v_{p0} of the local supporting vector field. Then we compute a tube $T^k_{p_0}$, centering at p_0, along the direction v_{p0} with a radius r and a height h (Fig. 3. (A)). For each voxel p_t within the tube $T^k_{p_0}$, we choose the single-peak component $F^k_{p_t}$ (Fig. 3. (B)) whose peak direction v_{pt} is closest to v_{p0}, and the peak direction v_{pt} is a vector at p_t (Fig. 3. (C)). We do not pick any vector for voxels without a valid peak direction whose angular difference is less than a threshold θ to the seeding vector v_{p0}. These peak vectors $\{v_{pt}\}$ form a vector field within this tube, and we use a second-order polynomial to fit each component of this vector field:

$$\arg\min_{a_0,a_1,\ldots,a_9} \sum_{t=0}^{card(T^k_{p_0})} ||v^d_{p_t} - a_0 - \sum_{i=1}^{3} a_i x_i - \sum_{l+h=2} a_n x^l_i x^h_j||^2_2 + \lambda_3 \sum_{n=4}^{9} ||a_n||^2 \quad (3)$$

where $v^d_{p_t}$ $(1 \leq d \leq 3)$ represent d^{th} component of the vector at voxel p_t. The second term regulates the second-order coefficients for smoothness. The polynomials are used to model the vector field $V^k_{p_0}$ within the tube $T^k_{p_0}$ that represents the kth underlying fiber bundle locally around the voxel p_0.

2.3 Rotation Calculation for SPHARM-Based FODs

For a target point q where we perform the interpolation, we choose the nearest voxel p_0, which has been augmented with a set of tubes $\{T^k_{p_0}\}$ and vector fields $\{V^k_{p_0}\}$ through the computation of Sect. 2.2. For each vector field $V^k_{p_0}$, we compute the vector v_q at point q using its polynomial representation. Each of the corresponding k^{th} single-peak FOD component $F^k_{p_t}$ from voxels within one voxel distance to q are used for interpolation. First, we rotate each single-peak component $F^k_{p_t}$ so that its peak direction is aligned with the vector v_q. An easy way to compute the rotation is $R_t = exp([\mathbf{r}]^\times)$, where $\mathbf{r}$ is a vector with its direction determined by the crossing product between the peak vectors v_{pt} and v_q, and its length is the angle between v_{pt} and v_q; $[\cdot]^\times$ is the cross-product matrix of a vector [8]. Since the rotated single-peak FOD components are now aligned in direction, we can compute the weighted mean of SPHARM coefficients, which is the interpolated FOD component corresponding to the kth peak around voxel p_O. The weights can be inverse distance or linear interpolation weights. After interpolating all the FOD single-peak components independently, we combine them into the complete interpolated FOD function at point q. The flowchart of the method is shown in Fig. 4. Our framework independently handles the single-peak components of different FODs and successfully obtains geometrically consistent interpolation of complicated crossing fiber geometry.

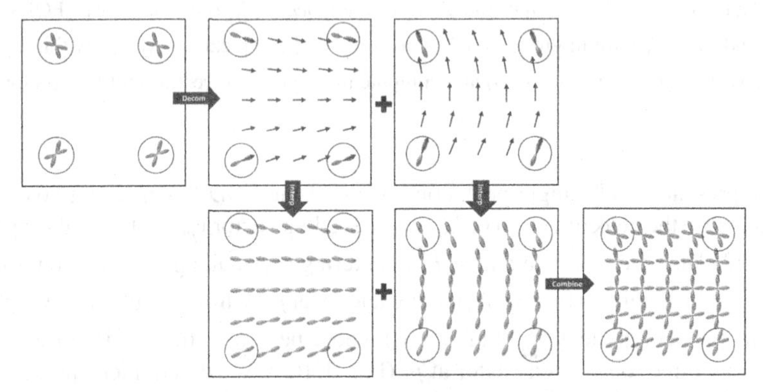

Fig. 4. In the flowchart, the original FODs are highlighted by red circles at the four corners. The FOD functions are decomposed into single-peak components in the first row and locally fitted by vector field flows. In the second row, we rotate the single-peak components at four corners to align with the vectors at each target point, and then we compute their weighted mean at each target point. Finally, we combine the single-peak components into complete FODs. (color figure online)

An essential step for the interpolation above is to transform the FOD function by a rotation *R*. A straightforward numerical way is to expand the FOD function on a spherical triangular mesh and rotate the mesh to rotate the function and compute the inner products between the rotated FOD function and each of the SPHARM basis to obtain the coefficients. However, the numerical method is computationally expensive. Instead, we propose a closed-form solution to derive a matrix from the rotation *R* that can be applied to the coefficients of the SPHARMs. Let $\mathrm{FOD_R}$ represent the FOD function after applying the rotation *R*. We have the following relation:

$$FOD_R(\theta, \varphi) = FOD(\theta_r, \varphi_r) \tag{4}$$

where (θ_r, φ_r) is the coordinate acquired by rotating the coordinate (θ, φ) with the inverse rotation R^{-1}. We represent (4) using the SPHARMs:

$$\sum_{n,m} v_n^m Y_n^m(\theta, \varphi) = \sum_{n,m} u_n^m Y_n^m(\theta_r, \varphi_r) \tag{5}$$

where v_n^m and u_n^m are coefficients for $\mathrm{FOD_R}$ and FOD, respectively. The key to computing coefficients v_n^m is representing the SPHARM function $Y_n^m(\theta_r, \varphi_r)$ by a linear combination of $Y_n^m(\theta, \varphi)$; namely, finding the transformation of SPHARMs under a coordinate system rotation. For rotation R^{-1}, we follow Wigner's work [10] to decompose it as three successive rotations around three axes:

$$R^{-1} = Z_\gamma Y_\beta Z_\alpha \tag{6}$$

where Z_γ and Z_α are the rotations around the current z-axis by angles γ and α, respectively; Y_β is the rotation around the current y-axis by an angle β.

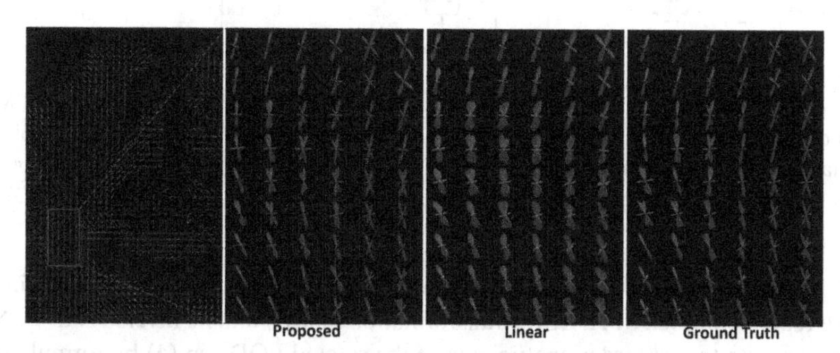

Fig. 5. One interpolated slice of the FODs. For FODs within the red box, a zoomed view of the proposed interpolation, the linear interpolation, and the ground truth are plotted.

We transform the real SPHARMs into complex SPHARMs for more straightforward computation. Based on a group symmetry argument [10], Wigner has proven that Wigner D-matrices can represent the transformation of the nth-order complex SPHARMs between two coordinate systems based on the decomposition in (6):

$$W_n(\theta_r, \varphi_r) = W_n(\theta, \varphi) D^n(\alpha, \beta, \gamma) \tag{7}$$

where W_n is a (2n + 1) complex vector that represents the nth-order complex SPHARMs; D^n is a (2n + 1)-by-(2n + 1) matrix whose elements are represented as:

$$D^n_{m,h}(\alpha, \beta, \gamma) = e^{im\alpha} d^n_{m,h}(\beta) e^{jh\gamma} \tag{8}$$

where the first and third terms correspond to the rotations Z_α and Z_γ in (6), and the rotations around the z-axis are trivial since they only change the azimuth angle φ in the complex SPHARMs. The middle term is induced from the rotation Y_β, which corresponds to a rotation around the y-axis, and is much more complicated:

$$\begin{aligned} d^n_{m,h}(\beta) &= N^h_m \sin^{h-m}(\beta)(1+\cos\beta)^m P^{(h-m,h+m)}_{n-h}(\cos\beta), \\ \text{with } N^h_m &= \frac{1}{2^h}\sqrt{\frac{(n-h)!(n+h)!}{(n-m)!(n+m)!}}, 0 \le m \le h \le n \end{aligned} \tag{9}$$

where P is the Jacobi polynomial, and other elements of this matrix can be induced by the symmetry property [11, 12]. Combining Eqs. (5) and (7), we have:

$$v_n = UD^n(\alpha, \beta, \gamma)U^{-1}u_n \tag{10}$$

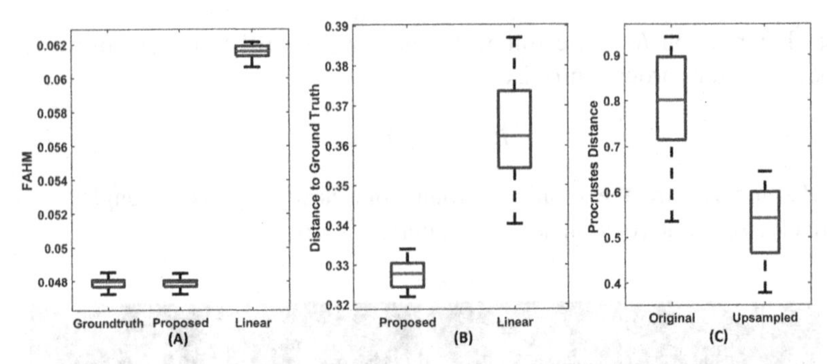

Fig. 6. Boxplots from quantitative comparisons using data from 40 HCP subjects. (A) Mean FAHM of the ground-truth FODs, upsampled FODs from the proposed and linear interpolation. (B) Relative error between up-sampled FODs and ground truth. (C) Procrustes distance of CST bundles from fiber tracking based on linear interpolation (original) and our up-sampled FODs.

where v_n and u_n are real (2n + 1) vectors whose m^{th} element is v_n^m and u_n^m in (5); U is the real to complex SPHARMs transformation matrix, and its inverse is U^{-1}. Now we can compute the nth-order coefficients of the rotated FOD_R in (4) by formula (10). The computation achieved by the closed-form representation is efficient because it only involves small-size matrix operations. For example, the largest matrix in coefficients computation for a FOD function represented by up to 16th-order SPHARMs is 33×33.

2.4 Evaluation Methods

We compare the proposed FOD interpolation method with the linear interpolation of SPHARM coefficients, the most used method in FOD-based tractography. We measure the quality of the interpolated FOD functions based on down-sampling; we down-simple a ground truth FOD volume data to half the resolution, interpolate the down-sampled data to the original resolution, and measure the interpolated FOD functions against the ground truth data based on two metrics. The first metric is to measure the sharpness of the interpolated FOD functions, which indicates the specificity and accuracy of the FOD function. Inspired by the full width at half maximum (FWHM) in signal processing, we define the full area at half maximum (FAHM) of a FOD function f as $FAHM(f) = area(\{x : f(x) > max(f/2)\})/4\pi$. The metric FAHM is more sensitive to boating effects than entropy [7] and generalized fractional anisotropy [8]. Another metric is the relative error between the interpolated FOD function and the ground truth FOD function. The relative error is the L^2 distance of two FOD functions divided by the L^2 norm of the ground truth FOD function.

We also evaluate the effectiveness of the proposed method on tractography. We up-sample the FOD volume images to super-resolution images, including the cortical spinal tract (CST) area that connects the cortical surface to the internal capsule. Then, we run the popular tractography from the MRTrix [13] on the original and super-resolution data. We use an evaluation called Topographic Regularity, an essential property widely presented in motor and sensory pathways [14–17], to show the improvements of the tractography on up-sampled FOD data. We measure the topographic regularity using an

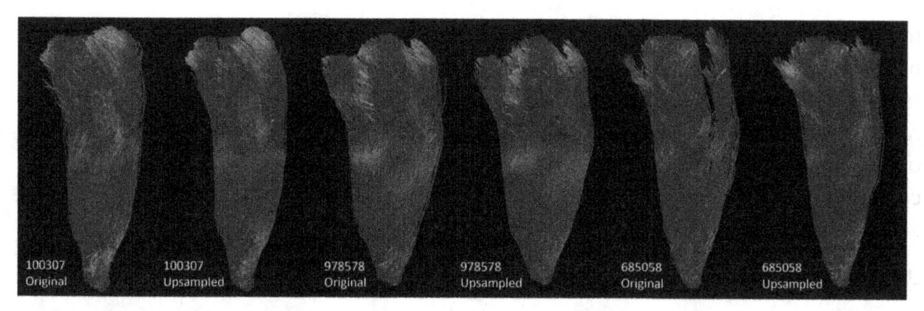

Fig. 7. Tractography of CST on original HCP data and upsampled data for three subjects, and for each bundle, we downsample the number of streamlines to 1000 for visualization.

intuitive metric proposed in [14], where the classical multidimensional scaling (MDS) is used to project both the beginning (cortical surface) and ending points (internal capsule) of the streamlines of each CST bundle to R^2. Then, the Procrustes distance between the projected beginning and ending points is computed to characterize how well topographic regularity is preserved during fiber tracking.

3 Experiment Results

We evaluated the FOD interpolation using 40 HCP subjects [18], including 20 females and 20 males. We reconstructed 16th-order SPHARM-based FODs [9] from the HCP data with an isotropic resolution of 1.25mm. For parameters in our method, we set λ_1, λ_2, and λ_3 in Eqs. (2) and (3) to be 1; the radius r, height h, and θ of the tubes to be three times of voxel size, five times of voxel size, and 10°.

The HCP FOD data is used as the ground truth for down-sampling-based evaluation. We show the FODs from one interpolated slice of a subject in Fig. 5 and highlight the FODs from an ROI (red box) where several fiber bundles cross. Contrasting to the proposed method, FODs from linear interpolation tend to lose their sharpness. For each subject, we computed the FAHM and relative L^2 error for each interpolated FOD, which was then used to compute the mean FAHM and mean relative L^2 error among all interpolated FODs. We show the boxplots of these measures from the 40 HCP subjects in Fig. 6. (A) and (B). The FAHM measurement shows our approach avoids the bloating effects and preserves a similar level of sharpness to the ground truth FODs; the lower mean relative L^2 error to the ground truth from our method further shows the proposed interpolation achieved more accurate interpolation.

We up-sampled the 40 HCP FOD volume images around the CST region to super-resolution images with an isotropic resolution of 0.25 mm. Then, we ran FOD-based probabilistic tractography on the original and up-sampled FOD data using the iFOD1 algorithm of the MRtrix software [13]. In each run, 10K seed points are randomly selected, and the main parameters of iFOD1 are set as: step size = 0.02 mm, which is around 0.1 times the voxel size of the up-sampled image, and angle threshold = 7°. The same parameters were used for the original HCP dataset to avoid the bias of parameters. Three representative examples of the reconstructed CST bundle from the motor cortex to the internal capsule are shown in Fig. 7, where we can easily see that the tracts from the super-resolution FOD by our proposed interpolation method are much smoother and better reflect the somatotopic organizational principles of the CST from neuroanatomy than the baseline tracking results. In Fig. 6. (C), boxplots of the results from Procrustes analyses further confirm this observation and demonstrate that our geometric FOD interpolation algorithm can significantly enhance the anatomical consistency of fiber tracking results.

4 Conclusion

We propose a novel interpolation method for FOD function with enhanced consistency of fiber geometry. The experiments show that our method provides a more accurate interpolation of FODs and can generate super-resolution FODs via upsampling to improve the tractography's performance significantly. In future work, we will integrate the proposed FOD interpolation with tractography algorithms and validate its performance in reducing false positives and negatives in challenging fiber bundles.

Acknowledgement. This work is supported by the National Institute of Health (NIH) under grants R01EB022744, RF1AG077578, RF1AG056573, RF1AG064584, R21AG064776, P41EB015922, U19AG078109.

References

1. Wandell, B.A.: Clarifying human white matter. Annu. Rev. Neurosci. **39**, 103–128 (2016)
2. Ugurbil, K., et al.: Pushing spatial and temporal resolution for functional and diffusion MRI in the human connectome project. Neuroimage **80**, 80–104 (2013)
3. Tournier, J.D., Calamante, F., Connelly, A.: Robust determination of the fibre orientation distribution in diffusion MRI: non-negativity constrained super-resolved spherical deconvolution. Neuroimage **35**, 1459–1472 (2007)
4. Thomas, C., et al.: Anatomical accuracy of brain connections derived from diffusion MRI tractography is inherently limited. Proc. Natl. Acad. Sci. USA **111**, 16574–16579 (2014)
5. Aydogan, D.B., Shi, Y.: Parallel transport tractography. IEEE Trans. Med. Imaging **40**, 635–647 (2021)
6. Goh, A., Lenglet, C., Thompson, P.M., Vidal, R.: A nonparametric Riemannian framework for processing high angular resolution diffusion images and its applications to ODF-based morphometry. Neuroimage **56**, 1181–1201 (2011)
7. Cetingul, H.E., Afsari, B., Wright, M.J., Thompson, P.M., Vidal, R.: Group action induced averaging for hardi processing. In: Proceedings of IEEE International Symposium on Biomedical Imaging, pp. 1389–1392 (2012)
8. Li, J., Shi, Y., Toga, A.W.: Diffusion of fiber orientation distribution functions with a rotation-induced riemannian metric. Med. Image Comput. Comput. Assist. Interv. **17**, 249–256 (2014)
9. Tran, G., Shi, Y.: Fiber orientation and compartment parameter estimation from multi-shell diffusion imaging. IEEE Trans. Med. Imaging **34**, 2320–2332 (2015)
10. Wigner, E.P.: Group theory and its application to the quantum mechanics of atomic spectra. Am. J. Phys. **28**, 408–409 (1960)
11. Aubert, G.: An alternative to Wigner d-matrices for rotating real spherical harmonics. AIP Adv. **3**, 62121–062121 (2013)

12. Lai, S.-T., Palting, P., Chiu, Y.-N.: On the closed form of Wigner rotation matrix elements. J. Math. Chem. **19**, 131–145 (1996)
13. Tournier, J.D., Calamante, F., Connelly, A.: MRtrix: diffusion tractography in crossing fiber regions. Int. J. Imag. Syst. Tech. **22**, 53–66 (2012)
14. Aydogan, D.B., Shi, Y.: Tracking and validation techniques for topographically organized tractography. Neuroimage **181**, 64–84 (2018)
15. Jbabdi, S., Sotiropoulos, S.N., Behrens, T.E.: The topographic connectome. Curr. Opin. Neurobiol. **23**, 207–215 (2013)
16. Nie, X., Shi, Y.: Topographic filtering of tractograms as vector field flows. In: Shen, D., et al. Medical Image Computing and Computer Assisted Intervention – MICCAI 2019. MICCAI 2019. Lecture Notes in Computer Science, vol 11766, pp. 564–572. Springer, Cham (2019). https://doi.org/10.1007/978-3-030-32248-9_63
17. Patel, G.H., Kaplan, D.M., Snyder, L.H.: Topographic organization in the brain: searching for general principles. Trends Cogn Sci **18**, 351–363 (2014)
18. Van Essen, D.C., et al.: The human connectome project: a data acquisition perspective. Neuroimage **62**, 2222–2231 (2012)

Learnable Subdivision Graph Neural Network for Functional Brain Network Analysis and Interpretable Cognitive Disorder Diagnosis

Dongdong Chen[1], Mengjun Liu[1], Zhenrong Shen[1], Xiangyu Zhao[1], Qian Wang[2], and Lichi Zhang[1(✉)]

[1] School of Biomedical Engineering, Shanghai Jiao Tong University, Shanghai, China
lichizhang@sjtu.edu.cn
[2] School of Biomedical Engineering, ShanghaiTech University, Shanghai, China

Abstract. Different functional configurations of the brain, also named as "brain states", reflect a continuous stream of brain cognitive activities. These distinct brain states can confer heterogeneous functions to brain networks. Recent studies have revealed that extracting information from functional brain networks is beneficial for neuroscience analysis and brain disorder diagnosis. Graph neural networks (GNNs) have been demonstrated to be superior in learning network representations. However, these GNN-based methods have few concerns about the heterogeneity of brain networks, especially the heterogeneous information of brain network functions induced by intrinsic brain states. To address this issue, we propose a learnable subdivision graph neural network (LSGNN) for brain network analysis. The core idea of LSGNN is to implement a learnable subdivision method to encode brain networks into multiple latent feature subspaces corresponding to functional configurations, and realize the feature extraction of brain networks in each subspace, respectively. Furthermore, considering the complex interactions among brain states, we also employ the self-attention mechanism to acquire a comprehensive brain network representation in a joint latent space. We conduct experiments on a publicly available dataset of cognitive disorders. The results affirm that our approach can achieve outstanding performance and also instill the interpretability of the brain network functions in the latent space. Our code is available at https://github.com/haijunkenan/LSGNN.

Keywords: Brain Network Analysis · Graph Neural Network · Brain States · Self-Attention Mechanism

---

Supplementary Information The online version contains supplementary material available at https://doi.org/10.1007/978-3-031-43993-3_6.

H. Greenspan et al. (Eds.): MICCAI 2023, LNCS 14227, pp. 56–66, 2023.
https://doi.org/10.1007/978-3-031-43993-3_6

1 Introduction

Studies on brain functional dynamics show that the brain network contains a variety of distinct functional configurations (brain states) during the course of brain cognition activities [5,11]. These distinct brain states depict the heterogeneous functional signature of the brain network (e.g., Visual, Attention, etc.) [1]. Generally, brain states can be characterized by discriminate connections constructed from brain networks, and various strategies have been developed to identify brain states with the objective of understanding how heterogeneous information is represented in the brain [9,24]. For example, the whole-brain functional connectivity patterns derived from independent component analysis (ICA) are employed for brain network analysis [13,18]. However, they typically struggle to perform well on high-dimensional data and need further design techniques for feature selection.

Recently, graph neural networks (GNNs) have been proven to be helpful in brain network analysis, due to their powerful ability in analyzing graph-structured data [21]. However, the existing GNN methods for heterogeneity mostly deal with the semantic heterogeneity from different modalities [22,23] or the connectivity heterogeneity among nodes [10], and rarely consider the functional heterogeneity of the whole brain network. Therefore, it may lead to a sub-optimal performance on disorder diagnosis. On the other hand, GNN as a deep learning model is typically poorly interpretable in brain network analysis [14]. Although several methods for GNN interpretation have been proposed, most of them concentrate on node-level or subject-level analysis. For instance, BrainGNN provides insights on the salient brain region of interest (ROIs) through specific node pooling operation [15]; IBGNN discusses the neural system mapping of subjects in different categories with an explanation generator mask [4]. However, few studies have been conducted on the interpretable analysis of brain states at network-level, especially the relationship between heterogeneous functional features of brain networks and corresponding disease diagnosis remains unexplored.

To address the above issues, we propose a learnable subdivision graph neural network to investigate brain networks with heterogeneous functional signatures, and the functional brain state can be combined with corresponding latent subspace for interpretable brain disorder diagnosis. The main contributions of this paper are as follows: 1)We propose a novel Learnable Subdivision Graph Neural Network (LSGNN) model for brain network analysis, which implements the extraction of heterogeneous features of brain networks under various functional configurations. 2) We develop a novel assignment method, that can encode brain networks into multiple latent feature subspaces in a learnable way. 3) Our method instills the interpretability of the latent space corresponding to brain states, which is beneficial to unveiling insights into the relationship between the signature of functional brain networks and cognitive disorder diagnosis.

2 Method

The framework of the LSGNN method is presented in Fig. 1, which is composed of three major components: the functional subdivision block (FSB), multiple GNN modules including the GNN layer (with shared weights) and graph pooling layer, and the functional aggregation block (FAB). Here, FSB is designed to automatically learn an assignment matrix to obtain the mask of each brain state. The inputs of FSB are acquired using two separate GNN layers with the same structure but different parameters. Furthermore, multiple GNN modules are designed to learn brain network representations under distinct brain states. Finally, FAB is developed to aggregate the information into a joint latent space, and acquire a comprehensive brain network representation for brain disorder diagnosis.

2.1 Preliminary

Problem Definition. The input of proposed model is a set of N weighted brain networks. For each network $\mathcal{G} = \{\mathcal{V}, \mathcal{E}, A\}$, $\mathcal{V} = \{v_i\}_{i=1}^{M}$ involves M nodes defined by the ROIs on a specific brain parcellation [7], and $\mathcal{E}$ records the distinct edge connections in each subject, which is represented with a weighted adjacency matrix $A \in \mathbb{R}^{M \times M}$ describing the correlation strengths among ROIs. The model finally outputs the prediction result $\hat{y}_n$ for each subject n.

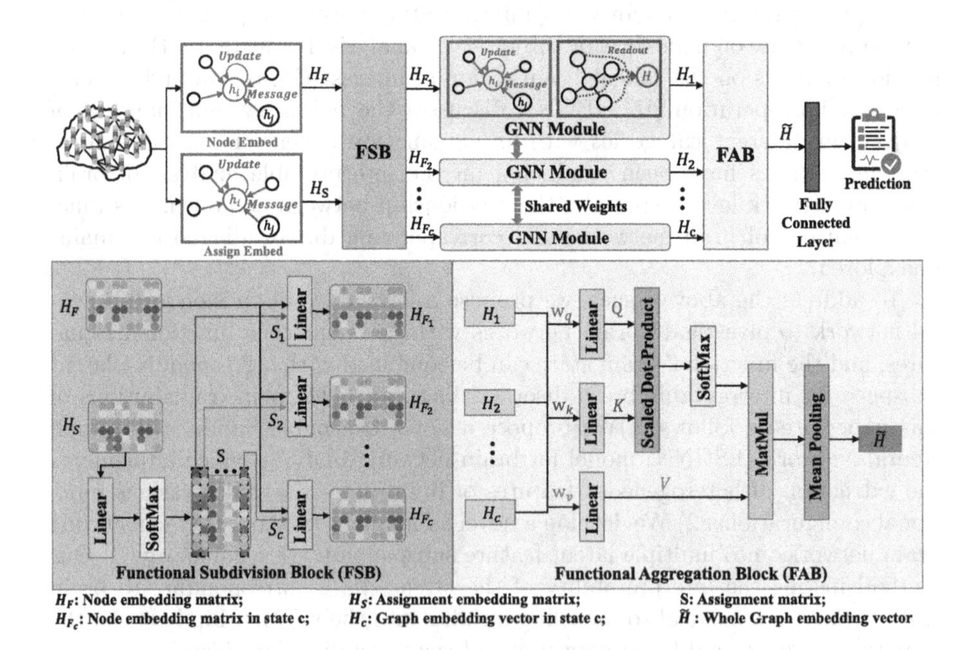

Fig. 1. Overview of the LSGNN framework which consists of FSB and FAB modules for learnable subdivision embedding and comprehensive brain network representation, respectively.

The GNN Module. The GNN module is superior in extracting structural information from the network, which is attributed by two kinds of layers [3]: 1) Message-passing-based GNN layers that extracts the embedding of ROI nodes through iteratively updating information with structural connections. The propagation rule can be formulated in matrix form as: $H^{(l+1)} = GNN\big(A, H^{(l)}; \theta^{(l)}\big)$, where $H^{(l+1)} \in \mathbb{R}^{M \times D}$ are the embeddings computed after l step of the GNN layers, and D is the embedding dimension. GNN is the message propagation function using a combination of linear transformations and ReLU non-linearities, which depends on the adjacency matrix A, trainable parameters $\theta^{(l)}$, and the node embeddings $H^{(l)}$ generated from the previous step. The input node embeddings H^0 are initialized using the node features on the graph. 2) The graph pooling layer. It realizes the aggregation of global information at the graph level using readout to convert all node embeddings into a graph embedding vector. Here, the readout function can be performed using the average pool.

2.2 Functional Subdivision Block

Considering the heterogeneity of brain networks induced by intrinsic functional brain states, we propose to encode brain networks into multiple latent feature subspaces corresponding to functional brain states. The key to achieve this goal is to build an assignment matrix that can allocate the feature representation of each dimension in the brain network to the latent subspace related to distinct brain states.

Here, we innovatively design a learnable assignment method, which is determined by the embedding matrix and automatically updated with the training of the whole model. Specifically, we first utilize two separate GNN layers to generate embeddings for preliminary node feature matrix $H_F = GNN_{\text{feat}}\left(A, H; \theta_{feat}\right)$ and input embeddings of assignment matrix $H_S = GNN_{\text{assig}}\left(A, H; \theta_{assig}\right)$, respectively. Note that these two GNN layers use the same data as input, but have distinct parameterizations and play separate roles. We continue to transpose the input embeddings of assignment matrix as $H_S^T \in \mathbb{R}^{D \times M}$, and the generation of assignment matrix $S \in \mathbb{R}^{D \times C}$ as follows:

$$S = \operatorname{softmax}\left(\mathrm{MLP}_1\left(H_S^T\right)\right), \tag{1}$$

where the softmax function is applied in a row-wise fashion, and the output dimension of MLP_1 corresponds to a pre-defined number C of brain states.

With the assignment matrix and node embedding matrix in hand, we further extract distinct node embeddings in different latent subspaces corresponding to brain states. For each column $\{S_c\}_{c=1}^C$ in the assignment matrix, it is essentially a mask vector where each element represents the probability that the feature is assigned to the brain state c. Therefore, we can obtain the brain network representation in each brain state through the product of feature and its corresponding assignment probability.

$$H_{F_c} = \mathrm{MLP}_2\left(H_F \odot \mathrm{R}\{S_c\}^T\right), \tag{2}$$

where R is a repeat function that extends the mask vector to the same dimension of the node embedding matrix, $\odot$ denotes element-wise multiplication, and MLP_2 maps each new node embedding matrix to distinct feature latent subspaces, which characterize heterogeneous information under different brain states.

2.3 Functional Aggregation Block

After assigning the node embedding matrix of brain network into distinct latent subspaces, we utilize multiple GNN modules (each consists of a GNN layer and a graph pooling layer) to obtain the graph embedding of brain network $H_c \in \mathbb{R}^{1\times D}$ corresponding to each brain state respectively. These GNN layers are performed with shared weight to reduce the complexity of the model. Considering that each brain state has different contributions to the final brain network representation, we propose a functional aggregation block based on the self-attention mechanism to aggregate the information into a joint latent space, and acquire a comprehensive brain network representation for brain disorder diagnosis.

In practice, the graph embedding in every brain state is first packed together into a matrix $\{H_C | \{H_c\}_{c=1}^{C}, H_C \in \mathbb{R}^{C\times D}\}$ and then mapped into three matrices, including query $Q = H_C W_q$, key $K = H_C W_k$, and value $V = H_C W_v$, where the weight matrices for W_q, W_k, W_v are the learned linear mappings. Therefore, we can calculate the self-attention by mapping a query and a set of key-value pairs, and combining with a mean pooling to obtain the whole graph embedding vector $\widetilde{H} \in \mathbb{R}^{1\times D}$ for the brain network.

$$\widetilde{H} = \frac{1}{C}\sum_{c=1}^{C} \text{softmax}(QK^T/\sqrt{D})V. \tag{3}$$

Here, the dot product is adopted to reduce computational and storage costs, softmax is used to normalize the self-attention and the scale $\sqrt{D}$ prevents the saturation led by softmax function. Finally, we fed the whole graph embedding vector into a fully connected layer to predict the diagnostic result $\hat{y}_n$.

2.4 Objective Function

In this work, we design an objective function composed of three components: First, we employ conventional supervised cross-entropy objective towards ground-truth y_n disorder prediction, defined as

$$\mathcal{L}_{\text{CLF}} = -\frac{1}{N}\sum_{n=1}^{N} \left(y_n \log\left(\hat{y}_n\right) + (1 - y_n)\log\left(1 - \hat{y}_n\right)\right). \tag{4}$$

Second, since each row of the assignment matrix S_i represents the probability of allocating the feature of this dimension to different latent subspaces, it should generally be close to a one-hot vector, i.e., each dimension feature is assigned to each latent subspace. Therefore, we design an entropy loss to reduce the uncertainty of mapping distribution by minimizing entropy H_{S_i}:

$$\mathcal{L}_{\mathrm{E}} = \frac{1}{D}\sum_{i=1}^{D} H(S_i). \tag{5}$$

Finally, to ensure the generalization ability of the model and reduce over-fitting, we add an $L2$ normalization term. To summarize, the total loss of the proposed model can be formulated as: $\mathcal{L} = \mathcal{L}_{\mathrm{CLF}} + \alpha * \mathcal{L}_{\mathrm{E}} + \beta * L_2$, where α, β are hyper-parameters that determine the relative importance of feature fusion loss items.

3 Experiments

3.1 Dataset and Experimental Settings

We evaluate our framework on publicly available Alzheimer's Disease Neuroimaging Initiative (ADNI) dataset [16], which includes 193 normal controls (NC), 240 early mild cognitive impairment (EMCI), and 149 late mild cognitive impairment (LMCI). We use them to form three binary classification tasks abbreviated as N-E, N-L, and E-L. The fMRI data is preprocessed in a standardized protocol including slice time correction, motion correction, spatial and temporal filtering, and covariates regression [2]. We follow the general process of the GNN-based method in brain network analysis and use the AAL atlas [19] to define 90 ROIs for every subject. We continue to construct brain networks using Pearson correlations.

We compare our proposed model with five different methods, including one conventional model (1) SVM [6] where functional brain networks are reshaped as feature vectors and are then put into models, two representative GNN models: (2) GCN [12], (3) GAT [20], and two state-of-the-art GNN-based models specifically designed for brain networks: (4) BrainGNN [15] and (5) IBGNN [4].

All deep learning experiments are conducted on NVIDIA GeForce GTX TITAN X GPUs with PyTorch [17] framework. We perform a grid search to determine the better choice for $\alpha = 10^{-3}$ and $\beta = 10^{-2}$, and set parameter $c = 7$ representing different configurations of the brain as suggested in [1,8]. We refer readers of interest to supplementary materials for detailed experimental settings. All reported results are averaged across 10 times ten-fold cross-validations. We finally adopt four commonly used metrics to evaluate all methods, including classification accuracy (ACC), sensitivity (SEN), specificity (SPE), and AUC.

3.2 Result Analysis

Table 1 shows the classification results of all methods on all tasks. We can have the following observations. First, compared with the conventional machine learning method (i.e., SVM), deep learning models generally achieve better performance on all tasks in terms of four evaluation metrics. It is indicated that the brain features obtained by automatic learning with neural networks may be better than the traditional handcrafted features in the diagnosis of brain diseases. Second, two GNN models specifically designed for brain networks (i.e.,

BrainGNN, and IBGNN) achieve better results than classical GNN models (i.e., GCN, and GAT), demonstrating the significance of considering the biomedical characteristics of brain networks when applying GNN-related models to brain network analysis. Finally, the effectiveness of our design for heterogeneous properties of the brain network is demonstrated by its superior performance compared with other SOTA models. Moreover, our method is statistically significantly better than other methods (with $p < 0.05$) based on pairwise t-test. For instance, in terms of ACC values, the proposed LSGNN obtains the improvement of at least 3% compared with the best alternatives (i.e., IBGNN) on all tasks. It is concluded that LSGNN is feasible to separate the heterogeneous feature of brain networks in latent space, which is beneficial to capture complex information through multiple GNN modules under distinct brain states, thus improving the ability of brain network representation.

Table 1. Classification results (mean ± std) of all methods on three tasks (%).

Tasks	Metrics	SVM	GCN	GAT	BrainGNN	IBGNN	LSGNN
N-E	ACC	$65.82_{\pm 1.23}$	$69.35_{\pm 1.39}$	$71.20_{\pm 1.11}$	$74.31_{\pm 1.48}$	$75.79_{\pm 1.39}$	$\mathbf{78.88}_{\pm 1.58}$
	SEN	$58.78_{\pm 1.34}$	$69.42_{\pm 1.85}$	$72.33_{\pm 1.75}$	$73.29_{\pm 1.80}$	$74.95_{\pm 2.13}$	$\mathbf{76.52}_{\pm 1.66}$
	SPE	$67.92_{\pm 1.28}$	$65.28_{\pm 1.44}$	$65.37_{\pm 1.28}$	$66.24_{\pm 1.92}$	$67.12_{\pm 1.03}$	$\mathbf{68.52}_{\pm 1.49}$
	AUC	$66.16_{\pm 1.33}$	$70.97_{\pm 1.71}$	$72.15_{\pm 1.29}$	$73.12_{\pm 1.30}$	$74.89_{\pm 1.86}$	$\mathbf{79.42}_{\pm 1.53}$
N-L	ACC	$69.14_{\pm 1.11}$	$74.52_{\pm 1.72}$	$75.49_{\pm 1.81}$	$82.34_{\pm 1.14}$	$82.90_{\pm 2.32}$	$\mathbf{87.47}_{\pm 1.71}$
	SEN	$62.48_{\pm 1.08}$	$79.77_{\pm 1.28}$	$78.34_{\pm 1.19}$	$79.21_{\pm 1.99}$	$78.10_{\pm 2.25}$	$\mathbf{81.58}_{\pm 1.52}$
	SPE	$72.92_{\pm 1.14}$	$75.98_{\pm 1.92}$	$75.48_{\pm 2.42}$	$78.12_{\pm 1.23}$	$77.85_{\pm 2.11}$	$\mathbf{79.86}_{\pm 1.33}$
	AUC	$71.45_{\pm 1.12}$	$76.59_{\pm 1.54}$	$74.29_{\pm 1.66}$	$78.74_{\pm 1.39}$	$79.12_{\pm 2.50}$	$\mathbf{83.15}_{\pm 1.55}$
E-L	ACC	$62.95_{\pm 1.14}$	$66.53_{\pm 2.20}$	$65.49_{\pm 2.30}$	$67.62_{\pm 2.25}$	$70.23_{\pm 1.95}$	$\mathbf{74.24}_{\pm 1.64}$
	SEN	$61.63_{\pm 1.09}$	$68.42_{\pm 1.94}$	$68.12_{\pm 2.89}$	$68.33_{\pm 1.93}$	$68.81_{\pm 1.69}$	$\mathbf{70.55}_{\pm 1.59}$
	SPE	$64.40_{\pm 1.05}$	$62.87_{\pm 2.37}$	$64.16_{\pm 1.97}$	$61.59_{\pm 2.61}$	$62.49_{\pm 1.30}$	$\mathbf{65.37}_{\pm 1.39}$
	AUC	$61.48_{\pm 1.10}$	$62.95_{\pm 1.98}$	$63.28_{\pm 2.21}$	$64.82_{\pm 2.10}$	$67.39_{\pm 1.51}$	$\mathbf{70.08}_{\pm 1.48}$

3.3 Ablation Study

We conduct ablation studies to verify the effectiveness of 1) the learnable assignment method in the FSB module, 2) the self-attention mechanism-based fusion method in the FAB module, and 3) entropy loss $\mathcal{L}_{\mathrm{E}}$. Specifically, In the FSB module, we compare the proposed learnable method with a fixed method, where we utilize k-means to cluster feature embedding into several latent subspaces with the same number as the learnable assignment matrix done. In the FAB module, we try a typically simple feature fusion method (i.e., feature concatenation) without self-attention. For the loss function, we conduct comparative experiments on whether to include entropy loss. The classification results of all six methods on the N-E task are listed in Table 2. It can be observed that the proposed learnable assignment method and self-attention mechanism-based fusion module are effective in cognitive disorder diagnosis. Furthermore, the results of entropy loss ablation prove that the introduction of entropy loss greatly enhances the robustness of the proposed model.

Table 2. Ablation study of LSGNN with different modules on N-E task.

FSB		FAB		Loss		Metrics		
Learnable	Fixed	Self-Att	Concat	$\mathcal{L}_{\mathrm{E}}$	Without	ACC	SEN	SPE
√		√		√		**78.88**$_{\pm\mathbf{1.58}}$	**76.52**$_{\pm\mathbf{1.66}}$	**68.52**$_{\pm 1.49}$
√		√			√	74.18$_{\pm 3.38}$	72.23$_{\pm 2.72}$	64.12$_{\pm 3.19}$
√			√	√		76.42$_{\pm 1.82}$	75.82$_{\pm 1.71}$	66.39$_{\pm\mathbf{1.13}}$
√			√		√	74.20$_{\pm 3.41}$	72.11$_{\pm 2.59}$	65.86$_{\pm 2.98}$
	√	√			√	72.30$_{\pm 2.19}$	71.12$_{\pm 1.74}$	64.23$_{\pm 2.45}$
	√		√		√	71.93$_{\pm 2.04}$	71.27$_{\pm 1.98}$	65.23$_{\pm 1.79}$

3.4 Interpretability of Brain States

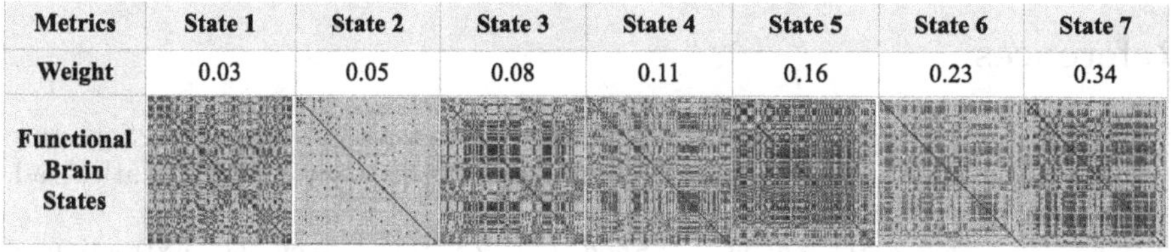

Metrics	State 1	State 2	State 3	State 4	State 5	State 6	State 7
Weight	0.03	0.05	0.08	0.11	0.16	0.23	0.34
Functional Brain States							

Fig. 2. Interpretability analysis of brain network in distinct brain states on N-E task.

To investigate brain states and revealing the impact of different brain states on diseases, we plot the functional brain states under distinct brain network feature subspaces on the N-E task. Specifically, we first calculate the assignment matrix using Eq. 2 and further compute the sum of each column combined with normalization. Each element is the probability weight of each brain state in the whole feature space, which also reflects the impact of each brain state on the final diagnosis task. The weight values of different brain states are sorted in the second line in Fig. 2. It is evident that the divergence of the impacts from distinct brain states is significant. For example, the weight of state 7 surpasses that of state 1 by over 10 times. The possible reason is that the heterogeneous functions (e.g. Visual, Attention, Memory, etc.) conferred from distinct brain states may have unequal contributions to the cognitive activities.

Then, for each brain state c, we extract node embedding matrix H_{F_c} and construct a corresponding functional connectivity matrix using Pearson correlation. Finally, we obtain distinct brain states by computing the average results of testing samples on the N-E task. From Fig. 2, we have an interesting observation that the brain state that has a great impact on diagnosis shows rich club regions, which reveals that there exist multiple subnetworks in the brain network that may correlate to different brain functions.

4 Conclusion

In this paper, we propose a novel learnable subdivision graph neural network method for functional brain network analysis and interpretable cognitive disorder diagnosis. Specifically, brain networks are embedded into multiple latent feature subspaces corresponding to functional configurations in a learnable way. Experimental results of the cognitive disorder diagnosis tasks verify the effectiveness of our proposed method. A direct future direction based on this work is to utilize heterogeneous graph construction techniques to describe brain network patterns. This allows for consideration of brain network heterogeneity from the initial step of brain network modeling, which could lead to a better understanding of brain networks.

Acknowledgement. This work was supported by the National Natural Science Foundation of China (No. 62001292).

References

1. Barttfeld, P., Uhrig, L., Sitt, J.D., Sigman, M., Jarraya, B., Dehaene, S.: Signature of consciousness in the dynamics of resting-state brain activity. Proc. Natl. Acad. Sci. **112**(3), 887–892 (2015)
2. Chen, X., Zhang, H., Zhang, L., Shen, C., Lee, S.W., Shen, D.: Extraction of dynamic functional connectivity from brain grey matter and white matter for MCI classification. Hum. Brain Mapp. **38**(10), 5019–5034 (2017)
3. Cui, H., et al.: Braingb: a benchmark for brain network analysis with graph neural networks. IEEE Trans. Med. Imaging (2022)
4. Cui, Hejie, Dai, Wei, Zhu, Yanqiao, Li, Xiaoxiao, He, Lifang, Yang, Carl: Interpretable Graph Neural Networks for Connectome-Based Brain Disorder Analysis. In: Wang, Linwei, Dou, Qi., Fletcher, P. Thomas., Speidel, Stefanie, Li, Shuo (eds.) Medical Image Computing and Computer Assisted Intervention – MICCAI 2022: 25th International Conference, Singapore, September 18–22, 2022, Proceedings, Part VIII, pp. 375–385. Springer, Cham (2022). https://doi.org/10.1007/978-3-031-16452-1_36
5. Deco, G., Jirsa, V.K., McIntosh, A.R.: Resting brains never rest: computational insights into potential cognitive architectures. Trends Neurosci. **36**(5), 268–274 (2013)
6. Dyrba, M., Grothe, M., Kirste, T., Teipel, S.J.: Multimodal analysis of functional and structural disconnection in a Alzheimer's disease using multiple kernel SVM. Hum. Brain Mapp. **36**(6), 2118–2131 (2015)
7. Figley, T.D., Mortazavi Moghadam, B., Bhullar, N., Kornelsen, J., Courtney, S.M., Figley, C.R.: Probabilistic white matter atlases of human auditory, basal ganglia, language, precuneus, sensorimotor, visual and visuospatial networks. Front. Hum. Neurosci. **11**, 306 (2017)

8. Gomez, Chloé, Grigis, Antoine, Uhrig, Lynn, Jarraya, Béchir.: Characterization of Brain Activity Patterns Across States of Consciousness Based on Variational Auto-Encoders. In: Wang, Linwei, Dou, Qi., Fletcher, P. Thomas., Speidel, Stefanie, Li, Shuo (eds.) Medical Image Computing and Computer Assisted Intervention – MICCAI 2022: 25th International Conference, Singapore, September 18–22, 2022, Proceedings, Part I, pp. 419–429. Springer, Cham (2022). https://doi.org/10.1007/978-3-031-16431-6_40
9. Grigis, Antoine, Gomez, Chloé, Frouin, Vincent, Uhrig, Lynn, Jarraya, Béchir.: Interpretable Signature of Consciousness in Resting-State Functional Network Brain Activity. In: Wang, Linwei, Dou, Qi., Fletcher, P. Thomas., Speidel, Stefanie, Li, Shuo (eds.) Medical Image Computing and Computer Assisted Intervention – MICCAI 2022: 25th International Conference, Singapore, September 18–22, 2022, Proceedings, Part I, pp. 261–270. Springer, Cham (2022). https://doi.org/10.1007/978-3-031-16431-6_25
10. Guo, J., Li, J., Leng, D., Pan, L.: Heterogeneous graph based deep learning for biomedical network link prediction. arXiv preprint arXiv:2102.01649 (2021)
11. Hudetz, A.G., Liu, X., Pillay, S.: Dynamic repertoire of intrinsic brain states is reduced in propofol-induced unconsciousness. Brain connectivity **5**(1), 10–22 (2015)
12. Kipf, T.N., Welling, M.: Semi-supervised classification with graph convolutional networks. arXiv preprint arXiv:1609.02907 (2016)
13. Leibovitz, Rotem, Osin, Jhonathan, Wolf, Lior, Gurevitch, Guy, Hendler, Talma: fMRI Neurofeedback Learning Patterns are Predictive of Personal and Clinical Traits. In: Wang, Linwei, Dou, Qi., Fletcher, P. Thomas., Speidel, Stefanie, Li, Shuo (eds.) Medical Image Computing and Computer Assisted Intervention – MICCAI 2022: 25th International Conference, Singapore, September 18–22, 2022, Proceedings, Part I, pp. 282–294. Springer, Cham (2022). https://doi.org/10.1007/978-3-031-16431-6_27
14. Li, X., Dvornek, N.C., Zhou, Y., Zhuang, J., Ventola, P., Duncan, J.S.: Graph Neural Network for Interpreting Task-fMRI Biomarkers. In: Shen, D., et al. (eds.) MICCAI 2019. LNCS, vol. 11768, pp. 485–493. Springer, Cham (2019). https://doi.org/10.1007/978-3-030-32254-0_54
15. Li, X., et al.: Braingnn: Interpretable brain graph neural network for fMRI analysis. Med. Image Anal. **74**, 102233 (2021)
16. Misra, C., Fan, Y., Davatzikos, C.: Baseline and longitudinal patterns of brain atrophy in MCI patients, and their use in prediction of short-term conversion to AD: results from ADNI. Neuroimage **44**(4), 1415–1422 (2009)
17. Paszke, A., et al.: Pytorch: An imperative style, high-performance deep learning library. Advances in neural information processing systems 32 (2019)
18. Shirer, W.R., Ryali, S., Rykhlevskaia, E., Menon, V., Greicius, M.D.: Decoding subject-driven cognitive states with whole-brain connectivity patterns. Cereb. Cortex **22**(1), 158–165 (2012)
19. Tzourio-Mazoyer, N., et al.: Automated anatomical labeling of activations in SPM using a macroscopic anatomical parcellation of the MNI MRI single-subject brain. Neuroimage **15**(1), 273–289 (2002)
20. Velickovic, P., Cucurull, G., Casanova, A., Romero, A., Lio, P., Bengio, Y., et al.: Graph attention networks. stat **1050**(20), 10–48550 (2017)
21. Xu, K., Hu, W., Leskovec, J., Jegelka, S.: How powerful are graph neural networks? In: International Conference on Learning Representations (2019)

22. Yao, Dongren, Yang, Erkun, Sun, Li., Sui, Jing, Liu, Mingxia: Integrating Multimodal MRIs for Adult ADHD Identification with Heterogeneous Graph Attention Convolutional Network. In: Rekik, Islem, Adeli, Ehsan, Park, Sang Hyun, Schnabel, Julia (eds.) PRIME 2021. LNCS, vol. 12928, pp. 157–167. Springer, Cham (2021). https://doi.org/10.1007/978-3-030-87602-9_15
23. Zhang, Y., Zhan, L., Cai, W., Thompson, P., Huang, H.: Integrating Heterogeneous Brain Networks for Predicting Brain Disease Conditions. In: Shen, D., et al. (eds.) MICCAI 2019. LNCS, vol. 11767, pp. 214–222. Springer, Cham (2019). https://doi.org/10.1007/978-3-030-32251-9_24
24. Zhao, Shijie, Fang, Long, Wu, Lin, Yang, Yang, Han, Junwei: Decoding Task Subtype States with Group Deep Bidirectional Recurrent Neural Network. In: Wang, Linwei, Dou, Qi., Fletcher, P. Thomas., Speidel, Stefanie, Li, Shuo (eds.) Medical Image Computing and Computer Assisted Intervention – MICCAI 2022: 25th International Conference, Singapore, September 18–22, 2022, Proceedings, Part I, pp. 241–250. Springer, Cham (2022). https://doi.org/10.1007/978-3-031-16431-6_23

FE-STGNN: Spatio-Temporal Graph Neural Network with Functional and Effective Connectivity Fusion for MCI Diagnosis

Dongdong Chen and Lichi Zhang(✉)

School of Biomedical Engineering, Shanghai Jiao Tong University, Shanghai, China
lichizhang@sjtu.edu.cn

Abstract. Brain connectivity patterns such as functional connectivity (FC) and effective connectivity (EC), describing complex spatio-temporal dynamic interactions in the brain network, are highly desirable for mild cognitive impairment (MCI) diagnosis. Major FC methods are based on statistical dependence, usually evaluated in terms of correlations, while EC generally focuses on directional causal influences between brain regions. Therefore, comprehensive integration of FC and EC with complementary information can further extract essential biomarkers for characterizing brain abnormality. This paper proposes Spatio-Temporal Graph Neural Network with Dynamic Functional and Effective Connectivity Fusion (FE-STGNN) for MCI diagnosis using resting-state fMRI (rs-fMRI). First, dynamic FC and EC networks are constructed to encode the functional brain networks into multiple graphs. Then, spatial graph convolution is employed to process spatial structural features and temporal dynamic characteristics. Finally, we design the position encoding-based cross-attention mechanism, which utilizes the causal linkage of EC during time evolution to guide the fusion of FC networks for MCI classification. Qualitative and quantitative experimental results demonstrate the significance of the proposed FE-STGNN method and the benefit of fusing FC and EC, which achieves 82% of MCI classification accuracy and outperforms state-of-the-art methods. Our code is available at https://github.com/haijunkenan/FE-STGNN.

Keywords: Resting-state MRI · Brain Connectivity Network · Graph Neural Network · Attention mechanism

1 Introduction

Mild cognitive impairment (MCI) is a prodromal stage of memory loss or other cognitive loss (e.g., language, visual and spatial perception) in individuals that may

---

Supplementary Information The online version contains supplementary material available at https://doi.org/10.1007/978-3-031-43993-3_7.

H. Greenspan et al. (Eds.): MICCAI 2023, LNCS 14227, pp. 67–76, 2023.
https://doi.org/10.1007/978-3-031-43993-3_7

progress to Alzheimer's disease (AD), which is a neurological brain disease that leaves individuals without the ability to live independently with daily activities [1]. Diagnosis of MCI would allow on-time intervention to delay or prevent the progress of the disease, which is crucial for advanced AD without effective treatment.

Resting-state functional magnetic resonance (rs-fMRI) as a non-invasive tool has demonstrated its potential to evaluate brain activity by measuring the blood oxygenation level-dependent (BOLD) signals over time [22]. Based on this, researchers model brain networks to analyze brain function and realize the diagnosis of brain diseases [13]. For example, compared with healthy controls, subjects with AD and MCI have shown changes in brain networks based on rs-fMRI data [14], manifested explicitly in the weakened connection between the right hippocampus and default mode network (DMN, a set of brain regions), and the enhanced connection between the left hippocampus with right lateral prefrontal lobe brain region, etc.

A brain network can be represented as a graph structure naturally, which consists of brain regions as nodes and connections among brain regions as edges [20]. Considering the characteristics of functional integration and separation in the brain, brain connectivity is typically modeled in three different patterns: namely, structural connectivity (SC) [18], functional connectivity (FC) [21], and effective connectivity (EC) [12]. Structural connectivity refers to a network of physical or anatomical connections linking sets of neurons, which characterize the associated structural biophysical attributes extracted from diffusion-weighted imaging (DWI) data. Differently, both functional and effective connectivity can be generated from fMRI data. FC describes brain activity from the perspective of statistical dependencies between brain regions, while EC emphasizes directional causal interactions of one brain region over another.

Currently, with the development of deep learning, various methods have been proposed to diagnose brain disease using rs-fMRI modeled with brain connectivity [7,11]. Among them, graph neural network-based (GNN) methods stand out from the others for their ability to capture network topology information [6]. However, there are three critical disadvantages to existing GNN-based methods for brain disease diagnosis. 1) Existing studies generally use a single brain connectivity pattern to model brain activities, which would restrict the following analysis to a single space of brain network characteristics. 2) Most of the GNN-based methods applied to brain diseases focus on extracting static topological information neglecting the dynamic nature of brain activity. 3) Many approaches seek to enhance the model's ability for disease diagnosis, while the underlying mechanism remains a black box, making it challenging to provide an explainable basis corresponding to brain neural activity.

Therefore, we propose Spatio-Temporal Graph Neural Network with Functional and Effective Connectivity Fusion (FE-STGNN), which mainly focuses on the fusion of local spatial structure information from FC and temporal causal evolution information from EC, for MCI diagnosis using rs-fMRI. As FC focuses on describing the strength of brain connections in each time slice, while EC for characterizing the dynamic information flow. Thus, our motivation is to utilize the causal linkage of EC during time evolution to guide the fusion of FC

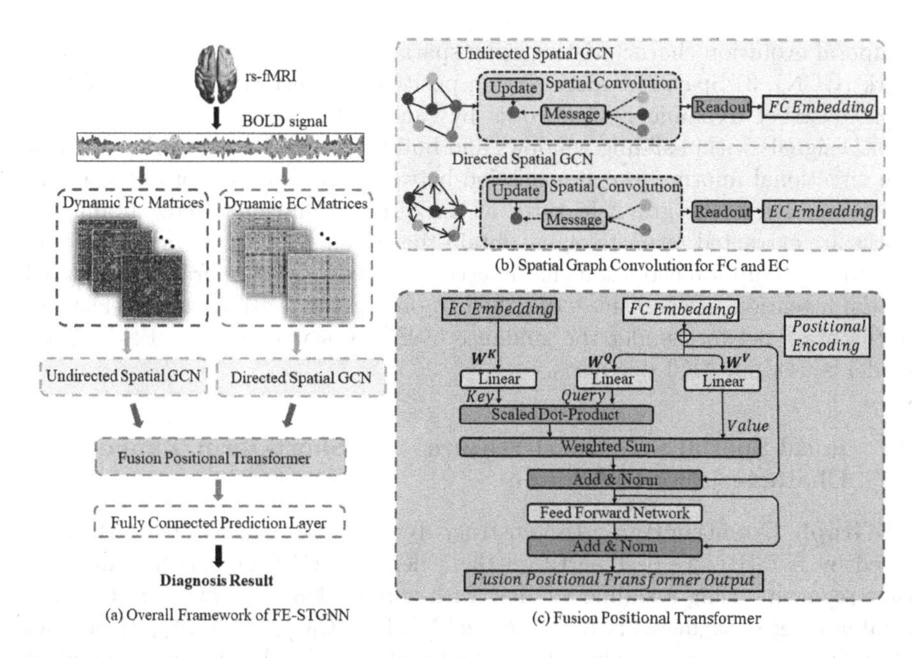

Fig. 1. Illustration of the proposed FE-STGNN model for MCI diagnosis based on rs-fMRI data. (a) is the overall framework of the proposed method. (b) describes the detailed process of spatial graph convolution for dynamic FC and EC networks from the perspective of the red node. (c) illustrates how FC aggregates information under the guidance of EC based on attention mechanism (Color figure online)

networks at discrete time slices. We summarize the novelty of the proposed method in three aspects:

1) We propose to model brain connectivity using FC and EC simultaneously, comprehensively considering the structural characteristics of the brain network and time-evolving properties of causal effects between brain regions.
2) Considering that FC networks focus on describing the strength of brain connections, while ECs are more detailed about the directional information flow among brain regions, we design a novel graph fusion framework based on a cross-attention mechanism that leads FCs to aggregate temporal structure information under EC guidance.
3) We track the model in the MCI diagnostic task and evaluate the importance of different brain regions for the impact of disease, which gives an explainable basis in combination with the background of biomedical knowledge.

2 Method

Figure 1 illustrates the framework of the proposed spatio-temporal graph neural network with functional and effective connectivity fusion, which consists of three main sequential components: 1) Dynamic FC and EC graph matrices construction; 2) Extraction module for local spatial structural features and short-term temporal evolution characteristics using spatial graph convolutional neural network (GCN); 3) Spatio-temporal fusion positional transformer. Specifically, we first construct dynamic FC networks by calculating the correlations between BOLD signals within sliding windows and build dynamic EC networks based on the directional information transmission between two brain regions at adjacent moments. Then, the graph features of the directed and undirected brain networks are extracted respectively to obtain the spatial structural correlation and short-term temporal evolution characteristics. Finally, a fusion framework based on the attention mechanism is designed to obtain the spatio-temporal features of the brain network under the guidance of EC networks, and the FE-STGNN model is used for MCI diagnosis.

2.1 Local Spatial Structural Features and Short-Term Temporal Characteristics Extraction

Bi-Graph Construction. To construct dynamic FC and EC networks captured by rs-fMRI, we first partition the T length BOLD signals into multiple time segments using a sliding window with window-length w and stride s. The total number of segments is $K = \lfloor (T-w)/s \rfloor$. Let $\{\mathcal{G}|\mathcal{G}_{FC}^t = (V, A_{FC}^t)\}$ be a set of undirected FC graphs, where V is a finite set of nodes $|V| = N$ corresponding to each brain Regions of Interests (ROI), t varies from 1 to K indicating the time segment, and the adjacency matrix $A_{FC}^t \in \mathbb{R}^{N \times N}$ consisting of edge element e_{vu} can be obtained by:

$$A_{FC}^t|e_{vu} = \begin{cases} \rho(v^t, u^t) & \text{if } v \neq u \\ 0 & \text{otherwise} \end{cases}, \tag{1}$$

where $\rho(v^t, u^t)$ measuring the Pearson Correlation [3] between v-th and u-th brain ROIs in t-th time segment.

Meanwhile, for directed EC networks we adopt Transfer Entropy [16] as a directional information-theoretic measure due to its advantages in detecting non-linear interactions between neurons. Therefore, we have EC graphs for each t-th segment with $\{\mathcal{G}|\mathcal{G}_{EC}^t = (V, A_{EC}^t)\}$, where the edge element e_{vu} can be computed through $\zeta(v^t \rightarrow u^{t+1})$, which measures the directed information of Transfer Entropy between two ROIs:

$$A_{EC}^t|e_{vu} = \begin{cases} \zeta(v^t \rightarrow u^{t+1}) & \text{if } v \neq u \\ 1 & \text{otherwise} \end{cases}. \tag{2}$$

Spatial Graph Convolution. The existing graph convolution methods are mainly divided into spectral methods and spatial methods. Here, we choose the spatial graph convolution for two reasons. First, the spatial method aggregates information based on the topological structure of the graph and pays close attention to the important spatial structural information in brain networks. Second, the spectral method cannot be applied to directed graphs because of the symmetry requirements of the spectral normalized Laplace matrix.

From the perspective of each node, spatial graph convolution can be described as a process of message passing and node updating [10], i.e., the aggregation of hidden features of local neighborhoods combined with transformations. Therefore, the spatial graph convolution for each layer can be obtained by two steps:

$$m_v^{l+1} = \sum_{u \in N(v)} M_l \left(h_v^l, h_u^l, e_{vu}\right), \quad h_v^{l+1} = U_l \left(h_v^l, m_v^{l+1}\right), \tag{3}$$

where $N(v)$ denotes the neighbors of v in the graph, h_v^l and h_u^l are hidden representations of node v and u in l-th layer respectively. e_{vu} as the edge weights are elements of the adjacent matrix, and m_v^{l+1} is the message aggregated from the target node v combined with local neighborhoods. The message functions M_l and node update functions U_l are all learned differentiable functions. We further aggregate the representations of all nodes in the graph $\mathcal{G}$ from the final convolution layer (i.e., L-th layer) to obtain FC and EC graph embeddings $\{H | H_{FC}^t \in \mathbb{R}^{1 \times d_H}, H_{EC}^t \in \mathbb{R}^{1 \times d_H}\}$ in each time segment respectively, where d_H is the feature dimension of the graph embedding.

$$H = R\left(\{h_v^L \mid v \in \mathcal{G}\}\right). \tag{4}$$

Here, the readout function R can be a simple permutation invariant function such as graph-level pooling function.

2.2 Spatio-Temporal Fusion with Dynamic FC and EC

Fusion Positional Transformer. After obtaining the graph representations of FC and EC in each time segment, we further design a graph fusion network based on attention mechanism, using the causal linkage of EC in time to guide the fusion of FC networks. Considering that the attention itself does not have position awareness, it is treated equally in attention at any time segment. We first introduce a learnable temporal positional encoding $P \in \mathbb{R}^{K \times d_H}$ to encode the time sequence properties of the brain network, which is randomly initialized and added to FC embeddings before attention.

The attention mechanism can be described as mapping a query and a set of key-value pairs to an output which is computed as a weighted sum of the values. In practice, we first conduct linear transforms on dynamic FCs, which are computed in the form of a matrix $\{H_{FC} | \{H_{FC}^t\}_{t=1}^K, H_{FC} \in \mathbb{R}^{K \times d_H}\}$, combined with the temporal positional encoding. The dynamic EC are also linear transformed in the form of matrix $\{H_{EC} | \{H_{EC}^t\}_{t=1}^K, H_{EC} \in \mathbb{R}^{K \times d_H}\}$. Therefore, both dynamic FC and EC are encoded into high-dimensional latent subspaces, including the query subspace $Q \in \mathbb{R}^{K \times d_k}$, the value subspace $V \in \mathbb{R}^{K \times d_k}$ from FC, and the key subspace $K \in \mathbb{R}^{N \times d_k}$ from EC simultaneously, where d_k is the dimension of latent subspace. The conducting progress can be formulated as:

$$\begin{aligned} Q &= (H_{FC} + P) * W_Q \\ K &= H_{EC} * W_K \\ V &= (H_{FC} + P) * W_V, \end{aligned} \tag{5}$$

where W_Q, W_K, and W_V are the weight matrices for Q, K, and V, respectively. We further calculate the spatio-temporal dependencies of the brain network with the dot product of FCs' query and ECs' key. Therefore, the attention of FC and EC fusion can be obtained by:

$$M = \text{softmax}\left(\frac{QK^T}{\sqrt{d_k}}\right)V. \tag{6}$$

Here, the softmax is used to normalize the spatio-temporal dependencies and the scale $\sqrt{d_k}$ prevents the saturation led by softmax function. To ensure the stable training, we adopt the residual connection formulated as $\hat{M} = M + (H_{FC}+P)$. Furthermore, a feed-forward neural network with nonlinear activation combined with another residual connection layer are applied to further improve the prediction conditioned on the embeddings $\hat{M}$. As a result, the output of the fusion positional transformer is $\hat{U} = \text{ReLU}\left(\hat{M}W_0 + b_0\right) + \hat{M}$, where W_0 and b_0 are learnable weight and bias.

Prediction Diagnosis. In the end, the prediction layer leverages a fully connected network to map the output embeddings of the fusion positional transformer into the corresponding label space, therefore we can obtain the prediction result $Y = \text{ReLU}\left(\hat{U}W_1 + b_1\right)$ with W_1 and b_1 are learnable weight and bias. In the training process, we design the loss function as shown in Eq. 7. The first term is used to minimize the error between the real diagnosis result $\hat{Y}$ and the prediction Y. The second term L_{reg} is the $L2$ regularization term that helps to avoid an overfitting problem with λ as a hyper-parameter.

$$\text{loss} = \left\|Y - \hat{Y}\right\| + \lambda L_{reg}. \tag{7}$$

3 Experiments

3.1 Dataset and Experimental Settings

In order to verify the performance of our method on Alzheimer's Disease Neuroimaging Initiative (ADNI) data [17], we adopt the individual rs-fMRI images including $T = 137$ time points consisting of 60 normal controls (NC) and 54 MCI subjects, following the standard pre-processing pipeline same to [4]. Considering that medical imaging data in the real world is often insufficient for deep learning, we randomly selected balanced data with small samples for experiments. We perform 10-fold cross-validation 10 times for all the experiments, a sliding window of $w = 37$ and stride $s = 10$ is used to construct dynamic brain networks for all the dynamic FC-based methods. We train our model with parameters: $\lambda = 0.01$ weight of the regularization term, $1e^{-3}$ learning rate, and a maximum number of 800 epoches. We performed a grid search to determine these hyper-parameters.

Table 1. Performance of ablation study on ADNI dataset (mean ± std %).

	Metric	LSTM			**Transformer**		
		FC	EC	FC+EC	FC	EC	**FC+EC**
GAT	ACC	74.6 ± 2.4	72.8 ± 2.1	76.3 ± 2.1	75.4 ± 2.8	73.7 ± 2.5	78.1 ± 2.3
	SEN	74.3 ± 2.6	70.4 ± 2.5	76.0 ± 2.7	75.4 ± 3.4	72.3 ± 3.1	77.8 ± 2.9
	SPE	71.7 ± 2.4	75.0 ± 2.2	81.7 ± 2.3	76.7 ± 3.1	75.0 ± 2.8	83.3 ± 2.7
GCN	ACC	74.6 ± 2.3	75.4 ± 2.2	79.0 ± 2.3	78.1 ± 2.7	77.2 ± 2.6	**82.5 ± 2.2**
	SEN	74.4 ± 2.6	74.8 ± 2.4	78.4 ± 2.5	78.1 ± 3.0	77.8 ± 3.0	**82.3 ± 2.9**
	SPE	78.3 ± 2.5	76.7 ± 2.3	88.3 ± 2.3	78.7 ± 2.8	76.7 ± 3.0	**86.7 ± 2.8**

3.2 Ablation Studies

Here we conduct ablation studies to verify the effectiveness of 1) the fusion of dynamic FC and EC networks and 2) each module in the proposed FE-STGNN. Specifically, we first replace graph convolutional neural network (GCN) with graph attention network (GAT) in the spatial graph convolution module (Fig. 1(b)). Both of them are a kind of graph neural network methods that have been widely used in brain network analysis [5]. Then, we replace the proposed fusion positional transformer with LSTM in the spatio-temporal fusion module (Fig. 1(c)). Finally, we compare the models with different modules: only FCs, only ECs, and fusion FCs and ECs. It is worth noting that, in order to fuse FC and EC in LSTM, we specifically design the high-dimensional feature vector product of FC and EC in the LSTM cell.

The result of ablation studies is presented in Table 1. These highlight modules (e.g. GCN, Transformer) are the modules constituting the proposed model. Comparing GAT and GCN methods, we find that GCN performs better than GAT in almost all models. The finding may support that GCN utilizes the Laplace matrix based on the adjacent matrix to aggregate structural information, while GAT uses attention coefficients based on node features, which are less important in brain network analysis. The Transformer also performs better than the LSTM method, which may be because compared with LSTM, the Transformer can not only extract the information of a period of time before the current time segment but also associate with the subsequent time.

To compare FC and FC+EC models, we further draw the ROC curves of comparison models in Fig. 2(a). It is shown that the ROC curves of fusion methods (i.e., green line and red line) almost wrap the curves (i.e., blue line and yellow line) of the models that only depend on the FC network. The result validate the merit of fusing FC and EC in processing and identifying brain networks.

3.3 Comparison with Other Methods

We also compare the proposed model against baseline algorithms including three static methods: (1) Support Vector Machine (SVM) [19]; (2) Two layers of the

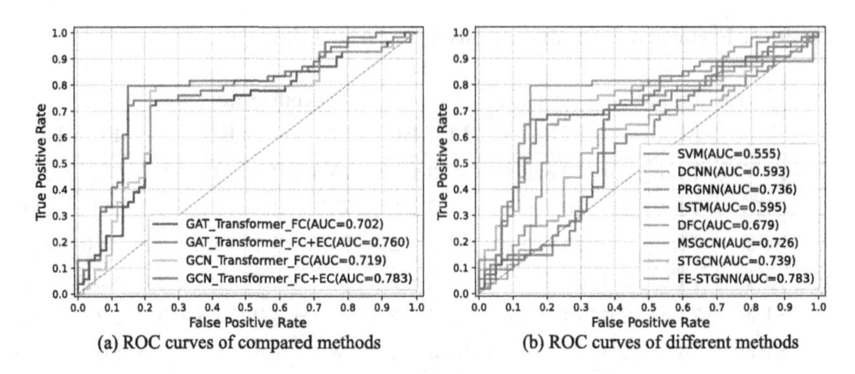

(a) ROC curves of compared methods (b) ROC curves of different methods

Fig. 2. Results of ROC curves and AUC values achieved by ablation studies and eight different methods

Table 2. Classification results for MCI Comparing with Different Methods (mean ± std %).

	Model	ACC	SEN	SPE	F-score
Static	SVM [19]	58.7 ± 1.0	58.5 ± 1.2	63.3 ± 1.4	58.5 ± 1.0
	DCNN [2]	64.1 ± 3.8	63.9 ± 4.8	65.0 ± 4.1	63.9 ± 3.7
	PRGNN [15]	76.3 ± 3.7	75.7 ± 5.2	86.7 ± 4.5	75.8 ± 4.1
Dynamic	LSTM [8]	65.8 ± 4.2	66.0 ± 4.1	61.7 ± 2.9	65.8 ± 3.8
	DFC [4]	72.8 ± 2.9	72.4 ± 2.1	80.0 ± 3.3	72.5 ± 3.2
	MSGCN [23]	75.4 ± 2.8	75.0 ± 3.1	83.3 ± 2.8	75.1 ± 2.7
	STGCN [9]	79.8 ± 3.2	79.5 ± 3.5	85.0 ± 2.7	79.6 ± 3.3
	FE-STGNN	$\mathbf{82.5 \pm 2.2}$	$\mathbf{82.3 \pm 2.9}$	$\mathbf{86.7 \pm 2.8}$	$\mathbf{82.4 \pm 2.1}$

Diffusion Convolutional Neural Networks (DCNN) [2]; (3) Graph Neural Network with regularized pooling layers (PRGNN) [15], and four dynamic methods: (4) Classical LSTM [8]; (5) Dynamic Functional Connectivity Matrix Graph Embedding Method (DFC) [4]; (6) MS-GCN with same normalized adjacency matrix (MS-GCN) [23]; (7) Spatio-Temporal Graph Convolution [9].

From Table 2 and Fig. 2(b), one can have the following observations. First, compared with three static models (i.e., SVM, DCNN, and PRGNN), almost all the methods based on dynamic FC networks obtain better performance. For example, in terms of ACC value, FE-STGNN has achieved at least 6% improvement, which indicates that dynamic functional network modeling is more effective in brain network analysis than static methods. Furthermore, our FE-STGNN yields significantly higher ACC and AUC compared to the other seven competing methods. These results validate the effectiveness of FE-STGNN in identifying MCI patients based on the fusion of FC and EC networks.

4 Conclusion

In this paper, we present a novel FE-STGNN framework for modeling spatio-temporal connectivity patterns of brain networks for MCI diagnosis. To our knowledge, it is one of the earliest studies to use the fusion of FC and EC in a deep-learning fashion to model both spatial and temporal patterns of brain networks at the same time. The ablation studies examine the efficacy of each module of the proposed method as well as the significant benefits of combining FC and EC. The proposed model's superiority is also demonstrated when compared to other methods. We plan to extend the proposed FE-STGNN to diagnose other brain functional diseases in the future.

Acknowledgement. This work was supported by the National Natural Science Foundation of China (No. 62001292).

References

1. Association, A., et al.: 2021 Alzheimer's disease facts and figures. Alzheimers Dement. **17**(3), 327–406 (2021)
2. Atwood, J., Towsley, D.: Diffusion-convolutional neural networks. Adv. Neural Inform. Process. Syst. **29** 1993–2001 (2016)
3. Benesty, J., Chen, J., Huang, Y., Cohen, I.: Pearson correlation coefficient. In: Cohen, I., Huang, Y., Chen, J., Benesty, J. (eds.) Noise Reduction in Speech Processing, pp. 1–4. Springer, Berlin, Heidelberg (2009)
4. Chen, X., Zhang, H., Zhang, L., Shen, C., Lee, S.W., Shen, D.: Extraction of dynamic functional connectivity from brain grey matter and white matter for MCI classification. Hum. Brain Mapp. **38**(10), 5019–5034 (2017)
5. Cui, H., et al.: Braingb: a benchmark for brain network analysis with graph neural networks. IEEE Trans. Med. Imaging **42**(2), 493–506 (2022)
6. DelEtoile, J., Adeli, H.: Graph theory and brain connectivity in Alzheimer's disease. Neuroscientist **23**(6), 616–626 (2017)
7. Dennis, E.L., Thompson, P.M.: Functional brain connectivity using fMRI in aging and Alzheimer's disease. Neuropsychol. Rev. **24**(1), 49–62 (2014)
8. Dvornek, N.C., Ventola, P., Pelphrey, K.A., Duncan, J.S.: Identifying Autism from resting-state fMRI using long short-term memory networks. In: Wang, Q., Shi, Y., Suk, H.-I., Suzuki, K. (eds.) Machine Learning in Medical Imaging, pp. 362–370. Springer, Cham (2017). https://doi.org/10.1007/978-3-319-67389-9_42
9. Gadgil, S., Zhao, Q., Pfefferbaum, A., Sullivan, E.V., Adeli, E., Pohl, K.M.: Spatio-temporal graph convolution for resting-state fMRI analysis. In: Martel, A.L., et al. (eds.) Medical Image Computing and Computer Assisted Intervention – MICCAI 2020: 23rd International Conference, Lima, Peru, October 4–8, 2020, Proceedings, Part VII, pp. 528–538. Springer, Cham (2020). https://doi.org/10.1007/978-3-030-59728-3_52
10. Gilmer, J., Schoenholz, S.S., Riley, P.F., Vinyals, O., Dahl, G.E.: Neural message passing for quantum chemistry. In: International Conference On Machine Learning, pp. 1263–1272. PMLR (2017)
11. Huang, S., et al.: Learning brain connectivity of Alzheimer's disease by sparse inverse covariance estimation. Neuroimage **50**(3), 935–949 (2010)

12. Ji, J., Zou, A., Liu, J., Yang, C., Zhang, X., Song, Y.: A survey on brain effective connectivity network learning. IEEE Trans. Neural Netw. Learn. Syst. (2021)
13. Langer, N., Pedroni, A., Gianotti, L.R., Hänggi, J., Knoch, D., Jäncke, L.: Functional brain network efficiency predicts intelligence. Hum. Brain Mapp. **33**(6), 1393–1406 (2012)
14. Lee, H., et al.: Harmonic holes as the submodules of brain network and network dissimilarity. In: Marfil, R., Calderón, M., Díaz del Río, F., Real, P., Bandera, A. (eds.) Computational Topology in Image Context: 7th International Workshop, CTIC 2019, Málaga, Spain, January 24-25, 2019, Proceedings, pp. 110–122. Springer I, Cham (2019). https://doi.org/10.1007/978-3-030-10828-1_9
15. Li, X., et al.: Pooling regularized graph neural network for fMRI biomarker analysis. In: Martel, A.L., et al. (eds.) Medical Image Computing and Computer Assisted Intervention – MICCAI 2020: 23rd International Conference, Lima, Peru, October 4–8, 2020, Proceedings, Part VII, pp. 625–635. Springer, Cham (2020). https://doi.org/10.1007/978-3-030-59728-3_61
16. Massey, J., et al.: Causality, feedback and directed information. In: Proc. Int. Symp. Inf. Theory Applic. (ISITA-90). pp. 303–305 (1990)
17. Misra, C., Fan, Y., Davatzikos, C.: Baseline and longitudinal patterns of brain atrophy in mci patients, and their use in prediction of short-term conversion to ad: results from adni. Neuroimage **44**(4), 1415–1422 (2009)
18. Sanchez, J.F.Q., Liu, X., Zhou, C., Hildebrandt, A.: Nature and nurture shape structural connectivity in the face processing brain network. Neuroimage **229**, 117736 (2021)
19. Suykens, J.A., Vandewalle, J.: Least squares support vector machine classifiers. Neural Process. Lett. **9**(3), 293–300 (1999)
20. Vecchio, F., Miraglia, F., Rossini, P.M.: Connectome: graph theory application in functional brain network architecture. Clin. Neurophysiol. Pract. **2**, 206–213 (2017)
21. Wang, H.E., Bénar, C.G., Quilichini, P.P., Friston, K.J., Jirsa, V.K., Bernard, C.: A systematic framework for functional connectivity measures. Front. Neurosci. **8**, 405 (2014)
22. Wee, C.Y., Yang, S., Yap, P.T., Shen, D.: Sparse temporally dynamic resting-state functional connectivity networks for early MCI identification. Brain Imaging Behav. **10**(2), 342–356 (2016)
23. Yu, S., Yue, G., Elazab, A., Song, X., Wang, T., Lei, B.: Multi-scale graph convolutional network for mild cognitive impairment detection. In: Zhang, D., Zhou, L., Jie, B., Liu, M. (eds.) Graph Learning in Medical Imaging: First International Workshop, GLMI 2019, Held in Conjunction with MICCAI 2019, Shenzhen, China, October 17, 2019, Proceedings, pp. 79–87. Springer, Cham (2019). https://doi.org/10.1007/978-3-030-35817-4_10

Learning Normal Asymmetry Representations for Homologous Brain Structures

Duilio Deangeli[1,2](✉), Emmanuel Iarussi[2,3], Juan Pablo Princich[4], Mariana Bendersky[3,4,5], Ignacio Larrabide[1,2], and José Ignacio Orlando[1,2]

[1] Yatiris, PLADEMA, UNICEN, Tandil, Buenos Aires, Argentina
ddeangeli@pladema.exa.unicen.edu.ar
[2] CONICET, Buenos Aires, Argentina
[3] Universidad Torcuato Di Tella, CABA, Buenos Aires, Argentina
[4] ENyS, CONICET-HEC-UNAJ, Florencio Varela, Buenos Aires, Argentina
[5] Normal Anatomy Department, UBA, CABA, Buenos Aires, Argentina

Abstract. Although normal homologous brain structures are approximately symmetrical by definition, they also have shape differences due to e.g. natural ageing. On the other hand, neurodegenerative conditions induce their own changes in this asymmetry, making them more pronounced or altering their location. Identifying when these alterations are due to a pathological deterioration is still challenging. Current clinical tools rely either on subjective evaluations, basic volume measurements or disease-specific deep learning models. This paper introduces a novel method to learn normal asymmetry patterns in homologous brain structures based on anomaly detection and representation learning. Our framework uses a Siamese architecture to map 3D segmentations of left and right hemispherical sides of a brain structure to a normal asymmetry embedding space, learned using a support vector data description objective. Being trained using healthy samples only, it can quantify deviations-from-normal-asymmetry patterns in unseen samples by measuring the distance of their embeddings to the center of the learned normal space. We demonstrate in public and in-house sets that our method can accurately characterize normal asymmetries and detect pathological alterations due to Alzheimer's disease and hippocampal sclerosis, even though no diseased cases were accessed for training. Our source code is available at https://github.com/duiliod/DeepNORHA.

Keywords: Normal asymmetry · Brain MRI · Anomaly detection

1 Introduction

(Sub)cortical brain structures are approximately symmetrical between the left and right hemispheres [24]. Although their appearance and size are similar,

Supplementary Information The online version contains supplementary material available at https://doi.org/10.1007/978-3-031-43993-3_8.

H. Greenspan et al. (Eds.): MICCAI 2023, LNCS 14227, pp. 77–87, 2023.
https://doi.org/10.1007/978-3-031-43993-3_8

they usually present difficult-to-characterize morphometric differences that vary among healthy populations [27], e.g. due to natural ageing [3]. Moreover, it has been studied that some neurological conditions, including Alzheimer's disease [10,16] (AD), schizophrenia [6], and epilepsy [4,17], are associated to asymmetry of the hippocampus or the amygdala [26]. In regular medical practice, radiologists detect pathological changes in asymmetry by manually inspecting the structure using brain MRIs [4]. They rely on their own subjective experience and knowledge, which varies among observers and lacks reproducibility, or standard quantitative measurements, e.g. volume differences in segmentations [3,9,12,16,17,20], which fail to capture morphological asymmetries beyond differences in size [10,21]. No tools automate quantifying normal asymmetry patterns beyond volume [16], e.g. to detect deviations caused by a neurodegenerative disease. Some deep learning tools approximate this goal in a binary classification setting to differentiate one particular condition from normal cases [5,7]. However, this form is heavily specialized to discern asymmetry alterations associated with one specific disease, requiring retraining for every new condition [11,12].

This paper introduces a novel framework for learning NORmal Asymmetries of Homologous cerebral structures (deep NORAH) based on anomaly detection and representation learning. Unlike previous methods that train Siamese neural networks with volume descriptors [12], our model takes 3D segmentations of left and right components from MRIs as inputs, and maps them into an embedding that summarizes their shape differences. Essentially, our Siamese architecture includes a shape characterization encoder that extracts morphological features directly from segmentations and an Asymmetry Projection Head (APH) that merges their differences to create a compact representation of asymmetries. To ensure this embedding learns the heterogeneity in normal individuals, our network is trained only with healthy samples, using a self-supervised pre-training stage based on a Contractive Autoencoder (CAE) and then fine-tuning using a Support Vector Data Description (SVDD) objective. Our experiments in the hippocampus show that our model can easily project new cases to the normal asymmetry space. Furthermore, we show that the distance between the embedding and the center of the normal space is a measure of deviation-from-normal-asymmetry, as we empirically observed increased distance in pathological cases. Hence, deep NORAH can be used to diagnose, e.g. AD, hippocampal sclerosis and even mild cognitive impairment (MCI) by simply detecting the differences in asymmetry regarding the normal set, without needing diseased cases for training.

In summary, our contributions are as follows: (i) ours is the first unsupervised deep learning model explicitly designed to learn normal asymmetries in homologous brain structures; (ii) although it is trained only with normal data, it can be used to detect diseased samples by quantifying the degree of deviation with respect to a healthy population , unlike existing methods that capture only disease-specific asymmetries [11]; and (iii) compared to other state-of-the-art anomaly detection approaches, our method demonstrates consistently better results for discriminating both synthetic and diseased-related asymmetries.

2 Methods

Figure 1 depicts a flowchart of our method as applied in test time. Our goal is to automatically measure the asymmetry of a given homologous brain structure $x = (x_L, x_R)$, with x_L and x_R being the 3D segmentations of its left and right lateral elements, and learn if these differences are typical for a normal population. To do so, we propose to learn a Siamese deep neural network $\mathbf{F}_\theta(x) = \mathbf{z}_{L-R}$ with θ parameters using an anomaly detection objective and samples from healthy individuals. $\mathbf{z}_{L-R}$ is a compact representation of the asymmetries in x, obtained by learning a hypersphere $\mathcal{S}$ with a center $\mathbf{c}$ and minimum radius. In test time, samples with normal asymmetries are projected to the vicinity of $\mathbf{c}$, while those with unexpected differences fall far from this point. As a result, the distance $d = \|\mathbf{F}_\theta(x) - \mathbf{c}\|_2^2$ can be used as a deviation-from-normal-asymmetry index.

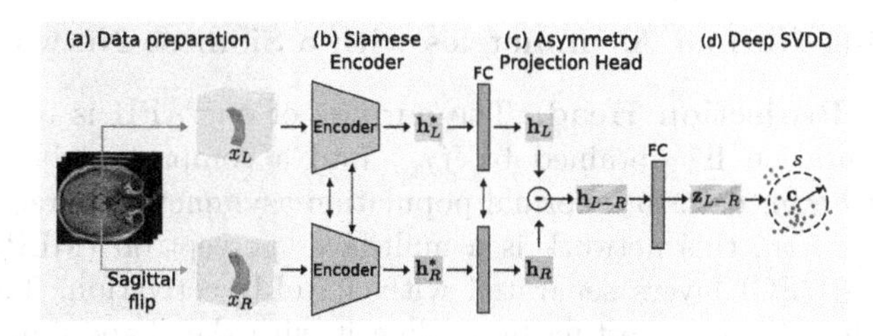

Fig. 1. Schematic of our framework for capturing abnormal asymmetries in homologous brain structures from 3D MRI.

To train this model, we first learn a 3D shape encoder $f_{\theta_{\mathrm{EN}}}(x)$ as part of a CAE, using normal samples (Sect. 2.1). This network can take any single segmentation of a lateral element $x_{(i)}$ as input, and map it to a high dimensional shape representation $\mathbf{h}^*_{(i)}$. We then add this encoder to a Siamese architecture by attaching it to an APH, which captures the differences in shape from $\mathbf{h}^*_L$ and $\mathbf{h}^*_R$, and project them into the unique asymmetry embedding $\mathbf{z}_{L-R}$. To this end, both the pre-trained encoder and the APH are trained using a deep SVDD objective (Sect. 2.2). This second learning phase not only trains the APH from scratch but fine-tunes the shape encoder to capture those morphological characteristics that are the most common source of asymmetry in normal individuals.

2.1 Pre-training the Shape Characterization Encoder as a CAE

Our shape encoder $f_{\theta_{\mathrm{EN}}}$ indistinctly map an arbitrarily left or right segmentation $\mathbf{x}_{(i)}$ of an homologous structure x, to a feature vector $\mathbf{h}^*_{(i)}$ that describes its shape. To this end, we apply a warm-up learning phase that trains $f_{\theta_{\mathrm{EN}}}$ as the encoding path of a CAE, using a self-supervised learning loss (Fig. 2a). Hence, the encoder is simultaneously trained with a decoder $g_{\theta_{\mathrm{DE}}}(i)$ using a reconstruction task. The encoder compresses the input into a lower-dimensional representation $\mathbf{h}^*_{(i)}$ by applying a series of convolutional and pooling operations, and the decoder tries to reconstruct it using upsampling operations and convolutions.

Formally, let $(g_{\theta_{\text{DE}}} \circ f_{\theta_{\text{EN}}})(x_{(i)})$ be a convolutional CAE with a decoding path $g_{\theta_{\text{DE}}}(\mathbf{h}^*_{(i)})$ with parameters θ_{DE} that outputs a reconstruction $\hat{x}_{(i)}$ of the input $x_{(i)}$ from its hidden representation $\mathbf{h}^*_{(i)}$. This is achieved by minimizing a mean square error objective (Fig. 2a). After it, the decoder is discarded and the shape encoder is used in the Siamese setting of our anomaly detection network.

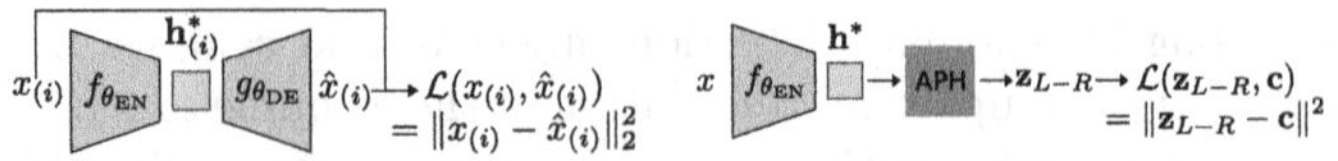

$x_{(i)}$ $f_{\theta_{\text{EN}}}$ $\mathbf{h}^*_{(i)}$ $g_{\theta_{\text{DE}}}$ $\hat{x}_{(i)} \rightarrow \mathcal{L}(x_{(i)}, \hat{x}_{(i)}) = \|x_{(i)} - \hat{x}_{(i)}\|_2^2$

(a) Pre-training

x $f_{\theta_{\text{EN}}}$ $\mathbf{h}^*$ $\rightarrow$ APH $\rightarrow \mathbf{z}_{L-R} \rightarrow \mathcal{L}(\mathbf{z}_{L-R}, \mathbf{c}) = \|\mathbf{z}_{L-R} - \mathbf{c}\|^2$

(b) Fine-tuning

Fig. 2. Training sequence of our method.

2.2 Learning Normal Asymmetries with a Siamese Network

Asymmetry Projection Head. The purpose of our APH is to project the shape representation $\mathbf{h}^*$ obtained by $f_{\theta_{\text{EN}}}$ into a compact embedding $\mathbf{z}_{L-R}$ (Fig. 2b) that better describes normal population asymmetry characteristics. In our implementation, this network is a multilayer perceptron (MLP) with two fully connected (FC) layers separated with a ReLU activation. The first FC layer is used in a Siamese setting by feeding it with the shape representations $\mathbf{h}^*_L$ and $\mathbf{h}^*_R$ of the left and right elements, respectively. Each of these inputs are projected into two new feature vectors $\mathbf{h}_L$ and $\mathbf{h}_R$ with a lower dimensionality. A merging operation (e.g. subtraction or concatenation) combines them into a joint representation $\mathbf{h}_{(L-R)}$, which is projected by the second FC layer into the asymmetry embedding $\mathbf{z}_{(L-R)}$. Notice that the main design choices to be made are the dimensionality of the outputs of each FC layer and the merging operation.

One-Class Deep SVDD. In order to enforce $\mathbf{z}_{(L-R)}$ to represent the asymmetry characteristics of normal individuals, we train the Siamese architecture $\mathbf{F}_\theta(x)$ in Fig. 2b using an anomaly detection objective. We adopted the one-class deep SVDD approach proposed in [22], which solves:

$$\min_{\theta} \frac{1}{n} \sum_{i=1}^{n} \|\mathbf{F}_\theta(x_{(i)}) - \mathbf{c}\|^2 + \frac{\lambda}{2} \|\theta\|_2^2. \quad (1)$$

The first term in Eq. 1 is a quadratic loss that penalizes the Euclidean distance of $\mathbf{z}_{(L-R)}$ from the center $\mathbf{c}$ of a hypersphere $\mathcal{S}$, that is implicitly determined by the distance itself in the representation space. The second term is a classic weight decay regularizer, controlled by λ. Notice that we do not contract $\mathcal{S}$ by explicitly penalizing its radius and samples lying outside its boundary, but by minimizing their mean Euclidean distance with respect to $\mathbf{c}$ [22]. To avoid convergence to a collapsed trivial solution with all zero weights, all layers in $\mathbf{F}_\theta$ do not use bias terms [22], and the center $\mathbf{c}$ was set to the average $\mathbf{z}_{(L-R)}$ obtained by feeding $\mathbf{F}_\theta$ with all training samples before fine-tuning, as in [28]. At that stage, f is already pre-trained using the self-supervised strategy described in Sect. 2.1, but the APH has random weights. Nevertheless, we experimentally observed that this center $\mathbf{c}$ is already enough to avoid a collapsed $\mathcal{S}$.

Deviation-from-Normal-Asymmetry Index. The distance between the asymmetry embedding of an unseen sample x and the center of the learned hypersphere, $s(x; \mathbf{c}) = \|\mathbf{F}_\theta(x) - \mathbf{c}\|^2$, can be used as a deviation-from-normal-asymmetry index: when the input x is a normal sample, its asymmetry embedding is expected to lie in the vicinity of $\mathbf{c}$, then associated to a small s value; on the other hand, if x is the segmentation of a subject with abnormal asymmetries, its associated $\mathbf{z}_{(L-R)}$ will lie afar from $\mathbf{c}$, reporting a higher s value.

3 Experimental Setup

We studied our method for hippocampal asymmetry characterization as a use case. First, we tested its ability to capture deviations in asymmetry using synthetically altered hippocampi with increased deformations, in a controlled setting. Then, we indirectly evaluated its performance as a diagnostic tool for neurodegenerative conditions, using s as a deviation-from-normal-asymmetry index. Finally, we performed an ablation study to understand the influence of design factors such as the shape encoder architecture, APH size, and merging operation.

Materials. We used a total of 3243 3D T1 brain MRIs, including 2945 from normal control (NC) individuals, 71 from patients with MCI, 179 with AD, and 16 and 32 with right (HSR) and left (HSL) hippocampal sclerosis, respectively. Samples were retrospectively collected from OASIS [14] (NC = 2217, AD = 33), IXI [2] (NC = 539) and ADNI [1] (NC = 53, AD = 33, MCI = 71) public sets, and from two in-house databases, ROFFO (NC = 83) and HEC (NC = 53, HSL = 32, HSR = 16), (see supp. mat. for demographics characteristics). All images were integrated in a single set, that we split into training, validation and test. The training set was used to learn patterns of normal asymmetry, with NC from ROFFO (63), IXI (539), and 70% of the NC from OASIS. Ages ranged from 19 to 95 years old, to ensure capturing normal variations due to natural aging. 60 NC images were kept aside for validation (see below). The test sets, were used to evaluate the diagnostic ability of our method on different cohorts. TEST-ADNI and TEST-HEC sets include all subjects from ADNI and HEC sets, while TEST-OASIS includes all AD and the remaining 30% of NC from OASIS.

Images from different devices were aligned and normalized to a standard reference MNI T1 template using SPM12 [19]. Hippocampal segmentations were obtained using HippMapp3r [8], a CNN model that is robust to atrophies and lesions. The resulting segmentations were cropped to create separate masks for each hippocampus, each with size 64 × 64 × 64 voxels. We created synthetic validation sets with abnormal hippocampal asymmetries to study the model's ability to identify asymmetries of different variations. We used 60 normal hippocampal segmentations (20 from ROFFO and 40 from OASIS) and applied elastic deformations [25] with $\sigma \in \{3,5,8\}$ to one of the two hippocampi of 20 individuals, resulting in 60 simulated abnormal pairs (see supp. mat. for qualitative examples). Four validation sets S_σ were created, with $\sigma = 3,5,8$ and all, including the original 60 normal pairs and the corresponding simulated cases in the first three and a mix of all of them in the last one. This allowed us to perform hyperparameter tuning and evaluate the model's performance.

Implementation Details. We studied two backbones for our Siamese shape encoder: a LeNet-based architecture similar to the one in [22] (but adapted to 3D inputs), with 3 convolutional layers with 16, 32 and 64 $5 \times 5 \times 5$ filters, each followed by a 3D batch normalization (BN) layer and a ReLU operation; and a deeper CNN equal to the encoding path of the CAE in [15] to learn anatomical shape variations from 3D organs. The size of the FC layers in the APH were adjusted based on the validation set. For CAE pre-training, the LeNet based encoder was attached to a bottleneck FC layer, followed by a symmetric decoder with 3 trilinear interpolation upsamplings, each followed by a convolutional layer with 3D BN and ReLU. For the deeper encoder, on the other hand, we used the exact same CAE from [15]. Further details about the architectures are provided in the supp. mat. We pre-trained the networks using the CAE approach for 250 epochs, and then fine-tuned them with SVDD for another 250 epochs. In all cases, we used Adam optimization with a learning rate of 10^{-4}, weight decay regularization with a factor of 10^{-6} and a batch size of 12 hippocampi pairs. We used PyTorch 1.12 and SciKit Learn for our experiments.

Baselines. We compared our model with respect to other multiple approaches. To account for the standard clinical methods, we included the absolute and normalized volume differences (AVD and NVD, respectively), used in [18] as scores for asymmetry. We also included shallow one-class support vector machines (OC-SVMs) [23] trained with the same NC subjects than ours but different feature sets. We used ShapeDNA [26] (ShapeDNA + OC-SVM), which was previously studied to characterize hippocampal asymmetries [21], and a combined large feature vector (LFV + OC-SVM) including volumetric differences, ShapeDNA and shape features obtained using PyRadiomics (sphericity, compactness, quadratic compactness, elongation, flatness, spherical disproportion, surface volume ratio, maximum 2D diameter, maximum 3D diameter and Major Axis). For standard deep practices in anomaly detection, we trained hybrid CAE + shallow OC-SVM using our LeNet (LeNet-CAE + OC-SVM) and deeper backbones (Oktay *et al.* [15] + OC-SVM). Finally, a binary network was trained to detect AD cases (AD classification), in order to have a supervised counterpart for comparison. We used a larger training set that, apart from the same NC subjects used for the anomaly detection models, included all samples in TEST-OASIS. The validation set had in this case the remaining NCs from OASIS and AD cases from ADNI.

The same backbone architecture was used, but with an additional FC layer that had softmax activation for classification, as in [11].

4 Results and Discussion

4.1 Characterization of Normal and Disease Related Asymmetries

To test our hypothesis that samples with abnormal hippocampal asymmetries deviate from the center of the normal hypersphere, we evaluated the distances $s(x; \mathbf{c})$ between all samples in the validation and test sets and grouped them by disease category. The distribution of these distances is shown in Fig. 3 (left), with statistical significance assessed using Mann-Whitney-Wilcoxon rank-sum tests ($\alpha = 0.05$) with Bonferroni correction applied for multiple comparisons. Figure 3 (right) represent the t-SNE projection of all representations. NC subjects are closely grouped in this plot, with the smallest distances to the center among all groups. These values are significantly lower than those obtained for synthetically altered samples ($p < 0.004$) and individuals with MCI ($p < 0.017$), AD ($p < 0.0083$), HSL ($p < 0.017$) and HSR ($p < 0.017$). Distances increase proportionally to σ for synthetic cases and conditions with unilateral hippocampus shrinkage such as HSL and HSR, which are located at the extremes in the t-SNE representation. Synthetic cases with $\sigma = 8$ group around one of the clusters, while those with $\sigma = 5$ and 3 are scattered closer but still far from the center. MCI subjects from ADNI are scattered similarly to NC samples from OASIS, which is consistent with their distances. This could be due to cognitive decline in MCI cases not necessarily associated with alterations in hippocampal asymmetry, which can resemble that of NC, but rather with changes in other brain areas such as amygdala or thalamus [13,26]. It is possible then that the MCI subjects in this study do not exhibit hippocampal asymmetry changes large enough to be distinguished from those in healthy controls. AD cases, on the other hand, are seen far from the hypersphere center as well, reporting distances higher than those from NC. Finally, when compared one another, we observed significant differences in the distances between NC from ADNI set ($p < 0.0055$), but not between OASIS and HEC sets ($p < 0.1241$).

4.2 Comparison with Other Approaches

The evaluation results of different methods are displayed in Table 1. Volumetric based approaches only detected unilateral atrophies in synthetic and HSL/R cases and poor performance for MCI and AD. Feature-based methods performed slightly better for AD and synthetic but dropped for HSL/R and MCI, lacking the required robustness to abnormal asymmetry changes. Hybrid CAE + OC-SVM models had good performance on synthetic data but not on real data tasks. Finally, the AD binary classifier was only able to detect AD but failed in any other cases, which is consistent with the disease specialization hypothesis. Our method, on the contrary, outperformed all other methods for detecting both synthetically induced and pathological changes in hippocampal asymmetry. Notice,

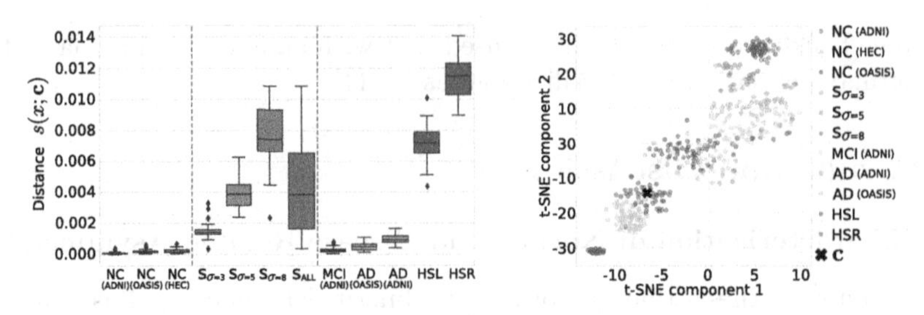

Fig. 3. Evaluation of our approach for characterizing different asymmetry patterns. Left: Distribution of distances $s(x; \mathbf{c})$ for each disease and NC group. Right: t-SNE representation of the normal asymmetry embeddings.

Table 1. AUC (95% CI) table for different approaches.

Method	Synthetic	MCI	AD (ADNI)	AD (OASIS)	HSL	HSR
AVD	0.72 (0.62–0.81)	0.56 (0.46–0.66)	0.55 (0.42–0.67)	0.62 (0.51–0.74)	0.95 (0.89–0.99)	0.94 (0.82–1.00)
NVD	0.82 (0.74–0.89)	0.58 (0.48–0.68)	0.62 (0.49–0.74)	0.64 (0.53–0.74)	0.95 (0.88–1.00)	0.94 (0.83–1.00)
ShapeDNA + OC-SVM	0.66 (0.56–0.74)	0.52 (0.42–0.60)	0.66 (0.55–0.76)	0.47 (0.35–0.60)	0.80 (0.69–0.90)	0.90 (0.81–0.98)
LFV + OC-SVM	0.93 (0.89–0.97)	0.58 (0.49–0.66)	0.67 (0.57–0.76)	0.65 (0.53–0.77)	0.92 (0.88–0.96)	0.98 (0.95–1.00)
LeNet-CAE + OC-SVM	0.95 (0.91–0.98)	0.48 (0.38–0.58)	0.50 (0.37–0.61)	0.46 (0.40–0.51)	0.77 (0.67–0.86)	0.69 (0.56–0.81)
Oktay *et al.* + OC-SVM	0.94 (0.90–0.98)	0.45 (0.35–0.56)	0.48 (0.35–0.59)	0.47 (0.42–0.53)	0.80 (0.69–0.88)	0.76 (0.65–0.86)
AD classification	0.49 (0.47–0.50)	0.64 (0.56–0.70)	0.73 (0.63–0.83)	(Used for training)	0.52 (0.48–0.56)	0.54 (0.48–0.61)
Deep NORAH (w/o CAE pretr.)	0.79 (0.71–0.87)	0.59 (0.48–0.69)	0.99 (0.99–1.00)	0.76 (0.67–0.84)	**1.00** (0.99–1.00)	**1.00** (0.99–1.00)
Deep NORAH (w/o FC dim. red.)	0.98 (0.94–1.00)	0.70 (0.60–0.79)	0.76 (0.65–0.87)	0.65 (0.54–0.76)	**1.00** (1.00–1.00)	**1.00** (1.00–1.00)
Deep NORAH (with CAE pretr.)	**0.99** (0.99–1.00)	**0.93** (0.87–0.97)	**1.00** (0.99–1.00)	**0.92** (0.86–0.96)	**1.00** (1.00–1.00)	**1.00** (1.00–1.00)

however, that the AD classification method is not a state-of-the-art approach but a comparable model with approximately the same backbone than ours. Other alternatives such as [10,12] might achieve much higher AUC values.

Ablation Analysis. Table 1 includes results with and without our CAE pretraining. This stage significantly improve performance in synthetic, MCI and AD (OASIS) cases, perhaps due to a better estimate of $\mathbf{c}$. Figure 4 illustrates the variations in AUC in $S_{\sigma=\text{all}}$ when changing the merge operation and the size of $h(\cdot)$. Using differences seems to be much more efficient in terms of capacity usage, with almost the same AUC obtained without a FC layer for dimensionality reduction (0.977) and with a FC layer with 512 outputs (0.997). Nevertheless, when applied on real cases, adding this additional component aids to improve the discrimination performance (see Table 1). Finally, Table 2 shows results in $S_{\sigma=\text{all}}$ for different encoder settings. Using a Siamese approach with difference as merge operation was in all cases superior than the other alternatives. In terms of architecture, LeNet and deep backbones showed similar performance, with LeNet being slightly better. This might be due to a higher number of FC parameters, as the output of the shape encoder is a vector with >32k features.

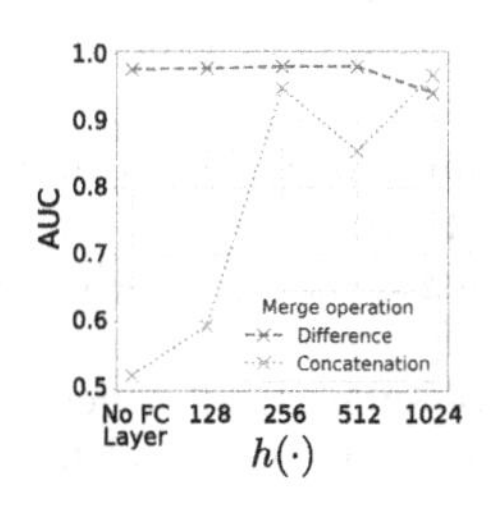

Fig. 4. APH ablation.

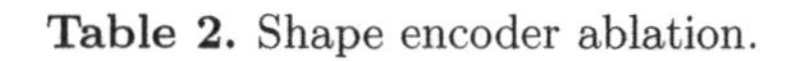

Table 2. Shape encoder ablation.

Backbone architecture	Siamese	Sagittal flip	Merge operation	Num. params	AUC ($S_{\sigma=\text{all}}$)
LeNet	✗	✗	N/A	5.4M	0.62
LeNet	✓	✗	Concat	2.7M	0.71
LeNet	✓	✗	Diff	2.7M	0.76
LeNet	✓	✓	Concat	2.7M	0.73
LeNet	✓	✓	Diff	2.7M	**0.95**
Deeper	✗	✗	N/A	1.1M	0.58
Deeper	✓	✗	Concat	555K	0.72
Deeper	✓	✗	Diff	555K	0.81
Deeper	✓	✓	Concat	555K	0.75
Deeper	✓	✓	Diff	555K	0.94

5 Conclusions

We presented a novel anomaly detection-based method for automatically characterizing normal asymmetry in homologous brain structures. Supervised alternatives restrict the definition of normal individuals due to explicitly learning their differences with respect to subjects with a specific condition. This implies that they ignore the asymmetry in control subjects, capturing only the asymmetries induced by the analyzed disease [12], and requiring retraining to detect new conditions unseen during training. Conversely, our unsupervised alternative leverages a recently introduced one-class-based objective to learn the space of normal asymmetries. Hence, it can detect diseased samples by quantifying their distance to the control space. Our experiments on hippocampus data showed that our approach could effectively use symmetry information to characterize normal populations and then identify disease presence by contrast, even though only NC subjects are used for training. Our model can potentially be applied to other homologous brain structures and diverse cohorts to aid radiologists in quantifying asymmetries of a normal brain better. In its current form, the model inherits the limitations of the segmentation approach, although it showed to achieve good performance using the outputs of HippMapp3r. Furthermore, it has the burden of not offering qualitative feedback, so future work should focus on bringing interpretability to this tool, e.g., by means of occlusion analysis.

Acknowledgments. This study was funded by PIP GI 2021-2023 0102472 (CONICET) and PICTs 2019-00070, 2020-00045 and 2021-00023 (Agencia I+D+i).

References

1. ADNI: Alzheimer's disease neuroimaging initiative. http://adni.loni.usc.edu/ Accessed Feb. 9 2023
2. IXI dataset website. http://brain-development.org/ixidataset/ Accessed Feb. 9 2023
3. Ardekani, B., et al.: Sexual dimorphism and hemispheric asymmetry of hippocampal volumetric integrity in normal aging and Alzheimer disease. Am. J. Neuroradiol. **40**(2), 276–282 (2019)

4. Bernasconi, N., et al.: Mesial temporal damage in temporal lobe epilepsy: a volumetric MRI study of the hippocampus, amygdala and parahippocampal region. Brain **126**(2), 462–469 (2003)
5. Borchert, R., et al.: Artificial intelligence for diagnosis and prognosis in neuroimaging for dementia; a systematic review. medRxiv 2021–12 (2021)
6. Csernansky, J.G., et al.: Abnormalities of thalamic volume and shape in schizophrenia. Am. J. Psychiatry **161**(5), 896–902 (2004)
7. Fu, Z., et al.: Altered neuroanatomical asymmetries of subcortical structures in subjective cognitive decline, amnestic mild cognitive impairment, and alzheimer's disease. J. Alzheimers Dis. **79**(3), 1121–1132 (2021)
8. Goubran, M., et al.: Hippocampal Segmentation for Brains with Extensive Atrophy Using Three-dimensional Convolutional Neural Networks. Tech. rep, Wiley Online Library (2020)
9. Herbert, M.R., et al.: Brain asymmetries in autism and developmental language disorder: a nested whole-brain analysis. Brain **128**(1), 213–226 (2005)
10. Herzog, N.J., Magoulas, G.D.: Brain asymmetry detection and machine learning classification for diagnosis of early dementia. Sensors **21**(3), 778 (2021)
11. Li, A., Li, F., Elahifasaee, F., Liu, M., Zhang, L.: Hippocampal shape and asymmetry analysis by cascaded convolutional neural networks for Alzheimer's disease diagnosis. Brain Imag. Behav. **15**(5), 2330–2339 (2021). https://doi.org/10.1007/s11682-020-00427-y
12. Liu, C.F., et al.: Using deep Siamese neural networks for detection of brain asymmetries associated with Alzheimer's disease and mild cognitive impairment. Magn. Reson. Imaging **64**, 190–199 (2019)
13. Low, A., et al.: Asymmetrical atrophy of thalamic Subnuclei in Alzheimer's disease and amyloid-positive mild cognitive impairment is associated with key clinical features. Alzheimer's Dementia: Diagnosis, Assess. Disease Monit. **11**(1), 690–699 (2019)
14. Marcus, D.S., et al.: Open access series of imaging studies: longitudinal MRI data in nondemented and demented older adults. J. Cogn. Neurosci. **22**(12), 2677–2684 (2010)
15. Oktay, O., et al.: Anatomically constrained neural networks (ACNNS): application to cardiac image enhancement and segmentation. IEEE Trans. Med. Imaging **37**(2), 384–395 (2017)
16. de Oliveira, A., et al.: Defining multivariate normative rules for healthy aging using neuroimaging and machine learning: an application to Alzheimer's disease. J. Alzheimers Dis. **43**(1), 201–212 (2015)
17. Park, B.y, et al.: Topographic divergence of atypical cortical asymmetry and atrophy patterns in temporal lobe epilepsy. Brain **145**(4), 1285–1298 (2022)
18. Pedraza, O., Bowers, D., Gilmore, R.: Asymmetry of the hippocampus and amygdala in MRI volumetric measurements of normal adults. JINS **10**(5), 664–678 (2004)
19. Penny, W.D., et al.: Statistical parametric mapping: the analysis of functional brain images. Elsevier (2011)
20. Princich, J.P., et al.: Diagnostic performance of MRI volumetry in epilepsy patients with hippocampal sclerosis supported through a random forest automatic classification algorithm. Front. Neurol. **12**, 613967 (2021)
21. Richards, R., et al.: Increased hippocampal shape asymmetry and volumetric ventricular asymmetry in autism spectrum disorder. NeuroImage: Clin. **26**, 102207 (2020)

22. Ruff, L., et al.: Deep one-class classification. In: ICML, pp. 4393–4402. PMLR (2018)
23. Schölkopf, B., et al.: Support vector method for novelty detection. Adv. Neural. Inform. Process. Syst. **12** (1999)
24. Tortora, G.J., Derrickson, B.H.: Principles of anatomy and physiology. John Wiley & Sons (2018)
25. van Tulder, G.: elasticdeform: Elastic deformations for n-dimensional images (2021)
26. Wachinger, C., et al.: Whole-brain analysis reveals increased neuroanatomical asymmetries in dementia for hippocampus and amygdala. Brain **139**(12), 3253–3266 (2016)
27. Woolard, A.A., Heckers, S.: Anatomical and functional correlates of human hippocampal volume asymmetry. Psych. Res.: Neuroimag. **201**(1), 48–53 (2012)
28. Zhang, Z., Deng, X.: Anomaly detection using improved deep SVDD model with data structure preservation. Pattern Recogn. Lett. **148**, 1–6 (2021)

Simulation of Arbitrary Level Contrast Dose in MRI Using an Iterative Global Transformer Model

Dayang Wang[1,2], Srivathsa Pasumarthi[1](✉), Greg Zaharchuk[3], and Ryan Chamberlain[1]

[1] Subtle Medical Inc., Menlo Park, CA 94025, USA
srivathsa@subtlemedical.com
[2] University of Massachusetts Lowell, Lowell, MA 01854, USA
[3] Stanford University, Stanford, CA 94035, USA

Abstract. Deep learning (DL) based contrast dose reduction and elimination in MRI imaging is gaining traction, given the detrimental effects of Gadolinium-based Contrast Agents (GBCAs). These DL algorithms are however limited by the availability of high quality low dose datasets. Additionally, different types of GBCAs and pathologies require different dose levels for the DL algorithms to work reliably. In this work, we formulate a novel transformer (Gformer) based iterative modelling approach for the synthesis of images with arbitrary contrast enhancement that corresponds to different dose levels. The proposed Gformer incorporates a sub-sampling based attention mechanism and a rotational shift module that captures the various contrast related features. Quantitative evaluation indicates that the proposed model performs better than other state-of-the-art methods. We further perform quantitative evaluation on downstream tasks such as dose reduction and tumor segmentation to demonstrate the clinical utility.

Keywords: Contrast-enhanced MRI · Iterative Model · Vision Transformers

1 Introduction

Gadolinium-Based Contrast Agents (GBCAs) are widely used in MRI scans owing to their capability of improving the border delineation and internal morphology of different pathologies and have extensive clinical applications [1]. However, GBCAs have several disadvantages like contraindications in patients with

D. Wang and S. Pasumarthi—The first two authors are equal contributors and co-first authors.

Supplementary Information The online version contains supplementary material available at https://doi.org/10.1007/978-3-031-43993-3_9.

H. Greenspan et al. (Eds.): MICCAI 2023, LNCS 14227, pp. 88–98, 2023.
https://doi.org/10.1007/978-3-031-43993-3_9

reduced renal function [2], patient inconvenience, high operation costs and environmental side effects [3]. Therefore, there is an increased emphasis on the paradigm of *"as low as reasonably achievable"* (ALARA) [4]. To tackle these concerns of GBCAs, several dose reduction [5,6] and elimination approaches [7] have been proposed. However, these deep learning(DL)-based dose reduction approaches require high quality low-dose contrast-enhanced (CE) images paired with pre-contrast and full-dose CE images. Acquiring such a dataset requires modification of the standard imaging protocol and involves additional training of the MR technicians. Therefore, it is important to simulate the process of T1w low-dose image acquisition, using images from the standard protocol. Moreover, it is crucial for these dose reduction approaches to establish the minimum dose level required for different pathologies as these are dependent on the scanning protocol and the GBCA compound injected. Therefore the simulation tool should also have the ability to synthesize images with multiple contrast enhancement levels, that correspond to multiple arbitrary dose levels.

Currently MRI dose simulation is done using physics-based models [8]. However, these physics-based methods are dependent on the protocol parameters and the type of GBCA and their relaxation parameters. Deep learning (DL) models have been widely used in medical imaging application due to their high capacity, generazibility, and transferability [9,10]. The performance of these DL models heavily depend on the availability of high quality data. There is a dearth of data-driven approaches to MRI dose-simulation given the lack of diverse ground truth data of the different dose levels. To this effect, we introduce a vision transformer based DL model[1] that can synthesize brain[2] MRI images that correspond to arbitrary dose levels, by training on a highly imbalanced dataset with only T1w pre-contrast, T1w 10% low-dose, and T1w CE standard dose images. The model backbone consists of a novel Global transformer (Gformer) with subsampling attention that can learn long-range dependencies of contrast uptake features. The proposed method also consists of a rotational shift operation that can further capture the shape irregularity of the contrast uptake regions. We performed extensive quantitative evaluation in comparison to other state-of-the art methods. Additionally, we show the clinical utility of the simulated T1w low-dose images using downstream tasks. To the best of our knowledge, this is the first DL based MRI dose simulation approach.

2 Methods

Iterative Learning Design: DL based models tend to perform poorly when the training data is highly imbalanced [11]. Furthermore, the problem of arbitrary dose simulation requires the interpolation of intermediate dose-levels using a minimum number of data points. Iterative models [12,13] are suitable for such

[1] A part of this work was presented as a poster in the conference of International Society for Magnetic Resonance in Medicine (ISMRM) 2023, held in Toronto.

[2] Refer Supplementary Material Fig. 1(b) for examples on Spine MRI.

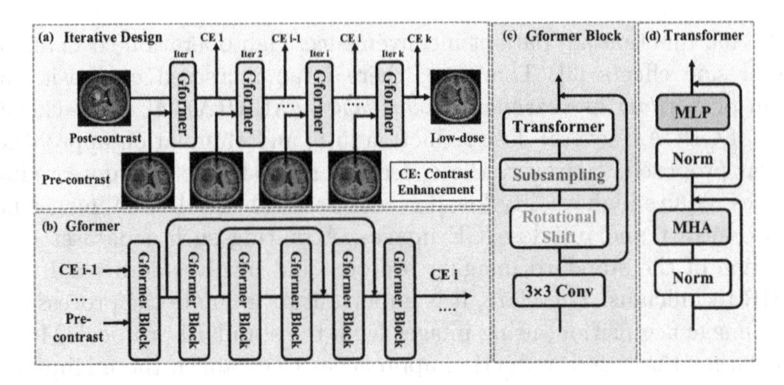

Fig. 1. (a) The proposed model based on iterative learning design. (b) The overall architecture of the proposed Gformer as the backbone network. (c) Layers inside the Gformer block. (d) Illustration of a typical vision transformer block.

applications as they work on the terminal images to generate step-wise intermediate solutions. We first utilize this design paradigm for the dose simulation task and train an end-to-end model on a highly imbalanced dataset where only T1w pre-contrast, T1w low-dose, and T1w post-contrast are available.

As shown in Fig. 1(a), the proposed iterative model $\mathcal{G} = \mathcal{F} \circ \mathcal{F} \circ \cdots \circ \mathcal{F}$ learns a transformation from the post-contrast to the low-dose image in k iterations, where $\mathcal{F}$ represents the base model. At each iteration i, the higher enhancement image from the previous step and the pre-contrast images are fed into $\mathcal{F}$ to predict the image with lower enhancement. The iterative model can be formulated as follows,

$$\begin{cases} \widehat{\mathbf{P_i}} = \mathcal{F}(\widehat{\mathbf{P_{i-1}}}, \mathbf{P_{pre}}) \\ \widehat{\mathbf{P_{low}}} = \overbrace{\mathcal{F} \circ \mathcal{F} \circ \cdots \circ \mathcal{F}}^{k}(\mathbf{P_{post}}, \mathbf{P_{pre}}), \end{cases} \tag{1}$$

where $\mathbf{P_{pre}}$, $\mathbf{P_{post}}$, and $\widehat{\mathbf{P_{low}}}$ denote the pre-contrast, post-contrast, and predicted low-dose images, respectively and $\widehat{\mathbf{P_{i-1}}}$ denotes the image with a higher enhancement than $\widehat{\mathbf{P_i}}$. This way, the intermediate outputs $\{\widehat{\mathbf{P_i}}\}_{i=1}^{k}$ having different enhancement levels, correspond to images with different contrast dose level with a uniform interval. This iterative model essentially learns a gradual dose reduction process, in which each iteration step removes a certain amount of contrast enhancement from the full-dose image.

Loss Functions and Model Convergence: The proposed iterative model aims to learn a mapping from the post-contrast & pre-contrast images to the synthesized low-dose images $\widehat{\mathbf{P_{low}}}$ and is trained with the true 10% low-dose image $\mathbf{P_{low}}$ as the ground truth. We used the $\mathbf{L_1}$ and structural similarity index measure (SSIM) losses. To tackle the problem of gradient explosion or vanishing, *"soft labels"* are generated using linear scaling. These *"soft labels"* serve as a reference to the intermediate outputs during the iterative training process and also aid model convergence, without which the model has to directly learn from post-contrast to low-dose. Given k iterations, the *soft label* $\{\mathcal{S}_i\}_{i=1}^{k-1}$ for iteration i is calculated as follows:

$$\mathcal{S}_i = \mathbf{P_{pre}} + [\gamma + (1-\gamma)\frac{k-i}{k}] \times \text{ReLU}(\widetilde{\mathbf{P_{post}}} - \widetilde{\mathbf{P_{pre}}} - \tau), \tag{2}$$

where $\widetilde{\mathbf{P_{post}}}$ and $\widetilde{\mathbf{P_{pre}}}$ denote the skull-stripped post-contrast and pre-contrast images. $\gamma = 0.1$ represents the dose level of the final prediction, and $\tau = 0.1$ denotes the threshold to extract the estimated contrast uptake $\mathcal{U} = \text{ReLU}(\widetilde{\mathbf{P_{post}}} - \widetilde{\mathbf{P_{pre}}} - \tau)$. Finally, the total losses are calculated as

$$\mathcal{L}_{\mathbf{total}} = \alpha \cdot \sum_{i=1}^{k-1} \mathcal{L}_{\mathbf{e}}(\widehat{\mathbf{P_i}}, \mathcal{S}_i) + \beta \cdot \mathcal{L}_{\mathbf{e}}(\widehat{\mathbf{P_{low}}}, \mathbf{P_{low}}). \tag{3}$$

where $\mathcal{L}_{\mathbf{e}} = \mathcal{L}_{\mathbf{L1}} + \mathcal{L}_{\mathbf{SSIM}}$ and $\alpha = 0.1$ and $\beta = 1$. The *"soft labels"* are assigned a small loss weight so that they do not overshadow the contribution of the real low-dose image. Additionally, in order to recover the high frequency texture information and to improve the overall perceptual quality, adversarial [14] and perceptual losses [15] are applied on $(\widehat{\mathbf{P_{low}}}, \mathbf{P_{low}})$ with a weight of 0.1.

Global Transformer (Gformer): Transformer models have risen to prominence in a wide range of computer vision applications [10,16]. Traditional Swin transformers compute attention on non-overlapping local window patches. To further exploit the global contrast information, we propose a hybrid global transformer (Gformer) as a backbone for the dose simulation task. As illustrated in Fig. 1(b), the proposed model design includes six sequential Gformer blocks as the backbone module with shortcuts. As shown in Fig. 1(c), the Gformer block contains a convolution block, a rotational shift module, a sub-sampling process, and a typical transformer module. The convolution layer extracts granular local information of the contrast uptake while the self-attention emphasizes more on the coarse global context, thereby paying attention to the overall contrast uptake structure.

Subsampling Attention: The sub-sampling is a key element in the Gformer block which generates a number of sub-images from the whole image as attention windows as shown in Fig. 2. Gformer performs self-attention on the sub-sampled images, which encompasses global contextual information with minimal self-attention overhead on small feature maps. Given the entire feature map $\mathbf{M_e} \in \mathbb{R}^{b \times c \times h \times w}$, where b, c, h, and w are the batch size, channel dimension, height, and width, respectively, the subsampling process aggregates the strided positions to the sub-feature maps as follows,

$$\{\mathbf{M_s}\}_{s=0}^{d^2-1} = \{\mathbf{M}[\,:\,,\,:\,,\ i:h:d\ ,\ j:w:d\ ]\}_{i=0,j=0}^{d-1,d-1}, \tag{4}$$

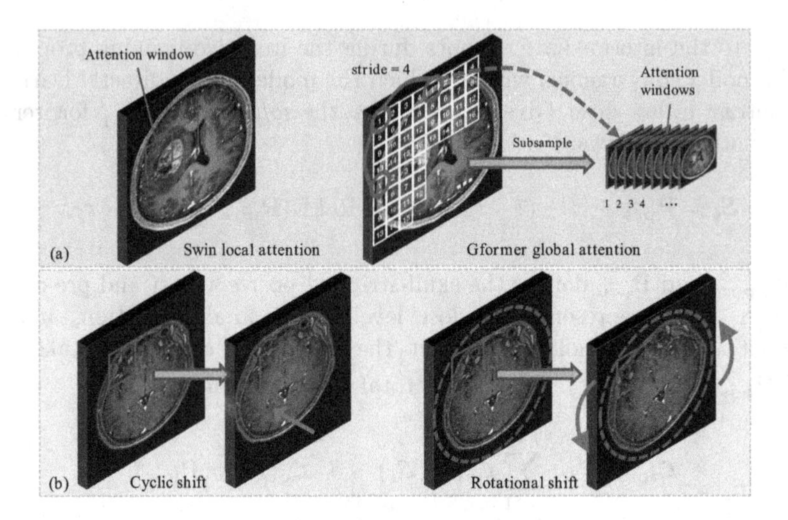

Fig. 2. (a) Illustration of the sub-sampling process with the stride of 4 as global attention window in Gformer block in comparison to local attention in Swin transformer. All pixels within the same number aggregate to the same sub-image. (b) Depiction of how rotational shift can enhance diverse contextual information fusion across layers compared to cyclic shift in Swin transformer.

where d denotes sampling a position every d pixels, and $\mathbf{M_s} \in \mathbb{R}^{b \times c \times \frac{h}{d} \times \frac{h}{d}}$ is the subsampled feature map. We set $h, d = 0$ to avoid any information loss during subsampling. These d^2 sub-feature maps are stacked onto the batch dimension as the attention windows for the transformer block shown in Fig. 1(c).

Rotational Shift: Image rotation has been widely used as a data augmentation technique in preprocessing and model training. Here, to further capture the heterogeneous nature of the contrast uptake areas, we employ the rotational shift as a module to facilitate the representation power of the Gformer. To prevent information loss on the edges due to rotation, only small angles (e.g., 10°, 20°) are used for rotation and residual shortcuts are also applied. Specifically, given the feature map $\mathbf{M_o} \in \mathbb{R}^{b \times c \times h \times w}$, rotational shift is performed around the vertical axis of height/width. The rotated feature map $\mathbf{M_r} \in \mathbb{R}^{b \times c \times h \times w}$ is obtained by the following equation:

$$\begin{bmatrix} \boldsymbol{p'} \\ \boldsymbol{q'} \\ \boldsymbol{x'} \\ \boldsymbol{y'} \end{bmatrix} = \begin{bmatrix} 1 & 0 & 0 & 0 \\ 0 & 1 & 0 & 0 \\ 0 & 0 & cos\boldsymbol{\lambda} & -sin\boldsymbol{\lambda} \\ 0 & 0 & sin\boldsymbol{\lambda} & cos\boldsymbol{\lambda} \end{bmatrix} \begin{bmatrix} \boldsymbol{p} \\ \boldsymbol{q} \\ \boldsymbol{x} - h//2 \\ \boldsymbol{y} - w//2 \end{bmatrix} + \begin{bmatrix} 0 \\ 0 \\ h//2 \\ w//2 \end{bmatrix}, \tag{5}$$

$$\mathbf{M_r}(\boldsymbol{p}, \boldsymbol{q}, \boldsymbol{x}, \boldsymbol{y}) = \begin{cases} \mathbf{M_o}(\boldsymbol{p'}, \boldsymbol{q'}, \lfloor \boldsymbol{x'} \rfloor, \lfloor \boldsymbol{y'} \rfloor), \text{ if } \boldsymbol{x'} \in [0, h) \text{ and } \boldsymbol{y'} \in [0, w) \\ 0, \text{ otherwise,} \end{cases} \tag{6}$$

where $\boldsymbol{\lambda}$ is the rotation angle. $(\boldsymbol{p}, \boldsymbol{q}, \boldsymbol{x}, \boldsymbol{y})$ and $(\boldsymbol{p'}, \boldsymbol{q'}, \boldsymbol{x'}, \boldsymbol{y'})$ denote the pixel index in the feature map tensor before and after rotational shift, respectively.

3 Experiments and Results

Dataset: With IRB approval and informed consent, we retrospectively used 126 clinical cases (113 training, 13 testing) from a internal private dataset[3] using Gadoterate meglumine contrast agent (Site A). For downstream task assessment we used 159 patient studies from another site (Site B) using Gadobenate dimeglumine. The detailed cohort description is given in Table 1. The clinical indications for both sites included suspected tumor, post-op tumor follow-up and routine brain. For each patient, 3D T1w MPRAGE scans were acquired for the pre-contrast, low-dose, and post-contrast images. These paired images were then mean normalized and affine co-registered (pre-contrast as the fixed image) using SimpleElastix [17]. The images were also skull-stripped, to account for differences in fat suppression, using the HD-BET brain extraction tool [18] for generating the *"soft labels"*.

Table 1. Dataset cohort description

Site	Total Cases	Gender	Age	Scanner	Field Strength	TE (sec)	TR (sec)	Flip Angle
Site A	126	55 Females 71 Males	48 ± 16	Philips Insignia	3T	2.97–3.11	6.41–6.70	8°
Site B	159	78 Females 81 Males	52 ± 17	GE Discovery	3T	2.99–5.17	7.73–12.25	8–20°

Implementation Details: All experiments were conducted with a single Tesla V100-SXM2 32GB GPU on a Intel(R) Xeon(R) CPU E5-2698 v4. The Subsampling stride for the six levels of Gformer block were {4,8,16,16,8,4}. The Rotational shift angles were {0,10,20,20,10,0} across all blocks. The model was optimized using the Adam optimizer with an initial learning rate of 1e-5 and a batch size of 4.

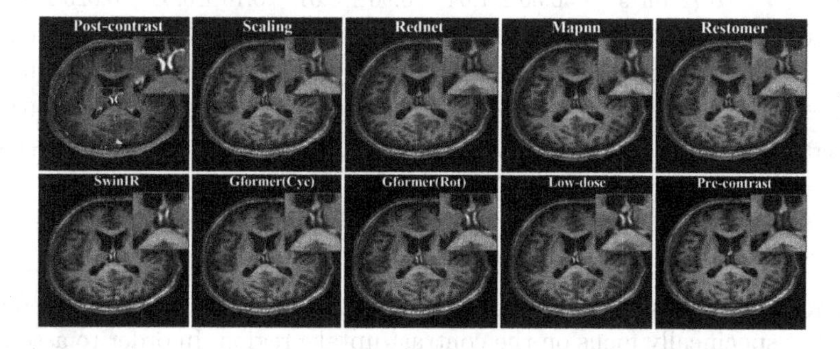

Fig. 3. The results of the synthesized 10% dose images from different methods. 'Rot' denotes rotational shift and 'Cyc' indicates cyclic shift.

[3] The dataset used in this study was obtained from Site A and B and are not available under a data sharing license.

Evaluation Settings: We quantitatively evaluated the proposed model using PSNR, SSIM, RMSE, and LPIPS perceptual metrics [19], between the synthesized and true low-dose images. We replaced the Gformer backbone with other state-of-the-art methods to compare the efficacy of the different methods. Particularly, the following backbone networks were studied: simple linear scaling (*"Scaling"*) approach, Rednet [20], Mapnn [13], Restormer [21], and SwinIR [22]. Unet [23] and Swin-Unet [24] models were not assessed due to their tendency to synthesize blurry artifacts in the iterative modelling. *throughput* metric (number of images generated per second) was also calculated to assess the inference efficiency.

Evaluation Results: Figure 4(a) shows that the proposed model is able to generate images that correspond to different dose levels. As shown in the zoomed inset, the hyperintensity of the contrast uptake in these images gradually reduces at each iteration. Figure 4(b) shows that the pathological structure in the synthesized low-dose image is similar to that of the ground truth. Figure 4(c) also shows that the model is robust to hyperintensities that are not related to contrast uptake. Figure 3 and Table 2 show that proposed model can synthesize enhancement patterns that look close to the true low-dose and that it performs better than the other competing methods with a reasonable inference throughput.

Table 2. Quantitative evaluation results of different base methods on the test cases. Bold-faced numbers indicate the best results.

Method	Throughput	PSNR (dB)↑	SSIM↑	RMSE↓	LPIPS↓
Post	–	33.93 ± 2.88	0.93 ± 0.03	0.34 ± 0.13	0.055 ± 0.016
Scaling	**0.79** Im/s	38.41 ± 2.22	0.94 ± 0.19	0.20 ± 0.05	0.027 ± 0.015
Rednet	0.71 Im/s	40.07 ± 2.72	0.97 ± 0.01	0.17 ± 0.05	0.029 ± 0.009
Mapnn	0.71 Im/s	40.56 ± 1.64	0.96 ± 0.01	0.16 ± 0.05	0.023 ± 0.012
Restormer	0.65 Im/s	40.04 ± 2.27	0.95 ± 0.01	0.16 ± 0.16	0.038 ± 0.016
SwinIR	0.58 Im/s	40.93 ± 2.25	0.96 ± 0.01	0.15 ± 0.06	0.028 ± 0.015
Gformer*(Cyc)	0.69 Im/s	41.46 ± 2.14	0.97 ± 0.02	0.14 ± 0.04	0.021 ± 0.007
Gformer*(Rot)	0.65 Im/s	$\mathbf{42.29 \pm 0.02}$	$\mathbf{0.98 \pm 0.01}$	$\mathbf{0.13 \pm 0.03}$	$\mathbf{0.017 \pm 0.005}$

Quantitative Assessment of Contrast Uptake: The above pixel-based metrics do not specifically focus on the contrast uptake region. In order to assess the contrast uptake patterns of the intermediate images, we used the following metrics as described in [25]: contrast to noise ratio(CNR), contrast to background ratio(CBR), and contrast enhancement percentage(CEP). The ROI for the contrast uptake was computed as the binary mask of the corresponding *"soft labels"* in Eq. 2. As shown in Fig. 5, the value of the contrast specific metrics increases in a non-linear fashion as the iteration step increases.

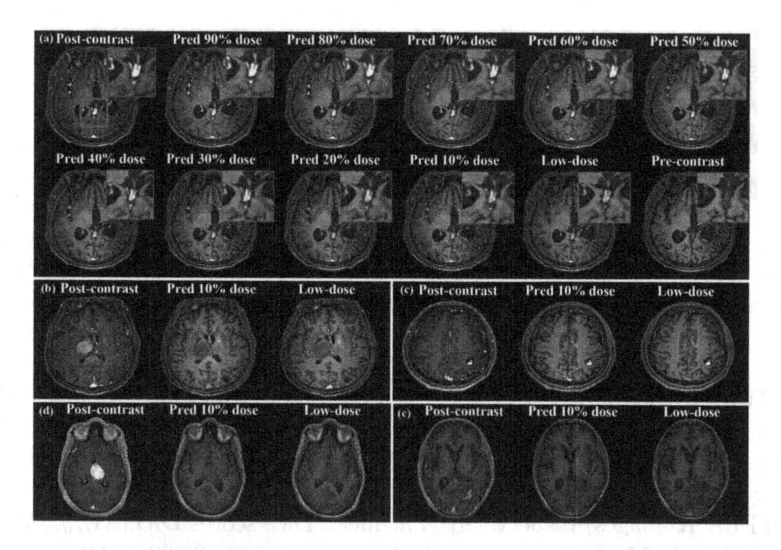

Fig. 4. (a) Model results showing images with different contrast enhancement corresponding to different dose levels along with the synthesized and true low-dose and pre-contrast. (b)–(c) Two representative slices of the synthesized 10% dose images. (d)–(e) Two representative slices using a different GBCA.

Downstream tasks: In order to demonstrate the clinical utility of the synthesized low-dose images, we performed two downstream tasks:

1) Low-dose to full-dose synthesis Using the DL-based algorithm to predict full-dose image from pre-contrast and low-dose images described in [5], we synthesized T1CE volumes using true low-dose (*T1CE-real-ldose*) and Gformer (rot) synthesized low-dose (*T1CE-synth-ldose*). We computed the PSNR and SSIM metrics of T1CE vs T1CE-synth/T1CE vs T1CE-synth-sim which are 29.82 ± 3.90 dB/28.10 ± 3.20 dB and $0.908 \pm 0.031/0.892 \pm 0.026$ respectively. This shows that the synthesized low-dose images perform similar[4] to that of the low-dose image in the dose reduction task. For this analysis we used the data from Site B.

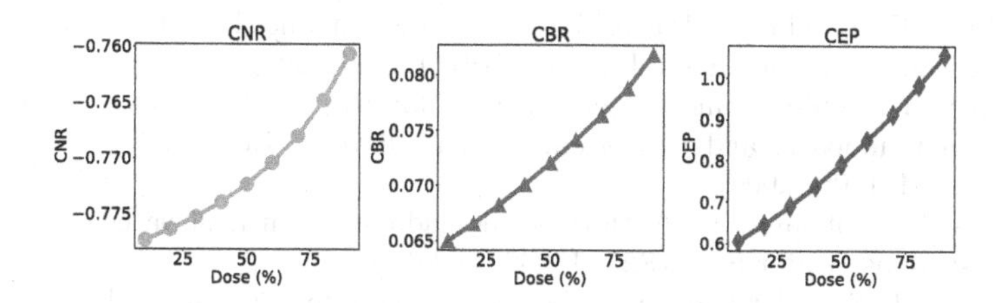

Fig. 5. Contrast uptake related quantitative metrics

[4] $p < 0.0001$ (Wilcoxon signed rank test).

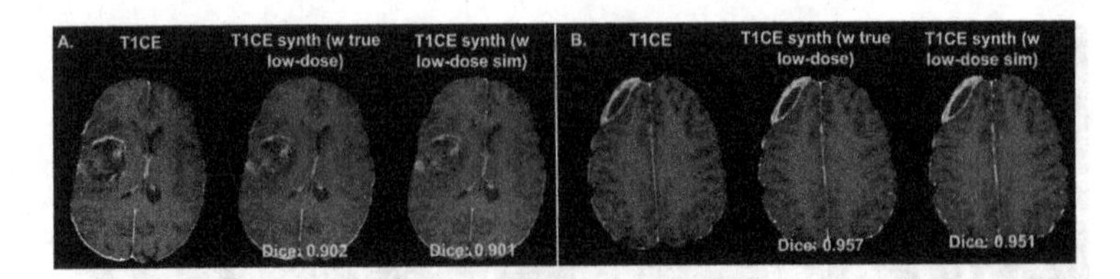

Fig. 6. Tumor segmentation (green overlay) on synthesized T1CE using real & simulated low-dose in comparison to the tumor segmentation on ground truth T1CE. The corresponding Dice scores are also shown in the bottom. (Color figure online)

2) Tumor segmentation Using the T1CE volumes synthesized in the above step, we perform tumor segmentation using the winning solution of BraTS 2018 challenge [26]. Let $\mathbf{M}_{true}$, $\mathbf{M}_{ldose}$ and $\mathbf{M}_{ldose-sim}$ be the whole tumor (WT) masks generated using T1CE, T1CE-real-ldose and T1CE-synth-ldose (+ T1, T2 and FLAIR images) respectively. The mean Dice scores $Dice(\mathbf{M}_{true}, \mathbf{M}_{ldose})$ and $Dice(\mathbf{M}_{true}, \mathbf{M}_{ldose-sim})$ on the test set were 0.889 ± 0.099 and 0.876 ± 0.092 respectively. Figure 6 shows visual examples of tumor segmentation performance. This shows that the clinical utility provided by the synthesized low-dose is similar[5] to that of the actual low-dose image.

4 Discussions and Conclusion

We have proposed a Gformer-based iterative model to simulate low-dose CE images. Extensive experiments and downstream task performance have verified the efficacy and clinical performance of the proposed model compared to other state-of-the art methods. In the future, further reader studies are required to assess the diagnostic equivalence of the simulated low-dose images. The model can be guided using physics-based models [27] that estimate contrast enhancement level using signal intensity. This simulation technique can easily be extended to other anatomies and contrast agents.

References

1. Minton, L.E., Pandit, R., Porter, K.K.: Contrast-enhanced MRI: history and current recommendations. Appl. Radiol. **50**(6), 15–19 (2021)
2. Grobner, T.: Gadolinium-a specific trigger for the development of nephrogenic fibrosing dermopathy and nephrogenic systemic fibrosis? Nephrol. Dial. Transplant. **21**(4), 1104–1108 (2006)
3. Brünjes, R., Hofmann, T.: Anthropogenic gadolinium in freshwater and drinking water systems. Water Res. **182**, 115966 (2020)
4. Uffmann, M., Schaefer-Prokop, C.: Digital radiography: the balance between image quality and required radiation dose. Eur. J. Radiol. **72**(2), 202–208 (2009)

[5] $p < 0.0001$ (Wilcoxon signed rank test).

5. Pasumarthi, S., Tamir, J.I., Christensen, S., Zaharchuk, G., Zhang, T., Gong, E.: A generic deep learning model for reduced gadolinium dose in contrast-enhanced brain MRI. Magn. Reson. Med. **86**(3), 1687–1700 (2021)
6. Gong, E., Pauly, J.M., Wintermark, M., Zaharchuk, G.: Deep learning enables reduced gadolinium dose for contrast-enhanced brain MRI. J. Magn. Reson. Imaging **48**(2), 330–340 (2018)
7. Kleesiek, J., et al.: Can virtual contrast enhancement in brain MRI replace gadolinium?: A feasibility study. Invest. Radiol. **54**(10), 653–660 (2019)
8. Sourbron, S., Buckley, D.L.: Tracer kinetic modelling in MRI: estimating perfusion and capillary permeability. Phys. Med. Biol. **57**(2), R1 (2011)
9. Wang, D., Wu, Z., Yu, H.: TED-Net: convolution-free T2T vision transformer-based encoder-decoder dilation network for low-dose CT denoising. In: Lian, C., Cao, X., Rekik, I., Xu, X., Yan, P. (eds.) MLMI 2021. LNCS, vol. 12966, pp. 416–425. Springer, Cham (2021). https://doi.org/10.1007/978-3-030-87589-3_43
10. Liu, J., Pasumarthi, S., Duffy, B., Gong, E., Zaharchuk, G., Datta, K.: One model to synthesize them all: Multi-contrast multi-scale transformer for missing data imputation. arXiv preprint arXiv:2204.13738 (2022)
11. Antipov, G., Baccouche, M., Dugelay, J.-L.: Face aging with conditional generative adversarial networks. In: 2017 IEEE International Conference on Image Processing (ICIP), pp. 2089–2093. IEEE (2017)
12. Liu, Y., Liu, Q., Zhang, M., Yang, Q., Wang, S., Liang, D.: IFR-Net: iterative feature refinement network for compressed sensing MRI. IEEE Trans. Comput. Imaging **6**, 434–446 (2019)
13. Shan, H., et al.: Competitive performance of a modularized deep neural network compared to commercial algorithms for low-dose CT image reconstruction. Nat. Mach. Intell. **1**(6), 269–276 (2019)
14. Goodfellow, I., et al.: Generative adversarial networks. Commun. ACM **63**(11), 139–144 (2020)
15. Johnson, J., Alahi, A., Fei-Fei, L.: Perceptual losses for real-time style transfer and super-resolution. In: Leibe, B., Matas, J., Sebe, N., Welling, M. (eds.) ECCV 2016. LNCS, vol. 9906, pp. 694–711. Springer, Cham (2016). https://doi.org/10.1007/978-3-319-46475-6_43
16. Liu, Z., et al.: Swin transformer: hierarchical vision transformer using shifted windows. In: Proceedings of the IEEE/CVF International Conference on Computer Vision, pp. 10012–10022 (2021)
17. Marstal, K., Berendsen, F., Staring, M., Klein, S.: SimpleElastix: a user-friendly, multi-lingual library for medical image registration. In: Proceedings of the IEEE Conference on Computer Vision and Pattern Recognition Workshops, pp. 134–142 (2016)
18. Schell, M., et al.: Automated brain extraction of multi-sequence MRI using artificial neural networks. In: European Congress of Radiology-ECR 2019 (2019)
19. Zhang, R., Isola, P., Efros, A.A., Shechtman, E., Wang, O.: The unreasonable effectiveness of deep features as a perceptual metric. In: Proceedings of the IEEE Conference on Computer Vision and Pattern Recognition, pp. 586–595 (2018)
20. Chen, H., et al.: Low-dose CT with a residual encoder-decoder convolutional neural network. IEEE Trans. Med. Imaging **36**(12), 2524–2535 (2017)
21. Zamir, S.W., Arora, A., Khan, S., Hayat, M., Khan, F.S., Yang, M.-H.: Restormer: efficient transformer for high-resolution image restoration. In: 2022 IEEE/CVF Conference on Computer Vision and Pattern Recognition (CVPR) (2022)

22. Liang, J., Cao, J., Sun, G., Zhang, K., Van Gool, L., Timofte, R.: SwiniR: image restoration using swin transformer. In: Proceedings of the IEEE/CVF International Conference on Computer Vision, pp. 1833–1844 (2021)
23. Ronneberger, O., Fischer, P., Brox, T.: U-Net: convolutional networks for biomedical image segmentation. In: Navab, N., Hornegger, J., Wells, W.M., Frangi, A.F. (eds.) MICCAI 2015. LNCS, vol. 9351, pp. 234–241. Springer, Cham (2015). https://doi.org/10.1007/978-3-319-24574-4_28
24. Cao, H., et al.: Swin-Unet: Unet-like pure transformer for medical image segmentation. In: Karlinsky, L., Michaeli, T., Nishino, K. (eds.) Computer Vision-ECCV 2022 Workshops. ECCV 2022. Lecture Notes in Computer Science. vol. 13803. Springer, Cham (2023). https://doi.org/10.1007/978-3-031-25066-8_9
25. Bône, A., et al.: From dose reduction to contrast maximization: can deep learning amplify the impact of contrast media on brain magnetic resonance image quality? A reader study. Invest. Radiol. **57**(8), 527–535 (2022)
26. Myronenko, A.: 3D MRI brain tumor segmentation using autoencoder regularization. In: Crimi, A., Bakas, S., Kuijf, H., Keyvan, F., Reyes, M., van Walsum, T. (eds.) BrainLes 2018. LNCS, vol. 11384, pp. 311–320. Springer, Cham (2019). https://doi.org/10.1007/978-3-030-11726-9_28
27. Mørkenborg, J., Pedersen, M., Jensen, F., Stødkilde-Jørgensen, H., Djurhuus, J., Frøkiær, J.: Quantitative assessment of GD-DTPA contrast agent from signal enhancement: an in-vitro study. Magn. Reson. Imaging **21**(6), 637–643 (2003)

Development and Fast Transferring of General Connectivity-Based Diagnosis Model to New Brain Disorders with Adaptive Graph Meta-Learner

Yuxiao Liu[1], Mianxin Liu[2], Yuanwang Zhang[1], and Dinggang Shen[1,3,4](✉)

[1] School of Biomedical Engineering, ShanghaiTech University, Shanghai 201210, China
[2] Shanghai Artificial Intelligence Laboratory, Shanghai 200232, China
[3] Shanghai United Imaging Intelligence Co., Ltd., Shanghai 200232, China
[4] Shanghai Clinical Research and Trial Center, Shanghai 201210, China
dgshen@shanghaitech.edu.cn

Abstract. The accurate and automatic diagnosis of new brain disorders (BDs) is crucial in the clinical stage. However, previous deep learning based methods require training new models with large data from new BDs, which is often not practical. Recent neuroscience studies suggested that BDs could share commonness from the perspective of functional connectivity derived from fMRI. This potentially enables developing a connectivity-based general model that can be transferred to new BDs to address the difficulty of training new models under data limitations. In this work, we demonstrate this possibility by employing the meta-learning algorithm to develop a general adaptive graph meta-learner and transfer it to new BDs. Specifically, we use an adaptive multi-view graph classifier to select the appropriate view for specific disease classification and a reinforcement-learning-based meta-controller to alleviate the over-fitting when adapting to new datasets with small sizes. Experiments on 4,114 fMRI data from multiple datasets covering a broad range of BDs demonstrate the effectiveness of modules in our framework and the advantages over other comparison methods. This work may pave the basis for fMRI-based deep learning models being widely used in clinical applications.

Keywords: Brain disorders · Graph convolutional network · Brain functional network · Few-shot adaptation · Meta-learning

1 Introduction

Brain disorders (BDs) pose severe challenges to public mental health in the global world. To understand the pathology, and for accurate diagnosis as well, functional MRI (fMRI), one of the MRI modalities, is widely studied for BDs. The fMRI provides assessments of the disease-induced changes in the brain functional connectivity networks (FCNs) among different brain regions of interest (ROIs). And a huge body of studies has successfully built effective classifiers for different BDs based on FCN and deep learning methods [6, 14, 15]. However, when facing new BDs, it is often needed to train a new classification model, which requires collections of large clinical data. During this process, the high costs in time, money, and labor prevent collecting sufficient

H. Greenspan et al. (Eds.): MICCAI 2023, LNCS 14227, pp. 99–108, 2023.
https://doi.org/10.1007/978-3-031-43993-3_10

data and thus the applications of deep learning models to new BDs with a small number of samples, especially for some rare BDs. Recently, there has been study [24] illustrating that BDs share significant commonness under the perspective of FCN alternations. Based on this knowledge, developing and transferring a general model to new BDs could be possible, which is promising to address the issues of building new classifiers for new BDs under data limitation.

Meta-learning based algorithm is one of the advanced methods to develop a general model based on heterogeneous information from data in different domains or for different tasks. It aims to learn optimal initial parameters for the model (**meta-learner**) which can be quickly generalized to new tasks, directly or with a few new training data for fine-tuning. There have been extensive discussions in the literature on developing more general models utilizing meta-learning [12,18,22,23] in medical fields.

Upon this evidence, it would be promising to develop a general BD diagnosis model based on FCN and further use meta-learning to solve the above-mentioned issues. However, there are still at least two challenges to achieve this goal. **First**, how to optimally extract the generalizable common knowledge (features) from the procedure of diagnosing various BDs? Previous methods focused on conventional FCNs (computed using simple linear correlations) and only treated FCN as a vector [12]. This manner of analysis *neither* properly explores topological features in the FCN, *nor* fully characterizes the complex, high-order functional interactions among brain regions, which are demonstrated to be associated with BDs [2]. **Second**, after developing a general model, how to optimally adapt the model to new datasets with different conditions? The best tuning of the model may exist only within a critical range of parameters, but previous studies blindly search for optimal parameters using *ad hoc* manual configurations, such as the adaption step size, which is easy to cause over-fitting and degrading the performance on small datasets. Theoretically, it would be beneficial to let the model adaptively configure the adaption step size and other parameters, according to the given dataset.

In this paper, we develop a novel framework (illustrated in Fig. 1) to explore the aforementioned issues. First, we assemble a large amount of data from both public datasets and in-house datasets (i.e., a total of 6 datasets, 4,114 subjects) to develop a general BD diagnosis model with meta-learning. Second, during the meta-learning procedure, we propose an adaptive multi-view graph classifier to mine topological information of low- and high-order FCNs, as different views of brain dynamics for classification. The attention mechanism is implemented to dynamically weigh and fuse different views of the information under given tasks, which collaborate with meta-learning and helps the model to learn to adapt to diagnoses for different BDs. Third, we apply a meta-controller driven by reinforcement learning [25] to choose the optimal adaption step size for the general model, for properly adapting to new BDs with small datasets and alleviating the over-fitting issue.

2 Methods

2.1 Notation and Problem Formulation

We define the entire set of BD diagnosis tasks using different datasets with $\mathcal{T}$. For a specific task $\mathcal{T}_j \in \mathcal{T}$ $(j = 1, 2, ..., 6)$, we have n pairs of FCN and labeled data $\mathcal{D} = \{(F_i, Y_i)\}_{i=1}^{n}$, where F is the fMRI data and Y is the label set of all subjects.

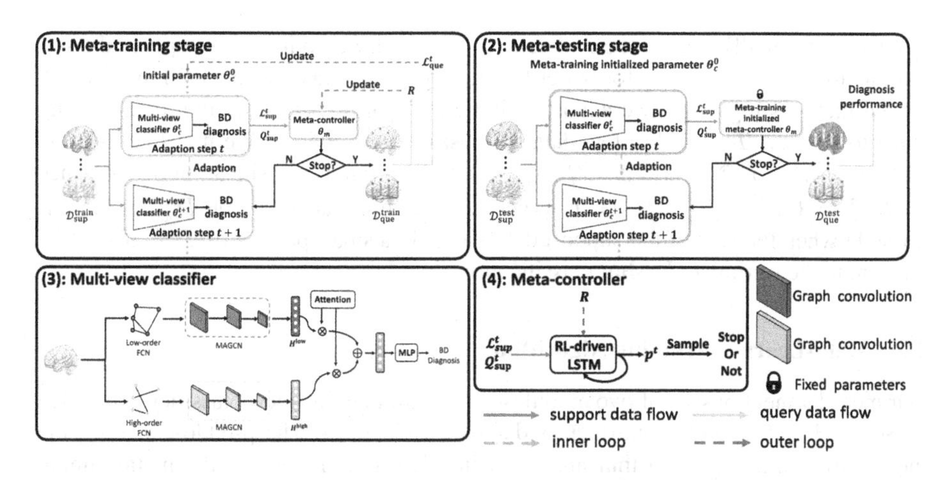

Fig. 1. The framework of our proposed method. During the meta-training stage (1), we first tune the initial parameters θ_c^0 for T steps on the support data $\mathcal{D}_{\text{sup}}^{\text{train}}$ as the inner loop when adapting to new meta-training tasks. At the step t, meta-controller (4) decides whether to stop based on graph embedding quality Q_{sup}^t and loss $\mathcal{L}_{\text{sup}}^t$ on the support data. If the controller decides to stop at step t, then we tune θ_m and θ_c^0 as the outer loop based on loss reward $R \in \mathbb{R}^{t\times 1}$ and $\mathcal{L}_{\text{que}}^t$ separately on the query data $\mathcal{D}_{\text{que}}^{\text{train}}$ for the next inner loop. Finally, we fix θ_m, and fine-tune the meta-training stage initialized θ_c^0 on $\mathcal{D}_{\text{sup}}^{\text{test}}$ in the meta-testing stage (2), predicting the labels on $\mathcal{D}_{\text{que}}^{\text{test}}$.

ROIs are defined in individual fMRI spaces based on the brain atlas [20] to extract ROI-based fMRI signals.

We extract three features from individual fMRI data for the graph deep learning analysis. First, we compute the low-order FCN A^{low}, where $A^{\text{low}} \in \mathbb{R}^{N_{\text{ROI}}\times N_{\text{ROI}}}$ is a graph adjacency matrix containing the pair-wise correlations among different ROI signals. Second, we compute the high-order FCN A^{high} based on the topological information in A^{low} (described in detail in Sect. 2.3). These two features will be used as *edge features*. Third, we use the corresponding order adjacency matrix, together with the mean and standard deviation value of ROI signals, as the *node features* for both low- and high-order graphs [14].

We have meta-training and meta-testing stages for our meta-learner. The corresponding tasks are named as meta-training task $\mathcal{T}^{\text{train}}$ and meta-testing task $\mathcal{T}^{\text{test}}$. The meta-training stage mimics cross-task adaptations, aiming to make the model *learning to adapt to* new task $\mathcal{T}^{\text{test}}$ with a small number of samples under initial parameters. To simulate the cross-task scenario, we utilized the episodic training mechanism [13], which samples small sets across all meta-training datasets. In detail, for each round the in meta-training stage, we randomly sample the non-overlapping support data $\mathcal{D}_{\text{sup}}^{\text{train}} = \left\{\left(F_i^{\text{train}}, Y_i^{\text{train}}\right)\right\}_{i=1}^{s}$ and query data $\mathcal{D}_{\text{que}}^{\text{train}} = \left\{\left(F_i^{\text{train}}, Y_i^{\text{train}}\right)\right\}_{i=1}^{q}$ across all meta-training datasets, based on which, the $\mathcal{T}^{\text{train}}$ is constructed. With constructed $\mathcal{T}^{\text{train}}$, we have two loops to update the initial parameters of the meta-learner, which are inner and outer loops. During inner loops, initial parameters update gradients are first estimated using the back-propagation learned from the $\mathcal{D}_{\text{sup}}^{\text{train}}$. Then, if initial parameters are updated by the first gradient, we further estimate the gradients based on $\mathcal{D}_{\text{que}}^{\text{train}}$ as the outer loop to finally update them. The two loops will be combined to update the initial meta-learner parameters for better fine-tuning on new tasks as detailed in Sect. 2.2. At the meta-testing stage, we adapt the initialized parameters from the meta-training stage to the new $\mathcal{T}^{\text{test}}$ with a few data and test its performance. We first fine-tune the meta-learner on $\mathcal{D}_{\text{sup}}^{\text{test}} = \{(F_i^{\text{test}}, Y_i^{\text{test}})\}_{i=1}^{s}$, and then report classification performance on $\mathcal{D}_{\text{que}}^{\text{test}} = \{(F_i^{\text{test}}, Y_i^{\text{test}})\}_{i=1}^{q}$. Our target is to make the accuracy on $\mathcal{D}_{\text{que}}^{\text{test}}$ as high as possible when the size of the support data s is only a small portion of $\mathcal{T}^{\text{test}}$. If $s = K$, we denote the setting as K**-shot** classification.

2.2 Meta-Learner Training Algorithm

Our meta-learner consists of two modules, which are 1) multi-view classifier as detailed in Sect. 2.3 and 2) meta-controller as detailed in Sect. 2.4. The pseudo-codes for the meta-learner training algorithm are given in Algorithm 1. During the meta-training stage, our meta-learner algorithm has two iterative parameter update loops, which are inner (lines 3–9) and outer loops (lines 10–14) [5]. The inner loop aims to fast adapt the multi-view graph classifier parameterized with θ_c to the new sampled $\mathcal{T}^{\text{train}}$ under given initial parameters (θ_c^0, θ_m). During the inner loop, the meta-controller parameterized with θ_m decides whether to stop at the adaption step t to avoid over-fitting. While the outer loop aims to improve the generality of the meta-learner by exploring the optimal initial parameters (θ_c^0, θ_m) according to the loss on $\mathcal{D}_{\text{que}}^{\text{train}}$ which can easily adapt to other tasks. The initial parameter updated in the outer loop will be set for the next sampled $\mathcal{T}^{\text{train}}$. At the meta-testing stage, the meta-learner will utilize the meta-training stage initialized (θ_c^0, θ_m), fixing θ_m and fine-tuning θ_c^0 on $\mathcal{D}_{\text{sup}}^{\text{test}}$, finally predicting the label $\mathcal{D}_{\text{que}}^{\text{test}}$.

2.3 Multi-view Graph Classifier θ_c

As mentioned above, exploring the topological information of the FCN is necessary for BD diagnosis. However, a single low-order FCN view can only illustrate the simple pair-wise relation. So, we additionally construct the high-order FCN, which reflects the correlation among ROIs in terms of their own FC patterns, as calculated below:

$$A_{ij}^{\text{high}} = \text{Pearson Correlation}(A_{i*}^{\text{low}}, A_{j*}^{\text{low}}). \tag{1}$$

After constructing high-order adjacency matrix A^{high} as a complementary view, we input them into the **multi-view graph classifier** parameterized with θ_c^t at the adaption step t. To extract the disease-related features of different-view graphs, we use convolution in GCN [11] to aggregate the neighboring node features.

Then, we use the pooling operation to further extract the global topological features of different graph views. Here, we opt for gPOOL operation [7] for its parameter-saving and ability to extract global topological features. For each pooling stage, it acquires the importance of each node by calculating the inner product between the node feature vectors and a learnable vector. The top-k important nodes and their corresponding subgraph will be sampled for the following graph convolutions. By iteratively repeating the node feature aggregation and pooling illustrated in Fig. 1 (3), we can finally acquire

Algorithm 1: Meta-training algorithm.

Input: learning rates: β_1, β_2, random initialization (θ_c^*, θ_m^*)
Output: Trained parameters (θ_c^0, θ_m).

1 $(\theta_c^0, \theta_m) = (\theta_c^*, \theta_m^*)$
2 **while** not done **do**
3 Sample $\mathcal{T}^{\text{train}}$ with $\mathcal{D}_{\text{sup}}^{\text{train}}$ and $\mathcal{D}_{\text{que}}^{\text{train}}$ from meta-training datasets
4 **while** $t \in [1, T_{\text{MAX}}]$ **do**
5 Calculate $\nabla_{\theta_c^{t-1}} \mathcal{L}_{\text{sup}}^t$ on $\mathcal{D}_{\text{sup}}^{\text{train}}$
6 $\theta_c^t \leftarrow \theta_c^{t-1} - \beta_1 \nabla_{\theta_c^{t-1}} \mathcal{L}_{\text{sup}}^t$
7 Calculate Q_{sup}^t by Eq. 3 on $\mathcal{D}_{\text{sup}}^{\text{train}}$
8 Calculate stop probability by Q_{sup}^t and $\mathcal{L}_{\text{sup}}^t$ and determine whether to stop at step t via Eq. 4, 5
9 Calculate reward R^t with loss changes on $\mathcal{D}_{\text{que}}^{\text{train}}$ via Eq. 7
10 **end**
11 $\theta_c^0 \leftarrow \theta_c^0 - \beta_1 \nabla_{\theta_c^0} \mathcal{L}_{\text{que}}^T$ on $D_{\text{que}}^{\text{train}}$
12 **for** $t = 0 : T$ **do**
13 $\theta_m \leftarrow \theta_m + \beta_2 R^t \nabla_{\theta_m} \ln p(t)$
14 **end**
15 **end**

the representation of the graph, denoted as H. Furthermore, to let the model adaptively choose the proper view for classification, we apply an attention-based mechanism to aggregate the learned embeddings by feeding the concatenation of the learned representations into an attention module as below:

$$\alpha = \text{Softmax}\left(\text{ReLU}\left(\left[H^{\text{low}} \| H^{\text{high}}\right] W_1\right) W_2\right). \tag{2}$$

The attention mechanism allows the model to decide which view should rely on for specific tasks more adaptively. Then, we forward attention-weighted features into an MLP and acquire the final prediction label. Here, we use the cross-entropy loss as a constraint to the network.

2.4 Meta-Controller θ_m

In machine learning, over-fitting is one of the critical problems that restrict the performance of models, especially in small datasets. Previous works utilize the early stopping to alleviate this problem, but the hand-crafted early stopping parameter is hard to choose. Here, we utilize a neural network to learn the stop policy adaptively, which we call **meta-controller** parameterized with θ_m depicted in Fig. 1 (4). The meta-controller uses a reinforcement learning based algorithm [17] to decide optimal adaption step sizes for cross-task adaptions. Considering the characteristics of the graph data, we let the model determine when to stop *not only* according to the classification loss, *but also* the graph embedding quality on the support data. For the embedding quality, we use the Average Node Information (ANI), denoted as Q, to measure it. It represents how a node can be measured by the neighboring nodes. A high ANI value indicates that the embedding module has learned the most information about the graph. If we keep aggregating the nodal features by graph convolution when the ANI is high, the over-smoothing will happen, making all nodes have similar features and thus degrading the performance. We define the ANI Q_{sup} of the support data within T^{train} with the L1 norm [9] as follows:

$$Q_i = \frac{1}{N_{\text{ROI}}} \sum_{j=1} \left\| \left[\left(I - D_i^{-1} A_i \right) X_i^L \right]_j \right\|_1, \; Q_{\text{sup}} = 1/s \times \sum_{i=1}^{s} \left(Q_i^{\text{high}} + Q_i^{\text{low}} \right), \quad (3)$$

where D_i X_i^L represent the degree matrix of A_i and the feature matrix in the last layer L, respectively; j denotes j-th node which is also the j-th row in those matrices, and $\|\cdot\|_1$ denotes the L1-norm of row vector.

For classification losses $\mathcal{L}_{\text{sup}}$ and ANI values Q_{sup} on $\mathcal{D}_{\text{sup}}^{\text{train}}$ across t adaption steps, we use them to compute the stop probability p^t at step t with an LSTM [8] model by considering the temporal information as follows:

$$o^t = \text{LSTM}\left(\left[Q_{\text{sup}}^t, \mathcal{L}_{\text{sup}}^t \right], o^{t-1} \right), p^t = \sigma\left(W o^t + b \right), \quad (4)$$

where o^t is the output of the LSTM model at step t and σ is the SoftMax function. Finally, we sample the choices c^t by Bernoulli distribution to decide whether we should stop at step t.

$$c^t \sim Bernouli(p^t) = \begin{cases} 1, & \text{stop the adaption} \\ 0, & \text{keep adaption} \end{cases}. \quad (5)$$

Since the relation between θ_m and θ_c is undifferentiable, it is impossible to take direct gradient descent on θ_m. We use stochastic policy gradient to optimize θ_m. Once we sample the stop choice at step T, we train the meta-controller according to loss changes on $\mathcal{D}_{\text{que}}^{\text{train}}$ across T steps. Given the parameter update trajectory of classifier $\{\theta_c^0, \theta_c^1, ...\theta_c^T\}$ during T steps, we calculate the corresponding loss change trajectory of $\mathcal{D}_{\text{que}}^{\text{train}}$. Based on that, we further define the controller immediate rewards r at step t as the loss change on $\mathcal{D}_{\text{que}}^{\text{train}}$ (caused by parameter update of step t):

$$r^{(t)} = \mathcal{L}\left(\mathcal{D}_{\text{que}}^{\text{train}}; \theta_c^{t-1} \right) - \mathcal{L}\left(\mathcal{D}_{\text{que}}^{\text{train}}; \theta_c^t \right) \quad (6)$$

Then, the accumulative reward R at step t is

$$R^t = \sum_{i=t}^{T} r^i = \mathcal{L}\left(\mathcal{D}_{\text{que}}^{\text{train}}; \theta_c^{t-1} \right) - \mathcal{L}\left(\mathcal{D}_{\text{que}}^{\text{train}}; \theta_c^T \right), \quad (7)$$

where T is the total number of steps and R^t is the change of classification loss on $\mathcal{D}_{\text{que}}^{\text{train}}$ from step t to the end of adaption. Then we update our meta-controller by policy gradients, which is a typical method in Reinforcement learning [25]:

$$\theta_m \leftarrow \theta_m + \beta_2 R^t \nabla_{\theta_m} \ln\left(p^t \right), \quad (8)$$

where∇_{θ_m} is the gradients over θ_m and β_2 is the learning rate.

3 Experiments

3.1 Dataset

We use fMRI meta-training data from five datasets including Alzheimer's Disease Neuroimaging Initiative (ADNI) [1,10], Open Access Series of Imaging Studies (OASIS) [19], and in-house dataset from Huashan Hospital (elder BDs datasets); ADHD-200 [3] and Autism Brain Imaging Data Exchange (ABIDE) [4] (youth BD datasets). For the meta-testing dataset, we use the in-house dataset from Zhongshan Hospital which is about vascular cognitive impairment (VCI). All datasets are shown in Table 1 The details of image acquisition parameters and processing procedures can be found in [16].

Table 1. Dataset for experiments

Dataset	ADNI	OASIS	ABIDE	ADHD-200	Huashan	Zhongshan
Disease	MCI, AD	AD	Autism	ADHD	MCI, AD	VCI
BD	785	83	499	280	100	151
HC	566	634	512	488	167	246
Total	1351	717	1011	768	267	397

3.2 Settings

To ensure a fair comparison, we use three graph convolutional layers, followed by corresponding pooling layers, for all GNN based methods. We use the ADAM optimizer with 1e-4 for learning rate and 5e-4 for weight decay, 100 for epochs, 0.001 for both β_1 and β_2, respectively. For the inner loop fast adaption, we set the minimum and maximum steps by 4 and 16. In the meta-training stage, we sample $\mathcal{D}_{\text{sup}}^{\text{train}}$ and $\mathcal{D}_{\text{que}}^{\text{train}}$ from all training datasets; and, in the meta-testing stage, we only randomly sample $\mathcal{D}_{\text{sup}}^{\text{test}}$ and $\mathcal{D}_{\text{que}}^{\text{test}}$ from the meta-testing dataset. The size of support data for both meta-training and meta-testing stages is depicted in the first line of Table 2, and we set the size of the query data as 256 for two stages. For different datasets, we only diagnose whether they are BD or not. We randomly select the support and the query data five times to report the mean and standard deviation value of the performance.

3.3 Results and Discussions

In Table 2, we compare our method with two SOTA meta-learning based methods [12,17] (in lines 3 and 4) under different K-shot configurations. It can be observed that, when the size of the support data increases, the performances of all methods increase and our proposed method outperforms the SOTA methods in terms of all performance metrics. We also validate the effectiveness of the use of multiple-BD datasets for training our model. When compared with the model without the meta-training stage (line

Table 2. Comparison of different methods, with bold denoting the highest performance. The proposed method shows statistically significant improvements (in level of p-value < 0.05) over all compared methods in terms of all metrics.

1		10-shot				30-shot				50-shot			
2	methods	Acc	Sen	Spe	AUC	Acc	Sen	Spe	AUC	Acc	Sen	Spe	AUC
3	Jaein	$52.6_{\pm 5.7}$	$52.7_{\pm 6.0}$	$55.2_{\pm 5.3}$	$55.8_{\pm 3.7}$	$58.8_{\pm 4.1}$	$59.3_{\pm 3.9}$	$56.8_{\pm 4.9}$	$59.7_{\pm 4.7}$	$63.3_{\pm 4.9}$	$65.2_{\pm 5.2}$	$61.3_{\pm 4.1}$	$63.9_{\pm 5.0}$
4	Ning	$54.3_{\pm 5.1}$	$53.1_{\pm 5.5}$	$56.7_{\pm 6.0}$	$55.4_{\pm 4.3}$	$60.7_{\pm 4.3}$	$61.7_{\pm 5.7}$	$60.3_{\pm 6.1}$	$61.4_{\pm 4.3}$	$65.1_{\pm 5.1}$	$64.7_{\pm 4.7}$	$66.3_{\pm 6.1}$	$65.7_{\pm 5.5}$
5	Ours (w/o pre-training)	$50.4_{\pm 6.3}$	$47.0_{\pm 5.7}$	$53.9_{\pm 6.2}$	$51.3_{\pm 4.9}$	$54.1_{\pm 4.3}$	$52.8_{\pm 4.4}$	$55.1_{\pm 5.3}$	$52.1_{\pm 4.9}$	$58.4_{\pm 4.7}$	$59.9_{\pm 4.9}$	$56.8_{\pm 5.2}$	$58.1_{\pm 6.3}$
6	Ours (w/o multi-view classifier)	$54.7_{\pm 5.8}$	$53.1_{\pm 5.5}$	$55.2_{\pm 6.1}$	$54.4_{\pm 3.0}$	$59.9_{\pm 5.7}$	$59.1_{\pm 4.7}$	$57.3_{\pm 5.4}$	$57.6_{\pm 3.7}$	$65.3_{\pm 4.9}$	$64.6_{\pm 6.0}$	$66.1_{\pm 7.1}$	$64.9_{\pm 3.8}$
7	Ours (w/o meta-controller)	$53.8_{\pm 6.7}$	$51.0_{\pm 7.7}$	$54.9_{\pm 5.2}$	$54.4_{\pm 3.9}$	$60.1_{\pm 7.1}$	$52.8_{\pm 6.2}$	$55.1_{\pm 8.3}$	$57.1_{\pm 4.4}$	$66.4_{\pm 7.7}$	$66.9_{\pm 4.3}$	$65.8_{\pm 7.6}$	$64.1_{\pm 6.6}$
8	Ours (w/o elder BD)	$53.3_{\pm 5.2}$	$52.7_{\pm 7.7}$	$54.9_{\pm 5.9}$	$55.7_{\pm 5.3}$	$57.8_{\pm 4.5}$	$57.6_{\pm 3.7}$	$58.1_{\pm 4.7}$	$58.8_{\pm 6.1}$	$62.4_{\pm 5.0}$	$59.8_{\pm 4.9}$	$66.9_{\pm 5.8}$	$64.4_{\pm 5.4}$
9	Ours (w/o youth BD)	$55.8_{\pm 4.9}$	$55.0_{\pm 5.7}$	$56.4_{\pm 6.3}$	$57.4_{\pm 4.8}$	$59.9_{\pm 5.7}$	$59.1_{\pm 5.7}$	$57.3_{\pm 6.0}$	$56.6_{\pm 5.5}$	$65.3_{\pm 5.3}$	$64.6_{\pm 4.9}$	$66.1_{\pm 6.5}$	$64.9_{\pm 6.3}$
10	Ours	$\mathbf{59.8}_{\pm 4.8}$	$\mathbf{58.0}_{\pm 4.3}$	$\mathbf{60.1}_{\pm 4.1}$	$\mathbf{59.4}_{\pm 4.0}$	$\mathbf{67.9}_{\pm 3.7}$	$\mathbf{66.1}_{\pm 4.2}$	$\mathbf{68.3}_{\pm 3.3}$	$\mathbf{67.6}_{\pm 3.7}$	$\mathbf{71.3}_{\pm 3.3}$	$\mathbf{70.7}_{\pm 3.2}$	$\mathbf{72.1}_{\pm 3.5}$	$\mathbf{70.0}_{\pm 3.6}$

5) or reducing either elder or youth BD datasets (lines 8 and 9), we find that the performances all drop significantly. This can validate that *not only* the information from MCI/AD, which is more similar to VCI, is useful, *but also* the general BD commonness suggested by data of youth BDs is beneficial. For the ablation study, we test the performance without a multi-view graph classifier or meta-controller as shown in Table 2 (lines 6 and 7) and Fig. 2 (a). Figure 2 (a) demonstrates that, when the epoch increases, the accuracy of the model assisted with both classifier and controller acquires steady and continuous improvement. In addition, we investigate the associations among ANI, support data accuracy and stop probability. As we can see from Figs. 2(b) and 2(c)), with the increment of ANI (red curve in Fig. 2(b)) and accuracy (green curve in Fig. 2(c)) on the support data, the stop probability also begins to increase (purple curves in Figs. 2(b) and 2(c)) to alleviate over-fitting in the query data, which supports the effectiveness of the meta-controller.

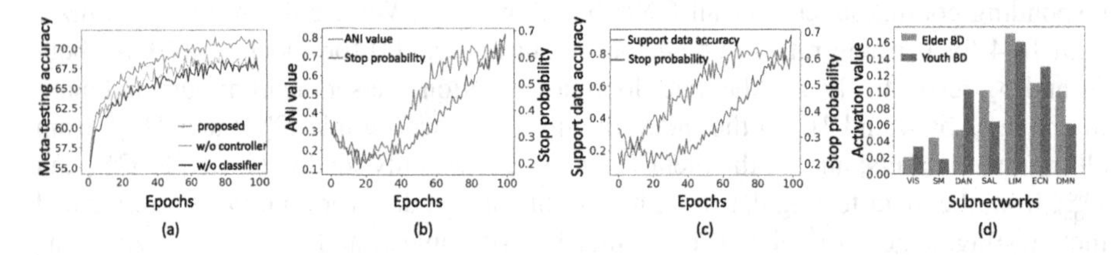

Fig. 2. (a) Accuracy curves; (b) ANI and stop probability curves; (c) Support data accuracy and stop probability curves; (d) Activation values of different subnetworks. (Color figure online)

Finally, we visualize the predictive importance of different resting-state networks, including visual network (VIS), somatomotor network (SM), dorsal attention network (DAN), salience network (SAL), limbic network (LIM), executive control network (ECN) and default mode network (DMN) when adapting to elder BD and younger BD as shown in Fig. 2 (d) by GradCAM [21]. In the results, the LIM consistently shows importance when diagnosing elder and younger BDs, which is in line with previous neuroscience studies [24]. This validates that our proposed method can properly detect meaningful common features among different BDs.

4 Conclusion

In this work, we focus on the issues of developing a classifier on new BD datasets with small samples and propose a novel framework. A broad of datasets covering elder and youth BDs are used to train the model to estimate the common features among BDs. An adaptive multi-view graph classifier is proposed to enable the model efficiently extract features for different BD diagnosis tasks. In addition, to avoid over-fitting during the adaptions to new data, we utilize a novel meta-controller driven by RL. Extensive experiments demonstrate the effectiveness and generalization of our proposed method. It is expected that advanced graph embedding methods can be integrated into our framework to improve performance. Our work is also promising to be extended to neuroscience studies to reveal both common and unique characteristics of different BDs.

Acknowledgment. This work was supported in part by National Natural Science Foundation of China (grant number 62131015), Science and Technology Commission of Shanghai Municipality (STCSM) (grant number 21010502600), and The Key R&D Program of Guangdong Province, China (grant number 2021B0101420006).

References

1. Aisen, P.S., et al.: Alzheimer's disease neuroimaging initiative 2 clinical core: progress and plans. Alzheimer's Dementia **11**(7), 734–739 (2015)
2. Chen, X., et al.: High-order resting-state functional connectivity network for MCI classification. Hum. Brain Mapp. **37**(9), 3282–3296 (2016)
3. ADHD-200 Consortium: The ADHD-200 consortium: a model to advance the translational potential of neuroimaging in clinical neuroscience. Front. Syst. Neurosci. **6**, 62 (2012)
4. Di Martino, A., et al.: The autism brain imaging data exchange: towards a large-scale evaluation of the intrinsic brain architecture in autism. Mol. Psychiatry **19**(6), 659–667 (2014)
5. Finn, C., Abbeel, P., Levine, S.: Model-agnostic meta-learning for fast adaptation of deep networks. In: International Conference on Machine Learning, pp. 1126–1135. PMLR (2017)
6. Gadgil, S., Zhao, Q., Pfefferbaum, A., Sullivan, E.V., Adeli, E., Pohl, K.M.: Spatio-temporal graph convolution for resting-state fMRI analysis. In: Martel, A.L., et al. (eds.) MICCAI 2020. LNCS, vol. 12267, pp. 528–538. Springer, Cham (2020). https://doi.org/10.1007/978-3-030-59728-3_52
7. Gao, H., Ji, S.: Graph U-nets. In: International Conference on Machine Learning, pp. 2083–2092. PMLR (2019)
8. Hochreiter, S., Schmidhuber, J.: Long short-term memory. Neural Comput. **9**(8), 1735–1780 (1997)
9. Hou, Y., et al.: Measuring and improving the use of graph information in graph neural networks. arXiv preprint arXiv:2206.13170 (2022)
10. Jack, C.R., Jr., et al.: Magnetic resonance imaging in Alzheimer's disease neuroimaging initiative 2. Alzheimer's Dementia **11**(7), 740–756 (2015)
11. Kipf, T.N., Welling, M.: Semi-supervised classification with graph convolutional networks (2017)
12. Lee, J., Kang, E., Jeon, E., Suk, H.-I.: Meta-modulation network for domain generalization in multi-site fMRI classification. In: MICCAI 2021. LNCS, vol. 12905, pp. 500–509. Springer, Cham (2021). https://doi.org/10.1007/978-3-030-87240-3_48

13. Li, D., Zhang, J., Yang, Y., Liu, C., Song, Y.Z., Hospedales, T.M.: Episodic training for domain generalization. In: Proceedings of the IEEE/CVF International Conference on Computer Vision, pp. 1446–1455 (2019)
14. Li, X., et al.: BrainGNN: interpretable brain graph neural network for fMRI analysis. Med. Image Anal. **74**, 102233 (2021)
15. Liu, M., et al.: Multiscale functional connectome abnormality predicts cognitive outcomes in subcortical ischemic vascular disease. Cereb. Cortex **32**(21), 4641–4656 (2022)
16. Liu, M., et al.: Deep learning reveals the common spectrum underlying multiple brain disorders in youth and elders from brain functional networks. arXiv preprint arXiv:2302.11871 (2023)
17. Ma, N., et al.: Adaptive-step graph meta-learner for few-shot graph classification. In: Proceedings of the 29th ACM International Conference on Information and Knowledge Management, pp. 1055–1064 (2020)
18. Mahajan, K., Sharma, M., Vig, L.: Meta-DermDiagnosis: few-shot skin disease identification using meta-learning. In: Proceedings of the IEEE/CVF Conference on Computer Vision and Pattern Recognition (CVPR) Workshops (2020)
19. Marcus, D.S., Wang, T.H., Parker, J., Csernansky, J.G., Morris, J.C., Buckner, R.L.: Open Access Series of Imaging Studies (OASIS): cross-sectional MRI data in young, middle aged, nondemented, and demented older adults. J. Cogn. Neurosci. **19**(9), 1498–1507 (2007)
20. Schaefer, A., et al.: Local-global parcellation of the human cerebral cortex from intrinsic functional connectivity MRI. Cereb. Cortex **28**(9), 3095–3114 (2018)
21. Selvaraju, R.R., Cogswell, M., Das, A., Vedantam, R., Parikh, D., Batra, D.: Grad-CAM: visual explanations from deep networks via gradient-based localization. In Proceedings of the IEEE International Conference on Computer Vision, pp. 618–626 (2017)
22. Singh, R., Bharti, V., Purohit, V., Kumar, A., Singh, A.K., Singh, S.K.: MetaMed: few-shot medical image classification using gradient-based meta-learning. Pattern Recognit. **120**, 108111 (2021)
23. Sun, L., et al.: Few-shot medical image segmentation using a global correlation network with discriminative embedding. Comput. Biol. Med. **140**, 105067 (2022)
24. Taylor, J.J., et al.: A transdiagnostic network for psychiatric illness derived from atrophy and lesions. Nat. Hum. Behav. **7**(3), 420–429 (2023)
25. Williams, R.J.: Simple statistical gradient-following algorithms for connectionist reinforcement learning. Mach. Learn. **8**, 229–256 (1992). https://doi.org/10.1007/BF00992696

Brain Anatomy-Guided MRI Analysis for Assessing Clinical Progression of Cognitive Impairment with Structural MRI

Lintao Zhang[1], Jinjian Wu[2], Lihong Wang[3], Li Wang[1], David C. Steffens[3], Shijun Qiu[2], Guy G. Potter[4(✉)], and Mingxia Liu[1(✉)]

[1] Department of Radiology and Biomedical Research Imaging Center, University of North Carolina at Chapel Hill, Chapel Hill, NC, USA
mxliu@med.unc.edu

[2] The First School of Clinical Medicine, Guangzhou University of Chinese Medicine, Guangzhou, Guangdong, China

[3] Department of Psychiatry, University of Connecticut School of Medicine, University of Connecticut, Farmington, CT, USA

[4] Department of Psychiatry and Behavioral Sciences, Duke University Medical Center, Durham, NC, USA
guy.potter@duke.edu

Abstract. Brain structural MRI has been widely used for assessing future progression of cognitive impairment (CI) based on learning-based methods. Previous studies generally suffer from the limited number of labeled training data, while there exists a huge amount of MRIs in large-scale public databases. Even without task-specific label information, brain anatomical structures provided by these MRIs can be used to boost learning performance intuitively. Unfortunately, existing research seldom takes advantage of such brain anatomy prior. To this end, this paper proposes a brain anatomy-guided representation (BAR) learning framework for assessing the clinical progression of cognitive impairment with T1-weighted MRIs. The BAR consists of a *pretext model* and a *downstream model*, with a shared brain anatomy-guided encoder for MRI feature extraction. The pretext model also contains a decoder for brain tissue segmentation, while the downstream model relies on a predictor for classification. We first train the pretext model through a brain tissue segmentation task on 9,544 auxiliary T1-weighted MRIs, yielding a generalizable encoder. The downstream model with the learned encoder is further fine-tuned on target MRIs for prediction tasks. We validate the proposed BAR on two CI-related studies with a total of 391 subjects with T1-weighted MRIs. Experimental results suggest that the BAR outperforms several state-of-the-art (SOTA) methods. The source code and pretrained models are available at https://github.com/goodaycoder/BAR.

Keywords: Structural MRI · Brain anatomy · Cognitive impairment

---

Supplementary Information The online version contains supplementary material available at https://doi.org/10.1007/978-3-031-43993-3_11.

H. Greenspan et al. (Eds.): MICCAI 2023, LNCS 14227, pp. 109–119, 2023.
https://doi.org/10.1007/978-3-031-43993-3_11

1 Introduction

Brain magnetic resonance imaging (MRI) has been increasingly used to assess future progression of cognitive impairment (CI) in various clinical and research fields by providing structural brain anatomy [1–6]. Many learning-based methods have been developed for automated MRI analysis and brain disorder prognosis, which usually heavily rely on labeled training data [7–10]. However, it is generally time-consuming and tedious to collect category labels for brain MRIs in practice, resulting in a limited number of labeled MRIs [11].

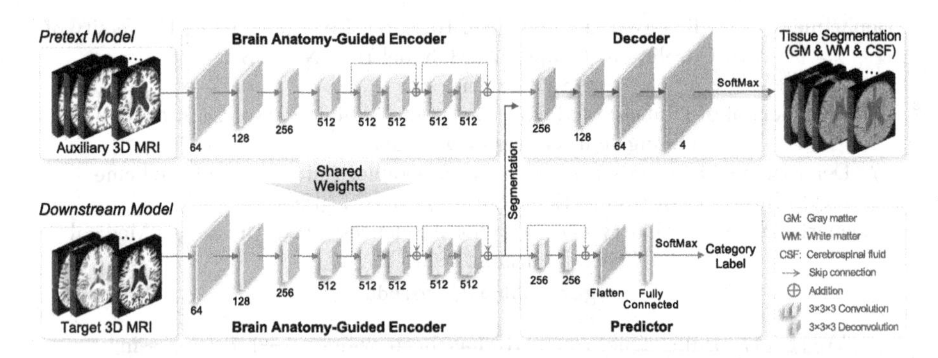

Fig. 1. Illustration of brain anatomy-guided representation (BAR) learning framework for assessing the clinical progression of cognitive impairment. The BAR consists of a *pretext model* and a *downstream model*, with a shared brain anatomy-guided encoder for MRI feature extraction. The pretext model also contains a decoder for brain tissue segmentation, while the downstream model relies on a predictor for prediction. The pretext model is trained on the large-scale ADNI [12] with 9,544 T1-weighted MRIs, yielding a generalizable encoder. With this learned encoder frozen, the downstream model is then fine-tuned on target MRIs for prediction tasks.

Even without task-specific category label information, brain anatomical structures provided by auxiliary MRIs can be employed as a prior to boost disease progression prediction performance. Considering that there are a large number of unlabeled MRIs in existing large-scale datasets [12,13], several deep learning methods propose to extract brain anatomical features from MRI without requiring specific category labels. For instance, Song *et al.* [14] suggest that the anatomy prior can be utilized to segment brain tumors, while Yamanakkanavar *et al.* [15] discuss how brain MRI segmentation improves disease diagnosis. Unfortunately, there are few studies that try to utilize such brain anatomy prior for assessing the clinical progression of cognitive impairment with structural MRIs.

To this end, we propose a brain anatomy-guided representation (BAR) learning framework for cognitive impairment prognosis with T1-weighted MRIs, incorporated with brain anatomy prior provided by a brain tissue segmentation task. As shown in Fig. 1, the BAR consists of a *pretext model* and a *downstream model*, with a shared anatomy-guided encoder for MRI feature extraction. These two models also use a decoder and a predictor for brain tissue segmentation and disease progression prediction, respectively. The pretext model is trained on 9,544 MRI scans from the public Alzheimer's Disease Neuroimaging Initiative (ADNI) [12] without any category label information, yielding a generalizable encoder. The downstream model is further fine-tuned on target MRIs for CI progression prediction. Experiments are performed on two CI-related studies with 391 subjects, with results suggesting the efficacy of BAR compared with state-of-the-art (SOTA) methods. The pretext model can also be used for brain tissue segmentation in other MRI-based studies. To the best of our knowledge, this is the first work that utilizes anatomy prior derived from large-scale T1-weighted MRIs for automated cognitive decline analysis. To promote reproducible research, the source code and trained models have been made publicly available to the research community (see https://github.com/goodaycoder/BAR).

2 Materials and Proposed Method

Data and Preprocessing. The *pretext model* is trained via a tissue segmentation task on auxiliary MRIs (without category label) from **ADNI**. A total of 9,544 T1-weighted MRIs from 2,370 ADNI subjects with multiple scans are used in this work. To provide accurate brain anatomy, we perform image preprocessing and brain tissue segmentation for these MRIs to generate ground-truth segmentation of three tissues, *i.e.*, white matter (WM), gray matter (GM) and cerebrospinal fluid (CSF), using an in-house toolbox iBEAT [16] with manual verification.

The *downstream model* is trained on 1) a late-life depression (**LLD**) study with 309 subjects from two sites [17,18], and 2) a type 2 diabetes mellitus (**DM**) study with 82 subjects from the First Affiliated Hospital of Guangzhou University of Chinese Medicine. Subjects in LLD are categorized into three groups: 1) 89 non-depressed cognitively normal (CN), 2) 179 depressed but cognitively normal (CND), 3) 41 depressed subjects (called CI) who developed cognitive impairment or even dementia in the follow-up years. Category labels in the LLD study are determined based on subjects' 5-year follow-up diagnostic information, while MRIs are acquired at baseline time. The DM contains 1) 45 health control (HC) subjects and 2) 37 diabetes mellitus patients with mild CI (MCI). Detailed image acquisition protocols are given in Table SI of *Supplementary Materials*. All MRIs are preprocessed via the following pipeline: 1) bias field correction, 2) skull stripping, 3) affine registration to the MNI space, 4) resampling to $1 \times 1 \times 1$ mm^3, 5) deformable registration to AAL3 [19] with SyN [20], and 6) warping 166 regions-of-interest (ROIs) of AAL3 back to MRI volumes.

Proposed Method. While it is often challenging to annotate MRIs in practice, there are a large number of MRIs (without task-specific category labels) in existing large-scale datasets. Even without category labels, previous studies propose to extract anatomical features (*e.g.*, ROI volumes of GM segmentation maps) to characterize brain anatomy [21,22]. Such brain anatomy prior learned via tissue segmentation can be employed to boost learning performance intuitively. Accordingly, we propose a brain anatomy-guided representation (BAR) learning framework for progression prediction of cognitive impairment, incorporated with brain anatomy prior provided by brain tissue segmentation. As shown in Fig. 1, the BAR consists of a *pretext model* for brain tissue segmentation and a *downstream model* for disease progression prediction, both equipped with brain anatomy-guided encoders (shared weights) for MRI feature learning.

(1) Pretext Model for Segmentation. To learn brain anatomical features from MRIs in a data-driven manner, we propose to employ a segmentation task for pretext model training. As shown in the top of Fig. 1, the pretext model consists of 1) a *brain anatomy-guided encoder* for MRI feature extraction and 2) a *decoder* for segmentation. The brain anatomy-guided encoder takes large-scale auxiliary 3D MRIs without category labels as input, and outputs 512 feature maps. It contains 8 convolution blocks, with each block containing a convolution layer (kernel size: $3 \times 3 \times 3$), followed by instance normalization and Parametric Rectified Linear Unit (PReLU) activation. The first 4 blocks downsample the input with a stride of $2 \times 2 \times 2$. The channel numbers of the eight blocks are [64, 128, 256, 512, 512, 512, 512, 512], respectively. A skip connection is applied to sum the input and output of every two of the last 4 blocks for residual learning.

The decoder takes the 512 feature maps as input and outputs segmentation maps of three tissues (*i.e.*, WM, GM, and CSF), thus guiding the encoder to learn brain anatomical features. The decoder contains four deconvolution blocks with 256, 128, 64 and 4 channels, respectively. Each deconvolution block shares the same architecture as the convolution block in the encoder. The output of the decoder is then fed into a SoftMax layer to get four probability maps that indicate the probability of a voxel belonging to a specific tissue (*i.e.*, background, WM, GM, and CSF). Besides, the reconstruction task can be used to train the pretext model instead of segmentation when lacking ground-truth segmentation maps. For *problems without ground-truth segmentation maps*, we can resort to an MRI reconstruction task to train the pretext model in an unsupervised manner.

(2) Downstream Model for Prediction. As shown in the bottom panel of Fig. 1, the downstream model takes target MRIs as input and outputs probabilities of category labels. It consists of 1) a *brain anatomy-guided encoder* and 2) a *predictor* for prognosis. This encoder shares the same architecture and parameters as that of the pre-trained pretext model. With the encoder frozen, predictor parameters will be updated when the downstream model is trained on target MRIs. The predictor has two convolution blocks (kernel size: $3 \times 3 \times 3$, stride: $2 \times 2 \times 2$, channel: 256) with a skip connection, followed by a flatten layer, an FC layer, and a SoftMax layer. The architecture of the predictor can be flexibly adjusted according to the requirements of different downstream tasks.

(3) Implementation. The proposed BAR is trained via two steps. 1) The pretext model is first trained on 9,544 MRIs from ADNI, with ground-truth segmentation as supervision. The Adam optimizer [23] with a learning rate of 10^{-4} and dice loss are used for training (batch size: 4, epoch: 30). 2) We then share

Table 1. Performance of fourteen methods in two MRI-based depression recognition tasks (*i.e.*, CND vs. CN and CI vs. CND classification) on the LLD study.

Method	CND vs. CN Classification					CI vs. CND Classification				
	AUC (%)	ACC (%)	SEN (%)	SPE (%)	F1s (%)	AUC (%)	ACC (%)	SEN (%)	SPE (%)	F1s(%)
SVM-GM	51.6(0.8)	52.3(3.8)*	48.7(12.2)	56.0(8.6)	49.9(8.4)	54.0(1.5)	50.8(1.1)*	51.6(2.4)	50.0(0.0)	50.5(1.8)
SVM-WM	57.9(2.6)	53.3(4.7)*	59.3(6.4)	47.3(6.4)	55.9(4.9)	38.7(3.5)	48.7(3.1)*	42.1(9.8)	55.0(7.1)	44.0(6.8)
XGB-GM	43.9(2.8)	42.3(1.5)*	54.0(2.8)	30.7(4.3)	48.3(1.4)	49.3(7.0)	44.1(9.5)*	36.8(12.9)	51.0(8.2)	38.7(12.5)
XGB-WM	50.8(2.5)	53.0(2.7)*	64.0(3.6)	42.0(3.8)	57.6(2.6)	61.7(5.0)	57.4(3.9)*	50.5(7.1)	64.0(6.5)	53.5(4.9)
ResNet18	65.4(7.2)	58.7(8.0)*	52.7(10.1)	64.7(9.0)	61.0(7.7)	61.3(8.1)	56.9(7.8)	**59.0**(14.6)	55.0(26.7)	**56.8**(4.9)
ResNet34	58.9(7.3)	57.7(6.1)*	56.0(8.3)	59.3(7.6)	58.3(6.2)	58.5(4.6)	56.4(8.1)	47.4(5.3)	65.0(13.7)	51.6(6.9)
ResNet50	63.0(5.4)	57.0(3.6)	67.3(8.6)	46.7(8.2)	60.9(4.5)	55.3(2.6)	50.8(4.6)	41.0(10.1)	60.0(13.7)	44.3(6.7)
Med3D18	57.9(2.4)	54.7(1.8)*	53.3(9.7)	56.0(9.3)	55.0(4.3)	59.6(9.1)	57.4(10.3)*	50.5(16.9)	64.0(10.8)	52.8(15.4)
Med3D34	58.7(7.2)	56.7(4.9)*	50.0(7.8)	63.3(11.1)	59.1(6.6)	57.6(4.6)	55.9(2.1)*	56.8(6.9)	55.0(7.9)	55.5(3.2)
Med3D50	60.2(2.5)	59.0(4.3)*	50.0(16.8)	**68.0**(13.9)	54.0(8.5)	47.3(9.2)	46.7(6.4)	45.3(24.6)	48.0(23.9)	43.0(13.6)
SEResNet	66.4(2.1)	57.7(4.3)	60.0(12.7)	55.3(14.6)	58.2(6.1)	59.5(5.5)	53.9(2.6)*	55.8(28.5)	52.0(28.0)	51.4(13.6)
EfficientNet	59.3(9.9)	58.0(5.9)*	64.7(25.7)	51.3(17.1)	54.1(6.7)	56.4(8.9)	53.3(5.2)*	56.8(31.9)	50.0(29.8)	49.4(22.9)
MobileNet	55.3(8.6)	54.3(3.5)*	44.7(31.2)	64.0(31.7)	53.9(20.6)	58.5(10.1)	53.3(5.8)*	56.8(40.0)	50.0(35.9)	47.2(27.5)
BAR (Ours)	**70.5**(4.1)	**65.0**(1.7)	**77.3**(5.5)	52.7(3.6)	**68.8**(2.4)	**64.5**(5.8)	**59.0**(5.1)	50.5(8.0)	**67.0**(6.7)	54.4(6.4)

the encoder of the pre-trained pretext model with the downstream model and fine-tune the predictor via a cross-entropy loss (batch size: 2, learning rate: 10^{-4}, epoch: 90). The learning rate of fine-tuning decays by 0.1 every 30 epochs. The BAR is implemented on PyTorch with NVIDIA TITAN Xp (memory: 12GB).

3 Experiment

Experimental Setting. Three classification tasks are performed: (1) depression recognition (*i.e.*, CND vs. CN classification) on LLD, (2) CI identification (*i.e.*, CI vs. CND classification) on LLD, and (3) MCI detection (*i.e.*, MCI vs. HC classification) on DM. The partition of training/test set is given in Table SII of *Supplementary Materials*. Such partition is repeated five times independently to avoid any bias introduced by random partition, and the mean and standard deviation results are recorded. The training data is duplicated and augmented using random affine transform. Five evaluation metrics are used, including area under ROC curve (AUC), accuracy (ACC), sensitivity (SEN), specificity (SPE), and F1-Score (F1s). Besides, we perform tissue segmentation by directly applying the trained pretext model to target MRIs from LLD and DM studies and visually compare the results of our BAR with those of FSL [24].

Competing Methods. We compare our BAR with two classic machine learning methods and five SOTA deep learning approaches, including (1) support vector machine (**SVM**) [25] with a radial basis function kernel (regularization: 1.0), (2) XGBoost (**XGB**) [26] (estimators: 300, tree depth: 4, learning rate: 0.2), (3) **ResNet**x with x convolution layers, (4) **Med3D**x [27] with x convolution layers, (5) **SEResNet** [28] that is an improved model by adding squeeze and excitation blocks to ResNet, (6) **EfficientNet** [29], and (7) **MobileNet** [30] that is an efficient lightweight CNN model. For SVM and XGB, we extract ROI-based WM and GM volumes of each MRI as input. All competing deep learning methods (with default architectures) take whole 3D MRIs as input and share the same training strategy as that used in the downstream model of the BAR. An early-stop training strategy (epoch: 90) is used in all deep learning models.

Table 2. Performance of fourteen methods in the MRI-based MCI detection task (*i.e.*, MCI vs. HC classification) on the DM study.

Method	AUC (%)	ACC (%)	SEN (%)	SPE (%)	F1s (%)
SVM-GM	52.5(1.7)	50.5(2.1)*	42.0(10.4)	59.0(9.6)	45.3(7.3)
SVM-WM	41.4(1.9)	47.5(1.8)*	43.0(13.0)	52.0(14.0)	44.0(9.2)
XGB-GM	39.1(5.2)	46.5(3.4)*	42.0(4.5)	51.0(4.2)	43.9(4.0)
XGB-WM	61.5(3.2)	55.5(2.1)*	48.0(2.7)	**63.0**(2.7)	51.9(2.5)
ResNet18	58.8(1.4)	55.5(2.7)*	60.0(7.1)	51.0(2.2)	57.3(4.5)
ResNet34	60.9(2.9)	55.5(1.1)*	59.0(5.5)	52.0(4.5)	56.9(2.7)
ResNet50	60.8(2.5)	58.0(2.1)	65.0(7.9)	51.0(4.2)	60.6(4.1)
Med3D18	56.0(3.1)	53.5(4.9)	58.0(22.5)	49.0(12.9)	53.9(11.7)
Med3D34	55.1(6.9)	51.5(5.8)	59.0(14.8)	44.0(20.7)	54.4(6.5)
Med3D50	61.3(5.4)	57.5(5.3)	63.0(6.7)	52.0(8.4)	59.7(5.0)
SEResNet	62.8(1.9)	59.0(2.9)	63.0(16.4)	55.0(17.0)	59.8(7.0)
EfficientNet	55.6(3.6)	53.5(8.0)*	64.0(19.8)	43.0(4.5)	56.9(12.5)
MobileNet	58.5(2.1)	53.0(4.1)*	59.0(26.1)	47.0(24.4)	53.3(13.5)
BAR (Ours)	**67.6**(1.7)	**60.5**(1.1)	**66.0**(10.8)	55.0(9.4)	**62.2**(4.5)

Results of Depression and CI Identification. In this task, we aim to recognize cognitively normal subjects with depression with a higher risk of progressing to CI than healthy subjects. The results of fourteen methods on the LLD study are reported in Table 1, where '*' denotes that the results of BAR and a competing method are statistically significantly different ($p < 0.05$ via paired t-test).

From the left of Table 1, we have the following observations on *CND vs. CN classification*. *First*, our BAR model generally outperforms thirteen competing methods in most cases. For instance, the BAR yields the results of AUC = 70.5% and SEN = 77.3%, which are 4.1% and 10.0% higher than those of the second-best methods (*i.e.*, SEResNet and ResNet50), respectively. This implies that the brain anatomical MRI features learned by our pretext model on large-scale datasets would be more discriminative, compared with those used in the competing methods. *Second*, among 10 deep models, our BAR produces the lowest standard deviation in most cases (especially on SEN and SPE), suggesting its robustness to bias introduced by random data partition in the downstream task. This could be due to the strong generalizability of the feature encoder guided by brain anatomy prior (derived from the auxiliary tissue segmentation task). *In addition*, our BAR significantly outperforms four machine learning methods and two lightweight deep models (*i.e.*, EfficientNet and MobileNet) with $p < 0.05$.

From the right of Table 1, we can see that the overall results of fourteen methods in *CI vs. CND classification* are usually worse than *CND vs. CN classification*. This suggests that the task of CI vs. CND classification is more challenging, which could be due to the more imbalanced training data in this task (as shown in Table SII of *Supplementary Materials*). On the other hand, the proposed BAR

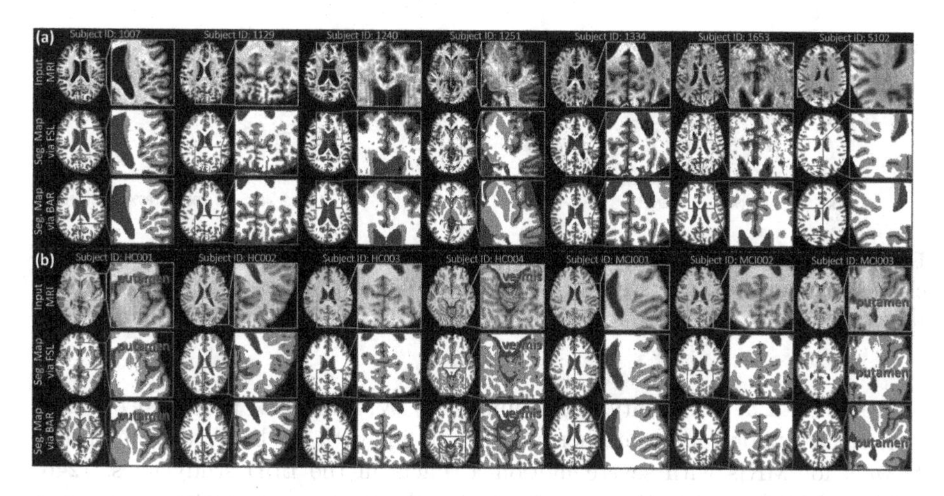

Fig. 2. Tissue segmentation (seg.) maps of white matter, gray matter and cerebrospinal fluid produced by FSL and our BAR on (a) LLD study and (b) DM study.

still performs best in terms of AUC=64.5% and SPE=67.0%, which are 2.8% and 2.0% higher than those of the second-best competing methods (*i.e.*, XGB-WM and ResNet34), respectively. These results further demonstrate the superiority of the BAR in MRI-based depression recognition.

Results of MCI Detection. The results of different methods in MCI detection (*i.e.*, MCI vs. HC classification) on the DM study are reported in Table 2. There are a total of 42 subjects (*i.e.*, 17 MCI and 25 HC) used for training in this task, which are fewer but more balanced than the two tasks in the LLD study (see Table SII). It can be observed from Tables 1 and 2 that the proposed BAR yields relatively lower standard deviations in terms of AUC and ACC in MCI vs. HC classification, compared with the two tasks on the LLD study. These results imply that data imbalance may be an important issue affecting the performance of deep learning models when the number of training samples is limited.

Segmentation Results. The pre-trained pretext model can also be used for brain tissue segmentation in downstream studies. Thus, we visualize brain segmentation maps generated by FSL and our BAR for target MRIs in both LLD and DM studies in Fig. 2. Note that T1-weighted MRIs in the LLD study are collected from 2 sites and have more inconsistent image quality when compared to those from DM. From Fig. 2, we have several interesting observations.

First, the segmentation results generated by the proposed BAR are generally better than those of FSL in most cases, especially for those *cortical surface areas* on the two studies. For instance, the WM region in segmentation maps generated by our BAR is much cleaner than that of FSL, indicating that our model is not sensitive to noise in MRI. Even for the LLD study with significant

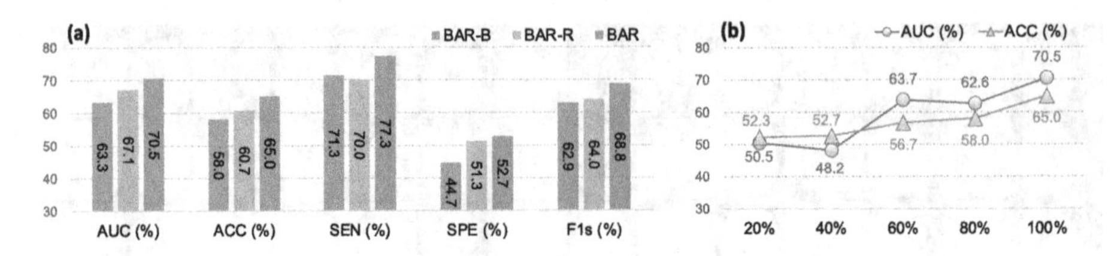

Fig. 3. (a) Results achieved by BAR with different pretext tasks in CND vs. CN classification on LLD. (b) Results of BAR with different numbers of MRIs from the target domain for downstream model training in CND vs. CN classification on LLD.

inter-site data heterogeneity, the boundary of WM and GM produced by BAR is more continuous and smoother, which is in line with the brain anatomy prior. *Second*, for MRIs with severe motion artifacts in the LLD study (IDs: 1240, 1334, and 1653), our method can produce high-quality segmentation maps, and the results are even comparable to those of MRIs without motion artifacts. This demonstrates that our model is robust to motion artifacts. The underlying reason could be that the pretext model is trained on large-scale MRIs, and thus, has good generalization ability when applied to MRIs with different image quality. *In addition*, both BAR and FSL often achieve better results in the DM study, since DM has relatively higher image quality than LLD. Still, the proposed BAR can achieve better segmentation results in many fine-grained brain regions, such as the *putamen region* (see HC001 and MCI003) and the *vermis region* (see HC004). These results demonstrate that our method has good adaptability when applied to classification and segmentation tasks in MRI-based studies.

Ablation Study. To validate the effectiveness of the learned brain anatomical MRI features, we further compare the BAR with its two variants (called **BAR-B** and **BAR-R**) that use anatomy prior derived from different pretext tasks in CND vs. CN classification on LLD. Specifically, the BAR-B is *trained from scratch* as a baseline on target data without any pre-trained encoder. The BAR-R trains the pretext model through an *MRI reconstruction task* in an unsupervised learning manner. As shown in Fig. 3(a), the BAR consistently performs better than its variants in terms of all five metrics. This implies that the learned MRI features guided by the segmentation task help promote prediction performance. Also, BAR and BAR-R outperform BAR-B in most cases, implying that brain anatomy prior derived from tissue segmentation or MRI reconstruction can help improve discriminative ability of MRI features and boost prediction performance.

Influence of Training Data Size. We also study the influence of training data size on BAR in CND vs. CN classification on LLD. With fixed test data, we randomly select a part of MRIs (*i.e.*, $[20\%, 40\%, \cdots, 100\%]$) from target training data to fine-tune the downstream prediction model. It can be observed from Fig. 3(b) that the overall performance in terms of AUC and ACC of our BAR increases with the increase of training data, and it produces the best results when using all training data for model fine-tuning. This suggests that using more data for downstream model fine-tuning helps promote learning performance.

4 Conclusion and Future Work

In this paper, we develop a brain anatomy-guided representation (BAR) learning framework for MRI-based progression prediction of cognitive impairment, incorporated by brain anatomy prior (derived from an auxiliary tissue segmentation task). We validate the proposed BAR on two CI-related studies with T1-weighted MRIs, and the experimental results demonstrate its effectiveness compared with SOTA methods. Besides, the pretext model trained on 9,544 MRIs from ADNI can be well adapted to tissue segmentation in the two CI-related studies. There is significant intra- and inter-site data heterogeneity in LLD with two sites. It is interesting to reduce such heterogeneity using domain adaptation [31,32], which will be our future work. Aside from tissue segmentation, one can use other auxiliary tasks to model brain anatomy, such as brain parcellation and brain MRI to CT translation. Besides, it is meaningful to compare our method with other model pre-training strategies [33,34], which will also be our future work.

Acknowledgment. L. Zhang, G.G. Potter, and M. Liu were supported by NIH grants RF1AG073297 and R01MH108560.

References

1. Ashtari-Majlan, M., Seifi, A., Dehshibi, M.M.: A multi-stream convolutional neural network for classification of progressive MCI in Alzheimer's disease using structural MCI images. IEEE J. Biomed. Health Inform. **26**(8), 3918–3926 (2022)
2. El-Gamal, F.E.Z.A., et al.: A personalized computer-aided diagnosis system for mild cognitive impairment (MCI) using structural MRI (sMRI). Sensors **21**(16), 5416 (2021)
3. Gonuguntla, V., Yang, E., Guan, Y., Koo, B.B., Kim, J.H.: Brain signatures based on structural MRI: classification for MCI, PMCI, and AD. Hum. Brain Mapp. **43**(9), 2845–2860 (2022)
4. Guo, M., et al.: A novel conversion prediction method of MCI to AD based on longitudinal dynamic morphological features using ADNI structural MRIs. J. Neurol. **267**(10), 2983–2997 (2020). https://doi.org/10.1007/s00415-020-09890-5
5. Lombardi, G., et al.: Structural magnetic resonance imaging for the early diagnosis of dementia due to Alzheimer's disease in people with mild cognitive impairment. Cochrane Database Syst. Rev. (3) (2020)
6. Yin, C., et al.: Anatomically interpretable deep learning of brain age captures domain-specific cognitive impairment. Proc. Natl. Acad. Sci. **120**(2), e2214634120 (2023)
7. Noor, M.B.T., Zenia, N.Z., Kaiser, M.S., Mahmud, M., Al Mamun, S.: Detecting neurodegenerative disease from MRI: A brief review on a deep learning perspective. In: Liang, P., Goel, V., Shan, C. (eds.) BI 2019. LNCS, vol. 11976, pp. 115–125. Springer, Cham (2019). https://doi.org/10.1007/978-3-030-37078-7_12

8. Chen, X., Tang, M., Liu, A., Wei, X.: Diagnostic accuracy study of automated stratification of Alzheimer's disease and mild cognitive impairment via deep learning based on MRI. Ann. Transl. Med. **10**(14) (2022)
9. Garg, N., Choudhry, M.S., Bodade, R.M.: A review on Alzheimer's disease classification from normal controls and mild cognitive impairment using structural MR images. J. Neurosci. Methods **384**, 109745 (2022)
10. Scarpazza, C., et al.: Translating research findings into clinical practice: a systematic and critical review of neuroimaging-based clinical tools for brain disorders. Transl. Psychiatry **10**(1), 107 (2020)
11. Nanni, L., et al.: Comparison of transfer learning and conventional machine learning applied to structural brain MRI for the early diagnosis and prognosis of Alzheimer's disease. Front. Neurol. **11**, 576194 (2020)
12. Jack Jr, C.R., et al.: The Alzheimer's disease neuroimaging initiative (ADNI): MRI methods. J. Magn. Reson. Imaging Off. J. Int. Soc. Magn. Reson. Med. **27**(4), 685–691 (2008)
13. LaMontagne, P.J., et al.: OASIS-3: longitudinal neuroimaging, clinical, and cognitive dataset for normal aging and Alzheimer disease. MedRxiv 2019-12 (2019)
14. Song, B., Chou, C.R., Chen, X., Huang, A., Liu, M.C.: Anatomy-guided brain tumor segmentation and classification. In: Crimi, A., Menze, B., Maier, O., Reyes, M., Winzeck, S., Handels, H. (eds.) BrainLes 2016. LNCS, vol. 10154, pp. 162–170. Springer, Cham (2016). https://doi.org/10.1007/978-3-319-55524-9_16
15. Yamanakkanavar, N., Choi, J.Y., Lee, B.: MRI segmentation and classification of human brain using deep learning for diagnosis of Alzheimer's disease: a survey. Sensors **20**(11), 3243 (2020)
16. Wang, L., Wu, Z., Chen, L., Sun, Y., Lin, W., Li, G.: iBEAT V2.0: a multisite-applicable, deep learning-based pipeline for infant cerebral cortical surface reconstruction. Nat. Protocols **18**(5), 1488–1509 (2023)
17. Steffens, D.C., et al.: Methodology and preliminary results from the neurocognitive outcomes of depression in the elderly study. J. Geriatr. Psychiatry Neurol. **17**(4), 202–211 (2004)
18. Steffens, D.C., Wang, L., Manning, K.J., Pearlson, G.D.: Negative affectivity, aging, and depression: results from the neurobiology of late-life depression (NBOLD) study. Am. J. Geriatr. Psychiatry **25**(10), 1135–1149 (2017)
19. Rolls, E.T., Huang, C.C., Lin, C.P., Feng, J., Joliot, M.: Automated anatomical labelling atlas 3. Neuroimage **206**, 116189 (2020)
20. Avants, B.B., Epstein, C.L., Grossman, M., Gee, J.C.: Symmetric diffeomorphic image registration with cross-correlation: evaluating automated labeling of elderly and neurodegenerative brain. Med. Image Anal. **12**(1), 26–41 (2008)
21. Elsayed, A.S.A.: Region of interest based image classification: a study in MRI brain scan categorization. The University of Liverpool (United Kingdom) (2011)
22. Magnin, B., et al.: Support vector machine-based classification of Alzheimer's disease from whole-brain anatomical MRI. Neuroradiology **51**, 73–83 (2009)
23. Kingma, D.P., Ba, J.: Adam: a method for stochastic optimization. arXiv preprint arXiv:1412.6980 (2014)
24. Jenkinson, M., Beckmann, C.F., Behrens, T.E., Woolrich, M.W., Smith, S.M.: FSL. NeuroImage **62**(2), 782–790 (2012)
25. Pisner, D.A., Schnyer, D.M.: Support vector machine. In: Machine Learning, pp. 101–121. Elsevier (2020)
26. Chen, T., Guestrin, C.: XGBoost: a scalable tree boosting system. In: Proceedings of the 22nd ACM SIGKDD International Conference on Knowledge Discovery and Data Mining, pp. 785–794 (2016)

27. Chen, S., Ma, K., Zheng, Y.: Med3D: transfer learning for 3D medical image analysis. arXiv preprint arXiv:1904.00625 (2019)
28. Hu, J., Shen, L., Sun, G.: Squeeze-and-excitation networks. In: Proceedings of the IEEE Conference on Computer Vision and Pattern Recognition, pp. 7132–7141 (2018)
29. Tan, M., Le, Q.: EfficientNet: rethinking model scaling for convolutional neural networks. In: International Conference on Machine Learning, pp. 6105–6114. PMLR (2019)
30. Howard, A.G., et al.: MobileNets: efficient convolutional neural networks for mobile vision applications. arXiv preprint arXiv:1704.04861 (2017)
31. Guan, H., Liu, M.: Domain adaptation for medical image analysis: a survey. IEEE Trans. Biomed. Eng. **69**(3), 1173–1185 (2021)
32. Guan, H., Liu, M.: DomainATM: domain adaptation toolbox for medical data analysis. NeuroImage **268**, 119863 (2023)
33. Zhou, Z., Sodha, V., Pang, J., Gotway, M.B., Liang, J.: Models genesis. Med. Image Anal. **67**, 101840 (2021)
34. Zhou, H.Y., Lu, C., Yang, S., Han, X., Yu, Y.: Preservational learning improves self-supervised medical image models by reconstructing diverse contexts. In: Proceedings of the IEEE/CVF International Conference on Computer Vision, pp. 3499–3509 (2021)

Dynamic Structural Brain Network Construction by Hierarchical Prototype Embedding GCN Using T1-MRI

Yilin Leng[1,2], Wenju Cui[2,3], Chen Bai[2,3], Zirui Chen[4], Yanyan Zheng[5], and Jian Zheng[2,3](✉)

[1] Institute of Biomedical Engineering, School of Communication and Information Engineering, Shanghai University, Shanghai, China
[2] Department of Medical Imaging, Suzhou Institute of Biomedical Engineering and Technology, Chinese Academy of Sciences, Suzhou, China
zhengj@sibet.ac.cn
[3] School of Biomedical Engineering (Suzhou), Division of Life Sciences and Medicine, University of Science and Technology of China, Hefei, China
[4] Mathematical and Physical Faculty, University College London, London, UK
[5] Wenzhou Medical University, Wenzhou, China

Abstract. Constructing structural brain networks using T1-weighted MRI (T1-MRI) presents a significant challenge due to the lack of direct regional connectivity. Current methods with T1-MRI rely on predefined regions or isolated pretrained modules to localize atrophy regions, which neglects individual specificity. Besides, existing methods capture global structural context only on the whole-image-level, which weaken correlation between regions and the hierarchical distribution nature of brain structure. We hereby propose a novel dynamic structural brain network construction method based on T1-MRI, which can dynamically localize critical regions and constrain the hierarchical distribution among them. Specifically, we first cluster spatially-correlated channel and generate several critical brain regions as prototypes. Then, we introduce a contrastive loss function to constrain the prototypes distribution, which embed the hierarchical brain semantic structure into the latent space. Self-attention and GCN are then used to dynamically construct hierarchical correlations of critical regions for brain network and explore the correlation, respectively. Our method is trained on ADNI-1 and tested on ADNI-2 databases for mild cognitive impairment (MCI) conversion prediction, and achieve the state-of-the-art (SOTA) performance.

Keywords: Dynamic Structural Brain Network · T1-MRI · Hierarchical Prototype Learning · GCN · Mild Cognitive Impairment

Y. Leng and W. Cui—Contribute equally to this work.

Supplementary Information The online version contains supplementary material available at https://doi.org/10.1007/978-3-031-43993-3_12.

H. Greenspan et al. (Eds.): MICCAI 2023, LNCS 14227, pp. 120–130, 2023.
https://doi.org/10.1007/978-3-031-43993-3_12

1 Introduction

T1-weighted magnetic resonance imaging (T1-MRI) is one of the indispensable medical imaging methods for noninvasive diagnosing neurological disorder [9]. Existing approaches [16,19] based on T1-MRI focus on extracting region of interests (ROIs) to analyze structural atrophy information associated with disease progression. However, some works [6,16,21] heavily rely on manual defined and selected ROIs, which have limitations in explaining the individual brain specificity. To address this issue, Lian et al. [15] localize discriminative regions by a pretrained module, where region localization and following feature learning cannot reinforce each other, resulting a coarse feature representation. Additionally, as inter-regional correlations are unavailable in T1-MRI directly, most related works [2,14] ignore inter-regional correlations or replace them with a generalized global information. These conventional modular approaches have limitations in explaining high-dimensional brain structure information [1,25].

Brain network is a vital method to analysis brain disease, which has been widely used in functional magnetic resonance imaging (fMRI) and diffusion tensor imaging (DTI). However, the structural brain network with T1-MRI is still underexplored due to the lack of direct regional connectivity. Recent advances [8,11,22,23] in graph convolution neural networks (GCNs) have optimized brain networks construction with fMRI and DTI. Given the successful application of GCN in these modalities, we think that it also has potential for construction of structural brain network using T1-MRI. Current approaches [4,8,12,13] to brain network construction involve the selection of ROIs and modeling inter-regional correlations, in which anatomical ROIs are employed as nodes, and inter-node correlations are modeled as edges. Some researches [18,31] have demonstrated that brain connectivity displays hierarchical structure distribution, yet most GCN-based methods [13,28] treat all nodes equally and ignore the hierarchical nature of brain connectivity. These structural brain networks are fixed and redundant, which may lead to coarse feature representation and suboptimal performance in downstream tasks.

To address these issues, we propose novel **d**ynamic structural brain network construction method named **h**ierarchical **pro**totypes embedding **GCN** (DH-ProGCN) to dynamically construct disease-related structural brain network based on T1-MRI. Firstly, a prototypes learning method is used to cluster spatially-correlated channel and generate several critical brain regions as prototypes. Then, we introduce a contrastive loss function to constrain the hierarchical distribution among prototypes to obtain the hierarchical brain semantic structure embedding in the latent space. After that, DH-ProGCN utilizes a self-attention mechanism to dynamically construct hierarchical correlations of critical regions for constructing structural brain network. GCN is applied to explore the correlation of the structural brain network for Mild Cognitive Impairment (MCI) conversion prediction. We verify the effectiveness of DH-ProGCN on the Alzheimer's Disease Neuroimaging Initiative-1 (ADNI-1) and ADNI-2 dataset. DH-ProGCN achieves state-of-the-art (SOTA) performance for the classification of progressive mild cognitive impairment (pMCI) and stable mild cognitive impairment (sMCI) based on T1-MRI.

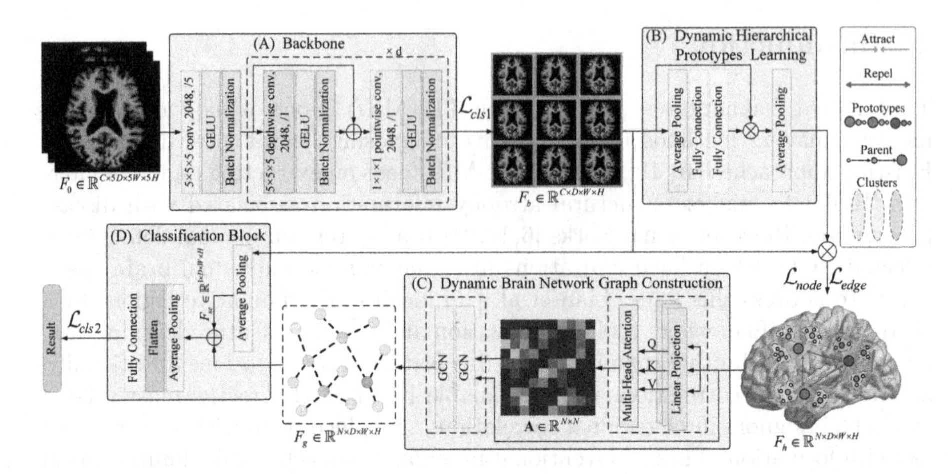

Fig. 1. The overall framework of the DH-ProGCN. (A) We first extract the feature F_b via backbone, and assume that the featuremap of each channel represents different discriminative regions which are showed as images with different colors in F_b. (B) The hierarchical feature F_h are then obtained by hierarchical clustering on the channel dimension. (C) We utilize a self-attention mechanism to model feature correlations matrix A and learn the feature graph F_g by a GCN. (D) F_g and the global representation F_b are concatenated for MCI conversion prediction.

2 Methods

2.1 Backbone

In this study, we utilize a Convmixer-like [24] block as the backbone to achieve primary discriminative brain regions localization, which could provide a large enough channel dimension for subsequent channel clustering with relatively low complexity. Specifically, depicted in Fig. 1(A), the backbone consists of a patch embedding layer followed by several full-convolution blocks. Patch embedding comprises a $5 \times 5 \times 5$ convolution, and the full convolution block comprises a $5 \times 5 \times 5$ depthwise convolution (grouped convolution with groups equal to the number of channels) and a pointwise convolution (kernel size is $1 \times 1 \times 1$) with 2048 channels. By the backbone, features of discriminative regions are finally extracted as $F_b \in \mathbb{R}^{C \times D \times H \times W}$, where D, H, W and C indicate depth, height, width and the number of channels, respectively.

2.2 Dynamic Hierarchical Prototype Learning

Prototypes Definition. In this study, we regard feature maps of each channel as corresponding to the response of distinct brain regions relevant to tasks, and cluster spatially-correlated subtle patterns as compact and discriminative parts from a group of channels whose peak responses appear in neighboring location following [30]. Intuitively, we utilize the location of each peak response as channel information, which can be represented as a position vector whose elements are coordinates from peak responses over all training images. The position vector of the same channel of all training images are combined as the candidate prototype, which can be obtained as following:

$$[t_x^1, t_y^1, t_z^1, t_x^2, t_y^2, t_z^2, \ldots, t_x^{\Omega}, t_y^{\Omega}, t_z^{\Omega}] \tag{1}$$

where $[t_x^i, t_y^i, t_z^i]$ represents the peak response coordinate of the *i-th* image and Ω represents the number of images in the training set. K-means [17] is used to achieve prototypes initialization. Specifically, vectors of all channels are clustered to obtain N sets of clusters $K = \{k_n\}_{n=1}^{N}$, and prototypes are defined as clustering centers $\Gamma = \{\gamma_n\}_{n=1}^{N}$ which are taken as N critical regions for the discriminative localization (i.e., ROIs). $F_h \in \mathbb{R}^{N \times D \times H \times W}$ represents features of clustering centers.

Dynamic Hierarchical Prototype Exploring. Inter-regional spatial connectivity is fixed, but the correlation between them is dynamic with disease progression. We argue that there are structural correlations between different regions, just as the complex hierarchical functional connectome in rich-clubs [25] organization with fMRI. We therefore explore the hierarchical semantic structure of critical brain regions by the hierarchical prototype clustering method. Specifically, we start by using the initial prototypes as the first hierarchy clustering prototypes, denoted as $\Gamma^0 = \{\gamma_n^0\}_{n=1}^{N_0}$. Then, K-means is applied iteratively to obtain parent prototypes of the lower-hierarchy prototypes $\Gamma^{i-1} = \{\gamma_n^{i-1}\}_{n=1}^{N_{i-1}}$, denoted as $\Gamma^i = \{\gamma_n^i\}_{n=1}^{N_i}$, where i represents the *i-th* hierarchy and N_i represents the number of clusters at *i-th* hierarchy, corresponding to the cluster $K^i = \{k_n^i\}_{n=1}^{N_i}$. In this paper, i is set as 2. The number of prototypes in the first, second and third hierarchy is set as 16, 8 and 4, respectively.

To facilitate optimal clustering of the network during training, we use two fully convolutional layers with two contrastive learning loss functions $\mathcal{L}_{node}$ and $\mathcal{L}_{edge}$ to approximate the clustering process. With $\mathcal{L}_{node}$, each channel clustering is enforced to become more compact inside and have significant inter-class differences with other clusterings, enabling all prototypes to be well separated:

$$\mathcal{L}_{node} = -\frac{1}{L}\sum_{l=1}^{L}\sum_{n=1}^{N_l}\sum_{u \in K_n^l} \log \frac{\exp\left(u \cdot \gamma_n^l / \phi_n^l\right)}{\sum_{i \neq n}^{N_l} \exp\left(u \cdot \gamma_i^l / \phi_n^l\right)} \tag{2}$$

$$\phi_n^l = \frac{\sum_{u \in K_n^l} \left\|u - \gamma_n^l\right\|_2}{|K_n^l| \cdot \log(|K_n^l| + \alpha)} \tag{3}$$

where L is the total number of layers, and N_l is the number of clusters in the *l-th* layer. K_n^l, γ_n^l, and ϕ_n^l denote the set of all elements, the cluster center (prototype), and the estimation of concentration of the *n-th* cluster in the *l-th* layer, respectively. α is a smoothing parameter to prevent small clusters from having overly-large ϕ.

The cluster concentration ϕ measures the closeness of elements in a cluster. A larger ϕ indicates more elements in the cluster or smaller total average distance between all elements and the cluster center. Ultimately, $\mathcal{L}_{node}$ compels all elements u in K_n^l to be close to their cluster center γ_n^l and away from other cluster center at the same level.

Similarly, $\mathcal{L}_{edge}$ aims to embed the hierarchical correlation between clustering prototypes, which can be expressed as:

$$\mathcal{L}_{edge} = -\frac{1}{L}\sum_{l=1}^{L-1}\sum_{n=1}^{N_l}\log\frac{\exp\left(\gamma_n^l \cdot Parent(\gamma_n^l)/\tau\right)}{\sum_{i\neq n}^{N^l}\exp\left(\gamma_n^l \cdot \gamma_i^l/\tau\right)} \tag{4}$$

$Parent(\gamma_n^l)$ represents the parent prototype of the prototype γ_n^l, and τ is a temperature hyper-parameter. $\mathcal{L}_{edge}$ forces all prototypes γ^l in the l-th layer to be close to their parent prototype and away from other prototypes within the same level.

2.3 Brain Network Graph Construction and Classification

Through Sect. 2.2, critical brain regions are clustered in a hierarchical semantic latent space. We hereby employ the prototypes regions as nodes and correlations between them as edges to construct structural brain network graphs.

We first apply a self-attention mechanism [26] to compute inter-region correlations to generate edges of the brain network. Then, the features F_h is input to three separate fully connected layers to obtain three vectors: query, key and value, which are used to compute attention scores $\mathbf{A} \in \mathbb{R}^{N\times N}$ between each pair of prototypes, followed by being used to weight the value vector and obtain the output of the self-attention layer as following operation:

$$\mathbf{A} = Attention(Q, K, V) = softmax(\frac{QK^T}{\sqrt{d_k}})V \tag{5}$$

where $Q \in \mathbb{R}^{N\times d_k}$, $K \in \mathbb{R}^{N\times d_k}$, $V \in \mathbb{R}^{N\times N}$ denote query, key, and value, respectively. d_k represents the dimension of Q, K. N represents the number of critical regions, which is set as 16 in this paper.

We then employ GCN to capture the topological interaction in the brain network graph and update features of nodes by performing the following operation:

$$GCN(\mathbf{X}) = \mathrm{ReLU}\left(\hat{\mathbf{D}}^{-1/2}\hat{\mathbf{A}}\hat{\mathbf{D}}^{-1/2}\mathbf{X}\Theta\right) \tag{6}$$

where $\hat{\mathbf{A}} = \mathbf{A}+\mathbf{I}$ is the adjacency matrix with inserted self-loops and $\mathbf{I}$ denotes an identity matrix. $\hat{\mathbf{D}}_{ii} = \sum_{j=0}\hat{\mathbf{A}}_{ij}$ is the diagonal degree matrix, and Θ represents learned weights. To prevent the network overfitting, we just use two GCN layers as the encoder to obtain the final graph feature $F_g \in \mathbb{R}^{N\times D\times H\times W}$. To achieve the classification, we perform channel squeezing on the backbone feature F_b to obtain global features $F_{se} \in \mathbb{R}^{1\times D\times H\times W}$, concatenate it with F_g and input them into the classification layer.

To this end, the information of critical brain regions are fully learned. Notably, as prototypes are dynamic, constructed brain network graphs are also dynamic, rather than predefined and fixed. This allows DH-ProGCN to model and explore the individual hierarchical information, providing a more personalise brain network representation for every subject.

3 Experiments

3.1 Dataset

The data we used are from two public databases: ADNI-1 (http://www.adni-info.org) and ADNI-2 [20]. The demographic information of the subjects and preprocessing steps are shown in the supplemental material. The images are finally resized to $91 \times 109 \times 91$ voxels. Through the quality checking, 305 images are left from ADNI-1 (197 for sMCI, 108 for pMCI), and 350 images are left from ADNI-2 (251 for sMCI, 99 for pMCI). Note that we only keep the earliest images for those who have more than two images at different times. Following [15], we train DH-ProGCN on ADNI-1 and perform independent testing on ADNI-2.

3.2 Implementation Details

We first train backbone with 2048 channels in all layers to extract the output features F_b with the cross-entropy loss $\mathcal{L}_{cls1}$. Then the clustering results are initialized by K-means and further optimized by CCL with $\mathcal{L}_{node}$ and $\mathcal{L}_{edge}$. Finally, The cross-entropy loss $\mathcal{L}_{cls2}$ is used for the final classification. The overall loss function is defined as:

$$\mathcal{L} = \mathcal{L}_{cls1} + \mathcal{L}_{cls2} + \mathcal{L}_{node} + \mathcal{L}_{edge} \tag{7}$$

where $\mathcal{L}_{node}$ and $\mathcal{L}_{edge}$ are explained in Sect. 2.2. Smooth parameter $\alpha = 10$ and temperature parameter $\tau = 0.2$ following [3].

All blocks are trained by SGD optimizer with a momentum of 0.9 and weight decay of 0.001. The model is trained for 300 epochs with an initial learning rate of 0.01 that is decreased by a factor of 10 every 100 epochs. Four metrics, namely accuracy (ACC), sensitivity (SEN), specificity (SPE), and area under the curve (AUC), are used to evaluate the performance of the proposed model. We use Python based on the PyTorch package and run the network on a single NVIDIA GeForce 3090 24 GB GPU. Source code is available at https://github.com/Leng-10/DH-ProGCN.

4 Results

4.1 Comparing with SOTA Methods

Six SOTA methods are used for comparison: 1) LDMIL [16] captured both local information conveyed by patches and global information; 2) H-FCN [15] implemented three levels of networks to obtain multi-scale feature representations

Table 1. Comparsion of our method with current SOTA methods for MCI conversion prediction on ADNI-2 obtained by the models trained on ADNI-1.

Method	ACC	SEN	SPE	AUC
LDMIL	0.769	0.421	0.824	0.776
H-FCN	0.809	0.526	0.854	0.781
HybNet	0.827	0.579	0.866	0.793
AD^{2}A	0.780	0.534	0.866	0.788
DSNet	0.762	**0.770**	0.742	0.818
MSA3D	0.801	0.520	0.856	0.789
DH-ProGCN	**0.849**	0.647	**0.928**	**0.845**

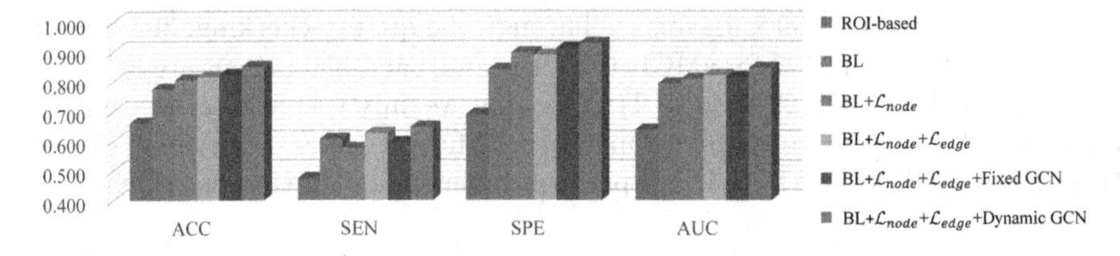

Fig. 2. Effects of each component of DH-ProGCN for MCI conversion prediction on ADNI-2 obtained by models trained on ADNI-1.

which are fused for the construction of hierarchical classifiers; 3) HybNet [14] assigned the subject-level label to patches for local feature learning by iterative network pruning; 4) AD^{2}A [10] located discriminative disease-related regions by an attention modules; 5) DSNet [19] provided disease-image specificity to an image synthesis network; 6) MSA3D [2] implemented a slice-level attention and a 3D CNN to capture subject-level structural changes.

Results in Table 1 show the superiority of DH-ProGCN over SOTA approaches for MCI conversion prediction. Specifically, DH-ProGCN achieves ACC of 0.849 and AUC of 0.845 tested on ADNI-2 by models trained on ADNI-1. It is worth noting that our method: 1) needs no predefined manual landmarks, but achieves better diagnostic results than existing deep-learning-based MCI diagnosis methods; 2) needs no pretrain network parameters from other tasks like AD diagnosis; 3) introduces hierarchical distribution structure to connect regions and form region-based specificity brain structure networks, rather than generalizing the correlations between regions with global information.

4.2 Ablation Study

Effect of Dynamic Prototype Learning. To verify the effect of dynamic prototype clustering, we compare 1) ROI-based approach [29], 2) backbone without channel clustering (BL), 3) backbone with dynamic prototypes clustering (BL+$\mathcal{L}_{node}$). As shown in Fig. 2, results indicate that dynamic prototype clus-

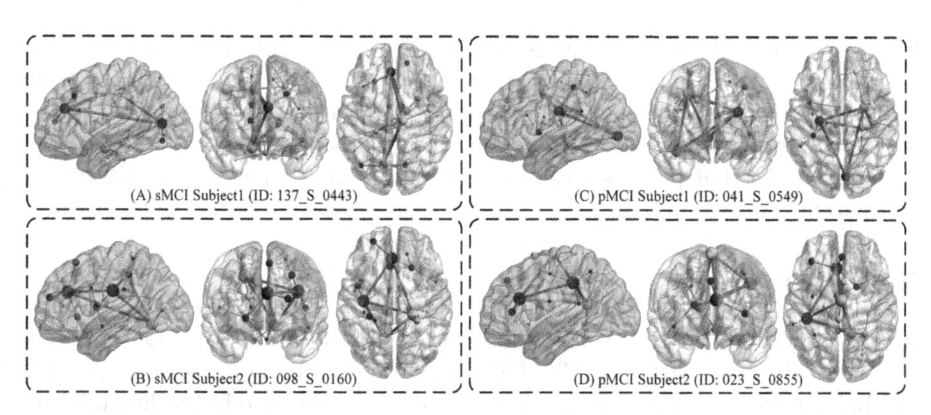

Fig. 3. Sagittal, coronal and axial views of connectome in hierarchical critical regions, (A)(B) represent brain network visualization of sMCI and (C)(D) represent pMCI subjects. BrainNet Viewer [27] is used to generate these figures with the peak response coordinates of cluster centers and the correlation matrix between them $mathbfA$ as node and edge features. The size of node increases with its hierarchy, and nodes with same color are clustered into the same parent prototype. Lower-hierarchy prototypes within cluster are closer to its parent prototypes, and higher-hierarchy prototypes between different clusters are closer than lower-hierarchy prototypes

tering outperforms the ROI-based and backbone on MCI conversion, and could generate better feature distributions for downstream brain images analysis tasks.

Effect of Hierarchical Prototype Learning. To evaluate the impact of hierarchical prototype learning, we compare backbone with flattened prototypes clustering (BL+$\mathcal{L}_{node}$), and hierarchical clustering (BL+$\mathcal{L}_{node}$+$\mathcal{L}_{edge}$). The results are presented in Fig. 2. With the constraint strengthened on the distribution of regions, the results are progressively improved. This implies that it makes sense to introduce hierarchical semantics into the construction of structure brain networks.

Effect of Dynamic Brain Network Construction. To verify whether our constructed dynamic brain network capability outperforms the fixed architecture, we obtained the fixed brain network graph by directly connecting all critical regions after obtaining hierarchical features and feeding them into the GCN for classification. The results are shown in Fig. 2, where the dynamic brain network structure performs better, suggesting that the correlation between regions needs to be measured dynamically to construct a better brain network.

In addition, we visualize the sagittal, coronal and axial views of hierarchical critical regions and their connectome in Fig. 3, which is based on graphs before GNN. The thickness of edges represents the correlation coefficient between nodes, i.e., the connected strength between brain regions. Localized regions are roughly distributed in anatomically defined parahippocampal gyrus, superior frontal gyrus, and cingulate gyrus for different sMCI subjects, lingual gyrus right, and superior longitudinal fasciculus for different pMCI subjects, which agree with previous studies. [5,7,9]. In general, critical regions and correlations are varied for different subjects, indicating the proposed network is a subject-wise dynamic model that processes each subject using data-dependent architectures and parameters.

5 Conclusion

In this paper, we propose a novel dynamic structural brain network construction method named DH-ProGCN. DH-ProGCN could dynamically cluster critical brain regions by the prototype learning, implicitly encode the hierarchical semantic structure of the brain into the latent space by hierarchical prototypes embedding, dynamically construct brain networks by self-attention and extract topology features in the brain network by GCN. Experimental results show that DH-ProGCN outperforms SOTA methods on the MCI conversion task. Essentially, DH-ProGCN has the potential to model hierarchical topological structures in other kinds of medical images. In our future work, we will apply this framework to other kinds of modalities and neurological disorders.

Acknowledgements. This study is partly supported by the Suzhou Science and Technology Plan Project (Grant NO. SZS2022008), and the Zhejiang Medical Health Science and Technology Project (Grant No. 2019RC276).

References

1. Bullmore, E., Sporns, O.: Complex brain networks: graph theoretical analysis of structural and functional systems. Nat. Rev. Neurosci. **10**(3), 186–198 (2009)
2. Chen, L., Qiao, H., Zhu, F.: Alzheimer's disease diagnosis with brain structural MRI using multiview-slice attention and 3D convolution neural network. Front. Aging Neurosci. **14**, 871706 (2022)
3. Chen, X., Fan, H., Girshick, R., He, K.: Improved baselines with momentum contrastive learning. arXiv preprint arXiv:2003.04297 (2020)
4. Chen, Y., et al.: Adversarial learning based node-edge graph attention networks for autism spectrum disorder identification. IEEE Trans. Neural Netw. Learn. Syst. (2022)
5. Chincarini, A., et al.: Local MRI analysis approach in the diagnosis of early and prodromal Alzheimer's disease. Neuroimage **58**(2), 469–480 (2011)
6. Cui, W., et al.: BMNet: a new region-based metric learning method for early Alzheimer's disease identification with FDG-PET images. Front. Neurosci. **16**, 831533 (2022)
7. Dickerson, B.C., et al.: MRI-derived entorhinal and hippocampal atrophy in incipient and very mild Alzheimer's disease. Neurobiol. Aging **22**(5), 747–754 (2001)
8. Duran, F.S., Beyaz, A., Rekik, I.: Dual-HINet: dual hierarchical integration network of multigraphs for connectional brain template learning. In: Wang, L., Dou, Q., Fletcher, P.T., Speidel, S., Li, S. (eds.) MICCAI 2022, Part I. LNCS, vol. 13431, pp. 305–314. Springer, Cham (2022). https://doi.org/10.1007/978-3-031-16431-6_29

9. Frisoni, G.B., Fox, N.C., Jack, C.R., Jr., Scheltens, P., Thompson, P.M.: The clinical use of structural MRI in Alzheimer disease. Nat. Rev. Neurol. **6**(2), 67–77 (2010)
10. Guan, H., Liu, Y., Yang, E., Yap, P.T., Shen, D., Liu, M.: Multi-site MRI harmonization via attention-guided deep domain adaptation for brain disorder identification. Med. Image Anal. **71**, 102076 (2021)
11. Kipf, T.N., Welling, M.: Semi-supervised classification with graph convolutional networks. arXiv preprint arXiv:1609.02907 (2016)
12. Lei, B., et al.: Self-calibrated brain network estimation and joint non-convex multitask learning for identification of early Alzheimer's disease. Med. Image Anal. **61**, 101652 (2020)
13. Li, Y., Wei, Q., Adeli, E., Pohl, K.M., Zhao, Q.: Joint graph convolution for analyzing brain structural and functional connectome. In: Wang, L., Dou, Q., Fletcher, P.T., Speidel, S., Li, S. (eds.) MICCAI 2022, Part I. LNCS, vol. 13431, pp. 231–240. Springer, Cham (2022). https://doi.org/10.1007/978-3-031-16431-6_22
14. Lian, C., Liu, M., Pan, Y., Shen, D.: Attention-guided hybrid network for dementia diagnosis with structural MR images. IEEE Trans. Cybern. **52**(4), 1992–2003 (2020)
15. Lian, C., Liu, M., Zhang, J., Shen, D.: Hierarchical fully convolutional network for joint atrophy localization and Alzheimer's disease diagnosis using structural MRI. IEEE Trans. Pattern Anal. Mach. Intell. **42**(4), 880–893 (2018)
16. Liu, M., Zhang, J., Adeli, E., Shen, D.: Landmark-based deep multi-instance learning for brain disease diagnosis. Med. Image Anal. **43**, 157–168 (2018)
17. Lloyd, S.: Least squares quantization in PCM. IEEE Trans. Inf. Theory **28**(2), 129–137 (1982)
18. Meunier, D., Lambiotte, R., Fornito, A., Ersche, K., Bullmore, E.T.: Hierarchical modularity in human brain functional networks. Front. Neuroinform. **3**, 37 (2009)
19. Pan, Y., Liu, M., Xia, Y., Shen, D.: Disease-image-specific learning for diagnosis-oriented neuroimage synthesis with incomplete multi-modality data. IEEE Trans. Pattern Anal. Mach. Intell. **44**(10), 6839–6853 (2021)
20. Petersen, R.C., et al.: Alzheimer's disease neuroimaging initiative (ADNI): clinical characterization. Neurology **74**(3), 201–209 (2010)
21. Shao, W., Peng, Y., Zu, C., Wang, M., Zhang, D., Initiative, A.D.N., et al.: Hypergraph based multi-task feature selection for multimodal classification of Alzheimer's disease. Comput. Med. Imaging Graph. **80**, 101663 (2020)
22. Song, X., et al.: Graph convolution network with similarity awareness and adaptive calibration for disease-induced deterioration prediction. Med. Image Anal. **69**, 101947 (2021)
23. Song, X., et al.: Multi-center and multi-channel pooling GCN for early AD diagnosis based on dual-modality fused brain network. IEEE Trans. Med. Imaging **42**(2), 354–367 (2022)
24. Trockman, A., Kolter, J.Z.: Patches are all you need? arXiv preprint arXiv:2201.09792 (2022)
25. Van Den Heuvel, M.P., Sporns, O.: Rich-club organization of the human connectome. J. Neurosci. **31**(44), 15775–15786 (2011)
26. Vaswani, A., et al.: Attention is all you need. In: Advances in Neural Information Processing Systems, vol. 30 (2017)
27. Xia, M., Wang, J., He, Y.: BrainNet viewer: a network visualization tool for human brain connectomics. PLoS ONE **8**(7), e68910 (2013)

28. Ye, J., He, J., Peng, X., Wu, W., Qiao, Yu.: Attention-driven dynamic graph convolutional network for multi-label image recognition. In: Vedaldi, A., Bischof, H., Brox, T., Frahm, J.-M. (eds.) ECCV 2020. LNCS, vol. 12366, pp. 649–665. Springer, Cham (2020). https://doi.org/10.1007/978-3-030-58589-1_39
29. Zhang, D., Wang, Y., Zhou, L., Yuan, H., Shen, D., Initiative, A.D.N., et al.: Multimodal classification of Alzheimer's disease and mild cognitive impairment. Neuroimage **55**(3), 856–867 (2011)
30. Zheng, H., Fu, J., Mei, T., Luo, J.: Learning multi-attention convolutional neural network for fine-grained image recognition. In: Proceedings of the IEEE International Conference on Computer Vision, pp. 5209–5217 (2017)
31. Zhou, C., Zemanová, L., Zamora, G., Hilgetag, C.C., Kurths, J.: Hierarchical organization unveiled by functional connectivity in complex brain networks. Phys. Rev. Lett. **97**(23), 238103 (2006)

Microstructure Fingerprinting for Heterogeneously Oriented Tissue Microenvironments

Khoi Minh Huynh[1,2], Ye Wu[1,2], Sahar Ahmad[1,2], and Pew-Thian Yap[1,2](✉)

[1] Department of Radiology, University of North Carolina, Chapel Hill, USA
[2] Biomedical Research Imaging Center (BRIC), University of North Carolina, Chapel Hill, USA
ptyap@med.unc.edu

Abstract. Most diffusion biophysical models capture basic properties of tissue microstructure, such as diffusivity and anisotropy. More realistic models that relate the diffusion-weighted signal to cell size and membrane permeability often require simplifying assumptions such as short gradient pulse and Gaussian phase distribution, leading to tissue features that are not necessarily quantitative. Here, we propose a method to quantify tissue microstructure without jeopardizing accuracy owing to unrealistic assumptions. Our method utilizes realistic signals simulated from the geometries of cellular microenvironments as fingerprints, which are then employed in a spherical mean estimation framework to disentangle the effects of orientation dispersion from microscopic tissue properties. We demonstrate the efficacy of microstructure fingerprinting in estimating intra-cellular, extra-cellular, and intra-soma volume fractions as well as axon radius, soma radius, and membrane permeability.

1 Introduction

In diffusion MRI, biophysical models offer a non-invasive means of probing the tissue micro-architecture of the human brain. Most models rely on closed-form formulas derived with simplifying assumptions such as short gradient pulse, Gaussian phase distribution, and the absence of compartmental exchange. The reliability and interpretability of these models diminish with deviation from these assumptions.

Monte Carlo (MC) simulations [1,2] and solving the Bloch-Torrey partial differential equation (BT-PDE) [3,4] are common methods for generating realistic signals associated with different gradient profiles and cellular geometries. However, the applications of these simulation techniques have been mostly limited to the validation of diffusion models rather than the estimation of microstructural properties.

This work was supported in part by the United States National Institutes of Health (NIH) through grants MH125479 and EB008374.

H. Greenspan et al. (Eds.): MICCAI 2023, LNCS 14227, pp. 131–141, 2023.
https://doi.org/10.1007/978-3-031-43993-3_13

Microstructure fingerprinting (MF) [5,6] exploits the representation accuracy and physical interpretability of simulated models to quantify microstructural properties. Diffusion signals are first simulated for a large collection of microstructural geometries, giving "fingerprints" of, for example, axons, somas, and the extra-cellular matrix. Tissue properties are inferred based on the relative contribution of each fingerprint to the voxel signal.

In this paper, we introduce a novel MF technique with the following key features:

1. MC simulation [5,6] is computationally expensive [1], limiting its ability in constructing a sufficient large dictionary of fingerprints for accurate tissue quantification. We will use SpinDoctor [3] to simulate diffusion signal by solving the BT-PDE. The significant speed-up of SpinDoctor over MC simulations allows fast construction of a comprehensive dictionary of fingerprints.
2. Inspired by [7], we will include fingerprints associated with different levels of membrane permeability to account for inter-cellular exchange. The utilization of simulated models has been shown to remove estimation bias associated with short exchange times [7] and to yield marked improvement and higher reproducibility over the widely used Kärger model [6].
3. Spherical Mean Spectrum Imaging (SMSI) [8] is used to eliminate confounding factors such as extra-axonal water and axonal orientation dispersion to improve sensitivity of the diffusion-weighted signal to axon radii [9].
4. MR-derived statistics are often biased toward large axons due to their dominant signal [10]. We introduce a method to correct for this bias based on signal fingerprints, offering radius and permeability measurements that are more biologically realistic.

Our method is able to efficiently and accurately probe microstructural properties such as cell size and membrane permeability without relying on assumptions associated with closed-form formulas.

2 Methods

2.1 Fingerprint Dictionary

The diffusion MRI signal at each voxel S is a combination of signals from multiple micro-environments, each represented by a signal fingerprint S_i:

$$S = \sum_i f[i] S_i(\mathcal{J}_i, \mathcal{P}), \tag{1}$$

where $f[i]$ is the volume fraction of the i-th fingerprint, $\mathcal{J}_i$ is a set of parameters characterizing the geometry of the corresponding micro-environment, and $\mathcal{P}$ is a set of acquisition parameters (e.g., pulse sequence, pulse duration, etc.). For brevity, we omit $\mathcal{P}$ as it is the same for all i's. In [8,11], $\mathcal{J}_i$ is based on a tensor model defined by longitudinal diffusivity $\lambda_{\parallel}[i]$ and radial diffusivity $\lambda_{\perp}[i]$, yielding only basic tissue properties such as diffusivity or anisotropy. Inferring geometrical properties such as the radius from diffusivity is not straight-forward [6]. To effectively quantify cell size and membrane permeability, we use Spin-Doctor [3,4] to generate realistic signal fingerprints for various microstructural geometries (called 'atoms') representing axons and somas with a range of radii and different levels of permeability in an extra-cellular matrix. Specifically, the dictionary of fingerprints covers four diffusion patterns:

1. Intra-axonal diffusion represented by packed cylinders with radii $r \in \{0.5, 2, 2.5, 3.0, 3.5, 4.0\}$ µm and permeabilities κ's from 0 (impermeable) to 50×10^{-6} µm µs^{-1}. For $b \leq 3000$s mm^{-2} (typical in most datasets), the diffusion signals of axons with radius from 0 to 2 µm are numerically indistinguishable. The atom for $r = 0.5$ µm summarizes the distribution in $r \in [0, 2]$ µm and the atoms for $r \in [2, 4]$ µm capture the tail of the distribution as in [12]. Cylinders are placed in an extra-cellular space to mimic realistic configurations.
2. Extra-axonal diffusion represented using a tensor model with $\frac{\lambda_{\parallel}}{\lambda_{\perp}} < \tau^2$, with geometric tortuosity $\tau = 2.6$. We choose $1.5 \times 10^{-3}\,\text{mm}^2\,\text{s}^{-1} \leq \lambda_{\parallel} \leq 2.0 \times 10^{-3}\,\text{mm}^2\,\text{s}^{-1}$.
3. Intra-soma diffusion represented using impermeable spheres with radii from 0 to 20 µm.
4. Free-water diffusion represented using a tensor model with $\lambda_{\parallel} = \lambda_{\perp} > \lambda_{\text{soma-max}}$, with $\lambda_{\text{soma-max}} = 1.0 \times 10^{-3}\,\text{mm}^2\,\text{s}^{-1}$ is the maximum apparent intra-soma diffusivity.

Parameters are chosen according to previous studies, covering the spectrum of biologically possible values in the human brain [6,8,13–20].

2.2 Solving the Bloch-Torrey Partial Differential Equation (BT-PDE)

We employ SpinDoctor [3] to numerically simulate the diffusion signal from a known geometry (Fig. 1) via solving the BT-PDE. SpinDoctor is typically 50 times faster than MC simulations and does not require GPUs. Our parallelized implementation generates a dictionary of fingerprints for Human Connectome Project (HCP) data with 48 intra-axonal fingerprints (6 radii and 8 permeabilities) and 20 intra-soma (20 radii) in 5 min. Note that the dictionary was generated once for the study, stored, and then resampled to match each subjects' gradient table as in [21].

2.3 Solving for Volume Fractions

From [8,11], the normalized spherical mean signal at each voxel $\bar{E}$ is a linear combination of the normalized spherical mean signals $\bar{E}_i$ from multiple atoms:

$$\bar{E} = \sum_i \nu[i]\bar{E}_i = \mathcal{A}\nu, \tag{2}$$

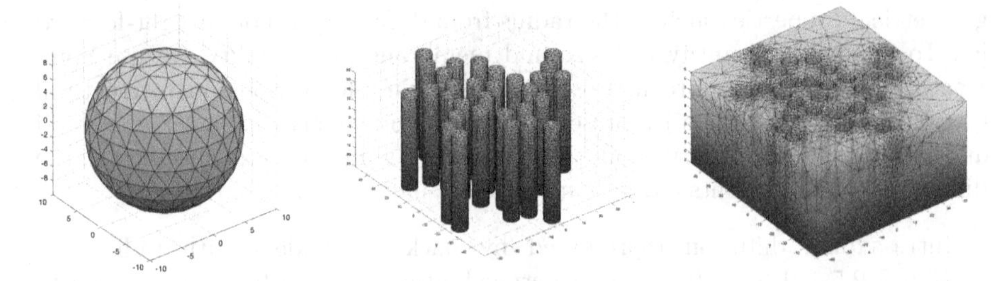

Fig. 1. Examples of SpinDoctor Geometric Configurations. From left to right: a sphere representing a soma with impermeable membrane, cylinders representing packed axons in a voxel, and extra-cellular space around axons allowing for compartmental exchange in case of non-zero axon permeability.

where $\mathcal{A}$ is a matrix containing $\bar{E}_i$ and ν is a vector consisting of $\nu[i]$. Note that $\bar{E}_i$ is a function of $\mathcal{J}_i$ but not the direction [22]. The direction-dependent signal S can be represented using rotational spherical harmonics (SHs) $\mathcal{R}(\mathcal{J}_i)$, the SHs $\mathcal{Y}_L$ of even orders up to L, and the SH coefficients φ_i of the fODF corresponding to the i-th fingerprint:

$$S \approx \sum_i \mathcal{R}(\mathcal{J}_i)\mathcal{Y}_L\varphi_i = \mathcal{B}\Phi. \tag{3}$$

From Φ, the volume fraction $\nu[i]$ of atom i is the 0-th order SH coefficient in φ_i [18,23].

Following [8], to remove fiber dispersion and degeneracy confounding effect, we solve for $\nu[i]$ by using both the spherical mean and the full signal:

1. Solve the full signal (FS) problem:

$$\min_{\Phi}\left\|\begin{pmatrix}\mathcal{B}\\ \sqrt{\gamma_1}I\end{pmatrix}\Phi-\begin{pmatrix}S\\ 0\end{pmatrix}\right\|_2^2, \tag{4}$$

   and estimate the FS problem volume fraction ν_{FS} from Φ.
2. Solve the mean signal (MS) problem:

$$\nu_{\mathrm{MS}} = \arg\min_{\nu\succeq 0}\left\|\begin{pmatrix}\mathcal{A}\\ \sqrt{\gamma_2}I\end{pmatrix}\nu-\begin{pmatrix}\bar{E}\\ 0\end{pmatrix}\right\|_2^2 + \gamma_3\|\nu\|_1. \tag{5}$$

3. Iterative reweighing until convergence:

$$\nu_j = \arg\min_{\nu_j\succeq 0}\left\|\begin{pmatrix}\mathcal{A}\\ \sqrt{\gamma_2}I\end{pmatrix}\nu_j-\begin{pmatrix}\bar{E}\\ 0\end{pmatrix}\right\|_2^2 + \gamma_3\|\mathrm{diag}(w_j)\nu_j\|_1, \tag{6}$$

   where $w_j[i] = \frac{1}{\xi+\nu_{j-1}[i]}$ with ξ being a constant and ν_0 the geometric mean of ν_{FS} and ν_{MS}.

We select the regularization parameters γ's using the Akaike information criterion (AIC) to balance the goodness of fit and complexity of the model. There is an empirical lower bound on the soma radius that can be detected via a mixture of somas (isotropic) and axons (anisotropic) when using the mean signal alone [24]. By using the full gradient-sensitized signal in Step 1, our method can distinguish between soma (isotropic signal) and axon (anisotropic signal). Our approach with both FS and MS allows for modeling a spectrum of diffusion from fine to coarse scales without fixing the number of compartments, e.g., one in [22], two (intra-cellular and extra-cellular) in [25], or three (intra-cellular, extra-cellular, and free-water) in [26].

2.4 Radius Bias Correction

The average axon radius for each voxel can be calculated by averaging the radii of the respective fingerprints, weighted by the volume fractions. Since the volume fractions ν are estimated from the normalized signal model [26], they are actually 'signal fractions'. Weighted averaging using signal fractions yields the axon radius index [17]. Since each axon's contribution to the voxel signal is approximately proportional to the square of its radius [27], using the signal fraction will create a bias toward axons with large radius [10]. This explains why the radius index is often in the range of 3 to 6 µm [13,17] while most axons have actual radii of 0.05 to 1.5 µm [12,28]. We correct for this bias by examining the microstructure model and the actual signal. The spherical mean signal $\bar{S}(b)$ for diffusion weighting b is

$$\frac{\bar{S}(b)}{\bar{S}(0)} = \sum_i \nu[i] \frac{\bar{S}_i(b)}{\bar{S}_i(0)}, \tag{7}$$

where $\nu[i]$ is the signal fraction and $\bar{S}_i(b)$ is the spherical mean signal of the i-th atom. The unbiased volume fraction $f[i]$ is given by

$$\bar{S}(b) = \sum_i f[i] \bar{S}_i(b). \tag{8}$$

Hence

$$\frac{\bar{S}(b)}{\bar{S}(0)} = \frac{1}{\bar{S}(0)} \sum_i f[i] \bar{S}_i(b) = \sum_i \nu[i] \frac{\bar{S}_i(b)}{\bar{S}_i(0)}, \tag{9}$$

allowing us to derive $f[i]$ from $\nu[i]$:

$$f[i] = \nu[i] \frac{\bar{S}(0)}{\bar{S}_i(0)}. \tag{10}$$

Computing f requires the non-diffusion-weighted signal of each compartment $\bar{S}_i(0)$. In our case, $\bar{S}_i(0)$, scaled by an arbitrary factor, is known from the SpinDoctor simulation. We define weight

$$w[i] = \frac{f[i]}{\sum_i f[i]} = \frac{\frac{\nu[i]}{\bar{S}_i(0)}}{\sum_i \frac{\nu[i]}{\bar{S}_i(0)}}, \tag{11}$$

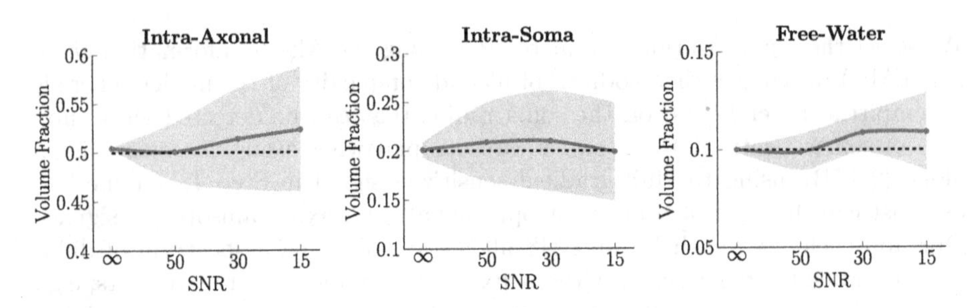

Fig. 2. Volume Fraction. Estimation of the volume fractions of various tissue compartments. The dashed lines represent the ground-truth values. The solid lines indicate mean values of 1000 instances at each noise level. The shaded regions indicate standard deviations.

which is not affected by the scaling factor and therefore can be used to compute the unbiased weighted-average radius.

3 Experiments

We validate our technique, called microstructure fingerprinting SMSI (MF-SMSI), using both in-silico and in-vivo data. The dictionary and synthetic data were generated with HCP-like parameters: 3 b-shells of 1000, 2000, 3000 $\mathrm{s\,mm^{-2}}$, 90 directions per shell, diffusion time $\Delta = 43.1$ ms, and pulse width $\delta = 10.6$ ms [29].

3.1 Volume Fraction

We evaluate the accuracy of MF-SMSI in volume fraction estimation by generating synthetic data following the model in [8,26]:

$$S = \nu_{\mathrm{FW}} S_{\mathrm{FW}} + \nu_{\mathrm{IS}} S_{\mathrm{IS}} + (1 - \nu_{\mathrm{FW}} - \nu_{\mathrm{IS}})(\nu_{\mathrm{IA}} S_{\mathrm{IA}} + (1 - \nu_{\mathrm{IA}}) S_{\mathrm{EC}}), \tag{12}$$

where ν and S are used to denote free-water (FW), intra-soma (IS), intra-axonal (IA), and extra-cellular (EC) volume fractions and signals. We set the ground truth volume fractions to $\nu_{\mathrm{FW}} = 0.1$, $\nu_{\mathrm{IS}} = 0.2$, and $\nu_{\mathrm{IA}} = 0.5$. We generated 1000 instances of the signal with SNR $= \infty$, 50, 30, and 15. MF-SMSI estimates accurately the volume fraction of each compartment (Fig. 2).

3.2 Cell Size and Membrane Permeability

To investigate the efficacy of our bias correction, we performed three experiments:

1. Axonal radius – We simulated signals for two impermeable axons with radii 2 μm and 4 μm, each with volume fraction 0.5. In Fig. 3 (left panel), the axon radius is biased toward the axon with higher baseline signal, giving a large average radius, similar to the observation in [26]. Our unbiased estimate of the radius is markedly closer to the ground truth of 3 μm.

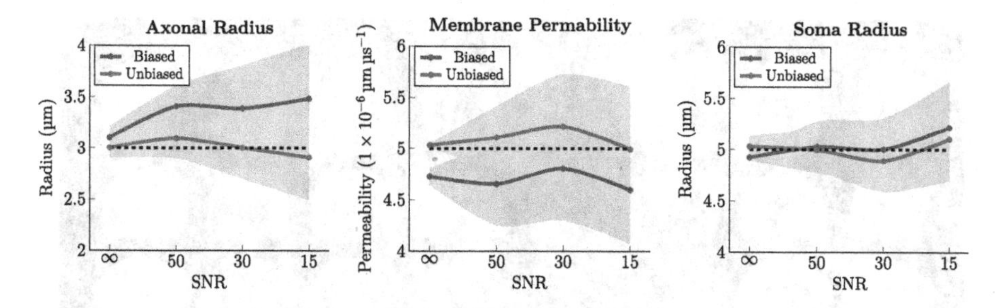

Fig. 3. Cell Size and Membrane Permeability. Estimation bias associated with axonal radius, membrane permeability, and soma radius. The dashed lines represent the ground truth. The solid lines indicate mean values of 1000 instances across noise levels. The shaded regions indicate standard deviations.

2. Membrane permeability – We simulated the signals for two axons with permeabilities 4 and 6 ($\times 1 \times 10^{-6}$ µm µs^{-1}), common radius 3 µm, and equal volume fraction 0.5. In Fig. 3 (middle panel), the biased permeability estimate is biased toward the axon with lower permeability and therefore higher baseline signal. MF-SMSI is able to estimate the correct permeability value matching the ground truth of $5 \times 1 \times 10^{-6}$ µm µs^{-1}.
3. Soma radius – We simulated signals for two somas with radii 4 µm and 6 µm with equal volume fraction 0.5. From Fig. 3 (right panel), the bias toward the soma with higher radius (higher baseline signal) is not severe. MF-SMSI yields estimate of the soma radius that is closer to the ground truth.

3.3 In-vivo Data

We compare our estimates with the axonal radius index $r_{\mathrm{axon}}^{\mathrm{index}}$ from ActiveAx [27]. Figure 4 presents the averaged MF-SMSI and ActiveAx maps of 35 HCP subjects. Intra-axonal (ν_{IA}), extra-cellular (ν_{EC}), free-water (ν_{FW}), and intra-soma (ν_{IS}) volume fractions are in great agreement with previous studies [8,19, 20,25]. Briefly, ν_{IA} is higher in white matter, ν_{FW} is high in CSF, and ν_{IS} is high in the cortical ribbon.

Axonal radius and permeability are lower in deep white matter, especially at the body of the corpus callosum and the forceps major, where axons are myelinated and densely packed [8]. Axonal radius r_{axon} ranges from 1.4 µm to 1.7 µm in most white matter areas with a small increase toward the cortex, similar to observations reported in histological studies [12,30,31]. Axonal radii are slightly lower in the anterior than the posterior part of the brain, in line with [6,13]. ActiveAx $r_{\mathrm{axon}}^{\mathrm{index}}$ is overestimated with values ranging from 3.5 to 7 µm with abrupt spatial changes: almost doubled from the corpus callosum to the cortex, which is unrealistic as the spatial variation of axonal radii was reported to be small [12,15,32].

The soma radii r_{soma} in the cortical ribbon have a mean value of 11 µm, similar to what was reported in [19].

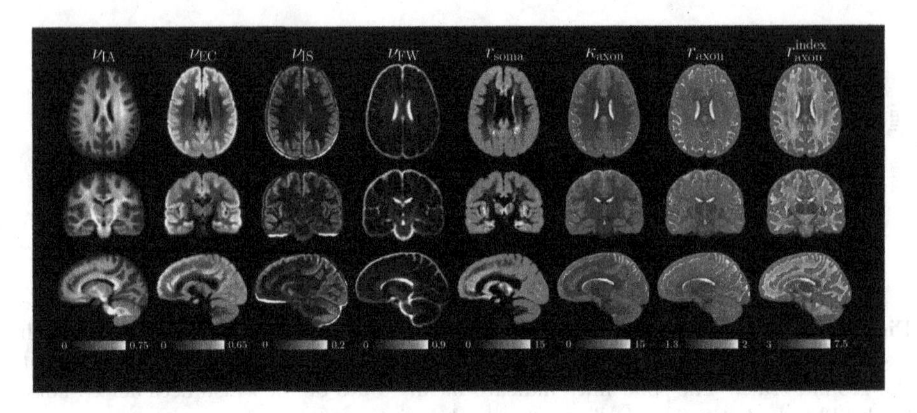

Fig. 4. Microstructural Measurements. Averaged microstructural indices from 35 HCP subjects. Top panel: Intra-axonal ν_{IA}, extra-cellular ν_{EC}, intra-soma ν_{IS}, and free-water ν_{FW} volume fractions. Bottom panel: soma radius r_{soma} (μm), axonal permeability κ_{axon} $(1 \times 10^{-6}$ μm μs$^{-1})$, axonal radius r_{axon} (μm), and ActiveAx axon radius index $r_{\text{axon}}^{\text{index}}$ (μm). Axonal measurements (overlaid on T2w images) are shown in white matter only.

3.4 Histological Corroboration

We compare our axonal radius estimate r_{axon}, ActiveAx axonal radius index $r_{\text{axon}}^{\text{index}}$, and the effective axon radius r_{eff} from [9], with histological samples

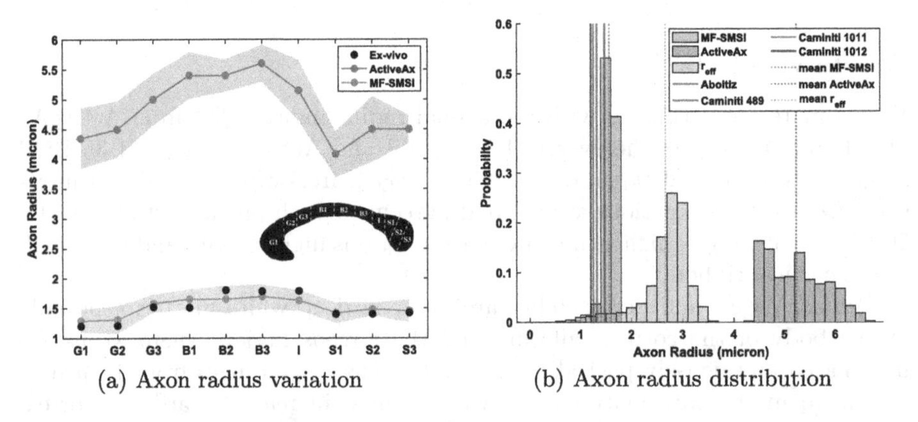

(a) Axon radius variation (b) Axon radius distribution

Fig. 5. Histological Evidence. Left: Mean axonal radii of 35 subjects in different regions of the corpus callosum given by MF-SMSI and ActiveAx. Shaded areas mark the standard deviations. Ex-vivo values are calculated from [12]. Right: Radius distribution in the corpus callosum. Histograms show the distributions and their mean values (marked with dotted lines) for MF-SMSI (unbiased estimate), ActiveAx (radii index), and effective axonal radii (Fig. 7 in [9]). The solid vertical lines show the mean values from histological studies [12,32] (accounted for shrinkage).

reported in [12,32]. Histological values were multiplied by 1.2 to correct for ex-vivo shrinkage [9].

Figure 5(a) shows the mean axonal radii estimated with MF-SMSI and ActiveAx for different areas of the corpus callosum from HCP subjects. Overall, there is a low-high-low trend going from the genu (anterior) through the mid-body to the splenium (posterior) of the corpus callosum. Ex-vivo values (after shrinkage correction) fall almost entirely in the range calculated with MF-SMSI, indicating strong agreement between our results and histological measurements.

Figure 5(b) shows the distribution of mean axonal radius in the corpus callosum from MF-SMSI, ActiveAx, and r_{eff} from [9]. Unlike Active Ax $r_{\mathrm{axon}}^{\mathrm{index}}$ and r_{eff}, which are indicators of axon radii, our estimates of r_{axon} are more realistic measurements of axonal radii, closer to ex-vivo samples.

4 Conclusion

We have presented a microstructure fingerprinting method that can provide accurate and reliable measurements of tissue properties beyond diffusivity and anisotropy, allowing quantification of cell size and permeability associated with axons and somas.

References

1. Hall, M.G., Alexander, D.C.: Convergence and parameter choice for Monte-Carlo simulations of diffusion MRI. IEEE Trans. Med. Imaging **28**(9), 1354–1364 (2009)
2. Balls, G., Frank, L.: A simulation environment for diffusion weighted MR experiments in complex media. Magn. Reson. Med. **62**(3), 771–778 (2009)
3. Li, J.R., et al.: SpinDoctor: a Matlab toolbox for diffusion MRI simulation. Neuroimage **202**, 116120 (2019)
4. Fang, C., Nguyen, V.D., Wassermann, D., Li, J.R.: Diffusion MRI simulation of realistic neurons with SpinDoctor and the Neuron Module. Neuroimage **222**, 117198 (2020)
5. Rensonnet, G., et al.: Towards microstructure fingerprinting: estimation of tissue properties from a dictionary of Monte Carlo diffusion MRI simulations. Neuroimage **184**, 964–980 (2019)
6. Nedjati-Gilani, G.L., et al.: Machine learning based compartment models with permeability for white matter microstructure imaging. Neuroimage **150**, 119–135 (2017)
7. Nilsson, M., Alerstam, E., Wirestam, R., Sta, F., Brockstedt, S., Lätt, J., et al.: Evaluating the accuracy and precision of a two-compartment Kärger model using Monte Carlo simulations. J. Magn. Reson. **206**(1), 59–67 (2010)
8. Huynh, K.M., et al.: Probing tissue microarchitecture of the baby brain via spherical mean spectrum imaging. IEEE Trans. Med. Imaging **39**(11), 3607–3618 (2020)
9. Veraart, J., et al.: Noninvasive quantification of axon radii using diffusion MRI. eLIFE **9**, e49855 (2020)
10. Frigo, M., Fick, R., Zucchelli, M., Deslauriers-Gauthier, S., Deriche, R.: Multi tissue modelling of diffusion MRI signal reveals volume fraction bias. In: International Symposium on Biomedical Imaging. (2020)

11. Huynh, K.M., et al.: Probing brain micro-architecture by orientation distribution invariant identification of diffusion compartments. In: International Conference on Medical Image Computing and Computer-Assisted Intervention, Springer, pp. 547–555 (2019)
12. Aboitiz, F., Scheibel, A.B., Fisher, R.S., Zaidel, E.: Fiber composition of the human corpus callosum. Brain Res. **598**(1–2), 143–153 (1992)
13. Fan, Q., et al.: Axon diameter index estimation independent of fiber orientation distribution using high-gradient diffusion MRI. NeuroImage **222** 117197 (2020)
14. Imae, T., et al.: Estimation of cell membrane permeability and intracellular diffusion coefficient of human gray matter. Magn. Reson. Med. Sci. **8**(1), 1–7 (2009)
15. Veraart, J., Raven, E.P., Edwards, L.J., Weiskopf, N., Jones, D.K.: The variability of MR axon radii estimates in the human white matter. Hum. Brain Mapp. **42**(7), 2201–2213 (2021)
16. Walter, A., Gutknecht, J.: Permeability of small nonelectrolytes through lipid bilayer membranes. J. Membr. Biol. **90**(3), 207–217 (1986)
17. Zhang, H., Hubbard, P.L., Parker, G.J., Alexander, D.C.: Axon diameter mapping in the presence of orientation dispersion with diffusion MRI. Neuroimage **56**(3), 1301–1315 (2011)
18. White, N.S., Leergaard, T.B., D'Arceuil, H., Bjaalie, J.G., Dale, A.M.: Probing tissue microstructure with restriction spectrum imaging: histological and theoretical validation. Hum. Brain Mapp. **34**(2), 327–346 (2013)
19. Palombo, M., et al.: SANDI: a compartment-based model for non-invasive apparent soma and neurite imaging by diffusion MRI. NeuroImage **215** 116835 (2020)
20. Huynh, K.M., et al.: Characterizing intra-soma diffusion with spherical mean spectrum imaging. In: International Conference on Medical Image Computing and Computer-Assisted Intervention, Springer, pp. 354–363 (2020)
21. Daducci, A., Canales-Rodríguez, E.J., Zhang, H., Dyrby, T.B., Alexander, D.C., Thiran, J.P.: Accelerated microstructure imaging via convex optimization (AMICO) from diffusion MRI data. Neuroimage **105**, 32–44 (2015)
22. Kaden, E., Kruggel, F., Alexander, D.C.: Quantitative mapping of the per-axon diffusion coefficients in brain white matter. Magn. Reson. Med. **75**(4), 1752–1763 (2016)
23. Tournier, J.D., Calamante, F., Gadian, D.G., Connelly, A.: Direct estimation of the fiber orientation density function from diffusion-weighted MRI data using spherical deconvolution. Neuroimage **23**(3), 1176–1185 (2004)
24. Afzali, M., Nilsson, M., Palombo, M., Jones, D.K.: Spheriously? the challenges of estimating sphere radius non-invasively in the human brain from diffusion MRI. Neuroimage **237**, 118183 (2021)
25. Kaden, E., Kelm, N.D., Carson, R.P., Does, M.D., Alexander, D.C.: Multi-compartment microscopic diffusion imaging. Neuroimage **139**, 346–359 (2016)
26. Zhang, H., Schneider, T., Wheeler-Kingshott, C.A., Alexander, D.C.: NODDI: practical in vivo neurite orientation dispersion and density imaging of the human brain. Neuroimage **61**(4), 1000–1016 (2012)
27. Alexander, D.C., et al.: Orientationally invariant indices of axon diameter and density from diffusion MRI. Neuroimage **52**(4), 1374–1389 (2010)
28. Drobnjak, I., Neher, P., Poupon, C., Sarwar, T.: Physical and digital phantoms for validating tractography and assessing artifacts. Neuroimage **245**, 118704 (2021)
29. Van Essen, D.C., et al.: The WU-Minn human connectome project: an overview. Neuroimage **80**, 62–79 (2013)

30. Liewald, D., Miller, R., Logothetis, N., Wagner, H.-J., Schüz, A.: Distribution of axon diameters in cortical white matter: an electron-microscopic study on three human brains and a macaque. Biol. Cybern. **108**(5), 541–557 (2014). https://doi.org/10.1007/s00422-014-0626-2
31. Sepehrband, F., Alexander, D.C., Kurniawan, N.D., Reutens, D.C., Yang, Z.: Towards higher sensitivity and stability of axon diameter estimation with diffusion-weighted MRI. NMR Biomed. **29**(3), 293–308 (2016)
32. Caminiti, R., Ghaziri, H., Galuske, R., Hof, P.R., Innocenti, G.M.: Evolution amplified processing with temporally dispersed slow neuronal connectivity in primates. Proc. Natl. Acad. Sci. **106**(46), 19551–19556 (2009)

AUA-dE: An Adaptive Uncertainty Guided Attention for Diffusion MRI Models Estimation

Tianshu Zheng[1], Ruicheng Ba[1], Xiaoli Wang[2], Chuyang Ye[3], and Dan Wu[1](✉)

[1] College of Biomedical Engineering and Instrument Science, Zhejiang University, Hangzhou, China
danwu.bme@zju.edu.cn

[2] School of Medical Imaging, Weifang Medical University, Weifang, China

[3] School of Integrated Circuits and Electronics, Beijing Institute of Technology, Beijing, China

Abstract. Diffusion MRI (dMRI) is a well-established tool for probing tissue microstructure properties. However, advanced dMRI models commonly have multiple compartments that are highly nonlinear and complex, and also require dense sampling in q-space. These problems have been investigated using deep learning based techniques. In existing approaches, the labels were calculated from the fully sampled q-space as the ground truth. However, for some of the dMRI models, dense sampling is hard to achieve due to the long scan time, and the low signal-to-noise ratio could lead to noisy labels that make it hard for the network to learn the relationship between the signals and labels. A good example is the time-dependent dMRI (TD-dMRI), which captures the microstructural size and transmembrane exchange by measuring the signal at varying diffusion times but requires dense sampling in both q-space and t-space. To overcome the noisy label problem and accelerate the acquisition, in this work, we proposed an *adaptive uncertainty guided attention for diffusion MRI models estimation* (AUA-dE) to estimate the microstructural parameters in the TD-dMRI model. We evaluated our proposed method with three different downsampling strategies, including q-space downsampling, t-space downsampling, and q-t space downsampling, on two different datasets: a simulation dataset and an experimental dataset from normal and injured rat brains. Our proposed method achieved the best performance compared to the previous q-space learning methods and the conventional optimization methods in terms of accuracy and robustness.

Keywords: Diffusion MRI · Noisy Data · Parameter Estimation · Uncertainty Attention

1 Introduction

Diffusion MRI (dMRI) is a powerful medical imaging tool for probing microstructural information based on the restricted diffusion assumption of the water molecules in biological tissues [7]. The conventional dMRI uses the apparent diffusion coefficient (ADC) to measure the diffusivity change, but it's not specific to

H. Greenspan et al. (Eds.): MICCAI 2023, LNCS 14227, pp. 142–151, 2023.
https://doi.org/10.1007/978-3-031-43993-3_14

microstructural properties. Recently, a series of advanced dMRI models, such as time-dependent dMRI (TD-dMRI) models [5,8,9], have been proposed to sample diffusion at different effective diffusion-times (so-called t-space), and related biophysical models have been developed to resolve the cellular microstructures.

Advanced dMRI models are typically multi-compartment models with highly non-linear and complex formulations. Accurate parameter estimation from such models requires dense q-space and/or t-space sampling. Deep learning techniques have been proposed to improve estimation accuracy from downsampled q-space data [2,13,15], by training networks to learn the relationship between downsampled diffusion signals and microstructural metrics from fully sampled q-space. However, TD-dMRI models require dense sampling in both q-space and t-space, which is challenging for clinical usage. To our best knowledge, previous works have not investigated downsampling models in t-space or joint q-t space.

The previous q-space learning networks [2,13,15], simply learned the mapping relationship between undersampled diffusion signals in q-space and diffusion model parameters. They neglected the "noisy label" problem, and thus, may suffer from training degradation due to failure to obtain valid information. Different from the annotated label, which is called ground-truth in the natural image, in the dMRI model estimation area, we used the parameters estimated from fully sampled q-space as the "gold standard" that actually suffer from estimation error due to noise in the data acquisition and also the limited number of samples in q-space. These two problems are particularly worth noting in TD-dMRI, which is known to have low SNR, and the errors could accumulate in the joint q-t space models. This imposed a vital problem for end-to-end t-space learning in practice. To address the challenge of learning from noisy data, we proposed an *adaptive uncertainty guided attention for diffusion MRI models estimation* (AUA-dE) based upon the previous AEME network [15] for estimating general dMRI model parameters.

In this work, we proposed a reweighting strategy to reduce the negative effects of noisy label based on uncertainty. Our contributions can be summarized below:

1. We brought up an important problem of the noisy label in dMRI model estimation which was not addressed before.
2. We proposed an attention-based sparse encoder to make the network focus on the key diffusion signal out of many q-space or t-space signals.
3. We developed an uncertainty-based reweighting strategy considering the uncertainty in both dMRI channels and spatial domain for microstructural estimation.
4. We proposed an end-to-end estimation strategy in both q-space and t-space with downsampled q-t space data.

In our work, we firstly demonstrated the effectiveness of our attention-based reweighting strategy on a simulation dataset, and then we evaluated our work with three different downsampling strategies (q-space, t-space, and q-t space) on a TD-dMRI dataset of normal and injured rat brains. We tested a TD-dMRI model which estimates the transmembrane water exchange time based on time-dependent diffusion kurtosis [10], here we named tDKI.

2 Method

2.1 *q-t* Space Sparsity

In this study, we extended the *q*-space learning model AEME [15] into the *q-t* space, and the signal can be represented as follows:

$$\mathbf{Y} = \mathbf{\Gamma}\mathbf{X}\mathbf{\Upsilon}^{\mathrm{T}} + \mathbf{H} \tag{1}$$

where $\mathbf{Y} = (y_1 \cdots y_V) \in \mathbb{R}^{K \times V}$($K$ is the number of diffusion signals including different b-values, gradients, and diffusion times (t_ds); and V is the size of patch), $y_v(v \in \{1, \cdots, V\})$ is the diffusion signal normalized by b0 at different t_d, and $\mathbf{X} \in \mathbb{R}^{N_\Gamma \times N_\Upsilon}$ is the matrix of the mixed sparse dictionary coefficients. $\mathbf{\Gamma} \in \mathbb{R}^{K \times N_\Gamma}$ and $\mathbf{\Upsilon} \in \mathbb{R}^{V \times N_\Upsilon}$ are decomposed dictionaries that encode the information in the mixed *q-t* domain and the spatial domain. $\mathbf{H}$ is the noise corresponding to $\mathbf{X}$. Then the sparse encoder can be formulated using the extragradient-based method similar to [15]:

$$\mathbf{X}^{n+\frac{1}{2}} = \mathcal{H}_M\left(AUA(\mathbf{X}^n, \mathbf{\Gamma}^{\mathrm{T}}\mathbf{Y}\mathbf{\Upsilon}, \mathbf{\Gamma}^{\mathrm{T}}\mathbf{\Gamma}\mathbf{X}^n\mathbf{\Upsilon}^{\mathrm{T}}\mathbf{\Upsilon})\right) \tag{2}$$

$$\mathbf{X}^{n+1} = \mathcal{H}_M\left(AUA(\mathbf{X}^n, \mathbf{A}_1\mathbf{\Gamma}^{\mathrm{T}}\mathbf{Y}\mathbf{\Upsilon}, \mathbf{A}_1\mathbf{\Gamma}^{\mathrm{T}}\mathbf{\Gamma}\mathbf{X}^{n+\frac{1}{2}}\mathbf{\Upsilon}^{\mathrm{T}}\mathbf{\Upsilon})\right) \tag{3}$$

where, $\mathbf{A}_1$ denotes a scalar matrix, $AUA(\cdot)$ is the adaptive uncertainty attention function, and $\mathcal{H}_M$ denotes a nonlinear operator corresponding to the threshold layer in Fig. 1(b):

$$\mathcal{H}_M(\mathbf{X}_{ij}) = \begin{cases} 0, & \text{if } \mathbf{X}_{ij} < \lambda \\ \mathbf{X}_{ij}, & \text{if } \mathbf{X}_{ij} \geq \lambda \end{cases} \tag{4}$$

So far, we can obtain the sparse representation of the signal.

2.2 Adaptive Uncertainty Attention Modelling

Uncertainty Attention Module. Inspired by the uncertainty modeling mechanism of Bayesian neural networks [6], we defined *uncertainty attention* (UA) to address the noisy label problem. For simplicity, we used Monte Carlo dropout to obtain the posterior distribution, other uncertainty quantification methods would also work. The basic attention module is adapted from CBAM [12], which includes channel and spatial attention. To better capture feature information, we employ mean and standard deviation pooling. Then the UA module is formulated according to the gray shaded box in Fig. 1(b). The top branch is the *channel uncertainty attention* (CUA) module, and the bottom branch is the *spatial uncertainty attention* (SUA) module.

Through the CUA module, the uncertainty reweighted sparse representation of the original signal $\mathbf{X}$ can be obtained: $\mathbf{X}_{\mathrm{CUA}} = CUA(\mathbf{X})$. $CUA(\cdot)$ is the channel-wise uncertainty attention function, which is used to model the uncertainty in the diffusion channels. The CUA of $\mathbf{X}$ can be obtained after the stochastic forward with dropout: $\mathbf{U}_{\mathrm{C}} = Var(\mathbf{X}^{k}_{\mathrm{CUA}})$. $\mathbf{U}_{\mathrm{C}}$ is the uncertainty of the channel-wise information, $\mathbf{X}^{k}_{\mathrm{CUA}}$ is a cluster of different sparse representations of $\mathbf{X}$ after dropout in the CUA module, and k is the number of stochastic forwards.

Similarly, the SUA module can be defined as: $\mathbf{X}_{\mathrm{SUA}} = SUA(\mathbf{X}_{\mathrm{CUA}})$. $SUA(\cdot)$ is the spatial-wise uncertainty attention, which is used to model the uncertainty of the spatial-wise information. The SUA of $\mathbf{X}_{\mathrm{CUA}}$ can be estimated as follows: $\mathbf{U}_{\mathrm{S}} = Var(\mathbf{X}^{k}_{\mathrm{SUA}})$. $\mathbf{U}_{\mathrm{S}}$ is the uncertainty of the spatial-wise information, $\mathbf{X}^{k}_{\mathrm{SUA}}$ is a cluster of different sparse representations of $\mathbf{X}_{\mathrm{CUA}}$ after dropout in the SUA module.

For simplicity and efficiency, in practice, we combined these two kinds of uncertainty together to reweight the whole sparse representation $\mathbf{X}$ as: $\mathbf{U} = Var(\mathbf{X}^{k})$. $\mathbf{U}$ is the uncertainty of the sparse representation, $\mathbf{X}^{k}$ is a cluster of different sparse representations of $\mathbf{X}$ after dropout in both CUA and SUA modules.

Adaptive Reweight Mechanism. In this work, we proposed an adaptive reweighting strategy by lowering the loss weight for a patch that may be corrupted by the noise. After the uncertainty $\mathbf{U}$ is approximated, We can set a weight tenor $\mathbf{U}_{\mathrm{w}}$ as below:

$$\mathbf{U}_{\mathrm{w}} = 1 - \mathbf{U} \tag{5}$$

Then, the impact of noise can be mitigated by an adaptive weight matrix $\mathbf{R}$:

$$\mathbf{R} = \begin{cases} t, & \text{if } \mathbf{U}_{\mathrm{w}} < t \\ \mathbf{U}_{\mathrm{w}}, & \text{if } \mathbf{U}_{\mathrm{w}} \geq t \end{cases} \tag{6}$$

where, t is a trainable parameter in the network, which can be modified adaptively. Then $\mathbf{X}$ will be reweighted by the $\mathbf{R}$ in the loss function as:

$$\mathcal{L} = ||M(\mathbf{R} \odot \mathbf{X}) - \widehat{P}||^{2} \tag{7}$$

where, $\odot$ denotes the element-wise multiplication, $M(\cdot)$ is a mapping function corresponding to the mapping networks, $\widehat{P}$ is the observed label of the estimated dMRI model parameters.

Network Construction. Following the *q-t* space sparsity analysis and the adaptive uncertainty mechanism mentioned above, we can incorporate historical information [15] into Eq. 2 and Eq. 3 to formulate the *adaptive uncertainty attention sparse encoder* (AUA-SE).

$$\tilde{\mathbf{C}}^{n+\frac{1}{2}} = \mathbf{W}^{\mathrm{m1}}\mathbf{Y}\mathbf{W}^{\mathrm{s1}} + \mathbf{X}^{n} - \mathbf{S}^{\mathrm{m1}}\mathbf{X}^{n}\mathbf{S}^{\mathrm{s1}} \tag{8}$$

$$\mathbf{C}^{n+\frac{1}{2}} = AUA(\mathbf{F}^{n+\frac{1}{2}} \odot \mathbf{C}^{n} + \mathbf{G}^{n+\frac{1}{2}} \odot \tilde{\mathbf{C}}^{n+\frac{1}{2}}) \tag{9}$$

$$\mathbf{X}_{\mathrm{UA}}{}^{n+\frac{1}{2}} = \mathcal{H}_{M}\left(\mathbf{C}^{n+\frac{1}{2}}\right) \tag{10}$$

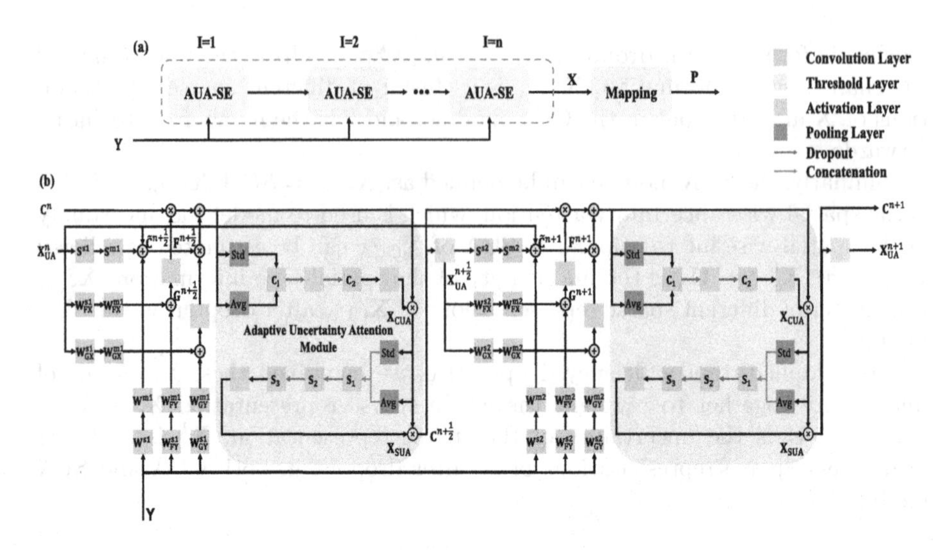

Fig. 1. **(a)** The overall structure of the proposed AUA-dE. The AUA-dE is made up of a number of AUA-SE units that are used to build the sparse representation $\mathbf{X}$ from the diffusion signals $\mathbf{Y}$, and then map the representation to the diffusion model parameters $\boldsymbol{P}$. **(b)** The AUA-SE unit is associated with the adaptive attention mechanism based on uncertainty (gray shaded box). $\mathbf{Y}$ is the input dMRI signal, and $\mathbf{X}_{\mathrm{UA}}^{n+1}$ is the sparse representation weighted by the uncertainty attention.

$$\tilde{\mathbf{C}}^{n+1} = \mathbf{W}^{\mathrm{m}2}\mathbf{Y}\mathbf{W}^{\mathrm{s}2} + \mathbf{X}^{n} - \mathbf{S}^{\mathrm{m}2}\mathbf{X}^{n+\frac{1}{2}}\mathbf{S}^{\mathrm{s}2} \tag{11}$$

$$\mathbf{C}^{n+1} = AUA(\mathbf{F}^{n+1} \odot \mathbf{C}^{n+\frac{1}{2}} + \mathbf{G}^{n+1} \odot \tilde{\mathbf{C}}^{n+1}) \tag{12}$$

$$\mathbf{X_{\mathrm{UA}}}^{n+1} = \mathcal{H}_{M}\left(\mathbf{C}^{n+1}\right) \tag{13}$$

where, $\mathbf{W}^{\mathrm{m}1} \in \mathbb{R}^{N_\Gamma \times K}$, $\mathbf{W}^{\mathrm{m}2} \in \mathbb{R}^{N_\Gamma \times K}$, $\mathbf{W}^{\mathrm{s}1} \in \mathbb{R}^{V \times N_\Upsilon}$, $\mathbf{W}^{\mathrm{s}2} \in \mathbb{R}^{V \times N_\Upsilon}$, $\mathbf{S}^{\mathrm{m}1} \in \mathbb{R}^{N_\Gamma \times N_\Gamma}$, $\mathbf{S}^{\mathrm{m}2} \in \mathbb{R}^{N_\Gamma \times N_\Gamma}$, $\mathbf{S}^{\mathrm{s}1} \in \mathbb{R}^{N_\Upsilon \times N_\Upsilon}$, and $\mathbf{S}^{\mathrm{s}2} \in \mathbb{R}^{N_\Upsilon \times N_\Upsilon}$ are the matrices that need to be learned through the neural network (convolution layers), where m and s correspond to the mixed domain and the spatial domain. $\tilde{\mathbf{C}}^{n+\frac{1}{2}}$, $\mathbf{C}^{n+\frac{1}{2}}$, $\tilde{\mathbf{C}}^{n+1}$, and $\mathbf{C}^{n+1}$ are four intermediate terms. The historical information is encoded through $\mathbf{F}^{n+\frac{1}{2}}$, $\mathbf{F}^{n+1}$, $\mathbf{G}^{n+\frac{1}{2}}$, and $\mathbf{G}^{n+1}$ and they can be defined as [15]. Further fundamental details on formulas and network architecture can be found in [14].

The overall network can be constructed by repeating the AUA-SE unit n times, and the output will be sent to the mapping networks for mapping the microstructure parameters, which consist of three fully connected layers of the feed-forward networks [15]. The overall structure is illustrated in Fig. 1 (a).

2.3 Dataset and Training

The *t*DKI model is defined as below [10]:

$$K(t) = \frac{2K_0\tau_m}{t} \times [1 - \frac{\tau_m}{t} \times (1 - e^{-\frac{t}{\tau_m}})] \tag{14}$$

where, $K(t)$ is the kurtosis at different t_d, K_0 is the kurtosis at $t_d = 0$, t is the t_d and τ_m is the water mixing time. $K(t)$ at individual t_d is obtained according to the DKE method [11].

Simulation Dataset was formed following the method of Barbieri et al., [1], where we plugged the varying parameters (K_0, τ_m) into Eq. 14 to obtain the kurtosis signal at different t_d (50, 100, and 200 ms). Parameter values were sampled uniformly from the following ranges according to the fitted values observed in rat brain data: K_0 between 0 and 3, τ_m between 2 and 200 ms, and a total of 409600 signals were generated. And 60% of them were used for training, 10% for validation, and 30% for testing.

In order to replicate the noisy label problem, we varied the noise level from SNR=10 to 30 in the t-space signal. In the training data, we used a Bayesian method modified from Gustafsson et al. [4] to obtain the label, and in the test data, we used "gold standard" label set from simulation.

Rat Brain Dataset was collected on a 7T Bruker scanner from 3 normal rats and 10 rats that underwent a model of ischemic injury by transient Middle Cerebral Artery Occlusion (MCAO). Diffusion gradients were applied in 18 directions per b-value at 3 b-values of 0.8, 1.5, and 2.5 ms/ μ m^2 and 5 t_d (50, 80, 100, 150, and 200 ms) with the following acquisition parameters: repetition time/echo time = 2207/18 ms, in-plane resolution = 0.3×0.3 mm^2, 10 slices with a slice thickness of 1 mm.

In order to get the gold standard, the DKE toolbox [11] was used to calculate the kurtosis at different t_d with the fully sampled q-space, and then used the Bayesian method mentioned above to estimate K_0 and τ_m with the fully sampled t-space. The dataset was downsampled with randomly selected 9 gradients at b = 0.8 and 1.5 ms/ μ m^2 in q-space and 3 t_d (50, 100, and 200 ms) in t-space. We mixed the 2 normal and 8 injured rats together for training (90%) and validation (10%), and 1 normal with 2 injured rats for testing.

Training. In this work, the dictionary size of our AUA-dE was set at 301, and the hidden size of the fully connected layer was 75. We used an early stopping strategy and a reducing learning rate with an initial learning rate of 1×10^{-4}. AdamW was selected as the optimizer with a batch size of 256. The dropout in the AUA-SE was 0.2 for 50 forward processes.

3 Experiments and Results

3.1 Ablation Study

We compared four different methods, including AEME [15] (baseline of the current network but without the attention or uncertainty mechanism), AEME with attention (the baseline with the attention), AEME with UA (the baseline with

UA but without adaptive mechanism), and AEME with AUA (our proposed AUA-dE) to evaluate the effectiveness of UA and AUA in mitigating the negative effect of the noisy label.

Here, we used the relative error (percentage of the gold standard) to compare different algorithms. Figure 2 showed that network structures with UA (AEME with UA and AUA-dE) achieved lower errors compared with other methods, especially in the lowest SNR environment. AUA-dE achieved the lowest estimation error because the threshold of uncertainty is a trainable parameter that does not need to be manually defined. Meanwhile, we performed paired t-tests for all comparative results, and AUA-dE showed significantly lower errors than all other methods.

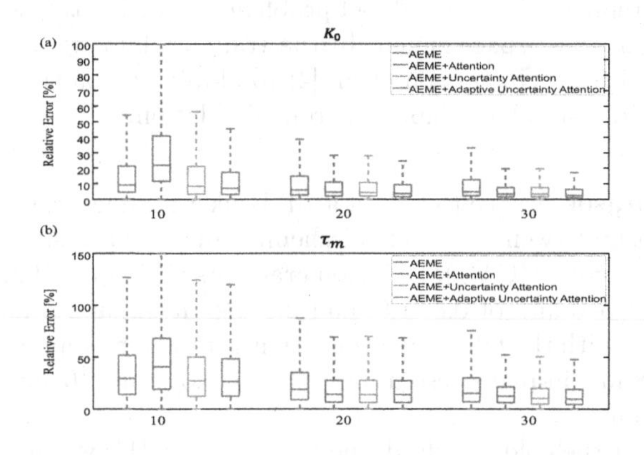

Fig. 2. Relative errors (percentage of the gold standard at the different SNR levels) of the estimated tDKI parameters at SNR levels from 10 to 30.

3.2 Performance Test

***q*-Space Downsampling.** We used our AUA-dE to estimate kurtosis at different t_d using the downsampled q-space, in comparison with an optimization based method DKE [11], a common q-space baseline q-DL [3], and the latest optimization based learning structure AEME [15]. Table 1 shows that our proposed AUA-dE achieved the lowest mean squared error (MSE) when compared to other methods for all t_d in both the normal and injured rat brain regions.

***t*-Space Downsampling.** In the t-space downsampling performance experiment, we used our AUA-dE to estimate the K_0 and τ_m based on $K(t)$ at varying t_d. We compared our method with the Bayesian method [4], q-DL, and AEME. From Table 2, it can be found our proposed AUA-dE achieved the lowest MSE compared with other methods in both normal and injured brain regions. Compared with previous methods, our error was only about 20% of the q-DL error in normal tissues and 40% in the injured regions.

Table 1. Evaluation of MSE on kurtosis using different methods on the downsampled q-space data with 9 different directions and 2 b-values (0.8 and 1.5 ms/μ m^2) under different t_d in both normal and injured rats brains. The lowest errors are in bold.

	Normal				Injured			
	DKE	q-DL	AEME	AUA-dE	DKE	q-DL	AEME	AUA-dE
50 ms	0.15	0.033	0.020	**0.017**	0.12	0.25	0.062	**0.052**
100 ms	0.18	0.050	0.027	**0.021**	0.15	0.65	0.078	**0.071**
200 ms	0.28	0.058	0.029	**0.027**	0.13	0.89	0.102	**0.087**

Table 2. Evaluation of MSE on tDKI parameters using different methods on the downsampled t-space data with 3 t_d (50, 100, and 200 ms) in both normal rats and injured regions. The lowest errors are in bold.

	Normal				Injured			
	Bayesian	q-DL	AEME	AUA-dE	Bayesian	q-DL	AEME	AUA-dE
K_0	0.122	0.078	0.062	**0.019**	0.138	0.065	0.063	**0.012**
τ_m	8567	1570	685	**303**	4569	763	537	**312**

q-t Space Downsampling. In q-t space downsampling, we used our proposed AUA-dE to estimate the K_0 and τ_m with jointly downsampled q-t space data with 5 times acceleration (3 folds in q-space and 1.7 folds in t-space). Figure 3 showed the estimated K_0 and τ_m maps of an injured rat brain, using DKE+Bayesian, q-DL, AEME, and AUA-dE. Only AUA-dE was capable of capturing the abnormal rise in the injured cortex in the τ_m map (denoted by the red arrow), indicating the clinical potential of this method for diagnosis of ischemic brain injury.

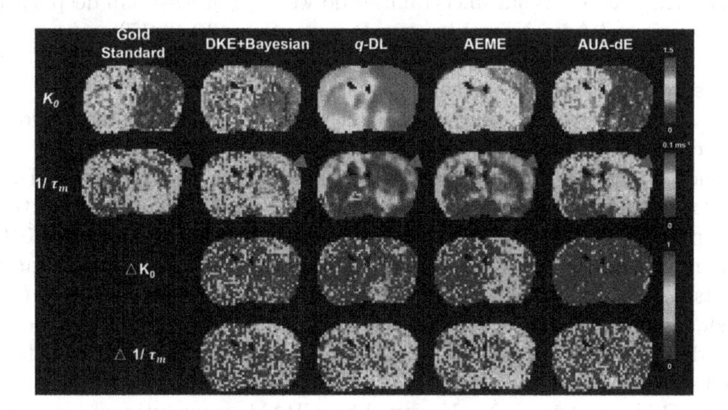

Fig. 3. The gold standard, estimated, and estimation errors of tDKI model parameters, based on DKE+Bayesian, q-DL, AEME, and AUA-dE (ours), in a injured rat brain with joint downsampled q-t space data. (Color figure online)

4 Conclusion

In this work, we proposed an *adaptive uncertainty guided attention for diffusion MRI models estimation* (AUA-dE) to address the noisy label problem in the estimation of TD-dMRI-based microstructural models. We tested our proposed network and module in a rat brain dataset and a simulation dataset. The proposed method showed the highest estimation accuracy in all of these datasets. Meanwhile, we demonstrated its performance on jointly downsampled q-t space data, for which previous algorithms did not work well with the highly accelerated setup (270/54). In the future, we will further investigate our proposed AUA module as a plug-in in different dMRI model estimation networks and also different dMRI models to test its generalizability and robustness.

References

1. Barbieri, S., Gurney-Champion, O.J., Klaassen, R., Thoeny, H.C.: Deep learning how to fit an intravoxel incoherent motion model to diffusion-weighted MRI. Magn. Reson. Med. **83**(1), 312–321 (2020)
2. Gibbons, E.K., et al.: Simultaneous NODDO and GFA parameter map generation from subsampled q-space imaging using deep learning. Magn. Reson. Med. **81**(4), 2399–2411 (2019)
3. Golkov, V., et al.: Q-space deep learning: twelve-fold shorter and model-free diffusion MRI scans. IEEE Trans. Med. Imaging **35**(5), 1344–1351 (2016)
4. Gustafsson, O., Montelius, M., Starck, G., Ljungberg, M.: Impact of prior distributions and central tendency measures on Bayesian intravoxel incoherent motion model fitting. Magn. Reson. Med. **79**(3), 1674–1683 (2018)
5. Jiang, X., et al.: In vivo imaging of cancer cell size and cellularity using temporal diffusion spectroscopy. Magn. Reson. Med. **78**(1), 156–164 (2017)
6. Kendall, A., Gal, Y.: What uncertainties do we need in Bayesian deep learning for computer vision? Adv. Neural Inform. Process. Syst. **30** (2017)
7. Mori, S., Zhang, J.: Principles of diffusion tensor imaging and its applications to basic neuroscience research. Neuron **51**(5), 527–539 (2006)
8. Panagiotaki, E., et al.: Noninvasive quantification of solid tumor microstructure using verdict MRI. Can. Res. **74**(7), 1902–1912 (2014)
9. Reynaud, O., Winters, K.V., Hoang, D.M., Wadghiri, Y.Z., Novikov, D.S., Kim, S.G.: Pulsed and oscillating gradient mri for assessment of cell size and extracellular space (pomace) in mouse gliomas. NMR Biomed. **29**(10), 1350–1363 (2016)
10. Solomon, E., et al.: Time-dependent diffusivity and kurtosis in phantoms and patients with head and neck cancer. Mag. Reson. Med. **89**(2), 522–535 (2022)
11. Tabesh, A., Jensen, J.H., Ardekani, B.A., Helpern, J.A.: Estimation of tensors and tensor-derived measures in diffusional kurtosis imaging. Magn. Reson. Med. **65**(3), 823–836 (2011)
12. Woo, S., Park, J., Lee, J.-Y., Kweon, I.S.: CBAM: convolutional block attention module. In: Ferrari, V., Hebert, M., Sminchisescu, C., Weiss, Y. (eds.) Computer Vision – ECCV 2018: 15th European Conference, Munich, Germany, September 8–14, 2018, Proceedings, Part VII, pp. 3–19. Springer, Cham (2018). https://doi.org/10.1007/978-3-030-01234-2_1

13. Ye, C.: Tissue microstructure estimation using a deep network inspired by a dictionary-based framework. Med. Image Anal. **42**, 288–299 (2017)
14. Ye, C., Li, X., Chen, J.: A deep network for tissue microstructure estimation using modified LSTM units. Med. Image Anal. **55**, 49–64 (2019)
15. Zheng, T., Zheng, W., Sun, Y., Zhang, Y., Ye, C., Wu, D.: An adaptive network with extragradient for diffusion MRI-based microstructure estimation. In: Wang, L., Dou, Q., Fletcher, P.T., Speidel, S., Li, S. (eds.) Medical Image Computing and Computer Assisted Intervention – MICCAI 2022: 25th International Conference, Singapore, September 18–22, 2022, Proceedings, Part I, pp. 153–162. Springer, Cham (2022). https://doi.org/10.1007/978-3-031-16431-6_15

Relaxation-Diffusion Spectrum Imaging for Probing Tissue Microarchitecture

Ye Wu[1(✉)], Xiaoming Liu[2], Xinyuan Zhang[3], Khoi Minh Huynh[4,5], Sahar Ahmad[4,5], and Pew-Thian Yap[4,5(✉)]

[1] School of Computer Science and Engineering, Nanjing University of Science and Technology, Nanjing, China
wuye@njust.edu.cn
[2] Union Hospital, Tongji Medical College, Huazhong University of Science and Technology, Wuhan, Hubei, China
[3] School of Biomedical Engineering, Southern Medical University, Guangzhou, China
[4] Department of Radiology, University of North Carolina, Chapel Hill, NC, USA
[5] Biomedical Research Imaging Center, University of North Carolina, Chapel Hill, NC, USA
ptyap@med.unc.edu

Abstract. Brain tissue microarchitecture is characterized by heterogeneous degrees of diffusivity and rates of transverse relaxation. Unlike standard diffusion MRI with a single echo time (TE), which provides information primarily on diffusivity, relaxation-diffusion MRI involves multiple TEs and multiple diffusion-weighting strengths for probing tissue-specific coupling between relaxation and diffusivity. Here, we introduce a relaxation-diffusion model that characterizes tissue apparent relaxation coefficients for a spectrum of diffusion length scales and at the same time factors out the effects of intra-voxel orientation heterogeneity. We examined the model with an in vivo dataset, acquired using a clinical scanner, involving different health conditions. Experimental results indicate that our model caters to heterogeneous tissue microstructure and can distinguish fiber bundles with similar diffusivities but different relaxation rates. Code with sample data is available at https://github.com/dryewu/RDSI.

Keywords: Diffusion MRI · Microstructure · Relaxation diffusion

1 Introduction

Recent advances in diffusion MRI (dMRI) and diffusion signal modeling equip brain researchers with an in vivo probe into microscopic tissue compositions [15,21]. Signal differences between water molecules in restricted, hindered, and free compartments

Y. Wu and X. Liu—Contributed equally to the paper.

This work was supported by the National Natural Science Foundation of China (No. 62201265, 61971214), and the Natural Science Foundation of Hubei Province of China (No. 2021CFB442). P.-T. Yap was supported in part by the United States National Institutes of Health (NIH) through grants MH125479 and EB008374.

H. Greenspan et al. (Eds.): MICCAI 2023, LNCS 14227, pp. 152–162, 2023.
https://doi.org/10.1007/978-3-031-43993-3_15

can be characterized by higher-order diffusion models for estimating the relative proportions of cell bodies, axonal fibers, and interstitial fluids within an imaging voxel. This allows for the detection of tissue compositional changes driven by development, degeneration, and disorders [13,22]. However, accurate characterization of tissue composition is not only affected by compartment-specific diffusivities but also transverse relaxation rates [4,27]. Several studies have shown that explicit consideration of the relaxation-diffusion coupling may improve the characterization of tissue microstructure [6,16,25].

Multi-compartment models are typically used to characterize signals from, for example, intra- and extra-neurite compartments [18,29]. However, due to the multitude of possible compartments and fiber configurations, solving for these models can be challenging. The problem can be simplified by considering per-axon diffusion models [8,10,28], which typically factor out orientation information and hence involve less parameters. However, existing models are typically constrained to data acquired with a single TE (STE) and do not account for compartment-specific T_2 relaxation. Several studies have shown that multi-TE (MTE) data can account better for intravoxel architectures and fiber orientation distribution functions (fODFs) [1,6,16,17,19].

Here, we propose a unified strategy to estimate using MTE diffusion data (i) compartment specific T_2 relaxation times; (ii) non-T_2-weighted (non-T_2w) parameters of multi-scale microstructure; and (iii) non-T_2w multi-scale fODFs. Our method, called relaxation-diffusion spectrum imaging (RDSI), allows for the direct estimation of non-T_2w volume fractions and T_2 relaxation times of tissue compartments. We evaluate RDSI using both ex vivo monkey and in vivo human brain MTE data, acquired with fixed diffusion times across multiple b-values. Using RDSI, we demonstrate the TE dependence of T_2w fODFs. Furthermore, we show the diagnostic potential of RDSI in differentiating tumors and normal tissues.

2 Methods

2.1 Multi-compartment Model

The diffusion-attenuated signal $S(\tau, b, \mathbf{g})$ acquired with TE τ, diffusion gradient vector $\mathbf{g}$, and gradient strength b can be modeled as

$$S(\tau, b, \mathbf{g}) \approx S(b, \mathbf{g}) e^{-\frac{\tau}{T_2(b)}}, \tag{1}$$

which can be expanded to a multi-compartment model:

$$S(\tau, b, \mathbf{g}) \approx S_r(b, \mathbf{g}) e^{-\tau r_r(b)} + S_h(b, \mathbf{g}) e^{-\tau r_h(b)} + S_f(b) e^{-\tau r_f(b)} \tag{2}$$

to account for signals $S_r(b, \mathbf{g})$, $S_h(b, \mathbf{g})$, and $S_f(b)$ and T_2 values of restricted, hindered, and free compartments. The apparent relaxation rates at different b-values, $r(b) = 1/T_2(b)$, can be estimated using single-shell data acquired with two or more TEs [14]. This model can be expressed using spherical deconvolution [9]:

$$S(\tau, b, \mathbf{g}) = e^{-\tau r_r(b)} \int_{D_r} R(b, \mathbf{g}, D_r) f(D_r) dD_r + e^{-\tau r_h(b)} \int_{D_h} R(b, \mathbf{g}, D_h) f(D_h) dD_h$$
$$+ e^{-\tau r_f(b)} \int_{D_f} R(b, D_f) f(D_f) dD_f$$
$$\text{s.t. } r_r(b),\ r_h(b),\ r_f(b) \geq 0\ \forall b,\ \mathcal{A}f(D_r),\ \mathcal{A}f(D_h),\ \mathcal{A}f(D_f) \geq 0, \tag{3}$$

where the compartment-specific response functions $R(b, \mathbf{g}, D_r)$, $R(b, \mathbf{g}, D_h)$, and $R(b, D_f)$ are associated with apparent diffusion coefficients D_r, D_h, and D_f, yielding compartment-specific multi-scale fODFs $f(D_r)$, $f(D_h)$, and $f(D_f)$. Operator $\mathcal{A}$ relates the spherical harmonics coefficients to fODF amplitudes.

2.2 Model Simplification via Spherical Mean

The spherical mean technique (SMT) [10] focuses on the direction-averaged signal to factor out the effects of the fiber orientation distribution. Taking the spherical mean, (3) can be written as

$$\bar{S}(\tau, b) = e^{-\tau r_r(b)} \int_{D_r} k(b, D_r) w(D_r) dD_r + e^{-\tau r_h(b)} \int_{D_h} k(b, D_h) w(D_h) dD_h$$
$$+ e^{-\tau r_f(b)} \int_{D_h} k(b, D_f) w(D_f) dD_f$$
$$\text{s.t. } r_r(b),\ r_h(b),\ r_f(b) \geq 0\ \forall b,\ w(D_r),\ w(D_h),\ w(D_f) \geq 0 \tag{4}$$

where $w(D_r)$, $w(D_h)$, and $w(D_f)$ are volume fractions and $k(b, D_r)$, $k(b, D_h)$, and $k(b, D_f)$ are spherical means of response functions $R(b, \mathbf{g}, D_r)$, $R(b, \mathbf{g}, D_h)$, and $R(b, D_f)$, respectively. Based on [8, 10], spherical means can be written as:

$$k(b, D_r) \equiv k(b, \{\lambda_\parallel, \lambda_\perp\} \in \Lambda_r) = e^{-b\lambda_\perp} \frac{\sqrt{\pi}\,\mathrm{erf}(\sqrt{b(\lambda_\parallel - \lambda_\perp)})}{2\sqrt{b(\lambda_\parallel - \lambda_\perp)}}, \quad \frac{\lambda_\parallel}{\lambda_\perp} \succeq \phi^2,$$
$$k(b, D_h) \equiv k(b, \{\lambda_\parallel, \lambda_\perp\} \in \Lambda_h) = e^{-b\lambda_\perp} \frac{\sqrt{\pi}\,\mathrm{erf}(\sqrt{b(\lambda_\parallel - \lambda_\perp)})}{2\sqrt{b(\lambda_\parallel - \lambda_\perp)}}, \quad 1 \prec \frac{\lambda_\parallel}{\lambda_\perp} \prec \phi^2,$$
$$k(b, D_f) \equiv k(b, \{\lambda_\parallel, \lambda_\perp\} \in \Lambda_f) = e^{-b\lambda_\perp}, \quad \lambda_\parallel = \lambda_\perp, \tag{5}$$

where D_r, D_h, and D_f are parameterized by parallel diffusivity $\lambda_\parallel$ and perpendicular diffusivity $\lambda_\perp$ for the restricted (Λ_r), hindered (Λ_h) and free (Λ_f) compartments. ϕ is the geometric tortuosity [28]. The spherical mean signal can thus be seen as the weighted combination of the spherical mean signals of spin packets. Similar to [8, 28], (4) allows us to probe the relaxation-diffusion coupling across a spectrum of diffusion scales. Anisotropic diffusion can be further separated as restricted or hindered.

2.3 Estimation of Relaxation and Diffusion Parameters

We first solve for the relaxation modulated spherical mean coefficients in (4). Next, we disentangle the relaxation terms from the spherical mean coefficients and solve for the relaxation rates. Finally, we estimate the fODFs using (3). Details are provided below:

(i) Relaxation modulated spherical mean coefficients. We rewrite (4) in matrix form as

$$\bar{\mathbf{S}} = \mathbf{K}\mathbf{E} \circ \mathbf{W} = \mathbf{K}\mathbf{X}, \tag{6}$$

where the mean signal $\bar{\mathbf{S}}$ is expressed as the product of the response function spherical mean matrix $\mathbf{K}$ and the Kronecker product ($\circ$) of relaxation matrix $\mathbf{E}$ and volume fraction matrix $\mathbf{W}$. We can solve for $\mathbf{X}$ in (6) via an augmented problem with the OSQP solver[1]:

$$\arg\min_{\mathbf{X}\succeq 0} \frac{1}{2} \left\|\mathbf{K}\mathbf{X} - \bar{\mathbf{S}}\right\|_2^2. \tag{7}$$

(ii) Relaxation times. With $\mathbf{X}$ solved, $\mathbf{E}$ and $\mathbf{W}$ can be determined by minimizing a constrained non-linear multivariate problem:

$$\min_{\mathbf{E},\mathbf{W}} \frac{1}{2} \left\|\mathbf{E} \circ \mathbf{W} - \bar{\mathbf{X}}\right\|_2^2 \text{ s.t. } \mathbf{E} \geq 0,\ \mathbf{W} \geq 0,\ \sum \mathbf{W} = 1, \tag{8}$$

which can be solved using a gradient based optimizer. Relaxation times can be determined based on $\mathbf{E}$.

(iii) fODFs. With $\mathbf{E}$ determined, (3) can be rewritten as a strictly convex quadratic programming (QP) problem:

$$\hat{\mathbf{f}} = \arg\min_{\mathbf{f}} \frac{1}{2}\mathbf{f}^\top \mathbf{P}\mathbf{f} + \mathbf{Q}^\top \mathbf{f}, \text{ s.t. } \mathbf{A}\mathbf{f}(D_r),\ \mathbf{A}\mathbf{f}(D_h),\ \mathbf{A}\mathbf{f}(D_f) \geq 0,\ \ \forall D_r, \forall D_h, \forall D_f, \tag{9}$$

which can be solved using the OSQP solver.

2.4 Microstructure Indices

Based on (3) and (4), various microstructure indices can be derived:

- Microscopic fractional anisotropy [20], per-axon axial and radial diffusivity [2], and free and restricted isotropic diffusivity.
- Axonal morphology indices derived based on [17,26] to compute the mean neurite radius (Mean NR), its internal deviation (Std. NR), and relative neurite radius (Cov. NR):
  - Mean NR $=$ mean$(\epsilon(\delta, \Delta) w_{D_r} \lambda_{\|} \lambda_{\perp})^{1/4}$, $\{\lambda_{\|}, \lambda_{\perp}\} \in D_r$, where $\epsilon \succ 0$ is a pulse scale that only depends on the pulse width δ and diffusion time Δ of the diffusion gradients.
  - Std. NR $=$ std$((\epsilon(\delta, \Delta) w_{D_r} \lambda_{\|} \lambda_{\perp})^{1/4})$.
  - Cov. NR $=$ cov$((\epsilon(\delta, \Delta) w_{D_r} \lambda_{\|} \lambda_{\perp})^{1/4}) \equiv$ cov$((w_{D_r} \lambda_{\|} \lambda_{\perp})^{1/4})$, which is independent on ϵ.

[1] https://osqp.org/.

2.5 Data Acquisition and Processing

Ex Vivo Data. We used an ex vivo monkey dMRI dataset[2] collected with a 7T MRI scanner [19]. A single-line readout PGSE sequence with pulse duration $\delta = 9.6$ ms and separation $\Delta = 17.5$ ms, echo times TE $= \{35.5, 45.5\}$ ms, TR $= 3500$ ms, and 0.5 mm isotropic resolution was utilized for acquisition across five shells, $b = \{4, 7, 23, 27, 31\} \times 10^3$ s/mm^2, each with a common set of 96 non-collinear gradient directions. A $b = 0$ s/mm^2 image was also acquired.

In Vivo Data. One healthy subject and three patients with gliomas were scanned using a Philips Ingenia CX 3T MRI scanner with a gradient strength of 80 mT/m and switching rates of 200 mT/m/ms. Diffusion data with seven TEs were obtained using a spin-echo echo-planar imaging sequence with fixed TR and diffusion time, $\{4, 4, 8, 8, 16\}$ diffusion-encoding directions at $b = \{0, 4, 8, 16, 32\} \times 10^2$ s/mm^2 respectively, TE $= \{75, 85, 95, 105, 115, 125, 135\}$ ms, TR $= 4000$ ms, 1.5 mm isotropic voxel size, image size = 160 $\times$ 160, 96 slices, whole-brain coverage, and acceleration factor = 3. The total imaging time was 21 min. Data processing includes (i) noise level estimation and removal; (ii) Rician unbiasing; (iii) removal of Gibbs ringing artifacts; and (iv) motion and geometric distortion corrections. To compensate for motion, all dMRIs were first preprocessed separately and then aligned using rigid registration based on the non-diffusion-weighted images. The lowest TE was set to minimize the contribution of the myelin water to the measured signal and the largest TE was chosen as a trade-off between image contrast and noise. Following previous studies [6, 16] and in-house testing, we used a spectrum of TE scales from 75 to 135 ms to cover tissue heterogeneity. We used MRtrix[3] to generate tissue segmentations (cortical and subcortical GM, WM, CSF, and pathological tissue) based on the T1w data.

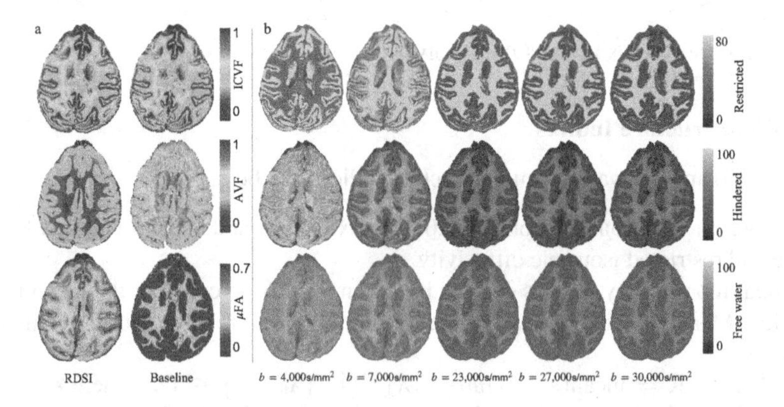

Fig. 1. Ex vivo data. (a) RDSI and REDIM parameter maps for tissue microstructure; (b) RDSI relaxation times for diffusion compartments.

[2] https://resources.drcmr.dk/MAPdata/axon-relaxation/.

[3] https://www.mrtrix.org/.

Implementation. To cover the whole diffusion spectrum, we set the diffusivity from 0 s/mm^2 (no diffusion) to 3×10^{-3} s/mm^2 (free diffusion). For the anisotropic compartment, $\lambda_{\parallel}$ was set from $1.5 \times 10^{-3}\,\text{mm}^2/\text{s}$ to $2 \times 10^{-3}\,\text{mm}^2/\text{s}$. Radial diffusivity $\lambda_{\perp}$ was set to satisfy $\lambda_{\parallel}/\lambda_{\perp} \geq 1.1$ as in [8,28]. For the isotropic compartment, we set the diffusivity $\lambda_{\parallel} = \lambda_{\perp}$ from $0\,\text{mm}^2/\text{s}$ to $3\times10^{-3}\,\text{mm}^2/\text{s}$ with step size $0.1\times10^{-3}\,\text{mm}^2/\text{s}$.

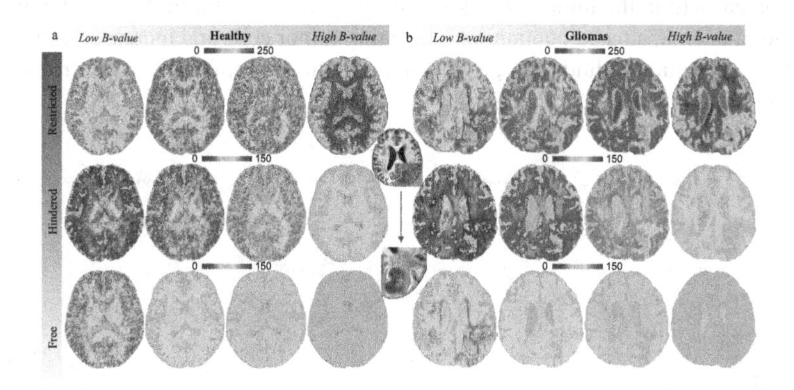

Fig. 2. In vivo data. Compartment-specific parameters for (a) healthy and (b) glioma subjects.

3 Results

3.1 Ex Vivo Data: Compartment-Specific Parameters

Figure 1(a) shows the estimated maps of T_2-independent parameters given by a baseline comparison method, called REDIM [16], and RDSI. We observe that the two methods yield similar intracellular volume fraction (ICVF) estimates. However, REDIM overestimates the anisotropic volume fraction (AVF) compared to RDSI, resulting in blurred boundaries between the gray matter and superficial white matter. RDSI yields consistent distribution between ICVF and μFA maps.

Figure 1(b) shows the RDSI T_2 relaxation maps of restricted, hindered, and free diffusion across b-values. As the b-value increases, the relaxation time increases for the restricted component but decreases for the hindered and free components. At lower b-values, the relaxation time for the extra-neurite compartment is substantially higher than that of the intra-neurite compartment.

3.2 In Vivo Data: Compartment-Specific Parameters

Figure 2 shows the RDSI T_2 relaxation maps of restricted, hindered, and free diffusion across b-values. The values are consistent between healthy and glioma subjects. The estimated relaxation times are in general in line with previous reports [6,11]. RDSI shows substantial differences between tumor and normal tissues in the relaxation maps (Fig. 2(b)).

Figure 3 shows the voxel distributions with respect to relaxation times and b-values. It is apparent that at higher b-values, a greater fraction of voxels in the restricted compartment have relaxation times within 100 to 200 ms, particularly for higher-grade gliomas. This might be related to prolonged transverse relaxation time due to increased water content within the tumor [5,7,24]. This property is useful in the visualization of peritumoral edema, an area containing infiltrating tumor cells and increased extracellular water due to plasma fluid leakage from aberrant tumor capillaries that surrounds the tumor core in higher-grade gliomas.

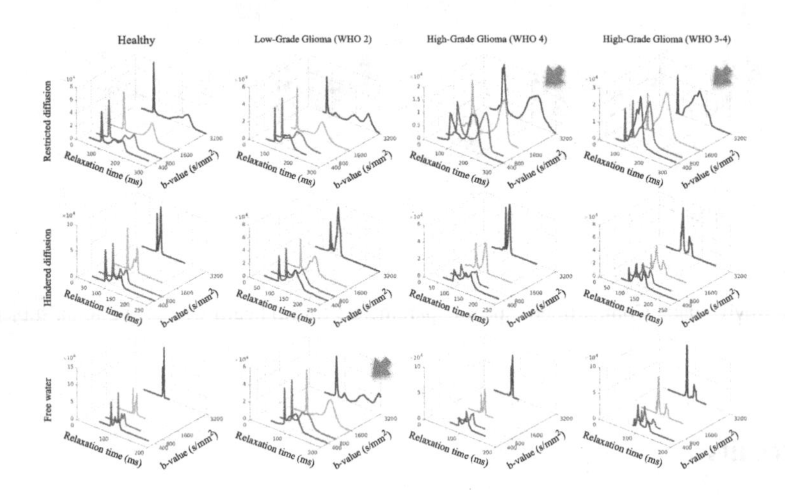

Fig. 3. Voxel distributions with respect to relaxation times and b-values.

3.3 In Vivo Data: Neurite Morphology

Figure 4(a) shows the relaxation times of the restricted compartment in white matter lesions, indicating that relaxation times are longer in gliomas than normal white matter tissue. The higher T_2 in grade 4 glioma is associated with changes in metabolite compositions, resulting in remarkable changes in neurite morphology in lesioned tissues (Fig. 4(c–d)), consistent with previous observations [12,23]. The rate of longitudinal relaxation time has been shown to be positively correlated with myelin content. Our results indicate that MTE dMRI is more sensitive to neurite morphology than STE dMRI (Fig. 4(b)).

Figures 4(c–d) show that the estimated Mean NR in the gray matter is approximately in the range of 10 μm, which is in good agreement with the sizes of somas in human brains, i.e., 11 ± 7 μm [26]. RDSI improves the detection of small metastases, delineation of tumor extent, and characterization of the intratumoral microenvironment when compared to conventional microstructure models (Fig. 4(c)). Our studies suggest that RDSI provides useful information on microvascularity and necrosis helpful for facilitating early stratification of patients with gliomas (Fig. 4(d)).

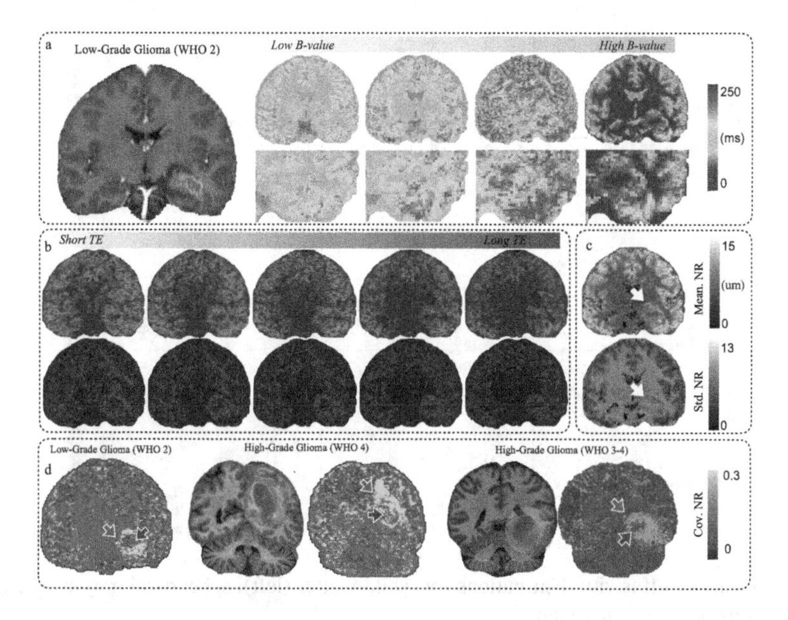

Fig. 4. Relaxation time of the restricted compartment of (a) a glioma patient, and (b) TE-dependent and (c–d) T_2-independent neurite morphology.

3.4 Relation Between Relaxation and Diffusivity

Figure 5 shows the relaxation-diffusivity distributions of white matter (WM), cortical gray matter (GM), and subcortical gray matter (SGM). The 2D plots show the contours of the joint distributions of the relaxation and diffusivity values across all voxels. The average diffusivity and relaxation in these regions indicate the existence of a single homogeneous region in WM and SGM. For GM, however, we observe a small peak for the relaxation rate arange 1e-3 to 1.5e-3.

3.5 fODFs

Figure 6 shows that the reconstructed fODFs are consistent with the expected WM arrangement of the healthy human brain. We provide a visual comparison of the fODFs estimated with and without the explicit consideration of relaxation. The two cases yield different fODFs. As expected, fiber populations are associated with different relaxation times, in line with [3, 16]. Our studies suggest that this difference could be caused by the spatially heterogeneous tissue microstructure, since fiber bundles with slower relaxation times contribute less to diffusion signals acquired with a longer TE. Explicitly taking into account relaxation in our model results in noteworthy contrast improvement in spatially heterogeneous superficial WM.

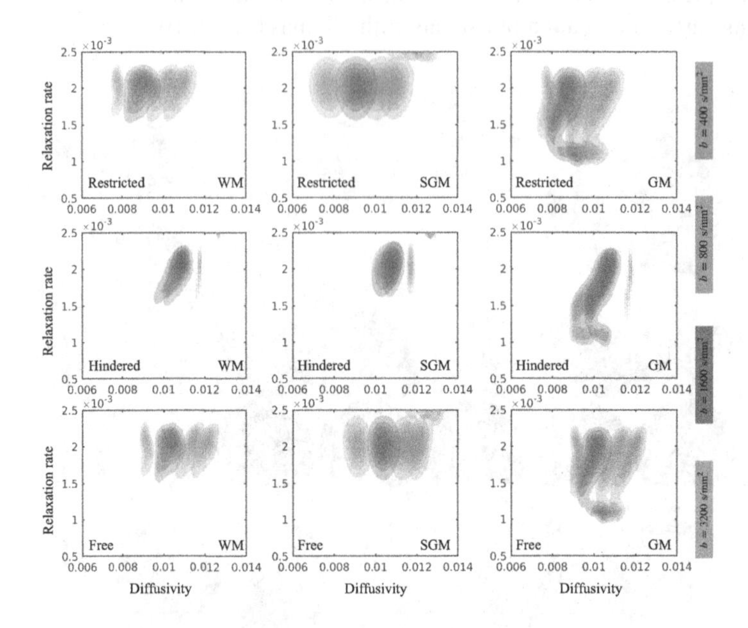

Fig. 5. Relaxation-diffusivity distributions for white matter (left), subcortical gray matter (middle), and cortical gray matter (right).

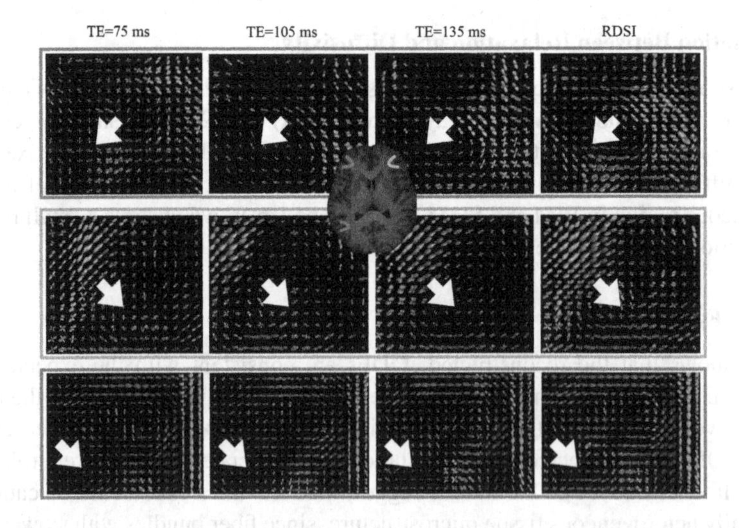

Fig. 6. TE-dependent and TE-independent fODFs in three superficial white matter regions.

4 Conclusion

RDSI provides a unified strategy for direct estimation of relaxation-independent volume fractions and compartment-specific relaxation times. Using MTE data, we demonstrated that RDSI can delineate heterogeneous tissue microstructure elusive to STE data. We also showed that RDSI provides information that is conducive to characterizing tissue abnormalities.

References

1. Anania, V., et al.: Improved diffusion parameter estimation by incorporating T2 relaxation properties into the DKI-FWE model. Neuroimage **256**, 119219 (2022)
2. Assaf, Y., Freidlin, R.Z., Rohde, G.K., Basser, P.J.: New modeling and experimental framework to characterize hindered and restricted water diffusion in brain white matter. Magn. Reson. Med. **52**(5), 965–978 (2004)
3. Barakovic, M., et al.: Bundle-specific axon diameter index as a new contrast to differentiate white matter tracts. Front. Neurosci. **15**, 646034 (2021)
4. Cowan, B., Cowan, B.P.: Nuclear Magnetic Resonance and Relaxation, vol. 427. Cambridge University Press, Cambridge (1997)
5. Ellingson, B.M., Wen, P.Y., Cloughesy, T.F.: Modified criteria for radiographic response assessment in glioblastoma clinical trials. Neurotherapeutics **14**(2), 307–320 (2017)
6. Gong, T., Tong, Q., He, H., Sun, Y., Zhong, J., Zhang, H.: MTE-NODDI: multi-TE NODDI for disentangling non-T2-weighted signal fractions from compartment-specific T2 relaxation times. Neuroimage **217**, 116906 (2020)
7. Hu, L.S., Hawkins-Daarud, A., Wang, L., Li, J., Swanson, K.R.: Imaging of intratumoral heterogeneity in high-grade glioma. Cancer Lett. **477**, 97–106 (2020)
8. Huynh, K.M., et al.: Probing tissue microarchitecture of the baby brain via spherical mean spectrum imaging. IEEE Trans. Med. Imaging **39**(11), 3607–3618 (2020)
9. Jeurissen, B., Tournier, J.D., Dhollander, T., Connelly, A., Sijbers, J.: Multi-tissue constrained spherical deconvolution for improved analysis of multi-shell diffusion MRI data. Neuroimage **103**, 411–426 (2014)
10. Kaden, E., Kelm, N.D., Carson, R.P., Does, M.D., Alexander, D.C.: Multi-compartment microscopic diffusion imaging. Neuroimage **139**, 346–359 (2016)
11. Lampinen, B., et al.: Towards unconstrained compartment modeling in white matter using diffusion-relaxation MRI with tensor-valued diffusion encoding. Magn. Reson. Med. **84**(3), 1605–1623 (2020)
12. Li, Y., Srinivasan, R., Ratiney, H., Lu, Y., Chang, S.M., Nelson, S.J.: Comparison of T1 and T2 metabolite relaxation times in glioma and normal brain at 3T. J. Magn. Reson. Imaging **28**(2), 342–350 (2008). https://onlinelibrary.wiley.com/doi/pdf/10.1002/jmri.21453
13. McDonald, E.S., et al.: Mean apparent diffusion coefficient is a sufficient conventional diffusion-weighted MRI metric to improve breast MRI diagnostic performance: results from the ECOG-ACRIN cancer research group A6702 diffusion imaging trial. Radiology **298**(1), 60 (2021)
14. McKinnon, E.T., Jensen, J.H.: Measuring intra-axonal T_2 in white matter with direction-averaged diffusion MRI. Magn. Reson. Med. **81**(5), 2985–2994 (2019)
15. Morozov, S., et al.: Diffusion processes modeling in magnetic resonance imaging. Insights Imaging **11**(1), 60 (2020)

16. Ning, L., Gagoski, B., Szczepankiewicz, F., Westin, C.F., Rathi, Y.: Joint relaxation-diffusion imaging moments to probe neurite microstructure. IEEE Trans. Med. Imaging **39**(3), 668–677 (2020)
17. Ning, L., Westin, C.F., Rathi, Y.: Characterization of b-value dependent T2 relaxation rates for probing neurite microstructure. bioRxiv. Cold Spring Harbor Laboratory (2022)
18. Palombo, M., et al.: SANDI: a compartment-based model for non-invasive apparent soma and neurite imaging by diffusion MRI. Neuroimage **215**, 116835 (2020)
19. Pizzolato, M., Andersson, M., Canales-Rodríguez, E.J., Thiran, J.P., Dyrby, T.B.: Axonal T2 estimation using the spherical variance of the strongly diffusion-weighted MRI signal. Magn. Reson. Imaging **86**, 118–134 (2022)
20. Reymbaut, A.: Diffusion anisotropy and tensor-valued encoding. In: Topgaard, D. (ed.) New Developments in NMR, pp. 68–102. Royal Society of Chemistry, Cambridge (2020)
21. Slator, P.J., et al.: Combined diffusion-relaxometry microstructure imaging: Current status and future prospects. Magn. Reson. Med. **86**(6), 2987–3011 (2021)
22. Sotardi, S., et al.: Voxelwise and regional brain apparent diffusion coefficient changes on MRI from birth to 6 years of age. Radiology **298**(2), 415 (2021)
23. Tavakoli, M.B., Khorasani, A., Jalilian, M.: Improvement grading brain glioma using T2 relaxation times and susceptibility-weighted images in MRI. Inform. Med. Unlocked **37**, 101201 (2023)
24. Upadhyay, N., Waldman, A.: Conventional MRI evaluation of gliomas. Br. J. Radiol. **84**(2), S107–S111 (2011)
25. Veraart, J., Novikov, D.S., Fieremans, E.: TE dependent Diffusion Imaging (TEdDI) distinguishes between compartmental T2 relaxation times. Neuroimage **182**, 360–369 (2018)
26. Veraart, J., et al.: Noninvasive quantification of axon radii using diffusion MRI. eLife **9**, e49855 (2020)
27. Weisskoff, R., Zuo, C.S., Boxerman, J.L., Rosen, B.R.: Microscopic susceptibility variation and transverse relaxation: theory and experiment. Magn. Reson. Med. **31**(6), 601–610 (1994)
28. White, N.S., Leergaard, T.B., D'Arceuil, H., Bjaalie, J.G., Dale, A.M.: Probing tissue microstructure with restriction spectrum imaging: histological and theoretical validation. Hum. Brain Mapp. **34**(2), 327–346 (2013)
29. Zhang, H., Schneider, T., Wheeler-Kingshott, C.A., Alexander, D.C.: NODDI: practical in vivo neurite orientation dispersion and density imaging of the human brain. Neuroimage **61**(4), 1000–1016 (2012)

Joint Representation of Functional and Structural Profiles for Identifying Common and Consistent 3-Hinge Gyral Folding Landmark

Shu Zhang(✉), Ruoyang Wang, Yanqing Kang, Sigang Yu, Huawen Hu, and Haiyang Zhang

Center for Brain and Brain-Inspired Computing Research, School of Computer Science, Northwestern Polytechnical University, Xi'an, China
Shu.zhang@nwpu.edu.cn

Abstract. The 3-hinge is a form of cortical fold, which is the intersection of the three gyri. And it has been proved to be unique anatomically, structurally, and functionally connective patterns. Compared with the normal gyri, the 3-hinge gyri have stronger structural connectivity and participate in more functional networks. Therefore, it is of great significance to further explore the 3-hinge regions, which could give more new insights on the study of mechanism of cortical folding patterns. However, for the large differences in brain across subjects, it is difficult to identify consistent 3-hinge regions across subjects and most previous studies on 3-hinges consistency merely focused on a single mode. In order to study the multi-modal consistency of 3-hinge regions, this paper proposes a joint representation of functional and structural profiles for identifying common and consistent 3-hinges. We use the representation of 3-hinge patterns in the functional network to obtain the functional consistency of 3-hinges cross subjects, then the distance between 3-hinge regions and Dense Individualized and Common Connectivity-Based Cortical Landmarks (DICCCOL) system to obtain the structural consistency. Combining these two sets of stability, 38 functionally and structurally consistent 3-hinge regions were successfully identified cross subjects. These consistent 3-hinge regions based on multi-modal data are more consistent than that merely based on structural data and experimental results elucidate those consistent 3-hinge regions are more correlated with visual function. This work deepens the understanding of the stability of 3-hinge regions and provides a basis for further inter-group analysis of 3-hinge gyral folding.

Keywords: 3-hinge Gyral Folding · Multimodal Data · Brain Landmark

1 Introduction

Cortical folds have been shown to be related to brain function, cognition, and behavior. Based on the research of the past decades, the cortex can be further decomposed into fine-grained basic morphological patterns, such as gyri and sulci. Gyri are more potential

H. Greenspan et al. (Eds.): MICCAI 2023, LNCS 14227, pp. 163–172, 2023.
https://doi.org/10.1007/978-3-031-43993-3_16

to be functional connection centers, which are responsible for exchanging information among remote gyri and nearby sulci; on the contrary, sulci exchange information directly with their nearby gyri [1].

Recent studies have shown that gyri can be further separated by the number of hinges it comprises, both for anatomical analysis and for functional timing analysis. Thus, the 3-hinge cyclotron fold pattern was gradually introduced and determined. It has been demonstrated that the 3-hinge regions have a thicker cortex [2], a stronger pattern of fiber connections [3], and a greater diversity of structural connections [4]. These studies revealed the salient features and potential value of the 3-hinge region. In a recent study, it was claimed that 3-hinges have been found as "connector" hubs in the brain [5].

Although these works have achieved great success, there remain several obstacles to studying and comprehending the role of 3-hinge gyral folding patterns. One of the most significant challenges is identifying common and consistent 3-hinge gyrus folding patterns across subjects, which has yet to be resolved and has impeded group-level 3-hinges analysis. The morphology of cortical folds varies greatly between individuals, so identifying stable 3-hinge regions between individuals is difficult. In 2017 Li *et al.* addressed this issue by manually labeling 3-hinges across subjects and species (macaques, chimpanzees, and humans), demonstrating that six 3-hinges have functional correspondence across subjects and even across species [4]. In 2020, Zhang *et al.* proposed a semi-automatic approach that combines the fold morphology of the cerebral cortex and the characteristics of white matter nerve fibers to estimate the correspondence of the 3-hinge regions across subjects [6]. In a recent study, by transferring Dense Individualized and Common Connectivity-Based Cortical Landmarks (DICCCOL) stability across subjects to the 3-hinge regions, a DICCCOL-based K-nearest landmark detection method was proposed, which automatically identified 79 consistent 3-hinge regions [7]. These studies have shown the possibility of automatically identifying consistent 3-hinge regions via data-driven approach. However, those identified consistent 3-hinge regions were still studied by single modality.

To identify the three-hinge regions with multi-modal stability via data-driven approach, we present a joint representation of functional and structural profiles for identifying consistent 3-hinges in this paper. We use functional network representation and fiber connectivity pattern of the DICCCOL system to obtain functional and structural consistency, respectively. And combine these two consistencies to identify 38 3-hinge regions that are consistent in both function and structure. We further compare these results with those based solely on structural data, which deepens our understanding of the 3-hinge region. Our work provides a basis for further inter-group analysis of the 3-hinge gyrus.

2 Method

2.1 Dataset and Preprocessing

We used the Q1 release of Human Connectome Project (HCP) [8] consortium and randomly selected 50 human brains from it in this study. The acquisition parameters of functional magnetic resonance imaging (fMRI) data are as follows: 90×104 matrix, 220 mm FOV, 72 slices, TR = 0.72 s, TE = 33.1 ms, flip angle = 52°, BW = 2290 Hz/Px, in-plane FOV = 208×180 mm, 2.0 mm isotropic voxels. For fMRI images, the preprocessing pipelines included skull removal, motion correction, slice time correction, spatial smoothing, global drift removal. All of these steps are implemented by FMRIB Software Library (FSL) FEAT [9]. We use resting state fMRI and task fMRI data. Among them, task fMRI data contains a total of seven tasks, which are EMOTION, GAMBLING, LANGUAGE, MOTOR, RELATION, SOCIAL and WM.

For diffusion weighted imaging (DWI) data, the parameters are as follows: Spin-echo EPI, TR = 5520 ms, TE = 89.5 ms, flip angle = 78°, refocusing flip angle = 160°, FOV 210 × 180 (RO × PE), matrix 168 × 144 (RO × PE), slice thickness 1.25 mm, 111 slices, 1.25 mm isotropic voxels, Multiband factor = 3, and Echo spacing = 0.78 ms. Fiber tracking and cortical surface can be reconstructed from DWI dataset. Please refer to [10, 11] for more pre-processing details. One subject is randomly selected as template and 10 subjects are randomly selected as referential subjects. The remaining subjects are used as predictive subjects to predict consistent 3-hinges.

2.2 Joint Representation of Functional and Structural Profiles

We introduce a joint representation of functional and structural profiles for identifying common and consistent 3-hinges. The method contains four steps, as shown in Fig. 1.

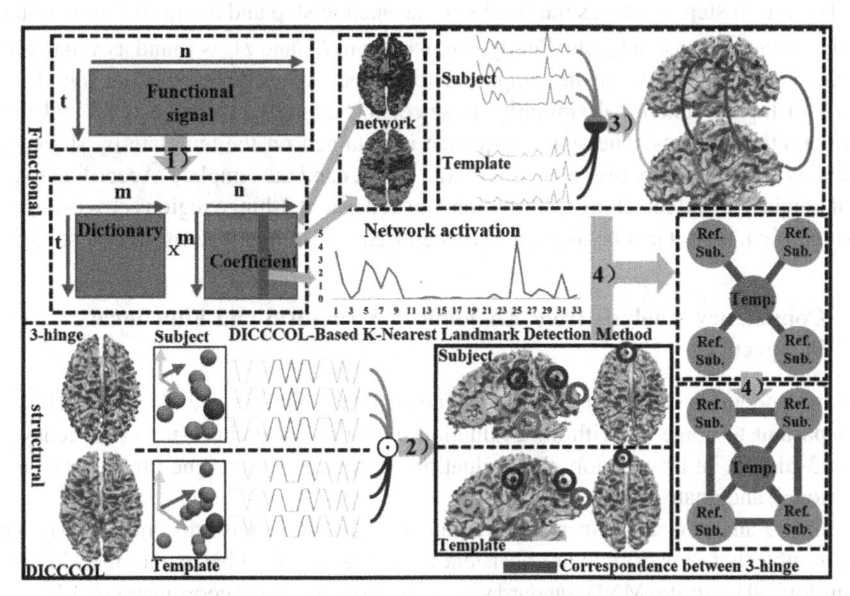

Fig. 1. Joint representation of functional and structural profiles. 1) Dictionary learning and sparse representation algorithm. 2) DICCCOL-based K-nearest landmark detection method. 3) Obtain functional consistency by functional network. 4) Combine the functional and structural consistency.

First step is to obtain the functional network. The 4-dimensional data of the resting state and the fMRI of 7 tasks are stretched into 2-dimensional data, which are input into dictionary learning and sparse representation algorithm [12]. 400 functional networks were obtained on the data of each modality. Based on prior knowledge, we select 10 functional networks from the resting state data and 23 functional networks from the task state data for this work. In the second step, we randomly select a subject as a template and use it as a bridge to study the consistency of 3-hinges across subjects. The second step is to obtain the corresponding of 3-hinges between the template and subjects using structural data. We improved the DICCCOL-based K-nearest landmark detection method [7] and retained the 3-hinges with the same spatial position and the same nearest DICCCOLs as the candidate matching set D_s. The third step is to obtain the corresponding of 3-hinges between the template and subjects using functional network. We register the functional network to all subjects, denote activation intensity vector of functional network with 3-hinge i on the template as A_i^t, and on other subjects as A_i^s. The candidate set D_j^f on subjects match 3-hinge j on the template based on functional network is given by formula 1:

$$D_j^f = |i|P\left(A_j^t, A_i^s\right) > maxP\left(A_j^t, A_i^s\right) - \text{th} \,| \tag{1}$$

where $P\left(A_j^t, A_i^s\right)$ represents the Pearson correlation coefficient of A_j^t and A_i^s, and the threshold is set to 0.1.

The fourth step combines the results of the second step and the third step to obtain stable 3-hinges across subjects. Firstly, the overlap in D_s and D_j^f is found as a matching result of 3-hinges between the template and subject. If multiple ones are found, the one with higher functional similarity is selected. Secondly, 10 subjects are selected as referential subjects. The stable sequence is obtained on the template by statistical matching information between the referential subject and the template. According to the stable sequence and matching information, the consistent 3-hinge regions cross subjects are identified. And it can be used to predict consistent 3-hinge regions on new subjects.

2.3 Consistency Analysis from Anatomical, Structural and Functional Perspective

From the previous section, a group of the consistent 3-hinges were identified. Then, it is important to come up with the evaluation operations and check the consistency of these 3-hinges. In this section, the evaluation operations are designed from structural, functional, and anatomical perspectives.

For the anatomical perspective, we calculate the voxel-wise distance of 3-hinges across subjects to measure their consistency. We register all subjects into the Montreal Neurological Institute (MNI) standard space. After obtaining the coordinates of 3-hinges, we calculated the voxel-level distance of 3-hinges between different subjects.

For the structural perspective, we use the similarity of fiber connection pattern passing through 3-hinges to evaluate the consistency of them. In detail, we count the nerve fibers passing through each 3-hinge region, and then used the trace-map method [13] to convert the nerve fiber bundles into vectors that could be quantified, and evaluated the consistency of the 3-hinges among groups by the Pearson correlation coefficient between these vectors. The trace-map is a computational model that transforms the directional information in the trajectory of an arbitrary bundle to a standard spherical surface, for quantitative comparison of structural connectivity patterns.

For the functional perspective, we count the activation intensity of the 3-hinge regions among different functional networks as vector and used the Pearson correlation coefficient of these vectors across different subjects to quantify the functional consistency of 3-hinge regions.

2.4 Comparative Analysis of Consistent 3-hinges for Structural Data and Multimodal Data

Both joint representation of functional and structural profiles and DICCCOL-based K-nearest landmark detection method are methods that identify consistent 3-hinges by using templates as bridges. We compare the results of the two methods and study the distribution and corresponding functions of the two groups of 3-hinges on the cortical surface. We register consistent 3-hinges on all subjects to the MNI standard space and calculate the distribution of consistent 3-hinges on the Automated anatomical labelling (AAL) template [14]. Since the 90 brain regions of the AAL template are symmetric between the left and right brain, we combine the left and right brain regions. For statistics, we count the distribution of consistent 3-hinges over 45 AAL brain regions on each subject and average across all subjects. We then investigated the functions with consistent 3-hinges distributions according to the functions corresponding to AAL regions.

3 Result

3.1 Visualization of the Identified Consistent 3-hinges

The consistent 3-hinges of subjects are shown in Fig. 2.

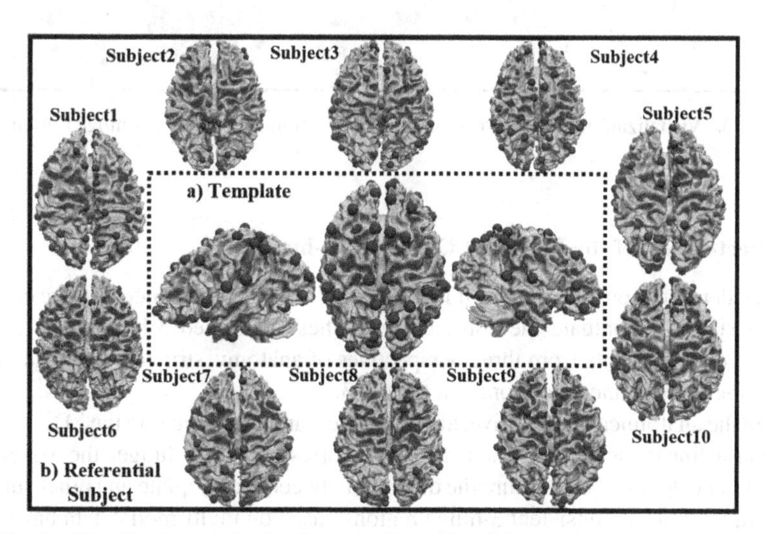

Fig. 2. Visualization of the distribution of identified 3-hinges in the cerebral cortex

Through adopting a joint representation of functional and structural profiles, stable sequences of length 65 are determined on the template and average 38 consistent 3-hinges can be successfully identified on subjects. Figure 2 shows the consistent 3-hinges of 10 referential subjects and the template subject. And we can see that those identified common and consistent 3-hinges are indeed consistent across the subjects.

In order to show the consistency of the identified 3-hinges among different subjects subjectively, in Fig. 3, we select 11 subjects (including a template, five referential subjects, and five predictive subjects) and mark the identified 3-hinges with the same color. As shown in Fig. 3, most of the 3-hinges can be found in the corresponding 3-hinges regions between different subjects, and these consistent 3-hinges regions are relatively close. We will quantitatively analyze the consistency of the 3-hinges identified in the following section.

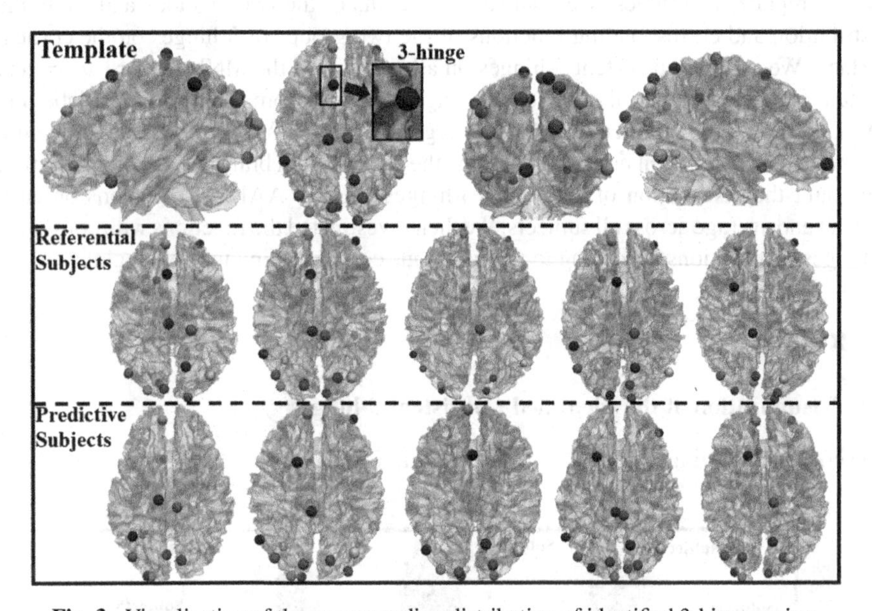

Fig. 3. Visualization of the corresponding distribution of identified 3-hinge region.

3.2 Effectiveness of the Proposed Consistent 3-hinges

After the identification of 3-hinges, it is important to evaluate whether they are common and consistent. To evaluate the consistency of these identified 3-hinges, we conduct quantitative experiments from three perspectives of anatomy, structure, and function to evaluate the performance mentioned in Sect. 2.3.

From the anatomical perspective, all the subjects are registered into the MNI standard space via a linear algorithm. Then, for each corresponding 3-hinge, the voxel-level distances are calculated to measure the distance between the template and other subjects. As shown in Table 1, consistent 3-hinge regions based on multi-modal data have better inter-population stability than three-hinge regions based on structural data.

Table 1. Voxel-level distances between the 3-hinges.

Consistent 3-hinge	Based on structural data	Based on multi-modal data
Referential subjects	6.03	5.31
Predictive subjects	6.10	5.31

From the structural perspective, the average Pearson correlation coefficient of fiber connection patterns of identified 3-hinges is as high as 0.44. As a comparison, the similarity is 0.37 for consistent 3-hinges based on structural data. This shows that the method of adding functional data to identify the consistent 3-hinges also has a great improvement in the structural stability of the identified 3-hinges.

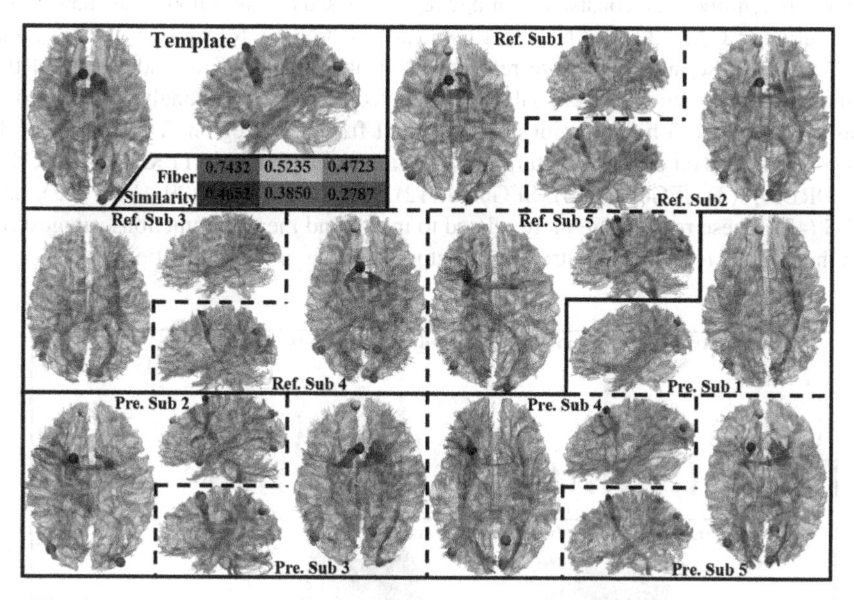

Fig. 4. Visualization of the corresponding fiber bundles of identified 3-hinge region.

In order to better demonstrate the structural consistency of these stable 3-hinges, in Fig. 4, we visualized six consistent 3-hinges on 11 selected subjects and the neural fiber modes across these 3-hinges. The stable 3-hinges and the corresponding nerve fibers are marked with the same color, and the similarity of the nerve fibers across the 3-hinge regions between the template and subjects is also marked with the same color. As shown in Fig. 4, the corresponding similarity of the 3-hinges in dark purple is close to the average similarity, and the corresponding nerve fibers are also similar between populations. This verifies the structural consistency of the identified 3-hinges.

From the functional perspective, the average similarity of activation intensity of consistent 3-hinges in a functional network based on multi-modal data between template and subject is 0.441, while the similarity based on structural data is 0.366. This indicates that the addition of functional data analysis does greatly improve the functional consistency of the identified 3-hinges.

In general, in this section, we verify that the consistency of the identified 3-hinges is significantly stronger than that of the methods based on structural data.

3.3 Comparative Analysis on the Consistent 3-hinges Based on Structural Data and Multimodal Data

As mentioned in Sect. 2.4, we analyze the distribution and corresponding functions of structural-based and functional-based consistent 3-hinge regions in the cerebral cortex. In Fig. 5, the distribution of these three hinges on the brain surface is measured by the AAL template. Next, we count the function of the brain regions with more 3-hinge regions in the two groups. In this step, the regions with fewer 3-hinge regions in both groups are ignored. The consistent 3-hinge regions based on multimodal data have more distribution in CAL (22), CUN (23), LING (24), SOG (25), MOG (26), IPL (31) and ANG (33). These areas are more relevant to visual function. This indicates that the introduction of functional data analysis has played a great role in analyzing the group consistency of the 3-hinge regions in the visual functional region. The consistent 3-hinge regions based on the structural data are more distributed in PreCG (1), SFGdor (2), ORBsup (3), IFGoperc (6), SFGmed (12), HIP (19), PoCG (29), PCL (35), and MTG (43). These regions are more related to motor and memory functions, suggesting that the 3-hinge regions have stronger structural stability in these functions.

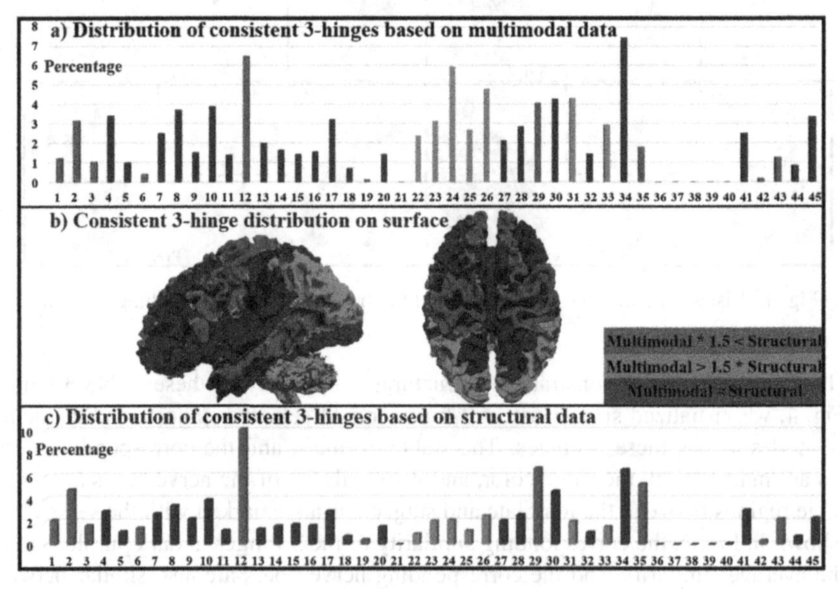

Fig. 5. Distribution of consistent 3-hinges in AAL brain regions based on structural and multimodal data

4 Conclusion

In this work, we propose a joint representation of functional and structural profiles in a data-driven approach for identifying consistent 3-hinges. We use functional network representation and fiber connectivity pattern of the DICCCOL system to obtain functional and structural consistency, respectively. And combine these two consistencies to identify 38 functionally and structurally consistent 3-hinge regions. Compared with the single-modal method with DICCCOL only, the results obtained by our proposed multimodal method have a more consistent 3-hinge pattern across subjects. And we further analyze that consistent 3-hinge regions based on multimodal data are closer to visual functions, while 3-hinge regions based on structural data are closer to memory and motor function.

Acknowledgements. This work was supported by the National Natural Science Foundation of China (62006194); High-level researcher start-up projects (06100-23SH0201228); Basic Research Projects of Characteristic Disciplines (G2023WD0146).

References

1. Deng, F., et al.: A functional model of cortical gyri and sulci. Brain Struct. Funct. **219**, 1473–1491 (2014)
2. Li, K., et al.: Gyral folding pattern analysis via surface profiling. Neuroimage **52**, 1202–1214 (2010)
3. Ge, F., et al.: Denser growing fiber connections induce 3-hinge gyral folding. Cereb. Cortex N. Y. N **1991**(28), 1064–1075 (2018)
4. Li, X., et al.: Commonly preserved and species-specific gyral folding patterns across primate brains. Brain Struct. Funct. **222**, 2127–2141 (2017)
5. Zhang, T., et al.: Cortical 3-hinges could serve as hubs in cortico-cortical connective network. Brain Imaging Behav. **14**, 2512–2529 (2020)
6. Zhang, T., et al.: Identifying Cross-individual Correspondences of 3-hinge Gyri. Med. Image Anal. **63**, 101700 (2020)
7. Zhang, S., et al.: A DICCCOL-based K-nearest landmark detection method for identifying common and consistent 3-hinge gyral folding landmarks. Chaos Solitons Fractals. **158**, 112018 (2022)
8. Van Essen, D.C., Smith, S.M., Barch, D.M., Behrens, T.E.J., Yacoub, E., Ugurbil, K.: WU-Minn HCP Consortium: the WU-Minn human connectome project: an overview. Neuroimage **80**, 62–79 (2013)
9. Woolrich, M.W., et al.: Bayesian analysis of neuroimaging data in FSL. Neuroimage **45**, S173-186 (2009)
10. Jiang, X., et al.: Modeling functional dynamics of cortical gyri and sulci. In: Ourselin, S., Joskowicz, L., Sabuncu, M.R., Unal, G., Wells, W. (eds.) MICCAI 2016. LNCS, vol. 9900, pp. 19–27. Springer, Cham (2016). https://doi.org/10.1007/978-3-319-46720-7_3
11. Liu, H., et al.: Elucidating functional differences between cortical gyri and sulci via sparse representation HCP grayordinate fMRI data. Brain Res. **1672**, 81–90 (2017)
12. Lv, J., et al.: Holistic atlases of functional networks and interactions reveal reciprocal organizational architecture of cortical function. IEEE Trans. Biomed. Eng. **62**, 1120–1131 (2015)

13. Zhu, D., et al.: DICCCOL: dense individualized and common connectivity-based cortical landmarks. Cereb. Cortex **23**, 786–800 (2013)
14. Rolls, E.T., Huang, C.-C., Lin, C.-P., Feng, J., Joliot, M.: Automated anatomical labelling atlas 3. Neuroimage **206**, 116189 (2020)

Single-subject Multi-contrast MRI Super-resolution via Implicit Neural Representations

Julian McGinnis[1,2,3](✉), Suprosanna Shit[1,4,5](✉), Hongwei Bran Li[5], Vasiliki Sideri-Lampretsa[1], Robert Graf[1,4], Maik Dannecker[1], Jiazhen Pan[1], Nil Stolt-Ansó[1], Mark Mühlau[2,3], Jan S. Kirschke[4], Daniel Rueckert[1,6], and Benedikt Wiestler[4]

[1] School of Computation, Information and Technology, TU Munich, Munich, Germany
{julian.mcginnis,suprosanna.shit}@tum.de
[2] TUM-Neuroimaging Center, TU Munich, Munich, Germany
[3] Department of Neurology, TU Munich, Munich, Germany
[4] Department of Neuroradiology, TU Munich, Munich, Germany
[5] Department of Quantitative Biomedicine, University of Zurich, Zurich, Switzerland
[6] Department of Computing, Imperial College London, London, UK

Abstract. Clinical routine and retrospective cohorts commonly include multi-parametric Magnetic Resonance Imaging; however, they are mostly acquired in different anisotropic 2D views due to signal-to-noise-ratio and scan-time constraints. Thus acquired views suffer from poor out-of-plane resolution and affect downstream volumetric image analysis that typically requires isotropic 3D scans. Combining different views of multi-contrast scans into high-resolution isotropic 3D scans is challenging due to the lack of a large training cohort, which calls for a subject-specific framework. This work proposes a novel solution to this problem leveraging Implicit Neural Representations (INR). Our proposed INR jointly learns two different contrasts of complementary views in a continuous spatial function and benefits from exchanging anatomical information between them. Trained within minutes on a single commodity GPU, our model provides realistic super-resolution across different pairs of contrasts in our experiments with three datasets. Using Mutual Information (MI) as a metric, we find that our model converges to an optimum MI amongst sequences, achieving anatomically faithful reconstruction. (Code is available at: https://github.com/jqmcginnis/multi_contrast_inr/).

Keywords: Multi-contrast Super-resolution · Implicit Neural Representations · Mutual Information

J. McGinnis and S. Shit—Equal contribution.

Supplementary Information The online version contains supplementary material available at https://doi.org/10.1007/978-3-031-43993-3_17.

H. Greenspan et al. (Eds.): MICCAI 2023, LNCS 14227, pp. 173–183, 2023.
https://doi.org/10.1007/978-3-031-43993-3_17

1 Introduction

In clinical practice, Magnetic Resonance Imaging (MRI) provides important information for diagnosing and monitoring patient conditions [4,16]. To capture the complex pathophysiological aspects during disease progression, multi-parametric MRI (such as T1w, T2w, DIR, FLAIR) is routinely acquired. Image acquisition inherently poses a trade-off between scan time, resolution, and signal-to-noise ratio (SNR) [19]. To maximize the source of information within a reasonable time budget, clinical protocol often combines anisotropic 2D scans of different contrasts in complementary viewing directions. Although acquired 2D scans offer an excellent in-plane resolution, they lack important details in the orthogonal out-of-plane. For a reliable pathological assessment, radiologists often resort to a second scan of a different contrast in the orthogonal viewing direction. Furthermore, poor out-of-plane resolution significantly affects the accuracy of volumetric downstream image analysis, such as radiomics and lesion volume estimation, which usually require isotropic 3D scans. As multi-parametric isotropic 3D scans are not always feasible to acquire due to time-constraints [19], motion [9], and patient's condition [10], super-resolution offers a convenient alternative to obtain the same from anisotropic 2D scans. Recently, it has been shown that acquiring three complementary 2D views of the *same contrast* may yield higher SNR at reduced scan time [19,29]. However, it remains under-explored if orthogonal anisotropic 2D views of *different contrasts* can benefit from each other based on the underlying anatomical consistency. Additionally, whether such strategies can further decrease scan times while preserving similar resolution and SNR remains unanswered. Moreover, unlike conventional super-resolution models trained on a cohort, a personalized model is of clinical relevance to avoid the danger of potential misdiagnosis caused by cohort-learned biases. In this work, we mitigate these gaps by proposing a novel multi-contrast super-resolution framework that only requires the patient-specific low-resolution MR scans of different sequences (and views) as supervision. As shown in various settings, our approach is not limited to specific contrasts or views but provides a generic framework for super-resolution. The contributions in this paper are three-fold:

1. To the best of our knowledge, our work is the first to enable subject-specific multi-contrast super-resolution from low-resolution scans without needing any high-resolution training data. We demonstrate that Implicit Neural Representations (INR) are good candidates to learn from complementary views of multi-parametric sequences and can efficiently fuse low-resolution images into anatomically faithful super-resolution.
2. We introduce Mutual Information (MI) [26] as an evaluation metric and find that our method preserves the MI between high-resolution ground truths in its predictions. Further observation of its convergence to the ground truth value during training motivates us to use MI as an early stopping criterion.
3. We extensively evaluate our method on multiple brain MRI datasets and show that it achieves high visual quality for different contrasts and views and preserves pathological details, highlighting its potential clinical usage.

Related Work. Single-image super-resolution (SISR) aims at restoring a high-resolution (HR) image from a low-resolution (LR) input from a single sequence and targets applications such as low-field MR upsampling or optimization of MRI acquisition [3]. Recent methods [3,8] incorporate priors learned from a training set [3], which is later combined with generative models [2]. On the other hand, multi-image super-resolution (MISR) relies on the information from complementary views of the same sequence [29] and is especially relevant to capturing temporal redundancy in motion-corrupted low-resolution MRI [9,27].

Multi-contrast Super-resolution (MCSR) targets using inter-contrast priors [20]. In conventional settings [15], an isotropic HR image of another contrast is used to guide the reconstruction of an anisotropic LR image. Zeng et al. [30] use a two-stage architecture for both SISR and MCSR. Utilizing a feature extraction network, Lyu et al. [14] learn multi-contrast information in a joint feature space. Later, multi-stage integration networks [6], separatable attention [7] and transformers [13] have been used to enhance joint feature space learning. However, all current MCSR approaches are limited by their need for a large training dataset. Consequently, this constrains their usage to specific resolutions and further harbors the danger of hallucination of features (e.g., lesions, artifacts) present in the training set and does not generalize well to unseen data.

Originating from shape reconstruction [18] and multi-view scene representations [17], Implicit Neural Representations (INR) have achieved state-of-the-art results by modeling a continuous function on a space from discrete measurements. Key reasons behind INR's success can be attributed to overcoming the low-frequency bias of Multi-Layer Perceptrons (MLP) [21,24,25]. Although MRI is a discrete measurement, the underlying anatomy is a continuous space. We find INR to be a good fit to model a continuous intensity function on the anatomical space. Once learned, it can be sampled at an arbitrary resolution to obtain the super-resolved MRI. Following this spirit, INRs have recently been successfully employed in medical imaging applications ranging from k-space reconstruction [11] to SISR [29]. Unlike [22,29], which learn anatomical priors in single contrasts, and [1,28], which leverage INR with latent embeddings learned over a cohort, we focus on employing INR in subject-specific, multi-contrast settings.

2 Methods

In this section, we first formally introduce the problem of joint super-resolution of multi-contrast MRI from only one image per contrast per patient. Next, we describe strategies for embedding information from two contrasts in a shared space. Subsequently, we detail our model architecture and training configuration.

Problem Statement. We denote the collection of all 3D coordinates of interest in this anatomical space as $\Omega = \{(x, y, z)\}$ with anatomical function $q : \Omega \rightarrow \mathbb{A}$. The image intensities are a function of the underlying anatomical properties $\mathbb{A}$. Two contrasts C_1 and C_2 can be scanned in a low-resolution subspace

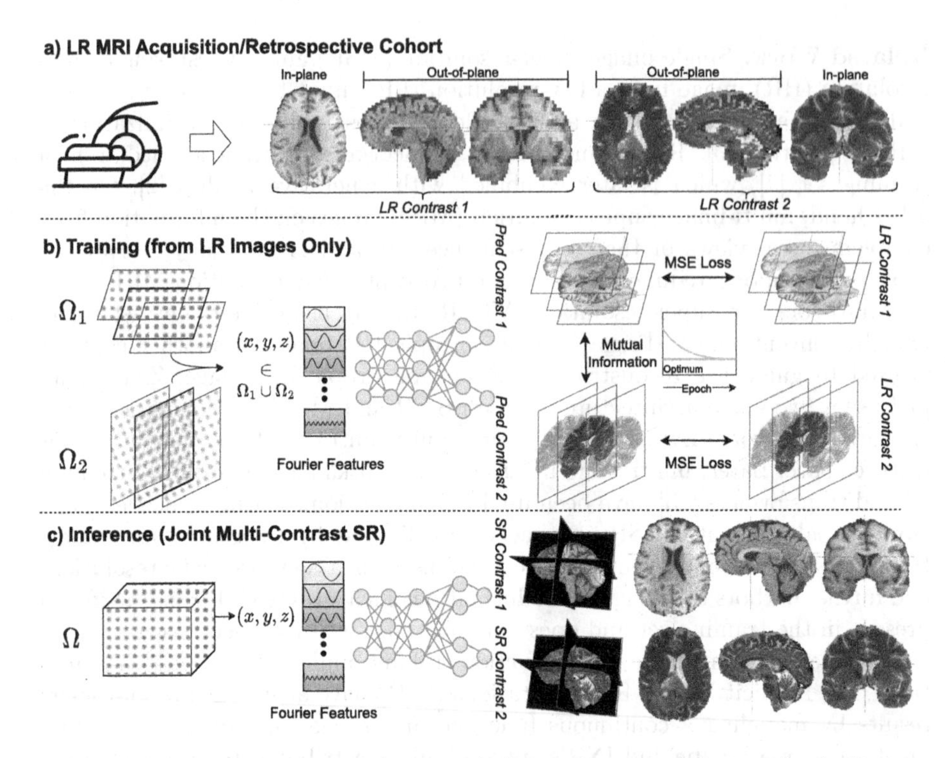

Fig. 1. Overview of our proposed approach (best viewed in full screen). a) Given a realistic clinical scenario, two MRI contrasts are acquired in complementary 2D views. b) Our proposed INR models both contrast from the supervision available in the 2D scans and, by doing so, learn to transfer knowledge from in-plane measurements to out-of-plane of the other contrast. Although our model is trained on MSELoss only for the observed coordinates, it constructs a continuous function space, converging to an optimum state of mutual information between the contrasts on the global space of Ω. c) Once learned, we can sample an isotropic grid and obtain the anatomically faithful and pathology-preserving super-resolution.

$\Omega_1, \Omega_2 \subset \Omega$. Let us consider $g_1, g_2 : \mathbb{A} \rightarrow \mathbb{R}$ that map from anatomical properties to contrast intensities C_1 and C_2, respectively. We obtain sparse observations $I_1 = \{g_1(q(\boldsymbol{x})) = f_1(\boldsymbol{x}); \forall \boldsymbol{x} \in \Omega_1\}$ and $I_2 = \{g_2(q(\boldsymbol{x})) = f_2(\boldsymbol{x}); \forall \boldsymbol{x} \in \Omega_2\}$, where f_i is composition of g_i and q. However, one can easily obtain the global anatomical space Ω by knowing Ω_1 and Ω_2, e.g., by rigid registration between the two images. In this paper, we aim to estimate $f_1, f_2 : \Omega \rightarrow \mathbb{R}$ given I_1 and I_2.

Joint Multi-contrast Modelling. Since both component-functions f_1 and f_2 operate on a subset of the same input space, we argue that it is beneficial to model them jointly as a single function $f : \Omega \rightarrow \mathbb{R}^2$ and optimize it based on their estimation error incurred in their respective subsets. This will enable information transfer from one contrast to another, thus improving the estimation and preventing over-fitting in single contrasts, bringing consistency to the prediction.

To this end, we propose to leverage INR to model a continuous multi-contrast function f from discretely sampled sparse observations I_1 and I_2.

MCSR Setup. Without loss of generalization, let us consider two LR input contrasts scanned in two orthogonal planes p_1 and p_2, where $p_1, p_2 \in$ {axial, sagittal, coronal}. We assume they are aligned by rigid registration requiring no coordinate transformation. Their corresponding in-plane resolutions are $(s_1 \times s_1)$ and $(s_2 \times s_2)$ and slice thickness is t_1 and t_2, respectively. Note that $s_1 < t_1$ and $s_2 < t_2$ imply high in-plane and low out-of-plane resolution. In the end, we aim to sample an isotropic $(s \times s \times s)$ grid for both contrasts where $s \leq s_1, s_2$.

Implicit Neural Representations for MCSR. We intend to project the information available in one contrast into another by embedding both in the shared weight space of a neural network. However, a high degree of weight sharing could hinder contrast-specific feature learning. Based on this reasoning, we aim to hit the sweet spot where maximum information exchange can be encouraged without impeding contrast-specific expressiveness. We propose a split-head architecture, as shown in Fig. 1, where the initial layers jointly learn the common anatomical features, and subsequently, two heads specialize in contrast-specific information. The model takes Fourier [25] Features $\boldsymbol{v} = [cos(2\pi B\boldsymbol{x}), sin(2\pi B\boldsymbol{x})]^T$ as input and predicts $[\hat{I}_1, \hat{I}_2] = f(\boldsymbol{v})$, where $\boldsymbol{x} = (x, y, z)$ and B is sampled from a Gaussian distribution $\mathcal{N}(\mu, \sigma^2)$. We use mean-squared error loss, $\mathcal{L}_{MSE}$, for training.

$$\mathcal{L}_{MSE} = \alpha \sum_{\boldsymbol{x} \in \Omega_1} (\hat{I}_1(\boldsymbol{x}) - I_1(\boldsymbol{x}))^2 + \beta \sum_{\boldsymbol{x} \in \Omega_2} (\hat{I}_2(\boldsymbol{x}) - I_2(\boldsymbol{x}))^2 \tag{1}$$

where α and β are coefficients for the reconstruction loss of two contrasts. Note that for points $\{(x, y, z)\} \in \Omega_2 \setminus \Omega_1$, there is no explicit supervision coming from low resolution C_1. For these points, one can interpret learning C_1 from the loss in C_2, and vice versa, to be a weakly supervised task.

Implementation and Training. Given the rigidly registered LR images, we compute Ω_1, $\Omega_2 \in \Omega$ in the scanner reference space using their affine matrices. Subsequently, we normalize Ω to the interval $[-1, 1]^3$ and independently normalize each contrast's intensities to $[0, 1]$. We use 512-dimensional Fourier Features in the input. Our model consists of a four-layer MLP with a hidden dimension of 1024 for the shared layers and two layers with a hidden dimension of 512 for the heads. We use Adam optimizer with a learning rate of 4e-4 and a Cosine annealing rate scheduler with a batch size of 1000. For the multi-contrast INR models, we use MI as in Eq. 2 for early stopping. Implemented in PyTorch, we train our model on a single A6000 GPU. Please refer to Table 3 in supplementary for an exhaustive hyper-parameter search.

Model Selection and Inference. Since our model is trained on sparse sets of coordinates, it is prone to overfitting them and has little incentive to generalize in out-of-plane predictions for single contrast settings. A remedy to this is to hold random points as a validation set. However, this will reduce the number of training samples and hinder the reconstruction of fine details. For multi-contrast settings, one can exploit the agreement between the two predicted contrasts. Ideally, the network should reach an equilibrium between the contrasts over the training period, where both contrasts optimally benefit from each other. We empirically show that Mutual Information (MI) [26] is a good candidate to capture such an equilibrium point without the need for ground truth data in its computation. For two predicted contrasts $\hat{I}_1$ and $\hat{I}_2$, MI can be expressed as:

$$\mathrm{MI}(\hat{I}_1, \hat{I}_2) = \sum_{y \in \hat{I}_2} \sum_{x \in \hat{I}_1} P_{(\hat{I}_1, \hat{I}_2)}(x, y) \log \left(\frac{P_{(\hat{I}_1, \hat{I}_2)}(x, y)}{P_{\hat{I}_1}(x) P_{\hat{I}_2}(y)} \right) \tag{2}$$

Compared to image registration, we do not use MI as a loss for aligning two images; instead, we use it as a quantitative assessment metric. Given two ground truth HR images for a subject, one can compute the optimum state of MI. We observe that the MI between our model predictions converges close to such an optimum state over the training period without any explicit knowledge about it, c.f. Fig. 3 in the supplementary. This observation motivates us to detect a plateau in MI between the predicted contrasts and use it as a stopping criterion for model selection in multi-contrast INR.

3 Experiments and Results

Datasets. To enable fair evaluation between our predictions and the reference HR ground truths, the in-plane SNR between the LR input scan and corresponding ground truth has to match. To synthetically create 2D LR images, it is necessary to downsample out-of-plane in the image domain anisotropically [32] while preserving in-plane resolution. Consequently, to mimic realistic 2D clinical protocol, which often has higher in-plane details than that of 3D scans, we use spline interpolation to model partial volume and downsampling. We demonstrate our network's modeling capabilities for different contrasts (T1w, T2w, FLAIR, DIR), views (axial, coronal, sagittal), and pathologies (MS, brain tumor). We conduct experiments on two public datasets, BraTS [16], and MSSEG [4], and an in-house clinical MS dataset (cMS). In each dataset, we select 25 patients that fulfill the isotropic acquisition criteria for both ground truth HR scans. Note that we only use the ground truth HR for evaluation, not anywhere in training. We optimize separate INRs for each subject with supervision from only its two LR scans. If required, we employ skull-stripping [12] and rigid registration to the MNI152 (MSSEG, cMS) or SRI24 (BraTS) templates. For details, we refer to Table 2 in the supplementary.

Metrics. We evaluate our results by employing common SR [5,14,29] quality metrics, namely PSNR and SSIM. To showcase perceptual image quality, we additionally compute the Learned Perceptual Image Patch Similarity (LPIPS)

Table 1. Quantitative results for MCSR on two public and one in-house datasets. All metrics consistently show that our split-head INR performs the best for MCSR.

Dataset	Contrasts	T1w			T2w			T1w & T2w		
BraTS 2019	Methods	PSNR ↑	SSIM ↑	LPIPS ↓	PSNR ↑	SSIM ↑	LPIPS ↓	$\varepsilon_{MI}^{T1}\downarrow$	$\varepsilon_{MI}^{T2}\downarrow$	$\hat{\varepsilon}_{MI}\downarrow$
	Cubic Spline	21.201	0.896	0.098	26.201	0.932	0.058	0.096	0.087	0.145
	LRTV	21.328	0.919	0.052	24.206	0.915	0.053	0.126	0.127	0.203
	SMORE	26.266	0.942	0.030	28.466	0.942	0.030	0.157	0.127	0.225
	Single Contrast INR	26.168	0.952	0.036	29.920	0.957	0.028	0.051	0.030	0.079
	Our vanilla INR	26.196	0.960	0.032	29.777	0.962	0.026	0.008	0.015	0.065
	Our split-head INR	**28.746**	**0.965**	**0.028**	**31.802**	**0.966**	**0.024**	**0.007**	**0.014**	**0.062**
MSSEG 2016	Contrasts	T1w			Flair			T1w & Flair		
	Methods	PSNR ↑	SSIM ↑	LPIPS ↓	PSNR ↑	SSIM ↑	LPIPS ↓	$\varepsilon_{MI}^{T1}\downarrow$	$\varepsilon_{MI}^{Flair}\downarrow$	$\hat{\varepsilon}_{MI}\downarrow$
	Cubic Spline	30.102	0.953	0.051	28.724	0.945	0.054	0.062	0.087	0.115
	LRTV	22.848	0.860	0.050	23.920	0.872	0.044	0.068	0.052	0.095
	SMORE	25.729	0.937	0.030	27.430	0.940	0.029	0.138	0.100	0.183
	Single Contrast INR	30.852	0.956	0.029	31.156	0.955	0.030	0.047	0.074	0.095
	Our vanilla INR	31.599	0.966	**0.019**	32.312	0.969	0.019	**0.008**	0.025	**0.024**
	Our split-head INR	**31.769**	**0.967**	**0.019**	**32.514**	**0.970**	**0.018**	**0.008**	**0.023**	**0.024**
cMS	Contrasts	DIR			Flair			DIR & Flair		
	Methods	PSNR ↑	SSIM ↑	LPIPS ↓	PSNR ↑	SSIM ↑	LPIPS ↓	$\varepsilon_{MI}^{DIR}\downarrow$	$\varepsilon_{MI}^{Flair}\downarrow$	$\hat{\varepsilon}_{MI}\downarrow$
	Cubic Spline	28.106	0.929	0.065	26.545	0.923	0.079	0.083	0.096	0.136
	LRTV	28.725	0.904	0.033	22.766	0.835	0.057	0.269	0.088	0.312
	SMORE	28.933	0.926	0.040	25.336	0.921	0.039	0.124	0.079	0.139
	Single Contrast INR	29.941	0.937	0.037	28.655	0.936	0.041	0.063	0.072	0.096
	Our vanilla INR	30.816	**0.956**	0.024	29.749	0.950	0.029	0.022	**0.033**	**0.009**
	Our split-head INR	**31.686**	**0.956**	**0.023**	**30.246**	**0.952**	**0.028**	**0.021**	**0.033**	**0.009**

[31] and measure the absolute error ϵ_{MI} in mutual information of two upsampled images to their ground truth counterparts as follows:

$$\varepsilon_{MI}^{C_i} = \frac{1}{N}\sum_{k=1}^{N} |\text{MI}(\hat{I}_i^k, I_j^k) - \text{MI}(I_i^k, I_j^k)|; \quad \hat{\varepsilon}_{MI} = \frac{1}{N}\sum_{k=1}^{N} |\text{MI}(\hat{I}_1^k, \hat{I}_2^k) - \text{MI}(I_1^k, I_2^k)|$$

Baselines and Ablation. To the best of our knowledge, there are no prior data-driven methods that can perform MCSR on a single-subject basis. Hence, we provide single-subject baselines that operate solely on single contrast and demonstrate the benefit of information transfer from other contrasts with our proposed models. In addition, we show ablations of our proposed split head model compared to our vanilla INR. Precisely, our experiments include:
***Baseline 1*:** Cubic-spline interpolation is applied on each contrast separately.
***Baseline 2*:** LRTV [23] applied on each contrast separately.
***Baseline 3*:** SMORE (v3.1.2) [32] applied on each contrast separately.
***Baseline 4*:** Two single-contrast INRs with one output channel each.
Our vanilla INR (ablation): Single INR with two output channels that jointly predicts the two contrast intensities.
Our proposed split-head INR: Single INR with two separate heads that jointly predicts the two contrast intensities (cf. Fig. 1).

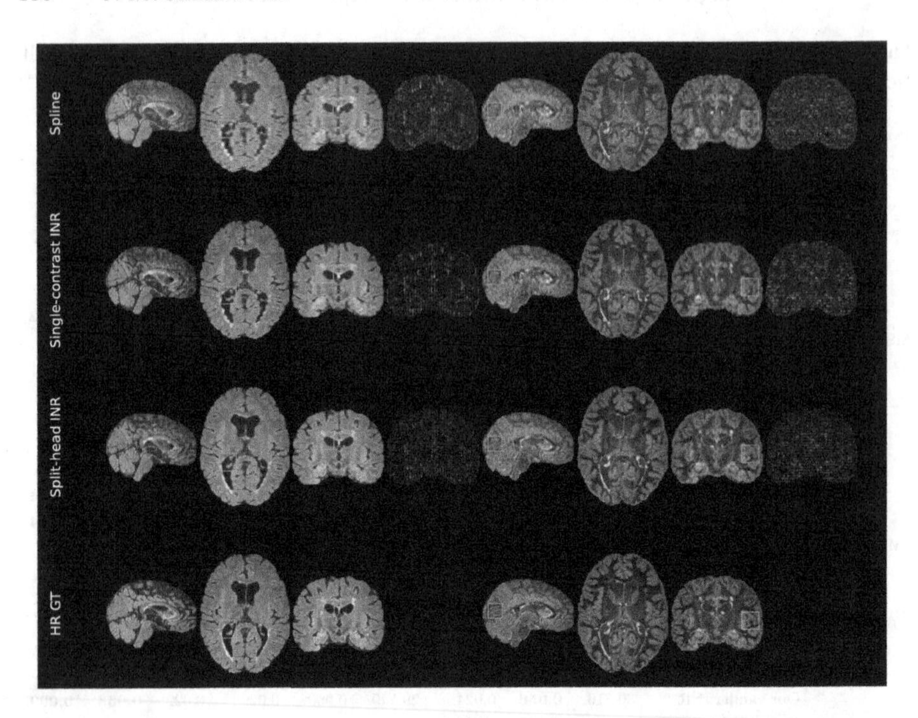

Fig. 2. Qualitative results for MCSR for cMS. The predictions of the split-head INR demonstrate the transfer of anatomical and lesion knowledge from complementing views and sequences. Yellow boxes highlight details recovered by the split-head INR in the out-of-plane reconstructions, where others struggle. (Color figure online)

Quantitative Analysis. Table 1 demonstrates that our proposed framework poses a trustworthy candidate for the task of MCSR. As observed in [32], LRTV struggles for anisotropic up-sampling while SMORE's overall performance is better than cubic-spline, but slightly worse to single-contrast INR. However, the benefit of single-contrast INR may be limited if not complemented by additional views as in [29]. For *MCSR* from single-subject scans, we achieve encouraging results across all metrics for all datasets, contrasts, and views. Since T1w and T2w both encode anatomical structures, the consistent improvement in BraTS for both sequences serves as a proof-of-concept for our approach. As FLAIR is the go-to-sequence for MS lesions, and T1w does not encode such information, the results are in line with the expectation that there could be a relatively higher transfer of anatomical information to pathologically more relevant FLAIR than vice-versa. Lastly, given their similar physical acquisition and lesion sensitivity, we note that DIR/FLAIR benefit to the same degree in the cMS dataset.

Qualitative Analysis. Figure 2 shows the typical behavior of our models on cMS dataset, where one can qualitatively observe that the split-head INR preserves the lesions and anatomical structures shown in the yellow boxes, which other models fail to capture. While our reconstruction is not identical to the GT HR, the coronal view confirms anatomically faithful reconstructions despite not receiving any in-plane supervision from any contrast during training. We refer to Fig. 4 in the supplementary for similar observations on BraTS and MSSEG.

4 Discussion and Conclusion

Given the importance and abundance of large multi-parametric retrospective cohorts [4,16], our proposed approach will allow the upscaling of LR scans with the help of other sequences. Deployment of such a model in clinical routine would likely reduce acquisition time for multi-parametric MRI protocols maintaining an acceptable level of image fidelity. Importantly, our model exhibits trustworthiness in its clinical applicability being 1) subject-specific, and 2) as its gain in information via super-resolution is validated by MI preservation and is not prone to hallucinations that often occur in a typical generative model.

In conclusion, we propose the first subject-specific deep learning solution for isotropic 3D super-resolution from anisotropic 2D scans of two different contrasts of complementary views. Our experiments provide evidence of inter-contrast information transfer with the help of INR. Given the supervision of only single subject data and trained within minutes on a single GPU, we believe our framework to be potentially suited for broad clinical applications. Future research will focus on prospectively acquired data, including other anatomies.

Acknowledgement. JM, MM and JSK are supported by Bavarian State Ministry for Science and Art (Collaborative Bilateral Research Program Bavaria - Québec: AI in medicine, grant F.4-V0134.K5.1/86/34). SS, RG and JSK are supported by European Research Council (ERC) under the European Union's Horizon 2020 research and innovation program (101045128-iBack-epic-ERC2021-COG). MD and DR are supported by ERC (Deep4MI - 884622) and ERA-NET NEURON Cofund (MULTI-FACT - 8810003808). HBL is supported by an Nvidia GPU grant.

References

1. Amiranashvili, T., Lüdke, D., Li, H.B., Menze, B., Zachow, S.: Learning shape reconstruction from sparse measurements with neural implicit functions. In: International Conference on Medical Imaging with Deep Learning (MIDL), pp. 22–34. PMLR (2022)
2. Chen, Y., Shi, F., Christodoulou, A.G., Xie, Y., Zhou, Z., Li, D.: Efficient and accurate MRI super-resolution using a generative adversarial network and 3D multi-level densely connected network. In: Frangi, A.F., Schnabel, J.A., Davatzikos, C., Alberola-López, C., Fichtinger, G. (eds.) MICCAI 2018. LNCS, vol. 11070, pp. 91–99. Springer, Cham (2018). https://doi.org/10.1007/978-3-030-00928-1_11

3. Chen, Y., Xie, Y., Zhou, Z., Shi, F., Christodoulou, A.G., Li, D.: Brain MRI super resolution using 3D deep densely connected neural networks. In: 2018 IEEE 15th International Symposium on Biomedical Imaging (ISBI), pp. 739–742. IEEE (2018)
4. Commowick, O., Cervenansky, F., Ameli, R.: MSSEG challenge proceedings: multiple sclerosis lesions segmentation challenge using a data management and processing infrastructure. In: International Conference on Medical Image Computing and Computer-Assisted Intervention (MICCAI) (2016)
5. Dong, C., Loy, C.C., He, K., Tang, X.: Image super-resolution using deep convolutional networks. IEEE TPAMI **38**(2), 295–307 (2015)
6. Feng, C.-M., Fu, H., Yuan, S., Xu, Y.: Multi-contrast MRI super-resolution via a multi-stage integration network. In: de Bruijne, M., et al. (eds.) MICCAI 2021. LNCS, vol. 12906, pp. 140–149. Springer, Cham (2021). https://doi.org/10.1007/978-3-030-87231-1_14
7. Feng, C.M., Yan, Y., Yu, K., Xu, Y., Shao, L., Fu, H.: Exploring separable attention for multi-contrast MR image super-resolution. arXiv preprint arXiv:2109.01664 (2021)
8. Georgescu, M.I., Ionescu, R.T., Verga, N.: Convolutional neural networks with intermediate loss for 3D super-resolution of CT and MRI scans. IEEE Access **8**, 49112–49124 (2020)
9. Gholipour, A., Estroff, J.A., Warfield, S.K.: Robust super-resolution volume reconstruction from slice acquisitions: application to fetal brain MRI. IEEE TMI **29**(10), 1739–1758 (2010)
10. Ha, J.Y., et al.: One-minute ultrafast brain MRI with full basic sequences: can it be a promising way forward for pediatric neuroimaging? AJR **215**(1), 198–205 (2020)
11. Huang, W., Li, H.B., Pan, J., Cruz, G., Rueckert, D., Hammernik, K.: Neural implicit k-space for binning-free non-cartesian cardiac MR imaging. In: Frangi, A., de Bruijne, M., Wassermann, D., Navab, N. (eds.) Information Processing in Medical Imaging. IPMI 2023. Lecture Notes in Computer Science. vol. 13939. Springer, Cham (2023). https://doi.org/10.1007/978-3-031-34048-2_42
12. Isensee, F., et al.: Automated brain extraction of multisequence MRI using artificial neural networks. Hum. Brain Mapp. **40**(17), 4952–4964 (2019)
13. Li, G., et al.: Transformer-empowered multi-scale contextual matching and aggregation for multi-contrast MRI super-resolution. In: 2022 IEEE/CVF Conference on Computer Vision and Pattern Recognition (CVPR), pp. 20636–20645. IEEE (2022)
14. Lyu, Q., et al.: Multi-contrast super-resolution MRI through a progressive network. IEEE TMI **39**(9), 2738–2749 (2020)
15. Manjón, J.V., Coupé, P., Buades, A., Collins, D.L., Robles, M.: MRI superresolution using self-similarity and image priors. J. Biomed. Imaging **2010**, 1–11 (2010)
16. Menze, B.H., et al.: The multimodal brain tumor image segmentation benchmark (brats). IEEE TMI **34**(10), 1993–2024 (2014)
17. Mildenhall, B., Srinivasan, P.P., Tancik, M., Barron, J.T., Ramamoorthi, R., Ng, R.: NeRF: representing scenes as neural radiance fields for view synthesis. In: Vedaldi, A., Bischof, H., Brox, T., Frahm, J.-M. (eds.) ECCV 2020. LNCS, vol. 12346, pp. 405–421. Springer, Cham (2020). https://doi.org/10.1007/978-3-030-58452-8_24
18. Park, J.J., Florence, P., Straub, J., Newcombe, R., Lovegrove, S.: DeepSDF: learning continuous signed distance functions for shape representation. In: 2019 IEEE/CVF Conference on Computer Vision and Pattern Recognition (CVPR), pp. 165–174. IEEE (2019)

19. Plenge, E.: Super-resolution methods in MRI: can they improve the trade-off between resolution, signal-to-noise ratio, and acquisition time? Magn. Reson. Med. **68**(6), 1983–1993 (2012)
20. Rousseau, F., Initiative, A.D.N., et al.: A non-local approach for image super-resolution using intermodality priors. Med. Image Anal. **14**(4), 594–605 (2010)
21. Saragadam, V., LeJeune, D., Tan, J., Balakrishnan, G., Veeraraghavan, A., Baraniuk, R.G.: WIRE: wavelet implicit neural representations. In: Proceedings of the IEEE/CVF Conference on Computer Vision and Pattern Recognition (CVPR), pp. 18507–18516. IEEE (2023)
22. Shen, L., Pauly, J., Xing, L.: NeRP: implicit neural representation learning with prior embedding for sparsely sampled image reconstruction. IEEE TNNLS (2022)
23. Shi, F., Cheng, J., Wang, L., Yap, P.T., Shen, D.: LRTV: MR image super-resolution with low-rank and total variation regularizations. IEEE TMI **34**(12), 2459–2466 (2015)
24. Sitzmann, V., Martel, J., Bergman, A., Lindell, D., Wetzstein, G.: Implicit neural representations with periodic activation functions. In: Advances in Neural Information Processing Systems (NeurIPS), pp. 7462–7473 (2020)
25. Tancik, M., et al.: Fourier features let networks learn high frequency functions in low dimensional domains. In: Advances in Neural Information Processing Systems (NeurIPS), pp. 7537–7547 (2020)
26. Wells, W.M., III., Viola, P., Atsumi, H., Nakajima, S., Kikinis, R.: Multi-modal volume registration by maximization of mutual information. Med. Image Anal. **1**(1), 35–51 (1996)
27. Wesarg, S., et al.: Combining short-axis and long-axis cardiac MR images by applying a super-resolution reconstruction algorithm. In: Medical Imaging 2010: Image Processing. vol. 7623, pp. 187–198. SPIE (2010)
28. Wu, Q., et al.: An arbitrary scale super-resolution approach for 3D MR images via implicit neural representation. IEEE JBHI **27**(2), 1004–1015 (2023)
29. Wu, Q., et al.: IREM: high-resolution magnetic resonance image reconstruction via implicit neural representation. In: de Bruijne, M., et al. (eds.) MICCAI 2021. LNCS, vol. 12906, pp. 65–74. Springer, Cham (2021). https://doi.org/10.1007/978-3-030-87231-1_7
30. Zeng, K., Zheng, H., Cai, C., Yang, Y., Zhang, K., Chen, Z.: Simultaneous single- and multi-contrast super-resolution for brain MRI images based on a convolutional neural network. Comput. Biol. Med. **99**, 133–141 (2018)
31. Zhang, R., Isola, P., Efros, A.A., Shechtman, E., Wang, O.: The unreasonable effectiveness of deep features as a perceptual metric. In: Proceedings of the IEEE Conference on Computer Vision and Pattern Recognition (CVPR), pp. 586–595. IEEE (2018)
32. Zhao, C., Dewey, B.E., Pham, D.L., Calabresi, P.A., Reich, D.S., Prince, J.L.: SMORE: a self-supervised anti-aliasing and super-resolution algorithm for MRI using deep learning. IEEE TMI **40**(3), 805–817 (2020)

DeepSOZ: A Robust Deep Model for Joint Temporal and Spatial Seizure Onset Localization from Multichannel EEG Data

Deeksha M. Shama[1(✉)], Jiasen Jing[1], and Archana Venkataraman[1,2]

[1] Department of Electrical and Computer Engineering, Johns Hopkins University, Baltimore, USA
{dshama1,jjing2}@jhu.edu, archanav@bu.edu

[2] Department of Electrical and Computer Engineering, Boston University, Boston, USA

Abstract. We propose a robust deep learning framework to simultaneously detect and localize seizure activity from multichannel scalp EEG. Our model, called DeepSOZ, consists of a transformer encoder to generate global and channel-wise encodings. The global branch is combined with an LSTM for temporal seizure detection. In parallel, we employ attention-weighted multi-instance pooling of channel-wise encodings to predict the seizure onset zone. DeepSOZ is trained in a supervised fashion and generates high-resolution predictions on the order of each second (temporal) and EEG channel (spatial). We validate DeepSOZ via bootstrapped nested cross-validation on a large dataset of 120 patients curated from the Temple University Hospital corpus. As compared to baseline approaches, DeepSOZ provides robust overall performance in our multi-task learning setup. We also evaluate the intra-seizure and intra-patient consistency of DeepSOZ as a first step to establishing its trustworthiness for integration into the clinical workflow for epilepsy.

Keywords: Epilepsy · EEG · Multi-instance learning · Trustworthy AI

1 Introduction

Epilepsy is a debilitating neurological disorder characterized by spontaneous and recurring seizures [17]. Roughly 30% of epilepsy patients are *drug resistant*, meaning they do not positively respond to anti-seizure medications. In such cases, the best alternative treatment is to identify and surgically resect the brain region responsible for triggering the seizures, i.e., the seizure onset zone (SOZ). Scalp electroencephalography (EEG) is the first and foremost modality used to monitor epileptic activity. However, seizure detection and SOZ localization from

Supplementary Information The online version contains supplementary material available at https://doi.org/10.1007/978-3-031-43993-3_18.

H. Greenspan et al. (Eds.): MICCAI 2023, LNCS 14227, pp. 184–194, 2023.
https://doi.org/10.1007/978-3-031-43993-3_18

scalp EEG are based on expert visual inspection, which is time consuming and heavily prone to the subjective biases of the clinicians [8].

Computer-aided tools for scalp EEG almost exclusively focus on the task of (temporal) seizure detection. Early works approached the problem via feature engineering and explored spectral [24,25], entropy-based [9], and graph-theoretic [1] features for the task. In general, these methods extract features from short time windows and use a machine learning classifier to discriminate between window-wise seizure and baseline activity [1,25]. More recently, deep learning models have shown promise in extracting generalizable information from noisy and heterogeneous datasets. Deep learning applications to EEG include convolutional neural networks (CNNs) [4,12,22,23], graph convolutional networks(GCNs) [21], and a combination of attention-based feature extraction [10] and recurrent layers to capture evolving dynamics [4,15,20]. Transformers have also been used for seizure detection, both in combination with CNNs [14] and directly on the EEG signals and their derived features [11,18]. While these methods have greatly advanced the problem of seizure detection, they provide little information about the SOZ, which is ultimately the more important clinical question.

A few works have explored the difficult task of localizing the SOZ via *post hoc* evaluations of deep networks trained for seizure detection. For example, the authors of [7,16] perform a cross-channel connectivity analysis of the learned representations to determine the SOZ. In contrast, the method of [2] identifies the SOZ by dropping out nodes of the trained GCN until the seizure detection performance degrades below a threshold. Finally, the SZTrack model of [6] jointly detects and tracks the spatio-temporal seizure spread by aggregating channel-wise detectors; the predictions of this model are seen to correlate with the SOZ. While valuable, the post hoc nature of these unsupervised analyses means that the results may not generalize to unseen patients. The first supervised approach for SOZ localization was proposed by [3] and uses probabilistic graphical models for simultaneous detection and localization. The more recent SZLoc model [5] proposes an end-to-end deep architecture for SOZ localization along with a set of novel loss functions to weakly supervise the localization task from coarse inexact labels. While these two methods represent seminal contributions to the field, they are difficult to train and only report the localization performance on short (i.e., < 2 min) EEG recordings around the time of seizure onset.

In this paper, we present DeepSOZ, a robust model for joint seizure detection and SOZ localization from multichannel scalp EEG. Our model consists of a spatial transformer encoder to combine cross-channel information and LSTM layers to capture dynamic activity for window-wise seizure detection. In parallel, we use a novel attention-weighted multi-instance pooling to supervise seizure-level SOZ localization at the single channel resolution. We curate a large evaluation dataset from the publicly available TUH seizure corpus by creating SOZ labels from the clinician notes for each patient. We perform extensive window-level, seizure-level, and patient-level evaluations of our model. Additionally, we analyze the consistency of predictions across seizure occurrences, which has not

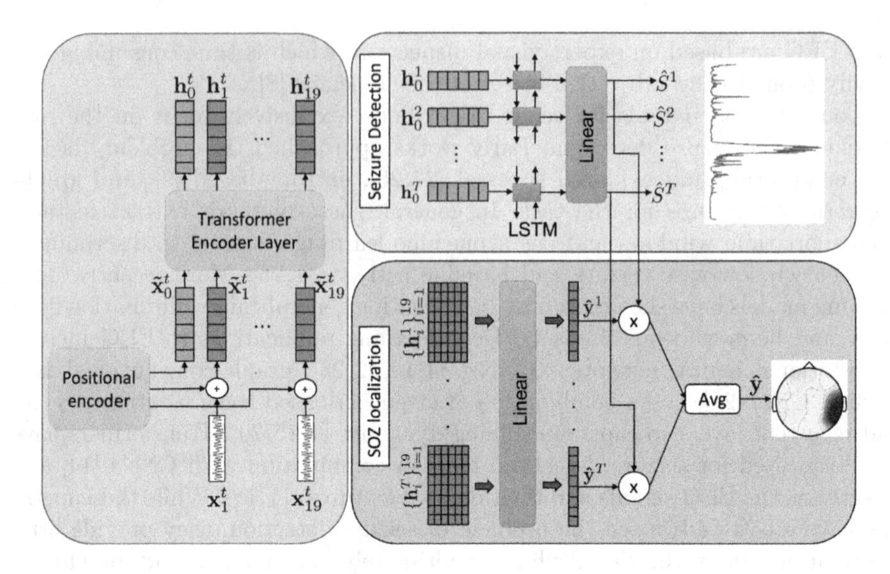

Fig. 1. Schematic of our DeepSOZ model. **Left:** Transformer encoder that uses positional encoding and self attention to generate hidden representations. **Top Right:** Bidirectional LSTM for seizure detection. **Bottom Right:** Attention weighted multi-instance pooling for SOZ localization.

previously been reported for SOZ localization. Quantifying the error variance is the first step in establishing trust in DeepSOZ for clinical translation.

2 Methodology

Figure 1 illustrates our DeepSOZ architecture. The inputs to DeepSOZ are multichannel EEG data for a single seizure recording segmented into one-second windows. The outputs are a temporal sequence of predicted seizure versus baseline activity (detection) and a channel-wise posterior distribution for the SOZ (localization). Formally, let $\mathbf{x}_i^t$ denote the EEG data for channel i and time window t. Clinical EEG is recorded in the 10–20 system, which consists of 19 channels distributed across the scalp. For training, let $S^t \in \{0, 1\}$ denote the seizure versus baseline activity label for time window t, and let $\mathbf{y} \in \{0, 1\}^{19\times1}$ be a vector representing the clinician annotated SOZ. Below, we describe each component of DeepSOZ, along with our training and validation strategy.

2.1 The DeepSOZ Model Architecture

Spatial Transformer Encoder: For each time window t, the multichannel EEG data $\{\mathbf{x}_i^t\}_{i=1}^{19}$ is passed to a transformer encoder consisting of multi-head attention (MHA) layers to generate both a global $\mathbf{h}_0^t$ and channel-wise $\{\mathbf{h}_i^t\}_{i=1}^{19}$ encodings. Since the spatial orientation of these channels is crucial for tracking seizure activity, we add a positional embedding generated by a trainable linear layer $\mathbf{W}_p$, resulting in the modified input $\tilde{\mathbf{x}}_i^t = \mathbf{x}_i^t + \mathbf{W}_p^T \mathbb{1}(i)$, where $\mathbb{1}(i)$ is the indicator function for a one hot encoding at element i.

The hidden representations are computed by the transformer encoder from the modified multichannel input $\tilde{\mathbf{X}}^t = [\tilde{\mathbf{x}}_1^t \quad \ldots \quad \tilde{\mathbf{x}}_{19}^t]$ as follows:

$$\mathbf{Z}^t = LN(MHA(\tilde{\mathbf{X}}^t) + \tilde{\mathbf{X}}^t) \qquad \mathbf{h}_i^t = LN(FF(\mathbf{z}_i^t) + \mathbf{z}_i^t), \tag{1}$$

where $LN(\cdot)$ denotes layer normalization, and $FF(\cdot)$ represents a learned two layer feed-forward network with ReLU activation.

The $MHA(\cdot)$ operation uses parallel self attentions to map the input data into a set of projections, as guided by the other channels in the montage. Formally, let n index the attention head. The attention weights $\mathbf{A}_n^t \in \mathbb{R}^{20\times 20}$ captures global (1) and cross-channel (19) similarities via the key matrix $\mathbf{K}_n^t = \mathbf{W}_n^K \tilde{\mathbf{X}}^t$ and query matrix $\mathbf{Q}_n^t = \mathbf{W}_n^Q \tilde{\mathbf{X}}^t$ as follows:

$$\mathbf{A}_n^t = \xi\left(\frac{\mathbf{Q}_n^t \mathbf{K}_n^{tT}}{\sqrt{d}}\right), \tag{2}$$

where $\xi(\cdot)$ represents the softmax function, and d is our model dimension. The attention $\mathbf{A}_n^t$ is multiplied by the value matrix $\mathbf{V}_n^t = \mathbf{W}_n^V \tilde{\mathbf{X}}^t$ to generate the output for head n. These outputs are concatenated and fed into a linear layer to produce $MHA(\cdot)$. Finally, these MHA outputs are passed into a two layer feed forward neural network with ReLU activation, post residual connections and layer normalization to generate the hidden encoding $\mathbf{H}^t \in \mathbb{R}^{20\times 200}$.

The matrices $\mathbf{W}_n^Q$, $\mathbf{W}_n^K$, and $\mathbf{W}_n^V$ are trained parameters of the encoder. For simplicity, we set the model dimension d to be the same as our input $\mathbf{x}_i^t$ ($d = 200$ in this work), and we specify 8 attention heads in the MHA operation.

LSTM for Temporal Seizure Detection: We use a bidirectional LSTM to capture evolving patterns in the global encodings of the one-second EEG windows, i.e., $\{\mathbf{h}_0^t\}_{t=1}^T$. We use a single LSTM layer with 100 hidden units to process the global encodings and capture both long-term and short-term dependencies. The output of the LSTM is passed into a linear layer, followed by a softmax function, to generate window-wise predictions $\hat{S}^t \in [0, 1]$. Here, $\hat{S}^t$ represents the posterior probability of seizure versus baseline activity at window t.

Attention-Weighted Multi-Instance Pooling for SOZ Localization: We treat the localization task as a multi-instance learning problem to predict a channel-wise posterior distribution for the SOZ vector $\{\mathbf{y}_i\}_{i=1}^{19}$ by computing a weighted average of the hidden representations from the transformer. We first map the channel-wise encodings $\{\mathbf{h}_i^t\}_{i=1}^{19} \in \mathbb{R}^{200}$ to scalars $\hat{\mathbf{y}}_i^t$ using the same linear layer across channels. We use the predicted seizure probability $\hat{S}^t$ as our attention to compute the final SOZ prediction as follows:

$$\hat{\mathbf{y}}_i = \sigma\left(\frac{\sum_{t=1}^T \hat{S}^t \cdot \hat{\mathbf{y}}_i^t}{\sum_{t=1}^T \hat{S}^t}\right), \qquad i = 1, \ldots, 19 \tag{3}$$

where $\sigma(\cdot)$ is the sigmoid function. The final patient-level predictions are obtained by averaging $\hat{\mathbf{y}}_i$ across all seizure recordings for that patient.

2.2 Loss Function and Model Training

We train DeepSOZ in two stages. First, the transformer and LSTM layers are trained for window-wise seizure detection using weighted cross entropy loss:

$$\mathcal{L}_{det} = -\frac{1}{T}\sum_{t=1}^{T}\Big((1-\delta)(1-S^t)\log(1-\hat{S}^t) + \delta \cdot S^t \log \hat{S}^t\Big), \tag{4}$$

where the weight $\delta = 0.8$ is fixed based on the ratio of non-seizure to seizure activity in the dataset. DeepSOZ is then finetuned for SOZ localization. To avoid catastrophic forgetting of the detection task, we freeze the LSTM layers and provide a weak supervision for detection via the loss function:

$$\mathcal{L}_{soz} = 0.1 \cdot \mathcal{L}_{det} - \frac{1}{19}\sum_{i=1}^{19}((1-\mathbf{y}_i)\log(1-\hat{\mathbf{y}}_i) + \mathbf{y}_i \log \hat{\mathbf{y}}_i) + \|\hat{\mathbf{y}}\|_1, \tag{5}$$

where the $||.||_1$ penalizes the L1 norm to encourage sparsity in predicted $\hat{\mathbf{y}}$

2.3 Model Validation

We evaluate DeepSOZ using bootstrapped 5-fold nested cross validation. Within each training fold, we select the learning rate and seizure detection threshold through a grid search with a fixed dropout of 0.15. We use PyTorch v1.9.0 with Adam [13] for training with a batch size of one patient; early stopping is implemented using a validation set drawn from the training data. We re-sample the original 5-fold split three times and report the results across all 15 models[1].

Seizure Detection: At the window level, we report sensitivity, specificity, and area under the receiver operating characteristic curve (AU-ROC). At the seizure level, we adopt the strategy of [4] and select a detection threshold that ensures no more than 2 min of false positive detections per hour in the validation dataset. To eliminate spikes, we smooth the output predictions using a 30 s window and count only the contiguous intervals beyond the calibrated detection threshold as seizure predictions. Following the standard of [4], we do not penalize post-ictal predictions. We report the false positive rate (FPR) per hour (min/hour), the sensitivity, and the latency (seconds) in seizure detection.

[1] Our code and data can be accessed at https://github.com/deeksha-ms/DeepSOZ.git.

Table 1. Description of patient demographics in the curated TUH dataset.

	curated TUH dataset
Number of patients	120
Male/Female	55/65
Average age	55.2 ± 16.6
Min/Max age	19/91
Seizures per patient	14.7 ± 25.2
Min/Max seizures per patient	1/152
Average EEG duration per patient	79.8 ± 135 min
Average seizure duration	88.0 ± 123.5 s
Min/Max seizure duration	7.5 ± 1121 s
Temporal/Extra-temporal Onset	72/48
Right/Left onset zone	59/61

SOZ Localization: By construction, DeepSOZ processes each seizure recording separately to find the SOZ. Patient-level SOZ predictions are obtained by averaging across all seizure recordings for that patient. The SOZ is correctly localized if the maximum channel-wise probability lies in the neighborhood determined by the clinician. We quantify the prediction variance at the seizure level by generating Monte Carlo samples during test via active dropout. At the patient level, we compute the prediction variance across all seizures for that patient.

Baseline Comparisons: We compare the performance of DeepSOZ with one model ablation and four state-of-the-art methods from the literature. Our ablation replaces the attention-weighted multi-instance pooling in DeepSOZ with a standard maxpool operation within the prediction seizure window (DeepSOZ-max). Our baselines consist of the CNN-BLSTM model for seizure detection developed by [4], the SZTrack model proposed by [6] that uses a convolutional-recurrent architecture for each channel, the SZLoc model by [5] consisting of CNN-transformer-LSTM layers, and the Temporal Graph Convolutional Network (TGCN) developed by [2]. SZTrack and SZLoc are trained and evaluated for localization via the approach published by the authors which uses *only 45 s of data around onset time*. We modify the TGCN slightly to extract channel-wise prediction for localization task but evaluate it on the full 10-minute recordings like DeepSOZ. Finally, we note that the CNN-BLSTM can only be used for seizure detection, and SZLoc is only trained for SOZ localization.

3 Experimental Results

Data and Preprocessing: We validate DeepSOZ on 642 EEG recordings from 120 adult epilepsy patients in the publicly available Temple University Hospital

Table 2. Temporal seizure detection performance on the TUH dataset. Window-level metrics are calculated for each one-second windows. Seizure-level metrics are aggregated over the duration of seizure after post-processing.

Model	Window-Level			Seizure-Level		
	AU-ROC	Sensitivity	Specificity	FPR	Sensitivity	Latency
DeepSOZ	.901 ± .027	.679 ± .100	.890 ± .030	.44 ± .23	.808 ± .106	-18.45 ± 15.67
DeepSOZ-max	.907 ± .032	.676 ± .079	.909 ± .029	.288 ± .153	.700 ± .105	-15.39 ± 9.91
TGCN [2]	.887 ± .032	.711 ± .148	.835 ± .085	.808 ± .591	.869 ± .085	-36.01 ± 29.49
SZTrack [6]	.5202 ± .045	.464 ± .300	.535 ± .303	2.06 ± .844	.799 ± .135	-50.5 ± 71.5
CNN-BLSTM [4]	.876 ± .044	.664 ± .135	.876 ± .055	.351 ± .45	.42 ± .281	28.89 ± 154.88

Table 3. SOZ localization metrics. The seizure-level results are calculated independently on all seizure recordings. Patient-level results are aggregated over multiple seizures of the patients. Number of model parameters is also given.

Model	Seizure-Level		Patient-Level		# Params
	Accuracy	Uncertainty	Accuracy	Uncertainty	
DeepSOZ	.731 ± .061	.009 ± .001	.744 ± .058	.142 ± .013	510K
DeepSOZ-max	.513 ± .154	.0 ± .0	.411 ± .076	.023 ± .007	510K
TGCN [2]	.479 ± .07	.0 ± .0	.486 ± .123	.153 ± .015	1.16M
SZTrack [6]	.454 ± .065	.003 ± .001	.450 ± .142	.017 ± .007	19K
SZLoc [5]	.682 ± .094	.008 ± .001	.740 ± .056	.074 ± .008	491K

(TUH) corpus [19] with a well characterized unifocal seizure onset. We use the clinical notes to localize the SOZ to a subset of the 19 EEG channels. Table 1 describes the seizure characteristics across patients in our curated subset.

Following [4], we re-sample the raw EEG to 200 Hz for uniformity, filter the signals between 1.6–30 Hz, and clip them at two standard deviations from mean to remove high intensity artifacts. All signals are normalized to have zero mean and unit variance. We standardize the input lengths by cropping the signals to 10 min around the seizure interval, while ensuring that the onset times are uniformly distributed within this period. We segment the EEG into one second non-overlapping windows to obtain the model inputs $\mathbf{x}_i^t$.

Seizure Detection Performance: Table 2 reports the seizure detection performance averaged over the 15 bootstrapped testing folds. At the window level, both aggregation strategies for DeepSOZ (weighted posterior and max pooling) perform similarly and achieve higher AU-ROC values than the other baselines. The TGCN and CNN-BLSTM baselines achieve notably worse AU-ROC values, establishing the power of a transformer encoder in extracting more meaningful features. SZTrack is trained using the published strategy in [6] and fails to detect seizures effectively. The differences in AU-ROC between DeepSOZ and TGCN, SZTrack, and CNN-BLSTM are statistically significant per a De Long's test at

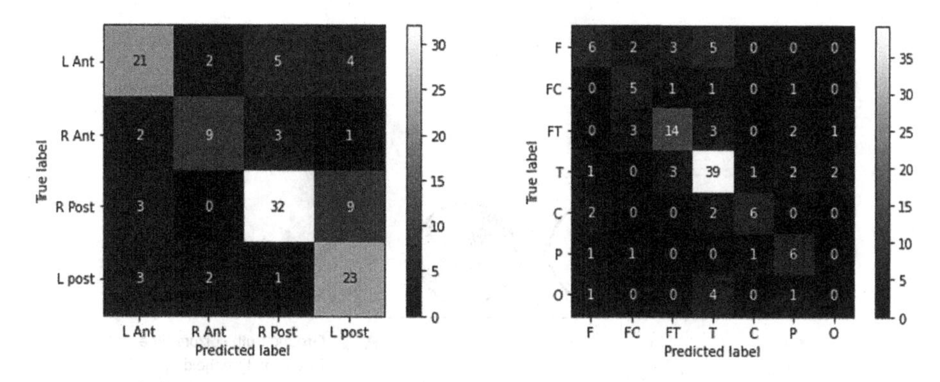

Fig. 2. Confusion matrices between the max channel-wise posterior and the true SOZ. **Left:** Quadrant-based aggregation (L: Left, R: Right, Ant: Anterior, Post: Posterior). **Right:** Functional region-based aggregation (F: Frontal, FC: Frontocentral, FT: Frontotemporal, T: Temporal, C: Central, P: Parietal, O: Occipital).

$p < 0.05$. At the seizure level, DeepSOZ achieves a good balance between sensitivity (0.81) and FPR (0.44 min/h). The negative latency of 18 s contributes towards the slightly elevated FPR. The TGCN and SZTrack have a high sensitivity, which comes at the cost of much higher FPR, while the CNN-BLSTM has a low detection sensitivity but comparable FPR.

SOZ Localization Performance: Table 3 summarizes the SOZ localization performance across models. DeepSOZ performs the best at both patient and seizure levels. In contrast, the SZTrack and TGCN baselines are confident in their predictions but more often incorrect, once again highlighting the value of a transformer encoder. While the SZLoc model performs the best of the baselines, we note that both it and SZTrack have an unfair advantage of being trained and evaluated on 45 s EEG recordings around the seizure onset time. In contrast, DeepSOZ processes full 10-minute recordings for both tasks.

Figure 2 aggregates the final predictions of DeepSOZ across the 120 patients into quadrants. As seen, DeepSOZ is adept at differentiating right- and left-hemisphere onsets but struggles to differentiate anterior and posterior SOZs. We hypothesize that this trend is due to the skew towards temporal epilepsy patients in the TUH dataset. A similar trend can be observed at the finer lobe-wise predictions. Figure 3 illustrate sample DeepSOZ outputs for two patients in the testing fold. As seen, DeepSOZ accurately detects the seizure interval in all cases but has two false positive detections for Patient 1. Nonetheless, DeepSOZ correctly localizes the seizure to the left frontal area. The localization for Patient 2 is more varied, which correlates with the patient notes that specify a right-posterior onset but epileptogenic activity quickly spreading to the left hemisphere. Overall, DeepSOZ is more uncertain about this patient.

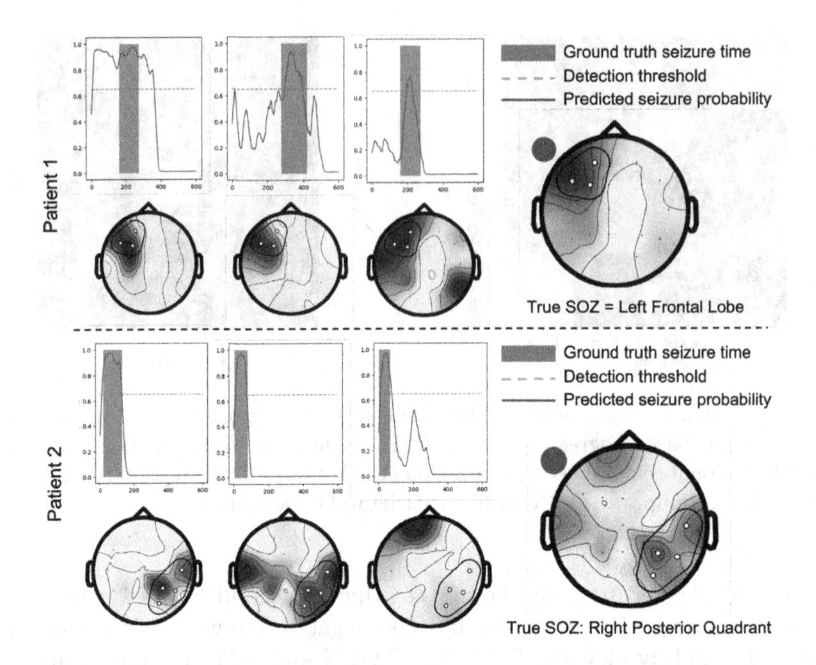

Fig. 3. Visualization for two testing patients. **Top:** Temporal seizure detection. Blue lines correspond to the DeepSOZ prediction; horizontal orange lines denote the seizure detection threshold from training; shaded region is the ground-truth seizure interval. **Bottom:** Predicted SOZ for the above seizure projected onto a topological scalp plot. **Side:** Patient-level SOZ with ground-truth below.

4 Conclusion

We have introduced DeepSOZ for joint seizure detection and SOZ localization from scalp EEG. DeepSOZ leverages a self-attention mechanism to generate informative global and channel-wise latent representations that strategically fuse multi-channel information. The subsequent recurrent layers and attention-weighted pooling allow DeepSOZ to generalize across a heterogeneous cohort. We validate DeepSOZ on data from 120 epilepsy patients and report improved detection and localization performance over numerous baselines. Finally, we quantify the prediction uncertainty as a first step towards building trust in the model.

Acknowledgements. This work was supported by the National Institutes of Health R01 EB029977 (PI Caffo), the National Institutes of Health R01 HD108790 (PI Venkataraman), and the National Institutes of Health R21 CA263804 (PI Venkataraman).

References

1. Akbarian, B., et al.: A framework for seizure detection using effective connectivity, graph theory, and multi-level modular network. Biomed. Sig. Process. Control **59**, 101878 (2020)
2. Covert, I.C., et al.: Temporal graph convolutional networks for automatic seizure detection. In: Machine Learning for Healthcare, pp. 160–180. PMLR (2019)
3. Craley, J., Johnson, E., Jouny, C., Venkataraman, A.: Automated noninvasive seizure detection and localization using switching markov models and convolutional neural networks. In: Shen, D., et al. (eds.) MICCAI 2019. LNCS, vol. 11767, pp. 253–261. Springer, Cham (2019). https://doi.org/10.1007/978-3-030-32251-9_28
4. Craley, J., et al.: Automated inter-patient seizure detection using multichannel convolutional and recurrent neural networks. Biomed. Sig. Process. Control **64**, 102360 (2021)
5. Craley, J., et al.: SZLoc: a multi-resolution architecture for automated epileptic seizure localization from scalp EEG. In: Medical Imaging with Deep Learning (2022)
6. Craley, J.E.A.: Automated seizure activity tracking and onset zone localization from scalp EEG using deep neural networks. PloS One **17**(2), e0264537 (2022)
7. Dissanayake, T., et al.: Geometric deep learning for subject independent epileptic seizure prediction using scalp EEG signals. IEEE J. Biomed. Health Inf. **26**(2), 527–538 (2021)
8. van Donselaar, C.A., et al.: Value of the electroencephalogram in adult patients with untreated idiopathic first seizures. Arch. Neurol. **49**(3), 231–237 (1992)
9. Güler, N.F., et al.: Recurrent neural networks employing Lyapunov exponents for EEG signals classification. Expert Syst. Appl. **29**(3), 506–514 (2005)
10. He, J., et al.: Spatial-temporal seizure detection with graph attention network and bi-directional LSTM architecture. Biomed. Sig. Process. Control **78**, 103908 (2022)
11. Hussein, R., et al.: Multi-channel vision transformer for epileptic seizure prediction. Biomedicines **10**(7), 1551 (2022)
12. Khan, H., Marcuse, L., Fields, M., Swann, K., Yener, B.: Focal onset seizure prediction using convolutional networks. IEEE Trans. Biomed. Eng. **65**(9), 2109–2118 (2017)
13. Kingma, D.P., et al.: Adam: A method for stochastic optimization. arXiv preprint arXiv:1412.6980 (2014)
14. Li, C., et al.: EEG-based seizure prediction via transformer guided CNN. Measurement **203**, 111948 (2022)
15. Liang, W., Pei, H., Cai, Q., Wang, Y.: Scalp EEG epileptogenic zone recognition and localization based on long-term recurrent convolutional network. Neurocomputing **396**, 569–576 (2020)
16. Mansouri, A., et al.: Online EEG seizure detection and localization. Algorithms **12**(9), 176 (2019)
17. Miller, J., et al.: Epilepsy. Hoboken (2014)
18. Pedoeem, J., Bar Yosef, G., Abittan, S., Keene, S.: TABS: transformer based seizure detection. In: Obeid, I., Picone, J., Selesnick, I. (eds.) Biomedical Sensing and Analysis. Springer, Cham (2022). https://doi.org/10.1007/978-3-030-99383-2_4
19. Shah, V., et al.: The temple university hospital seizure detection corpus. Front. Neuroinform. **12**, 83 (2018). https://isip.piconepress.com/projects/tuh_eeg/html/downloads.shtml

20. Vidyaratne, L., Glandon, A., Alam, M., Iftekharuddin, K.M.: Deep recurrent neural network for seizure detection. In: 2016 International Joint Conference on Neural Networks (IJCNN), pp. 1202–1207. IEEE (2016)
21. Wagh, N., et al.: EEG-GCNN: augmenting electroencephalogram-based neurological disease diagnosis using a domain-guided graph convolutional neural network. In: Machine Learning for Health, pp. 367–378. PMLR (2020)
22. Wei, Z., Zou, J., Zhang, J., Xu, J.: Automatic epileptic EEG detection using convolutional neural network with improvements in time-domain. Biomed. Sig. Process. Control **53**, 101551 (2019)
23. Yuan, Y., Xun, G., Jia, K., Zhang, A.: A multi-view deep learning framework for EEG seizure detection. IEEE J. Biomed. Health Inf. **23**(1), 83–94 (2018)
24. Zandi, A.S., et al.: Automated real-time epileptic seizure detection in scalp EEG recordings using an algorithm based on wavelet packet transform. IEEE Trans. Biomed. Eng. **57**(7), 1639–1651 (2010)
25. Zhang, Y., et al.: Integration of 24 feature types to accurately detect and predict seizures using scalp EEG signals. Sensors **18**(5), 1372 (2018)

Multimodal Brain Age Estimation Using Interpretable Adaptive Population-Graph Learning

Kyriaki-Margarita Bintsi[1(✉)], Vasileios Baltatzis[1,2], Rolandos Alexandros Potamias[1], Alexander Hammers[2], and Daniel Rueckert[1,3]

[1] Department of Computing, Imperial College London, London, UK
m.bintsi19@imperial.ac.uk
[2] Biomedical Engineering and Imaging Sciences, King's College London, London, UK
[3] Technical University of Munich, Munich, Germany

Abstract. Brain age estimation is clinically important as it can provide valuable information in the context of neurodegenerative diseases such as Alzheimer's. Population graphs, which include multimodal imaging information of the subjects along with the relationships among the population, have been used in literature along with Graph Convolutional Networks (GCNs) and have proved beneficial for a variety of medical imaging tasks. A population graph is usually static and constructed manually using non-imaging information. However, graph construction is not a trivial task and might significantly affect the performance of the GCN, which is inherently very sensitive to the graph structure. In this work, we propose a framework that learns a population graph structure optimized for the downstream task. An attention mechanism assigns weights to a set of imaging and non-imaging features (phenotypes), which are then used for edge extraction. The resulting graph is used to train the GCN. The entire pipeline can be trained end-to-end. Additionally, by visualizing the attention weights that were the most important for the graph construction, we increase the interpretability of the graph. We use the UK Biobank, which provides a large variety of neuroimaging and non-imaging phenotypes, to evaluate our method on brain age regression and classification. The proposed method outperforms competing static graph approaches and other state-of-the-art adaptive methods. We further show that the assigned attention scores indicate that there are both imaging and non-imaging phenotypes that are informative for brain age estimation and are in agreement with the relevant literature.

Keywords: Brain age regression · Interpretability · Graph Convolutional Networks · Adaptive graph learning

---

Supplementary Information The online version contains supplementary material available at https://doi.org/10.1007/978-3-031-43993-3_19.

H. Greenspan et al. (Eds.): MICCAI 2023, LNCS 14227, pp. 195–204, 2023.
https://doi.org/10.1007/978-3-031-43993-3_19

1 Introduction

Healthy brain aging follows specific patterns [2]. However, various neurological diseases, such as Alzheimer's disease [8], Parkinson's disease [23], and schizophrenia [18], are accompanied by an abnormal accelerated aging of the human brain. Thus, the difference between the biological brain age of a person and their chronological age can show the deviation from the healthy aging trajectory, and may prove to be an important biomarker for neurodegenerative diseases [6,9].

Recently, graph-based methods have been explored for brain age estimation as graphs can inherently combine multimodal information by integrating the subjects' neuroimaging information as node features and, through a similarity metric, the associations among subjects through as edges that connect these nodes [21]. However, in medical applications, the construction of a population-graph is not always simple as there are various ways subjects could be considered similar.

GCNs [16] have been extensively [1] used in the medical domain for node classification tasks, such as Alzheimer's prediction [15,21] and Autism classification [4]. They take graphs as input and, in most cases, the graph structure is predefined and static. GCN performance is highly dependent on the graph structure. This has been correlated in related literature with the heterophily of graphs in the semi-supervised node classification tasks, which refers to the case when the nodes of a graph are connected to nodes that have dissimilar features and different class labels [29]. It has been shown that if the homophily ratio is very low, a simple Multi-Layer Perceptron (MLP) that completely ignores the structure, can outperform a GCN [30].

A way to address this problem is through adaptive graph learning [27], which learns the graph structure through training. In the medical domain, there is little ongoing research on the topic [13,28]. However, the adaptive graphs in [7,13,14]

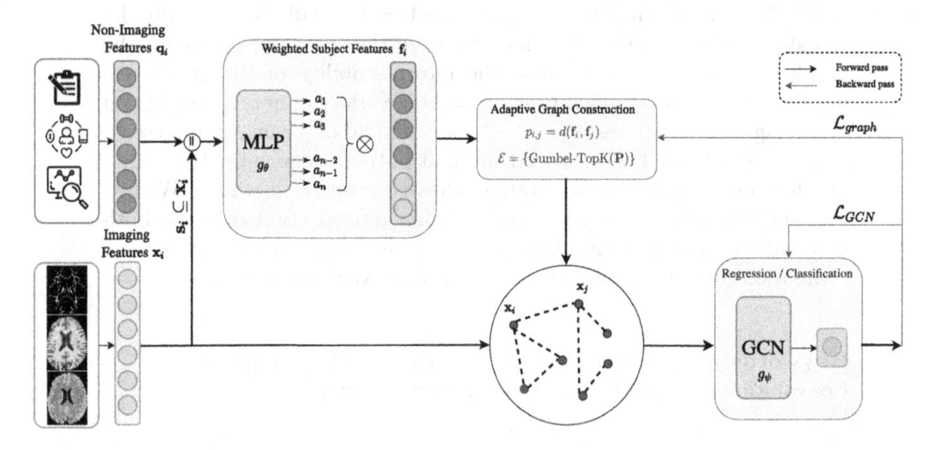

Fig. 1. Proposed methodology: The non-imaging features $\mathbf{q}_i$ and a subset of the imaging features $\mathbf{s}_i \subseteq \mathbf{x}_i$ per subject i are used as input to a MLP, which produces attention weights for each one of these features. Based on these, the edges of the graph are stochastically sampled using the Gumbel-Top-k trick. The constructed graph, which uses the imaging features $\mathbf{X}$ as node features is used to train a GCN for brain age estimation tasks.

are connected based on the imaging features and do not take advantage of the associations of the non-imaging information for the edges. In [11], non-imaging information is used for graph connectivity, but the edges are already pre-pruned similar to [21], and only the weights of the edges can be modified. In [26], even though the graph is dynamic, it is not being learnt. While the extracted graphs in these studies are optimized for the task at hand, there is no explanation for the node connections that were proposed. Given that interpretability is essential in the medical domain [24], since the tools need to be trusted by the clinicians, an adaptive graph learnt during the training, whose connections are also interpretable and clear can prove important. Additionally, most existing works focus on brain age classification in four bins, usually classifying the subjects per decade [7,13,14]. Age regression, which is a more challenging task, has not been extensively explored with published results not reaching sufficient levels of performance [25].

Contributions. This paper has the following contributions: 1) We combine imaging and non-imaging information in an attention-based framework to learn adaptively an optimized graph structure for brain age estimation. 2) We propose a novel graph loss that enables end-to-end training for the task of brain age regression. 3) Our framework is inherently interpretable as the attention mechanism allows us to rank all imaging and non-imaging phenotypes according to their significance for the task. 4) We evaluate our method on the UK Biobank (UKBB) and achieve state-of-the-art results on the tasks of brain age regression and classification. The code can be found on GitHub at: https://github.com/bintsi/adaptive-graph-learning.

2 Methods

Given a set of N subjects with M features $\mathbf{X} = [\mathbf{x}_1, ..., \mathbf{x}_N] \in \mathbb{R}^{N \times M}$ and labels $\mathbf{y} \in \mathbb{R}^N$, a population graph is defined as $\mathcal{G} = \{\mathcal{V}, \mathcal{E}\}$, where $\mathcal{V}$ is a set of nodes, one per subject, and $\mathcal{E}$ is a set of paired nodes that specifies the connectivity of the graph, meaning the edges of the graph. To create an optimized set of edges for the task of brain age estimation, we leverage a set of non-imaging phenotypes $\mathbf{q}_i \in \mathbb{R}^Q$ and a set of imaging phenotypes $\mathbf{s}_i \in \mathbb{R}^S$ per subject i through an attention-based framework. The imaging phenotypes are a subset of the imaging features $\mathbf{s}_i \subseteq \mathbf{x}_i$. The phenotypes are selected according to [5]. The resulting graph is used as input in a GCN that performs node-level prediction tasks. Here, we give a detailed description of the proposed architecture, in which the connectivity of the graph $\mathcal{E}$, is learnt through end-to-end training in order to find the optimal graph structure for the task of brain age estimation. An outline of the proposed pipeline may be found in Fig. 1.

Attention Weights Extraction. Based on the assumption that not all phenotypes are equally important for the construction of the graph, we train a MLP g_θ, with parameters θ, which takes as input both the non-imaging features $\mathbf{q}_i$ and the imaging features $\mathbf{s}_i$ for every subject i and outputs an attention weight vector $\mathbf{a} \in \mathbb{R}^{Q+S}$, where every element of $\mathbf{a}$ corresponds to a specific phenotype. Intuitively, we expect that the features that are relevant to brain age estimation will get attention weights close to 1, and close to 0 otherwise. Since we are interested in the overall relevance of the phenotypes for the task, the output weights need to be global and apply to all of the nodes. To do so, we average the attention weights across subjects and normalize them between 0 and 1.

Edge Extraction. The weighted phenotypes for each subject i are calculated as $\mathbf{f}_i = \mathbf{a} \odot (\mathbf{q}_i \parallel \mathbf{s}_i)$, where $\mathbf{f}_i \in \mathbb{R}^{Q+S}$, $\mathbf{a}$ are the attention weights produced by the MLP g_θ, $(\cdot \parallel \cdot)$ denotes the concatenation function between two vectors and $\odot$ denotes the Hadamard product. We further define the probability $p_{ij}(\mathbf{f}_i, \mathbf{f}_j; \theta, t)$ of a pair of nodes $(i, j) \in \mathcal{V}$ to be connected in Eq. (1):

$$p_{ij}(\mathbf{f}_i, \mathbf{f}_j; \theta, t) = e^{-td(\mathbf{f}_i, \mathbf{f}_j)^2} \tag{1}$$

where t is a learnable parameter and d is a distance function that calculates the distance between the weighted phenotypes of two nodes. To keep the memory cost low, we create a sparse k-degree graph. We use the Gumbel-Top-k trick [17], which acts as a stochastic relaxation of the kNN rule, in order to sample k edges for every node according to the probability matrix $\mathbf{P} \in \mathbb{R}^{N \times N}$. Since there is stochasticity in the sampling scheme, multiple runs are performed at inference time and the predictions are averaged.

Optimization. The extracted graph is used as input, along with the imaging features $\mathbf{X}$, which are used as node features, to a GCN g_ψ, with parameters ψ, which comprises of a number of graph convolutional layers, followed by fully connected layers. The pipeline is trained end-to-end with a loss function $\mathcal{L}$ that consists of two components and is defined as in Eq. (2).

$$\mathcal{L} = \mathcal{L}_{GCN} + \mathcal{L}_{graph}. \tag{2}$$

The first component, $\mathcal{L}_{GCN}$, is optimizing the GCN, g_ψ. For regression we use the Huber loss [12], while for classification we use the Cross Entropy loss function. The second component, $\mathcal{L}_{graph}$, optimizes the MLP g_θ, whose output are the phenotypes' attention weights. However, the graph is sparse with discrete edges and hence the network cannot be trained with backpropagation as is. To alleviate this issue, we formulate our graph loss in a way that rewards edges that lead to correct predictions and penalize edges that lead to wrong predictions. Inspired by [13], where a similar approach was used for classification, the proposed graph loss function is designed for regression instead and is defined in Eq. (3):

$$\mathcal{L}_{graph} = \sum_{(i,j) \in \mathcal{E}} \rho(y_i, g_\psi(\mathbf{x}_i))\, log(p_{ij}), \tag{3}$$

where $\rho(\cdot, \cdot)$ is the reward function which is defined in Eq. (4):

$$\rho(y_i, g_\psi(\mathbf{x}_i)) = |y_i - g_\psi(\mathbf{x}_i)| - \varepsilon \tag{4}$$

Here ε is the null model's prediction (i.e. the average brain age of the training set). Intuitively, when the prediction error $|y_i - g_\psi(\mathbf{x}_i)|$ is smaller than the null

Table 1. Performance of the proposed approach in comparison with a traditional machine learning model, static graph baselines, and the state-of-the-art for age regression (left), and 4-class age classification (right). (MAE in years)

Method	MAE	r score
Linear Regression	3.82	0.75
Static (Node features)	3.87	0.75
Static (Phenotypes)	3.98	0.74
DGM	3.72	0.75
Proposed	**3.61**	**0.79**

Method	Accuracy	AUC	F1
Logistic Regression	0.54	0.79	0.54
Static (Node features)	0.53	0.75	0.52
Static (Phenotypes)	0.52	0.75	0.51
DGM	0.55	**0.80**	0.55
Proposed	**0.58**	**0.80**	**0.56**

Table 2. Ablation studies. Left: Effect of the choice of phenotypes used for the connection of the edges. Right: Performance of the proposed using different distances on the phenotypes to connect the edges. (MAE in years)

No. of phenotypes	MAE	r score
20 (non-imaging only)	4.63	0.68
35 (non-imaging & imaging)	**3.61**	**0.79**
50 (non-imaging & imaging)	3.65	0.78
68 (imaging only)	3.73	0.77

Similarity Metric	MAE	r score
Random	5.59	0.61
Euclidean	**3.61**	**0.79**
Cosine	3.7	0.78
Hyperbolic	4.08	0.72

model's prediction then the reward function will be negative. In turn, this will encourage the maximization of p_{ij} so that $\mathcal{L}_{graph}$ is minimized.

3 Experiments

Dataset. The proposed framework is evaluated on the UKBB [3], which provides not only a wide collection of images of vital organs, including brain scans, but also various non-imaging information, such as demographics, and biomedical, lifestyle, and cognitive performance measurements. Hence, it is perfectly suitable for brain age estimation tasks that incorporate the integration of imaging and non-imaging information. Here, we use 68 neuroimaging phenotypes and 20 non-imaging phenotypes proposed by [5] as the ones most relevant to brain age in UKBB. The neuroimaging features are provided by UKBB, and include measurements extracted from structural MRI and diffusion weighted MRI. All phenotypes are normalized from 0 to 1. The age range of the subjects is 47–81 years. We only keep the subjects that have available the necessary phenotypes ending up with about 6500 subjects. We split the dataset into 75% for the training set, 5% for the validation set, and 20% for the test set. Our pipeline has been primarily designed to tackle the challenging regression task and therefore the main experiment is brain age regression. We also evaluate our framework on the 4-class classification task that has been used in other related papers.

Baselines. Given that a GCN trained on a meaningless graph can perform even worse than a simple regressor/classifier, our first baseline is a Linear/Logistic

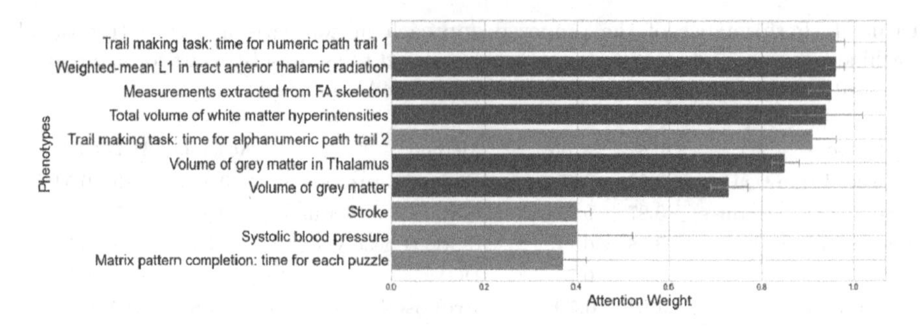

Fig. 2. Attention weights for the top 5 imaging (blue) and top 5 non-imaging phenotypes (pink) averaged across 10 runs of the pipeline. (Color figure online)

Regression model. For the second baseline, we construct a static graph based on a similarity metric, more specifically cosine similarity, of a set of features (either node features or non-imaging and imaging phenotypes) using the kNN rule with $k = 5$ and train a GCN on that graph. Using euclidean distance as the similarity metric leads to worse performance and is therefore not explored for the baselines. We also compare our method with DGM, which is the state-of-the-art on graph learning for medical applications [13]. Since this work is only applicable for classification tasks, we extend it to regression, implementing the graph loss function we used for our pipeline as well.

Implementation Details. The GCN architecture uses ReLU activations and consists of one graph convolutional layer with 512 units and one fully connected layer with 128 units before the regression/classification layer. The number and the dimensions of the layers are determined through hyperparameter search based on validation performance. The networks are trained with the AdamW optimizer [20], with a learning rate of 0.005 for 300 epochs and utilize an early stopping scheme. The average brain age of the training set, which is used in Eq. (3), is $\varepsilon = 6$. We use PyTorch [22] and a Titan RTX GPU. The reported results are computed by averaging 10 runs with different initializations.

3.1 Results

Regression. For the evaluation of the pipeline in the regression task, we use Mean Absolute Error (MAE), and the Pearson Correlation Coefficient (r score). A summary of the performance of the proposed and competing methods is available at Table 1. Linear regression (MAE = 3.82) outperforms GCNs trained on static graphs whether these are based on the node features (MAE = 3.87) or leverage the phenotypes (MAE = 3.98). The DGM outperforms the baselines (MAE = 3.72). The proposed method outperforms all others (MAE = 3.61).

Classification. For the classification task, we divide the subjects into four balanced classes. The metrics used for the classification task to evaluate the performance of the model are accuracy, the area under the ROC curve (AUC), which is

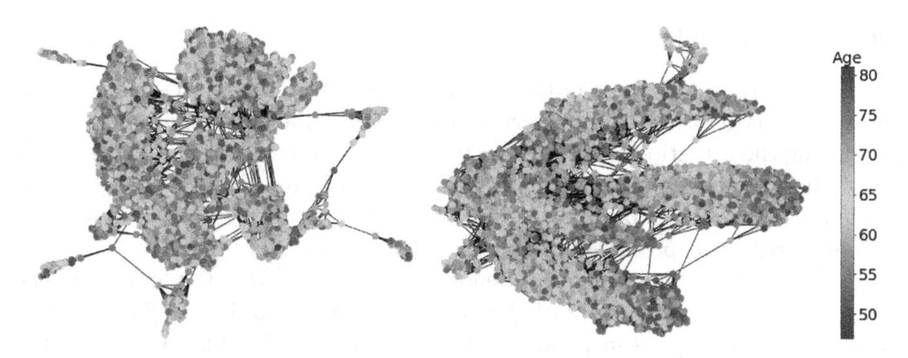

Fig. 3. Population-graph visualizations. Left: graph constructed based on the cosine similarity of the phenotypes. Right: graph constructed based on the cosine similarity of the weighted phenotypes, where the attention weights are the ones extracted from the proposed pipeline. The color of the nodes corresponds to the subject's age. It is evident that using the attention weights increases the homophily of the graph.

defined as the average of the AUC curve of every class, and the Macro F1-score (Table 1). A similar trend to the regression task appears here, with the proposed method reaching top performance with 58% accuracy.

Our hypothesis that the construction of a pre-defined graph structure is sub-optimal, and might even hurt performance, is confirmed. Both for the classification and regression tasks, GCNs trained on static graphs do not outperform a simple linear model that ignores completely the structure of the graph. The proposed approach proves that not all phenotypes are equally important for the construction of the graph, and that giving attention weights accordingly increases the performance for both tasks.

3.2 Ablation Studies

Number of Phenotypes. We perform an ablation test to investigate the effect of the number of features used for the extraction of the edges. We used only non-imaging phenotypes, only imaging phenotypes, or a combination of both (Table 2). The combination of imaging and non-imaging phenotypes (MAE = 3.61) performs better than using either one of them, while adding more imaging phenotypes does not necessarily improve performance.

Distance Metrics. Moreover, we explore how different distance metrics affect performance. Here, euclidean, cosine, and hyperbolic [19] distances are explored. We also include a random edge selection to examine whether using the phenotypic information improves the results. The results (Table 2) indicate that euclidean distance performs the best, closely followed by cosine similarity. Hyperbolic performed a bit worse (MAE = 4 years), while random edges gave a MAE of 5.59. Regardless of the distance metric, phenotypes do incorporate valuable information as performance is considerably better than using random edges.

3.3 Interpretability

A very important advantage of the pipeline is that the graph extracted through the training is interpretable, in terms of why two nodes are connected or not. Visualizing the attention weights given to the phenotypes, we get an understanding of the features that are the most relevant to brain aging. The regression problem is clinically more important, thus we will focus on this in this section. A similar trend was presented for the classification task as well.

The imaging and non-imaging phenotypes that were given the highest attention scores in the construction of the graph can be seen in Fig. 2. We color the non-imaging phenotypes in pink, and the imaging ones in blue. A detailed list of the names of the phenotypes along with the attention weights given by the pipeline can be found in the Supplementary Material. The non-imaging phenotypes that were given the highest attention weights, were two cognitive tasks, the numeric and the alphanumeric trail making tasks. Systolic blood pressure, and stroke, were the next most important non-imaging phenotypes, even though they were not as important as some of the imaging features. Various neuroimaging phenotypes were considered important, such as information regarding the tract anterior thalamic radiaton, the volume of white matter hyperintensities, gray matter volumes, as well as measurements extracted from the FA (fractional anisotropy, a measure of the connectivity of the brain) skeleton. Our findings are in agreement with the relevant literature [5].

Apart from the attention weights, we also visualize the population graph that was used as the static graph of the baseline (Fig. 3(left)). This population graph is constructed based on the cosine similarity of the phenotypes, with all the phenotypes playing an equally important role for the connectivity. In addition, we visualize the population graph using the cosine similarity of the weighted phenotypes (Fig. 3(right)), with the attention weights provided by our trained pipeline. In both of the graph visualizations, the color of each node corresponds to the subject's age. It is clear that learning the graph through our pipeline results in a population graph where subjects with similar ages end up in more compact clusters, whereas the static graph does not demonstrate any form of organization. Since GCNs are affected by a graph construction with low homophily, it is reasonable that the static graphs perform worse than a simple machine learning method, and why the proposed approach manages to produce state-of-the-art results. Using a different GNN architecture that is not as dependent on the constructed graph's homophily could prove beneficial [10,29].

4 Conclusion

In this paper, we propose an end-to-end pipeline for adaptive population-graph learning that is optimized for brain age estimation. We integrate multimodal information and use attention scores for the construction of the graph, while also increasing interpretability. The graph is sparse, which minimizes the computational costs. We implement the approach both for node regression and node classification and show that we outperform both the static graph-based and state-of-the-art approaches. We finally provide an insight into the most important phenotypes for graph construction, which are in agreement with related neurobiological literature.

In future work, we plan to extract node features from the latent space of a CNN. Training end-to-end will focus on latent imaging features that are also important for the GCN. Such features would potentially be more expressive and improve the overall performance. Finally, we leverage the UKBB because of the wide variety of multimodal data it contains, which makes it a perfect fit for brain age estimation but as a next step we also plan to evaluate our framework on different tasks and datasets.

Acknowledgements. KMB would like to acknowledge funding from the EPSRC Centre for Doctoral Training in Medical Imaging (EP/L015226/1).

References

1. Ahmedt-Aristizabal, D., Armin, M.A., Denman, S., Fookes, C., Petersson, L.: Graph-based deep learning for medical diagnosis and analysis: past, present and future. Sensors **21**(14), 4758 (2021)
2. Alam, S.B., Nakano, R., Kamiura, N., Kobashi, S.: Morphological changes of aging brain structure in MRI analysis. In: 2014 Joint 7th International Conference on Soft Computing and Intelligent Systems (SCIS) and 15th International Symposium on Advanced Intelligent Systems (ISIS), pp. 683–687. IEEE (2014)
3. Alfaro-Almagro, F., et al.: Image processing and quality control for the first 10,000 brain imaging datasets from UK biobank. NeuroImage **166**, 400–424 (2018)
4. Anirudh, R., Thiagarajan, J.J.: Bootstrapping graph convolutional neural networks for autism spectrum disorder classification. In: ICASSP 2019-2019 IEEE International Conference on Acoustics, Speech and Signal Processing (ICASSP), pp. 3197–3201. IEEE (2019)
5. Cole, J.H.: Multimodality neuroimaging brain-age in UK biobank: relationship to biomedical, lifestyle, and cognitive factors. Neurobiol. Aging **92**, 34–42 (2020)
6. Cole, J.H., et al.: Predicting brain age with deep learning from raw imaging data results in a reliable and heritable biomarker. Neuroimage **163**, 115–124 (2017)
7. Cosmo, L., Kazi, A., Ahmadi, S.-A., Navab, N., Bronstein, M.: Latent-graph learning for disease prediction. In: Martel, A.L., et al. (eds.) MICCAI 2020. LNCS, vol. 12262, pp. 643–653. Springer, Cham (2020). https://doi.org/10.1007/978-3-030-59713-9_62
8. Davatzikos, C., Bhatt, P., Shaw, L.M., Batmanghelich, K.N., Trojanowski, J.Q.: Prediction of MCI to ad conversion, via MRI, CSF biomarkers, and pattern classification. Neurobiol. Aging **32**(12), 2322-e19 (2011)
9. Franke, K., Gaser, C.: Ten years of brainage as a neuroimaging biomarker of brain aging: what insights have we gained? Front. Neurol. 789 (2019)
10. Hamilton, W., Ying, Z., Leskovec, J.: Inductive representation learning on large graphs. In: Advances in Neural Information Processing Systems, vol. 30 (2017)
11. Huang, Y., Chung, A.C.S.: Edge-variational graph convolutional networks for uncertainty-aware disease prediction. In: Martel, A.L., et al. (eds.) MICCAI 2020. LNCS, vol. 12267, pp. 562–572. Springer, Cham (2020). https://doi.org/10.1007/978-3-030-59728-3_55

12. Huber, P.J.: Robust estimation of a location parameter. In: Kotz, S., Johnson, N.L. (eds.) Breakthroughs in Statistics: Methodology and Distribution, pp. 492–518. Springer, New York (1992). https://doi.org/10.1007/978-1-4612-4380-9_35
13. Kazi, A., Cosmo, L., Ahmadi, S.A., Navab, N., Bronstein, M.: Differentiable graph module (DGM) for graph convolutional networks. IEEE Trans. Pattern Anal. Mach. Intell. **45**(2), 1606–1617 (2022)
14. Kazi, A., Farghadani, S., Navab, N.: IA-GCN: interpretable attention based graph convolutional network for disease prediction. arXiv preprint arXiv:2103.15587 (2021)
15. Kazi, A., et al.: InceptionGCN: receptive field aware graph convolutional network for disease prediction. In: Chung, A.C.S., Gee, J.C., Yushkevich, P.A., Bao, S. (eds.) IPMI 2019. LNCS, vol. 11492, pp. 73–85. Springer, Cham (2019). https://doi.org/10.1007/978-3-030-20351-1_6
16. Kipf, T.N., Welling, M.: Semi-supervised classification with graph convolutional networks. arXiv preprint arXiv:1609.02907 (2016)
17. Kool, W., Van Hoof, H., Welling, M.: Stochastic beams and where to find them: the gumbel-top-k trick for sampling sequences without replacement. In: International Conference on Machine Learning, pp. 3499–3508. PMLR (2019)
18. Koutsouleris, N., et al.: Accelerated brain aging in schizophrenia and beyond: a neuroanatomical marker of psychiatric disorders. Schizophr. Bull. **40**(5), 1140–1153 (2014)
19. Krioukov, D., Papadopoulos, F., Kitsak, M., Vahdat, A., Boguná, M.: Hyperbolic geometry of complex networks. Phys. Rev. E **82**(3), 036106 (2010)
20. Loshchilov, I., Hutter, F.: Decoupled weight decay regularization. arXiv preprint arXiv:1711.05101 (2017)
21. Parisot, S., et al.: Disease prediction using graph convolutional networks: application to autism spectrum disorder and Alzheimer's disease. Med. Image Anal. **48**, 117–130 (2018)
22. Paszke, A., et al.: PyTorch: an imperative style, high-performance deep learning library. In: Advances in Neural Information Processing Systems, vol. 32 (2019)
23. Reeve, A., Simcox, E., Turnbull, D.: Ageing and Parkinson's disease: why is advancing age the biggest risk factor? Ageing Res. Rev. **14**, 19–30 (2014)
24. Shaban-Nejad, A., Michalowski, M., Buckeridge, D.L.: Explainability and interpretability: keys to deep medicine. In: Shaban-Nejad, A., Michalowski, M., Buckeridge, D.L. (eds.) Explainable AI in Healthcare and Medicine. SCI, vol. 914, pp. 1–10. Springer, Cham (2021). https://doi.org/10.1007/978-3-030-53352-6_1
25. Stankeviciute, K., Azevedo, T., Campbell, A., Bethlehem, R., Lio, P.: Population graph GNNs for brain age prediction. In: ICML Workshop on Graph Representation Learning and Beyond (GRL+), pp. 17–83 (2020)
26. Wang, Y., Sun, Y., Liu, Z., Sarma, S.E., Bronstein, M.M., Solomon, J.M.: Dynamic graph CNN for learning on point clouds. ACM Trans. Graph. (tog) **38**(5), 1–12 (2019)
27. Wei, S., Zhao, Y.: Graph learning: a comprehensive survey and future directions. arXiv preprint arXiv:2212.08966 (2022)
28. Zheng, S., et al.: Multi-modal graph learning for disease prediction. IEEE Trans. Med. Imaging **41**(9), 2207–2216 (2022)
29. Zheng, X., Liu, Y., Pan, S., Zhang, M., Jin, D., Yu, P.S.: Graph neural networks for graphs with heterophily: a survey. arXiv preprint arXiv:2202.07082 (2022)
30. Zhu, J., Yan, Y., Zhao, L., Heimann, M., Akoglu, L., Koutra, D.: Beyond homophily in graph neural networks: current limitations and effective designs. Adv. Neural. Inf. Process. Syst. **33**, 7793–7804 (2020)

BrainUSL: Unsupervised Graph Structure Learning for Functional Brain Network Analysis

Pengshuai Zhang[1,2], Guangqi Wen[1,2], Peng Cao[1,2,3](✉), Jinzhu Yang[1,2,3], Jinyu Zhang[1,2], Xizhe Zhang[4], Xinrong Zhu[5], Osmar R. Zaiane[6], and Fei Wang[5](✉)

[1] Computer Science and Engineering, Northeastern University, Shenyang, China
caopeng@cse.neu.edu.cn
[2] Key Laboratory of Intelligent Computing in Medical Image of Ministry of Education, Northeastern University, Shenyang, China
[3] National Frontiers Science Center for Industrial Intelligence and Systems Optimization, Shenyang, China
[4] Biomedical Engineering and Informatics of Nanjing Medical University, Nanjing, China
[5] Early Intervention Unit, Department of Psychiatry, Affiliated Nanjing Brain Hospital, Nanjing Medical University, Nanjing, China
fei.wang@yale.edu
[6] Amii, University of Alberta, Edmonton, AB, Canada

Abstract. The functional connectivity (FC) between brain regions is usually estimated through a statistical dependency method with functional magnetic resonance imaging (fMRI) data. It inevitably yields redundant and noise connections, limiting the performance of deep supervised models in brain disease diagnosis. Besides, the supervised signals of fMRI data are insufficient due to the shortage of labeled data. To address these issues, we propose an end-to-end unsupervised graph structure learning method for sufficiently capturing the structure or characteristics of the functional brain network itself without relying on manual labels. More specifically, the proposed method incorporates a graph generation module for automatically learning the discriminative graph structures of functional brain networks and a topology-aware encoding module for sufficiently capturing the structure information. Furthermore, we also design view consistency and correlation-guided contrastive regularizations. We evaluated our model on two real medical clinical applications: the diagnosis of Bipolar Disorder (BD) and Major Depressive Disorder (MDD). The results suggest that the proposed method outperforms state-of-the-art methods. In addition, our model is capable of identifying associated

P. Zhang and G. Wen—Contribute equally to this work.

Supplementary Information The online version contains supplementary material available at https://doi.org/10.1007/978-3-031-43993-3_20.

H. Greenspan et al. (Eds.): MICCAI 2023, LNCS 14227, pp. 205–214, 2023.
https://doi.org/10.1007/978-3-031-43993-3_20

biomarkers and providing evidence of disease association. To the best of our knowledge, our work attempts to construct learnable functional brain networks with unsupervised graph structure learning. Our code is available at https://github.com/IntelliDAL/Graph/tree/main/BrainUSL.

Keywords: Functional connectivity analysis · Graph structure learning · Unsupervised learning · fMRI

1 Introduction

Recent studies have shown that rs-fMRI based analysis for brain functional connectivity (FC) is effective in helping understand the pathology of brain diseases [8,17,25]. The functional connectivity in the brain network can be modeled as the graph where nodes denote the brain regions and the edges represent the correlations between those regions [6]. Hence, the brain disease identification can be seen as the graph classification with the refined graph structures [28].

The representation learning of brain network heavily relies on the graph structure quality. The existing brain network construction methods [16,23] are often noisy or incomplete due to the inevitably error-prone data measurement or collection. The noisy or incomplete graphs often lead to unsatisfactory representations and prevent us from fully understanding the mechanism underlying the disease. In pursuit of an optimal graph structure for graph classification, recent studies have sparked an effort around the central theme of Graph Structure Learning (GSL), which aims to learn an optimized graph structure and corresponding graph representations. However, most works for GSL rely on the human annotation, which plays an important role in providing supervision signals for structure improvement. Since the fMRI data is expensive and limited, unsupervised graph structure learning is urgently required [1,4,14]. Moreover, disease interpretability is essential as it can help decision-making during diagnosis.

Considering the above issues, we aim to discover useful graph structures via a learnable graph structure from the BOLD signals instead of measuring the associations between brain regions by a similarity estimation. In this paper, we propose an end-to-end unsupervised graph structure learning framework for functional brain network analysis (BrainUSL) directly from the BOLD signals. The unsupervised graph structure learning consists of a graph generation module and the topology-aware encoder. We propose three loss functions to constrain the graph structure learning, including the sparsity-inducing norm, the view consistency regularization and the correlation-guided contrastive loss. Finally, the generated graph structures are used for the graph classification. We evaluate our model on two real medical clinical applications: Bipolar Disorder diagnosis and Major Depressive Disorder diagnosis. The results demonstrate that our BrainUSL achieves remarkable improvements and outperforms state-of-the-art methods. The main contributions of this paper are summarized below:

- We propose an end-to-end unsupervised graph structure learning method for functional brain network analysis.
- We propose the correlation-guided contrastive loss to model the correlations between graphs by defining the sample correlation estimation matrix.

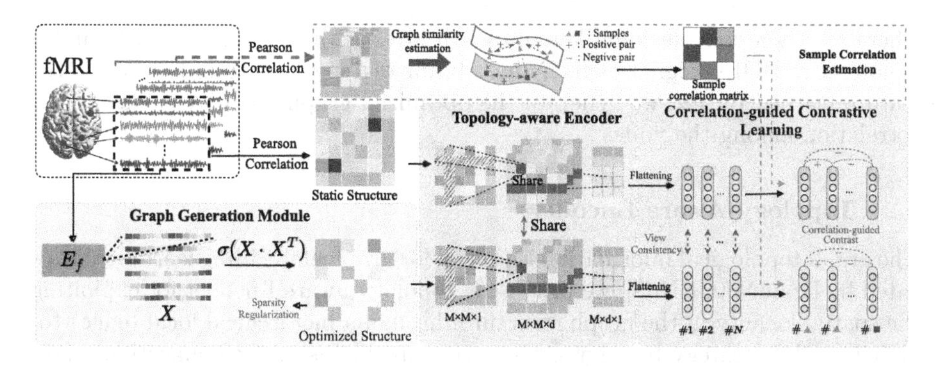

Fig. 1. Illustration of the proposed BrainUSL. The unsupervised graph structure learning module consists of a graph generation module for generating optimized sparsity-induced graph structures, a topology-aware encoder for capturing the essential topological representation and a correlation-guided contrastive learning for exploiting discriminative and sparse graph structures. Then, three loss functions are proposed for guiding the procedure of graph structure learning. Finally, the generated graph structure for each subject and the topology-aware encoder are further used for the downstream classification.

- Our method provides a perspective for disease interpretable analysis and association analysis between BD and MDD.
- The experimental results demonstrate the advantage of the proposed method in brain disorder diagnosis.

2 Method

The constructed graph structure of brain network in existing works are often noisy or incomplete. To address this issue, we propose a novel unsupervised graph structure learning method, including the graph generation module for generating optimized sparsity-induced graphs and the topology-aware encoder for capturing the topological information in graphs, as illustrated in Fig. 1. Then, we propose a new objective function for unsupervised graph structure learning from the perspectives of constraining structure sparsity as well as view consistency and preserving discriminative patterns at the same time.

2.1 Graph Generation Module

To exploit the information in fMRI signals for generating the optimized sparsity-induced graph structure, we propose a graph generation module, which containing a graph generation module contains BOLD signal feature aggregation (E_f) with a stack of convolutional layers [7] for learning the low-dimensional BOLD signal features. The feature is learned as follows:

$$E_f^{(l+1)}(u) = \sum_{s=0}^{U-1} E_f^{(l)}(u-s) * \mathcal{K}^{(l)}(s), \tag{1}$$

where $\mathcal{K}^{(l)}$ is a convolutional kernel of l-th layer with a kernel size of U and u denotes the BOLD signal element in a brain region of the input x. With the feature learned by E_f, we generate the optimized graph A_G by calculating the correlation among the nodes.

2.2 Topology-Aware Encoder

The graph topological information is crucial for graph embedding learning. Motivated by BrainNetCNN [5], we propose a Topology-aware Encoder for exploiting the spatial locality in the graph structure through a hierarchical local (edge)-to-global (node) strategy by aggregating the embeddings of the connections associated with the nodes at the two ends of each edge. The topology-aware encoder involves an operator of *edge aggregation* (E_g) with multiple cross-shaped filters for capturing the spatial locality in the graph and *node aggregation* (E_n) for aggregating the associated edge. The *cross-shaped filters* in edge aggregation involve a combination of $1 \times M$ and $M \times 1$ basis filters with horizontal and vertical orientations, which are defined as:

$$H_g = E_g(A) = \sum_{i=0}^{M} \sum_{j=0}^{M} A^{(i,\cdot)} w^r + A^{(\cdot,j)} w^h, \tag{2}$$

where $w^r \in \mathbb{R}^{1 \times M}$ and $w^h \in \mathbb{R}^{M \times 1}$ denote the learned vectors of the horizontal and vertical convolution kernel, M denotes the number of ROIs, A and H_g denote the adjacency matrix and the edge embeddings. With the learned edge embeddings, we further learn the node embeddings by aggregating the associated edges with the nodes with a learnable layer. More specifically, the node aggregation takes the edge embedding as the inputs and obtains the node embedding from a node-wise view by a 1D convolutional filter. The node aggregation is defined as:

$$H_n = E_n(H_g) = \sum_{i=1}^{M} H_g^{(i,\cdot)} w^n, \tag{3}$$

where $H_n \in \mathbb{R}^{M \times d}$ is the node embedding, and $w^n \in \mathbb{R}^{1 \times M}$ is the learned vector of the filter, and d is the dimensionality of the node embeddings.

2.3 Objective Functions

To better exploit the graph structure, we design three loss functions including a sparsity-inducing norm, a view consistency regularization and a correlation-guided contrastive loss. We assume that the sparsity of the generated graphs allows for the preservation of the important edges and removal of noise. To achieve this, we utilize an l_1 norm to remove the irrelevant connections and preserve the sparsity of the generated graphs.

Furthermore, we introduce a view consistency regularization to ensure the consistency of two views by maximizing the agreement between the node embeddings learned from the fixed graph structure A_P and learnable graph structure

A_G. The view consistency regularization is defined as $L_{vc} = \sum_{i=1}^{N} sim(e_i, \hat{e}_i)$, where $sim(\cdot, \cdot)$ is a cosine similarity measure, N denotes the number of samples, $\hat{e}_i$ and e_i represent the i-th graph embeddings from A_P and A_G.

The motivation of contrastive learning is to capture the graph embeddings by modeling the correlations between graphs [24]. However, it produces bias depending on the simple data augmentations, which can degrade performance on downstream tasks. Hence, we introduce the graph correlation estimation by graph kernel [27] to construct the correlation matrix $S \in N \times N$. Then, we binarize the matrix by a simple thresholding with a threshold value θ. If $S_{ij} \geq \theta$, we set S_{ij} to 1, which indicates that the i-th and j-th samples are regarded as a positive sample pair. Otherwise, S_{ij} is set to 0, indicating they are considered as a negative sample pair. With the estimated positive and negative pairs, the correlation-guided contrastive loss is defined as:

$$L_{cc} = -\sum_{i=1}^{N} \log \frac{\sum_{S_{ij}=1} \exp(sim(e_i, e_j)/\tau)}{\sum_{j=1}^{N} \mathbb{1}_{i \neq j} \exp(sim(e_i, e_j)/\tau)}, \tag{4}$$

where $\mathbb{1}(\cdot) = \{0, 1\}$ is an indicator function, and τ is a temperature factor to control the desired attractiveness strength.

The final objective function is formulated as:

$$L = L_{vc} + \alpha L_{cc} + \beta \left\| A_G \right\|_1, \tag{5}$$

where α and β are the trade-off hyper-parameters. Finally, based on the generated graphs and pre-trained topology-aware encoder, we leverage the multi-layer perceptron (MLP) with the cross-entropy loss for the graph classification.

3 Experiments and Results

3.1 Dataset and Experimental Details

We evaluated our BrainUSL on a private dataset constructed from Nanjing Medical University (NMU) for BD and MDD diagnosis by repeating the 5-fold cross-validation 5 times with different random seeds. We deal with the original fMRI data by dpabi [20] and divide the whole brain into 116 brain regions based on Automated Anatomical Labeling (AAL) for analysis, which included spatial normalization to Echo Planar Imaging (EPI) template of standard Montreal Neurological Institute (MNI) space (spatial resolution 3 mm × 3 mm × 3 mm), spatial and temporal smoothing with a 6 mm × 6 mm × 6 mm Gaussian kernel and filter processing with adopting 0.01–0.08 Hz low-frequency fluctuations to remove interference signals. The dataset includes 172 health controls (104 females and 68 males, aged 24.89 ± 7.14 years, range 18–43 years), 127 MDDs (90 females and 37 males, age range 17–34 years) and 102 BDs (76 females and 26 males, age range 16–32 years), who were scanned at a single site with identical inclusion and exclusion criteria.

Table 1. Classification results on the diagnosis of BD and MDD. L_{vc}: view consistency regularization; L_{cc}: correlation-guided contrastive loss.

Methods	HC vs. BD				HC vs. MDD			
	ACC (%)	AUC (%)	SEN (%)	SPEC (%)	ACC (%)	AUC (%)	SEN(%)	SPEC (%)
FC + SVM [21]	73.2	70.8	65.3	72.9	68.1	67.5	66.5	75.4
FC + RF [21]	67.7	60.2	60.1	65.0	63.4	59.8	63.8	66.1
GroupINN [22]	67.9	63.3	62.8	67.1	66.8	65.3	63.1	65.8
ASD-DiagNet [2]	73.3	70.1	58.5	79.6	68.2	66.7	60.2	74.3
MVS-GCN [18]	66.9	64.2	62.1	73.5	68.3	68.2	63.8	61.2
ST-GCN [3]	67.1	57.5	56.6	73.5	58.1	52.3	53.3	55.1
BrainNetCNN [5]	73.3	71.7	64.1	79.3	69.4	68.4	61.5	75.2
BrainUSL - w/o l_1	75.5	73.6	**65.2**	82.0	75.4	74.2	67.5	81.0
BrainUSL - w/o L_{vc}	75.6	72.7	62.8	82.7	73.8	73.1	**69.5**	76.7
BrainUSL - w/o L_{cc}	75.0	71.7	59.1	84.3	74.1	73.1	65.4	80.8
BrainUSL (Ours)	**77.3**	**74.4**	63.0	**85.7**	**76.7**	**75.3**	67.6	**82.6**
BrainUSL - BD	–	–	–	–	75.2	74.1	67.8	80.5
BrainUSL - MDD	76.4	73.7	63.2	84.2	–	–	–	–

3.2 Classification Results

We compare our BrainUSL with state-of-the-art models in terms of Accuracy (ACC), Area Under the Curve (AUC), Sensitive (SEN) and Specificity (SPEC). The comparable methods can be grouped into two categories: traditional methods including FC+SVM/RF [21] and deep learning methods including GroupINN [22], ASD-DiagNet [2], MVS-GCN [18], ST-GCN [3] and BrainNetCNN [5].

Comparison with SOTA. Compared with state-of-the-art methods, the proposed BrainUSL generally achieves the best performance on MDD and BD identification, the results are shown in Table 1. Specifically, the results show that our BrainUSL yields the best ACC and AUC results on MDD (ACC = 76.7% and AUC = 75.3%) and BD (ACC = 77.3% and AUC = 74.4%), compared to the existing brain disease diagnosis approaches. Moreover, Fig. 2 illustrates the influence of the pre-training epochs of BrainUSL on the classification performance. It can be found that the performance improves with the pre-training epochs increasing until 40/60 epochs for BD/MDD, which demonstrates that more pre-training epochs help capture accurate structural representation. In addition, the similar sparsity patterns are observed for both the diagnosis of two disorders. The results demonstrate that our generated graphs are more discriminative than the graphs constructed by pearson correlation coefficient, which confirms that the quality of the graph structure is critical for functional brain network representation learning, and noisy or redundant connections in brain network impede understanding of disease mechanisms.

Ablation Study. There are three parts in our final objective function. Next, we perform a sequence of ablation studies on the three parts of our model. As

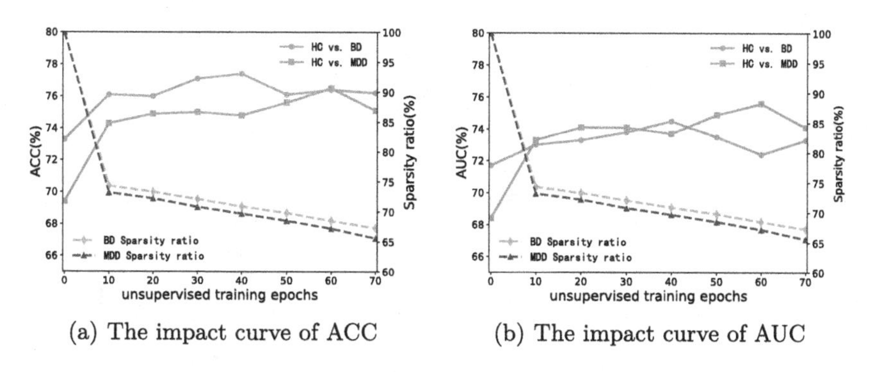

(a) The impact curve of ACC (b) The impact curve of AUC

Fig. 2. The impact of unsupervised training epochs.

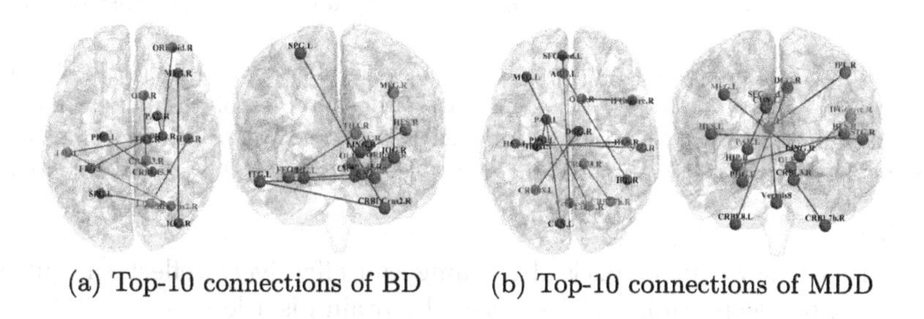

(a) Top-10 connections of BD (b) Top-10 connections of MDD

Fig. 3. Illustration of discriminative connections for brain disease diagnosis.

shown the third part in Table 1, all the proposed loss functions obviously improve the classification performance, showing the crucial role of the each component and the complementary one another. Therefore, our results demonstrate that the graph structure constructed in an unsupervised manner can provide the potential correlations and discriminative information between brain regions precisely.

3.3 Functional Connectivity Analysis

We use BrainNetViewer [19] to illustrate the discriminative top-10 connections identified by our method for brain disease diagnosis in Fig. 3. Neuroimaging studies have demonstrated that the subnetworks of SN, CEM, and DMN are often co-activated or deactivated during emotional expression task. We find that some identified connections in MDD and BD such as Frontal-Mid-R between Frontal-Sup-Medial-L and Angular-L between Thalamus-L are the key connections in DMN, CEN and SN, which demonstrates that our model can generate the discriminative brain structure and facilitate the identification of biomarkers [10–12]. Furthermore, as shown in Fig. 4, by comparing the graphs constructed by PCC and BrainUSL, it can be observed that our method produces a sparser structure for the brain network, indicating that only a small portion of the functional connections are relevant to the outcome. The results indicate that the generated

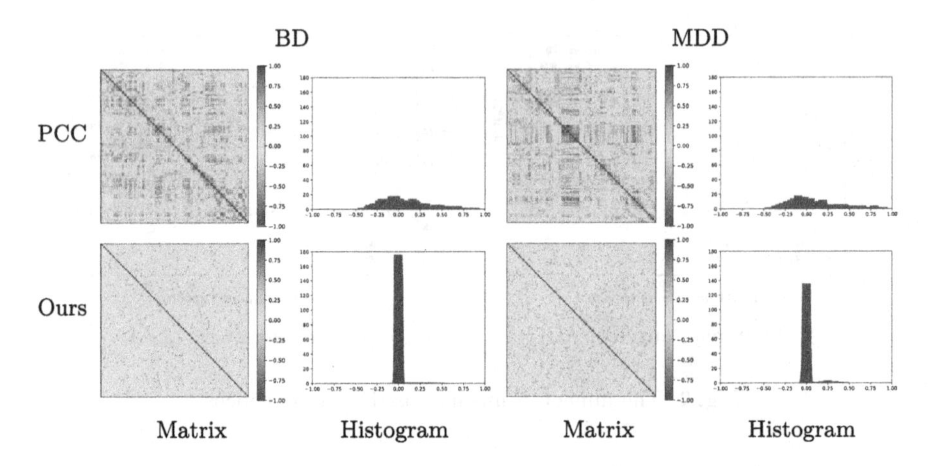

Fig. 4. Differences in connectivity patterns constructed by PCC (Top) and BrainUSL (Bottom). The 1st and 3rd columns indicate two examples of the generated graph structures. The 2nd and 4th columns indicate the histograms of connection strengths excluding the diagonal elements for all the subjects by PCC and our learnable manner.

graph structures via unsupervised learning can effectively reflect the intrinsic connections between brain regions caused by brain disorders.

3.4 Association of Brain Diseases

A number of studies [9,15] have demonstrated there exists associations between different psychiatric disorders [13], patients with one psychiatric disorder are more susceptible to other psychiatric disorders. We evaluate the association between the disorders and the transfer learning ability of our model. Specifically, We pre-train our model on one dataset, then fine-tune and evaluate it on the other dataset. The results are illustrated in the last part of Table 1. We observed that the transfer learning between two different brain disorder disease also achieve a better results compared with other methods. The result indicates that the two diseases are correlated, which is consistent with the existing study results [26]. Moreover, the results also indicate that our model learns more transferable representations and provides a perspective for the study of disease associations through transfer learning on the functional brain network analysis.

4 Conclusion

Due to the inevitably error-prone data measurement or collection, the functional brain networks constructed by existing works are often noisy and incomplete. To address this issue, we propose the unsupervised graph structure learning framework for functional brain network analysis (BrainUSL), which generates the sparse and discriminative graph structures according to the characteristics of the graph data itself. We conducted extensive experiments on the NMU dataset which indicate that our BrainUSL achieves promising performance with the SOTA methods. In addition, we discuss the interpretability of our model and find discriminative correlations in functional brain networks for diagnosis.

Acknowledgment. This paper is supported by the National Natural Science Foundation of China (No. 62076059), the Science Project of Liaoning province (2021-MS-105) and the 111 Project (B16009).

References

1. Chavas, J., Guillon, L., Pascucci, M., Dufumier, B., Rivière, D., Mangin, J.F.: Unsupervised representation learning of cingulate cortical folding patterns. In: Wang, L., Dou, Q., Fletcher, P.T., Speidel, S., Li, S. (eds.) MICCAI 2022, Part I. LNCS, vol. 13431, pp. 77–87. Springer, Cham (2022). https://doi.org/10.1007/978-3-031-16431-6_8
2. Eslami, T., Mirjalili, V., Fong, A., Laird, A.R., Saeed, F.: ASD-DiagNet: a hybrid learning approach for detection of autism spectrum disorder using fMRI data. Front. Neuroinform. **13**, 70 (2019)
3. Gadgil, S., Zhao, Q., Pfefferbaum, A., Sullivan, E.V., Adeli, E., Pohl, K.M.: Spatio-temporal graph convolution for resting-state fMRI analysis. In: Martel, A.L., et al. (eds.) MICCAI 2020. LNCS, vol. 12267, pp. 528–538. Springer, Cham (2020). https://doi.org/10.1007/978-3-030-59728-3_52
4. Imran, A.A.Z., Wang, S., Pal, D., Dutta, S., Zucker, E., Wang, A.: Multimodal contrastive learning for prospective personalized estimation of CT organ dose. In: Wang, L., Dou, Q., Fletcher, P.T., Speidel, S., Li, S. (eds.) MICCAI 2022, Part I. LNCS, vol. 13431, pp. 634–643. Springer, Cham (2022). https://doi.org/10.1007/978-3-031-16431-6_60
5. Kawahara, J., et al.: BrainNetCNN: convolutional neural networks for brain networks; towards predicting neurodevelopment. Neuroimage **146**, 1038–1049 (2017)
6. Khosla, M., Jamison, K., Ngo, G.H., Kuceyeski, A., Sabuncu, M.R.: Machine learning in resting-state fMRI analysis. Magn. Reson. Imaging **64**, 101–121 (2019)
7. Kiranyaz, S., Avci, O., Abdeljaber, O., Ince, T., Gabbouj, M., Inman, D.J.: 1D convolutional neural networks and applications: a survey. Mech. Syst. Sig. Process. **151**, 107398 (2021)
8. Kumar, V., Garg, R.: Resting state functional connectivity alterations in individuals with autism spectrum disorders: a systematic review. medRxiv (2021)
9. Lawry Aguila, A., Chapman, J., Janahi, M., Altmann, A.: Conditional VAEs for confound removal and normative modelling of neurodegenerative diseases. In: Wang, L., Dou, Q., Fletcher, P.T., Speidel, S., Li, S. (eds.) MICCAI 2022, Part I. LNCS, vol. 13431, pp. 430–440. Springer, Cham (2022). https://doi.org/10.1007/978-3-031-16431-6_41
10. Lee, W.H., Frangou, S.: Linking functional connectivity and dynamic properties of resting-state networks. Sci. Rep. **7**(1), 16610 (2017)
11. Lynch, C.J., Uddin, L.Q., Supekar, K., Khouzam, A., Phillips, J., Menon, V.: Default mode network in childhood autism: posteromedial cortex heterogeneity and relationship with social deficits. Biol. Psychiatry **74**(3), 212–219 (2013)
12. Nebel, M.B., et al.: Intrinsic visual-motor synchrony correlates with social deficits in autism. Biol. Psychiatry **79**(8), 633–641 (2016)

13. Radonjić, N.V., et al.: Structural brain imaging studies offer clues about the effects of the shared genetic etiology among neuropsychiatric disorders. Mol. Psychiatry **26**(6), 2101–2110 (2021)
14. Sauty, B., Durrleman, S.: Progression models for imaging data with longitudinal variational auto encoders. In: Wang, L., Dou, Q., Fletcher, P.T., Speidel, S., Li, S. (eds.) MICCAI 2022, Part I. LNCS, vol. 13431, pp. 3–13. Springer, Cham (2022). https://doi.org/10.1007/978-3-031-16431-6_1
15. Seyfioğlu, M.S., et al.: Brain-aware replacements for supervised contrastive learning in detection of Alzheimer's disease. In: Wang, L., Dou, Q., Fletcher, P.T., Speidel, S., Li, S. (eds.) MICCAI 2022, Part I. LNCS, vol. 13431, pp. 461–470. Springer, Cham (2022). https://doi.org/10.1007/978-3-031-16431-6_44
16. Wang, Y., Kang, J., Kemmer, P.B., Guo, Y.: An efficient and reliable statistical method for estimating functional connectivity in large scale brain networks using partial correlation. Front. Neurosci. **10**, 123 (2016)
17. Wang, Z., et al.: Distribution-guided network thresholding for functional connectivity analysis in fMRI-based brain disorder identification. IEEE J. Biomed. Health Inform. **26**(4), 1602–1613 (2021)
18. Wen, G., Cao, P., Bao, H., Yang, W., Zheng, T., Zaiane, O.: MVS-GCN: a prior brain structure learning-guided multi-view graph convolution network for autism spectrum disorder diagnosis. Comput. Biol. Med. **142**, 105239 (2022)
19. Xia, M., Wang, J., He, Y.: BrainNet Viewer: a network visualization tool for human brain connectomics. PLoS ONE **8**(7), e68910 (2013)
20. Yan, C.G., Wang, X.D., Zuo, X.N., Zang, Y.F.: DPABI: data processing & analysis for (resting-state) brain imaging. Neuroinformatics **14**(3), 339–351 (2016). https://doi.org/10.1007/s12021-016-9299-4
21. Yan, K., Zhang, D.: Feature selection and analysis on correlated gas sensor data with recursive feature elimination. Sens. Actuators, B Chem. **212**, 353–363 (2015)
22. Yan, Y., Zhu, J., Duda, M., Solarz, E., Sripada, C., Koutra, D.: GroupINN: grouping-based interpretable neural network for classification of limited, noisy brain data. In: Proceedings of the 25th ACM SIGKDD International Conference on Knowledge Discovery & Data Mining, pp. 772–782 (2019)
23. Yin, W., Li, L., Wu, F.X.: Deep learning for brain disorder diagnosis based on fMRI images. Neurocomputing **469**, 332–345 (2022)
24. You, Y., Chen, T., Sui, Y., Chen, T., Wang, Z., Shen, Y.: Graph contrastive learning with augmentations. In: Advances in Neural Information Processing Systems, vol. 33, pp. 5812–5823 (2020)
25. Zhang, Z., Ding, J., Xu, J., Tang, J., Guo, F.: Multi-scale time-series kernel-based learning method for brain disease diagnosis. IEEE J. Biomed. Health Inform. **25**(1), 209–217 (2020)
26. Zhao, H., Nyholt, D.R.: Gene-based analyses reveal novel genetic overlap and allelic heterogeneity across five major psychiatric disorders. Hum. Genet. **136**, 263–274 (2017). https://doi.org/10.1007/s00439-016-1755-6
27. Zhou, Z.H., Sun, Y.Y., Li, Y.F.: Multi-instance learning by treating instances as non-IID samples. In: Proceedings of the 26th Annual International Conference on Machine Learning, pp. 1249–1256 (2009)
28. Zhu, Y., Xu, W., Zhang, J., Liu, Q., Wu, S., Wang, L.: Deep graph structure learning for robust representations: a survey. arXiv preprint arXiv:2103.03036 (2021)

Learning Asynchronous Common and Individual Functional Brain Network for AD Diagnosis

Xiang Tang[1], Xiaocai Zhang[2], Mengting Liu[1], and Jianjia Zhang[1(✉)]

[1] School of Biomedical Engineering, Shenzhen Campus of Sun Yat-sen University, Shenzhen 518107, China
zhangjj225@mail.sysu.edu.cn

[2] Institute of High Performance Computing, Agency for Science, Technology and Research (A*STAR), Singapore 138632, Singapore

Abstract. Construction and analysis of functional brain network (FBN) with rs-fMRI is a promising method to diagnose functional brain diseases. Traditional methods usually construct FBNs at the individual level for feature extraction and classification. There are several issues with these approaches. Firstly, due to the unpredictable interferences of noises and artifacts in rs-fMRI, these individual-level FBNs have large variability, leading to instability and unsatisfactory diagnosis accuracy. Secondly, the construction and analysis of FBNs are conducted in two successive steps without negotiation with or joint alignment for the target task. In this case, the two steps may not cooperate well. To address these issues, we propose to learn common and individual FBNs adaptively within the Transformer framework. The common FBN is shared, and it would regularize the FBN construction as prior knowledge, alleviating the variability and enabling the network to focus on these disease-specific individual functional connectivities (FCs). Both the common and individual FBNs are built by specially designed modules, whose parameters are jointly optimized with the rest of the network for FBN analysis in an end-to-end manner, improving the flexibility and discriminability of the model. Another limitation of the current methods is that the FCs are only measured with synchronous rs-fMRI signals of brain regions and ignore their possible asynchronous functional interactions. To better capture the actual FCs, the rs-fMRI signals are divided into short segments to enable modeling cross-spatiotemporal interactions. The superior performance of the proposed method is consistently demonstrated in early AD diagnosis tasks on ADNI2 and ADNI3 data sets.

Keywords: Functional brain network · Alzheimer's disease · rs-fMRI · Computer-aided diagnosis · Transformer

1 Introduction

Building and analyzing functional brain network (FBN) based on resting-state magnetic resonance imaging (rs-fMRI) have become a promising approach to

H. Greenspan et al. (Eds.): MICCAI 2023, LNCS 14227, pp. 215–225, 2023.
https://doi.org/10.1007/978-3-031-43993-3_21

functional brain disease diagnosis, e.g., Alzheimer's disease (AD) and Parkinson's disease. Rs-fMRI probes neural activity by the fluctuations in the blood-oxygen-level-dependent (BOLD) signals. The strength of functional connectivities (FCs) between brain regions is measured by the correlation between pairwise BOLD signals. FBN represents the interaction patterns between brain regions during brain functioning and can be used to identify abnormal changes caused by brain diseases [9,28,32]. Such abnormal alterations can be used as diagnostic biomarkers of brain diseases [3,11,14,16,21,22].

Traditional FBNs are constructed at the individual level using Pearson's correlation [26]or sparse representation (SR) [12]. Although this approach is popular due to its simplicity and efficiency, it suffers from multiple limitations. Firstly, due to the unpredictable interferences of noises and artifacts, a potential issue is that the constructing of FBNs at the individual level inevitably leads to large variability in the topographical structure of the networks. So the subtle disease-specific FCs change in the FBNs of patients in comparison with the healthy are likely to be overwhelmed by such large variability [31]. Such variability results in unsatisfactory diagnosis accuracy and poor generality for the classifier. Another issue with the traditional FBN-based methods is that the construction and analysis of FBNs are conducted in two successive steps. Such a two-step approach may not be optimal considering the lack of communication between the two steps. Specifically, the FBNs constructed in the first step do not necessarily work well with the feature extraction methods in the second step since there are no interactions and there is no aligned objective towards solving the target task.

It has been demonstrated that the construction of the common FBN at population level as prior knowledge is beneficial to reduce the effects of the first issue of large variability aforementioned [24]. For example, G. Varoquaux [24] adds a common sparse structure across all subjects as the prior knowledge to construct group consistent individual FBN. Many methods based on Group Sparse Representation (GSR) [15,27] estimate the FBN for all subjects using Group Lasso with $l_{1,2}$-norm constraint to alleviate the variability. These results indicate that a typical whole FBN can be decomposed into a common and an individual FBN components. The common FBN component is shared by all subjects and can be used as prior knowledge regularization to increase the stability of FBN for a subject and facilitate the detection of disease-specific FCs. Inspired by this finding, we propose to learn common and individual FBNs adaptively within Transformer framework to increase the model's stability and efficacy. Unlike the traditional methods, both the common and individual FBNs in the proposed method are built by specially designed modules, whose parameters are jointly optimized within the FBN analysis modules of the network in an end-to-end manner, and this could improve the model's flexibility and discriminability.

Another equally important issue with the existing methods is that the current measures of FC strength only take the synchronous functional interaction into account and ignore the potential asynchronous interactions, as illustrated in Fig. 1(a). Specifically, the current FC measures, e.g., correlation or SR, implicitly assume that the activities of different brain regions start exactly at the same

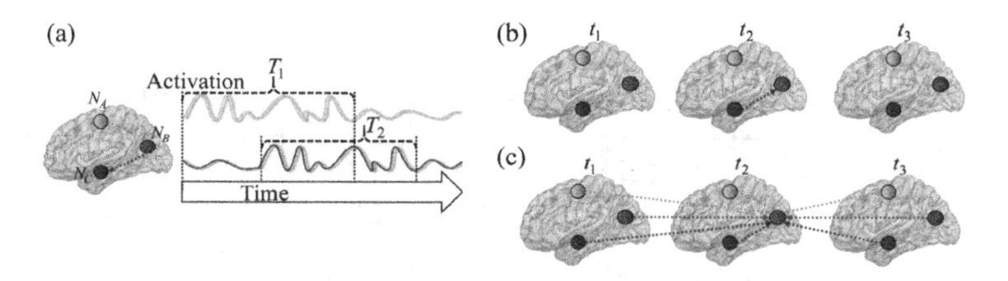

Fig. 1. Illustration of asynchronous functional connectivity. (a) Time-lagged response of N_C in T_2 for N_A' activity in T_1. Direct calculation of the correlation between their BOLD signals during T_1 cannot reflect their actual functional connectivity. (b) Traditional measures of synchronous correlation. (c) The proposed asynchronous cross-spatiotemporal correlation measure.

time, as shown in Fig. 1(b). However, this is not necessarily always the actual case in brain functioning, and certain brain regions may not be synchronized. The time-lagged effects could originate from two-fold: on the one hand, the information flow within the brain takes time, which would cause asynchronous rs-fMRI observations [7,19]; on the other hand, even when various brain regions get activated at the same time, their individual activities may have different patterns and durations, which will also lead to asynchronous time-lagged interactions [4,5]. In this case, only measuring the FCs synchronicity cannot fully capture the underlying functional interactions between brain regions, and this would degrade the model performance. To reflect the asynchronous FCs, we propose to divide the whole BOLD signal into short-term segments with the sliding windows [20]. Then their asynchronous interactions are measured in a specially designed attention module in the Transformer framework, as shown in Fig. 1(c).

To validate the effectiveness of the proposed method, we conducted experimental studies on rs-fMRI data sets of ADNI2 and ADNI3 for early AD diagnosis, i.e., classification of mild cognitive impairment (MCI) and normal control (NC). The experimental results show that the proposed method consistently achieves state-of-the-art performance.

2 Proposed Method

The overall network of our method is shown in Fig. 2(a). Firstly, the raw rs-fMRI BOLD signal $\boldsymbol{X}_{raw} \in R^{N \times T \times D_{raw}}$ is transformed into $\boldsymbol{X}_0 \in R^{N \times T \times D_0}$ by applying 1D temporal convolution for input mapping, where N denotes the number of brain nodes, T denotes frames, and D denotes the feature dimension. Then, before feeding into the L-th Transformer encoder, $\boldsymbol{X}_0$ needs to be position encoded for incorporating node index as spatial identity. The position encoding matrix $\boldsymbol{P}$ is defined as: $\boldsymbol{P}(pos,:,2dim) = sin(pos/10000^{2dim/D})$; $\boldsymbol{P}(pos,:,2dim+1) = cos(pos/10000^{2dim/D})$, where $pos \in [0,\cdots,N-1]$ is the node index, the symbol ':' refers to all indexes in the corresponding dimension, and $dim \in [0,\cdots,\frac{D}{2}-1]$ indicates the index of the feature dimension. $\boldsymbol{P}$ is added to $\boldsymbol{X}_0$, i.e., $\boldsymbol{X_0} = \boldsymbol{X}_0 + \boldsymbol{P}$. If we measure the interactions between brain

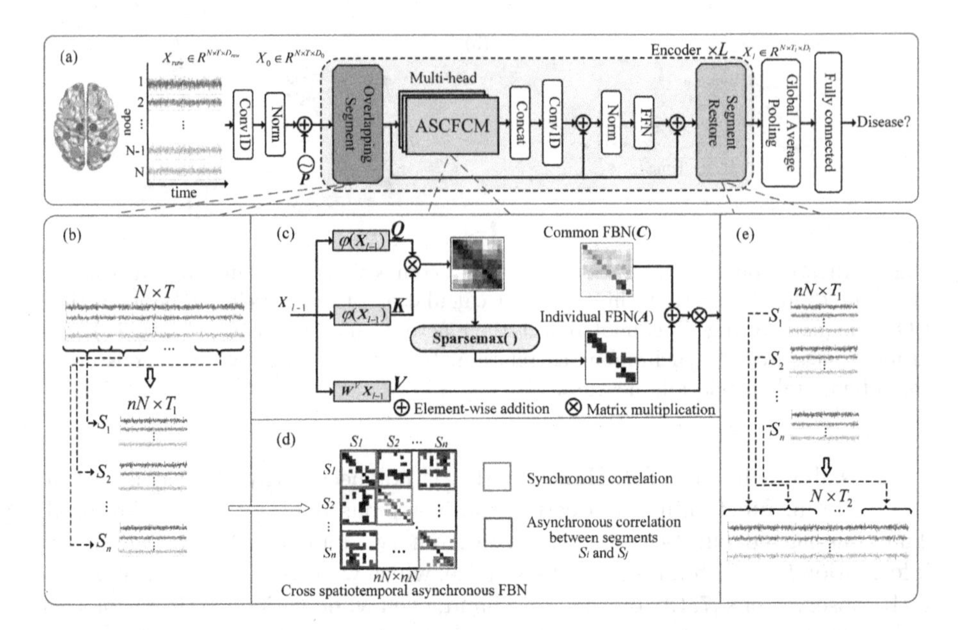

Fig. 2. Illustration of the proposed method. (a) Overall framework. (b) Dividing rs-fMRI signals into segments by sliding window. (c) Attention-based Sparse Common-and-individual FBN Construction Module (ASCFCM). (d) Asynchronous cross spatiotemporal FCs. (e) Restore the rs-fMRI segments.

regions directly, we can obtain synchronous FCs. In order to consider the time-lagged interaction between brain regions, we use a sliding window to segment the signals into short-time segments, the details are illustrated in Fig. 2(b) and introduced in detail later. Then, an improved version of the multi-head self-attention module is proposed to measure the dependencies between brain regions to obtain FBN. Conceptually, each encoder module can be formulated as a transformation function as:

$$\boldsymbol{X}_l = f_l(a_l(\boldsymbol{X}_{l-1}), \boldsymbol{X}_{l-1}), l \in 1, \cdots, L. \tag{1}$$

where $f_l(\cdot)$ denotes the transformation function of the l-th encoder layer with input from $\boldsymbol{X}_{l-1}$, which is the output of the (l-1)-th layer, and $a_l(\cdot)$ denotes the attention module to construct FBN. The details of $a_l(\cdot)$ are illustrated in Fig. 2(c) and introduced later. The final feature $\boldsymbol{X}_L$ of the L-th layer is fed into the fully connected layer for classification.

2.1 Attention-Based Sparse Common-and-Individual FBN Construction Module (ASCFCM)

In order to respect the sparsity property of the human brain and introduce population-level constraint, we redesign the kernel self-attention module in [30] and propose an attention-based sparse common-and-individual FBN construction module (ASCFCM), which will be introduced in detail below.

Revisiting Kernel Self-attention Mechanism: Here, we first review the kernel attention module proposed in [30] and then introduce how it is redesigned to fit the target task. In order to measure the FCs between N nodes, the l-th layer takes $\boldsymbol{X}_{l-1} \in R^{N \times T \times D_{l-1}}$ as input data and generates queries $\boldsymbol{Q}$, keys $\boldsymbol{K}$, and values $\boldsymbol{V}$ through the kernel embedding $\varphi(\cdot)$ and linear projection matrix $\boldsymbol{W}^V \in R^{D_{l-1} \times D_v}$, i.e., $\boldsymbol{Q} = \varphi(\boldsymbol{X}_{l-1}); \boldsymbol{K} = \varphi(\boldsymbol{X}_{l-1}); \boldsymbol{V} = \boldsymbol{X}_{l-1}\boldsymbol{W}^V$, where D_v is set as D_{l-1}/H, and H is the number of heads. The resulting attention matrix $\boldsymbol{A} = \boldsymbol{Q}\boldsymbol{K}^T$ is $N \times N$, and $\boldsymbol{A}_{i,j}$ is actually a Gaussian RBF kernel function [30] with the input features of the i-th and j-th brain regions from $\boldsymbol{X}_{l-1}$, denoted as x_i and x_j, i.e., $\boldsymbol{A}_{i,j} = \langle \varphi(x_i), \varphi(x_j) \rangle = \exp(-\beta \|x_i - x_j\|^2)$. The kernel parameter β can be adaptively learned during the model training process. One advantage of this method over the conventional methods is that it significantly reduces the number of parameters in calculating $\boldsymbol{Q}$ and $\boldsymbol{K}$. Another advantage is that it can effectively model the non-linearity between the features of brain nodes and can better model the complex interactions between brain regions. With the obtained attention matrix $\boldsymbol{A}$, the feature mapping function of the l-th layer $a_l(\cdot)$ can be defined as:

$$a_l(\boldsymbol{Q}, \boldsymbol{K}, \boldsymbol{V}) = \text{softmax}(\frac{\boldsymbol{A}}{\sqrt{TD_v}})\boldsymbol{V} \tag{2}$$

where $\text{softmax}(\frac{\boldsymbol{A}}{\sqrt{TD_v}})$ is a $N \times N$ matrix and it can be viewed as an adaptively built FBN, presenting the dependencies between N brain regions.

Sparsity: From Eq. (2), the original attention scores are transformed by $\text{softmax}(\cdot)$. Due to the non-negative nature, i.e., $\text{softmax}(z_i) = \frac{\exp(z_i)}{\sum_j \exp(z_j)}$, all the FCs between brain regions are positive, leading to a dense FBN. However, the brain is a sparse network [2], which is characterized by a limited number of FCs for most brain regions. In this case, the resulting FBN built with $\text{softmax}(\cdot)$ does not respect the nature of the human brain, and this may degrade the effectiveness of detecting disease-specific patterns. To address this issue, we propose to build a sparse FBN by using $\text{sparsemax}(\cdot)$ [17]. Let $\Delta^K := \{p \in R^K | 1^T p = 1, p \geq 0\}$ be the K-dimensional probability distribution simplex. The key idea of sparse transformation is mapping an input z into the nearest simplex from Δ^K by $\text{sparsemax}(z) := \underset{p \in \Delta^K}{\text{argmin}} \|p - z\|^2$. The optimization could produce a sparse distribution with proper regularization, as in [17], leading to a sparse FBN.

Population-Level Common Pattern: As mentioned previously, the above FBN has large variability since it is constructed with the rs-fMRI data of each subject. Previous research indicates that human brains have similar topologies, which can be modeled as prior knowledge separately to regularize the model, helping to alleviate the issue of large variability. Therefore, we parameterize a data-driven common FBN module as a regularization, as shown in Fig. 2(c). The resulting common FBN, denoted as $\boldsymbol{C}$, is shared by all the subjects and integrated with FBN at the individual level for joint diagnosis. The final $a_l(\cdot)$ in

Eq. 1 is defined as:

$$a_l(\boldsymbol{Q}, \boldsymbol{K}, \boldsymbol{V}) = (\text{sparsemax}(\frac{\boldsymbol{A}}{\sqrt{TD_v}}) + \boldsymbol{C})\boldsymbol{V} \tag{3}$$

2.2 Cross Spatiotemporal Asynchronous FCs

To simulate the time-lagged asynchronous FCs, we segment the BOLD signals $\boldsymbol{X}_l \in R^{N \times T \times D_l}$ into $\boldsymbol{X}_l \in R^{nN \times T_1 \times D_l}$ by using a sliding window, as shown in Fig. 2(b). The length and number of segments, denoted as T_1 and n, respectively, are determined by the length and steps of the sliding window. Measuring the FCs between these new segments breaks the limitations of synchronous FCs, yielding cross-spatiotemporal interaction patterns, as shown in Fig. 2(d). These segments will be concatenated to restore the original number of nodes, i.e., $nN \rightarrow N$, to avoid exponential expansion of the network complexity (see Fig. 2(e)). By doing this, both the synchronous FCs in the traditional methods and the asynchronous ones targeted in this paper can be simultaneously modeled, providing more clues for disease diagnosis.

3 Experiments

3.1 Data and Preprocessing

Both rs-fMRI data sets of ADNI2 and ADNI3 from the Alzheimer's Disease Neuroimaging Initiative (ADNI) (https://adni.loni.usc.edu/) are used to evaluate the effectiveness of the proposed method with the most challenging task of differentiating MCI from NC. The subject number is 410 (147 MCI vs. 263 NC) for ADNI2 and 425 (168 MCI vs. 257 NC) for ADNI3. The ADNI2 data are acquired on a 3 Tesla (Philips) scanner with TR/TE set at 3000/30 ms and a flip angle of 80°. The ADNI3 data are acquired with a flip angle of 90°. The preprocessing is carried out using DPARSFA toolbox (http://rfmri.org/DPARSF) and SPM-12 (ehttps://www.fil.ion.ucl.ac.uk/spm/software/). We perform a standard approach to processing rs-fMRI by following [6], including discarding the first 10 volumes of each time series, slice timing, head motion correction, and Montreal Neurological Institute spatial normalization. Empirically, the AAL [23] atlas is used to extract all voxels of time series values in the 90 regions of interest (ROI).

3.2 Experimental Settings

We conduct a 5-fold cross validation in our evaluation to obtain stable results, and all the methods involved in the comparison use the same partitions of data for fairness. The performance of the model is measured by four metrics: Area Under the receiver operating characteristic Curve (AUC), accuracy (ACC), sensitivity (SEN), and specificity (SPE). The encoder layers of the proposed network L are set to 5. Asynchronous FCs are explored in the first two layers. The sliding

Table 1. The NC-vs-MCI classification performance by different methods on ADNI2 and ADNI3.

Method	ADNI2				ADNI3			
	ACC	SEN	SPE	AUC	ACC	SEN	SPE	AUC
Compact Representation [29]	66.8 ± 1.7	98.5 ± 0.9	10.2 ± 4.8	54.3 ± 2.7	65.1 ± 4.1	52.3 ± 16.3	73.8 ± 16.2	63.0 ± 1.8
Kernel Transformer [30]	68.3 ± 1.3	98.1 ± 1.7	15.0 ± 6.0	56.5 ± 2.2	69.4 ± 3.7	47.9 ± 6.2	83.3 ± 5.5	65.6 ± 3.7
LCC [10]	69.7 ± 2.9	88.5 ± 10.1	36.1 ± 19.9	62.2 ± 5.9	71.0 ± 3.2	88.8 ± 6.3	43.7 ± 6.0	66.2 ± 3.2
Transformer [25]	70.3 ± 2.6	94.3 ± 2.8	29.7 ± 4.9	60.8 ± 2.7	72.9 ± 3.2	51.6 ± 10.0	86.8 ± 6.2	69.2 ± 3.8
BrainGNN [13]	71.2 ± 5.2	88.6 ± 6.3	41.3 ± 26.6	64.2 ± 9.5	68.4 ± 9.1	41.9 ± 32.6	85.6 ± 6.6	63.8 ± 13.2
STGCN [8]	74.2 ± 2.2	84.6 ± 8.8	48.2 ± 9.8	68.4 ± 3.0	71.3 ± 4.6	84.4 ± 7.4	50.8 ± 18.2	67.7 ± 6.6
BrainNetCNN [11]	78.5 ± 3.3	84.8 ± 5.5	67.3 ± 9.1	74.6 ± 5.8	72.4 ± 2.0	46.8 ± 8.9	89.1 ± 6.2	71.8 ± 2.4
Proposed network	83.2 ± 2.4	91.3 ± 4.4	68.8 ± 9.2	80.0 ± 5.3	78.1 ± 1.8	63.0 ± 7.3	87.9 ± 2.9	75.5 ± 5.6

window of each layer has a length of 32 and a step size of 16 while the number of heads H in each layer is set as 2. The model input $\boldsymbol{X}_{raw} \in R^{90\times128\times1}$ is transformed into $\boldsymbol{X}_0 \in R^{90\times128\times6}$ by a mapping with a kernel size of 3×1. The output feature dimension D_l of each layer is set as 12. The drop rate of the attention score is 0.2, and the batch size is 16. We train 250 epochs with stochastic gradient descent (SGD), whose learning rate is 0.1 and momentum is 0.9. The code is publicly available at https://github.com/seuzjj/ACIFBN.

3.3 Experimental Results

The average performance of the proposed method and the competing methods is summarized in Table 1, and their ROC curves are shown in Fig. 3(a). As seen, our proposed method achieves the highest accuracy compared with other classical methods, i.e., 83.2% on ADNI2 and 78.1% on ADNI3, obtaining an improvement of 4.7% and 5.7%, respectively, over the current state-of-the art methods. Our result increases by 14.9% on ADNI2 over the base model–Kernel Transformer. These results are encouraging, indicating the significant efficacy and stability of our method.

Table 2. Different model components affect classification performance. The backbone structure is Kernel Transformer, and except for the settings available in the table, all other settings are the same.

FC pattern		activation		asynchrony	ACC of ADNI2	ACC of ADNI3
individual	common	softmax	sparsemax			
✓		✓			68.3 ± 1.3	69.4 ± 3.7
	✓				79.2 ± 2.9	76.7 ± 5.1
✓	✓	✓			79.5 ± 4.0	77.1 ± 3.5
✓	✓		✓		82.3 ± 1.1	76.7 ± 1.3
✓	✓		✓	✓	83.2 ± 2.4	78.1 ± 1.8

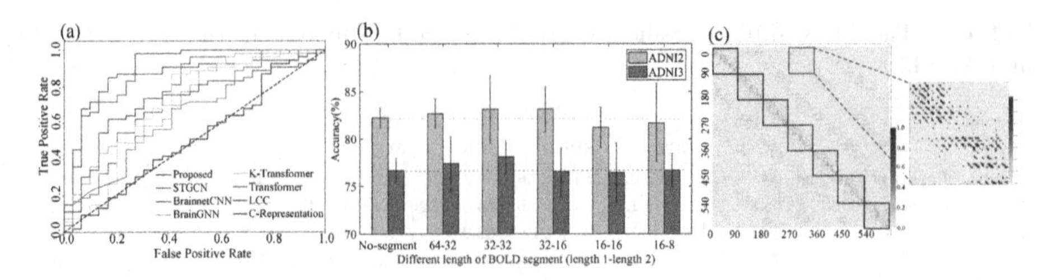

Fig. 3. Visualization of experimental results. (a) ROC on ADNI2. (b) The effect of different T_1. (c) Asynchronous FCs.

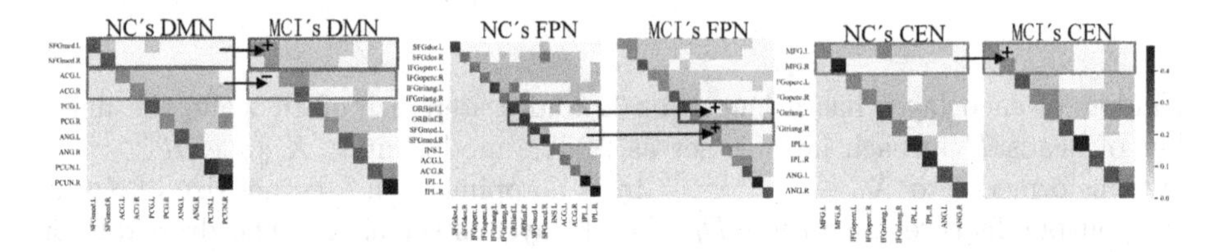

Fig. 4. Difference of DMN, FPN, CEN between NC and MCI.

3.4 Ablation Study

In order to further verify the effectiveness of the proposed method and the contribution of each component, we conduct ablation studies and the results are shown in Table 2. As seen, introducing common FBN improves the accuracy significantly from 68.3% to 79.2%, and this is probably attributed to the efficacy of it as prior knowledge regularizer. On top of it, considering the sparsity property of FBN further increases the accuracy to 82.3%. Modeling asynchronous FCs, denoted as 'asynchrony', further boost the model's accuracy, reaching 83.2%.

To verify the effects of the settings during dividing rs-fMRI signals into segments, we set the window lengths as different values in the two asynchronous FCs modeling layers, seen in Fig. 3(b). The step of windows is half of the length. It can be consistently observed on ADNI2 and ADNI3 that the model performance first improves with the decreasing segment lengths, and then start to decrease. This implies that modeling asynchronous FCs in a wide range is beneficial, however, too short segments would harm the performance due to increased model complexity or weak of temporal information within short segments.

4 Visualization and Conclusion

The learned individual FBNs carry diagnostic clues, which are visualized in Fig. 4 for MCI and NC comparison. As seen, increased connectivities could be identified in the superior frontal gyrus (SFGmed) of the default mode network (DMN) and central executive network (CEN), the middle frontal gyrus (MFG) of the frontal parietal network (FPN), and the part inferior frontal gyrus (ORBinf) of CEN, while decreased connectivities could be found in the anterior cingulate and paracingulate gyri (ACG) of the DMN. These results are consistent with the literature [1,18]. The effects of modeling asynchronous FCs are shown in Fig. 3(c). As seen, the dominant FCs are mainly synchronous from the diagonal blocks (inside the black block), while certain asynchronous FCs (inside the red block) are also well captured. This is reasonable since most brain regions interact synchronously, while a portion of them would have asynchronous time-lagged functional responses.

In sum, this paper proposes to adaptively construct and analyze asynchronous common and individual functional brain networks within a specially designed Transformer-based network, which is more flexible, discriminative and capable of capturing both synchronous and asynchronous FCs. The experimental results on two ADNI data sets well demonstrate the effectiveness of the proposed method.

Acknowledgments. This work was supported in part by National Natural Science Foundation of China (grant number 62101611), Guangdong Basic and Applied Basic Research Foundation (grant number 2022A1515011375, 2023A1515012278) and Shenzhen Science and Technology Program (grant number JCYJ20220530145411027, JCYJ20220818102414031).

References

1. Ahmadi, H., Fatemizadeh, E., Motie-Nasrabadi, A.: Identifying brain functional connectivity alterations during different stages of Alzheimer's disease. Int. J. Neurosci. **132**(10), 1005–1013 (2022)
2. Bassett, D.S., Bullmore, E.T.: Small-world brain networks revisited. Neuroscientist **23**(5), 499–516 (2017)
3. Chen, H., Zhang, Y., Zhang, L., Qiao, L., Shen, D.: Estimating brain functional networks based on adaptively-weighted fMRI signals for MCI identification. Front. Aging Neurosci. **12**, 595322 (2021)
4. Deshpande, G., Santhanam, P., Hu, X.: Instantaneous and causal connectivity in resting state brain networks derived from functional MRI data. Neuroimage **54**(2), 1043–1052 (2011)
5. Deshpande, G., Sathian, K., Hu, X.: Assessing and compensating for zero-lag correlation effects in time-lagged Granger causality analysis of fMRI. IEEE Trans. Biomed. Eng. **57**(6), 1446–1456 (2010)
6. Esteban, O., et al.: fMRIPrep: a robust preprocessing pipeline for functional MRI. Nat. Methods **16**(1), 111–116 (2019)
7. Friston, K., Moran, R., Seth, A.K.: Analysing connectivity with Granger causality and dynamic causal modelling. Curr. Opin. Neurobiol. **23**(2), 172–178 (2013)
8. Gadgil, S., Zhao, Q., Pfefferbaum, A., Sullivan, E.V., Adeli, E., Pohl, K.M.: Spatio-temporal graph convolution for resting-state fMRI analysis. In: Martel, A.L., et al. (eds.) MICCAI 2020. LNCS, vol. 12267, pp. 528–538. Springer, Cham (2020). https://doi.org/10.1007/978-3-030-59728-3_52
9. Ghanbari, M., et al.: Alterations of dynamic redundancy of functional brain subnetworks in Alzheimer's disease and major depression disorders. NeuroImage Clin. **33**, 102917 (2022)

10. Kaiser, M.: A tutorial in connectome analysis: topological and spatial features of brain networks. Neuroimage **57**(3), 892–907 (2011)
11. Kawahara, J., et al.: BrainNetCNN: convolutional neural networks for brain networks; towards predicting neurodevelopment. Neuroimage **146**, 1038–1049 (2017)
12. Lee, H., Lee, D.S., Kang, H., Kim, B.N., Chung, M.K.: Sparse brain network recovery under compressed sensing. IEEE Trans. Med. Imaging **30**(5), 1154–1165 (2011)
13. Li, X., et al.: Braingnn: interpretable brain graph neural network for fMRI analysis. Med. Image Anal. **74**, 102233 (2021)
14. Li, Y., Liu, J., Tang, Z., Lei, B.: Deep spatial-temporal feature fusion from adaptive dynamic functional connectivity for MCI identification. IEEE Trans. Med. Imaging **39**(9), 2818–2830 (2020)
15. Li, Y., Yu, Z.L., Bi, N., Xu, Y., Gu, Z., Amari, S.I.: Sparse representation for brain signal processing: a tutorial on methods and applications. IEEE Signal Process. Mag. **31**(3), 96–106 (2014)
16. Liu, M., Zhang, H., Shi, F., Shen, D.: Building dynamic hierarchical brain networks and capturing transient meta-states for early mild cognitive impairment diagnosis. In: de Bruijne, M., et al. (eds.) MICCAI 2021. LNCS, vol. 12907, pp. 574–583. Springer, Cham (2021). https://doi.org/10.1007/978-3-030-87234-2_54
17. Martins, A., Astudillo, R.: From softmax to sparsemax: a sparse model of attention and multi-label classification. In: International Conference on Machine Learning, pp. 1614–1623. PMLR (2016)
18. Metmer, H., Lu, J., Zhao, Q., Li, W., Lu, H.: Evaluating functional connectivity of executive control network and frontoparietal network in Alzheimer's disease (2013)
19. Mitra, A., Snyder, A.Z., Hacker, C.D., Raichle, M.E.: Lag structure in resting-state fMRI. J. Neurophysiol. **111**(11), 2374–2391 (2014)
20. Qiu, H., Hou, B., Ren, B., Zhang, X.: Spatio-temporal tuples transformer for skeleton-based action recognition. arXiv preprint arXiv:2201.02849 (2022)
21. Riaz, A., et al.: FCNet: a convolutional neural network for calculating functional connectivity from functional MRI. In: Wu, G., Laurienti, P., Bonilha, L., Munsell, B.C. (eds.) CNI 2017. LNCS, vol. 10511, pp. 70–78. Springer, Cham (2017). https://doi.org/10.1007/978-3-319-67159-8_9
22. Shi, Y., et al.: ASMFS: adaptive-similarity-based multi-modality feature selection for classification of Alzheimer's disease. Pattern Recogn. **126**, 108566 (2022)
23. Tzourio-Mazoyer, N., et al.: Automated anatomical labeling of activations in SPM using a macroscopic anatomical parcellation of the MNI MRI single-subject brain. Neuroimage **15**(1), 273–289 (2002)
24. Varoquaux, G., Gramfort, A., Poline, J.B., Thirion, B.: Brain covariance selection: better individual functional connectivity models using population prior. In: Advances in Neural Information Processing Systems, vol. 23 (2010)
25. Vaswani, A., et al.: Attention is all you need. In: Advances in Neural Information Processing Systems, vol. 30 (2017)
26. Wang, J., et al.: Disrupted functional brain connectome in individuals at risk for Alzheimer's disease. Biol. Psychiat. **73**(5), 472–481 (2013)
27. Wee, C.Y., Yap, P.T., Zhang, D., Wang, L., Shen, D.: Group-constrained sparse fMRI connectivity modeling for mild cognitive impairment identification. Brain Struct. Funct. **219**, 641–656 (2014)
28. Zhang, H.Y., et al.: Detection of PCC functional connectivity characteristics in resting-state fMRI in mild Alzheimer's disease. Behav. Brain Res. **197**(1), 103–108 (2009)

29. Zhang, J., Zhou, L., Wang, L., Li, W.: Functional brain network classification with compact representation of SICE matrices. IEEE Trans. Biomed. Eng. **62**(6), 1623–1634 (2015)
30. Zhang, J., Zhou, L., Wang, L., Liu, M., Shen, D.: Diffusion kernel attention network for brain disorder classification. IEEE Trans. Med. Imaging **41**(10), 2814–2827 (2022)
31. Zhang, Y., et al.: Strength and similarity guided group-level brain functional network construction for MCI diagnosis. Pattern Recogn. **88**, 421–430 (2019)
32. Zhu, X., Cortes, C.R., Mathur, K., Tomasi, D., Momenan, R.: Model-free functional connectivity and impulsivity correlates of alcohol dependence: a resting-state study. Addict. Biol. **22**(1), 206–217 (2017)

Self-pruning Graph Neural Network for Predicting Inflammatory Disease Activity in Multiple Sclerosis from Brain MR Images

Chinmay Prabhakar[1(✉)], Hongwei Bran Li[1,2,5], Johannes C. Paetzold[3], Timo Loehr[2], Chen Niu[4], Mark Mühlau[5], Daniel Rueckert[2,3], Benedikt Wiestler[5], and Bjoern Menze[1,2]

[1] Department of Quantitative Biomedicine, University of Zurich, Zürich, Switzerland
chinmay.prabhakar@uzh.ch

[2] Department of Computer Science, Technical University of Munich, Munich, Germany

[3] Department of Computing, Imperial College London, London, UK

[4] The First Affiliated Hospital of Xi'an Jiaotong University, Xi'an, China

[5] Klinikum rechts der Isar, Technical University of Munich, Munich, Germany

Abstract. Multiple Sclerosis (MS) is a severe neurological disease characterized by inflammatory lesions in the central nervous system. Hence, predicting inflammatory disease activity is crucial for disease assessment and treatment. However, MS lesions can occur throughout the brain and vary in shape, size and total count among patients. The high variance in lesion load and locations makes it challenging for machine learning methods to learn a globally effective representation of whole-brain MRI scans to assess and predict disease. Technically it is non-trivial to incorporate essential biomarkers such as lesion load or spatial proximity. Our work represents the first attempt to utilize graph neural networks (GNN) to aggregate these biomarkers for a novel global representation. We propose a two-stage MS inflammatory disease activity prediction approach. First, a 3D segmentation network detects lesions, and a self-supervised algorithm extracts their image features. Second, the detected lesions are used to build a patient graph. The lesions act as nodes in the graph and are initialized with image features extracted in the first stage. Finally, the lesions are connected based on their spatial proximity and the inflammatory disease activity prediction is formulated as a graph classification task. Furthermore, we propose a self-pruning strategy to auto-select the most critical lesions for prediction. Our proposed method outperforms the existing baseline by a large margin (AUCs of 0.67 *vs.* 0.61 and 0.66 *vs.* 0.60 for one-year and two-year inflammatory disease activity, respectively). Finally, our proposed method enjoys inherent explainability by

B. Wiestler and B. Menze—Contributed equally as senior authors.

Supplementary Information The online version contains supplementary material available at https://doi.org/10.1007/978-3-031-43993-3_22.

H. Greenspan et al. (Eds.): MICCAI 2023, LNCS 14227, pp. 226–236, 2023.
https://doi.org/10.1007/978-3-031-43993-3_22

assigning an importance score to each lesion for the overall prediction. Code is available at https://github.com/chinmay5/ms_ida.git.

1 Introduction

Multiple Sclerosis (MS) is a severe central nervous system disease with a highly nonlinear disease course where periodic relapses impair the patient's quality of life. Clinical studies show that relapses co-occur with the appearance of new inflammatory MS lesions in MR images [19,25], making MR imaging a central element for the clinical management of MS patients. Further, assessing new MS lesions is crucial for disease assessment and therapy monitoring [5,8]. Unfortunately, prevailing therapies often involve highly active immunomodulatory drugs with potentially severe side effects. Hence, it necessitates developing machine learning models capable of predicting the future disease activity of individual patients to select the best therapy.

While recent approaches applied convolutional neural networks (CNN) to directly learn features from MR image space [3,4,21,26,28], there remain challenges to obtain an effective global representation to characterize disease status. We attribute the difficulty of MS inflammatory disease activity prediction to a set of distinct disease characteristics that are well observable in MR images. First, while lesions have a sufficient signal-to-noise profile in images, their variation in shape, size, and number of occurrences amongst patients make it challenging for existing CNN-based methods that process the whole MRI scan in one go. Second, with advanced age, small areas appearing in the MRI of healthy individuals may resemble MS. As such, it is crucial to not only predict MS but also to identify the lesions deemed consequential for the final prediction.

To solve this problem, we use concepts from geometric deep learning. Specifically, we propose a two-stage pipeline. First, the lesions in the MRI scans are segmented using a state-of-the-art 3D segmentation algorithm [9], and their image features are extracted with a self-supervised method [14]. Second, the extracted lesions are converted into a patient graph. The lesions act as nodes of the graph, while the edge connectivity is determined using the spatial proximity of the lesions. By this representation, we can solve the MS inflammatory disease activity prediction task as a graph-level classification problem. We argue that formulating the MS inflammatory disease activity prediction in our two-stage pipeline has the following advantages: (1) Graph neural networks can easily handle different numbers of nodes (lesions) and efficiently incorporate their spatial locations. (2) Modern segmentation [9,16,23] and representation learning methods [2,11,14] are effective tools for lesion detection and allow us to extract pathology-specific features. (3) By operating at the lesion level, it is possible to discover the lesions that contribute most to the eventual prediction, making the decisions more interpretable. Thus, our proposed solution can be a viable methodology for MS inflammatory disease activity prediction to handle the associated challenges.

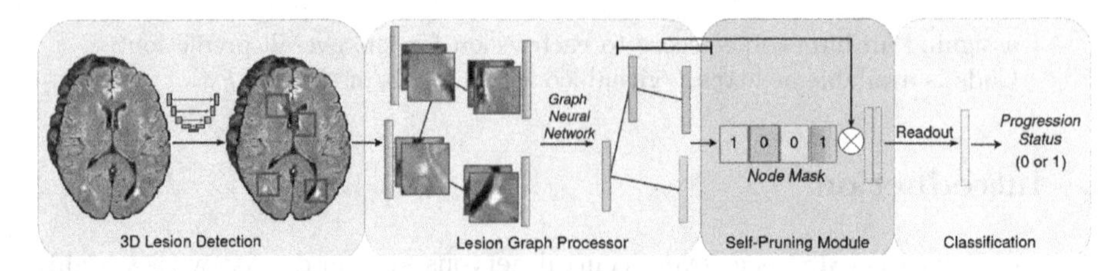

Fig. 1. Our proposed Multiple Sclerosis (MS) inflammatory disease activity prediction framework. We first detect lesions in the MRI scan using nn-Unet [9]. A crop centered at the detection is extracted and used to learn self-supervised lesion features. Next, we build a graph from these detected lesions, where each lesion becomes a node, with the connections (edges) between lesions (nodes) defined by spatial proximity. This graph is processed using a graph neural network to generate enriched lesion features. Next, our self-pruning module (SPM) processes these enriched lesion features to determine an importance score for each lesion. The least scoring lesions are pruned-off and the highest scoring lesions are passed to the readout layer to obtain a graph-level feature vector. This graph-level vector is used for the final prediction.

Contributions. Our contribution is threefold: (1) we are the first to formulate the MS inflammatory disease activity prediction task as a graph classification problem, thereby bringing a new set of methods to a significant clinical problem. (2) We propose a two-stage pipeline that effectively captures inherent MS variations in MRI scans, thus generating an effective global representation. (3) We develop a self-pruning module, which assigns an importance score to each lesion and reduces the task complexity by prioritizing the critical lesions. Additionally, the assigned per-lesion importance score improves our model's explainability.

2 Methodology

Overview. The objective is to predict MS inflammatory disease activity, i.e., to classify if new or significantly enlarged inflammatory lesions appear in the follow-up after the initial MRI scans. We denote the dataset as $D(X, y)$, where X is the set of lesion patches extracted from MR scans, and y $\in$ {0, 1} is the inflammatory disease activity status. For patient i, multiple lesion patches $\{x_1^i, x_2^i, ...x_n^i\}$ can exist, where n is the total number of lesions. We aim to learn a mapping function f: $\{x_1^i, x_2^i, ...x_n^i\} \rightarrow \{0, 1\}$. Please note that our formulation *differs* from existing methods [21,26,28], which aim to learn a direct mapping from the MR image to the inflammatory disease activity label. As shown in Fig. 1, our proposed method consists of four distinct components. We describe each component in the following sections.

Lesion Detection and Feature Extraction. We focus on the individual lesions instead of processing whole-brain MRI scans. This is important because MS lesions typically comprise less than 1% of voxels in the MRI scan. With this strategy, the graph model can aggregate lesion-level features for an effective patient-specific representation. First, we detect the lesions using a state-of-the-art nn-Unet [9] pre-trained with MR images and their consensus annotations from two neuro-radiologists. Then, for each detected lesion, we extract a small fixed-size patch centered at it. Finally, we learn self-supervised features z for the lesion using a recent Transformer-based approach [14].

Lesion Graph Processor. In the second stage, we generate a patient-specific graph $G(V, E, Z)$ from the detected lesions. The lesions act as vertices V of this graph and are initialized with the crop-derived features Z. The spatial location s of the lesions is used to determine their connectivity E using a k-Nearest Neighbor (kNN) graph method [27]. Furthermore, the connected edges are weighted based on their spatial proximity. Specifically, given two lesions v_i and v_j, with spatial locations s_i and s_j respectively, the edge weight $w(v_i, v_j) = exp(-\frac{||s_i - s_j||^2}{\tau^2})$. τ is a scalar that controls the contribution of distant lesions. Hence the final graph connectivity can be represented as:

$$E_{i,j} = \begin{cases} w(v_i, v_j), \text{ if } j \in N(v_i) \text{ } or \text{ } i \in N(v_j) \\ 0, \qquad\qquad\quad \text{otherwise} \end{cases} \tag{1}$$

where $N(v_i)$ are the nodes directly connected to the node v_i. Constructing a graph from the lesions is instrumental in two aspects: (1) The framework allows us to work with varying numbers of lesions in different patients. Alternatively, sequential models could be employed. However, since the lesions lack a canonical ordering, such models would not achieve an effective global representation [12]. (2) It is possible to incorporate meaningful lesion properties such as spatial proximity and the number of lesions.

Please note that separate graphs are created for individual patients. Thus, MS inflammatory disease activity prediction is formulated as a graph-level classification task. The graph $G(V, E, Z)$ can be processed using message-passing neural networks (such as GCN [10], GAT [1], EdgeConv [24], GraphSage [6], to name a few) to learn enriched lesion features $\hat{Z}$. These enriched features are passed through to the self-pruning module.

Self-pruning Module. The number of lesions can vary substantially among the patients, including the possibility of false positives in the segmentation stage. As such, it is crucial to recognize the most relevant lesions for the final prediction. In addition, this will bring inherent explainability and make it easier for a doctor to validate model predictions. To accomplish this, the enriched lesion features $\hat{Z}$ are passed through a self-pruning module (SPM). The SPM produces a binary mask M for each lesion to determine whether a lesion contributes to the classification. The SPM uses a learnable projection vector $\vec{p}$ to compute importance scores ($\hat{Z}\vec{p}/||\vec{p}||$) for the lesions. These scores are scaled with a sigmoid layer. We retain the high-scoring lesions and discard the rest, which is formulated as:

$$M = top\text{-}r(\sigma(\frac{\hat{Z}\vec{p}}{||\vec{p}||}), r) \tag{2}$$

where $\sigma(x) = 1/(1 + e^{-x})$ is the sigmoid function and $top\text{-}r(\cdot)$ is an operator which selects a fraction r of the lesions based on high importance score. r is a hyper-parameter in our setup. Since the masking process is part of the forward pass through the model during both training and inference stages and not a post hoc modification, we refer to it as *self-pruning* of nodes. Features of the remaining nodes ($\hat{Z}' = \hat{Z} \otimes M$) are passed to the classification head.

It should be noted that the existence of multiple lesions is a typical characteristic of MS. Hence, a crucial aspect of MS management is that clinicians must identify signs of inflammatory disease activity in MR images to make treatment decisions [19]. Therefore, along with predicting inflammatory disease activity, interpreting the contribution of individual lesions to the prediction is essential. By assigning an importance score to each lesion, the SPM can provide explainability to clinicians at a lesion level, while existing CNN methods can not.

Classification Head. The classification head consists of a readout layer aggregating all the remaining node's features to produce a single feature vector $\hat{z}'$ for the entire graph. This graph-level feature is passed through an MLP to obtain the final prediction $\hat{y}$. We train our model using a binary cross-entropy loss.

$$\mathcal{L}_{clf} = -\frac{1}{N}\sum_{i=1}^{N}(y_i \log(\hat{y_i}) + (1 - y_i)\log(1 - \hat{y_i})) \tag{3}$$

where N is the number of patients, y_i is the ground truth inflammatory disease activity information and $\hat{y_i}$ is the model prediction.

3 Experiments

Datasets and Image Preprocessing. Our approach is evaluated on a cohort of 430 MS patients collected following approval from the local IRB [7]. Patients included in this analysis were diagnosed with relapsing-remitting MS, with a maximum disease duration of three years at the time of baseline scan. We collect the FLAIR and T1w MR scans for each patient. The scans have a uniform voxel size of $1 \times 1 \times 1\,\text{mm}^3$, were rigidly co-registered to the MNI152 atlas and skull-stripped using HD-BET [17]. Three neuro-radiologists independently read longitudinal subtraction imaging, where FLAIR images from two time points were co-registered and subtracted. In this vein, new and significantly enlarged lesions are identified as positive inflammatory disease activity.

The dataset contains MS inflammatory disease activity information for clinically relevant one-year and two-year intervals [20]. At the end of the first year, we have the inflammatory disease activity status of 430 patients, with 303 showing activity and 127 not. Similarly, at the end of two years, we have data available for 347 patients, with 287 showing activity and 60 not. Thus, the dataset shows a slight imbalance in favor of inflammatory disease activity. This imbalance is a typical property in the MS patients cohort that impairs algorithm development.

Feature Extraction and Training Configuration. We use an nn-Unet [9] for lesion segmentation and detection. Then a uniform crop of size $24 \times 24 \times 24$ mm^3 is extracted centered at each lesion. The cropped patches are passed through a transformer-based masked autoencoder [14] to extract self-supervised lesion features. The encoder produces a 768-dimensional feature vector for each patch. We also append normalized lesion coordinates to the encoder output to get the final lesion features.

The lesions are connected using a k-nearest neighbor algorithm with $k = 5$. Further, these connections (edges) are weighted using $\tau = 0.01$ (Eq. 1). Two message-passing layers with hidden dimensions of 64 and 8, respectively, process the generated graph to enrich lesion features. Next, the enriched features are passed through the SPM. The SPM uses a learnable projection vector $\vec{p} \in R^8$ and sigmoid activation to learn the importance score. Based on this importance score, a mask is produced to select $r = 0.5$ (i.e., 50%) of the highest-scoring lesions and discard the rest. Next, a sum aggregation is used as the readout function. These aggregated features are passed through 2 feed-forward layers with hidden dimensions of size 8. Finally, the features are passed to a sigmoid function to obtain the final prediction.

The model is trained for 300 epochs using *AdamW* optimizer [13] with 0.0001 weight decay. The base learning rate is 1e-4. The batch size is set to 16. A dropout layer with $p = 0.5$ is used between different feed-forward blocks. Since the dataset is imbalanced in favor of patients experiencing inflammatory disease activity, we use a balanced batch sampler to load approximately the same number of positive and negative samples in each mini-batch.

Evaluation Strategy, Classifier, and Metrics. We report our results on MS inflammatory disease activity prediction for the clinically relevant one and two-year intervals [20]. The Area Under the Receiver Operating Characteristic Curve (AUC) is used as the evaluation metric. We use 80% of the samples as the training set and 10% as the validation. The validation set is used to select the best model which is then applied to the remaining 10% cases. This procedure is iterated until all cases have been assigned to a test set once (ten-fold cross-validation). The same folds are used for the proposed model and baseline algorithms.

4 Results

Quantitative Comparison. Table 1 shows the classification performance for the ten folds on one-year and two-year lesion inflammatory disease activity prediction. The $\pm$ indicates the corresponding standard deviations. We compare our method against two existing approaches for MS inflammatory disease activity prediction baselines, a 3D Res-Net [28], and a multi-resolution CNN architecture [21]. These methods learn a direct mapping from the MR image to the inflammatory disease activity label. Our graph model outperforms the baseline methods on one (**0.67** *vs.* 0.61 AUC) and two-year inflammatory disease activity prediction (**0.66** *vs.* 0.60 AUC). In the following, we analyze and discuss each component of our established framework.

Table 1. Comparison of our method against the existing CNN-based solutions. We report the AUC score for MS Inflammatory Disease Activity (IDA) prediction at the end of one and two years. Our proposed two-stage solution outperforms the existing baselines, achieving the best AUC score on both prediction tasks.

Methods	One year IDA	Two year IDA
	AUC	AUC
3D ResNet	0.595 ± 0.104	0.575 ± 0.099
CNN multi-res [21]	0.610 ± 0.053	0.600 ± 0.059
Ours	**0.671 ± 0.062**	**0.664 ± 0.063**

Ablation Study. In this section, we analyze the importance of different components of our proposed method. We defer the analysis of the lesion feature representation to the appendix owing to space constraints.

The Effectiveness of Graph Structure. Since the lesion feature extractor generates rich lesion features, one may argue that the graph structure is unwarranted. There are two alternatives to using a graph, (i) completely discard the graph structure, use a feed-forward layer to enrich the lesion features further, and aggregate them to perform classification [15] (Since this formulation regards the input as a set, we call this *Set-Proc* model). (ii) Aggregate all the lesion features for a patient using a mean aggregation and process the aggregated feature by traditional machine learning algorithms such as random forest (RF), support vector machine (SVM) with the RBF-kernel, and logistic regression (LR). Table 2 compares our model's performance against these alternatives. Our proposed solution obtains better AUC than the alternatives, indicating that incorporating the graph structure is beneficial for eventual prediction.

Table 2. Effectiveness of the graph structure. We compare our method to traditional ML algorithms and a set-based aggregation baseline, both of which discard the graph structure. The incorporation of the graph structure is beneficial for downstream prediction.

Methods	One year	Two year
	AUC	AUC
SVM	0.635 ±0.05	0.513 ±0.09
RF	0.639 ±0.04	0.650 ±0.09
LR	0.658 ±0.05	0.655 ±0.07
Set-Proc	0.654 ±0.06	0.658 ±0.08
Ours	**0.671** ±0.06	**0.664** ±0.06

Table 3. Importance of spatial proximity. GAT (partially) and Transformer-Conv (completely) ignore spatial proximity in the graph while GCN, Edge-Conv, and GraphSAGE incorporate it. The incorporation of spatial proximity is beneficial for downstream prediction.

Methods	One year	Two year
	AUC	AUC
Tra. Conv [18]	0.632 ± 0.06	0.627 ± 0.06
GAT	0.624 ±0.08	0.631 ±0.09
Edge	0.640 ±0.09	0.657 ±0.09
SAGE	0.650 ±0.05	0.634 ±0.06
GCN(Ours)	**0.671** ±0.06	**0.664** ±0.06

Table 4. Comparison of the performance of different message passing layers with and without the self-pruning module. Incorporating the self-pruning module is beneficial for most message passing layers.

Methods	One year		Two year	
	w/o SPM	w/ SPM	w/o SPM	w/ SPM
GAT	**0.628** ±0.07	0.624 ±0.08	0.607 ±0.06	**0.631**±0.09
Edge	0.629 ±0.07	**0.640** ±0.09	0.651 ±0.07	**0.657** ±0.09
Sage	0.647 ±0.06	**0.650** ±0.05	0.630 ±0.07	**0.634** ±0.06
GCN(Ours)	0.640 ±0.04	**0.671** ±0.06	0.643 ± 0.07	**0.664**±0.06

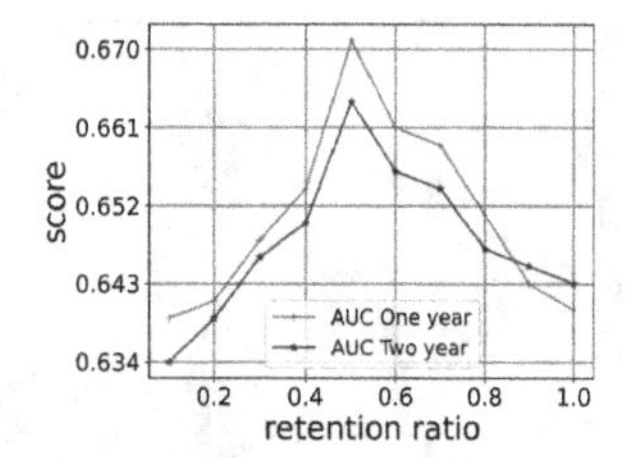

Fig. 2. Model performance against different retention ratio r. *Best performance observed for* $r = 0.5$.

The Importance of Encoding Spatial Proximity. The spatial proximity in our model is encoded at two levels. First, lesion connectivity is determined using a NN graph, and second, we weigh the edges based on their distance. Graph convolution layers such as EdgeConv, GCN, and GraphSAGE take the edge weights into account. (EdgeConv does it implicitly by taking a difference of lesion features that already contain spatial information).

On the other hand, the GAT model learns an attention weight and ignores the pre-defined edge weights. However, it still computes these coefficients on only the connected nodes. We can go further, completely ignore the distances and instead use a fully connected graph. The TransformerConv [18] on such a graph is equivalent to applying the well-known transformer encoder [22] on the inputs. Table 3 shows that the methods that ignore spatial proximity (TransformerConv) or do not use distance-based weighting (GAT) struggle. On the other hand, EdgeConv, GCN, and GraphSAGE work better. We use GCN in our model owing to its superior performance.

The Contribution of the Self-Pruning Module (SPM). The SPM selects a subset of lesions for the final prediction during the training and evaluation phases. However, the proposed classification method can work without it. In this case, none of the lesions is discarded during the readout operation. Table 4 shows the classification results with and without the SPM. We observe that including SPM leads to better outcomes across different message-passing networks. An explanation could be that the SPM can better handle patient variations (in terms of the total number of lesions) by operating on a subset of lesions (Fig. 3).

Analysis of Hyperparameters. The retention ratio r and the number of neighbors k used for building the graph are the two critical hyperparameters in our proposed framework. We discuss the effect of the retention ratio r here and defer discussion about k to the appendix.

Effect of Retention Ratio r. The retention ratio $r \in (0, 1]$ controls the fraction of lesions retained after the self-pruning module. If we set its value to 1, all the lesions are retained for the final prediction and thus, bypassing the self-pruning

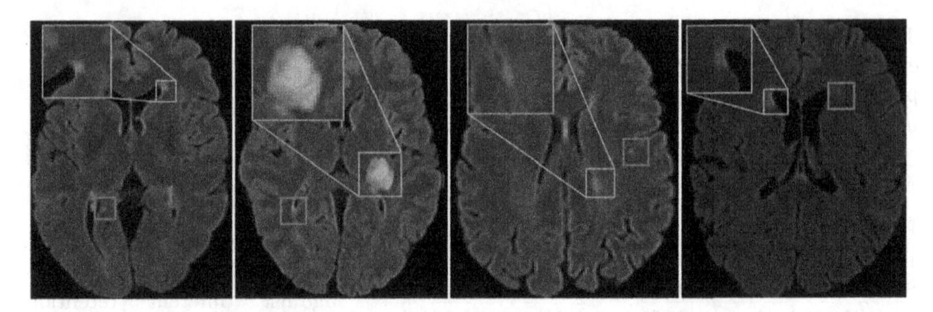

Fig. 3. Lesions selected by the *SPM* for two-year inflammatory disease activity prediction are highlighted with a green bounding box. We also show the zoomed-in view of the lesion. A concurrent lesion in the scan ignored by the *SPM* is shown with a blue bounding box. According to their size and location, the selected lesions are most likely to be relevant to the prediction. (Color figure online)

module. Any other value implies that we ignore at least a few lesions in the readout layer. Since the number of lesions can vary across graphs, we retain $\lceil (N.r) \rceil$ lesions after the self-pruning layer. To find the optimal r, we test our model with r between 0.1 and 1.0. The results are summarized in Fig. 2. We set r to 0.5 for both tasks.

5 Conclusion

Predicting MS inflammatory disease activity is a clinically relevant, albeit challenging task. In this work, we propose a two-stage graph-based pipeline that surpasses existing CNN-based methods by decoupling the tasks of detecting and learning rich semantic features for lesions. We also propose a self-pruning module that further improves model generalizability by handling variations in the number of lesions within patients. Most importantly, we frame the MS inflammatory disease activity prediction as a graph classification problem. We hope our work provides a new perspective and leads to cutting-edge research at the intersection of graph processing and MS inflammatory disease activity prediction.

Acknowledgement. This work was supported by Helmut Horten Foundation. B.W., M.M., and B.M. were supported through the DFG, SPP Radiomics. H.B.L. is supported by an Nvidia GPU research grant.

References

1. Brody, S., Alon, U., Yahav, E.: How attentive are graph attention networks? arXiv preprint arXiv:2105.14491 (2021)
2. Chen, T., Kornblith, S., Norouzi, M., Hinton, G.: A simple framework for contrastive learning of visual representations. In: International Conference on Machine Learning, pp. 1597–1607. PMLR (2020)

3. Durso-Finley, J., Falet, J.P., Nichyporuk, B., Douglas, A., Arbel, T.: Personalized prediction of future lesion activity and treatment effect in multiple sclerosis from baseline MRI. In: International Conference on Medical Imaging with Deep Learning, pp. 387–406. PMLR (2022)
4. Falet, J.P.R., et al.: Estimating individual treatment effect on disability progression in multiple sclerosis using deep learning. Nat. Commun. **13**(1), 5645 (2022)
5. Filippi, M., et al.: Identifying progression in multiple sclerosis: new perspectives. Ann. Neurol. **88**(3), 438–452 (2020)
6. Hamilton, W., Ying, Z., Leskovec, J.: Inductive representation learning on large graphs. In: Advances in Neural Information Processing Systems, vol. 30 (2017)
7. Hapfelmeier, A., et al.: Retrospective cohort study to devise a treatment decision score predicting adverse 24-month radiological activity in early multiple sclerosis. Ther. Adv. Neurol. Disord. **16**, 17562864231161892 (2023)
8. Hauser, S.L., Cree, B.A.: Treatment of multiple sclerosis: a review. Am. J. Med. **133**(12), 1380–1390 (2020)
9. Isensee, F., Jaeger, P.F., Kohl, S.A., Petersen, J., Maier-Hein, K.H.: nnU-Net: a self-configuring method for deep learning-based biomedical image segmentation. Nat. Methods **18**(2), 203–211 (2021)
10. Kipf, T.N., Welling, M.: Semi-supervised classification with graph convolutional networks. arXiv preprint arXiv:1609.02907 (2016)
11. Li, H., et al.: Imbalance-aware self-supervised learning for 3D radiomic representations. In: de Bruijne, M., et al. (eds.) MICCAI 2021. LNCS, vol. 12902, pp. 36–46. Springer, Cham (2021). https://doi.org/10.1007/978-3-030-87196-3_4
12. Liu, C.M., Ta, V.D., Le, N.Q.K., Tadesse, D.A., Shi, C.: Deep neural network framework based on word embedding for protein glutarylation sites prediction. Life **12**(8), 1213 (2022)
13. Loshchilov, I., Hutter, F.: Decoupled weight decay regularization. arXiv preprint arXiv:1711.05101 (2017)
14. Prabhakar, C., Li, H.B., Yang, J., Shit, S., Wiestler, B., Menze, B.: ViT-AE++: improving vision transformer autoencoder for self-supervised medical image representations. arXiv preprint arXiv:2301.07382 (2023)
15. Qi, C.R., Su, H., Mo, K., Guibas, L.J.: Pointnet: deep learning on point sets for 3D classification and segmentation. In: Proceedings of the IEEE Conference on Computer Vision and Pattern Recognition, pp. 652–660 (2017)
16. Ronneberger, O., Fischer, P., Brox, T.: U-Net: convolutional networks for biomedical image segmentation. In: Navab, N., Hornegger, J., Wells, W.M., Frangi, A.F. (eds.) MICCAI 2015. LNCS, vol. 9351, pp. 234–241. Springer, Cham (2015). https://doi.org/10.1007/978-3-319-24574-4_28
17. Schell, M., et al.: Automated brain extraction of multi-sequence MRI using artificial neural networks. European Congress of Radiology-ECR 2019 (2019)
18. Shi, Y., Huang, Z., Feng, S., Zhong, H., Wang, W., Sun, Y.: Masked label prediction: unified message passing model for semi-supervised classification. arXiv preprint arXiv:2009.03509 (2020)
19. Sormani, M.P., Bruzzi, P.: MRI lesions as a surrogate for relapses in multiple sclerosis: a meta-analysis of randomised trials. Lancet Neurol. **12**(7), 669–676 (2013)
20. Sormani, M.P., De Stefano, N.: Defining and scoring response to IFN-β in multiple sclerosis. Nat. Rev. Neurol. **9**(9), 504–512 (2013)
21. Tousignant, A., Lemaître, P., Precup, D., Arnold, D.L., Arbel, T.: Prediction of disease progression in multiple sclerosis patients using deep learning analysis of MRI data. In: International Conference on Medical Imaging with Deep Learning, pp. 483–492. PMLR (2019)

22. Vaswani, A., et al.: Attention is all you need. In: Advances in Neural Information Processing Systems, vol. 30 (2017)
23. Wang, H., et al.: Mixed transformer U-Net for medical image segmentation. In: ICASSP 2022-2022 IEEE International Conference on Acoustics, Speech and Signal Processing (ICASSP), pp. 2390–2394. IEEE (2022)
24. Wang, Y., Sun, Y., Liu, Z., Sarma, S.E., Bronstein, M.M., Solomon, J.M.: Dynamic graph CNN for learning on point clouds. ACM Trans. Graph. (TOG) **38**(5), 1–12 (2019)
25. Wattjes, M.P., et al.: 2021 MAGNIMS-CMSC-NAIMS consensus recommendations on the use of MRI in patients with multiple sclerosis. Lancet Neurol. **20**(8), 653–670 (2021)
26. Yoo, Y., et al.: Deep learning of brain lesion patterns for predicting future disease activity in patients with early symptoms of multiple sclerosis. In: Carneiro, G., et al. (eds.) LABELS/DLMIA -2016. LNCS, vol. 10008, pp. 86–94. Springer, Cham (2016). https://doi.org/10.1007/978-3-319-46976-8_10
27. Zhang, X., He, L., Chen, K., Luo, Y., Zhou, J., Wang, F.: Multi-view graph convolutional network and its applications on neuroimage analysis for Parkinson's disease. In: AMIA Annual Symposium Proceedings, vol. 2018, p. 1147. American Medical Informatics Association (2018)
28. Zhang, Y.D., Pan, C., Sun, J., Tang, C.: Multiple sclerosis identification by convolutional neural network with dropout and parametric ReLU. J. Comput. Sci. **28**, 1–10 (2018)

atTRACTive: Semi-automatic White Matter Tract Segmentation Using Active Learning

Robin Peretzke[1,2(✉)], Klaus H. Maier-Hein[1,2,3,4,5,6,7], Jonas Bohn[1,4,8], Yannick Kirchhoff[1,6,7], Saikat Roy[1,7], Sabrina Oberli-Palma[1], Daniela Becker[9,10], Pavlina Lenga[9], and Peter Neher[1,4]

[1] Division of Medical Image Computing (MIC), German Cancer Research Center (DKFZ), Heidelberg, Germany
robin.peretzke@dkfz-heidelberg.de
[2] Medical Faculty Heidelberg, Heidelberg University, Heidelberg, Germany
[3] German Cancer Consortium (DKTK), Partner Site Heidelberg, Heidelberg, Germany
[4] National Center for Tumor Diseases (NCT), NCT Heidelberg, a Partnership Between DKFZ and University Medical Center Heidelberg, Heidelberg, Germany
[5] Pattern Analysis and Learning Group, Heidelberg University Hospital, Heidelberg, Germany
[6] HIDSS4Health - Helmholtz Information and Data Science School for Health, Karlsruhe/Heidelberg, Germany
[7] Faculty of Mathematics and Computer Science, Heidelberg University, Heidelberg, Germany
[8] Faculty of Bioscience, Heidelberg University, Heidelberg, Germany
[9] Department of Neurosurgery, Heidelberg University Hospital, Heidelberg, Germany
[10] IU, International University of Applied Sciences, Erfurt, Germany

Abstract. Accurately identifying white matter tracts in medical images is essential for various applications, including surgery planning and tract-specific analysis. Supervised machine learning models have reached state-of-the-art solving this task automatically. However, these models are primarily trained on healthy subjects and struggle with strong anatomical aberrations, e.g. caused by brain tumors. This limitation makes them unsuitable for tasks such as preoperative planning, wherefore time-consuming and challenging manual delineation of the target tract is typically employed. We propose semi-automatic entropy-based active learning for quick and intuitive segmentation of white matter tracts from whole-brain tractography consisting of millions of streamlines. The method is evaluated on 21 openly available healthy subjects from the Human Connectome Project and an internal dataset of ten neurosurgical cases. With only a few annotations, the proposed approach enables segmenting tracts on tumor cases comparable to healthy subjects (dice = 0.71), while the performance of automatic methods, like TractSeg dropped substantially (dice = 0.34) in comparison to healthy subjects. The method is implemented as a prototype named atTRACTive in the freely available software MITK Diffusion. Manual experiments on tumor data showed higher efficiency due to lower segmentation times compared to traditional ROI-based segmentation.

H. Greenspan et al. (Eds.): MICCAI 2023, LNCS 14227, pp. 237–246, 2023.
https://doi.org/10.1007/978-3-031-43993-3_23

Keywords: DWI and tractography · Active learning · Tract segmentation · Fiber tracking

1 Introduction

Diffusion-weighted MRI enables visualization of brain white matter structures. It can be used to generate tractography data consisting of millions of synthetic fibers or streamlines for a single subject stored in a tractogram that approximate groups of biological axons [1]. Many applications require streamlines to be segmented into individual tracts corresponding to known anatomy. Tract segmentations are used for a variety of tasks, including surgery planning or tract-specific analysis of psychiatric and neurodegenerative diseases [2,11,12,17].

Automated methods built on supervised machine learning algorithms have attained the current state-of-the-art in segmenting tracts [3,18,21]. Those are trained using various features, either directly from diffusion data in voxel space or from tractography data. Models may output binary masks containing the target white matter tract, or perform a classification on streamline level. However, such algorithms are commonly trained on healthy subjects and have shown issues in processing cases with anatomical abnormalities, e.g. brain tumors [20]. Consequently, they are unsuitable for tasks such as preoperative planning of neurosurgical patients, as they may produce incomplete or false segmentations, which could have harmful consequences during surgery [19]. Additionally, supervised techniques are restricted to fixed sets of predetermined tracts and are trained on substantial volumes of hard-to-generate pre-annotated reference data.

Manual methods are still frequently used for all cases not yet covered by automatic methods, such as certain populations like children, animal species, new acquisition schemes or special tracts of interests. Experts determine regions of interest (ROI) in areas where a particular tract is supposed to traverse or through which it must not pass, and segmentations can be accomplished either (1) by virtually excluding and maintaining streamlines from tractography according to the defined ROI or (2) by using these regions for tract-specific ROI-based tractography. Both approaches require a similar effort, although the latter is more commonly used. The correct definition of ROIs can be time-consuming and challenging, especially for inexperienced users. Despite these limitations, ROI-based techniques are currently without vivid alternatives for segmenting tracts that automated methods cannot handle.

Methods to simplify tract segmentation have been proposed before. Clustering approaches were developed to reduce complexity of large amounts of streamlines in the input data [4,6]. Tractome is a tool that allows interactive segmentation of such clusters by representing them as single streamlines that can interactively be included or excluded from the target tract [14]. Although the approach has shown promise, it has not yet supplanted conventional ROI-based techniques.

We propose a novel semi-automated tract segmentation method for efficient and intuitive identification of arbitrary white matter tracts. The method employs entropy-based active learning of a random forest classifier trained on features of the dissimilarity representation [13]. Active learning has been utilized for several cases in the medical domain, while it has never been applied in the context of tract segmentation [7,9,16]. It reduces manual efforts by iteratively identifying the most informative or ambiguous samples, here, streamlines, during classifier training, to be annotated by a human expert. The method is implemented as the tool atTRACTive in MITK Diffusion[1], enabling researchers to quickly and intuitively segment tracts in pathological datasets or other situations not covered by automatic techniques, simply by annotating a few but informative streamlines.

2 Methods

2.1 Binary Classification for Tract Segmentation

To create a segmentation of a white matter tract from an individual whole-brain tractogram T, streamlines which not belong to this tract must be excluded from the tractography data. This is formulated as a binary classification of a streamline $s \in T$, depending on whether it belongs to the target tract t (see Fig. 1 for a brief summary of the nomenclature of this work)

$$e(s) = \begin{cases} 1, & s \in t \\ 0, & else \end{cases}. \tag{1}$$

To perform the classification, supervised models have been trained on various features representing the data. We choose the dissimilarity representation proposed by Olivetti to classify streamlines, which has shown well performance and can be computed quickly for arbitrary data [3,13].

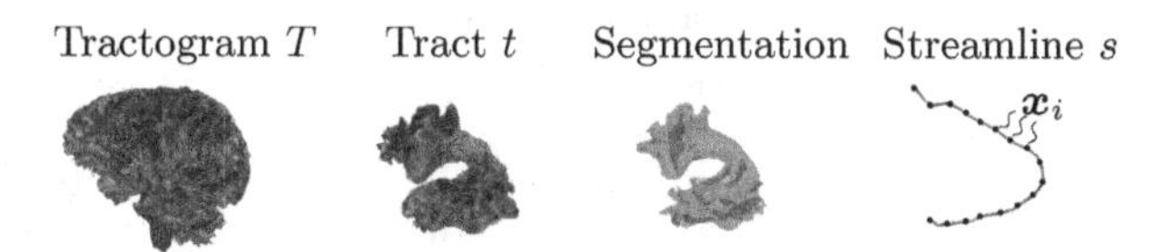

Fig. 1. Visualization of a tractogram T, a white matter tract t (arcuate fasciculus), its binary segmentation mask and a single streamline s containing 3D points $\boldsymbol{x}_i$.

A number of n streamlines, in this case, $n = 100$, are used as prototypes forming a reference system of the entire tractogram. A streamline is expressed through a feature vector relative to this reference system. Briefly, a single streamline $s = [\boldsymbol{x}_1, ..., \boldsymbol{x}_m]$, i.e. a polyline containing varying numbers of ordered 3D points $\boldsymbol{x}_i = [x_i, y_i, z_i] \in \mathbb{R}^3, i = 1...m$, is described by its minimum average direct flip distance d_{MDF} to each prototype [5]. The d_{MDF} of a streamline s_a to a prototype p_a is defined as

$$d_{MDF}(p_a, s_a) = \min((d_{direct}(p_a, s_a), d_{flipped}(p_a, s_a)) \tag{2}$$

where $d_{direct}(p_a, s_a) = \frac{1}{m}\sum_{i=1}^{m} ||\mathbf{x}_i^{p_a} - \mathbf{x}_i^{s_a}||_2$ with m being the number of 3D points of the streamlines and $d_{flip}(p_a, s_a) = \frac{1}{m}\sum_{i=1}^{m} ||\mathbf{x}_i^{p_a} - \mathbf{x}_{m-i+1}^{s_a}||_2$.

[1] https://github.com/MIC-DKFZ/MITK-Diffusion.

Additionally to d_{MDF}, the endpoint distance d_{END} between a streamline and a prototype is calculated, which is equal to d_{MDF}, besides, only the start points x_1 and endpoints x_m of the streamline and prototype are respected for the calculation [3]. Hence, features for a single streamline are represented by a vector twice the number of prototypes. In order to calculate these, all streamlines must have the same number of 3D points and are thus resampled to $m = 40$ points.

2.2 Active Learning for Tract Selection

Commonly, for training classifiers, large amounts of annotated and potentially redundant data are used, leading to high annotation efforts and long training times. Active learning reduces both by training machine learning models with only small and iteratively updated labeled subsets of the originally unlabeled data. The proposed workflow is initialized, as shown in Fig. 2(a), by presenting a randomly selected subset $S_{rand} = [s_1, ..., s_n]$ of $n =$ streamlines from an individual whole-brain tractogram to an expert for annotation, where $S_{rand} \subset T$. Subsequently, the dissimilarity representation is calculated using initially 100 prototypes (Fig. 2(b)), and a classifier is trained, in this case, a random forest. Within completing the training, which takes only a few seconds, the classifier predicts whether the remaining unlabeled streamlines belong to the target tract. Based on the predicted class labels, the target tract is presented (Fig. 2(c)). Furthermore, the class probabilities $p(s)$ determined by the random forest are

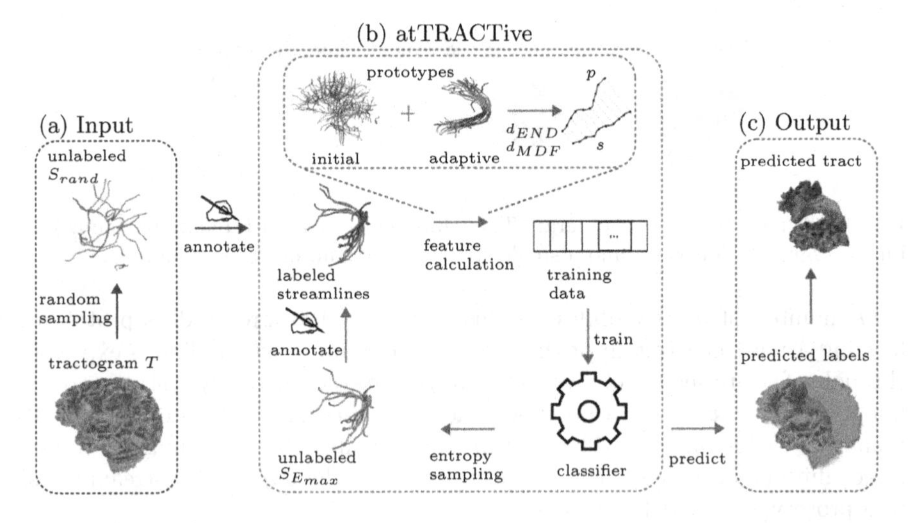

Fig. 2. Active Learning workflow: Extraction of subset S_{rand} of unlabeled streamlines from the whole-brain tractogram T for annotation (a). Initial and adaptive prototypes are used to compute d_{MDF} and d_{END} for each streamline, followed by training the random forest classifier (b), which predicts on remaining unlabeled streamlines (c). Entropy reflecting the uncertainty is used to select a new subset $S_{E_{max}}$ to annotate in the subsequent iterations until the human expert is satisfied with the prediction.

used to estimate its uncertainty with respect to each sample by calculating the entropy E

$$E(s_i) = \sum_{i=1}^{n} p(s_i) \log p(s_i). \tag{3}$$

Next, a subset $S_{E_{max}}$ of streamlines with the highest entropy or uncertainty is selected to be labeled by the expert and is added to the training data (Fig. 2 (c)) [8]. Additionally, these streamlines are utilized as adaptive prototypes until a threshold of $n = 100$ adaptive prototype streamlines is reached. Since the model selects ambiguous streamlines in the target tract region, utilizing them as supplementary prototypes improves feature expressiveness in this region of interest.

This process is repeated iteratively until the expert accepts the prediction.

3 Experiments

3.1 Data

The proposed technique was tested on a healthy-subject dataset and on a dataset containing tumor cases. The first comprises 21 subjects of the human connectome project (HCP) that were used for testing the automated methods TractSeg and Classifyber [3,18]. Visual examinations revealed false-negatives in the reference tracts, meaning that some streamlines that belong to the target tract were not included in the reference. These false-negatives did not affect the generation of accurate segmentation masks, since most false-negatives are occupied by true-positive streamlines, but negatively influenced our experiments. To reduce false-negatives, the reference segmentation mask as well as start- and end-region segmentations were used to reassign streamlines from the tractogram using two criteria: Streamlines must be inside the binary reference segmentation (1) and start and end in the assigned regions (2). As the initial size of ten million streamlines is computationally challenging and unsuitable for most tools, all tractograms were randomly down-sampled to one million streamlines. We focused on the left optic radiation (OR), the left cortico-spinal tract (CST), and the left arcuate-fasciculus (AF), representing a variety of established tracts.

To test the proposed method on pathological data, we used an in-house dataset containing ten presurgical scans of patients with brain tumors. Tractography was performed using probabilistic streamline tractography in MITK Diffusion. To reduce computational costs, we retained one million streamlines that passed through a manually inserted ROI located in an area traversed by the OR [15]. Subjects have tumor appearance with varying sizes ($(17.87 \pm 12.73\,\mathrm{cm}^3)$) in temporoloccipital, temporal, and occipital regions, that cause deformations around the OR and lead to deviations of the tract from the normative model.

3.2 Experimental Setup

To evaluate the proposed method, we conducted two types of experiments. Manual segmentation experiments using an interactive prototype of atTRACTive were initiated on the tumor data (holistic evaluation). Additionally, reproducible simulations on the freely available HCP and the internal tumor dataset were created (algorithmic evaluation). In order to mimic expert annotation during algorithmic evaluation, class labels were assigned to streamlines using previously generated references. The quality of the predictions was measured by calculating the dice score of the binary mask. The code used for these experiments is publicly available[2].

For the algorithmic evaluation, the initial training dataset was created with 20 randomly selected streamlines from the whole-brain tractogram, which have been shown to be a decent number to start training. Since some tracts contain only a fraction of streamlines from the entire tractogram, it might be unlikely that the training dataset will contain any streamline belonging to the target tract. Therefore, two streamlines of the specific tract were further added to the training dataset, and class weights were used to compensate for the class unbalance. According to Fig. 2, the dissimilarity representation was determined, the random forest classifier was trained and the converged model was used to predict on the unlabeled streamlines and to calculate the entropy. In each iteration, the ten streamlines with the highest entropy are added to the training dataset, which has been determined to be a good trade-off between annotation effort and prediction improvement. The process was terminated after 20 iterations, increasing the size of the training data from 22 to 222 out of one million streamlines.

The holistic evaluation was conducted with equal settings, except that the workflow was terminated when the prediction matched the expectation of the expert. To ensure that the initial dataset S_{rand} contained streamlines from the target tract, the expert initiated the active learning workflow by defining a small ROI that included fibers of the tract. S_{rand} was created by randomly sampling only those streamlines that pass through this ROI. To allow comparison between the proposed and traditional ROI-based techniques, the OR of subjects from the tumor dataset were segmented using both approaches by an expert familiar with the respective tool, and the time required was reported to measure efficiency.

Note, in all experiments, the classifier is trained from scratch every iteration, prototypes are generated for each subject individually, and the classifier predicts on data from the same subject it is trained with, as it performs subject-individual tract segmentation and is not used as a fully automated method. To ensure a stable active learning setup that generalizes across different datasets, the whole method was developed on the HCP and applied with fixed settings to the tumor data [10].

3.3 Results

In Table 1, the dice score of the active learning simulation on the HCP and tumor data after the fifth, tenth, and twentieth iterations are shown and compared with outcomes of Classifyber and TractSeg. Results for the HCP data were already on par with the benchmark of automatic methods between the fifth and tenth

[2] https://github.com/MIC-DKFZ/atTRACTive_simulations.

iterations. On the tumor data, the performance of the proposed method remains above 0.7 while the performance of TractSeg drops substantially. Furthermore, Classifyber does not support the OR and is therefore not listed in Table 1.

Table 1. Dice score of TractSeg and Classifyber on both datasets compared to entropy-based active learning after the fifth, tenth, and twentieth iteration.

Tract	HCP dataset			Tumor dataset
	CST	AF	OR	OR
Classifyber [3]	0.86 ± 0.01	0.84 ± 0.03	–	–
TractSeg [18]	0.86 ± 0.02	0.86 ± 0.03	0.83 ± 0.02	0.34 ± 0.19
Active Learning$_{it=5}$	0.83 ± 0.05	0.86 ± 0.03	0.79 ± 0.11	0.71 ± 0.06
Active Learning$_{it=10}$	0.88 ± 0.03	0.88 ± 0.02	0.85 ± 0.05	0.72 ± 0.05
Active Learning$_{it=20}$	$\mathbf{0.90 \pm 0.03}$	$\mathbf{0.90 \pm 0.02}$	$\mathbf{0.88 \pm 0.02}$	$\mathbf{0.73 \pm 0.08}$

Figure 3 depicts the quantitative gain of active learning on the three tracts of the HCP data and compares it to pure random sampling by displaying the dice score depending on annotated streamlines. While active learning leads to an increase in the metric until the predictions at around five to ten iterations show no meaningful improvements, the random selection does not improve overall.

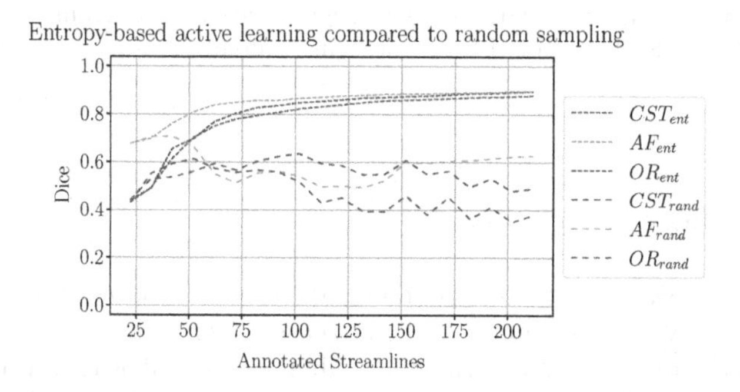

Fig. 3. Dice score of entropy-based active learning of the HCP dataset of the OR, CST, and AF ($[Tract]_{ent}$) compared to random sampling ($[Tract]_{rand}$)

Qualitative results of the algorithmic evaluation of the AF of a randomly chosen subject of the HCP dataset are shown in Fig. 4(a). Initially, the randomly sampled streamlines in the training data are distributed throughout the brain, while entropy-based selected streamlines from subsequent iterations cluster around the AF. The prediction improves iteratively, as indicated by a rising dice score. When accessing qualitative results of the pathological dataset visual inspection revealed particularly poorly performance of TractSeg in cases where

OR fibers were in close proximity to tumor tissue, leading to fragmented segmentations, while complete segmentations were reached with active learning even for these challenging tracts after a few iterations, as shown in Fig. 4(b).

The initial manual experiments with atTRACTive were consistent with the simulations. The prediction aligned with the expectations of the expert at around five to seven iterations taking a mean of 4,5 min, while it took seven minutes on average to delineate the tract with ROI-based segmentation. During the iterations, streamlines around the target tract were suggested for labeling, and the prediction improved. Visual comparison yielded more false-positive streamlines with the ROI-based approach while atTRACTive created more compact tracts.

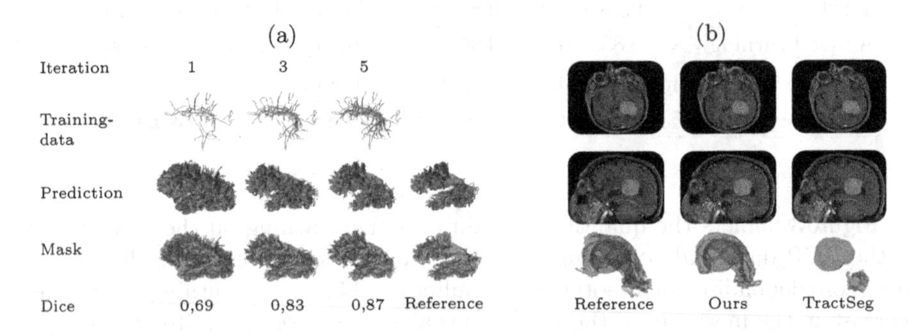

Fig. 4. Training data, prediction, segmentation mask, and dice-score after first, third, and fifth iteration of active learning with the reference of the AF (a) and tumor (blue) and OR segmentation (red) with reference, active learning outcome after the third iteration and TractSegs output (b). (Color figure online)

4 Discussion

Active learning-based white matter tract segmentation enables the identification of arbitrary pathways and can be applied to cases where fully automated methods are unfeasible. In this work, algorithmic evaluation as well as the implementation of the technique into the GUI-based tool atTRACTive including further holistic manual experiments were conducted.

The algorithmic evaluation yielded consistent results from the fifth to the tenth iterations on both the HCP and tumor datasets. As expected, outcomes obtained from the tumor dataset were not quite as good as those of the HCP dataset. This trend is generally observed in clinical datasets, which tend to exhibit lower performance levels compared to high-quality datasets, which could be responsible for the decline in the results. Preliminary manual experiments with atTRACTive indicated active learning to have shorter segmentation times compared to traditional ROI-based techniques. These experiments are in line with the simulations as the generated tracts matched the expectations of the expert after around five to seven iterations, meaning that less than a hundred out of million annotated streamlines are required to train the model. Enhancements to the usability of the prototype are expected to further improve efficiency. A current limitation of atTRACTive is the selection of the initial subset, based on randomly sampling streamlines passing through a manually inserted ROI. This approach does not guarantee that streamlines of the target tract are included in the subset. In that case, the ROI has to be replaced or S_{rand} needs to be regenerated.

Future analyses, evaluating the inter- and intra-rater variability compared to other interactive approaches, will be conducted on further tracts. For selected scenarios, the ability of the classifier to generalize by learning from previously annotated subjects will be investigated, which may even allow to train a fully automatic classifier for new tracts once enough data is annotated. To further optimize the method, the feature representation or sampling procedure could be improved. Uncertainty sampling may select redundant streamlines due to similar high entropy values. Instead, annotating samples with high entropy values being highly diverse or correcting false classifications could convey more information.

By introducing active learning into tract segmentation, we provide an efficient and intuitive alternative compared to traditional ROI-based approaches.

atTRACTive has the potential to interactively assist researchers in identifying arbitrary white matter tracts not captured by existing automated approaches.

References

1. Basser, P.J., Pajevic, S., Pierpaoli, C., Duda, J., Aldroubi, A.: In vivo fiber tractography using DT-MRI data. Magn. Reson. Med. **44**(4), 625–632 (2000)
2. Berman, J.: Diffusion MR tractography as a tool for surgical planning. Magn. Reson. Imaging Clin. N. Am. **17**(2), 205–214 (2009)
3. Bertò, G., et al.: Classifyber, a robust streamline-based linear classifier for white matter bundle segmentation. Neuroimage **224**, 117402 (2021)
4. Garyfallidis, E., Brett, M., Correia, M.M., Williams, G.B., Nimmo-Smith, I.: QuickBundles, a method for tractography simplification. Front. Neurosci. **6**, 175 (2012)
5. Garyfallidis, E., Ocegueda, O., Wassermann, D., Descoteaux, M.: Robust and efficient linear registration of white-matter fascicles in the space of streamlines. Neuroimage **117**, 124–140 (2015)
6. Guevara, P., et al.: Robust clustering of massive tractography datasets. Neuroimage **54**(3), 1975–1993 (2011)
7. Hao, R., Namdar, K., Liu, L., Khalvati, F.: A transfer learning-based active learning framework for brain tumor classification. Front. Artif. Intell. **4**, 635766 (2021)
8. Holub, A., Perona, P., Burl, M.C.: Entropy-based active learning for object recognition. In: 2008 IEEE Computer Society Conference on Computer Vision and Pattern Recognition Workshops, pp. 1–8. IEEE (2008)
9. Liu, L., Lei, W., Wan, X., Liu, L., Luo, Y., Feng, C.: Semi-supervised active learning for covid-19 lung ultrasound multi-symptom classification. In: 2020 IEEE 32nd International Conference on Tools with Artificial Intelligence (ICTAI), pp. 1268–1273. IEEE (2020)

10. Lüth, C.T., Bungert, T.J., Klein, L., Jaeger, P.F.: Toward realistic evaluation of deep active learning algorithms in image classification. arXiv preprint arXiv:2301.10625 (2023)
11. McIntosh, A.M., et al.: White matter tractography in bipolar disorder and schizophrenia. Biol. Psychiat. **64**(12), 1088–1092 (2008)
12. Mukherjee, P., McKinstry, R.C.: Diffusion tensor imaging and tractography of human brain development. Neuroimaging Clin. **16**(1), 19–43 (2006)
13. Olivetti, E., Avesani, P.: Supervised segmentation of fiber tracts. In: Pelillo, M., Hancock, E.R. (eds.) SIMBAD 2011. LNCS, vol. 7005, pp. 261–274. Springer, Heidelberg (2011). https://doi.org/10.1007/978-3-642-24471-1_19
14. Porro-Muñoz, D., Olivetti, E., Sharmin, N., Nguyen, T.B., Garyfallidis, E., Avesani, P.: Tractome: a visual data mining tool for brain connectivity analysis. Data Min. Knowl. Disc. **29**(5), 1258–1279 (2015). https://doi.org/10.1007/s10618-015-0408-z
15. Tuch, D.S.: Q-ball imaging. Magn. Reson. Med.: Off. J. Int. Soc. Magn. Reson. Med. **52**(6), 1358–1372 (2004)
16. Wang, J., Yan, Y., Zhang, Y., Cao, G., Yang, M., Ng, M.K.: Deep reinforcement active learning for medical image classification. In: Martel, A.L., et al. (eds.) MICCAI 2020, Part I. LNCS, vol. 12261, pp. 33–42. Springer, Cham (2020). https://doi.org/10.1007/978-3-030-59710-8_4
17. Wasserthal, J., et al.: Multiparametric mapping of white matter microstructure in catatonia. Neuropsychopharmacology **45**(10), 1750–1757 (2020)
18. Wasserthal, J., Neher, P., Maier-Hein, K.H.: TractSeg-fast and accurate white matter tract segmentation. Neuroimage **183**, 239–253 (2018)
19. Yang, J.Y.M., Yeh, C.H., Poupon, C., Calamante, F.: Diffusion MRI tractography for neurosurgery: the basics, current state, technical reliability and challenges. Phys. Med. Biol. **66**(15), 15TR01 (2021)
20. Young, F., Aquilina, K., A Clark, C., D Clayden, J.: Fibre tract segmentation for intraoperative diffusion MRI in neurosurgical patients using tract-specific orientation atlas and tumour deformation modelling. Int. J. Comput. Assist. Radiol. Surg. **17**, 1–9 (2022)
21. Zhang, F., Karayumak, S.C., Hoffmann, N., Rathi, Y., Golby, A.J., O'Donnell, L.J.: Deep white matter analysis (DeepWMA): fast and consistent tractography segmentation. Med. Image Anal. **65**, 101761 (2020)

Domain-Agnostic Segmentation of Thalamic Nuclei from Joint Structural and Diffusion MRI

Henry F. J. Tregidgo[1(✉)], Sonja Soskic[1], Mark D. Olchanyi[2,3], Juri Althonayan[1], Benjamin Billot[1,4], Chiara Maffei[5], Polina Golland[4], Anastasia Yendiki[5], Daniel C. Alexander[1], Martina Bocchetta[6,7], Jonathan D. Rohrer[7], and Juan Eugenio Iglesias[1,4,5]

[1] Centre for Medical Image Computing, UCL, London, UK
h.tregidgo@ucl.ac.uk
[2] Neurostatistics Research Laboratory, MIT, Cambridge, USA
[3] Neurotechnology and Neurorecovery, MGH and Harvard Medical School, Boston, USA
[4] Computer Science and Artificial Intelligence Laboratory, MIT, Cambridge, USA
[5] Martinos Center for Biomedical Imaging, MGH and Harvard Medical School, Boston, USA
[6] Centre for Cognitive and Clinical Neuroscience, Brunel University, London, UK
[7] Dementia Research Centre, UCL, London, UK

Abstract. The human thalamus is a subcortical brain structure that comprises dozens of nuclei with different function and connectivity, which are affected differently by disease. For this reason, there is growing interest in studying the thalamic nuclei *in vivo* with MRI. Tools are available to segment the thalamus from 1 mm T1 scans, but the image contrast is too faint to produce reliable segmentations. Some tools have attempted to refine these boundaries using diffusion MRI information, but do not generalise well across diffusion MRI acquisitions. Here we present the first CNN that can segment thalamic nuclei from T1 and diffusion data of any resolution without retraining or fine tuning. Our method builds on our histological atlas of the thalamic nuclei and silver standard segmentations on high-quality diffusion data obtained with our recent Bayesian adaptive segmentation tool. We combine these with an approximate degradation model for fast domain randomisation during training. Our CNN produces a segmentation at 0.7 mm isotropic resolution, irrespective of the resolution of the input. Moreover, it uses a parsimonious model of the diffusion signal (fractional anisotropy and principal eigenvector) that is compatible with virtually any set of directions and b-values, including huge amounts of legacy data. We show results of our proposed method on three heterogeneous datasets acquired on dozens of different scanners. The method is publicly available at freesurfer.net/fswiki/ThalamicNucleiDTI.

1 Introduction

The human thalamus is a brain region with connections to the whole cortex [6,30]. It comprises dozens of nuclei that are involved in diverse

H. Greenspan et al. (Eds.): MICCAI 2023, LNCS 14227, pp. 247–257, 2023.
https://doi.org/10.1007/978-3-031-43993-3_24

functions like cognition, memory, sensory, motor, consciousness, language, and others [14,30,32]. Crucially, these nuclei are differently affected by diseases such as Parkinson's [17], Alzheimer's [9,10], or frontotemporal dementia [40]. Such differentiation has sparked interest in studying the thalamic nuclei *in vivo* with MRI. This requires automated segmentation methods at the subregion level, as opposed to the whole thalamus provided by neuromaging packages like FreeSurfer [15] or FSL [27], or by convolutional neural networks (CNNs) like DeepNAT [41] or SynthSeg [7].

Different approaches have been used to segment thalamic nuclei. Some methods have attempted to register manually labelled histology to MRI [20,23,29], but accuracy is limited by the difficulty in registering two modalities with such different contrasts, resolutions, and artifacts. Diffusion MRI (dMRI) has been used to spatially cluster voxels into subregions, based on similarity in diffusion signal [5,25,31] or connectivity to cortical regions [6,21]. Clustering based on functional MRI connectivity has also been explored [42]. However, such clusters are not guaranteed to correspond to anatomically defined nuclei.

Other methods have relied on specialised MRI sequences to highlight the anatomical boundaries of the thalamus, typically at 7T [24,36] or with advanced dMRI acquisitions [4]. A popular method within this category is "THOMAS", a labelled dataset of 7T white-matter-nulled scans that has been used to segment the thalamic nuclei with multi-atlas segmentation [35] and CNNs [38]. Its disadvantage is requiring such advanced acquisitions at test time, which precludes its application to legacy data or at sites without the required expertise or resources.

One approach that supports training and test data of different modalities is Bayesian segmentation, which combines a probabilistic atlas (derived from one modality) with a likelihood model to compute adaptive segmentations on other modalities using Bayesian inference [3]. A probabilistic atlas of thalamic nuclei built from 3D reconstructed histology is available on FreeSurfer, along with a companion Bayesian segmentation method to segment the nuclei from structural scans [18]. We have recently released an improved version of this method that incorporates dMRI into the likelihood model [37], but it inherits the well-known problems of Bayesian segmentation with partial voluming (PV) [39]. While this tool works well with high-resolution dMRI data (like the Human Connectome Project, or HCP [33]), the lack of PV modelling is detrimental at resolutions much lower than ~1 mm isotropic. This is the case of virtually every legacy dataset, and many modern datasets that use lower resolutions to keep acquisition time short or for consistency with older timepoints (e.g., ADNI [19]).

Finally, there are also supervised discriminative methods that label the thalamus from dMRI. An early approach by Stough et al. [34] used a random forest to segment the thalamus into six groups of nuclei. As features, they used local measures like fractional anisotropy (FA) and the principal eigenvector ($V1$), and connectivity with cortical regions. A recent approach [13] segmented the whole thalamus using the six unique elements of the diffusion tensor image (DTI) as inputs. While these supervised approaches can provide excellent performance on the training domain, they falter on datasets with different resolution.

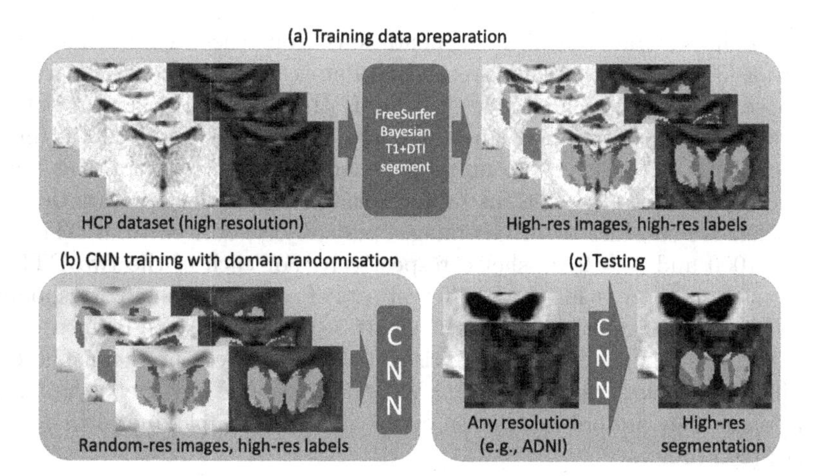

Fig. 1. Overview of the proposed method. (a) Generation of labelled training data. (b) Training with domain randomisation. (c) Testing. Images are in coronal view.

Here, we present the first CNN for joint dMRI and T1-weighted (*T1w*) images that can segment the thalamic nuclei without retraining or fine-tuning. We use domain randomisation to model resolution during training, which enables the CNN to produce super-resolved 0.7 mm isotropic segmentations, independently of the resolution of the input images. Aggressive data augmentation is used to ensure robustness against variations in contrast, shape and artifacts. Finally, our CNN uses a parsimonious representation of the dMRI data (FA+V1), which makes our publicly available tool compatible with virtually every dMRI dataset.

2 Methods

A summary of our method is shown in Fig. 1. We use our joint T1/DTI Bayesian method in FreeSurfer to segment the thalamic nuclei from a large number of modern, high-quality scans. These segmentations are used as silver standard to train a CNN, thus circumventing the need for manual segmentations. Our approach uses a hybrid domain randomisation and augmentation strategy that enables the network to generalise to virtually any diffusion dataset.

2.1 Training Dataset, Preprocessing, and Data Representation

To make the CNN compatible with legacy datasets, we choose a simple representation based on the FA and V1 of the DTI fit at each voxel. DTI only requires 7 measurements and is thus compatible with even the oldest datasets. As in many DTI visualisation tools, we combine the FA and V1 into a single 3×1 red-green-blue vector at every voxel. This RGB vector has brightness proportional to the FA, and its colour encodes the direction of V1 as shown in Fig. 1a.

To obtain accurate training segmentations from the Bayesian method in FreeSurfer [37] we require a high-resolution dataset with reduced PV artifacts. We choose the HCP, which includes 0.7 mm isotropic T1 and 1.25 mm isotropic dMRI with 90 directions and three b-values (1000, 2000, and 3000 s/mm^2). We use the HCP to generate our targets and then generate training images at a wide spectrum of resolutions by increasing the voxel size with a degradation model. We consider two RGB images per subject, derived from DTI fits of the b = 1000 and b = 2000 shells, respectively. For each of the two DTI fits, the Bayesian method yields three different sets of segmentations (corresponding to three available likelihood models). All six segmentations are defined on the .7 mm grid (Fig. 1a, right), and comprise 23 thalamic nuclei per hemisphere (46 total) [37]. We exclude the Pc, Pt and VM nuclei provided by the Bayesian segmentation as they are not labelled in every training example due to their small volumes (2–3 mm^3).

2.2 Domain Randomisation and Data Augmentation

We employ domain randomisation and aggressive data augmentation for both our T1 and diffusion data in order to model: *(i)* the degradation in quality from HCP to more standard acquisition protocols, and *(ii)* the variability in appearance due to differences in acquisitions and scanners at test time.

Domain Randomisation for Resolution: at the crux of our method is the domain randomisation of input resolutions. At every iteration, we randomly sample the voxel dimensions for the T1 and DTI (independently) in two steps. First, we sample a "coarse" scalar voxel size from a uniform distribution between 1 and 3 mm. Then, we sample the voxel side length in each direction from a normal distribution centred on this "coarse" mean with $\sigma = 0.2$ mm.

Next, we resample the T1 and RGB channels to the sampled resolution. For the T1, we use a publicly available PV-aware degradation model [8], which accounts for variability in slice thickness and slice spacing. For the RGB, one should theoretically downsample the original diffusion-weighted images, and recompute the DTI at the target resolution. However, the exact characteristics of the blurring depend on the set of directions and b-values, which will not be the same for the training and test datasets. Moreover, recomputing the DTI is too slow for on-the-fly augmentation. Instead, we apply the degradation model to the RGB image directly, which can be done very efficiently. This is only an approximation to the actual degradation, but in practice, the domain randomisation strategy minimises the effects of the domain gap created by the approximation.

Data Augmentations: we also apply a number of geometric and intensity augmentations, some standard, and some specific to our dMRI representation.

- *Global geometric augmentation:* we use random uniform scaling (between 0.85 and 1.15) and rotations about each axis (between −15 and 15°). Rotations are applied to the images and also used to reorient the V1 vectors.

- *Local geometric augmentation:* we deform the scans with a piecewise linear deformation field, obtained by linear interpolation on a $5 \times 5 \times 5 \times 3$ grid. V1 vectors are reoriented with the PPD method ("preservation of principle direction" [1]).
- *Local orientation augmentation:* we generate a smooth grid of random rotations between $[-15°, 15°]$ around each axis using piecewise linear interpolation on a $5 \times 5 \times 5 \times 3$ grid. These are applied to V1 to simulate noise in principle direction
- *DTI "speckles":* To account for infeasible FA and V1 voxels generated by potentially unconstrained DTI fitting, we select random voxels in the low resolution images (with probability $p = 1 \times 10^{-4}$), randomise their RGB values, and renormalise them so that their effective FA is between 0.5 and 1.
- *Noise, brightness, contrast, and gamma:* we apply random Gaussian noise to both the T1 and FA; randomly stretch the contrast and modify the brightness of the T1; and apply a random gamma transform to the T1 and FA volumes.

These augmentations are applied to the downsampled images. After that, the augmented images are upscaled back to 0.7 mm isotropic resolution. This ensures that all the channels (including the target segmentations) are defined on the same grid, independently of the intrinsic resolution of the inputs (Fig. 1b). At test time, this enables us to produce .7 mm segmentations for scans of any resolution (prior upscaling to the .7 mm grid).

2.3 Loss

We build on the soft dice loss [26]. Since some labels in the atlas are very small, we implemented a grouped soft dice by combining nuclei into 10 larger groupings [37]. We then combined this Dice with the average Dice of the individual nuclei and the Dice of the whole thalamus into the following composite loss:

$$\mathcal{L} = -\sum_{l} SDC(X_l, Y_l) - \sum_{g} SDC\left(\{X_l\}_{G_g}, \{Y_l\}_{G_g}\right) - SDC\left(\{X_l\}_{l \neq 0}, \{Y_l\}_{l \neq 0}\right), \quad (1)$$

where, $X_l = \{x_i^l\}$ and $Y_l = \{y_i^l\}$ are the predicted and ground truth probability maps for label $l \in [0, \ldots, L]$; G_g is the set of label indices in nuclear group $g \in [1, \ldots, 10]$, label $l = 0$ corresponds to the background and SDC is the soft Dice coefficient: $SDC(X, Y) = (2 \times \sum_i x_i y_i)/(\sum_i x_i^2 + \sum_i y_i^2)$.

2.4 Architecture and Implementation Details

Our CNN is a 3D U-net [11,28] with 5 levels (2 layers each), $3 \times 3 \times 3$ kernels and ELU activations [12]. The first level has 24 features, and every level has twice as many features as the previous one. The last layer has a softmax activation. The loss in Eq. 1 was optimised for 200,000 iterations with Adam [22]. A random crop of size $128 \times 128 \times 128$ voxels (guaranteed to contain the thalami) was used at every iteration. The T1 scans are normalised by scaling the median white matter

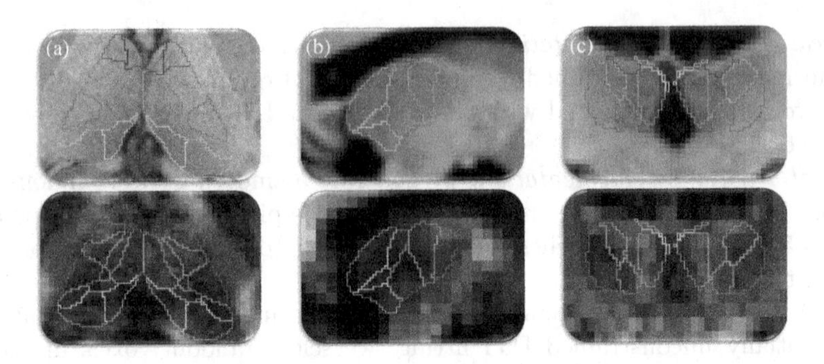

Fig. 2. Example segmentations from: (a) HCP (axial); (b) LOCAL (sagittal); and (c) ADNI (coronal). Top: T1. Bottom: RGB encoding. The CNN produces accurate segmentations at high isotropic resolution despite the heterogeneous acquisitions.

intensity to 0.75 and clipping at [0, 1]. The DTI volumes are upsampled to the 0.7 mm space of the T1s (using the log domain [2]) prior to the RGB computation. To generate a training target we combine all six segmentation candidates (three likelihood models times two shells) in a two step process: *(i)* averaging the one-hot encodings of each segmentation, and *(ii)* coarsely segmenting into 10 label groups and renormalising the soft target.

For validation purposes, we used the Bayesian segmentations of 50 withheld HCP subjects and 14 withheld ADNI subjects. Even though the Bayesian segmentation of ADNI data is not reliable enough to be used as ground truth for evaluation (due to the PV problems described in the Introduction), it is still informative for validation purposes – particularly when combined with HCP data. The final model for each network is chosen based on the validation loss averaged between the HCP and ADNI validation sets.

3 Experiments and Results

3.1 MRI Data

We trained on 600 HCP subjects (see Sect. 2.1). For evaluation, we used:

HCP: 10 randomly selected subjects (not overlapping with the training data), with manual segmentations of 10 groups of labels (the same as in Sect. 2.3).

LOCAL: 21 healthy subjects (9 males, ages 53–80), each with a 1.1 mm isotropic T1 and a test-retest pair of 2.5 mm isotropic dMRI (64 directions, b=1,000).

ADNI: 90 subjects from the ADNI, 45 with with Alzheimer's disease (AD) and 45 healthy controls (73.8$\pm$7.7 years; 44 females), each with a T1w (1.2$\times$1$\times$1 mm, sagittal) and dMRI (1.35$\times$1.35$\times$2.7 mm, axial, 41 directions, b=1,000) scan.

3.2 Competing Methods and Ablations

To the best of our knowledge, the only available tool that can segment T1/dMRI of any resolution is Freesurfer [37]. We therefore compare our method with this

Table 1. Mean Dice for ground truth comparison (left columns) and test re-test (right columns). Dice is shown for labels grouped into: histological labels ("hist", 23 labels), manual protocol ("manual", 10 labels), nuclear groups [37] ("nuclear", 5 labels), and whole thalamus. CNNs are sorted in descending order of average Dice across columns.

Model	Dice with manual tracing:			Test re-test Dice score:			
	Manual	Nuclear	Whole	Hist	Manual	Nuclear	Whole
Bayesian	0.547	0.639	0.885	0.716	0.826	0.880	0.946
CNN No defm.	0.585	0.678	0.895	0.824	0.885	0.917	0.963
CNN single target	0.582	0.676	0.905	0.824	0.883	0.911	0.962
CNN full model	0.576	0.674	0.903	0.819	0.876	0.910	0.961
CNN majority vote	0.579	0.676	0.897	0.816	0.875	0.909	0.961
CNN no rotation	0.580	0.672	0.902	0.808	0.873	0.909	0.961
CNN no speckle	0.583	0.680	0.903	0.799	0.869	0.907	0.958
CNN average target	0.575	0.669	0.898	0.799	0.867	0.906	0.959
CNN Dice Loss	0.544	0.618	0.806	0.824	0.867	0.899	0.963

algorithm, along with seven ablations of model options: using only the Dice loss; three variations on the way of merging the candidate Bayesian segmentations into a training target (average one-hot; majority voting; and selecting a segmentation at random); and three ablations on the augmentation (omitting the "speckle" DTI voxels, the random V1 rotations, and the piecewise linear deformation).

3.3 Results

Qualitative results are shown in Fig. 2. Our CNN successfully segments all scans at 0.7 mm resolution, despite the different input voxel sizes. Quantitative results are presented below for three experimental setups, one with each dataset.

Direct Evaluation with Manual Ground truth Using HCP: We first evaluated all competing methods and ablations on the 10 manually labelled subjects. Table 1 (left columns) shows mean Dice scores at different levels of granularity. Thanks to the ground truth aggregation, domain randomisation and aggressive augmentation, most of the CNNs produce higher accuracy than the Bayesian method at every level of detail – despite having been trained on automated segmentations from the Bayesian tool. The only ablation with noticeable lower performance is the one using the Dice of only the fine histological labels (i.e., no Dice of groupings), which highlights the importance of our composite Dice loss.

Test-Retest Using LOCAL: Table 1 (right columns) shows Dice scores between the segmentations of the two available sets of images for the LOCAL dataset, for different levels of granularity. All the networks are more stable than the Bayesian method, with considerably higher test-retest dice scores.

Best-Performing CNN: Considering Table 1 as a whole, the CNN with the highest mean Dice is the one without local geometric augmentation. We hypothesise that this is because the benefit of this augmentation is negligible due to

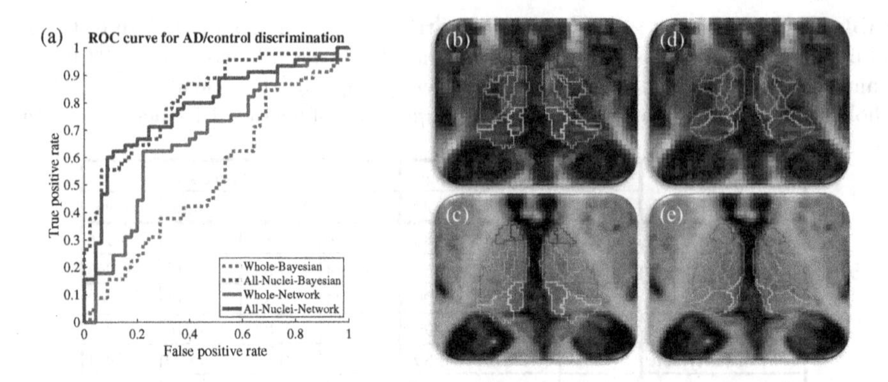

Fig. 3. Comparison of Bayesian and CNN segmentation on ADNI. (a) ROC curves for AD discrimination. (b) Axial view of RGB encoding of a subject with Bayesian segmentation overlaid. (c) Corresponding view of T1w scan. (d–e) Corresponding plane with CNN segmentation. Red arrows point at oversegmentation by the Bayesian method. (Color figure online)

Table 2. Area under the curve and accuracy at elbow for AD discrimination.

Method	Whole thal AUC	Whole thal acc. at elbow	Nuclei AUC	Nuclei acc. at elbow
Bayesian	53.3	57.8	81.9	74.4
CNN	67.1	70.0	78.3	75.6

the large number of training cases, and thus does not compensate for the loss of performance due to the approximations that are required to augment on the fly.

Group Study Using ADNI: We segmented the ADNI subjects with the best-performing CNN, and computed volumes of the thalamic nuclei normalised by the intracranial volume (estimated with FreeSurfer). We then computed ROC curves for AD discrimination using a threshold on: *(i)* the whole thalamic volume; and *(ii)* the likelihood ratio given by a linear discriminant analysis (LDA, [16]) on the volumes of the 23 nuclei (left-right averaged). The ROC curves are shown in Fig. 3(a). The area under the curve (AUC) and accuracy at the elbow are shown in Table 2. The LDA from the CNN and Bayesian methods show no significant difference in overall discriminative power. However, the atrophy detected by the CNN shows greater significance with the AV and VA reaching significance after correction for multiple comparisons ($p < 0.002$ Wilcoxon rank-sum). Additionally, there is a significant increase in the discriminative power of the whole thalamus from the CNN compared to the Bayesian method ($p < 0.005$ paired DeLong test). This indicates the external boundary of our method may be more useful than that provided by the Bayesian method and often corresponds to a reduction of oversegmentation into the pulvinar (Figs. 3b–e).

4 Discussion and Conclusion

We have presented the first method that can segment the thalamic nuclei from T1 and dMRI data obtained with virtually any acquisition, solving the problems posed by PV to Bayesian methods. Using Bayesian segmentations generated from multiple diffusion models while applying hybrid domain randomisation and augmentation methods, we remarkably improve upon both the accuracy and reliability of our source segmentations. Our tool is robust against misregistration from geometric distortion, which is generally more problematic in frontal and occipital regions. Nuclei volumes resulting from the tool show similar discriminative power to those provided by the Bayesian tool, while improving the utility of whole thalamus measurements and increasing segmentation resolution. Crucially, our use of the FA and V1 representation of dMRI data as input means that our tool is compatible with virtually every dMRI dataset. Publicly sharing this ready-to-use tool as part of FreeSurfer 7.4 will enable neuroimaging studies of the thalamic nuclei without requiring any expertise in neuroanatomy or machine learning, and without any specialised computational resources.

Acknowledgments. Work primarily funded by ARUK (IRG2019A003). Additional support by the NIH (RF1MH123195, R01AG070988, P41EB015902, R01EB021265, R56MH121426, R01NS112161), EPSRC (EP/R006032/1), Wellcome Trust (221915/Z/20/Z), Alzheimer's Society (AS-JF-19a-004-517), Brain Research UK, Wolfson; UK NIHR (BRC149/NS/MH), UK MRC (MR/M008525/1), Marie Curie (765148), ERC (677697), and Miriam Marks Brain Research.

References

1. Alexander, D., Pierpaoli, C., Basser, P., Gee, J.: Spatial transformations of diffusion tensor magnetic resonance images. IEEE Trans. Med. Imaging **20**(11), 1131–1139 (2001)
2. Arsigny, V., Fillard, P., Pennec, X., Ayache, N.: Log-Euclidean metrics for fast and simple calculus on diffusion tensors. Magn. Reson. Med. **56**(2), 411–421 (2006)
3. Ashburner, J., Friston, K.: Unified segmentation. Neuroimage **26**(3), 839–51 (2005)
4. Basile, G., Bertino, S., Bramanti, A., Ciurleo, R., et al.: In vivo super-resolution track-density imaging for thalamic nuclei identification. Cereb Cortex **31**, 5613–36 (2021)
5. Battistella, G., Najdenovska, E., Maeder, P., Ghazaleh, N., et al.: Robust thalamic nuclei segmentation method based on local diffusion magnetic resonance properties. Brain Struct. Funct. **222**(5), 2203–16 (2017)
6. Behrens, T.E., Johansen-Berg, H., Woolrich, M., Smith, S., et al.: Non-invasive mapping of connections between human thalamus and cortex using diffusion imaging. Nat. Neurosci. **6**(7), 750–57 (2003)
7. Billot, B., Greve, D.N., Puonti, O., Thielscher, A., et al.: SynthSeg: segmentation of brain MRI scans of any contrast and resolution without retraining. Med Image Anal. **86**, 102789 (2023)
8. Billot, B., Robinson, E., Dalca, A.V., Iglesias, J.E.: Partial volume segmentation of brain MRI scans of any resolution and contrast. In: Martel, A.L., et al. (eds.) MICCAI 2020. LNCS, vol. 12267, pp. 177–187. Springer, Cham (2020). https://doi.org/10.1007/978-3-030-59728-3_18

9. Braak, H., Braak, E.: Alzheimer's disease affects limbic nuclei of the thalamus. Acta Neuropathol. **81**(3), 261–268 (1991)
10. Braak, H., Braak, E.: Neuropathological stageing of Alzheimer-related changes. Acta Neuropathol. **82**(4), 239–259 (1991)
11. Çiçek, Ö., Abdulkadir, A., Lienkamp, S.S., Brox, T., Ronneberger, O.: 3D U-net: learning dense volumetric segmentation from sparse annotation. In: Ourselin, S., Joskowicz, L., Sabuncu, M.R., Unal, G., Wells, W. (eds.) MICCAI 2016. LNCS, vol. 9901, pp. 424–432. Springer, Cham (2016). https://doi.org/10.1007/978-3-319-46723-8_49
12. Clevert, D.A., Unterthiner, T., Hochreiter, S.: Fast and accurate deep network learning by exponential linear units. arXiv preprint arXiv:1511.07289 (2015)
13. Ewert, C., Kügler, D., Yendiki, A., Reuter, M.: Learning anatomical segmentations for tractography from diffusion MRI. In: Computing dMRI Workshop 2020, pp. 81–93
14. Fama, R., Sullivan, E.V.: Thalamic structures and associated cognitive functions: relations with age and aging. Neurosci. Biobehav. R **54**, 29–37 (2015)
15. Fischl, B., Salat, D.H., Busa, E., Albert, M., et al.: Whole brain segmentation: automated labeling of neuroanatomical structures in the human brain. Neuron **33**(3), 341–355 (2002)
16. Fisher, R.A.: The use of multiple measurements in taxonomic problems. Ann. Eugenic **7**(2), 179–88 (1936)
17. Henderson, J.M., Carpenter, K., Cartwright, H., Halliday, G.M.: Loss of thalamic intralaminar nuclei in progressive supranuclear palsy and Parkinson's disease: clinical and therapeutic implications. Brain **123**(7), 1410–21 (2000)
18. Iglesias, J.E., Insausti, R., Lerma-Usabiaga, G., Bocchetta, M., et al.: A probabilistic atlas of the human thalamic nuclei combining ex vivo MRI and histology. Neuroimage **183**, 314–26 (2018)
19. Jack, C.R., Jr., Bernstein, M.A., Fox, N.C., Thompson, P., et al.: The Alzheimer's disease neuroimaging initiative: MRI methods. J. Magn. Reson. Imaging **27**, 685–91 (2008)
20. Jakab, A., Blanc, R., Berényi, E.L., Székely, G.: Generation of individualized thalamus target maps by using statistical shape models and thalamocortical tractography. Am. J. Neuroradiol. **33**(11), 2110 (2012)
21. Johansen-Berg, H., Behrens, T., Sillery, E., Ciccarelli, O., et al.: Functional-anatomical validation and individual variation of diffusion tractography-based segmentation of the human thalamus. Cereb. Cortex **15**(1), 31–39 (2005)
22. Kingma, D., Ba, J.: Adam: a method for stochastic optimization. arXiv:1412.6980 (2014)
23. Krauth, A., Blanc, R., Poveda, A., Jeanmonod, D., et al.: A mean three-dimensional atlas of the human thalamus: generation from multiple histological data. Neuroimage **49**(3), 2053–62 (2010)
24. Liu, Y., D'Haese, P.F., Newton, A.T., Dawant, B.M.: Generation of human thalamus atlases from 7 T data and application to intrathalamic nuclei segmentation in clinical 3 T T1-weighted images. Magn. Reson. Med. **65**, 114–128 (2020)
25. Mang, S., Busza, A., Reiterer, S., Grodd, W.: Klose: Thalamus segmentation based on the local diffusion direction: a group study. Magn. Reson. Med. **67**, 118–26 (2012)
26. Milletari, F., Navab, N., Ahmadi, S.A.: V-net: fully convolutional neural networks for volumetric medical image segmentation. In: 3DV Conference 2016, pp. 565–571 (2016)

27. Patenaude, B., Smith, S., Kennedy, D., Jenkinson, M.: A Bayesian model of shape and appearance for subcortical brain segmentation. Neuroimage **56**, 907–22 (2011)
28. Ronneberger, O., Fischer, P., Brox, T.: U-net: convolutional networks for biomedical image segmentation. In: Navab, N., Hornegger, J., Wells, W.M., Frangi, A.F. (eds.) MICCAI 2015. LNCS, vol. 9351, pp. 234–241. Springer, Cham (2015). https://doi.org/10.1007/978-3-319-24574-4_28
29. Sadikot, A.F., Chakravarty, M., Bertrand, G., Rymar, V.V., et al.: Creation of computerized 3D MRI-integrated atlases of the human basal ganglia and thalamus. Front. Syst. Neurosci. **5**, 71 (2011)
30. Schmahmann, J.: Vascular syndromes of the thalamus. Stroke **34**, 2264–2278 (2003)
31. Semedo, C., et al.: Thalamic nuclei segmentation using tractography, population-specific priors and local fibre orientation. In: Frangi, A.F., Schnabel, J.A., Davatzikos, C., Alberola-López, C., Fichtinger, G. (eds.) MICCAI 2018. LNCS, vol. 11072, pp. 383–391. Springer, Cham (2018). https://doi.org/10.1007/978-3-030-00931-1_44
32. Sherman, S.M., Guillery, R.W.: Exploring the Thalamus. Elsevier, Amsterdam (2001)
33. Sotiropoulos, S.N., Jbabdi, S., Xu, J., Andersson, J.L., et al.: Advances in diffusion MRI acquisition and processing in the human connectome project. Neuroimage **80**, 125–143 (2013)
34. Stough, J.V., Glaister, J., Ye, C., Ying, S.H., Prince, J.L., Carass, A.: Automatic method for thalamus parcellation using multi-modal feature classification. In: Golland, P., Hata, N., Barillot, C., Hornegger, J., Howe, R. (eds.) MICCAI 2014. LNCS, vol. 8675, pp. 169–176. Springer, Cham (2014). https://doi.org/10.1007/978-3-319-10443-0_22
35. Su, J.H., Thomas, F.T., Kasoff, W.S., Tourdias, T., et al.: Thalamus optimized multi atlas segmentation (THOMAS): fast, fully automated segmentation of thalamic nuclei from structural MRI. Neuroimage **194**, 272–82 (2019)
36. Tourdias, T., Saranathan, M., Levesque, I.R., Su, J., Rutt, B.K.: Visualization of intra-thalamic nuclei with optimized white-matter-nulled MPRAGE at 7T. Neuroimage **84**, 534–545 (2014)
37. Tregidgo, H.F.J., Soskic, S., Althonayan, J., Maffei, C., et al.: Accurate Bayesian segmentation of thalamic nuclei using diffusion MRI and an improved histological atlas. Neuroimage **274**, 120129 (2023)
38. Umapathy, L., Keerthivasan, M.B., Zahr, N.M., Bilgin, A., Saranathan, M.: Convolutional neural network based frameworks for fast automatic segmentation of thalamic nuclei from native and synthesized contrast structural MRI. Neuroinformatics 1–14 (2021)
39. Van Leemput, K., Maes, F., Vandermeulen, D., Suetens, P.: A unifying framework for partial volume segmentation of brain MR images. IEEE Trans. Med. Imaging **22**(1), 105–119 (2003)
40. Vatsavayai, S.C., Yoon, S.J., Gardner, R.C., Gendron, T.F., et al.: Timing and significance of pathological features in C9orf72 expansion-associated frontotemporal dementia. Brain **139**(12), 3202–16 (2016)
41. Wachinger, C., Reuter, M., Klein, T.: DeepNAT: deep convolutional neural network for segmenting neuroanatomy. Neuroimage **170**, 434–445 (2018)
42. Zhang, D., Snyder, A., Fox, M., Sansbury, M., et al.: Intrinsic functional relations between human cerebral cortex and thalamus. J. Neurophysiol. **100**, 1740–48 (2008)

Neural Pre-processing: A Learning Framework for End-to-End Brain MRI Pre-processing

Xinzi He[1(✉)], Alan Q. Wang[2], and Mert R. Sabuncu[1,2]

[1] School of Biomedical Engineering, Cornell University, Ithaca, USA
xh278@cornell.edu

[2] School of Electrical and Computer Engineering, Cornell University, Ithaca, USA

Abstract. Head MRI pre-processing involves converting raw images to an intensity-normalized, skull-stripped brain in a standard coordinate space. In this paper, we propose an end-to-end weakly supervised learning approach, called Neural Pre-processing (NPP), for solving all three sub-tasks simultaneously via a neural network, trained on a large dataset without individual sub-task supervision. Because the overall objective is highly under-constrained, we explicitly disentangle geometric-preserving intensity mapping (skull-stripping and intensity normalization) and spatial transformation (spatial normalization). Quantitative results show that our model outperforms state-of-the-art methods which tackle only a single sub-task. Our ablation experiments demonstrate the importance of the architecture design we chose for NPP. Furthermore, NPP affords the user the flexibility to control each of these tasks at inference time. The code and model are freely-available at https://github.com/Novestars/Neural-Pre-processing.

Keywords: Neural network · Pre-processing · Brain MRI

1 Introduction

Brain magnetic resonance imaging (MRI) is widely-used in clinical practice and neuroscience. Many popular toolkits for pre-processing brain MRI scans exist, e.g., FreeSurfer [9], FSL [26], AFNI [5], and ANTs [3]. These toolkits divide up the pre-processing pipeline into sub-tasks, such as skull-stripping, intensity normalization, and spatial normalization/registration, which often rely on computationally-intensive optimization algorithms.

Recent works have turned to machine learning-based methods to improve pre-processing efficiency. These methods, however, are designed to solve individual sub-tasks, such as SynthStrip [15] for skull-stripping and Voxelmorph [4] for registration. Learning-based methods have advantages in terms of inference time

Supplementary Information The online version contains supplementary material available at https://doi.org/10.1007/978-3-031-43993-3_25.

H. Greenspan et al. (Eds.): MICCAI 2023, LNCS 14227, pp. 258–267, 2023.
https://doi.org/10.1007/978-3-031-43993-3_25

and performance. However, solving sub-tasks independently and serially has the drawback that each step's performance depends on the previous step. In this paper, we propose a neural network-based approach, which we term Neural Pre-Processing (NPP), to solve three basic tasks of pre-processing simultaneously.

NPP first translates a head MRI scan into a skull-stripped and intensity-normalized brain using a translation module, and then spatially transforms to the standard coordinate space with a spatial transform module. As we demonstrate in our experiments, the design of the architecture is critical for solving these tasks together. Furthermore, NPP offers the flexibility to turn on/off different pre-processing steps at inference time. Our experiments demonstrate that NPP achieves state-of-the-art accuracy in all the sub-tasks we consider.

2 Methods

2.1 Model

As shown in Fig. 1, our model contains two modules: a geometry-preserving translation module, and a spatial-transform module.

Geometry-Preserving Translation Module. This module converts a brain MRI scan to a skull-stripped and intensity normalized brain. We implement it using a U-Net style [24] f_θ architecture (see Fig. 1), where θ denotes the model weights. We operationalize skull stripping and intensity normalization as a pixel-wise multiplication of the input image with a scalar multiplier field χ, which is the output of the U-Net f_θ:

$$T_\theta(x) = f_\theta(x) \otimes x, \tag{1}$$

where $\otimes$ denotes the element-wise (Hadamard) product.

Such a parameterization allows us to impose constraints on χ. In this work, we penalize high-frequencies in χ, via the total variation loss described below. Another advantage of χ is that it can be computed at a lower resolution to boost both training and inference speed, and then up-sampled to the full resolution grid before being multiplied with the input image. This is possible because the multiplier χ is spatially smooth. In contrast, if we have f_θ directly compute the output image, doing this at a lower resolution means we will inevitably lose high frequency information. In our experiments, we take advantage of thibass by having the model output the multiplicative field at a grid size that is 1/2 of the original input grid size along each dimension. The scalar field, which solves both skull stripping and intensity normalization, is not range restricted by design. The appropriate values will be learned from the data. In practice, we found that thresholding it at 0.2 yields a good brain mask.

Spatial Transformation Module. Spatial normalization is implemented as a variant of the Spatial Transformer Network (STN) [17]; in our implementation, the STN outputs the 12 parameters of an affine matrix Φ_{aff}. The STN takes

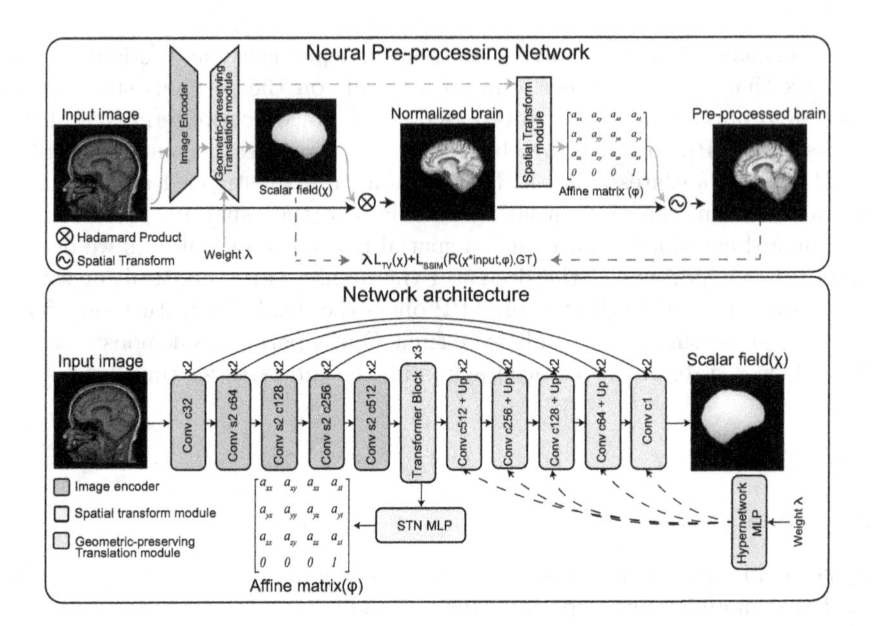

Fig. 1. (Top) An overview of Neural Pre-processing. (Bottom) The network architecture of Neural Pre-processing

as input the bottleneck features from the image translation network f_θ and feeds it through a multi-layer perceptron (MLP) that projects the features to a 12-dimensional vector encoding the affine transformation matrix. This affine transformation is in turn applied to the output of the image translation module $T_\theta(x)$ via a differentiable resampling layer [17].

2.2 Loss Function

The objective to minimize is composed of two terms. The first term is a reconstruction loss L_{rec}. In this paper, we use SSIM [31] for L_{rec}. The second term penalizes T_θ from making high-frequency intensity changes to the input image, encapsulating our prior knowledge that skull-stripping and MR bias field correction involve a pixel-wise product with a spatially smooth field. In this work, we use a total variation penalty [23] L_{TV} on the multiplier field χ, which promotes sparsity of spatial gradients in χ. The final objective is:

$$\arg\min_\theta L_{rec}(T_\theta(x) \circ \Phi_{aff}, x_{gt}) + \lambda L_{TV}(\chi), \tag{2}$$

where x_{gt} is the pre-processed ground truth images, $\lambda \geq 0$ controls the trade-off between the two loss terms, and $\circ$ denotes a spatial transformation.

Conditioning on λ. Classically, hyperparameters like λ are tuned on a held-out validation set - a computationally-intensive task which requires training multiple models corresponding to different values of λ. To avoid this, we condition on λ

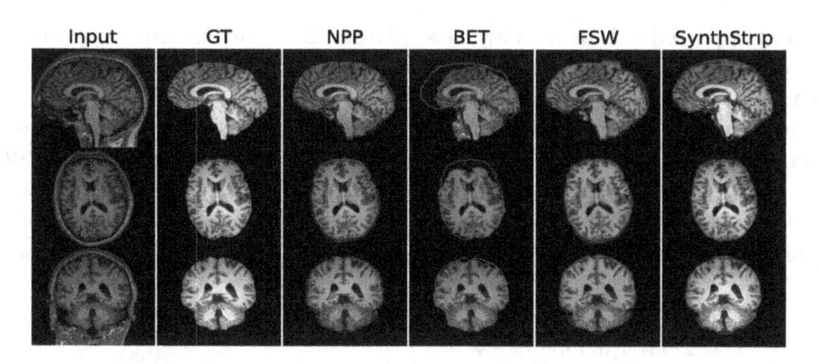

Fig. 2. Representative slices for skull-stripping. From top to bottom: coronal, axial and sagittal views. Green and red contours depict ground truth and estimated brain masks, respectively. (Color figure online)

in f_θ by passing in λ as an input to a separate MLP $h_\phi(\lambda)$ (see Fig. 1), which generates a λ-conditional scale and bias for each channel of the decoder layers. h_ϕ can be interpreted as a hypernetwork [12,14,30] which generates a conditional scale and bias similar to adaptive instance normalization (AdaIN) [16].

Specifically, for a given decoder layer with C intermediate feature maps $\{z_1, ..., z_C\}$, $h_\phi(\lambda)$ generates the parameters to scale and bias each channel z_c such that the the channel values are computed as:

$$\hat{z}_c = \alpha_c z_c + \beta_c, \tag{3}$$

for $c \in \{1, ..., C\}$. Here, α_c and β_c denote the scale and bias of channel c, conditioned on λ. This is repeated for every decoder layer, except the final layer.

3 Experiments

Training Details. We created a large-scale dataset of 3D T1-weighted (T1w) brain MRI volumes by aggregating 7 datasets: GSP [13], ADNI [22], OASIS [20], ADHD [25], ABIDE [32], MCIC [11], and COBRE [1]. The whole training set contains 10,083 scans. As ground-truth target images, we used FreeSurfer generated skull-stripped, intensity-normalized and affine-registered (to so-called MNI atlas coordinates) images.

We train NPP with ADAM [19] and a batch size of 2, for a maximum of 60 epochs. The initial learning rate is 1e−4 and then decreases by half after 30 epochs. We use random gamma transformation as a data augmentation technique, with parameter log gamma (−0.3, 0.3). We randomly sampled λ from a log-uniform distribution on $(-3, 1)$ for each mini-batch.

Architecture Details. f_θ is a U-Net-style architecture containing an encoder and decoder with skip connections in between. The encoder and decoder have 5

Table 1. Supported sub-tasks and average runtime for each method. Skull-stripping (SS), intensity normalization (IN) and spatial normalization (SN). Units are sec, **bold** is best.

Method	SS	IN	SN	GPU	CPU
SynthStrip	16.5	–	–	✓	
BET	9.1	262.2	–		✓
C2F	–	–	5.6	✓	
Freesurfer	747.2	481.5	671.6		✓
NPP (Ours)		**2.94**		✓	

Table 2. Performance on intensity normalization. Higher is better, **bold** is best.

Method	Rec SSIM	Rec PSNR	Bias SSIM	Bias PSNR
FSL	98.5	34.2	92.5	39.2
FS	98.9	35.7	92.1	39.3
NPP	**99.1**	**36.2**	**92.7**	**39.4**

levels and each level consists of 2 consecutive 3D convolutional layers. Specifically, each 3D convolutional layer is followed by an instance normalization layer and LeakyReLU (negative slope of 0.01). In the bottleneck, we use three transformer blocks to enhance the ability of capturing global information [29]. Each transformer block contains a self-attention layer and a MLP layer. For the transformer block, we use patch size $1 \times 1 \times 1$, 8 attention heads, and an MLP expansion ratio of 1. We perform tokenization by flattening the 3D CNN feature maps into a 1D sequence.

The hypernetwork, h_ϕ, is a 3-layer MLP with hidden layers 512, 2048 and 496. The STN MLP is composed of a global average pooling layer and a 2-layer MLP with hidden layers of size 256 and 12. The 2-layer MLP contains: linear (256 channels); ReLU; linear (12 channels); Tanh. Note an identity matrix is added to the output affine matrix to make sure the initial transformation is close to identity. It's widely used in the affine registration literature to improve convergence and efficiency.

Baselines. We chose three popular and widely-used tools, SynthStrip [15], C2F [21], FSL [26], and FreeSurfer [9], as baselines. SynthStrip (SS) is a learning-based skull-stripping method, while FSL and FreeSurfer (FS) is a cross-platform brain processing package containing multiple tools. FSL's Brain Extraction Tool (BET) and FMRIB's Automated Segmentation Tool are for skull stripping and MR bias field correction, respectively. FS uses a watershed method for skull-stripping, a model-based tissue segmentation (N4biasfieldcorrection [28]) for intensity normalization and bias field correction.

3.1 Runtime Analyses

The primary advantage of NPP is runtime. As shown in Table 1, for images with resolution $256 \times 256 \times 256$, NPP requires less than 3 s on a GPU and less than 8 s on a CPU for all three pre-processing tasks. This is in part due to the fact that the multiplier field can be computed at a lower resolution (in

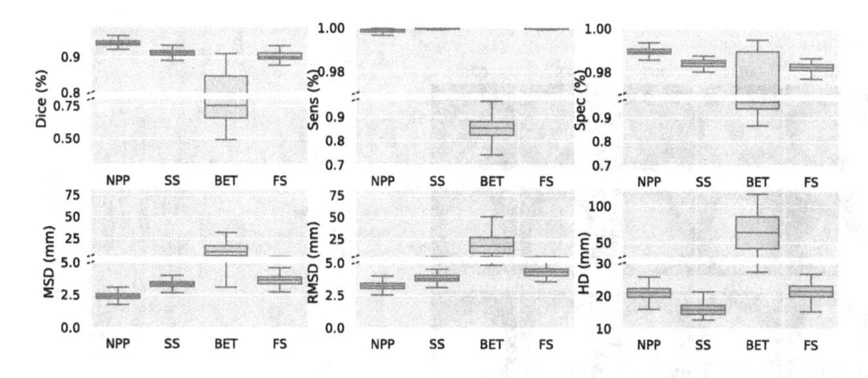

Fig. 3. Skull-stripping performance on various metrics. (Top) Higher is better. (Bottom) Lower is better.

our case, on a grid of $128 \times 128 \times 128$). The output field is then up-sampled with trilinear interpolation before being multiplied with the input image. In contrast, SynthStrip needs 16.5 s on a GPU for skull stripping and C2F needs 5.6 on a GPU for spatial normalization. FSL's optimized implementation takes about 271.3 s per image for skull stripping and intensity normalization, whereas FreeSurfer needs more than 10 min.

3.2 Pre-processing Performance

We empirically validate the performance of NPP for the three tasks we consider: skull-stripping, intensity normalization, and spatial transformation.

Evaluation Datasets. For skull-stripping, we evaluate on the Neuralfeedback skull-stripped repository (NFSR) [7] dataset. NFSR contains 125 manually skull-stripped T1w images from individuals aged 21 to 45, and are diagnosed with a wide range of psychiatric diseases. The definition of the brain mask used in NFSR follows that of FS. For intensity normalization, we evaluate on the test set (N = 856) from the Human Connectome Project (HCP). The HCP dataset includes T1w and T2w brain MRI scans which can be combined to obtain a high quality estimate of the bias field [10,27]. For spatial normalization, we use T1w MRI scans from the Parkinson's Progression Markers Initiative (PPMI). These images were automatically segmented using Freesurfer into anatomical regions of interest (ROIs)[1] [6].

Metrics. For skull-stripping, we quantify performance using the Dice overlap coefficient (Dice), Sensitivity (Sens), Specificity (Spec), mean surface distance

[1] In this work, the following ROIs were used to evaluate performance: brain stem (Bs), thalamus (Th), cerebellum cortex (Cbmlc), cerebellum white matter (Wm), cerebral white matter (Cblw), putamen (Pu), ventral DC (Vt), pallidum (Pa), caudate (Ca), lateral ventricle (LV), and hippocampus (Hi).

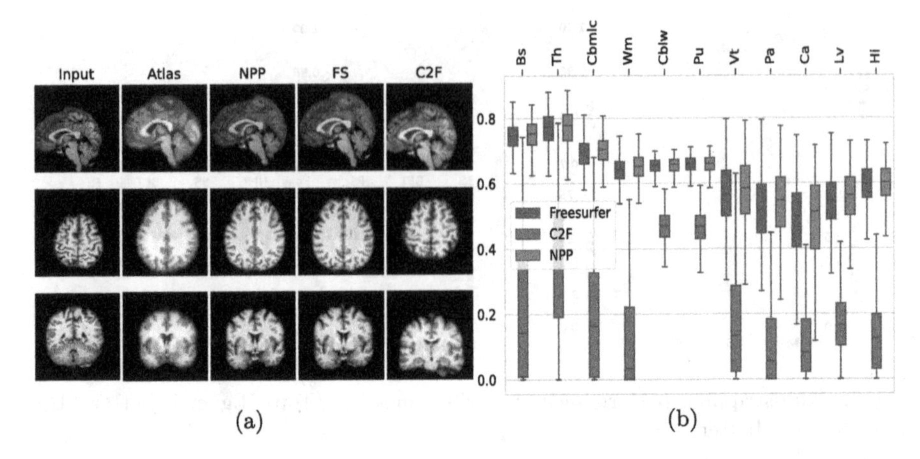

Fig. 4. (a) Representative examples for spatial normalization. Rows depict sagittal, axial and coronal view. For each view, from left to right: input image, atlas, NPP results, and FreeSurfer results. (b) Boxplots illustrating Dice scores of each anatomical structure for NPP and FreeSurfer in the atlas-based registration with the PPMI dataset. We combine left and right hemispheres of the brain into one structure for visualization.

(MSD), residual mean surface distance (RMSD), and Hausdorff distance (HD), as defined elsewhere [18]. For intensity normalization, we evaluate the intensity-normalized reconstruction (Rec) and estimated bias image (Bias, which is equal to the multiplier field χ) to the ground truth images, using PSNR and SSIM. We quantify registration between ROIs in an individual MRI and the atlas ROI for the assessment of spatial normalization by calculating the Dice score between the spatially transformed segmentations (resampled using the estimated affine transformation) and the probabilistic labels of the target atlas.

Results. Figure 2 and Fig. 3 shows skull-stripping performance for all models. We observe that the proposed method outperforms all traditional and learning-based baselines on Dice, Spec, and MSD/RMSD. Importantly, NPP achieved 93.8% accuracy and 2.7% improvement on Dice and 2.39mm MSD. Especially for MSD, NPP is 28% better than the second-best method, SynthStrip. We further observe that BET commonly fails, which has also been noted in the literature [8].

Table 2 shows the quantitative results of FSL, FS and NPP(see visualization results in Supplementary S1). NPP outperforms the baselines on all metrics. From Table 2, we see that FreeSurfer's reconstruction is better than BET's, but the bias field estimates are relatively worse. We can appreciate this in the figure in Supplementary S1, as we observe that FS's bias field estimate (f) contains too much high-frequency anatomical detail.

Figure 4(b) shows boxplots of Dice scores for NPP and FreeSurfer and C2F, for each ROI. Compared to FS and C2F, NPP achieves consistent improvement

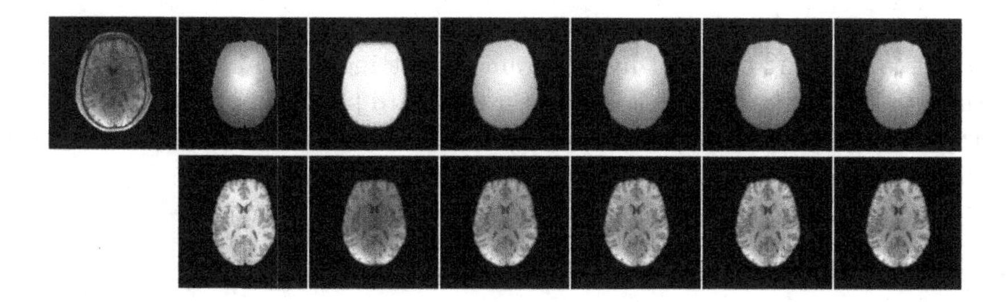

Fig. 5. (Top) From left to right, raw input image, ground truth bias field, estimated multiplier fields from NPP for different values of $\lambda = 10, 1, 0.1, 0.01, 0.001$. (Bottom) From left to right, ground truth, outputs from NPP for different values of λ.

on all ROIs measured. Figure 4(a) shows representative slices for spatial normalization.

3.3 Ablation

As ablations, we compare the specialized architecture of NPP against a naive U-Net trained to solve all three tasks at once. Additionally, we implemented a different version of our model where the U-Net directly outputs the skull-stripped and intensity-normalized image, which is in turn re-sampled with the STN. In this version, we did not have the scalar multiplication field and thus our loss function did not include the total variation term. We call this version U-Net+STN. As another alternative, we trained the U-Net+STN architecture via UMIRGPIT [2], which encourages the translation network (U-Net) to be geometry-preserving by alternating the order of the translation and registration. We note again that for all these baselines, we used the same architecture as f_θ, but instead of computing the multiplier field χ, f_θ computes the final intensity-normalized and spatially transformed image directly. The objective is the reconstruction loss L_{rec}. All other implementation details were the same as NPP. For evaluation, we use the test images from the HCP dataset.

Results: Tables 3 and 4 lists the SSIM values for the estimated reconstruction and bias fields, for different ablations and NPP with a range of λ values. We

Table 3. Ablation study results of different λ. **Bold** is best.

Method	RecSSIM	Bias SSIM
NPP, $\lambda = 10$	96.38 ± 1.29	96.02 ± 1.29
NPP, $\lambda = 1$	99.07 ± 0.61	98.09 ± 0.62
NPP, $\lambda = 0.1$	**99.25 ± 0.52**	**98.40 ± 0.40**
NPP, $\lambda = 0.01$	99.24 ± 0.51	98.22 ±0.39
NPP, $\lambda = 0.001$	99.22 ± 0.52	98.13 ± 0.40

Table 4. Comparison with ablated models. **Bold** is best.

Method	*Rec SSIM*
Naive U-Net	84.12 ± 3.34
U-Net+STN	84.87 ± 3.04
UMIRGPIT	84.26 ± 3.02
NPP, $\lambda = 0.1$	**99.25 ± 0.52**

observe that there is a sweet spot around $\lambda = 0.1$, which underscores the importance of considering different hyperparameter settings and affording the user to optimize this at test time. All ablation results are poor, supporting the importance of our architectural design. Figure 5 shows some representative results for a range of λ values.

4 Conclusion

In this paper, we propose a novel neural network approach for brain MRI pre-processing. The proposed model, called NPP, disentangles geometry-preserving translation mapping (which includes skull stripping and bias field correction) and spatial transformation. Our experiments demonstrate that NPP can achieve state-of-the-art results for the major tasks of brain MRI pre-processing.

Funding. Funding for this project was in part provided by the NIH grant R01AG053949, and the NSF CAREER 1748377 grant.

References

1. Aine, C.J., et al.: Multimodal neuroimaging in schizophrenia: description and dissemination. Neuroinformatics **15**, 343–364 (2017)
2. Arar, M., Ginger, Y., Danon, D., Bermano, A.H., Cohen-Or, D.: Unsupervised multi-modal image registration via geometry preserving image-to-image translation. In: Proceedings of the IEEE/CVF Conference on Computer Vision and Pattern Recognition, pp. 13410–13419 (2020)
3. Avants, B.B., Epstein, C.L., Grossman, M., Gee, J.C.: Symmetric diffeomorphic image registration with cross-correlation: evaluating automated labeling of elderly and neurodegenerative brain. Med. Image Anal. **12**, 26–41 (2008)
4. Balakrishnan, G., Zhao, A., Sabuncu, M.R., Guttag, J., Dalca, A.V.: VoxelMorph: a learning framework for deformable medical image registration. IEEE Trans. Med. Imaging **38**, 1788–1800 (2019)
5. Cox, R.W.: AFNI: software for analysis and visualization of functional magnetic resonance neuroimages. Comput. Biomed. Res. **29**(3), 162–173 (1996)
6. Dalca, A.V., Guttag, J., Sabuncu, M.R.: Anatomical priors in convolutional networks for unsupervised biomedical segmentation. In: Proceedings of the IEEE Conference on Computer Vision and Pattern Recognition, pp. 9290–9299 (2018)
7. Eskildsen, S.F., et al.: BEaST: brain extraction based on nonlocal segmentation technique. NeuroImage **59**, 2362–2373 (2012)
8. Ezhilarasan, K., et al.: Automatic brain extraction from MRI of human head scans using Helmholtz free energy principle and morphological operations. Biomed. Signal Process. Control **64**, 102270 (2021)
9. Fischl, B.: FreeSurfer. Neuroimage **62**(2), 774–781 (2012)
10. Glasser, M.F., et al.: The minimal preprocessing pipelines for the human connectome project. NeuroImage **80**, 105–124 (2013)
11. Gollub, R.L., et al.: The MCIC collection: a shared repository of multi-modal, multi-site brain image data from a clinical investigation of schizophrenia. Neuroinformatics **11**, 367–388 (2013)
12. Ha, D., Dai, A., Le, Q.V.: Hypernetworks (2016). http://arxiv.org/abs/1609.09106

13. Holmes, A.J., et al.: Brain genomics superstruct project initial data release with structural, functional, and behavioral measures. Sci. Data **2**, 1–16 (2015)
14. Hoopes, A., Hoffmann, M., Fischl, B., Guttag, J., Dalca, A.V.: HyperMorph: amortized hyperparameter learning for image registration (2021). http://arxiv.org/abs/2101.01035
15. Hoopes, A., Mora, J.S., Dalca, A.V., Fischl, B., Hoffmann, M.: SynthStrip: skull-stripping for any brain image. Neuroimage **260**, 119474 (2022)
16. Huang, X., Belongie, S.: Arbitrary style transfer in real-time with adaptive instance normalization. In: Proceedings of the IEEE International Conference on Computer Vision, pp. 1501–1510 (2017)
17. Jaderberg, M., Simonyan, K., Zisserman, A., et al.: Spatial transformer networks. In: Advances in Neural Information Processing Systems, vol. 28 (2015)
18. Jadon, S.: A survey of loss functions for semantic segmentation. In: 2020 IEEE Conference on Computational Intelligence in Bioinformatics and Computational Biology (CIBCB), pp. 1–7. IEEE (2020)
19. Kingma, D.P., Ba, J.: Adam: a method for stochastic optimization (2014). http://arxiv.org/abs/1412.6980
20. Marcus, D.S., Fotenos, A.F., Csernansky, J.G., Morris, J.C., Buckner, R.L.: Open access series of imaging studies: longitudinal MRI data in nondemented and demented older adults. J. Cogn. Neurosci. **22**(12), 2677–2684 (2010)
21. Mok, T.C., Chung, A.: Affine medical image registration with coarse-to-fine vision transformer. In: Proceedings of the IEEE/CVF Conference on Computer Vision and Pattern Recognition, pp. 20835–20844 (2022)
22. Mueller, S.G., et al.: The Alzheimer's disease neuroimaging initiative. Neuroimaging Clin. **15**(4), 869–877 (2005)
23. Osher, S., Burger, M., Goldfarb, D., Xu, J., Yin, W.: An iterative regularization method for total variation-based image restoration. Multiscale Model. Simul. **4**, 460–489 (2005)
24. Ronneberger, O., Fischer, P., Brox, T.: U-net: convolutional networks for biomedical image segmentation. In: Navab, N., Hornegger, J., Wells, W.M., Frangi, A.F. (eds.) MICCAI 2015, Part III. LNCS, vol. 9351, pp. 234–241. Springer, Cham (2015). https://doi.org/10.1007/978-3-319-24574-4_28
25. Sibley, M.H., Coxe, S.J.: The ADHD teen integrative data analysis longitudinal (TIDAL) dataset: background, methodology, and aims. BMC Psychiatry **20**, 1–12 (2020)
26. Smith, S.M., et al.: Advances in functional and structural MR image analysis and implementation as FSL, vol. 23 (2004)
27. Song, W., et al.: Jointly estimating bias field and reconstructing uniform MRI image by deep learning. J. Magn. Reson. **343**, 107301 (2022)
28. Tustison, N.J., et al.: N4ITK: improved N3 bias correction. IEEE Trans. Med. Imaging **29**(6), 1310–1320 (2010)
29. Vaswani, A., et al.: Attention is all you need. In: Advances in Neural Information Processing Systems, vol. 30 (2017)
30. Wang, A.Q., Dalca, A.V., Sabuncu, M.R.: Computing multiple image reconstructions with a single hypernetwork. Mach. Learn. Biomed. Imaging **1** (2022)
31. Wang, Z., Bovik, A.C., Sheikh, H.R., Simoncelli, E.P.: Image quality assessment: from error visibility to structural similarity. IEEE Trans. Image Process. **13**, 600–612 (2004)
32. de Wilde, A., et al.: Alzheimer's biomarkers in daily practice (abide) project: rationale and design. Alzheimer's & Dementia: Diagnosis, Assessment & Disease Monitoring **6**, 143–151 (2017)

Dynamic Functional Connectome Harmonics

Hoyt Patrick Taylor IV[1] and Pew-Thian Yap[2,3](✉)

[1] Department of Computer Science, University of North Carolina, Chapel Hill, NC, USA
[2] Department of Radiology, University of North Carolina, Chapel Hill, NC, USA
ptyap@med.unc.edu
[3] Biomedical Research Imaging Center, University of North Carolina, Chapel Hill, NC, USA

Abstract. Functional connectivity (FC) "gradients" enable investigation of connection topography in relation to cognitive hierarchy, and yield the primary axes along which FC is organized. In this work, we employ a variant of the "gradient" approach wherein we solve for the normal modes of FC, yielding functional connectome harmonics. Until now, research in this vein has only considered static FC, neglecting the possibility that the principal axes of FC may depend on the timescale at which they are computed. Recent work suggests that momentary activation patterns, or brain states, mediate the dominant components of functional connectivity, suggesting that the principal axes may be invariant to change in timescale. In light of this, we compute functional connectome harmonics using time windows of varying lengths and demonstrate that they are stable across timescales. Our connectome harmonics correspond to meaningful brain states. The activation strength of the brain states, as well as their inter-relationships, are found to be reproducible for individuals. Further, we utilize our time-varying functional connectome harmonics to formulate a simple and elegant method for computing cortical flexibility at vertex resolution and demonstrate qualitative similarity between flexibility maps from our method and a method standard in the literature.

Keywords: dynamic functional connectivity · harmonics · flexibility

1 Introduction

Nonlinear dimensionality reduction techniques such as diffusion embedding and principal component analysis applied to FC data provide the primary axes along which FC is organized [11]. Laplacian embedding is a related nonlinear dimensionality reduction technique, which yields the normal modes or functional connectome harmonics, which are the focus of this study. Connectome harmonics have a straightforward interpretation as the brain analog of the Fourier basis and can be used to filter the underlying fMRI signal.

A growing body of evidence suggests that the strongest components of FC are underpinned by brief but strong patterns of activation, or brain states, resembling the dominant components of FC (e.g., the default mode network) rather than sustained,

This work was supported in part by the United States National Institutes of Health (NIH) through grants MH125479 and EB008374.

H. Greenspan et al. (Eds.): MICCAI 2023, LNCS 14227, pp. 268–276, 2023.
https://doi.org/10.1007/978-3-031-43993-3_26

low-magnitude signal coherence between brain regions implicated in those strongest components of FC [3,10,14]. Simultaneously, few studies have analyzed how functional connectome harmonics depend on the timescale on which they are computed. If briefly activated brain states mediate the primary axes of FC, we would expect momentary functional connectome harmonics to be stable with respect to timescale.

The robustness of connectivity patterns across populations and the reliability and reproducibility of FC gradients across subjects and choices of parameters have been well studied [1,9,17]. However, it is not well understood to what degree the magnitude of and relationship between the activation timeseries of certain brain states is reproducible within and across subjects. Further, the reliability of time-varying FC harmonics has yet to be studied. Given both our hypothesis that functional connectome harmonics represent discrete brain states and evidence that discrete brain states underpin FC, we expect that the activation energy of principal functional connectome harmonics with respect to the underlying fMRI signal will be reproducible within individuals.

Previous work has shown that network configuration is dynamic during task and rest, with the community allegiance of brain regions changing over time [4]. The propensity of brain regions to change network allegiance, or flexibility, has been implicated in learning, development, and psychiatric disorders [2,8,19]. Previous studies have defined cortical flexibility in terms of the frequency of community allegiance change and demonstrated relationships between flexibility, learning, and development. Functional connectome harmonics correspond to axes along which connectivity is organized, yielding both a convenient basis to represent underlying functional activation and a spectral representation of each brain region. The spectral coordinates define the position of a brain region in terms of the primary axes of FC, with similar spectral coordinates indicating similar network allegiance profiles. By computing time-varying functional connectome harmonics, we can study the temporal dynamics of spectral coordinates, which provides a new lens through which to examine cognitive flexibility. Previous studies have relied on community detection algorithms to compute the degree of change in network allegiance of brain regions, which are hindered by free parameters governing the number of communities, the strength of connection between adjacent time points, and the choice of parcellation due to computational restraints on community detection algorithms. Dynamic functional connectome harmonics present a simpler method for computing flexibility: change in network allegiance—and therefore flexibility—is proportional to change in spectral coordinates given by connectome harmonics.

2 Methods

We used the resting-state fMRI time series of 44 subjects from the HCP minimal preprocessed test-retest dataset mapped to the 32k MSM-All surface atlas [7]. For each subject, the test and retest datasets contain data from two rs-fMRI sessions for a total of 176 sessions. To investigate the variation in of functional connectome harmonics at different time scales, we computed FC matrices at 4 window lengths: $w = 0.5$, 1, 2, and 3.5 min. For each window length, each 28 min time series was divided into $N_w = \frac{28}{w}$ windows, and an FC matrix was computed for each window. FC between two surface vertices was computed as the correlation between the z-normalized time series. In light of studies demonstrating that thresholding FC matrices increases the reliability of FC gradients [9], the top 10% of the correlations for each row in each FC matrix was retained, and the FC matrices were made symmetric by taking the average of each matrix and its transpose.

For each FC matrix $\hat{A}(t)$ at timepoint t, the normalized graph Laplacian matrix, $\hat{L}(t) = \hat{D}(t)^{\frac{1}{2}}\hat{A}(t)\hat{D}(t)^{\frac{1}{2}}$, with $\hat{D}(t)_{i,i} = \sum_{j=0}^{N_v} \hat{A}(t)_{i,j}$ was computed. The first N_h eigenvectors of $\hat{L}(t)$, solutions to $\hat{L}(t)\boldsymbol{\psi}(t)_k = \lambda(t)_k\boldsymbol{\psi}(t)_k$, were computed, yielding the functional connectome harmonics of window t, $\hat{\psi}(t) = \{\boldsymbol{\psi}_k(t)|k = 1,\ldots,N_h\}$. Note that $\hat{\psi}(t)$ is a matrix of size $N_v \times N_h$, where each column is a harmonic $\boldsymbol{\psi}(t)_k$, N_v = 52,924 is the number of vertices, and N_h = 7 is the number of harmonics computed. This process is carried out for all windows for each acquisition, yielding the dynamic functional connectome harmonics for a particular subject and window length: $\{\hat{\psi}(t)|t = 1,\ldots,N_w\}$.

To extract a single set of harmonics for each acquisition at each window length, the window-wise harmonics from all windows for that acquisition were stacked horizontally into a large matrix of size $N_v \times (N_h N_w)$ on which PCA was performed, yielding one set of principal components for each acquisition and each subject. We refer to the components obtained via this approach as the most-prevalent harmonics, $\bar{\hat{\psi}}$, as they encapsulate orthogonal patterns of maximum variation across all window-wise harmonics. This process was also repeated on the group level for the test and retest cohorts separately at each window length, yielding the most prevalent harmonics $\bar{\hat{\Psi}}$ for the test and retest groups. For computation of $\bar{\hat{\Psi}}$, the matrix on which PCA was performed is of size $N_v \times (N_h N_w N_a N_s)$, where N_a = 2 is the number of acquisitions per subject for either test or retest, and N_s = 44 is the number of subjects.

To investigate the reproducibility of dynamic functional connectome harmonics as a basis for reconstructing the underlying fMRI signal, we utilize the harmonics as operators on functional timeseries. For a vertex-wise functional time series $\hat{X} = [\boldsymbol{x}(1),\ldots,\boldsymbol{x}(N_T)]$, where $\boldsymbol{x}(t)$ is the fMRI activation at timepoint t, we define its $N_h \times N_T$ spectral coefficient timeseries matrix $\hat{\alpha}$ with respect to harmonics $\hat{\psi}$ as

$$\hat{\alpha} = \hat{\psi}^{\top}\hat{X}, \tag{1}$$

where each row $\boldsymbol{\alpha}_k$ of $\hat{\alpha}$ is the spectral coefficient timeseries of timeseries $\hat{X}$ for harmonic $\boldsymbol{\psi}_k$. Each $\boldsymbol{\alpha}_k$ of $\hat{\alpha}$ gives the activation power of harmonic k at each time point, and the L2 norm of $\boldsymbol{\alpha}_k$ yields the activation energy $E(\boldsymbol{\psi}_k, \hat{X}) = ||\boldsymbol{\alpha}_k||$ of harmonic k across the entire timeseries. Further, we define the harmonic-filtered timeseries $\tilde{\hat{X}}$ and the representation efficiency ratio γ as

$$\tilde{\hat{X}} = (\hat{\alpha}^{\top}\hat{\psi}^{\top})^{\top}; \quad \gamma = \frac{||\tilde{\hat{X}}||}{||\hat{X}||}. \tag{2}$$

A set of harmonics $\{\boldsymbol{\psi}_k|k = 1,\ldots,N_h\}$ defines a spectral coordinate vector $\boldsymbol{\xi}^i = [\psi_1^i,\ldots,\psi_{N_h}^i]$ for each vertex i on the surface, specifying a position in the harmonic embedding space. To allow for comparison of spectral coordinates across a series of consecutive time windows, we first compute the most-prevalent harmonics $\bar{\hat{\psi}}$ of the time series, and perform Procrustes alignment [18] between each set of window-wise harmonics $\hat{\psi}(t)$ and the most-prevalent harmonics $\bar{\hat{\psi}}$ from that acquisition using the implementation from [17]. These aligned harmonic time series define a trajectory in spectral space for each vertex. Vertices that have similar network allegiance have similar spectral coordinates, taking on similar values in each harmonic. A large change in spectral coordinates between timepoints t and $t+1$ indicates a change in FC topology relative to that vertex. We therefore define flexibility in terms of the distance between spectral coordinates at consecutive time points. For N_w sets of aligned harmonics computed at consecutive windows, we define the flexibility of vertex i as

$$f^i = \left(\sum_{t=1}^{N_w-1} ||\boldsymbol{\xi}^i(t) - \boldsymbol{\xi}^i(t+1)||^2 \right)^{\frac{1}{2}} . \tag{3}$$

In order to evaluate our method of computing flexibility, we also employ a standard community detection based technique for comparison with our method wherein a graph modularity metric is optimized to obtain a community partition using a multi-layer temporal graph [2, 13]. Due to computational complexity of the modularity optimization process involved, we apply this standard flexibility metric to FC data parcellated using the FreeSurfer Destrieux Atlas [5].

3 Results

Our most-prevalent harmonics comprise meaningful patterns of activation on the cortex (Fig. 1). On the group level, $\bar{\boldsymbol{\Psi}}_1$ reflects the default mode network (DMN) and $\bar{\boldsymbol{\Psi}}_2$ the task positive network (TPN) [6]. Our harmonics largely resemble FC gradients computed in the literature, although we do not observe the visual vs somatosensory component, which has been observed as the second gradient in static FC studies [11, 17]. The absence of this harmonic at shorter timescales may indicate that it is mediated by longer-term, lower magnitude signal co-fluctuations and within visual and somatosensory cortices.

We evaluated the stability of the group level most-prevalent harmonics across repeated measurements (test and retest group cohorts) and across varying window length using cosine similarity and found high similarity ($\geq$0.70) between all pairings, indicating high stability (Fig. 1). Further, we evaluated the intra-subject stability of the individual most-prevalent harmonics using cosine similarity and found high stability across repeated measurements (4 scans per subject) for the first 3 harmonics. Table 1 displays the mean $\pm$ the standard deviation of the intra-subject cosine similarity between individual most-prevalent harmonics, and of the inter-window cosine similarity between dynamic harmonics of a given subject for $k = 1, 2, 3$. Note that individual most-prevalent and dynamic harmonics were Procrustes aligned before comparison in Table 1.

We investigated how the harmonics relate to functional timeseries by examining the stability of the harmonic activation energy of the first two harmonics, $||\boldsymbol{\alpha}_1||$ and $||\boldsymbol{\alpha}_2||$, the representation efficiency ratio γ, and the correlation between $\boldsymbol{\alpha}_1$ and $\boldsymbol{\alpha}_2$. For

Table 1. Intra-subject and inter-window cosine similarity in harmonics

	intra-subject $\bar{\psi}_k$			inter-window $\psi(t)_k$		
w	$k=1$	$k=2$	$k=3$	$k=1$	$k=2$	$k=3$
0.5	0.81 ± 0.11	0.79 ± 0.10	0.72 ± 0.13	0.22 ± 0.10	0.17 ± 0.08	0.12 ± 0.06
1	0.83 ± 0.09	0.80 ± 0.09	0.74 ± 0.11	0.41 ± 0.12	0.33 ± 0.11	0.25 ± 0.09
2	0.84 ± 0.09	0.82 ± 0.09	0.75 ± 0.11	0.58 ± 0.11	0.50 ± 0.12	0.40 ± 0.12
3.5	0.86 ± 0.07	0.82 ± 0.09	0.76 ± 0.11	0.70 ± 0.09	0.61 ± 0.11	0.52 ± 0.12

Table 2. Reliability of harmonic decomposition metrics

	$\|\|\alpha_1\|\|$	$\|\|\alpha_2\|\|$	$\rho(\alpha_1, \alpha_2)$	γ
ICC	$0.68 \pm .12$	$0.59 \pm .13$	$0.64 \pm .12$	$0.58 \pm .13$
p	1.1×10^{-16}	5.85×10^{-14}	2.22×10^{-16}	1.3×10^{-15}
F-stat	7.85	5.59	6.60	6.20

each session and each subject, we computed each quantity using the most-prevalent harmonics $\bar{\bar{\psi}}$ of that subject and session and investigated the reliability of these harmonic-derived metrics using the intraclass correlation coefficient (ICC), with results summarized in Table 2. ICC results indicate larger variability in each metric across subjects than across sessions with high significance ($p \leq 1 \times 10^{-13}$ for all metrics), with moderate to good reliability for $||\alpha_1||$ and $\rho(\alpha_1, \alpha_2)$, and moderate reliability for $||\alpha_2||$ and γ. We also found that $\rho(\alpha_1, \alpha_2)$ had a mean of –0.153 and standard deviation of .07, indicating that the activation timecourses of the harmonics most closely resembling the DMN and TPN are anticorrelated.

We computed flexibility using our dynamic functional connectome harmonic formulation (Eq. 3) for each subject and session for each window length. Average flexibility computed using $w = 2$ is displayed in Fig. 2a. Importantly, there were not significant differences in the qualitative map of flexibility with varying window length. As a sanity-check on our definition of flexibility, we also computed flexibility using traditional methods from the literature [2,8,14,19], using the Leidenalg community detection algorithm [15] with interlayer connection strength $\omega = 1$ and resolution parameter $r = 1$, and window length $w = 2$. Flexibility was computed as the ratio of the total number of community assignment changes to the total number of possible transitions for each node (parcel). Using the community-detection based flexibility approach, we find very similar distribution of flexibility values, but with higher values in the prefrontal cortex, and slightly lower values in the occipital cortex. Flexibility results from both methods with the exception of the high values in the occipital cortex are in keeping with previous studies on flexibility, indicating that our definition of flexibility is an effective method for measuring the propensity of brain regions to change community allegiance [12]. Importantly, our definition of flexibility involves fewer parameters than conventional methods, and yields a vertex-wise map of flexibility.

4 Discussion

Individual most-prevalent harmonics are found to be similar across acquisitions after Procrustes alignment, while there is notably lower average similarity between dynamic harmonics, particularly at shorter timescales (Table 1). This is not unexpected, as we anticipate variability in FC between windows. Higher intra-subject similarity in $\bar{\psi}_k$ compared with inter-window $\psi(t)_k$ similarity indicates that our PCA-based method of extracting the most-prevalent harmonics from the dynamic harmonics achieves its goal of extracting the common brain states for a given acquisition and that those states are reproducible within subjects. Further, our most-prevalent harmonics are advantageous over naive averaging approaches, as they preserve the orthogonality constraint intrinsic to connectome harmonics.

Our most-prevalent functional connectome harmonics show invariance under change in timescale and strongly resemble FC "gradients" in the literature computed on static FC [11]. As connectome harmonics represent the primary axes along which FC is organized, their invariance to timescale is significant: persistence of the same harmonics at long, medium, and short timescales supports the hypothesis that the harmonics themselves are brain states whose activation fluctuates. Notably, at the group level, our second harmonic $\bar{\boldsymbol{\Psi}}_2$ resembles the TPN, whereas in the literature the second gradient is found to be an axis spanning from visual to somatosensory areas. Although our $\bar{\boldsymbol{\Psi}}_3$ has high intensity values in the visual cortex, it is not a clear axis between visual and somatosensory regions. This difference between our results and static FC gradient results may indicate that the second FC gradient from the literature is the result of longer-timescale, lower-magnitude signal variations that differentiate the visual and somatosensory cortices.

Our analysis indicates that our individual most-prevalent harmonics $\bar{\bar{\psi}}$ are a reproducible basis with which to decompose the underlying fMRI signal. The projection coefficient timeseries of a given harmonic, $\boldsymbol{\alpha}_k$, represents the timecourse of relative activation of $\bar{\psi}_k$. Importantly, we found that the magnitude of these activation timecourses for the first and second most-prevalent harmonic (corresponding to the DMN and TPN), as well as the correlation between them were reliable across repeated measurements in individual subjects, with significantly higher variability across subjects than across scans (Table 2).This indicates that the strength of activation and the temporal relationship between the activation of $\bar{\psi}_1$ and $\bar{\psi}_2$ are reliable, and suggests that our method of extracting most-prevalent harmonics provides a reliable subject-specific method for analyzing the manifestation and relationship between different connectivity patterns. It is also worth noting that we found that $\rho(\boldsymbol{\alpha}_1, \boldsymbol{\alpha}_2)$ was negative in general, indicating that activation of the DMN and TPN is anticorrelated, in keeping with previous studies [6, 16]. In light of this, further analysis of the strength of activation of meaningful harmonics and their interrelationships could provide meaningful insights into differences in network dynamics in diseased populations.

Our dynamic functional connectome harmonic definition of flexibility yields a vertex-wise map of cortical flexibility for each subject, requires fewer parameters, and is less computationally expensive than standard definitions of flexibility. Our group-average flexibility result (Fig. 2a) shows strong resemblance to the result computed using the community-detection based flexibility definition (Fig. 2b). Transmodal

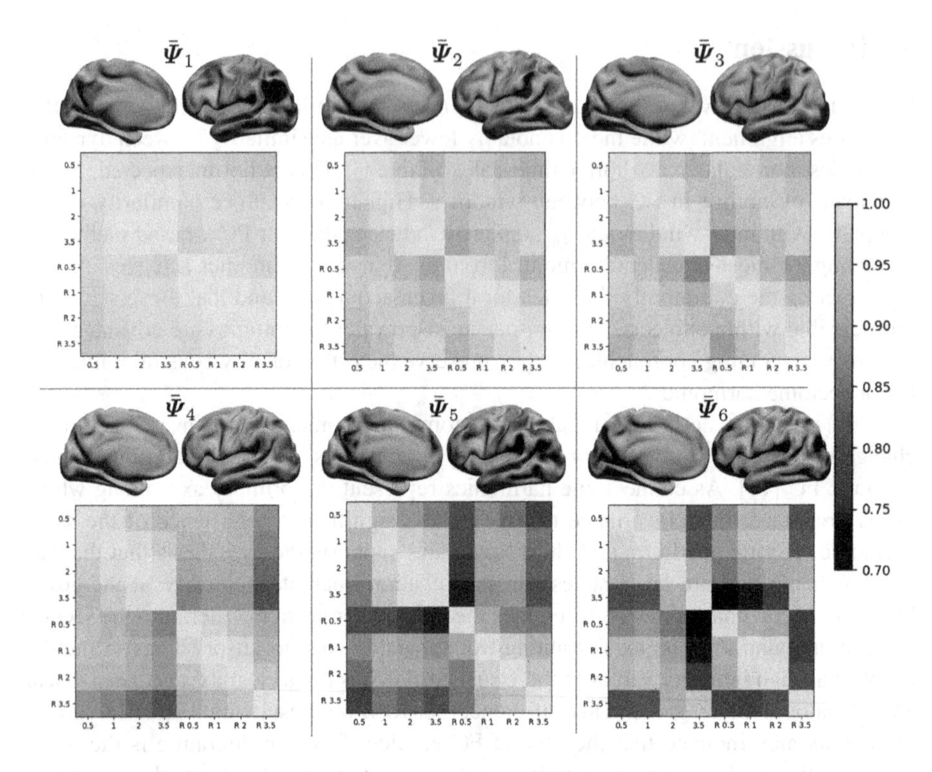

Fig. 1. Group-level most-prevalent harmonics $\bar{\Psi}_k$ derived from 2 min window for test cohort mapped to the cortical surface. Matrix below each $\bar{\Psi}_k$ is cosine similarity between group-level harmonics computed for different window lengths and test or retest cohort. Matrix labels are window length in minutes, with R denoting retest cohort. $\bar{\Psi}_1$ differentiates the default mode network from the rest of the brain, $\bar{\Psi}_2$ differentiates regions in the task positive network from the rest of the brain. Color bar ranges from cosine similarity of 0.70 to 1.00.

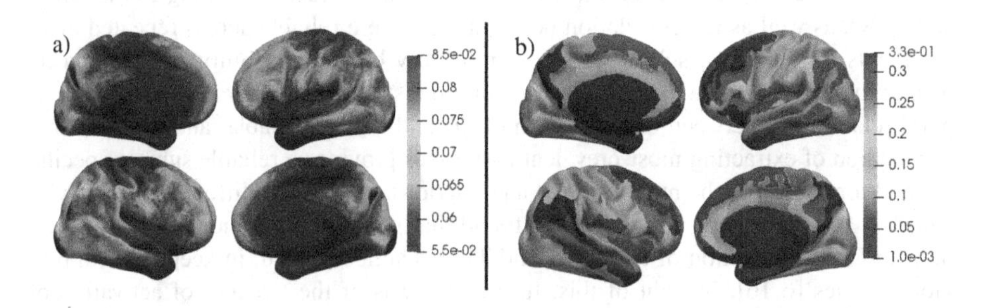

Fig. 2. Group average flexibility computed using **a)** dynamic functional connectome harmonic method (Eq. 3) and **b)** conventional community detection based method. Note that both flexibility maps are computed using 2 min window length.

regions such as the angular gyrus, temporal-parietal junction, prefrontal and medial prefrontal cortices, precuneus, and middle temporal gyrus display high flexibility due to their involvement in complex high order cognitive functions. In contrast, unimodal regions such as motor, somatosensory, limbic, and auditory cortex demonstrate low flexibility. An interesting feature of our flexibility result is high flexibility in the visual cortex. Although this has not been consistently found in previous studies, it is worth noting that many of the seminal works on flexibility deal with task-based fMRI data [2, 14], wherein flexibility in the visual cortex may be lower than at rest. One possible explanation is that during rest, daydreaming and imagination lead to rapid switching in the recruitment of the visual cortex, leading to high flexibility values.

5 Conclusion

Our most-prevalent functional connectome harmonics and their dynamic counterparts provide a powerful lens through which to study the organization of FC as well as its dynamics. We demonstrate that the most-prevalent harmonics are invariant when the timescale at which their underlying connectivity matrices are computed is changed, and that they are highly stable across repeated measurements on the group and individual level. Further, we show that our harmonics provide a reliable basis with which to filter fMRI timeseries and to extract the activation timecourses of individual harmonic brain states. Importantly, the reproducibility of these harmonic decompositions indicates that they provide a subject-specific method with which to study the dynamics of specific brain states and their interrelationships. We also present a novel formulation of cortical flexibility defined in terms of dynamic functional connectome harmonics and show that it gives vertex resolution flexibility maps that qualitatively similar results to previous definitions of flexibility which are constrained by free parameters and computational complexity.

References

1. Abrol, A., et al.: Replicability of time-varying connectivity patterns in large resting state fMRI samples. Neuroimage **163**, 160–176 (2017). https://doi.org/10.1016/j.neuroimage.2017.09.020
2. Bassett, D.S., Wymbs, N.F., Porter, M.A., Mucha, P.J., Carlson, J.M., Grafton, S.T.: Dynamic reconfiguration of human brain networks during learning. Proc. Natl. Acad. Sci. U.S.A. **108**, 7641–7646 (2011). https://doi.org/10.1073/pnas.1018985108
3. Betzel, R.F., Faskowitz, J., Sporns, O.: High-amplitude co-fluctuations in cortical activity drive resting-state functional connectivity (2019). https://doi.org/10.1101/800045
4. Cohen, J.R.: The behavioral and cognitive relevance of time-varying, dynamic changes in functional connectivity (2018). https://doi.org/10.1016/j.neuroimage.2017.09.036
5. Destrieux, C., Fischl, B., Dale, A., Halgren, E.: Automatic parcellation of human cortical gyri and sulci using standard anatomical nomenclature. Neuroimage **53**, 1–15 (2010). https://doi.org/10.1016/j.neuroimage.2010.06.010
6. Fox, M.D., Snyder, A.Z., Vincent, J.L., Corbetta, M., Van Essen, D.C., Raichle, M.E.: The human brain is intrinsically organized into dynamic, anticorrelated functional networks. Proc. Natl. Acad. Sci. U.S.A. **102**, 9673–9678 (2005). https://doi.org/10.1073/pnas.0504136102

7. Glasser, M.F., et al.: The minimal preprocessing pipelines for the human connectome project. Neuroimage **80**, 105–124 (2013). https://doi.org/10.1016/j.neuroimage.2013.04.127
8. Harlalka, V., Bapi, R.S., Vinod, P.K., Roy, D.: Atypical flexibility in dynamic functional connectivity quantifies the severity in autism spectrum disorder. Front. Hum. Neurosci. **13**, 6 (2019). https://doi.org/10.3389/fnhum.2019.00006
9. Hong, S.J., et al.: Toward a connectivity gradient-based framework for reproducible biomarker discovery. Neuroimage **223**, 117322 (2020). https://doi.org/10.1016/j.neuroimage.2020.117322
10. Liu, X., Zhang, N., Chang, C., Duyn, J.H.: Co-activation patterns in resting-state fMRI signals (2018). https://doi.org/10.1016/j.neuroimage.2018.01.041
11. Margulies, D.S., et al.: Situating the default-mode network along a principal gradient of macroscale cortical organization. Proc. Natl. Acad. Sci. U.S.A. **113**, 12574–12579 (2016). https://doi.org/10.1073/pnas.1608282113
12. Mattar, M.G., Betzel, R.F., Bassett, D.S.: The flexible brain (2016). https://doi.org/10.1093/brain/aww151
13. Mucha, P.J., Richardson, T., Macon, K., Porter, M.A., Onnela, J.P.: Community structure in time-dependent, multiscale, and multiplex networks. Science **328**, 876–878 (2010). https://doi.org/10.1126/science.1184819
14. Reddy, P.G., et al.: Brain state flexibility accompanies motor-skill acquisition. Neuroimage **171**, 135–147 (2018). https://doi.org/10.1016/j.neuroimage.2017.12.093
15. Traag, V.A., Waltman, L., van Eck, N.J.: From Louvain to Leiden: guaranteeing well-connected communities. Sci. Rep. **9**, 5233 (2019). https://doi.org/10.1038/s41598-019-41695-z
16. Uddin, L.Q., Kelly, A.M., Biswal, B.B., Castellanos, F.X., Milham, M.P.: Functional connectivity of default mode network components: correlation, anticorrelation, and causality. Hum. Brain Mapp. **30**, 625–637 (2009). https://doi.org/10.1002/hbm.20531
17. Vos de Wael, R., et al.: BrainSpace: a toolbox for the analysis of macroscale gradients in neuroimaging and connectomics datasets. Commun. Biol. **3**, 103 (2020). https://doi.org/10.1038/s42003-020-0794-7
18. Wang, C., Mahadevan, S.: Manifold alignment using procrustes analysis. In: Proceedings of the 25th International Conference on Machine Learning (2008). https://doi.org/10.1145/1390156.1390297
19. Yin, W., et al.: The emergence of a functionally flexible brain during early infancy. Proc. Natl. Acad. Sci. U.S.A. **117**, 23904–23913 (2020). https://doi.org/10.1073/pnas.2002645117

Wasserstein Distance-Preserving Vector Space of Persistent Homology

Tananun Songdechakraiwut[1,2(✉)], Bryan M. Krause[2], Matthew I. Banks[2], Kirill V. Nourski[3], and Barry D. Van Veen[2]

[1] Duke University, Durham, USA
t.song@duke.edu
[2] University of Wisconsin–Madison, Madison, USA
[3] University of Iowa, Iowa City, USA

Abstract. Analysis of large and dense networks based on topology is very difficult due to the computational challenges of extracting meaningful topological features from networks. In this paper, we present a computationally tractable approach to topological data analysis of large and dense networks. The approach utilizes principled theory from persistent homology and optimal transport to define a novel vector space representation for topological features. The feature vectors are based on persistence diagrams of connected components and cycles and are computed very efficiently. The associated vector space preserves the Wasserstein distance between persistence diagrams and fully leverages the Wasserstein stability properties. This vector space representation enables the application of a rich collection of vector-based models from statistics and machine learning to topological analyses. The effectiveness of the proposed representation is demonstrated using support vector machines to classify measured functional brain networks. Code for the topological vector space is available at https://github.com/topolearn.

Keywords: Persistent homology · Wasserstein distance · Graph embedding · Functional brain networks · Consciousness

1 Introduction

Networks are ubiquitous representations for describing complex, highly interconnected systems that capture intricate patterns of relationships between nodes. [3]. Finding meaningful, computationally tractable characterizations of network structure is very difficult, especially for large and dense networks with node degrees ranging over multiple orders of magnitude [5].

Persistent homology [10] is an emerging tool for understanding, characterizing and quantifying the topology of complex networks [19,21]. Connected components and cycles are the most dominant and fundamental topological features of real networks. For example, many networks naturally organize into modules or connected components [5]. Similarly, cycle structure is ubiquitous and is often interpreted in terms of information propagation, redundancy and feedback loops

H. Greenspan et al. (Eds.): MICCAI 2023, LNCS 14227, pp. 277–286, 2023.
https://doi.org/10.1007/978-3-031-43993-3_27

[14]. Topological features are represented in persistent homology using descriptors called *persistence diagrams* [10]. Effective use of such topological descriptors in machine learning requires a notion of proximity. Wasserstein distance is often used to quantify the distance between persistence diagrams, motivated by its central stability properties [18]. However, the integration of persistence diagrams and Wasserstein distance with standard learning methods from statistics and machine learning has been a challenging open problem due to the differences between Wasserstein distance and standard Euclidean-based metrics [8].

Approaches that embed persistence diagrams into vector spaces [1] or Hilbert spaces [7,13] have recently been proposed to address this challenge. None of the embedding methods proposed thus far preserve Wasserstein distance in the original space of persistence diagrams [6]. Thus, these approaches do not inherit the stability properties of Wasserstein distance.

Recently, it was shown that persistence diagrams are inherently 1-dimensional if the topological features of networks are limited to connected components and cycles, and that the Wasserstein distance between these diagrams has a closed form expression [19]. Consequently, the work in [20] provides a computationally tractable, topological clustering approach for complex networks. However, significant limitations of the result in [19,20] are that it is unclear how this approach can be incorporated with standard Euclidean-based learning methods from statistics and machine learning, and that the approach is limited to evaluating networks with the *identical* number of nodes. There are many opportunities for applications of topological analysis of networks of *different* size, such as studies of the human brain when different subjects are sampled at different resolutions.

In this work, we present a novel *topological vector space* (TopVS) that embeds 1-dimensional persistence diagrams representing connected components and cycles for networks of *different* sizes. Thus, TopVS enables topological machine learning with networks of different sizes and greatly expands the applicability of previous work. Importantly, TopVS preserves the Wasserstein distance in the original space of persistence diagrams. Preservation of the Wasserstein distance ensures the theoretical stability properties of persistence diagrams carry over to the proposed embedding. In addition to the robustness benefits, TopVS also enables the application of a wide variety of Euclidean metric-based learning methods to topological data analysis. Particularly, the utility of TopVS is demonstrated in topology-based classification problems using support vector machines. TopVS is illustrated by classifying measured functional brain networks based on data obtained from subjects with *different* numbers of electrodes. The results show that TopVS performs very well compared to other competing approaches.

2 Wasserstein Distance-Preserving Vector Space

2.1 One-Dimensional Persistence Diagrams

Define a network as an undirected weighted graph $G = (V, \boldsymbol{w})$ with a set of nodes V and a weighted adjacency matrix $\boldsymbol{w} = (w_{ij})$. Define a binary graph G_ϵ with the identical node set V by thresholding the edge weights so that an edge between nodes i and j exists if $w_{ij} > \epsilon$. The binary graph is viewed as a 1-skeleton [15]. As ϵ increases, more and more edges are removed from the network G. Thus, we have a graph filtration: $G_{\epsilon_0} \supseteq G_{\epsilon_1} \supseteq \cdots \supseteq G_{\epsilon_k}$, where $\epsilon_0 \leq \epsilon_1 \leq \cdots \leq \epsilon_k$ are called filtration values. Persistent homology keeps track of the birth and death of topological features over filtration values ϵ. A topological feature that is born at a filtration b_i and persists up to a filtration d_i, is represented by a point (b_i, d_i) in a 2D plane. A set of all the points $\{(b_i, d_i)\}$ is called *persistence diagram* [10]. In the 1-skeleton, the only non-trivial topological features are connected components and cycles [22]. As ϵ increases, the number of connected components $\beta_0(G_\epsilon)$ and cycles $\beta_1(G_\epsilon)$ are monotonically increasing and decreasing, respectively [19]. Thus, the representation of the connected components and cycles can be simplified to a collection of sorted birth values $B(G) = \{b_i\}_{i=1}^{|V|-1}$ and a collection of sorted death values $D(G) = \{d_i\}_{i=1}^{1+|V|(|V|-3)/2}$, respectively [19]. $B(G)$ comprises edge weights in the *maximum spanning tree* (MST) of G. Once $B(G)$ is identified, $D(G)$ is given as the remaining edge weights that are not in the MST. Thus $B(G)$ and $D(G)$ are computed very efficiently in $O(n \log n)$ operations with n number of edges in networks.

2.2 Closed-Form Wasserstein Distance for Different-Size Networks

The Wasserstein distance between the 1-dimensional persistence diagrams can be obtained using a closed-form solution. Let G_1 and G_2 be two given networks possibly with different node sizes, i.e., their birth and death sets may *differ* in size. Their underlying empirical distributions on the persistence diagrams for connected components are defined in the form of Dirac masses [23]: $f_{G_1,B}(x) := \frac{1}{|B(G_1)|} \sum_{b \in B(G_1)} \delta(x-b)$ and $f_{G_2,B}(x) := \frac{1}{|B(G_2)|} \sum_{b \in B(G_2)} \delta(x-b)$, where $\delta(x-b)$ is a Dirac delta centered at the point b. Then the empirical distribution functions are the integration of $f_{G_1,B}$ and $f_{G_2,B}$ as $F_{G_1,B}(x) = \frac{1}{|B(G_1)|} \sum_{b \in B(G_1)} \mathbb{1}_{b \leq x}$ and $F_{G_2,B}(x) = \frac{1}{|B(G_2)|} \sum_{b \in B(G_2)} \mathbb{1}_{b \leq x}$, where $\mathbb{1}_{b \leq x}$ is an indicator function taking the value 1 if $b \leq x$, and 0 otherwise. A pseudoinverse of $F_{G_1,B}$ is defined as $F^{-1}_{G_1,B}(z) = \inf\{b \in \mathbb{R} \,|\, F_{G_1,B}(b) \geq z\}$, i.e., $F^{-1}_{G_1,B}(z)$ is the smallest b for which $F_{G_1,B}(b) \geq z$. Similarly, we define a pseudoinverse of $F_{G_2,B}$ as $F^{-1}_{G_2,B}(z) = \inf\{b \in \mathbb{R} \,|\, F_{G_2,B}(b) \geq z\}$. Then the empirical Wasserstein distance for connected components has a closed-form solution in terms of these pseudoinverses as

$$W_{p,B}(G_1, G_2) = \Big(\int_0^1 |F^{-1}_{G_1,B}(z) - F^{-1}_{G_2,B}(z)|^p \, dz \Big)^{1/p}. \tag{1}$$

Similarly, the Wasserstein distance for cycles $W_{p,D}(G_1, G_2)$ is defined in terms of empirical distributions for death sets $D(G_1)$ and $D(G_2)$.

The empirical Wasserstein distances $W_{p,B}$ and $W_{p,D}$ are approximated by computing the Lebesgue integration in (1) numerically as follows. Let $\widehat{B}(G_1) = \{F^{-1}_{G_1,B}(1/m), F^{-1}_{G_1,B}(2/m), ..., F^{-1}_{G_1,B}(m/m)\}$ and $\widehat{D}(G_1) = \{F^{-1}_{G_1,D}(1/n), ..., F^{-1}_{G_1,D}(n/n)\}$ be pseudoinverses for network G_1 sampled with partitions of equal intervals. Let $\widehat{B}(G_2)$ and $\widehat{D}(G_2)$ be sampled pseudoinverses for network G_2 with the same partitions of m and n, respectively. Then the approximated Wasserstein distances are given by

$$\widehat{W}_{p,B}(G_1, G_2) = \Big(\frac{1}{m^p}\sum_{k=1}^{m} |F^{-1}_{G_1,B}(k/m) - F^{-1}_{G_2,B}(k/m)|^p\Big)^{1/p}, \qquad (2)$$

$$\widehat{W}_{p,D}(G_1, G_2) = \Big(\frac{1}{n^p}\sum_{k=1}^{n} |F^{-1}_{G_1,D}(k/n) - F^{-1}_{G_2,D}(k/n)|^p\Big)^{1/p}. \qquad (3)$$

For a special case when networks G_1 and G_2 have the same number of nodes, i.e., $|B(G_1)| = |B(G_2)|$ and $|D(G_1)| = |D(G_2)|$, then exact computation of the Wasserstein distance is achieved using those birth and death sets, and setting m to the cardinality of the birth sets and n to that of the death sets.

2.3 Vector Representation of Persistence Diagrams

A collection of 1-dimensional persistence diagrams together with the Wasserstein distance is a metric space. 1-dimensional persistence diagrams can be embedded into a vector space that preserves the Wasserstein metric on the original space of persistence diagrams as follows. Let $G_1, G_2, ..., G_N$ be N observed networks possibly with different node sizes. Let $F^{-1}_{G_i,B}$ be a pseudoinverse of network G_i. The vector representation of a persistence diagram for connected components in network G_i is defined as a vector of the pseudoinverse sampled at $1/m, 2/m, ..., m/m$, i.e., $\boldsymbol{v}_{B,i} := \big(F^{-1}_{G_i,B}(1/m), F^{-1}_{G_i,B}(2/m), ..., F^{-1}_{G_i,B}(m/m)\big)^\top$. A collection of these vectors $M_B = \{\boldsymbol{v}_{B,i}\}_{i=1}^N$ with the p-norm $||\cdot||_p$ induces the p-norm metric $d_{p,B}$ given by $d_{p,B}(\boldsymbol{v}_{B,i}, \boldsymbol{v}_{B,j}) = ||\boldsymbol{v}_{B,i} - \boldsymbol{v}_{B,j}||_p = m\widehat{W}_{p,B}$. Thus, for $p = 1$ the proposed vector space describes Manhattan distance, $p = 2$ Euclidean distance, and $p \to \infty$ the maximum metric, which in turn correspond to the earth mover's distance (W_1), 2-Wasserstein distance (W_2), and the bottleneck distance (W_∞), respectively, in the original space of persistence diagrams. Similarly, we can define a vector space of persistence diagrams for cycles $M_D = \{\boldsymbol{v}_{D,i}\}_{i=1}^N$ with the p-norm metric $d_{p,D}$. The normed vector space $(M_B, d_{p,B})$ describes topological space of connected components in networks, while $(M_D, d_{p,D})$ describes topological space of cycles in networks.

The topology of a network viewed as a 1-skeleton is *completely* characterized by connected components and cycles. Thus, we can fully describe the network topology using both M_B and M_D as follows. Let $M_B \times M_D = \{(\boldsymbol{v}_{B,i}, \boldsymbol{v}_{D,i}) \,|\, \boldsymbol{v}_{B,i} \in M_B, \boldsymbol{v}_{D,i} \in M_D\}$ be the Cartesian product between M_B and M_D so the vectors in $M_B \times M_D$ are the concatenations of $\boldsymbol{v}_{B,i}$ and $\boldsymbol{v}_{D,i}$. For this product space to represent meaningful topology of network G_i, the vectors $\boldsymbol{v}_{B,i}$ and $\boldsymbol{v}_{D,i}$ must be a network decomposition, as discussed in Sect. 2.1. Thus $\boldsymbol{v}_{B,i}$ and $\boldsymbol{v}_{D,i}$ are constructed by sampling their psudoinverses with $m = \mathcal{V} - 1$ and $n = 1 + \frac{\mathcal{V}(\mathcal{V}-3)}{2}$, respectively, where $\mathcal{V}$ is a free parameter indicating a reference network size. The metrics $d_{p,B}$ and $d_{p,D}$ can be put together to form a

p-product metric $d_{p,\times}$ on $M_B \times M_D$ as [9]

$$d_{p,\times}\big((\boldsymbol{v}_{B,i},\boldsymbol{v}_{D,i}),(\boldsymbol{v}_{B,j},\boldsymbol{v}_{D,j})\big) = \big([d_{p,B}(\boldsymbol{v}_{B,i},\boldsymbol{v}_{B,j})]^p + [d_{p,D}(\boldsymbol{v}_{D,i},\boldsymbol{v}_{D,j})]^p\big)^{1/p}$$
$$= \big([m\widehat{W}_{p,B}]^p + [n\widehat{W}_{p,D}]^p\big)^{1/p}, \tag{4}$$

where $(\boldsymbol{v}_{B,i},\boldsymbol{v}_{D,i}),(\boldsymbol{v}_{B,j},\boldsymbol{v}_{D,j}) \in M_B \times M_D$, $m = \mathcal{V} - 1$ and $n = 1 + \frac{\mathcal{V}(\mathcal{V}-3)}{2}$. Thus, $d_{p,\times}$ is a weighted combination of p-Wasserstein distances, and is simply the p-norm metric between vectors constructed by concatenating $\boldsymbol{v}_{B,i}$ and $\boldsymbol{v}_{D,i}$. The normed vector space $(M_B \times M_D, d_{p,\times})$ is termed *topological vector space* (TopVS). Note the form of $d_{p,\times}$ given in (4) results in an unnormalized mass after multiplying m and n by their reciprocals given in (2) and (3). This unnormalized variant of Wasserstein distance is widely used in both theory [8,18] and application [7,19,21] of persistent homology. A direct consequence of the equality given in (4) is that the mean of persistence diagrams under the approximated Wasserstein distance is equivalent to the sample mean vector in TopVS. In addition, the proposed vector representation is highly interpretable because persistence diagrams can be easily reconstructed from vectors by separating sorted births and deaths.

For a special case in which networks $G_1, G_2, ..., G_N$ have the same number of nodes, the vectors $\boldsymbol{v}_{B,i}$ and $\boldsymbol{v}_{D,i}$ are simply the original birth set $B(G_i)$ and death set $D(G_i)$, respectively, and the p-norm metric $d_{p,\times}$ is expressed in terms of exact Wasserstein distances as $d_{p,\times} = ([mW_{p,B}]^p + [nW_{p,D}]^p)^{1/p}$.

3 Application to Functional Brain Networks

Dataset. We evaluate our method using functional brain networks from the anesthesia study reported by [2]. The brain networks are based on alpha band (8–12 Hz) weighted phase lag index applied to 10-second segments of resting state intracranial electroencephalography recordings. These recordings were made from eleven neurosurgical patients during administration of increasing doses of the general anesthetic propofol just prior to surgery. Each segment is labeled as one of the three arousal states: pre-drug *wake* (W), *sedated* but responsive to command (S), or *unresponsive* (U). The number of brain networks belonging to each subject varies from 71 to 119, resulting in the total of 977 networks from all the subjects. The network size varies from 89 to 199 nodes across subjects.

Classification Performance Evaluation. We are interested in whether candidate methods 1) can differentiate arousal states within individual subjects, and 2) generalize their learned knowledge to unknown subjects afterwards. As a result, we consider two different nested cross validation (CV) tasks as follows.

1. For the first task, we classify a collection of brain networks belonging to each subject separately. Specifically, we apply a nested CV comprising an outer loop of stratified 2-fold CV and an inner loop of stratified 3-fold CV. Since we may get a different split of data folds each time, we perform the nested CV

for 100 trials and report an average accuracy score and standard deviation for each subject. We also average these individual accuracy scores across subjects (11×100 scores) to obtain an overall accuracy.
2. For the second task, we use a different nested CV comprising both outer and inner loops with a leave-one-subject-out scheme. That is, a classifier is trained using all but one test subject. The inner loop is used to determine optimal hyperparameters, while the outer loop is used to assess generalization capacity of the candidate methods to unknown subjects in the population.

Method Comparison. Brain networks are used to compare the classification performance of the proposed TopVS relative to that of five state-of-the-art kernel methods and two well-established graph neural network methods. Three of these kernel methods are based on conventional 2-dimensional persistence diagrams for connected components and cycles: the *persistence image* (PI) vectorization [1], the *sliced Wasserstein kernel* (SWK) [7] and the *persistence weighted Gaussian kernel* (PWGK) [13]. The other two kernel methods are based on graph kernels: the *propagation kernel* (Prop) [16] and the *GraphHopper kernel* (GHK) [11]. The PI method embeds persistence diagrams into a vector space in which classification is performed using linear support vector machines (SVMs). The non-linear SWK, PWGK, Prop and GHK methods are combined with SVMs to perform classification. While nearly any classifier may be used with TopVS, here we illustrate results using the SVM with the linear kernel, which maximizes Wasserstein distance-based margin. When the TopVS method is applied to different-size networks, we upsample birth and death sets of smaller networks to match that of the largest network in size. Hyperparameters are tuned using grid search. SVMs have a regularization parameter $\mathcal{C} = \{0.01, 1, 100\}$. Thus, a grid search trains TopVS and PI methods with each $C \in \mathcal{C}$. The SWK and WGK methods have a bandwidth parameter $\Sigma = \{0.1, 1, 10\}$, and thus grid search trains both methods with each pair $(C, \sigma) \in \mathcal{C} \times \Sigma$. The Prop method has a maximum number of propagation iterations $T_{max} = \{1, 5, 10\}$, and thus is trained with each pair $(C, t_{max}) \in \mathcal{C} \times T_{max}$. GHK method uses the RBF kernel with a parameter $\Gamma = \{0.1, 1, 10\}$ between node attributes, and thus is trained with each pair $(C, \gamma) \in \mathcal{C} \times \Gamma$.

In addition, we also evaluate two well-established graph neural network methods including *graph convolutional networks* (GCN) [12] and *graph isomorphism network* (GIN) [25]. GCN and GIN are based on configurations and choices of hyperparameter values used in [25] as follows. Five graph neural network layers are applied, and the Adam optimizer with initial learning rate and weight decay of 0.01 are employed. We tune the following hyperparameters: the number of hidden units in $\{16, 32\}$, the batch size in $\{32, 128\}$ and the dropout ratio in $\{0, 0.5\}$. The number of epochs is set to 100 to train both methods.

Results. Results for the first task are summarized in Fig. 1, in which classification accuracy for individual subjects is shown. There is variability in individual subject performance because a different subject's network has a different number of electrodes, different electrode locations and different effective signal to

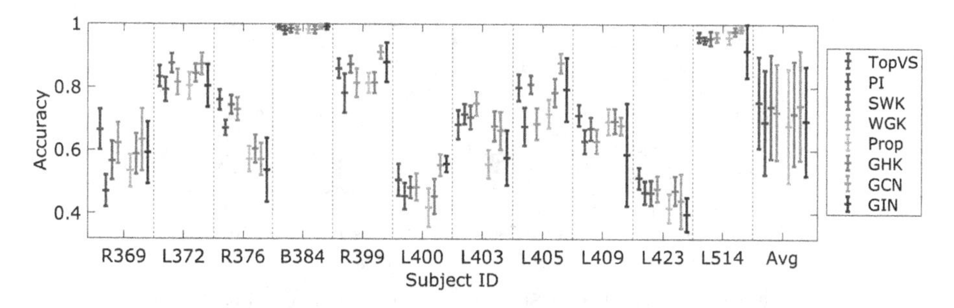

Fig. 1. Accuracy classifying brain networks within individual subjects. The last column displays the average accuracy obtained across all subjects. The center markers and bars depict the means and standard deviations obtained over 100 different trials.

noise ratio. So we expect these subjects to exhibit a diverse set of topological features across spatial resolutions. In most subjects all methods perform relatively well. The consistently poorer performance of PI, Prop and GIN is evident in the lower overall performance. On the other hand, our TopVS method is consistently among the best performing classifiers, resulting in the higher overall performance. For classification accuracy across subjects from the second task, we have 0.65 ± 0.21 for TopVS, 0.58 ± 0.22 for PI, 0.57 ± 0.20 for SWK, 0.60 ± 0.21 for WGK, 0.36 ± 0.12 for Prop, 0.43 ± 0.14 for GHK, 0.53 ± 0.20 for GCN and 0.48 ± 0.19 for GIN. TopVS is also among the best methods for classifying across subjects, while the performance of all the graph neural networks and graph kernels is significantly weaker. These results suggest that the use of computationally demanding and complex classification methods, such as GCN and GIN, does not result in significant increase in generalizability when classifying brain networks.

In addition, we compute confusion matrices to gain insights into the across-subject predictions for the second task, as displayed in Fig. 2. The persistent homology based methods, including TopVS, PI, SWK and WGK, are generally effective for separating unresponsive (U) from the other two states, and the majority of classification errors are associated with the differentiation between wake (W) and sedated (S) states. Prior work [2] demonstrated that wake and sedated brains are expected to have a great deal of similarity in comparison to the less similar unresponsive brains. However, the work in [2] performed the analysis on each individual subject separately while the results presented here are based on the analysis across subjects. Thus, not only the results here are consistent with the previous work [2] but also suggest that such biological expectation carries over to brains across subjects and that topology based methods can potentially derive biomarkers of changes in arousal states in the population, which underlie transitions into and out of consciousness, informing our understanding of the neural correlates of consciousness in clinical settings. TopVS shows clear advantages over all other topological baseline methods for differentiating wake and sedated states, suggesting that the proposed vector representation is an effective choice for representing subtle topological structure in brain networks.

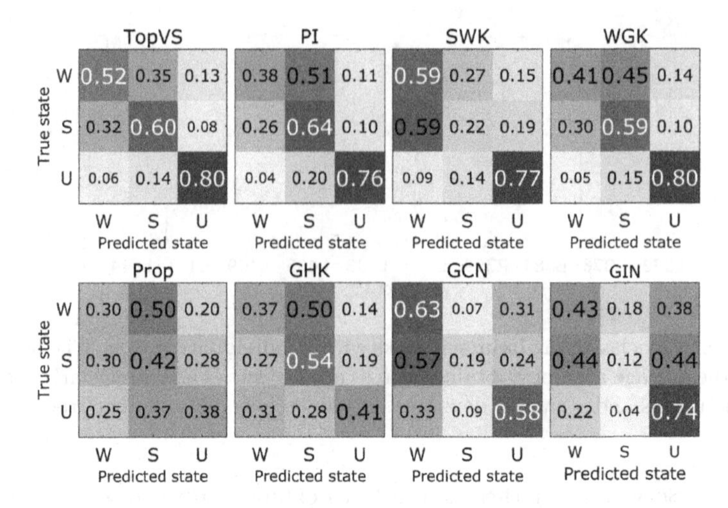

Fig. 2. Confusion matrices illustrating method performance for classifying across subjects. The numbers represent the fraction of brain networks in the test subjects being predicted as one of the three states: wake (W), sedated (S), and unresponsive (U). The confusion matrices are normalized with the entries in each row summing to 1.

Runtime Experiment. The kernel candidate methods are evaluated for a runtime experiment based on Intel Core i7 CPU with 16 GB of RAM. Figure 3 displays the runtime vs input size plot. The result clearly shows that all three persistent homology based kernels (PI, SWK and WGK) are limited to dense networks with a few hundred nodes, representing the current scaling limit of persistent homology embedding methods. On the other hand, TopVS is able to compute a kernel between 2000-node networks each with approx. two million edges in about one second. The computational practicality of TopVS extends its applicability to the large-scale analyses of brain networks that cannot be analyzed using prior methods based on conventional 2-dimensional persistence diagrams. Note that the time complexity of Prop is linear while TopVS has the slightly higher complexity as linearithmic. While Prop is the most efficient among all the methods, it has the lowest average accuracy when classifying the brain network data.

Potential Impact and Limitation. An open problem in neuroscience is identifying an algorithm that reliably extracts a patient's level of consciousness from passively recorded brain signals (i.e., biomarkers) and is robust to inter-patient variability, including where the signals are recorded in the brain. Conveniently, the anesthesia dataset is labeled according to consciousness state, and electrode placement (node location) was dictated solely by clinical considerations and thus varied across patients. Importantly, the relatively robust performance across patients suggests there are reliable topological signatures of consciousness captured by TopVS. The distinction between Wake and Sedated states involves relatively nuanced differences in connectivity, yet TopVS exploits the subtle differences in topology that differentiate these states better than the com-

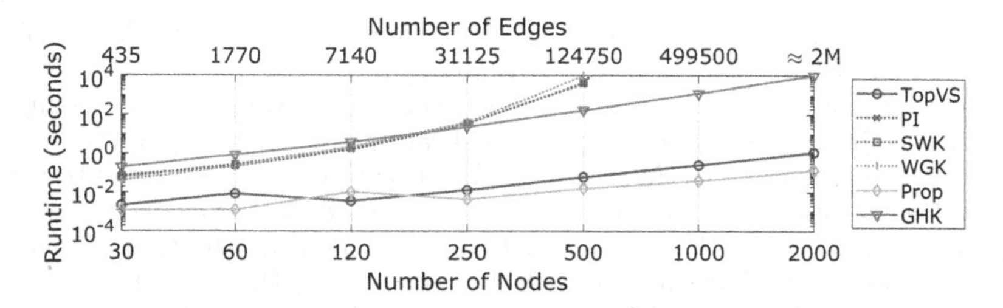

Fig. 3. Runtime experiment. We measured the runtime as the average amount of time each algorithm takes to compute its kernel between two complete graphs starting from edge weights as a given input. The runtime is plotted with respect to network size in terms of both the number of nodes and edges.

peting methods. Our results suggest that the neural correlates of consciousness can be captured in measurements of brain network topology, a longstanding problem of great significance. Additionally, TopVS is a principled framework that connects persistent homology theory with practical applications. Our versatile vector representation can be used with various vector-based statistical and machine learning models, expanding the potential for analyzing extensive and intricate networks beyond the scope of this paper. While TopVS is limited to representing connected components and cycles, assessment of higher-order topological features beyond cycles is of limited value due to their relative rarity and interpretive challenges, and consequent minimal discriminitive power [4,17,24].

References

1. Adams, H., et al.: Persistence images: a stable vector representation of persistent homology. JMLR **18** (2017)
2. Banks, M.I., et al.: Cortical functional connectivity indexes arousal state during sleep and anesthesia. NeuroImage **211**, 116627 (2020)
3. Barrat, A., Barthelemy, M., Pastor-Satorras, R., Vespignani, A.: The architecture of complex weighted networks. Proc. Natl. Acad. Sci. **101**(11), 3747–3752 (2004)
4. Biagetti, M., Cole, A., Shiu, G.: The persistence of large scale structures. Part I. Primordial non-Gaussianity. J. Cosmol. Astropart. Phys. **2021**(04), 061 (2021)
5. Bullmore, E., Sporns, O.: Complex brain networks: graph theoretical analysis of structural and functional systems. Nat. Rev. Neurosci. **10**(3), 186–198 (2009)
6. Carrière, M., Bauer, U.: On the metric distortion of embedding persistence diagrams into separable Hilbert spaces. In: Symposium on Computational Geometry (2019)
7. Carriere, M., Cuturi, M., Oudot, S.: Sliced Wasserstein kernel for persistence diagrams. In: International Conference on Machine Learning (2017)
8. Cohen-Steiner, D., Edelsbrunner, H., Harer, J., Mileyko, Y.: Lipschitz functions have LP-stable persistence. Found. Comput. Math. **10**, 127–139 (2010)
9. Deza, M.M., Deza, E.: Encyclopedia of Distances. Springer, Heidelberg (2009)
10. Edelsbrunner, H., Harer, J.L.: Computational Topology: An Introduction. American Mathematical Society (2022)

11. Feragen, A., Kasenburg, N., Petersen, J., de Bruijne, M., Borgwardt, K.M.: Scalable kernels for graphs with continuous attributes. In: NIPS (2013)
12. Kipf, T.N., Welling, M.: Semi-supervised classification with graph convolutional networks. In: International Conference on Learning Representations (2017)
13. Kusano, G., Hiraoka, Y., Fukumizu, K.: Persistence weighted Gaussian kernel for topological data analysis. In: International Conference on Machine Learning (2016)
14. Kwon, Y.K., Cho, K.H.: Analysis of feedback loops and robustness in network evolution based on Boolean models. BMC Bioinform. **8**(1), 1–9 (2007)
15. Munkres, J.R.: Elements of Algebraic Topology. CRC Press, Boca Raton (2018)
16. Neumann, M., Garnett, R., Bauckhage, C., Kersting, K.: Propagation kernels: efficient graph kernels from propagated information. Mach. Learn. **102**(2), 209–245 (2016)
17. Sizemore, A.E., Giusti, C., Kahn, A., Vettel, J.M., Betzel, R.F., Bassett, D.S.: Cliques and cavities in the human connectome. J. Comput. Neurosci. **44**, 115–145 (2018)
18. Skraba, P., Turner, K.: Wasserstein stability for persistence diagrams. arXiv preprint arXiv:2006.16824 (2020)
19. Songdechakraiwut, T., Chung, M.K.: Topological learning for brain networks. Ann. Appl. Stat. **17**(1), 403–433 (2023)
20. Songdechakraiwut, T., Krause, B.M., Banks, M.I., Nourski, K.V., Veen, B.D.V.: Fast topological clustering with Wasserstein distance. In: International Conference on Learning Representations (ICLR) (2022)
21. Songdechakraiwut, T., Shen, L., Chung, M.: Topological learning and its application to multimodal brain network integration. In: de Bruijne, M., et al. (eds.) MICCAI 2021. LNCS, vol. 12902, pp. 166–176. Springer, Cham (2021). https://doi.org/10.1007/978-3-030-87196-3_16
22. Sunada, T.: Homology groups of graphs. In: Topological Crystallography. Surveys and Tutorials in the Applied Mathematical Sciences, vol. 6, pp. 37–51. Springer, Tokyo (2013). https://doi.org/10.1007/978-4-431-54177-6_4
23. Turner, K., Mileyko, Y., Mukherjee, S., Harer, J.: Fréchet means for distributions of persistence diagrams. Discrete Comput. Geom. **52**(1), 44–70 (2014)
24. Xia, K., Wei, G.W.: Persistent homology analysis of protein structure, flexibility, and folding. Int. J. Numer. Methods Biomed. Eng. **30**(8), 814–844 (2014)
25. Xu, K., Hu, W., Leskovec, J., Jegelka, S.: How powerful are graph neural networks? In: International Conference on Learning Representations (2019)

Community-Aware Transformer for Autism Prediction in fMRI Connectome

Anushree Bannadabhavi[1], Soojin Lee[2], Wenlong Deng[1], Rex Ying[3], and Xiaoxiao Li[1](✉)

[1] Department of Electrical and Computer Engineering, The University of British Columbia, Vancouver, BC V6Z 1Z4, Canada
xiaoxiao.li@ece.ubc.ca
[2] Department of Medicine, The University of British Columbia Vancouver, Vancouver, BC V6Z 1Z4, Canada
[3] The Department of Computer Science, Yale University, New Haven, CT 06510, USA

Abstract. Autism spectrum disorder(ASD) is a lifelong neurodevelopmental condition that affects social communication and behavior. Investigating functional magnetic resonance imaging (fMRI)-based brain functional connectome can aid in the understanding and diagnosis of ASD, leading to more effective treatments. The brain is modeled as a network of brain Regions of Interest (ROIs), and ROIs form communities and knowledge of these communities is crucial for ASD diagnosis. On the one hand, Transformer-based models have proven to be highly effective across several tasks, including fMRI connectome analysis to learn useful representations of ROIs. On the other hand, existing transformer-based models treat all ROIs equally and overlook the impact of community-specific associations when learning node embeddings. To fill this gap, we propose a novel method, `Com-BrainTF`, a hierarchical local-global transformer architecture that learns intra and inter-community aware node embeddings for ASD prediction task. Furthermore, we avoid over-parameterization by sharing the local transformer parameters for different communities but optimize unique learnable prompt tokens for each community. Our model outperforms state-of-the-art (SOTA) architecture on ABIDE dataset and has high interpretability, evident from the attention module. Our code is available at https://github.com/ubc-tea/Com-BrainTF.

Keywords: Autism · fMRI · Transformers

This work is supported in part by the Natural Sciences and Engineering Research Council of Canada (NSERC) and Compute Canada.

Supplementary Information The online version contains supplementary material available at https://doi.org/10.1007/978-3-031-43993-3_28.

H. Greenspan et al. (Eds.): MICCAI 2023, LNCS 14227, pp. 287–297, 2023.
https://doi.org/10.1007/978-3-031-43993-3_28

1 Introduction

Autism spectrum disorder (ASD) is a developmental disorder that affects communication, social interaction, and behaviour [18]. ASD diagnosis is challenging as there are no definitive medical tests such as a blood tests, and clinicians rely on behavioural and developmental assessments to accurately diagnose ASD. As a more objective alternative measurement, neuroimaging tools, such as fMRI has been widely used to derive and validate biomarkers associated with ASD [9]. Conventionally, fMRI data is modelled as a functional connectivity (FC) matrix, *e.g.*, by calculating Pearson correlation coefficients of pairwise brain ROIs to depict neural connections in the brain. FC-based brain connectome analysis is a powerful tool to study connectivity between ROIs and their impact on cognitive processes, facilitating diagnosis and treatment of neurological disorders including ASD [14,28].

Deep learning (DL) models have led to significant advances in various fields including fMRI-based brain connectome analysis. For instance, BrainNetCNN [17] proposes a unique convolutional neural network (CNN) architecture with special edge-to-edge, edge-to-node, and node-to-graph convolutional filters. Considering brain network's non-euclidean nature, graph neural networks (GNNs) [30] have emerged as a promising method for analyzing and modelling brain connectome by constructing graph representation from FC matrices [4,16,19,30]. More recently, transformers [29] have been utilized for brain connectome analysis as they can learn long-range interaction between ROIs without a predefined graph structure. Brain Network Transformer (BNT) [15] has transformer encoders that learn node embeddings using a pearson correlation matrix and a readout layer that learns brain clusters. BNT [15] shows that transformer-based methods outperform CNN and GNN models for fMRI-based classification.

It is worth noting that our brain is comprised of functional communities [12] (often referred to as functional networks), which are groups of ROIs that perform similar functions as an integrated network [25,26]. These communities are highly reproducible across different studies [3,5] and are essential in understanding the functional organization of the brain [11]. Their importance extends beyond basic neuroscience, as functional communities have been shown to be relevant in understanding cognitive behaviours [12], mental states [8], and neurological and psychiatric disorders [2,24]. For example, numerous studies have found that individuals with ASD exhibit abnormalities in default mode network (DMN) [22], which is characterized by a combination of hypo- and hyper-connectivity with other functional networks. ASD is also associated with lifelong impairments in attention across multiple domains, and significant alterations in the dorsal attention network (DAN) and DMN have been linked to these deficits [7]. Such prior knowledge of functional relationships between functional communities, can be highly beneficial in predicting ASD. However, existing DL models often overlook this information, resulting in sub-optimal brain network analysis.

To address this limitation, we propose a novel community-aware transformer for brain network analysis, dubbed `Com-BrainTF`, which integrates functional community information into the transformer encoder. `Com-BrainTF` consists of a hierarchical transformer with a local transformer that learns community-specific embeddings, and a global transformer that fuses the whole brain information. The local transformer takes FC matrices as input with its parameters shared across all communities, while, personalized prompt tokens are learnt to differentiate the local transformer embedding functions. The local transformer's output class tokens and node embeddings are passed to the global transformer, and a pooling layer summarizes the final prediction. Our approach enhances the accuracy of fMRI brain connectome analysis and improves understanding of the brain's functional organization. Our key contributions are:

1. We propose a novel local-global hierarchical transformer architecture to efficiently learn and integrate community-aware ROI embeddings for brain connectome analysis by utilizing both ROI-level and community-level information.
2. We avoid over-parameterization by sharing the local transformer parameters and design personalized learnable prompt tokens for each community.
3. We prove the efficacy of `Com-BrainTF` with quantitative and qualitative experiment results. Our visualization demonstrates the ability of our model to capture functional community patterns that are crucial for ASD vs. Healthy Control (HC) classification.

2 Method

2.1 Overview

Problem Definition. In brain connectome analysis, we first parcellate the brain into N ROIs based on a given atlas. FC matrix is constructed by calculating the Pearson correlation coefficients between pairs of brain ROIs based on the strength of their fMRI activations. Formally, given a brain graph with N number of nodes, we have a symmetric FC matrix $X \in \mathbb{R}^{N \times N}$. Node feature vector of ROI j is defined as the jth row or column of this matrix. Given K functional communities and the membership of ROIs, we rearrange the rows and columns of the FC matrix, resulting in K input matrices $\{X_1, X_2, \ldots, X_K\}$. $X_k \in \mathbb{R}^{N_k \times N}$ is viewed as a N_k length sequence with N dimensional tokens and $\sum_k N_k = N$ (Fig. 1(1)). This process helps in grouping together regions with similar functional connectivity patterns, facilitating the analysis of inter-community and intra-community connections. `Com-BrainTF` inputs X_k to the community k-specific local transformer and outputs N dimensional tokens $H_k \in \mathbb{R}^{N_k \times N}$. A global transformer learns embedding $H = [H_1, \ldots, H_k] \mapsto Z_L \in \mathbb{R}^{N \times N}$, followed by a pooling layer and multi-layer perceptrons (MLPs) to predict the output.

Overview of Our Pipeline. Human brain connectome is a hierarchical structure with ROIs in the same community having greater similarities compared to inter-community similarities. Therefore, we designed a local-global transformer architecture(Fig. 1(2)) that mimics this hierarchy and efficiently leverages community labels to learn community-specific node embeddings. This approach allows the model to effectively capture both local and global information. Our model has three main components: one local transformer, one global transformer and a pooling layer details of which are discussed in the following subsections.

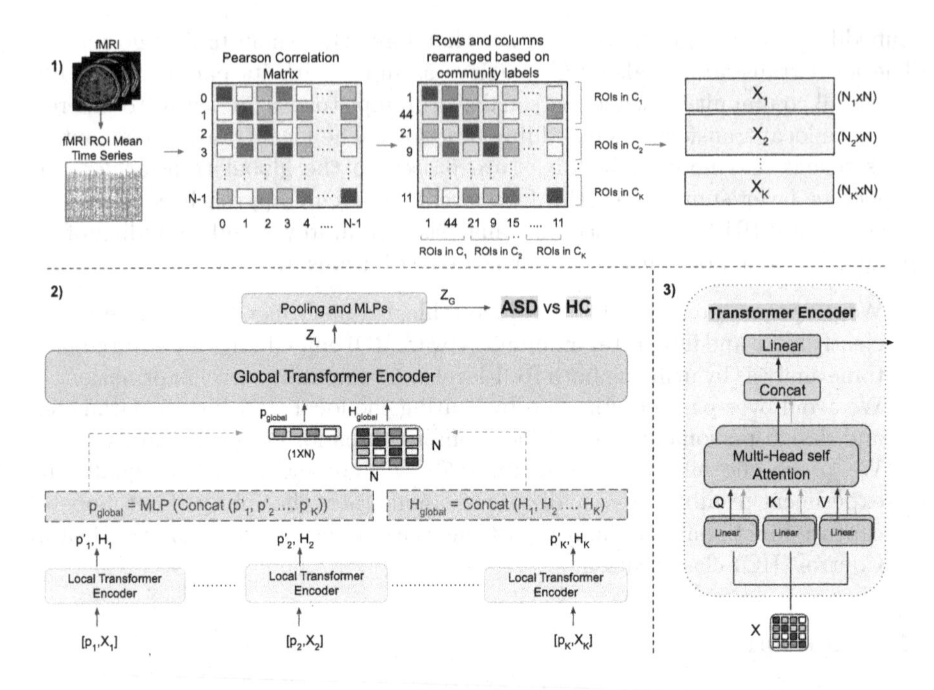

Fig. 1. 1) fMRI images parcellated by an atlas to obtain the Functional connectivity matrix for 'N' ROIs. Rows and columns are rearranged based on community labels of each ROI to obtain matrices $X_1, X_2 ... X_K$, inputs to local transformer 2) Overview of our local-global transformer architecture 3) Transformer encoder module

2.2 Local-Global Transformer Encoder

The transformer encoder [29] (Fig. 1(3)) is a crucial component of both local and global transformers. The transformer encoder takes as input an FC matrix (no position encodings) and has a multi-head-attention module to capture interdependencies between nodes using attention mechanism. The learnt node feature matrix H_i is given by

$$H_i = (\|_{m=1}^{M} h^m) W_O, \tag{1}$$

$$h^m = \text{softmax}\left(\frac{Q^m (K^m)^T}{\sqrt{d_k^m}}\right) V^m, \text{where,} \tag{2}$$

$Q^m = W_Q X_i', K^m = W_K X_i', V^m = W_V X_i', X_i' = [p_i, X_i]$, M = number of attention heads, $\|$ is concatenation operator, W_Q, W_K, W_V, W_O are model parameters.

Local Transformer: For optimal analysis of fMRI brain connectomes, it is important to incorporate both functional community information and node features. Therefore, we introduce the knowledge of community labels to the network by grouping node features based on community labels producing inputs

$\{X_1, X_2, \ldots, X_K\}$. However, separate local transformers for each input would result in a significant increase in model parameters. Hence, we use the same local transformer for all inputs, but introduce unique learnable personalized 1D prompt tokens $\{p_1, p_2, \ldots, p_K\}$, where $p_i \in \mathbb{R}^{1\times N}$, that learn to distinguish between node feature matrices of each community and therefore avoid over-parameterization of the model. Previous transformer-based models like BNT [15] do not have any prompt tokens and only use the FC matrix as input. Using attention mechanism, pairwise connection strength between ROIs is learnt, producing community-specific node embeddings H_i and prompt tokens $p_i^{'}$,

$$p_i^{'}, H_i = \text{LocalTransformer}([p_i, X_i]) \text{ where, } i \in [1, 2...K]. \tag{3}$$

Global Transformer: On obtaining community-specific node embeddings and prompt tokens from the local transformer, it is essential to combine this information and design a module to learn the brain network at a global level. Therefore, we introduce a global transformer encoder to learn inter-community dependencies. Input to the global transformer is the concatenated, learnt node feature matrices from the local transformer and a prompt token (Fig. 1(2)). Prompt tokens learnt by the local transformer contain valuable information to distinguish different communities and thus are concatenated and passed through an MLP to obtain a prompt token input for the global transformer as follows:

$$p_{global} = \text{MLP (Concat } (p_1^{'}, p_2^{'} \ldots p_K^{'})), \tag{4}$$

$$H_{global} = \text{Concat}(H_1, H_2, \ldots, H_K), \tag{5}$$

$$p^{'}, Z_L = \text{GlobalTransformer}([p_{global}, H_{global}]). \tag{6}$$

The resulting attention-enhanced, node embedding matrix Z_L is then passed to a pooling layer for further coarsening of the graph. Extensive ablation studies are presented in Sect. 3.3 to justify the choice of inputs and output.

2.3 Graph Readout Layer

The final step involves aggregating global-level node embeddings to obtain a high-level representation of the brain graph. We use OCRead layer [15] for aggregating the learnt node embeddings. OCRead initializes orthonormal cluster centers $E \in \mathbb{R}^{K\times N}$ and softly assigns nodes to these centers. Graph level embedding Z_G is then obtained by $Z_G = A^\top Z^L$, where $A \in \mathbb{R}^{K\times N}$is a learnable assignment matrix computed by OCRead. Z_G is then flattened and passed to an MLP for a graph-level prediction. The whole process is supervised with Cross-Entropy (CE) loss.

3 Experiments

3.1 Datasets and Experimental Settings

ABIDE is an open-source collection of resting-state functional MRI (rs-fMRI) data from 17 international sites [1] with Configurable Pipeline for the Analysis of

Table 1. Performance comparison with baselines (Mean ± standard deviation)

Model	AUROC	Accuracy	Sensitivity	Specificity
BrainNetTF [15]	78.3 ± 4.4	68.1 ± 3.1	78.1 ± 10.0	58.9 ± 12.0
BrainNetCNN [17]	74.1 ± 5.1	67.5 ± 3.1	65.3 ± 4.3	**69.6 ± 4.1**
FBNETGNN [16]	72.5 ± 8.3	64.9 ± 8.9	60.9 ± 11.3	67.5 ± 13.1
Com-BrainTF	**79.6 ± 3.8**	**72.5 ± 4.4**	**80.1 ± 5.8**	65.7 ± 6.4

Connectomes (CPAC), band-pass filtering (0.01–0.1 Hz), no global signal regression, parcellated by Craddock 200 atlas [23]. ROIs of ABIDE dataset belong to either of the eight functional communities namely, cerebellum and subcortical structures (CS & SB), visual network (V), somatomotor network (SMN), DAN, ventral attention network (VAN), limbic network (L), frontoparietal network (FPN) and DMN, following the assignments in Yeo 7 network template [31]. ABIDE has fMRI data of 1009 subjects, 51.14% of whom were diagnosed with ASD. Pearson correlation matrix is computed using mean time series of ROIs. Stratified sampling strategy [15] is used for train-validation-test data split.

Experimental Settings: All models are implemented in PyTorch and trained on V100D-8C (C-Series Virtual GPU for Tesla V100) with 8GB memory. For both local and global transformers, the number of attention heads is equal to the number of communities. We used Adam optimizer with an initial learning rate of 10^{-4} and a weight decay of 10^{-4}. The train/validation/test data spilt ratio is 70:10:20, and the batch size is 64. Prediction of ASD vs. HC is the binary classification task. We use AUROC, accuracy, sensitivity, and specificity on the test set to evaluate performance. Models are trained for 50 epochs with an average runtime of around 1 minute and we use an early stopping strategy. The mean and standard deviation values were obtained from five random runs with five different seeds. The epoch with the highest AUROC performance on the validation set is used for performance comparison on the test set.

3.2 Quantitative and Qualitative Results

Comparison with Baselines (Quantitative Results). Following the comparison method mentioned in BNT [15], we compare the performance of our model with three types of baselines (i) Comparison with transformer-based models - BNT (ii) Comparison with neural network model on fixed brain network - BrainNetCNN [17] (iii) Comparison with neural network models on learnable brain network - FBNETGEN [16]. Note that the selected baselines are reported to be the best among other alternatives in the same category on ABIDE dataset by BNT [15]. As seen in Table 1, our model outperforms all the other architectures on three evaluation metrics. In contrast to other models, Com-BrainTF is hierarchical, similar to the brain's functional organization and therefore is capable of learning relationships within and between communities.

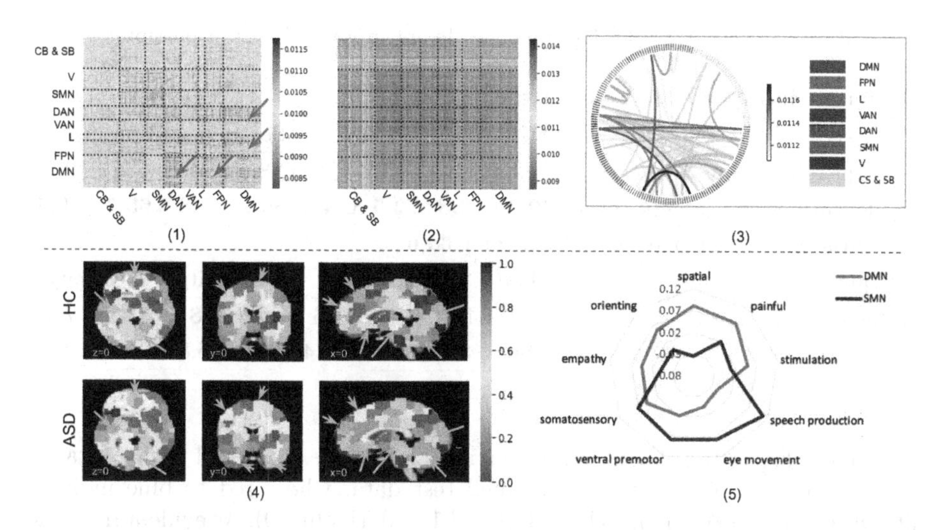

Fig. 2. (1) Attention matrix of the global transformer of Com-BrainTF highlighting prominent communities that influence ASD prediction (2) Attention matrix of the first transformer encoder module of BNT (3) Chord plot of the top 1% of attention values (4) Neurosynth results generated from averaged prompt vectors (5) Correlation between functional communities and functional keywords decoded by Neurosynth [27].

Interpretibility of Com-BrainTF (Qualitative Results). We perform extensive experiments to analyze the interpretability of `Com-BrainTF`. The following results are discussed in detail:

1. Interpretation of the learned attention matrix: Fig. 2(1) and Fig. 2(2) show the averaged attention matrices of `Com-BrainTF` and BNT [15] respectively. These attention matrices are obtained by averaging attention scores over correctly classified test data. In comparison to BNT, our learned attention scores clearly highlight the communities important for ASD prediction. Specifically, attention scores in the SMN region (green arrow) are high indicating that within SMN, connectivity plays an essential role in ASD prediction. This aligns with prior studies on autism which have shown reduced FC within SMN [13], reflecting altered sensory and motor processing [10,20,21]. Additionally, attention scores in DMN, DAN, and FPN regions were found to be high (magenta arrows) suggesting that the functional connectivity between DMN and DAN and between DMN and FPN are also crucial for ASD prediction which is in line with previous studies that report abnormalities in DMN [22] and its functional connectivity with other functional networks including DAN and FPN. These connections can be better visualized from the chord plot in Fig. 2(3). To summarize, `Com-BrainTF` is able to correctly identify functional communities associated with ASD and the results are consistent with neuroscience findings.

2. Ability of the prompt's attention vector to differentiate between ASD and HC subjects: We investigate the first row of the learned attention matrix, namely the attention of ROIs corresponding to the prompt. The ROI-

Table 2. Ablation study: Examining different input and output possibilities

Different inputs to the global transformer				
Inputs	AUROC	Accuracy	Sensitivity	Specificity
Prompt tokens only	73.2 ± 4.5	65.7 ± 5.8	**88.8 ± 4.1**	43.7 ± 9.1
Node features and prompt token	**79.6 ± 3.8**	**72.5 ± 4.4**	80.1 ± 5.8	**65.7 ± 6.4**
CE loss on node features vs prompt token				
Outputs	AUROC	Accuracy	Sensitivity	Specificity
Prompt token	71.9 ± 4.6	66.1 ± 3.6	**80.4 ± 8.2**	51.5 ± 7.6
Node features	**79.6 ± 3.8**	**72.5 ± 4.4**	80.1 ± 5.8	**65.7 ± 6.4**

wise normalized attention scores shown in Fig. 2(4) are generated using averaged prompt vectors over correctly classified test data, where red to blue indicate attention scores from high to low (dark red:1 to dark blue: 0). As evident from the figure, clear differences can be seen in the important brain regions among ASD and HC subjects. Additionally, most differences in the prompt vector embeddings are seen in DMN and SMN (supporting figures in the appendix), consistent with previous neuroscience literature.
3. Meta-analysis of the important functional networks that influence ASD prediction: DMN and SMN networks have been found to be crucial for ASD prediction based on Fig. 2(4). Using the difference of the prompt vector values in those regions between ASD and HC subjects, Fig. 2(5) was generated based on Meta-analysis using Neurosynth[1] database [27]. Regions in DMN were found to be associated with trait keywords such as empathy, painful, stimulation, orienting and spatial, whereas regions in the SMN showed higher correlations with eye movement, ventral premotor, somatosensory and speech production.

3.3 Ablation Studies

We conduct ablation studies to justify the designs of input and output pairing of the global transformer that facilitates effective learning of node embeddings, consequently resulting in superior performance. The input global transformer prompt token is intentionally retained in all of the experiments because of (i) its capability to identify important communities for the prediction task (Sect. 3.2) and (ii) its capability to learn relationships within and between communities (Sect. 3.2 Fig. 2 (3)).

Input: Node Features vs. Prompt Tokens of Local Transformers. We evaluated two input possibilities for the global transformer: (i) use only prompt tokens from the local transformer (ii) incorporate both prompt tokens and

[1] The meta-analytic framework (www.neurosynth.org) provides the posterior probability $P(\text{Feature}|\text{Coordinate})$ for psychological features (i.e., word or phrase) at a given spatial location based on neuroscience literature.

updated node features. Table 2 reveals that prompt tokens alone are unable to capture the complete intra-functional connectivity and negatively affect performance.

Output: Cross Entropy Loss on the Learned Node Features vs. Prompt Token. In comparison to vision models like ViT [6], which only use the class token (similar to the prompt token in our structure) for classification, we use output node embeddings from the global transformer for further processing steps. We justify this design by conducting a performance comparison experiment: (i) use learnt node features as the global transformer output (ii) use only the prompt token as output. The results (Table 2) demonstrated a significant decrease in performance when prompt token alone was used. This is expected since brain networks are non-euclidean structures and the learnt node features capture more information about the underlying graph, making them essential for accurately capturing the inter-community dependencies.

4 Conclusion

In this work, we introduce `Com-BrainTF`, a hierarchical, community-aware local-global transformer architecture for brain network analysis. Our model learns intra- and inter-community aware node embeddings for ASD prediction tasks. With built-in interpretability, `Com-BrainTF` not only outperforms SOTA on the ABIDE dataset but also detects salient functional networks associated with ASD. We believe that this is the first work leveraging functional community information for brain network analysis using transformer architecture. Our framework is generalizable for the analysis of other neuroimaging modalities, ultimately benefiting neuroimaging research. Our future work includes investigating alternate variants for choosing different atlases and community network parcellations.

References

1. Craddock, C., et al.: The neuro bureau preprocessing initiative: open sharing of preprocessed neuroimaging data and derivatives (2013)
2. Canario, E., Chen, D., Biswal, B.: A review of resting-state fMRI and its use to examine psychiatric disorders. Psychoradiology **1**(1), 42–53 (2021)
3. Chen, S., et al.: Group independent component analysis reveals consistent resting-state networks across multiple sessions. Brain Res. **1239**, 141–151 (2008)
4. Cui, H., et al.: Interpretable graph neural networks for connectome-based brain disorder analysis. In: Medical Image Computing and Computer Assisted Intervention-MICCAI 2022: 25th International Conference, Singapore, September 18–22, 2022, Proceedings, Part VIII (2022)
5. Damoiseaux, J.S., et al.: Consistent resting-state networks across healthy subjects. Proc. National Acad. Sci. **103**(37), 138484–13853 (2006)
6. Dosovitskiy, A., et al.: An image is worth 16x16 words: Transformers for image recognition at scale. CoRR (2020)

7. Farrant, K., Uddin, L.Q.: Atypical developmental of dorsal and ventral attention networks in autism. Develop. Sci. **19**(4),550–563 (2015)
8. Geerligs, L., et al.: State and trait components of functional connectivity: individual differences vary with mental state. J. Neurosci. **34**(41), 13949–13961 (2015)
9. Goldani, A.A.S., et al.: Biomarkers in autism. Front. Psych. **5**, 100 (2014)
10. Gowen, E., Hamilton, A.: Motor abilities in autism: a review using a computational context. J. Autism Develop. Disorders **43**(2), 323–344 (2013). https://doi.org/10.1007/s10803-012-1574-0
11. van den Heuvel, M.P., et al.: Functionally linked resting-state networks reflect the underlying structural connectivity architecture of the human brain. Hum. Brain Mapp. **30**(10), 3127–3141 (2009)
12. den Heuvel, M.P.V., Pol, H.E.H.: Exploring the brain network: A review on resting-state fMRI functional connectivity. Europ. Neuropsychopharmacol. **20**(8), 519–534 (2010)
13. Ilioska, I., et al.: Connectome-wide mega-analysis reveals robust patterns of atypical functional connectivity in autism. Biol. Psych. **94**(1), 29–39 (2022)
14. Kaiser, M.D., et al.: Severe impairments of social interaction and associated abnormalities in children: epidemiology and classification. Proc. Natl. Acad. Sci. **107**(49), 21223–21228 (2010)
15. Kan, X., et al.: Brain network transformer. Adv. Neural Inform. Process. Syst. **35**, 25586–25599 (2022)
16. Kan, X., et al.: FBNETGEN: Task-aware GNN-based fMRI analysis via functional brain network generation. In: Medical Imaging with Deep Learning, pp. 618–637 (2022)
17. Kawahara, J., et al.: Brainnetcnn: convolutional neural networks for brain networks; towards predicting neurodevelopment. NeuroImage **146**, 1038–1049 (2017)
18. L, W., J, G.: Severe impairments of social interaction and associated abnormalities in children: epidemiology and classification. J. Autism Dev. Disorders **146**, 1038–1049 (1979)
19. Li, X., et al.: Braingnn: Interpretable brain graph neural network for fMRI analysis. Med. Image Anal. **74**, 102233 (2021)
20. Marco, E.J., et al.: Sensory processing in autism: a review of neurophysiologic findings. Pediatr. Res. **69**(8), 48–54 (2011)
21. Mostofsky, S.H., Ewen, J.B.: Altered connectivity and action model formation in autism is autism. The Neuroscientist **17**(4), 437–448 (2011)
22. Padmanabhan, A., et al.: The default mode network in autism. Biol. Psych.: Cogn. Neurosci. Neuroimaging **2**(6), 476–486 (2017)
23. RC, C., et al.: A whole brain fMRI atlas generated via spatially constrained spectral clustering. Hum Brain Mapp. **33**(8), 1914–1928 (2012)
24. Rosazza, Cristina, Minati, Ludovico: Resting-state brain networks: literature review and clinical applications. Neurol. Sci. **32**(5), 773–785 (2011)
25. Smith, S.M., et al.: Correspondence of the brain's functional architecture during activation and rest. Proc. National Acad. Sci. **106**(31), 13040–13045 (2009)
26. Smith, S.M., et al.: Functional connectomics from resting-state fMRI. Trends Cogn. Sci. **7**(3), 141–144 (2013)
27. T, Y., et al.: Large-scale automated synthesis of human functional neuroimaging data. Nature Methods **8**(8) 665–70 (2011)
28. True, P., et al.: Multiple-network classification of childhood autism using functional connectivity dynamics. In: Medical Image Computing and Computer-Assisted Intervention - MICCAI 2014 (2014)

29. Vaswani, A., et al.: Attention is all you need. Adv. Neural Inform. Process. Syst. **30** (2017)
30. Veličković, P., et al.: Graph attention networks. In: International Conference on Learning Representations (2018)
31. Thomas Yeo, B.T., et al.: The organization of the human cerebral cortex estimated by intrinsic functional connectivity. J. Neurophysiol. **106**(3), 1125–1165 (2011). https://doi.org/10.1152/jn.00338.2011

Overall Survival Time Prediction of Glioblastoma on Preoperative MRI Using Lesion Network Mapping

Xingcan Hu[1,2], Li Xiao[1,2(✉)], Xiaoyan Sun[1,2], and Feng Wu[1,2]

[1] Department of Electronic Engineering and Information Science, University of Science and Technology of China, Hefei 230052, China
[2] Institute of Artificial Intelligence, Hefei Comprehensive National Science Center, Hefei 230088, China
xiaoli11@ustc.edu.cn

Abstract. Glioblastoma (GBM) is the most aggressive malignant brain tumor. Its poor survival rate highlights the pressing need to adopt easily accessible, non-invasive neuroimaging techniques to preoperatively predict GBM survival, which can benefit treatment planning and patient care. MRI and MRI-based radiomics, although effective for survival prediction, do not consider brain's functional alternations caused by tumors, which are clinically significant for guiding therapeutic strategies aimed at inhibiting tumor-brain communication. In this paper, we propose an augmented lesion network mapping (A-LNM) based survival prediction framework, where a novel neuroimaging feature family, called functional lesion network (FLN) maps generated by the A-LNM, is achieved from patients' structural MRI, and thus are more readily available than functional connections measured with functional MRI of patients. Specifically, for each patient, the A-LNM first estimates functional disconnection (FDC) maps by embedding the lesion (the whole tumor) into an atlas of functional connections in a large cohort of healthy subjects, and many FLN maps are then obtained by averaging subsets of the FDC maps such that we can artificially boost data volume (i.e., FLN maps), which helps to mitigate over-fitting and improve survival prediction performance when learning a deep neural network from a small sized dataset. The augmented FLN maps are finally fed to a 3D ResNet-based backbone followed by the average pooling operation and fully-connected layers for GBM survival prediction. Experimental results on the BraTS 2020 training dataset demonstrate the effectiveness of our proposed framework with the A-LNM derived FLN maps for GBM survival classification. Moreover, we identify the survival-relevant brain regions that can be traced back with biological interpretability.

Keywords: Brain tumor · functional connections · lesion network mapping · survival prediction

H. Greenspan et al. (Eds.): MICCAI 2023, LNCS 14227, pp. 298–307, 2023.
https://doi.org/10.1007/978-3-031-43993-3_29

1 Introduction

Glioblastomas (GBMs, known as grade IV gliomas) are the most common primary malignant brain tumors with high spatial heterogeneity and varying degrees of aggressiveness [22]. Patients with GBM generally have a very poor survival rate; the median overall survival time is about 14 months [17]; and the overall survival time is affected by many factors, including patient characteristics (e.g., age and physical status), tissue histopathology (e.g., cellular density and nuclear atypia), and molecular pathology (e.g., mutations and gene expression levels) [1,14,15]. Although these factors, particularly molecular information, have usually proved to be strong predictors of survival in GBM, there remain substantial challenges and unmet clinical needs to exploit easily accessible, non-invasive neuroimaging data acquired preoperatively to predict overall survival time of GBM patients, which can benefit treatment planning.

To do so, magnetic resonance imaging (MRI) and its derived radiomics have been widely used to study GBM preoperative prognosis over the last few decades. For example, Anand et al. [2] first applied a forest of trees to assign an importance value to each of the 1022 radiomic features extracted from T1 MRI, and then the 32 most important features were fed to the random forest regressor for predicting overall survival time of a GBM patient. Based on patches from multi-modal MRI images, Nie et al. [19] trained a 3D convolutional neural network (CNN) to learn the high-level semantic features, which were eventually input to a support vector machine (SVM) for classifying long- and short-term GBM survivors. In addition, an integrated model by fusing radiomics features, MRI-based CNN features, and clinical features, was presented for GBM survival group classification, resulting in better performance than using any single type of features [12].

Although both MRI and its derived radiomics features have been demonstrated to have predictive power for survival analysis in the aforementioned literature, they do not account for brain's functional alternations caused by tumors, which are clinically significant as biologically-interpretable biomarkers of recovery and therapy. These alternations can be reflected by changes in resting-state functional MRI (fMRI)-derived functional connectivities/connections (FCs) between the blood oxygenation level-dependence (BOLD) time series of paired brain regions. Therefore, the use of FCs to predict overall survival time for GBM has recently attracted increasing attention [7,16,24], and more importantly, survival-related FC patterns or brain regions were found to guide therapeutic solutions aimed at inhibiting tumor-brain communication.

Nevertheless, current FC-based survival prediction still suffers from two main deficiencies when applied to GBM prognosis. First, due to mass effect and physical infiltration of GBM in the brain, FCs estimated directly from GBM patients' resting-state fMRI might be inaccurate, especially when the tumors are near or in the regions of interest. Second, resting-state fMRI data are not routinely collected for GBM clinical practices, which restricts the size of annotated datasets such that it is infeasible to train a reliable prediction model based on deep learning for survival prediction. In order to circumvent these issues, in this paper we introduce a novel neuroimaging feature family, namely functional lesion network

(FLN) maps that are generated by our augmented lesion network mapping (A-LNM), for overall survival time prediction of GBM patients. Our A-LNM is motivated by lesion network mapping (LNM) [8] which can localize neurological deficits to functional brain networks and identify regions relate to a clinical syndrome. By embedding the lesion into a normative functional connectome and computing functional connectivity between the lesion and the rest of the brain using fMRI of all healthy subjects in the normative cohort, LNM has been successfully employed to the identification of the brain network underlying particular symptoms or behavioral deficits in stoke [4,13].

The details of our workflow are described as follows.

1) We first manually segment the whole tumor (regarded as lesion in this paper) on structural MRI for all GBM patients, and the resulting lesion masks are mapped onto a reference brain template, e.g., the MNI152 $2mm^3$ template.
2) The proposed A-LNM is next used to generate FLN maps for each GBM patient by using resting-state fMRI from a large cohort of healthy subjects. Specifically, for each patient, we correlate the mean BOLD time series of all voxels within the lesion with the BOLD time series of every voxel in the whole brain for all N subjects in the normative cohort, producing N functional disconnection (FDC) maps of voxel-wise correlation values (transformed to z-scores). These resulting N FDC maps are partitioned into M disjoint subsets of equal size, and M FLN maps are separately obtained by averaging the FDC maps in each of the M subsets. Similar to data augmentation schemes, we can artificially boost data volume (i.e., FLN maps) up to M times through producing M FLN maps for each patient in the A-LNM, which helps to mitigate the risk of over-fitting and improve the performance of overall survival time prediction when learning a deep neural network from a small sized dataset. For this reason, we propose the name "augmented LNM (A-LNM)", compared to the traditional LNM where only one FLN map is generated per patient by averaging all the N FDC maps.
3) Finally, these augmented FLN maps are fed to a 3D ResNet-based backbone network followed by the average pooling operation and fully-connected layers for GBM survival prediction.

To our knowledge, this paper is the first to demonstrate a successful extension of LNM for survival prediction in GBM. To evaluate the predictive power of the FLN maps generated by our A-LNM, we conduct extensive experiments on 235 GBM patients in the training dataset of BraTS 2020 [18] to classify the patients into three overall survival time groups viz. long, mid, and short. Experimental results show that our A-LNM based survival prediction framework outperforms previous state-of-the-art methods. In addition, an explainable analysis driven by the Gradient-weighted Class Activation Mapping (Grad-CAM) [10] for survival-related brain regions is fulfilled.

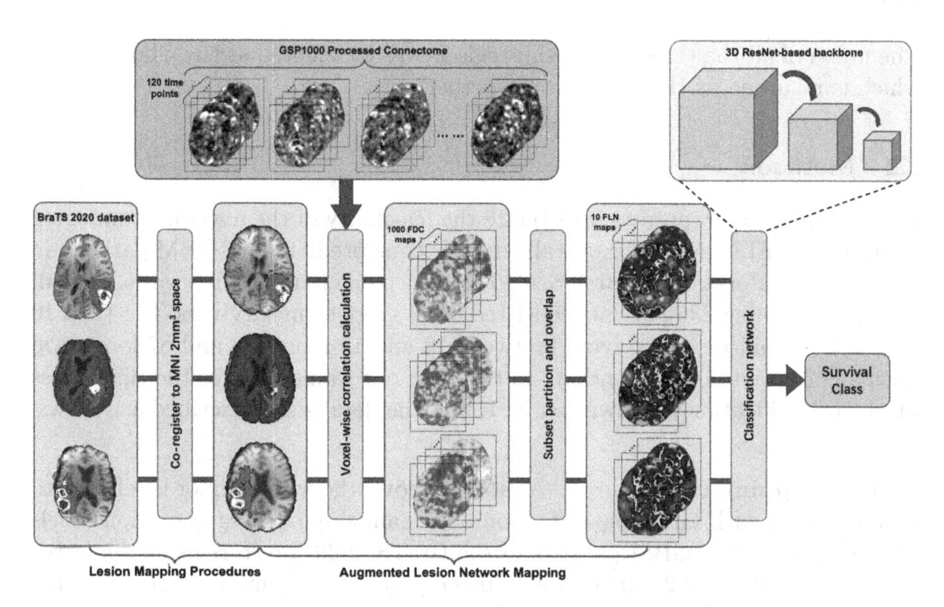

Fig. 1. The proposed framework for GBM survival prediction.

2 Materials and Methods

2.1 Materials

GSP1000 Processed Connectome. It publicly released preprocessed resting-state fMRI data of 1000 healthy right-handed subjects with an average age 21.5 ± 2.9 years and approximately equal numbers of males and females from the Brain Genomics Superstruct Project (GSP) [5], where the concrete image acquisition parameters and preprocessing procedures can be found as well. Specifically, a slightly modified version of Yeo's Computational Brain Imaging Group (CBIG) fMRI preprocessing pipeline (https://github.com/bchcohenlab/CBIG) was employed to obtain either one or two preprocessed resting-state fMRI runs of each subject that had 120 time points per run and were spatially normalized into the MNI152 template with 2mm^3 voxel size. We downloaded and used the first-run preprocessed resting-state fMRI of each subject for the following analysis.

BraTS 2020. It provided an open-access pre-operative imaging training dataset to segment brain tumors of glioblastoma (GBM, belonging to high grade glioma) and low grade glioma (LGG) patients, as well as to predict overall survival time of GBM patients [18]. This training dataset contained 133 LGG and 236 GBM patients, and each patient had four MRI modalities, including T1, post-contrast T1-weighted, T2-weighted, and T2 Fluid Attenuated Inversion Recovery. Manual expert segmentation delineated three tumor sub-regions, i.e., the GD-enhancing tumor, the peritumoral edema, and the necrotic and non-enhancing tumor core.

The union of all the three tumor sub-regions was considered as the whole tumor, which is regarded as the lesion in this paper.

2.2 Methods

In this paper, we propose to investigate the feasibility of the novel neuroimaging features, i.e., FLN maps, for overall survival time prediction of GBM patients in the training dataset of the BraTS 2020, in which one patient alive was excluded, and the remaining 235 patients consisted of 89 short-term survivors (less than 10 months), 59 mid-term survivors (between 10 and 15 months), and 87 long-term survivors (more than 15 months). To this end, our framework for the three-class survival classification is shown in Fig. 1, and the details are described as follows.

Lesion Mapping Procedures. As stated above, the whole tumor is referred to as a lesion for each GBM patient. From the manual expert segmentation labels of lesions in the 235 GBM patients of the BraTS 2020, we co-register the lesion masks to the MNI152 2mm^3 template by employing a Symmetric Normalization algorithm in ANTsPy [3].

Augmented Lesion Network Mapping (A-LNM). After lesion mapping, we introduce a modified LNM (called augmented LNM (A-LNM) in this paper) to generate FLN maps for each GBM patient by using resting-state fMRI of all 1000 GSP healthy subjects, as described below. *i*) For each patient, the lesion is viewed as a seed region to calculate FDC in the healthy subjects with resting-state fMRI. Specifically, to compute FDC, the mean BOLD time series of voxels within each lesion is correlated with the BOLD time series of every voxel in the whole brain for all the 1000 healthy subjects, yielding 1000 FDC maps of voxel-wise correlation values (transformed to z-scores), where an FDC map is actually a three-dimensional voxel-wise matrix of size $91 \times 109 \times 91$ (spatial resolution: 2mm^3 voxel size). *ii*) Different from the commonly used LNM where the resulting 1000 FDC maps are thresholded or averaged to obtain a single FLN map for each patient, the A-LNM generates many FLN maps for each patient in a manner that partitions all the 1000 FDC maps into disjoint subsets of equal size and averages each subset to produce one FLN map. One can clearly see that similar to data augmentation schemes, we artificially boost the number of training samples (i.e., FLN maps) by our A-LNM, which helps to mitigate the risk of over-fitting and improve the performance of overall survival time prediction when learning a deep neural network from such a small sized training set used in this paper. Note that in Sect. 3 of this paper, according to experimental results, we divided the 1000 FDC maps into 100 subsets, and randomly chose 10 out of the resulting 100 FLN maps for each patient as input to the downstream prediction model.

Deep Neural Network for Overall Survival Time Prediction. By taking the obtained FLN maps as input, we apply a 3D ResNet-based backbone network transferred from the encoder of MedicalNet [6] to extract CNN features from each FLN map. The features are then combined using the average pooling operation and fed to a fully-connected layer with kernel size $(1, 1, 1)$ to classify each GBM patient into one of the three overall survival time groups (i.e., short-term survival, mid-term survival, and long-term survival).

3 Experiments and Results

3.1 Experimental Settings

Implementation Details. Our proposed method was implemented in PyTorch 1.13.1 on NVIDIA A100 Tensor Core GPUs. The loss function was the standard cross-entropy loss. The Adam optimizer with the weight decay of 10^{-5} was adopted. Three 3D ResNet-based backbones with different numbers of layers (10, 50, and 101) were performed, where the initial learning rates were set as 10^{-4}, 10^{-4}, and 10^{-5}, respectively, and would decrease by a factor of 5 if the classification performance is not improved within 5 epochs. The number of epochs for training was 50, and the batch size was fixed as 64.

Performance Evaluation. We evaluated the classification performance of our proposed method using 235 GBM patients in the BraTS 2020 training dataset, because only these 235 patients had both overall survival time and manual expert segmentation labels of lesions. In all experiments, we conducted five-fold cross-validation ten times in order to reduce the effect of sampling bias. Moreover, the A-LNM was performed ten times randomly to avoid particular data distribution and obtain more reliable results. The classification results were reported in terms of accuracy, macro precision (macro-P), macro recall (macro-R), and macro F1 score (macro-F1), respectively.

3.2 Comparison Studies

Quantitative Comparison of Different Prediction Models. As this paper is the first application of FLN maps in the overall survival time prediction for GBM, comparison among the classification performance of different models using the same type of features, i.e., the A-LNM or the LNM derived FLN maps, is demanded for model selection. To validate the effectiveness of the 3D ResNet-based backbones for GBM survival prediction, we made quantitative comparison of a ridge classifier (RC) with PCA [23], a support vector classifier (SVC) with PCA, a logistic regression (LR) with PCA, and three 3D ResNet-based backbones with different numbers of layers (10, 50, and 101). As presented in Table 1, all the 3D ResNet-based backbones outperformed the other three machine learning-based models. In addition, all the 3D ResNet-based backbones achieved better classification results by using the A-LNM derived FLN maps than the LNM derived FLN maps, which implies that the A-LNM derived FLN maps have stronger predictive power than the LNM derived FLN maps.

Table 1. Quantitative comparison of different prediction models on the BraTS 2020 training dataset. (mean±sd)

Feature	Model	Accuracy	macro-P	macro-R	macro-F1
FLN maps (LNM)	RC with PCA	0.475 ± 0.030	0.320 ± 0.035	0.426 ± 0.014	0.365 ± 0.024
	LR with PCA	0.487 ± 0.029	0.428 ± 0.047	0.441 ± 0.034	0.434 ± 0.039
	SVC with PCA	0.483 ± 0.039	0.328 ± 0.070	0.438 ± 0.050	0.375 ± 0.041
	3D ResNet-10	0.553 ± 0.065	0.460 ± 0.039	0.548 ± 0.042	0.500 ± 0.043
	3D ResNet-50	0.511 ± 0.046	0.359 ± 0.037	0.455 ± 0.034	0.401 ± 0.041
	3D ResNet-101	0.468 ± 0.076	0.319 ± 0.054	0.437 ± 0.040	0.369 ± 0.046
FLN maps (A-LNM)	3D ResNet-10	$\mathbf{0.658} \pm 0.047$	0.604 ± 0.032	$\mathbf{0.639} \pm 0.037$	0.621 ± 0.032
	3D ResNet-50	0.632 ± 0.048	$\mathbf{0.622} \pm 0.054$	0.623 ± 0.038	$\mathbf{0.622} \pm 0.036$
	3D ResNet-101	0.532 ± 0.051	0.379 ± 0.125	0.464 ± 0.037	0.417 ± 0.044

Quantitative Comparison of Different Types of Features. Subsequently, we compared the predictive power of FLN maps and the other four widely used types of features, i.e., clinical features with an LR [27], biophysics features with an SVC [9], radiomics features with a gradient boosting classifier (GBC) [21], and MRI-based features with a 3D CNN model [9]. These competing models were executed following the instructions in their respective papers to achieve the best performance. Quantitative results of different types of features are displayed in Table 2. One can see that FLN maps showed the strongest predictive power on all the four metrics. Specifically, the 3D ResNet-10 backbone with the A-LNM derived FLN maps improved the classification performance by 10.5% to 18.5% in terms of accuracy, which again demonstrates the superiority of the A-LNM derived FLN maps for GBM survival prediction.

Table 2. Quantitative comparison of different types of features on the BraTS 2020 training dataset. (mean±sd; †: p-value< 0.05)

Feature & Model	Accuracy	macro-P	macro-R	macro-F1
clinical features + LR [27]	$0.473 \pm 0.080^{\dagger}$	$0.319 \pm 0.053^{\dagger}$	$0.424 \pm 0.034^{\dagger}$	$0.364 \pm 0.042^{\dagger}$
biophysics features + SVC [9]	$0.498 \pm 0.064^{\dagger}$	$0.338 \pm 0.038^{\dagger}$	$0.446 \pm 0.053^{\dagger}$	$0.385 \pm 0.044^{\dagger}$
radiomics features + GBC [21]	$0.549 \pm 0.050^{\dagger}$	$0.533 \pm 0.036^{\dagger}$	$0.522 \pm 0.027^{\dagger}$	$0.527 \pm 0.031^{\dagger}$
MRI images + 3D DenseNet-121 [9]	$0.545 \pm 0.047^{\dagger}$	$0.525 \pm 0.031^{\dagger}$	$0.513 \pm 0.035^{\dagger}$	$0.519 \pm 0.032^{\dagger}$
FLN maps (LNM) + 3D ResNet-10	$0.553 \pm 0.065^{\dagger}$	$0.460 \pm 0.039^{\dagger}$	$0.548 \pm 0.042^{\dagger}$	$0.500 \pm 0.043^{\dagger}$
FLN maps (A-LNM) + 3D ResNet-10	$\mathbf{0.658} \pm 0.047$	$\mathbf{0.604} \pm 0.032$	$\mathbf{0.639} \pm 0.037$	$\mathbf{0.621} \pm 0.032$

3.3 Brain Regions in Relation to GBM Survival

To identify the most discriminative brain regions associated with overall survival time in GBM, we estimated the relative contribution of each voxel to the classification performance in our proposed method by using the Grad-CAM [10]. To obtain steady results, as shown in Fig. 2(A), the voxels with top 5% weights in the class activation maps (CAMs) of all candidate models were overlapped by class, and the position covered by more than half of the models is displayed. The CAMs of three classes of survivors overlapped in Fig. 2(B) where both coincident and non-coincident areas exist.The association of an increased degree of invasion within the frontal lobe with decreased survival time can be observed, which is in concordance with a previous study [20]. Patients whose frontal lobe is affected by tumors showed more executive dysfunction, apathy, and disinhibition [11]. On the dominant left hemisphere, the CAMs of long-term survivors and mid-term survivors overlapped at the superior temporal gyrus and Wernicke's area which are involved in the sensation of sound and language comprehension respectively, and have been associated with decreased survival in patients with high-grade glioma [26]. In addition, the CAM of mid-term survivors covered more areas of the middle and inferior temporal gyri which were considered as one of the higher level ventral streams of visual processing linked to facial recognition [25].

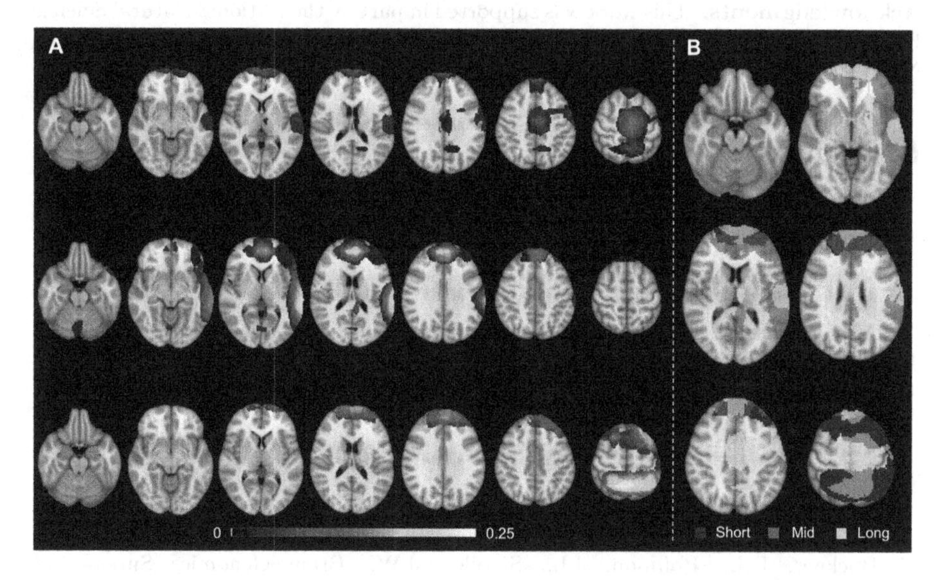

Fig. 2. (A) The processed class activation maps of long-term, mid-term, and short-term survivors from the top down. The color scale indicates the average weights of class features. (B) The overlapped class activation map of long-term, mid-term, and short-term survivors where the purple color, orange color, and yellow color represent the map of short-term, mid-term, and long-term survivors respectively. (Color figure online)

4 Conclusion

In this paper, we introduce a novel neuroimaging feature family, called A-LNM derived FLN maps, for overall survival time prediction of GBM patients. A-LNM was presented to generate plenty of FLN maps for each GBM patient by partitioning the FDC maps obtained from resting-state fMRI of 1000 GSP healthy subjects into disjoint subsets of equal size and averaging each subset. We applied a 3D ResNet-based backbone network to extract features from the generated FLN maps and classify GBM patients into three overall survival time groups. Experimental results on the BraTS 2020 training dataset validated the effectiveness of the A-LNM derived FLN maps for GBM survival prediction. Moreover, a visualization analysis implemented by the Grad-CAM revealed the brain regions associated with GBM survival. In future work, we will try to fuse the FLN maps and MRI-based radiomics features to study their combined predictive power for GBM survival analysis.

Acknowledgments. This work was supported in part by the National Natural Science Foundation of China under Grant 62202442, and in part by the Anhui Provincial Natural Science Foundation under Grant 2208085QF188.

References

1. Adeberg, S., Bostel, T., König, L., Welzel, T., Debus, J., Combs, S.E.: A comparison of long-term survivors and short-term survivors with glioblastoma, subventricular zone involvement: a predictive factor for survival? Radiat. Oncol. **9**, 1–6 (2014)
2. Anand, V.K., et al.: Brain tumor segmentation and survival prediction using automatic hard mining in 3D CNN architecture. In: Crimi, A., Bakas, S. (eds.) BrainLes 2020. LNCS, vol. 12659, pp. 310–319. Springer, Cham (2021). https://doi.org/10.1007/978-3-030-72087-2_27
3. Avants, B.B., Tustison, N., Song, G., et al.: Advanced normalization tools (ants). Insight J **2**(365), 1–35 (2009)
4. Bowren, M., Jr., et al.: Post-stroke outcomes predicted from multivariate lesion-behaviour and lesion network mapping. Brain **145**(4), 1338–1353 (2022)
5. Buckner, R.L., Roffman, J.L., Smoller, J.W.: Brain Genomics Superstruct Project (GSP) (2014). https://doi.org/10.7910/DVN/25833,https://doi.org/10.7910/DVN/25833
6. Chen, S., Ma, K., Zheng, Y.: Med3D: Transfer learning for 3D medical image analysis. arXiv preprint arXiv:1904.00625 (2019)
7. Daniel, A.G., et al.: Functional connectivity within glioblastoma impacts overall survival. Neuro-oncology **23**(3), 412–421 (2021)
8. Fox, M.D.: Mapping symptoms to brain networks with the human connectome. N. Engl. J. Med. **379**(23), 2237–2245 (2018)
9. González, S.R., Zemmoura, I., Tauber, C.: 3D brain tumor segmentation and survival prediction using ensembles of convolutional neural networks. In: Crimi, A., Bakas, S. (eds.) BrainLes 2020. LNCS, vol. 12659, pp. 241–254. Springer, Cham (2021). https://doi.org/10.1007/978-3-030-72087-2_21

10. Gotkowski, K., Gonzalez, C., Bucher, A., Mukhopadhyay, A.: M3d-CAM: A pyTorch library to generate 3D data attention maps for medical deep learning (2020)
11. Gregg, N., et al.: Neurobehavioural changes in patients following brain tumour: patients and relatives perspective. Support. Care Cancer **22**, 2965–2972 (2014)
12. Hu, Z., et al.: A deep learning model with radiomics analysis integration for glioblastoma post-resection survival prediction. arXiv preprint arXiv:2203.05891 (2022)
13. Kim, N.Y., et al.: Network effects of brain lesions causing central poststroke pain. Ann. Neurol. **92**(5), 834–845 (2022)
14. Kraus, J.A., et al.: Molecular analysis of the PTEN, TP53 and CDKN2A tumor suppressor genes in long-term survivors of glioblastoma multiforme. J. Neurooncol. **48**, 89–94 (2000)
15. Kraus, J.A., et al.: Long-term survival of glioblastoma multiforme: importance of histopathological reevaluation. J. Neurol. **247**, 455–460 (2000)
16. Lamichhane, B., Daniel, A.G., Lee, J.J., Marcus, D.S., Shimony, J.S., Leuthardt, E.C.: Machine learning analytics of resting-state functional connectivity predicts survival outcomes of glioblastoma multiforme patients. Front. Neurol. **12**, 642241 (2021)
17. Liu, L., et al.: Overall survival time prediction for high-grade glioma patients based on large-scale brain functional networks. Brain Imaging Behav. **13**, 1333–1351 (2019)
18. Menze, B.H., et al.: The multimodal brain tumor image segmentation benchmark (BRATS). IEEE Trans. Med. Imaging **34**(10), 1993–2024 (2014)
19. Nie, D., et al.: Multi-channel 3D deep feature learning for survival time prediction of brain tumor patients using multi-modal neuroimages. Sci. Rep. **9**(1), 1103 (2019)
20. Prasanna, P., et al.: Mass effect deformation heterogeneity (MEDH) on gadolinium-contrast T1-weighted MRI is associated with decreased survival in patients with right cerebral hemisphere glioblastoma: a feasibility study. Sci. Rep. **9**(1), 1145 (2019)
21. Rajput, S., Agravat, R., Roy, M., Raval, M.S.: Glioblastoma multiforme patient survival prediction. In: Su, R., Zhang, Y.-D., Liu, H. (eds.) MICAD 2021. LNEE, vol. 784, pp. 47–58. Springer, Singapore (2022). https://doi.org/10.1007/978-981-16-3880-0_6
22. Ricard, D., Idbaih, A., Ducray, F., Lahutte, M., Hoang-Xuan, K., Delattre, J.Y.: Primary brain tumours in adults. Lancet **379**(9830), 1984–1996 (2012)
23. Salvalaggio, A., De Filippo De Grazia, M., Zorzi, M., Thiebaut de Schotten, M., Corbetta, M.: Post-stroke deficit prediction from lesion and indirect structural and functional disconnection. Brain **143**(7), 2173–2188 (2020)
24. Sprugnoli, G., et al.: Tumor bold connectivity profile correlates with glioma patients' survival. Neuro-Oncol. Adv. **4**(1), vdac153 (2022)
25. Talacchi, A., Santini, B., Savazzi, S., Gerosa, M.: Cognitive effects of tumour and surgical treatment in glioma patients. J. Neurooncol. **103**, 541–549 (2011)
26. Thomas, R., O'Connor, A.M., Ashley, S.: Speech and language disorders in patients with high grade glioma and its influence on prognosis. J. Neurooncol. **23**, 265–270 (1995)
27. Wang, S., Dai, C., Mo, Y., Angelini, E., Guo, Y., Bai, W.: Automatic brain tumour segmentation and biophysics-guided survival prediction. In: Crimi, A., Bakas, S. (eds.) BrainLes 2019. LNCS, vol. 11993, pp. 61–72. Springer, Cham (2020). https://doi.org/10.1007/978-3-030-46643-5_6

Exploring Brain Function-Structure Connectome Skeleton via Self-supervised Graph-Transformer Approach

Yanqing Kang, Ruoyang Wang, Enze Shi, Jinru Wu, Sigang Yu, and Shu Zhang(✉)

Center for Brain and Brain-Inspired Computing Research, School of Computer Science, Northwestern Polytechnical University, Xi'an, China
shu.zhang@nwpu.edu.cn

Abstract. Understanding the relationship between brain functional connectivity and structural connectivity is important in the field of brain imaging, and it can help us better comprehend the working mechanisms of the brain. Much effort has been made on this issue, but it is still far from satisfactory. The brain transmits information through a network architecture, which means that the regions and connections of the brain are significant. The main difficulties with this issue are currently at least two aspects. On the one hand, the importance of different brain regions in structural and functional integration has not been fully addressed; on the other hand, the connectome skeleton of the brain, plays the role in common and key connections in the brain network, has not been clearly studied. To alleviate the above problems, this paper proposes a transformer-based self-supervised graph reconstruction framework (TSGR). The framework uses the graph neural network (GNN) to fuse functional and structural information of the brain, reconstructs the brain graph through a self-supervised model and identifies the regions that are important to the reconstruction task. These regions are considered as key connectome regions which play an essential role in the communication connectivity of the brain network. Based on key brain regions, the connectome skeleton can be obtained. Experimental results demonstrate the effectiveness of the proposed method, which obtains key regions and connectome skeleton in the brain network. This provides a new angle of view to explore the relationship between brain function and structure. Our code is available at https://github.com/kang105/TSGR.

Keywords: Brain Function · Brain Structure · Self-supervised · Graph Neural Network · Transformer

1 Introduction

Understanding the relationship between brain functional connectivity and structural connectivity is a key issue in studying the working mechanisms of the brain [1, 2]. In the early stage, due to the dynamic variability nature of functional connectivity and the unique stability of structural connectivity, the two were often analyzed separately [3, 4]. Later,

H. Greenspan et al. (Eds.): MICCAI 2023, LNCS 14227, pp. 308–317, 2023.
https://doi.org/10.1007/978-3-031-43993-3_30

attention was paid to their relationship, and a large number of studies were proposed to analyze this issue [5–7]. For example, Greicius *et al.* [5] found that higher structural connectivity tended to be accompanied by higher functional connectivity; MIŠIĆ *et al.* [6] used a multimodal approach to correlate structural and functional connectivity with each other using partial least squares analysis while searching for the best covariance pattern of both; Sarwa *et al.* [7] adopted a deep learning framework for structural connectivity to functional connectivity prediction.

Despite the great success of the above methods, they are far from satisfactory for studying the relationship between brain functional and structural connectivity. Connections in brain regions make up networks, and it is crucial to identify key nodes and connections. Therefore, there are currently at least two difficulties. On the one hand, key brain regions, act as hubs for information transmission in the brain network, have not been completely identified in the joint analysis of brain functional and structural profiles. On the other hand, it has not been clearly studied for the connectome skeleton of brain, which plays a key role in both functional and structural connections of brain.

To overcome the above limitations, we propose a transformer-based graph self-supervised graph reconstruction framework. It can obtain the key connectome regions and skeleton of brain by combining brain function and structure. The method has two main characteristics. For one thing, graph neural networks (GNN) [8] is applied to integrate brain function and structure. We represent the brain as a graph, which its nodes are the regions of interest (ROIs) defined by the atlas. The edge information and node features of the graph come from the structural connectivity and the functional connectivity of the ROIs, respectively. The graph features are propagated through the self-attention mechanism [9] and graph convolution network (GCN) [10]. For the other thing, a self-supervised model is adopted to reconstruct the brain graph and obtain the contribution scores of ROIs for the reconstruction task. ROIs with higher scores are more important for reconstruction and are considered as key connectome regions of brain. We used several functional magnetic resonance imaging (fMRI) from the Human Connectome Project (HCP) 900 datasets [11], combined them with structure connectivity from diffusion-weighted MRI (dMRI) separately, and acquired the corresponding key connectome regions. Based on key regions, we obtained the connectome skeleton of the brain.

Experimental results demonstrate the effectiveness of proposed method. First, we obtained low loss values, which indicated that the model achieved the reconstruction task. Second, we obtained the contribution scores of brain ROIs. ROIs with high scores play a role in the transmission of information in the network, are regarded as key connectome regions. Finally, we obtained the connectome skeleton of the brain, which expresses the most key connections of the brain both in functional and structural networks.

2 Method

2.1 Overview

The pipeline of the transformer-based self-supervised graph reconstruction (TSGR) framework proposed in this paper is shown in Fig. 1. The method analyzes key connectome regions of brain in the joint analysis of brain functional and structural profiles.

The brain is represented as a graph and used as input for the ScorePool-AE module, then the reconstructed graph and contribution scores of the nodes can be obtained. The framework consists of three main parts, i.e., graph generation, ScorePool-AE module, and reconstruction target. In the graph generation part, we adopt Destrieux Atlas to initialize the brain surface into 148 ROIs and use them as nodes of the graph. Based on the ROIs, functional and structural information are utilized to generate the node features and edge features of the graph, respectively. In the ScorePool-AE module, we design ScorePool to get the contribution score of each node in the graph. In the reconstruction target part, the mean square error (MSE) between the node features of the input graph and the reconstructed graph is applied as the loss function.

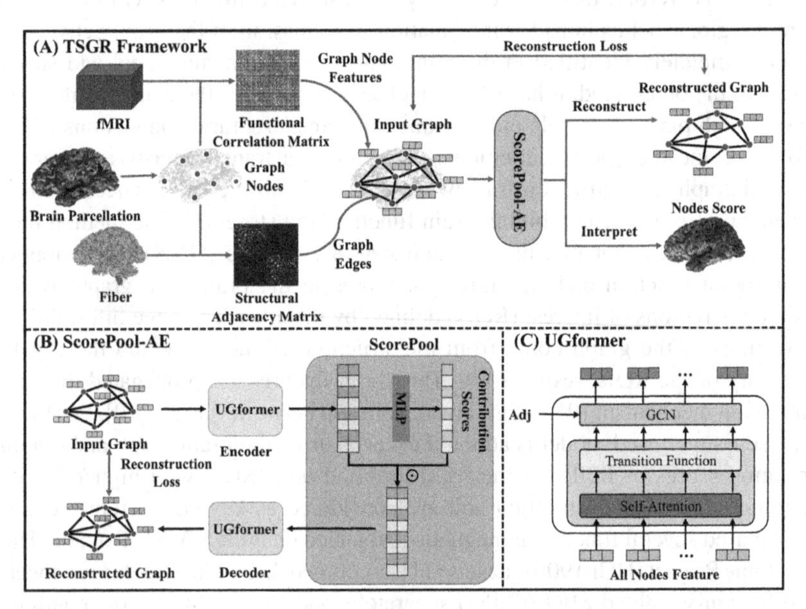

Fig. 1. The pipeline of the proposed TSGR framework. (A) represents the process of the framework, including three parts, i.e., graph generation, ScorePool-AE module, and reconstruction target. (B) represents the ScorePool-AE module architecture. (C) represents the UGformer.

2.2 Data and Preprocess

We used the HCP 900 dataset and randomly selected 98 subjects from it. T1-weighted MRI data were used to reconstruct the brain cortical surface, dMRI were utilized to reconstruct the fiber bundles from white matter, resting-state fMRI (rs-fMRI) and task fMRI showed the functional changes of the brain. Among them, task fMRI data contains a total of seven tasks, which are EMOTION, GAMBLING, LANGUAGE, MOTOR, RELATION, SOCIAL and WM.

Standard Freesurfer pipeline including tissue-segmentation and white matter surface (inner surface) reconstruction [12] is proposed to preprocess the T1-weighted MRI. For rs-fMRI, we adopted the graycoordinate system [13] as the platform to extract rs-fMRI time sequence for each surface vertex. For dMRI, we followed the method in Van den Heuvel *et al.* [14] to use deterministic tractography to derive white matter fibers, and reconstructed 5×10^4 fiber tracts for each subject.

To jointly use these three modalities, we aligned them into the same space. A linear image registration method (FLIRT) [15] and a nonlinear one (FNIRT) [16] were cascaded to transform and warp T1-weighted MRI to the dMRI space. The preprocessed rs-fMRI uses graycoordinate system, which is located at the same space as T1-weighted MRI, and the vertex-wise correspondence between them can be directly established.

2.3 TSGR Framework

Graph Generation. We define the brain as an undirected graph $G = \{V, E\}$. $V = \{v_i | i \in 1, 2, \ldots, n\}$ represents the set of nodes of the graph. $F \in \mathbb{R}^{n \times D}$ represents the feature matrix of graph nodes. $E \in \mathbb{R}^{n \times n}$ represents the adjacency matrix of the graph.

Graph Nodes with Generated Features. The brain surface is divided into 148 ROIs and used as nodes of the graph. fMRI signals are selected to express node features. We compute the Pearson correlation coefficients of the signals among all ROIs to obtain the functional similarity matrix and use it as the feature matrix F of the graph nodes. The length of each node feature D is 148.

Graph Edge. For each pair of ROIs, we calculate the total number of fibers directly connected of them and then divide it by the geometric mean of their areas, thus obtaining the structural connectivity matrix S. We set the threshold t_s for S to make it sparse and do the binarization to obtain E. $E_{i,j} = 1$ means that nodes i and j are connected, otherwise $E_{i,j} = 0$.

ScorePool-AE Module. In ScorePool-AE module, we adopt the encoder-decoder structure to implement the graph reconstruction task. The encoder is applied to extract node representations of the graph, then the ScorePool is used to obtain the contribution scores of nodes, and finally the decoder is used to reconstruct the node features of the graph.

The encoder is based on the Ugformer [17], which implements the self-attention mechanism on the nodes of the graph via transformer. For one of the layers of UGformer, the self-attention mechanism is adopted on all nodes of the graph rather than on neighboring nodes, then GCNs is applied to exploit the structural information of the graph. The process is shown as Eq. (1) and (2).

$$H'^{(k)} = Attention_V\left(H^{(k)}Q^{(k)}, H^{(k)}K^{(k)}, H^{(k)}V^{(k)}\right) \tag{1}$$

$$H^{(k+1)} = GCN\left(E, H'^{(k)}\right) \tag{2}$$

where $H^{(k)}$ is the node representations of the graph at the k-th layer of UGformer. V is the set of all nodes of the graph. $Q^{(k)}$, $K^{(k)}$, $V^{(k)} \in \mathbb{R}^{d_0 \times d}$ are the projection matrices.

Inspired by the pooling operation of the GNN [18], we design the ScorePool to get the contribution score of each node in the graph for the reconstruction task. ScorePool comes after encoder and before decoder. The output of encoder is the graph node representations $X \in \mathbb{R}^{n \times d}$, which is used as the input of the ScorePool. After an MLP layer, X is mapped to $Y \in \mathbb{R}^{n\times 1}$. Y is seen as the scores of graph nodes. Finally, we do the element-wise product operation of X and Y, and use it as the input of the decoder. It is calculated as Eqs. (3) and (4). The decoder has the same structure as the encoder and is utilized to reconstruct the node features.

$$Y = softmax(MLP(X)) \tag{3}$$

$$X\prime = X \odot Y \tag{4}$$

where $\odot$ denotes the element-wise product and $X\prime$ denotes the input of the decoder.

Reconstruction Target. Our TSGR framework reconstructs the graph by predicting the nodes features of the graph. The loss function is MSE between the feature matrix $F\prime$ of reconstructed graph nodes and the feature matrix F of input graph nodes.

2.4 Analyzing Brain Key Connectome ROIs and Hierarchical Networks

In this work, we propose a method that identifies key connectome ROIs and can be applied to the hierarchical analysis of brain networks. First, we obtained the average contribution scores of brain ROIs for all individuals. Second, we verify whether the ROIs with high scores are key connectome ROIs in the functional and structural networks. The functional network is obtained by averaging F over all individuals and setting a threshold. The structural network is the same operation done for S. We use three network centrality metrics, i.e., degree centrality, closeness centrality and PageRank centrality, to measure key connectome ROIs. In addition, we calculate the number of all ROIs participating in the functional network, where the functional networks are obtained by dictionary learning [19]. Finally, we divide all brain ROIs into 3 scales and analyze the characteristics of networks consisting of ROIs at different scales, namely, the connectivity strength and global efficiency of the networks.

2.5 Exploring the Connectome Skeleton of Brain ROIs

This study explores the connectome skeleton of the brain based on important ROIs. The connectome skeleton plays the role in common connections in the brain network and is a core connectivity pattern. First, we obtain the key functional connections of eight (one resting-state and seven tasks) functional networks. Specifically, we calculate the eight functional network connections consisting of Scale-1 and Scale-2 ROIs, and a connection is selected if it exists simultaneously in six and more functional networks. Then, we obtain the intersection between the key functional connections and the structural network connection as the connectome skeleton. Finally, we analyze the connectome skeleton, namely, counting the strength of connectivity between brain regions and the length of fibers in the connectome skeleton.

3 Experiments and Results

3.1 Experimental Performance

Experimental Setup. A total of 98 individuals is evaluated by the proposed method. Each of individual has eight brain graphs, which have the same topology corresponding to the structural information and different node features corresponding to eight kinds of functional information. We set up eight sets of experiments, each of which is trained independently. Each set of experiments use one of the brain graphs from 98 individuals, which corresponded to the same kind of functional information. These eight sets of experiments are called REST, EMOTION, GAMBLING, LANGUAGE, MOTOR, RELATION, SOCIAL and WM. We set the number of layers of the UGformer in the encoder and decoder to 1, and the number of heads in the self-attention to 4. The model is trained for 200 epochs using the Adam optimizer to update parameters.

Reconstruction Effect. The final loss values of experiments obtained are low. To verify the effect of reconstruction graph, we calculated Correlation and SSIM [20] between the input node features and the reconstructed node features for all individuals, as shown in Fig. 2. We find out that for all eight sets of experiments, the average Correlation and SSIM of all individuals are around 0.9997. This indicates that our method achieves the reconstruction task and we can further analyze based on experimental results.

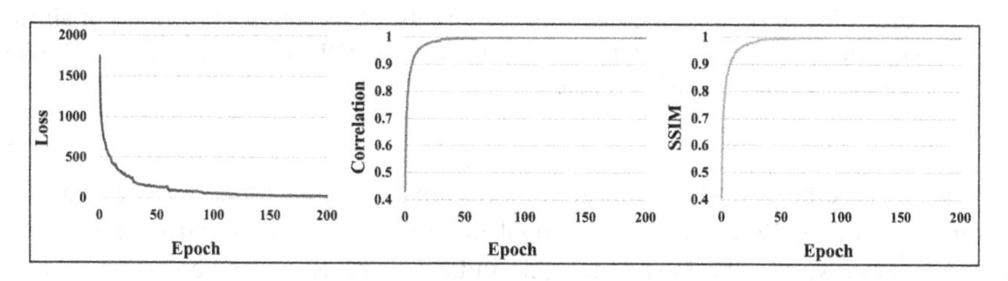

Fig. 2. Experimental performance of REST

3.2 Analysis of Key Brain ROIs and Network Hierarchy

For each set of experiments, the contribution scores of brain ROIs are obtained by averaging the scores of all individuals, as shown in Fig. 3. According to the scores from high to low, we divide the ROIs into 3 scales. Each scale contains 20%, 30%, and 50% of the ROIs, and the corresponding number of ROIs is 30, 44, and 74 respectively. Scale-1 ROIs are considered as key connectome ROIs. Scale-2 and Scale-3 are in descending order of importance in brain connectivity.

Figure 3(B) shows Scale-1 ROIs in eight sets of experiments, which are basically distributed in the superior frontal, parietal, and occipital of the lateral brain, with little distribution in the medial and bottom parts of the brain. These regions have important biological significance, for example, Heuvel *et al.* confirmed that the parietal and prefrontal cortex contain multiple hubs in almost all species [21]. Figure 3(D) lists the Scale-1 specific ROIs for all experiments. Some ROIs exist in the Scale-1 of most experiments, such as the F_mid, Precentral, and O_mid, indicating that these ROIs are involved in the functional network of multiple tasks; some ROIs are unique to one experiment, such as the fronto-margin only in the SOCIAL, indicating that this ROI is more important in this task. Figure 3(C) shows the Scale-1 IoU in all experiments, and it can be seen that EMOTION and LANGUAGE have the highest value, indicating that the functional activation of these two tasks is similar.

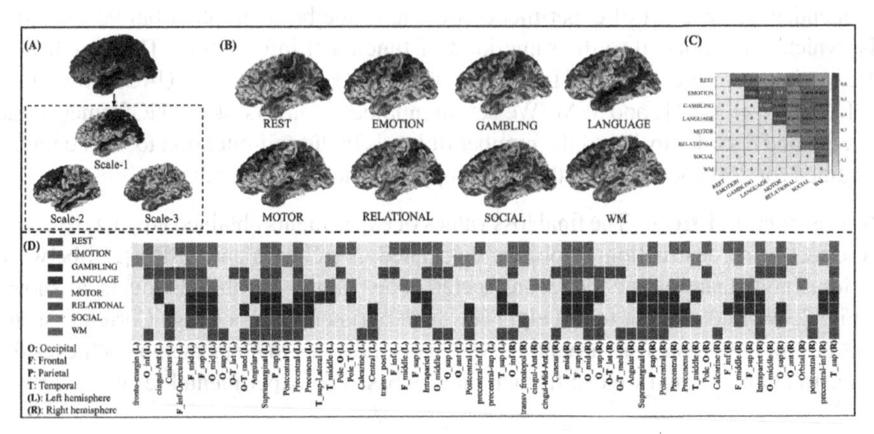

Fig. 3. Visualization of contribution scores of brain ROIs. (A) represents the scores of all ROIs and division of ROIs at REST. (B) and (D) show the Scale-1 ROIs in all experiments. (C) shows the IoU of the Scale-1 ROIs in all experiments.

To verify whether the ROIs with high scores are key connectome ROIs, we calculate the centrality of all ROIs in the functional network (orange) and structural network (green) separately, and the horizontal axis indicates the ROIs ranked from lowest to highest scores, as shown in Fig. 4(A). Here we use three centrality metrics, namely degree centrality, closeness centrality, and PageRank centrality. The results show that the metrics are positively correlated with scores of ROIs. We also calculate the number of all ROIs participating in the functional network (purple), and the results are positively correlated with scores of ROIs. This indicates that ROIs with high scores are located in key positions at the network and participate in more functional networks.

In addition, we also investigate the networks composed among the scales. As shown in Fig. 4(B), the functional and structural connectivity strength is relatively similar overall, with decreasing strength of connections in the Scale-1, Scale-2 and Scale-3. We also analyze the global efficiency of the network and the coupling between the functional and structural networks, as shown in Fig. 4(C). Among the functional and structural networks, S1_1 has the highest global efficiency for most of the experiments, followed by S1_2 and S2_2. It can be seen at the bottom of Fig. 4(C), S1_1 has the highest functional-structural coupling. Therefore, we consider the Scale-1 ROIs as key connectome ROIs, and the network they form assumes the main information transfer function in the functional and structural network.

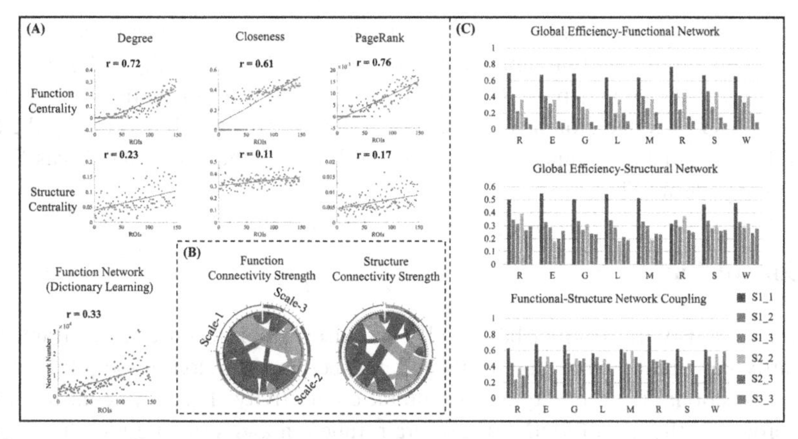

Fig. 4. Analysis of key ROIs and hierarchical network. (A) represents the centrality of all ROIs in the functional and structural networks, and the number of ROIs participating in the functional network in the REST. (B) represents the functional and structural connectivity strength among the 3 scales in the REST. (C) represents the analysis of the network composed among the 3 scales.

3.3 Analysis of Connectome Skeleton in Brain Networks

We obtain the key connection for the functional network of all eight functional networks, where the functional network consists of connections within the ROIs including Scale-1 and Scale-2. Then we combine the key functional connections and the structural connections to obtain the connectome skeleton, as shown in Fig. 5(A).

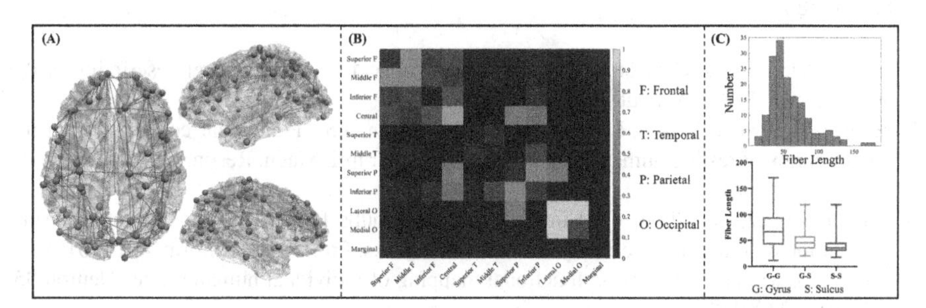

Fig. 5. Analysis of connectome skeleton. (A) represents the visualization of the connectome skeleton. (B) and (C) represents the distribution of connectome skeleton in brain regions and fiber length, respectively.

It shows that the connections are distributed more evenly over the whole brain rather than concentrated in a particular region. From Fig. 5(B), it can be concluded that the intra-occipital and intra-central connections are the most numerous, while the frontal and parietal connections are more evenly distributed. This indicates that the occipital region plays an important role in the connectome skeleton. We count the fiber lengths in the connectome skeleton, as shown in the upper of Fig. 5(C). The fiber lengths are mostly concentrated around 50 mm, and long fibers (fiber lengths greater than 80 mm) account for about 20%, which are more distributed in the frontal region. The high percentage of short fibers improves communication efficiency. The bottom of Fig. 5(C) counts the fiber lengths of the connectome skeleton in the Gyrus-Sulcus connection, in which the fibers of the G-G connection are the longest, followed by the G-S and finally the S-S, which indicates that the connectome skeleton still follows the traditional Gyrus-Sulcus connection pattern [22, 23].

4 Conclusion

We propose a new transformer-based self-supervised graph reconstruction framework to identify key brain connectome ROIs and connectome skeleton in the joint analysis of brain functional connectivity and structural connectivity. The main contribution of the approach is the use of GNN to fuse brain function and structure and the use of a self-supervised model to identify the ROIs that important for graph reconstruction. The experimental results validate the effectiveness of the method. First, we obtain the scores of ROIs. Second, we verify that the ROIs with high scores are key connectome ROIs. Finally, we obtain the connectome skeleton of the brain. It provides a new approach for analyzing the relationship between functional and structural connectivity, and further analysis will be settled in future work.

Acknowledgement. This work was supported by the National Natural Science Foundation of China (62006194).

References

1. Sporns, O., Tononi, G., Kötter, R.: The human connectome: a structural description of the human brain. PLoS Comput. Biol. **1**, e42 (2005)
2. Biswal, B., Zerrin Yetkin, F., Haughton, V.M., Hyde, J.S.: Functional connectivity in the motor cortex of resting human brain using echo-planar mri. Magn. Reson. Med. **34**, 537–541 (1995)
3. Zamora-López, G., Zhou, C., Kurths, J.: Cortical hubs form a module for multisensory integration on top of the hierarchy of cortical networks. Front. Neuroinform. **4**, 1 (2010)
4. Gordon, E.M., et al.: Precision functional mapping of individual human brains. Neuron **95**, 791-807.e7 (2017)
5. Greicius, M.D., Supekar, K., Menon, V., Dougherty, R.F.: Resting-state functional connectivity reflects structural connectivity in the default mode network. Cereb. Cortex **19**, 72–78 (2009)
6. Mišić, B., et al.: Network-level structure-function relationships in human neocortex. Cereb. Cortex **26**, 3285–3296 (2016)
7. Sarwar, T., Tian, Y., Yeo, B.T.T., Ramamohanarao, K., Zalesky, A.: Structure-function coupling in the human connectome: a machine learning approach. Neuroimage **226**, 117609 (2021)
8. Scarselli, F., Gori, M., Tsoi, A.C., Hagenbuchner, M., Monfardini, G.: The graph neural network model. IEEE Trans. Neural Netw. **20**, 61–80 (2009)
9. Dosovitskiy, A., et al.: An Image is Worth 16x16 Words: Transformers for Image Recognition at Scale. arXiv preprint http://arxiv.org/abs/2010.11929 (2021)

10. Kipf, T.N., Welling, M.: Semi-Supervised Classification with Graph Convolutional Networks. arXiv preprint http://arxiv.org/abs/1609.02907 (2017)
11. Van Essen, D.C., et al.: The human connectome project: a data acquisition perspective. Neuroimage **62**, 2222–2231 (2012)
12. Fischl, B., et al.: Whole brain segmentation: automated labeling of neuroanatomical structures in the human brain. Neuron **33**, 341–355 (2002)
13. Glasser, M.F., et al.: The minimal preprocessing pipelines for the human connectome project. Neuroimage **80**, 105–124 (2013)
14. van den Heuvel, M.P., Kahn, R.S., Goñi, J., Sporns, O.: High-cost, high-capacity backbone for global brain communication. Proc. Natl. Acad. Sci. **109**, 11372–11377 (2012)
15. Jenkinson, M., Bannister, P., Brady, M., Smith, S.: Improved optimization for the robust and accurate linear registration and motion correction of brain images. Neuroimage **17**, 825–841 (2002)
16. Jenkinson, M., Beckmann, C.F., Behrens, T.E.J., Woolrich, M.W., Smith, S.M.: FSL. NeuroImage **62**, 782–790 (2012)
17. Nguyen, D.Q., Nguyen, T.D., Phung, D.: Universal Graph Transformer Self-Attention Networks. arXiv preprint http://arxiv.org/abs/1909.11855 (2022)
18. Gao, H., Ji, S.: Graph U-Nets. In: Proceedings of the 36th International Conference on Machine Learning, pp. 2083–2092. PMLR (2019)
19. Mallat, S.G., Zhang, Z.: Matching pursuits with time-frequency dictionaries. IEEE Trans. Signal Process. **41**, 3397–3415 (1993)
20. Wang, Z., Bovik, A.C., Sheikh, H.R., Simoncelli, E.P.: Image quality assessment: from error visibility to structural similarity. IEEE Trans. Image Process. **13**, 600–612 (2004)
21. van den Heuvel, M.P., Sporns, O.: Network hubs in the human brain. Trends Cogn. Sci. **17**, 683–696 (2013)
22. Deng, F., et al.: A functional model of cortical gyri and sulci. Brain Struct. Funct. **219**, 1473–1491 (2014)
23. Nie, J., et al.: Axonal fiber terminations concentrate on gyri. Cereb Cortex. **22**, 2831–2839 (2012)

Vertex Correspondence in Cortical Surface Reconstruction

Anne-Marie Rickmann[1,2(✉)], Fabian Bongratz[2], and Christian Wachinger[1,2]

[1] Lab for Artificial Intelligence in Medical Imaging, Ludwig Maximilians University, Munich, Germany
arickman@med.lmu.de

[2] Department of Radiology, Technical University Munich, Munich, Germany

Abstract. Mesh-based cortical surface reconstruction is a fundamental task in neuroimaging that enables highly accurate measurements of brain morphology. Vertex correspondence between a patient's cortical mesh and a group template is necessary for comparing cortical thickness and other measures at the vertex level. However, post-processing methods for generating vertex correspondence are time-consuming and involve registering and remeshing a patient's surfaces to an atlas. Recent deep learning methods for cortex reconstruction have neither been optimized for generating vertex correspondence nor have they analyzed the quality of such correspondence. In this work, we propose to learn vertex correspondence by optimizing an L1 loss on registered surfaces instead of the commonly used Chamfer loss. This results in improved inter- and intra-subject correspondence suitable for direct group comparison and atlas-based parcellation. We demonstrate that state-of-the-art methods provide insufficient correspondence for mapping parcellations, highlighting the importance of optimizing for accurate vertex correspondence.

1 Introduction

The reconstruction of cortical surfaces from brain MRI scans, a fundamental process in neuroimaging, involves extracting the pial surface (outer cerebral cortex layer) and the white matter surface (white-gray matter boundary). Various methods, including FreeSurfer [8] and CAT12 [5], have been widely employed for cortical surface reconstruction. While a single patient's surface can be used for computing metrics such as cortical thickness, curvature, and gyrification, one of the main objectives of reconstructing cortical surfaces is to perform group comparisons, which are essential for detecting differences in brain structures between patients and healthy control groups. To enable such comparisons, it is necessary to establish point-to-point correspondence between the vertices of a patient's cortical mesh and a group template. This allows for measures such as cortical thickness to be compared at the vertex level. In addition, vertex correspondence enables the mapping of an atlas parcellation from the template onto

Supplementary Information The online version contains supplementary material available at https://doi.org/10.1007/978-3-031-43993-3_31.

H. Greenspan et al. (Eds.): MICCAI 2023, LNCS 14227, pp. 318–327, 2023.
https://doi.org/10.1007/978-3-031-43993-3_31

individual surfaces, which is useful for comparing measures at a regional level. This includes computing cortical thickness for specific parcels, which has wide applications for studying cortex maturation, as well as, aging- and disease-related cortical atrophy [14,17,20,21]. Currently, vertex correspondence is generated in a post-processing step, which is a time-consuming process that typically involves registering and remeshing a patient's surfaces to an atlas [7,9]. Therefore, directly generating surfaces with vertex correspondence would be valuable for fast, reliable, and accurate cortical surface comparison.

Recently, several deep learning methods for cortex reconstruction have emerged, including DeepCSR [4], Vox2Cortex [3], CorticalFlow [13,19], CortexODE [15], and Topofit [11]. These methods can be divided into two categories: (i) *implicit surface reconstruction* methods, and (ii) *explicit template deformation* methods. Implicit surface reconstruction methods represent a 2D surface as a signed distance function and rely on mesh extraction and topology correction, which can be computationally demanding and result in geometric artifacts [3,13].

Explicit deformation approaches for cortical surface reconstruction take a template mesh as input to the deep learning model, where a vertex-wise deformation field is learned conditioned on 3D MRI scans. One advantage of mesh-based methods in cortical surface reconstruction is that the sphere-like topology of neural tissue boundaries can already be incorporated into the template mesh. As a result, there is no need for topology correction during the reconstruction process. The main challenge with these methods is generating smooth and watertight output meshes, i.e., ensuring a diffeomorphic mapping from the input to the output mesh. Researchers have addressed this issue through regularization losses [3,11] or numerical integration of a deformation-describing ODE [13,15,19]. The connectivity of the output mesh is mostly determined by the template mesh, with potential up- or down-sampling of the mesh resolution. Keeping the mesh resolution constant facilitates comparisons between reconstructed surfaces as the output mesh has the same number of vertices and vertex connectivity as the input mesh. Despite maintaining constant mesh resolution, Vox2Cortex [3] (V2C), CorticalFlow++ [19] (CFPP), and Topofit [11] have not focused on optimizing or evaluating the accuracy of vertex correspondence.

In this work, we propose a novel surface reconstruction approach that natively provides correspondence to a template without the need for spherical registration, see Fig. 1. We achieve this by training on meshes with vertex correspondence instead of meshes that vary in the number of vertices and vertex connectivity. For the network to learn these correspondences, we replace the commonly used Chamfer loss with the L1 loss, which has not yet been used for cortical surface reconstruction. We use V2C as our backbone network as it is fast to train and provides white and pial surfaces for both hemispheres with one network, but our approach is generic and can also be integrated in other surface reconstruction methods. We term our method Vox2Cortex with Correspondence (V2CC). We demonstrate that template deformation methods trained with the Chamfer loss, such as V2C, Topofit, and CFPP, provide vertex correspondences that are insufficient for mapping parcellations. Instead, our approach results in improved inter- and intra-subject vertex correspondence, making it suitable for direct group comparison and atlas-based parcellation.

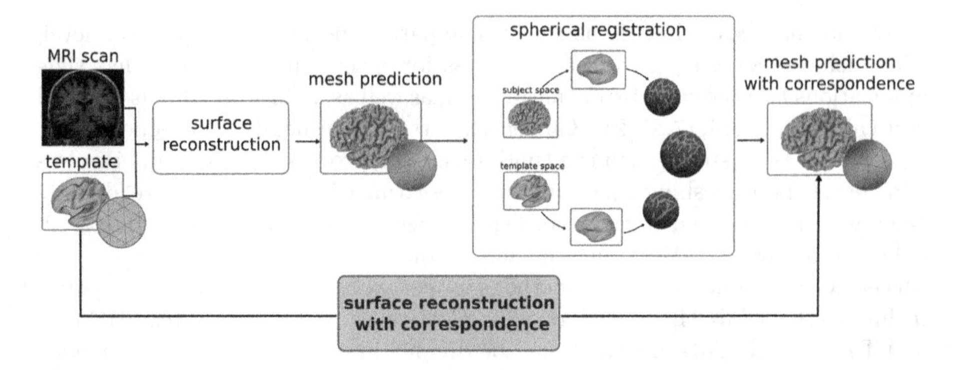

Fig. 1. Top: Overview of existing cortical surface reconstruction approaches, that are dependent on a cumbersome spherical registration as post-processing to obtain vertex correspondence to a template. Bottom: Our approach directly yields surface predictions with correspondence to the input template and does not require any registration.

2 Methods

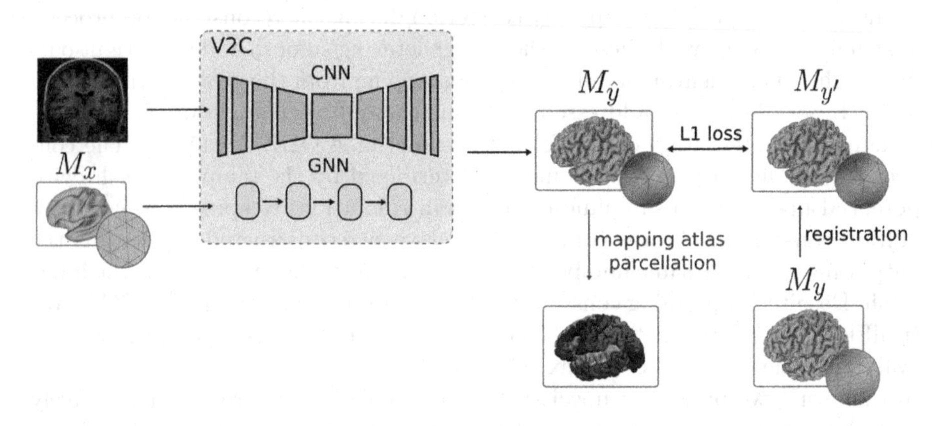

Fig. 2. Overview of our V2CC method. The ground truth mesh is registered to the template mesh in a pre-processing step, allowing to compute the L1 loss on the vertex locations. We use V2C [3] as the surface reconstruction network. Vertex correspondence to the template enables direct mapping of an atlas parcellation at inference.

In cortical surface reconstruction, template deformation methods transform a mesh template to match the neuroanatomy of the given patient. Let M_x be the triangular mesh template, where $M_x = \{\mathbf{V}_x, \mathbf{F}_x, \mathbf{E}_x\}$, contains n vertices, represented as $\mathbf{V}_x \in \mathbb{R}^{n\times 3}$, o faces $\mathbf{F}_x \in \mathbb{R}^{o\times 3}$, storing the indices of the respective vertices that make up the triangles, and r edges $E \in \mathbb{R}^{r\times 2}$, storing the indices of two adjacent faces that share a common edge. The surface reconstruction algorithm computes the displacement $f : \mathbb{R}^{n\times 3} \rightarrow \mathbb{R}^{n\times 3}$ for the set of vertices $\mathbf{V}_x$. In V2C, this displacement is computed by a graph convolutional network, which is conditioned on image features from a convolutional neural network that takes the MRI scan as input. The two sub-networks are connected via feature-sampling modules that map features extracted by the CNN to vertex locations of the meshes. V2C addresses the issue of self-intersections in explicit surface reconstruction methods by incorporating multiple regularization terms into the loss function. Let $M_y = \{\mathbf{V}_y, \mathbf{F}_y, \mathbf{E}_y\}$ be the ground truth mesh, and $M_{\hat{y}} = \{\mathbf{V}_{\hat{y}}, \mathbf{F}_{\hat{y}}, \mathbf{E}_{\hat{y}}\}$ the predicted deformed mesh, where $\hat{\mathbf{V}}_y \in \mathbb{R}^{n\times 3}$, and $\mathbf{V}_y \in \mathbb{R}^{m\times 3}$. Note that $n \neq m$ and therefore there exists no one-to-one mapping for vertex correspondence. The full loss function of V2C consists of a loss term for the CNN and a loss term for the mesh reconstruction, with further details in [3]. Here, we focus on the mesh reconstruction loss, which contains a geometric consistency loss and several regularization terms. The geometric consistency loss $\mathcal{L}_C$ is a curvature weighted Chamfer loss, and is defined as:

$$\mathcal{L}_C(M_y, M_{\hat{y}}) = \frac{1}{|P_y|}\sum_{\mathbf{u}\in P_y} \min_{\mathbf{v}\in P_{\hat{y}}} \|\mathbf{u}-\mathbf{v}\|^2 + \frac{1}{|P_{\hat{y}}|}\sum_{\mathbf{v}\in P_{\hat{y}}} \min_{\mathbf{u}\in P_y} \|\mathbf{v}-\mathbf{u}\|^2, \tag{1}$$

where $P_y \in \mathbb{R}^{q\times 3}$, and $P_{\hat{y}} \in \mathbb{R}^{q\times 3}$ are point clouds sampled from the surfaces of M_y and $M_{\hat{y}}$ respectively. For simplicity, we have omitted the curvature weights. In order to optimize for vertex correspondence, we propose to use a pre-processing step that registers the Mesh M_y to the template mesh M_x, resulting in a resampled ground truth mesh $M_{y'} = \{\mathbf{V}_{y'}, \mathbf{F}_{y'}, \mathbf{E}_{y'}\}$, with $\mathbf{V}_{y'} \in \mathbb{R}^{n\times 3}$ and $\mathbf{F}_{y'} \in \mathbb{R}^{o\times 3}$, where each index $i \in 1, \dots, n$ represents the same anatomical location in both $\mathbf{V}_x$ and $\mathbf{V}_{y'}$. Instead of the Chamfer loss in Eq. (1), we propose the loss function of V2CC as $\mathcal{L}(M_y, M_{y'}) = \mathcal{L}1(M_y, M_{y'}) + \lambda\mathcal{L}_{reg}(M_{\hat{y}})$, where $\mathcal{L}_1$ is the mean absolute distance between corresponding vertices in M_y and My', and $\mathcal{L}_{reg}$ is the normal consistency regularization to avoid self-intersections in $M_{\hat{y}}$. $\mathcal{L}_1$ and $\mathcal{L}_{reg}$ are defined as:

$$\mathcal{L}_1(M_y, M_{y'}) = \frac{1}{n}\sum_{i}^{n}|\mathbf{v}_i - \mathbf{u}_i|, \tag{2}$$

$$\mathcal{L}_{reg}(M_{\hat{y}}) = \frac{1}{|E_{\hat{y}}|}\sum_{a,b\in E_{\hat{y}}}(1-(\hat{\mathbf{n}}_a\cdot\hat{\mathbf{n}}_b))^2, \tag{3}$$

where $\hat{\mathbf{n}}_i$ is the unit normal of the i-th face of $M_{\hat{y}}$. Our method relies on only one regularization term, compared to three in V2C. The regularization factor λ needs to be tuned as a hyperparameter. Our proposed V2CC method and the pre-processing step are presented in Fig. 2.

3 Experiments and Results

Evaluation Metrics: To assess the quality of reconstructed cortical surfaces, we employ four metrics. With the average symmetric Chamfer distance (*cdist*) and the percentage of self-intersecting faces ($\%SIF$), we evaluate the reconstructed surfaces' quality. For evaluating vertex correspondence, we use two different approaches for intra- and inter-subject cases. In *intra-subject cases*, we measure whether the same template vertex moves to the same location when provided with different scans of the same subject. In this case, we use scans that were acquired within a brief period of time to avoid structural changes. We calculate the consistency of vertex locations using the root-mean-square deviation (RMSD) of vertex positions. In *inter-subject cases*, we assess the ability of our method to map pre-defined parcellation atlases, such as the Desikan-Killany-Tourville (DKT) atlas [1,12], onto cortical surfaces. This mapping allows for the assessment of morphological measurements, such as cortical thickness in cortical regions. To evaluate inter-subject vertex correspondence, we directly map vertex classes from the template atlas onto the predicted mesh and calculate the Dice overlap (Dice) to FreeSurfer's silver standard parcellation.

Data: For evaluation, we used the ADNI dataset (available at http://adni.loni.usc.edu), which provides MRI T1 scans for subjects with Alzheimer's disease, mild cognitive impairment, and cognitively normal. After removing scans with processing artifacts, we split the data into training, validation, and testing sets, balanced according to diagnosis, age, and sex. As ADNI is a longitudinal study, only the initial (baseline) scan for each subject was used. Our ADNI subset contains 1,155 subjects for training, 169 for validation, and 323 for testing. We used the TRT dataset [16] to evaluate intra-subject correspondence, which contains 120 MRI T1 scans from 3 subjects, where each subject was scanned twice in 20 days. We further tested generalization to the Mindboggle-101 dataset [12] (101 scans) and the Japanese ADNI (J-ADNI, https://www.j-adni.org) (502 baseline scans). All three datasets contain scans from various scanner vendors, with different field strengths (1.5 and 3T).

Implementation Details: To prepare for training, we pre-processed the data using FreeSurfer v7.2 [8], generating orig.mgz files and white and pial surfaces to use as silver standard ground truth surfaces. We use FreeSurfer's *mri_surf2surf* tool to register surfaces to fsaverage6 (40,962 vertices) and fsaverage (163,842 vertices) template surfaces. We further followed the pipeline of [3,4], registering MRI scans to the MNI152 space. We used public implementations of baseline methods [2,6,18] and made adaptations so that all methods use the same template for training and testing. All models were trained on NVIDIA Titan-RTX or A100 GPUs. The hyperparameter λ was set to 0.003 for white matter surfaces and 0.007 for pial surfaces after grid search. Our code is publicly available[1].

Results and Discussion: We compare the proposed V2CC method to state-of-the-art models V2C, CFPP, and Topofit on the ADNI dataset with FreeSurfer's fsaverage6 right hemisphere templates as an input mesh. All methods were trained using the resampled ground truth meshes. For baseline models we used hyperparameters proposed by the original method. We show the results in the top part of Table 1. Topofit achieves the highest surface reconstruction accuracy

[1] https://github.com/ai-med/V2CC.

Table 1. Comparison of mesh quality of right hemisphere surfaces by average symmetric chamfer distance (cdist) in mm ±std and mean of percentage of self-intersecting faces (% SIF), and vertex correspondences by mean RMSD ±std of vertex positions and Dice overlap of mapped atlas parcellation. All models were trained on the ADNI data. RMSD values were computed on the TRT dataset, all other metrics on the data specified by the data column. Bold numbers indicate best performing methods.

		fsaverage6 template						
		Right Pial			Right WM			Average
Method	data	RMSD↓	cdist↓	% SIF↓	RMSD↓	cdist↓	% SIF↓	Dice↑
V2C [3]	ADNI	1.015 ±0.496	0.437 ±0.0311	1.123	0.961 ±0.447	0.372 ± 0.030	0.185	0.762
CFPP [19]	ADNI	0.884 ±0.353	0.3314 ±0.029	**0.052**	0.778 ±0.294	0.337 ±0.031	**0.013**	0.813
Topofit [11]	ADNI	–	–	–	1.271 ±0.410	**0.180 ±0.030**	0.022	0.838
V2CC only $\mathcal{L}_1$	ADNI	**0.816 ±0.337**	**0.268 ±0.036**	2.880	**0.739 ±0.268**	**0.228 ±0.036**	0.073	**0.921**
V2CC	ADNI	0.825 ±0.360	0.285 ±0.040	1.335	0.748 ±0.285	0.231 ±0.036	0.073	**0.921**
		fsaverage template						
V2C [3]	ADNI	1.139 ±0.569	0.210 ±0.030	3.174	1.010 ±0.485	0.185 ±0.032	0.727	0.823
Topofit [11]	ADNI	–	–	–	1.326 ±0.406	**0.137 ±0.033**	**0.020**	0.871
V2CC	ADNI	**0.911 ±0.404**	**0.192 ±0.029**	**2.981**	**0.821 ±0.326**	0.186 ±0.035	0.110	**0.920**
V2C [3]	Mindb	–	0.305 ±0.045	5.372	–	0.196 ±0.023	1.272	0.780
V2CC	Mindb	-	0.305 ±0.048	4.453	–	0.204 ±0.030	0.157	0.865
V2C [3]	J-ADNI	–	0.262 ±0.046	3.578	–	0.222 ±0.078	1.063	0.803
V2CC	J-ADNI	-	0.262 ±0.048	3.614	–	0.230 ±0.079	0.140	0.913

on the white matter surface, and CFPP has the lowest number of self-intersecting faces pial surfaces. V2C achieves lower surface accuracy compared to CFPP and Topofit and has a higher number of self-intersections. We believe this could be due to longer training time for Topofit and CFPP, 2600 and 1000 epochs, compared to 100 epochs for V2C. When replacing the loss function in V2C with $\mathcal{L}_1$, we interestingly observe an immense boost in surface accuracy, as well as improved inter- and intra-subject correspondence. The disadvantage of using the $\mathcal{L}_1$ alone, is seen in an increase of self-intersecting faces, especially on pial surfaces. Self-intersections of pial surfaces can be reduced by introducing the normal consistency regularizer in Eq. (2).

Next, we trained the baseline V2C model, Topofit, and our V2CC model on higher resolution templates (fsaverage) and images. We present the results on the right hemisphere in the lower section of Table 1. Results for the left hemisphere can be found in the supplementary material. We did not train CFPP on high resolution because of the long training process (about four weeks). We can observe that all models achieve lower surface reconstruction error (cdist) when trained with higher resolution, but also more self-intersections. We believe this is partly due to already existing self-intersections in the fsaverage templates and the resampled ground truth meshes. For the vertex correspondence metrics, we can observe that both baselines, V2C and Topofit, achieve higher parcellation scores than in the low-resolution experiment, but are still outperformed by V2CC. We have further trained a state-of-the-art parcellation model (FastSurfer [10]) on the same dataset, which yields a Dice score of 0.88 ± 0.022, so we

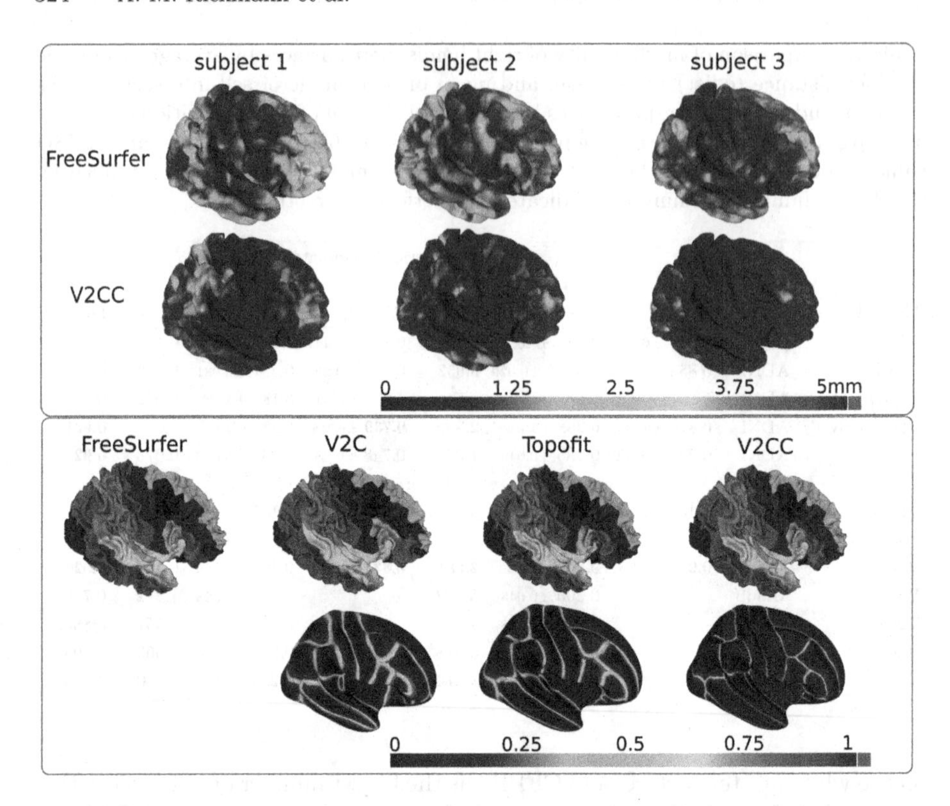

Fig. 3. Top box: vertex RMSD on the TRT dataset. Bottom box: Top: Parcellation examples on a white matter surface of the right hemisphere of an example subject from the ADNI test set. Bottom: Fraction of misclassified vertices over the test set, displayed on the smoothed fsaverage template.

can conclude that our atlas-based parcellation can even outperform dedicated parcellation models. We visualize the quality of intra-subject correspondence of V2CC and FreeSurfer in the top box of Fig. 3, where we display the per-vertex RMSD on each subject's white matter surface of the right hemisphere. We can observe that for all three subjects, our method leads to less variance of vertex positions than FreeSurfer. This is interesting, as FreeSurfer results were registered and resampled to obtain vertex correspondence and our predictions were not. Further, this shows that even though FreeSurfer surfaces have been used as ground truth to train our model, V2CC generalizes well and is more robust to subtle changes in the images. The bottom box in Fig. 3 visualizes the parcellation result of one example subject and the average parcellation error over the whole test set. We observe that parcellation errors occur mainly in boundary regions for all methods, but these boundary regions are much finer in V2CC. To test the generalization ability of our method, we tested V2CC and V2C on two unseen datasets J-ADNI and Mindboggle. We observe, that V2CC achieves better parcellation Dice scores, while the surface reconstruction accuracy is similar for both methods.

Downstream Applications: We hypothesize that our meshes with vertex correspondence to the template can be directly used for downstream applications such as group comparisons or disease classification, without the need for post-processing steps. We performed a group comparison of subjects with Alzheimer's disease (AD) and healthy controls of the ADNI test set, where we compare cortical thickness measures on a per-vertex level. We present a visualization of p-values in Fig. 4. We observe that meshes generated with V2CC highlight similar regions to FreeSurfer meshes. The visualization shows significant atrophy throughout the cortex, with stronger amount of thinning in the left hemisphere which matches findings from studies on cortical thinning in Alzheimer's disease [17,21].We further performed an AD classification study based on thickness measures on the ADNI test set. We computed mean thickness values per parcel (DKT atlas parcellation) for V2CC, FreeSurfer, and the V2C [3] baseline. We show the classification results for AD and controls using a gradient-boosted regression tree, trained on thickness measurements from the ADNI training set. The classifiers achieved 0.810 balanced accuracy (bacc) for Freesurfer, 0.804 bacc for V2CC, and 0.776 bacc, for V2C on the ADNI test set. Demonstrating that V2CC achieves comparable results to FS and outperforms V2C.

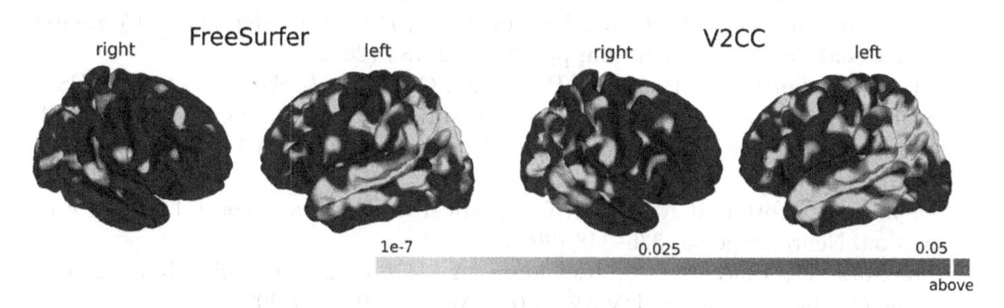

Fig. 4. Group study of per-vertex cortical thickness measures in patients with Alzheimer's disease and healthy controls on the ADNI test set. Colors indicate regions with significantly lower cortical thickness in AD subjects (t-test, one-sided). Note that our predicted meshes can be directly compared on a per-vertex basis while FreeSurfer meshes need to be inflated to a sphere and registered.

4 Conclusion

In this work, we proposed V2CC, a novel approach for cortical surface reconstruction that directly provides vertx correspondence. V2CC utilizes a pre-processing step, where ground truth meshes are registered to a template, and directly learns the correspondences by optimizing an L1 loss instead of the commonly used Chamfer loss. We evaluated V2CC on several datasets, including ADNI,

TRT, Mindboggle-101, and J-ADNI, and compared it to state-of-the-art methods. Our experimental results show that V2CC achieves comparable performance to previous methods in terms of surface reconstruction accuracy. However, V2CC improves intra- and inter-subject correspondence and disease classification based on cortical thickness. We have evaluated our proposed pre-processing step and loss function with V2C as the backbone network, but the underlying concepts are generic and could also be integrated in other methods like Topofit or CFPP.

Acknowledgments. This research was supported by the German Research Foundation and the Federal Ministry of Education and Research in the call for Computational Life Sciences (DeepMentia, 031L0200A). We gratefully acknowledge the computational resources provided by the Leibniz Supercomputing Centre (www.lrz.de).

References

1. Desikan, R.S., et al.: An automated labeling system for subdividing the human cerebral cortex on MRI scans into gyral based regions of interest. NeuroImage **31**(3), 968–980 (2006)
2. Hoopes, A., Greve, D.: TopoFit GitHub repository. https://github.com/ahoopes/topofit. Accessed 04 Mar 2023
3. Bongratz, F., Rickmann, A.M., Pölsterl, S., Wachinger, C.: Vox2Cortex: fast explicit reconstruction of cortical surfaces from 3D MRI scans with geometric deep neural networks. In: Proceedings of the IEEE/CVF Conference on Computer Vision and Pattern Recognition, pp. 20773–20783 (2022)
4. Cruz, R.S., Lebrat, L., Bourgeat, P., Fookes, C., Fripp, J., Salvado, O.: DeepCSR: a 3D deep learning approach for cortical surface reconstruction. In: 2021 IEEE Winter Conference on Applications of Computer Vision (WACV), pp. 806–815 (2021)
5. Dahnke, R., Yotter, R.A., Gaser, C.: Cortical thickness and central surface estimation. Neuroimage **65**, 336–348 (2013)
6. Bongratz, F., Bongratz, F., Rickmann, A. M.: Vox2Cortex github repository. https://github.com/ai-med/Vox2Cortex. Accessed 04 Mar 2023
7. Fischl, B., Sereno, M., Tootell, R., Dale, A.: High-resolution inter-subject averaging and a coordinate system for the cortical surface. Hum. Brain Mapp. **8**(4), 272–284 (1999).https://surfer.nmr.mgh.harvard.edu/ftp/articles/fischl99-morphing.pdf
8. Fischl, B.: FreeSurfer. Neuroimage **62**(2), 774–781 (2012)
9. Fischl, B., Sereno, M.I., Dale, A.M.: Cortical surface-based analysis: II: inflation, flattening, and a surface-based coordinate system. Neuroimage **9**(2), 195–207 (1999)
10. Henschel, L., Conjeti, S., Estrada, S., Diers, K., Fischl, B., Reuter, M.: FastSurfer - a fast and accurate deep learning based neuroimaging pipeline. Neuroimage **219**, 117012 (2020)
11. Hoopes, A., Iglesias, J.E., Fischl, B., Greve, D., Dalca, A.V.: TopoFit: rapid reconstruction of topologically-correct cortical surfaces. In: Medical Imaging with Deep Learning (2021)
12. Klein, A., Tourville, J.: 101 labeled brain images and a consistent human cortical labeling protocol. Front. Neurosci. **6**, 171 (2012). https://doi.org/10.3389/fnins.2012.00171

13. Lebrat, L., et al.: CorticalFlow: a diffeomorphic mesh transformer network for cortical surface reconstruction. In: Advances in Neural Information Processing Systems, vol. 34 (2021)
14. Lerch, J.P., Pruessner, J.C., Zijdenbos, A.P., Hampel, H., Teipel, S.J., Evans, A.C.: Focal decline of cortical thickness in Alzheimer's disease identified by computational neuroanatomy. Cereb. Cortex **15**(7), 995–1001 (2005)
15. Ma, Q., Li, L., Robinson, E.C., Kainz, B., Rueckert, D., Alansary, A.: CortexODE: learning cortical surface reconstruction by neural odes. IEEE Trans. Med. Imaging **42**, 430–443 (2022)
16. Maclaren, J.R., Han, Z., Vos, S.B., Fischbein, N.J., Bammer, R.: Reliability of brain volume measurements: a test-retest dataset. Sci. Data **1**, 1–9 (2014)
17. Roe, J.M., et al.: Asymmetric thinning of the cerebral cortex across the adult lifespan is accelerated in Alzheimer's disease. Nat. Commun. **12**(1), 721 (2021)
18. Santa Cruz, R., et al.: CorticalFlow++. https://bitbucket.csiro.au/projects/CRCPMAX/repos/corticalflow/browse. Accessed 04 Mar 2023
19. Santa Cruz, R., et al.: CorticalFlow++: boosting cortical surface reconstruction accuracy, regularity, and interoperability. In: Wang, L., Dou, Q., Fletcher, P.T., Speidel, S., Li, S. (eds.) MICCAI 2022. LNCS, vol. 13435, pp. 496–505. Springer, Cham (2022). https://doi.org/10.1007/978-3-031-16443-9_48
20. Shaw, M.E., Sachdev, P.S., Anstey, K.J., Cherbuin, N.: Age-related cortical thinning in cognitively healthy individuals in their 60s: the path through life study. Neurobiol. Aging **39**, 202–209 (2016). https://doi.org/10.1016/j.neurobiolaging.2015.12.009
21. Singh, V., Chertkow, H., Lerch, J.P., Evans, A.C., Dorr, A.E., Kabani, N.J.: Spatial patterns of cortical thinning in mild cognitive impairment and Alzheimer's disease. Brain **129**(11), 2885–2893 (2006)

Path-Based Heterogeneous Brain Transformer Network for Resting-State Functional Connectivity Analysis

Ruiyan Fang[1], Yu Li[1], Xin Zhang[1,2(✉)], Shengxian Chen[1], Jiale Cheng[1,3], Xiangmin Xu[1,2], Jieling Wu[4], Weili Lin[3], Li Wang[3], Zhengwang Wu[3], and Gang Li[3(✉)]

[1] School of Electronic and Information Engineering, South China University of Technology, Guangzhou, Guangdong, China
eexinzhang@scut.edu.cn
[2] Pazhou Lab, Guangzhou, China
[3] Department of Radiology and Biomedical Research Imaging Center, University of North Carolina at Chapel Hill, Chapel Hill, NC, USA
gang_li@med.unc.edu
[4] Guangdong Maternal and Child Health Center, Guangzhou, China

Abstract. Brain functional connectivity analysis is important for understanding brain development, aging, sexual distinction and brain disorders. Existing methods typically adopt the resting-state functional connectivity (rs-FC) measured by functional MRI as an effective tool, while they either neglect the importance of information exchange between different brain regions or the heterogeneity of brain activities. To address these issues, we propose a Path-based Heterogeneous Brain Transformer Network (PH-BTN) for analyzing rs-FC. Specifically, to integrate the path importance and heterogeneity of rs-FC for a comprehensive description of the brain, we first construct the brain functional network as a path-based heterogeneous graph using prior knowledge and gain initial edge features from rs-FC. Then, considering the constraints of graph convolution in aggregating long-distance and global information, we design a Heterogeneous Path Graph Transformer Convolution (HP-GTC) module to extract edge features by aggregating different paths' information. Furthermore, we adopt Squeeze-and-Excitation (SE) with HP-GTC modules, which can alleviate the over-smoothing problem and enhance influential features. Finally, we apply a readout layer to generate the final graph embedding to estimate brain age and gender, and thoroughly evaluate the PH-BTN on the Baby Connectome Project (BCP) dataset. Experimental results demonstrate the superiority of PH-BTN over other state-of-the-art methods. The proposed PH-BTN offers a powerful tool to investigate and explore brain functional connectivity.

Keywords: Path-based heterogeneous network · Heterogeneous path graph transformer convolution · Brain functional connectivity analysis

Supplementary Information The online version contains supplementary material available at https://doi.org/10.1007/978-3-031-43993-3_32.

H. Greenspan et al. (Eds.): MICCAI 2023, LNCS 14227, pp. 328–337, 2023.
https://doi.org/10.1007/978-3-031-43993-3_32

1 Introduction

Brain functional network refers to the integrator of information exchange between different neurons, neuron clusters or brain regions. It can not only reveal the working mechanism and developmental changes of the brain [24], but also reflect anatomical connectivity of brain structure [12], making it a hot topic of neuroscience research in recent years. Current research [15,19,32] based on brain functional network mainly focus on two directions: brain physiological basis and brain diseases. Brain diseases are often associated with abnormal connections and have been shown to be related to physiological basis, especially age and sex [2,4,9]. In neuroscience, age and gender prediction based on brain functional network would be the basis for better studying brain diseases, and understanding and exploring the operating mechanisms of baby brains.

As a powerful neuroimaging tool, the resting-sate functional Magnetic Resonance Imaging (rs-fMRI) constructs brain functional networks by capturing the changes of blood oxygen level-dependent (BOLD) signals and computing their correlation between different regions of interest (ROIs) [5]. Owing to the benefits of non-invasive and high-resolution of rs-fMRI, resting-state functional connectivity (rs-FC) derived from BOLD signals is increasingly used to analyze brain age and gender [6]. Up to now, rs-FC analysis methods have primarily included correlation-based methods and graph-based approaches [26]. Compared with correlation-based methods, representing rs-FC data as a graph can preserve the natural topological properties of brain networks, where the nodes are defined as ROIs by an atlas, and the edges are calculated as pairwise correlations between ROIs. However, the assumptions of most existing graph-based methods are still far from the reality of the human brain with the following limitations:

Ignoring Path Importance. Some studies have indicated that connectivity is the core of the brain and no neuron is an island [1,21]. Similar as in the graph theory, these connections can be defined as paths in a brain graph. However, popular methods PR-GNN [18] and BrainGNN [17] mainly focused on node features, which ignored the significance of path features. Despite the fact that BrainNetCNN [14] built an edge-based brain network which can be viewed as a special path-based network, the edge-based description can only depict the direct connections, but cannot account for a variety of indirect connections between brain regions.

Neglecting Heterogeneity. Although BC-GCN [19] formulated brain network as a path-based graph, the graph was homogeneous with only one type of path. In fact, brain functional network is heterogeneous, as proved by abundant studies [7, 22,30].

Overlooking Global Structures. Attention mechanism can help the model focus on crucial connections. Inspired by this, GAT model [25] was employed to analyze brain networks in [31], but limited by graph convolution, it only considered the local structures of neighboring nodes. According to emerging research, global and local structures provide different views for brain network analysis [8,28], and global brain information concerned by Graph Transformer can further improve predictive performance [3,13].

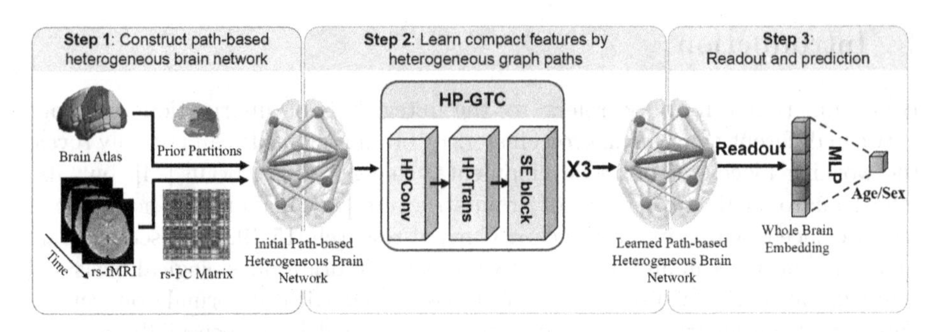

Fig. 1. An overview of PH-BTN for brain age or gender prediction. The color only represents type of paths and the thickness of edges/paths reflects the change of features.

To address these issues, we propose to construct a **P**ath-based **H**eterogeneous **B**rain **T**ransformer **N**etwork (PH-BTN) from rs-fMRI data, learn compact edge features by heterogeneous graph paths and finally readout for brain rs-FC analysis. The proposed PH-BTN compensates the graph neural network (GNN) with the capacity of modeling heterogeneous and global information of the brain graph by designing the **H**eterogeneous **P**ath **G**raph **T**ransformer **C**onvolution (HP-GTC) module with multi-path hypothesis. The major contributions of this work are highlighted below:

- We offer a new perspective on modeling the brain network as a heterogeneous graph with multiple types of path-based features under the prior knowledge of brain partitions, which takes into account the path significance and heterogeneity of the brain, and better simulates the brain network.
- We develop a novel Graph Transformer Network, namely PH-BTN, which is unprecedently able to encode path-based heterogeneous brain network with Transformer to enforce neighbors while incorporating global information. The core component in PH-BTN, namely HP-GTC module, is combined with heterogeneous graph convolution and attention, which can aggregate rich and crucial path information to generate compact brain representation. Furthermore, the Squeeze-and-Excitation (SE) block [11] adopted in HP-GTC module can alleviate the over-smoothing problem of GNN.
- We conduct extensive experiments on the Baby Connectome Project (BCP) dataset to verify the superiority of our proposed method compared with other state-of-the-art methods, and explore the age and gender relevance to the brain functional network.

2 Methodology

An overview of the proposed PH-BTN is illustrated in Fig. 1. Below, we first introduce the construction of the path-based heterogeneous brain network. Then, we focus on the HP-GTC module and elaborate on the novel design and its two components with SE block, i.e., heterogeneous graph path convolution and transformer layer. Finally, we briefly describe the readout and prediction stages.

2.1 Path-Based Heterogeneous Graph Generation

Path-Based Heterogeneous Brain Network. To retain the heterogeneity and path-based structure, we encode the brain functional network captured by rs-fMRI as a path-based heterogeneous graph $\mathcal{G} = (\mathcal{V}, \mathcal{E}, \mathcal{P})$, where $\mathcal{V} = \{v_i\}_{i=1}^{N} \in \mathbb{R}^N$ is the node set of size N defined by ROIs on a specific brain atlas, $\mathcal{E} = [e_{ij}] \in \mathbb{R}^{N \times N \times D}$ is the edge set constructed by using the Pearson's correlation coefficient (PCC) between a sub-series of BOLD signals between nodes, with each edge e_{ij} initialized with D-dimensional edge features h_{ij}, and $\mathcal{P}$ is the path set along with a path type mapping function $\psi : \mathcal{P} \rightarrow \mathcal{R}$, where $\mathcal{R}$ denotes the graph path types, $|\mathcal{R}| \geq 2$.

Graph Path. In graph theory, a graph path p is composed of a finite sequence of n edges, where n denotes that this path is a n-hop path p^n . For example, a 2-hop path p^2_{ikj} between node v_i and v_j can be represented as $p^2_{ikj} = \{e_{ik}, e_{kj}\}$, where e_{ij} is the edge between v_i and v_j and $i \neq j \neq k$. In particular, the 0-hop path indicates the self-loop of the node. According to ablation results of multi-hops [20], we can see that graph paths with finite hops contain enough effective information. To simplify, we limit the highest hop to 2. Then, the multi-hop paths between node v_i and v_j can be defined as $P_{ij} := (P^0_{ij}, P^1_{ij}, P^2_{ij})$, where P^n_{ij} means the set of all n-hop paths between node v_i and v_j, $P^0_{ij} = \{e_{ij} | i = j\}$ and $P^1_{ij} = \{e_{ij} | i \neq j\}$. Particularly, when combining P^0_{ij} and P^1_{ij}, we can gain the sequence $\{e_{ik}, e_{kj}\}$, where $i = k \neq j \ or \ i \neq k = j$. This conjunctive sequence of P^0_{ij} and P^1_{ij} can be regarded as a special 2-hop paths $\{p_{ikj} | i = k \neq j \ or \ i \neq k = j\}$, where k denotes the index of intermediate node. Thus, the multi-hop paths between node v_i and v_j can be recorded as $P_{ij} := P_{ikj} = \{p_{ikj} | \forall k \in N\}$.

Heterogeneous Graph Path. In order to obtain the type of paths, we introduce a brain partition (e.g., frontal, parietal, temporal, occipital and insular) as prior to define heterogeneous paths. It is intuitive to define the types of graph path by edges. However, with the increase of edges, the complexity of the path type definition will increase greatly. Thus, we directly regard the type of intermediate node v_k as the type of graph path p_{ikj}. For instance, if v_k in the frontal lobe, we have $\psi(p_{ikj}) = frontal$. Considering the fact that the feature spaces of different types of paths are not completely irrelevant (as shown in Fig. 3-(a)), we have additionally defined all graph path as the base type, which can reflect the common information shared by different types of graph paths. For example, when using the above-mentioned partition as prior to define path type, we have $|\mathcal{R}| = |\{base, frontal, parietal, temporal, occipital, insular\}| = 6$.

2.2 HP-GTC: Heterogeneous Path Graph Transformer Convolution Module to Learn Compact Features

After brain network construction, we present the HP-GTC module to extract and aggregate specific and common features from different types of graph paths. As illustrated in Fig. 2, the HP-GTC module consists of two layers, which are **H**eterogeneous Graph **P**ath **Conv**olution (HPConv) layer and **H**eterogeneous Graph **P**ath **Trans**former (HPTrans) layer respectively.

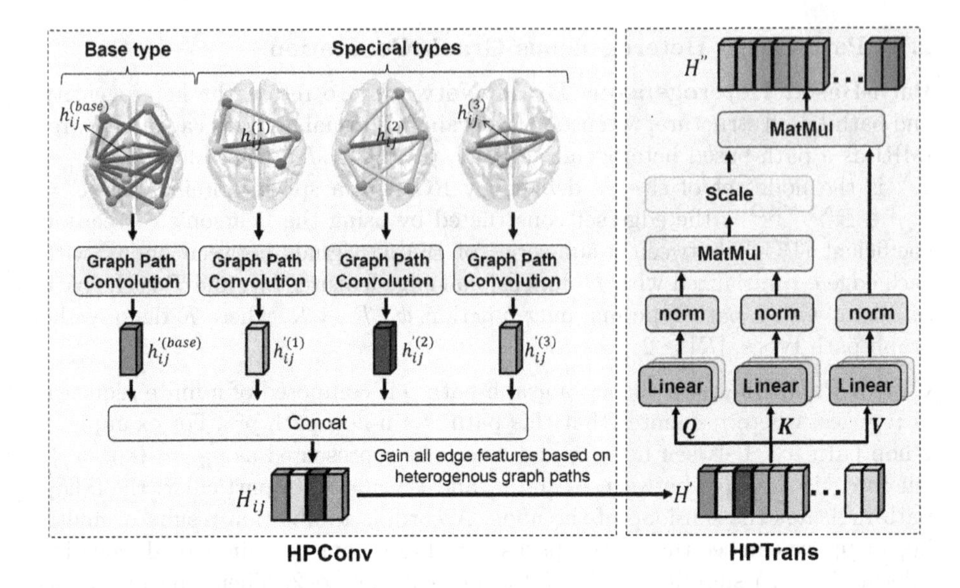

Fig. 2. Diagram for the details of HPConv and HPTrans layers for feature learning. For the sake of simplicity and clarity, we here only show the above four paths' types as an example, which does not mean that we only use four types.

HPConv. Considering that different heterogeneous paths have different distributions and contain different information, we design a novel graph convolution to learn the edge features of each path type independently. Under each path type, the brain network can be regarded as a homogeneous graph. Inspired by Li et al. [19], we utilize the same graph path convolution for each path type here. Within each path-based heterogeneous brain network $\mathcal{G}$, we define the following simple propagation model as HPConv layer for the forward-pass update of an edge denoted by e_{ij} under the graph type $r \in \mathcal{R}$:

$$\boldsymbol{h}_{ij}^{'(r)} = \sum_{k:\ \psi(p_{ikj})=r} p_{ikj}^{(r)} \boldsymbol{w}_k^{(r)} = \sum_{k:\ \psi(p_{ikj})=r} \left(h_{ik}^{(r)} + h_{kj}^{(r)} \right) \boldsymbol{w}_k^{(r)} \tag{1}$$

where $\boldsymbol{h}_{ij}^{(r)} \in \mathbb{R}^d$ and $\boldsymbol{h}_{ij}^{'(r)} \in \mathbb{R}^{d'}$ are the input and output features of HPConv layer in the lth HP-GTC module under the graph type r, d and d' are features' dimension, $p_{ikj}^{(r)}$ is the path feature and $\boldsymbol{w}_k^{(r)}$ denotes the learnable parameters of transformation. Thus, we finally have the intermediate representation feature matrix $\boldsymbol{H}^{'(r)} = \left[\boldsymbol{h}_{ij}^{'(r)} \right]$ under the graph path type r. Intuitively, the HPConv layer extracts edge features in specific and common path spaces, so as to ensure that the edge features learned in different specific feature spaces are independent and those learned in common spaces can retain global and shared information.

HPTrans. Note that different types of graph paths would have different impacts on a specific edge, and the same type of graph paths may similarly affect the specific edge. To model these characteristics, we propose an attention mechanism to capture these heterogeneous and common information to learn more effective edge features. Inspired by Kan et al. [13], to further incorporate global information and inconsistent contributions from different types of paths, we design a novel Graph Transformer layer to further aggregate output features $\boldsymbol{H}^{'} = \left[\boldsymbol{h}_{ij}^{'(r \in \mathcal{R})}\right]$ of HPConv layer for learning more compact edge features. The HPTrans layer is formulated as follows:

$$\boldsymbol{Q} = norm(\boldsymbol{H}^{'}\boldsymbol{W}_Q),\ \boldsymbol{K} = norm(\boldsymbol{H}^{'}\boldsymbol{W}_K),\ \boldsymbol{V} = norm(\boldsymbol{H}^{'}\boldsymbol{W}_V) \quad (2)$$

$$\boldsymbol{A} = (\boldsymbol{Q}\boldsymbol{K}^{\top})/\sqrt{|R|},\ \boldsymbol{H}^{''} = HPTrans(\boldsymbol{H}^{'}) = \boldsymbol{A}\boldsymbol{V} \quad (3)$$

where $\boldsymbol{H}^{''}$ denotes the output features, and the weight parameters $\boldsymbol{W}_Q$, $\boldsymbol{W}_K$, $\boldsymbol{W}_V \in \mathbb{R}^{d' \times \frac{d'}{2}}$ linearly map edge features to different feature spaces, and then adopt L2-normalized operator $norm(\cdot)$ to generate the corresponding representations $\boldsymbol{Q}$, $\boldsymbol{K}$, $\boldsymbol{V}$ for subsequent attention matrix $\boldsymbol{A}$ calculation and feature aggregation. For simplicity of illustration, in this paper, we only consider the single-head self-attention and assume $\boldsymbol{V} = \boldsymbol{H}^{'}$. The extension to the multi-head attention is straightforward, and we omit bias terms for simplicity.

To further enhance influential features and alleviate the over-smoothing problem of GNN, we adopt SE block in the HP-GTC module. Therefore, the final formulation of HP-GTC module can be represented as below:

$$\boldsymbol{H}^{l+1} = SE\left(HPTrans\left(HPConv\left(\boldsymbol{H}^{l}\right)\right)\right) \quad (4)$$

where $\boldsymbol{H}^{l}$ is the edge feature matrix of lth HP-GTC module.

2.3 Readout and Prediction

Lastly, inspired by Li et al. [19], a readout layer is adopted to transform the edge features learned by heterogeneous paths into the final graph embedding $\boldsymbol{H}_{G_m}$. Then $\boldsymbol{H}_{G_m}$ is sent to a multi-layer perceptron (MLP) to give the final prediction $\hat{y}_m$, where G_m denotes the mth graph sampled in subject s. Specifically, the final prediction of subject s is the average or weighted voting result of its' all samples during inferring.

3 Experimental Results

Dataset and Implementation Details. We validated our PH-BTN on the Baby Connectome Project (BCP) dataset [10] including 612 longitudinal rs-fMRI scans from 248 subjects (ages 6-811days, 106 boys vs. 142 girls). We here adopted Harvard-Oxford atlas (N = 112 ROIs) [23] and the mainstream brain functional partition as prior, where $|\mathcal{R}|$ =

Table 1. Brain age prediction and sex classification results (mean ± std) of all comparison methods on BCP dataset. (**bold**: best; underline: runner-up)

Graph mode		Method	Age Prediction		Sex Classification	
			MAE(days) ↓	PCC(%) ↑	ACC(%) ↑	F1(%) ↑
Non-graph		MLP	82.16 ± 0.46	86.16 ± 1.08	74.57 ± 5.51	74.14 ± 6.06
		CNN	89.90 ± 1.72	84.49 ± 4.23	76.62 ± 7.28	75.88 ± 6.85
Homogeneous	Node-based	GCN [16]	81.21 ± 1.67	86.27 ± 4.12	77.42 ± 7.86	75.62 ± 9.29
		PR-GNN [18]	79.02 ± 2.51	87.91 ± 5.65	78.65 ± 4.92	78.29 ± 4.87
		BrainGNN [17]	75.55 ± 1.94	88.69 ± 4.54	77.43 ± 6.53	76.65 ± 7.03
	Edge-based	BrainNetCNN [14]	70.57 ± 1.25	89.65 ± 3.82	79.03 ± 6.21	78.23 ± 6.95
	Path-based	BC-GCN$_{SE}$ [19]	61.40 ± 1.02	90.32 ± 3.64	79.43 ± 5.82	78.63 ± 5.69
Heterogeneous		PH-BTN	**58.91 ± 1.64**	**92.56 ± 2.16**	**81.87 ± 5.06**	**81.50 ± 5.18**

$|\{base,\ frontal,\ parietal,\ temporal,\ occipital, insular\}| = 6$. For experiments, we evaluated PH-BTN and related competing models (i.e., four brain backbone models - BrainNetCNN, BrainGNN, PR-GNN and BC-GCN, and three classical deep learning models - MLP, CNN and GCN) through 10-fold cross-validation. We chose the classification accuracy (ACC) and weighted F1-score (F1) as evaluation metrics for sex classification, and evaluated the Mean Absolute Error (MAE) and Pearson's Correlation Coefficient (PCC) between predicted and ground truth ages. To guarantee training is fair, we used the identical parameters across all experimental baselines. See *Supplementary Materials* for more details of dataset and implementation.

Performance Analysis. As shown in Table 1, we first report the overall results of age prediction and sex classification achieved by the proposed PH-BTN and related alternative methods on the BCP dataset. From Table 1, we can see that the proposed PH-BTN outperformed alternative models both on age prediction with MAE of 58.91 days (7.32% of the whole age distribution) and 92.56% PCC, and sex classification with 81.87% ACC and 81.50% F1. Furthermore, we have following interesting observations when comparing these statistics. First, graph-based methods are superior to other non-graph methods when utilizing rs-FC features, which demonstrates the rationality and effectiveness of formulating the brain as a graph. Second, we found that the performance of node-based methods is worse than that of edge-based/path-based methods, especially at least 5.28 days in age prediction. This confirms the view that the connectivity is the core of the brain. Third, we also observed that brain networks with heterogeneous design have higher ACC and lower MAE than the homogeneous design, which implies that the heterogeneity of brain connections. All above results indicate that our formulation is more in line with brain functional network by considering brain graph topology, path information and connective heterogeneity.

Ablation Studies. As shown in *Supplementary Materials*, we have done several ablation experiments to further evaluate the effectiveness of the proposed method, e.g., different path types by prior or random division, whether to use

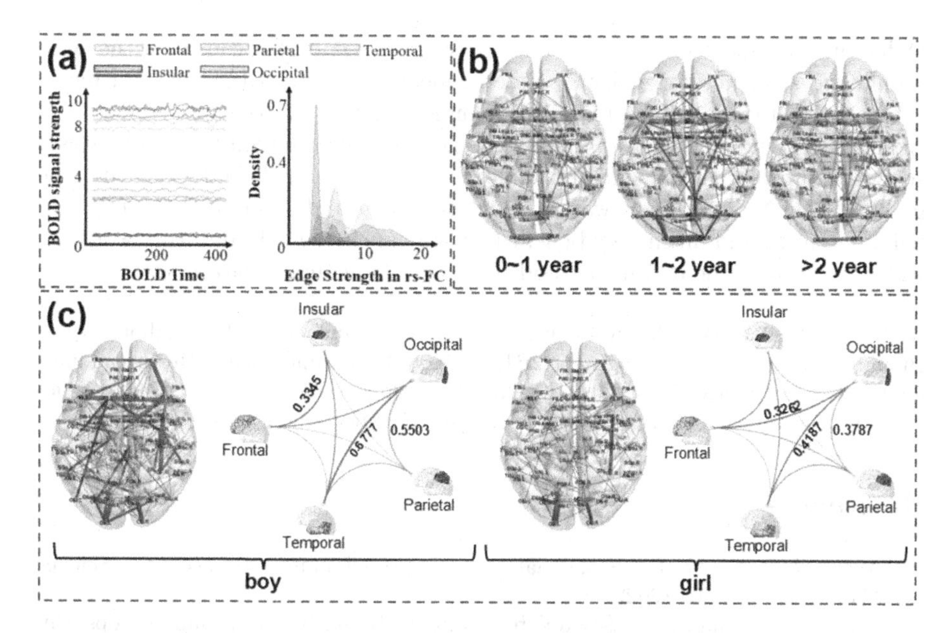

Fig. 3. (a) Heterogeneous analysis of five functional partitions before experiments. (b)–(c) Visualization of important connections of different age and sex learned by PH-BTN.

HPTrans or not, etc. After comparison, we can draw the following conclusions: (1) Reasonable path-type division is the key to learning brain heterogeneity. (2) Regardless of special types, the base type representing shared information is essential for brain functional connectivity analysis (w/o base loses at least 8.44 days). (3) The transformer-based mechanism can improve the performance (e.g., MAE 3.94 days) of PH-BTN in global and heterogeneous views.

Visualization and Discussion. In this section, we utilized the gradient back-tracking method [19] to visualize brain functional networks learned by the proposed PH-BTN. For visualizing age-related connections, we divided all scans into three typical groups, i.e., 0–1 year, 1–2 year and >2 year. All of our visual findings illustrated in Fig. 3(b)–(c) are consistent with those of many research [7,27,29], which have showed that frontal, parietal and occipital lobes are deeply related to brain age and gender, and well correspond to our statistical analysis (Fig. 3(a)) of different brain regions in the original data. We refer the readers to the *Supplementary Materials* for additional results of visualization.

4 Conclusion

In this paper, we propose a novel network, namely PH-BTN, for encoding path-based heterogeneous brain networks for analyzing brain functional connectivity.

Different from most existing methods, our proposed model considers the path significance and heterogeneity by heterogeneous graph convolution, and incorporates global brain structure and key connections by Transformer mechanism. Experiments and visualization of age and gender prediction on the BCP dataset show the superiority and effectivity of our PH-BTN model. Moreover, the proposed PH-BTN offers a new way to understand neural development, explore sexual differences, and ultimately benefit neuroimaging research. In future work, we will extend and validate our methods on larger benchmark datasets.

Acknowledgements. This work was supported by Guangdong Key Laboratory of Human Digital Twin Technology (2022B1212010004) and Fundamental Research Funds for the Central Universities (2022ZYGXZR104). This work also utilized approaches developed by the efforts of the UNC/UMN Baby Connectome Project Consortium.

References

1. Axer, M., Amunts, K.: Scale matters: the nested human connectome. Science **378**(6619), 500–504 (2022)
2. Bao, A.M., Swaab, D.F.: Sex differences in the brain, behavior, and neuropsychiatric disorders. Neuroscientist **16**(5), 550–565 (2010)
3. Cai, H., Gao, Y., Liu, M.: Graph transformer geometric learning of brain networks using multimodal MR images for brain age estimation. IEEE Trans. Med. Imaging **42**(2), 456–466 (2023)
4. Cole, J.H., Franke, K.: Predicting age using neuroimaging: innovative brain ageing biomarkers. Trends Neurosci. **40**(12), 681–690 (2017)
5. Friston, K.J.: Functional and effective connectivity in neuroimaging: a synthesis. Hum. Brain Mapp. **2**(1–2), 56–78 (1994)
6. Gadgil, S., Zhao, Q., Pfefferbaum, A., Sullivan, E.V., Adeli, E., Pohl, K.M.: Spatio-temporal graph convolution for resting-state fMRI analysis. In: Martel, A.L., et al. (eds.) MICCAI 2020. LNCS, vol. 12267, pp. 528–538. Springer, Cham (2020). https://doi.org/10.1007/978-3-030-59728-3_52
7. Gao, W., Alcauter, S., Smith, J.K., Gilmore, J.H., Lin, W.: Development of human brain cortical network architecture during infancy. Brain Struct. Funct. **220**, 1173–1186 (2015)
8. He, S., Grant, P.E., Ou, Y.: Global-local transformer for brain age estimation. IEEE Trans. Med. Imaging **41**(1), 213–224 (2021)
9. Hou, Y., et al.: Ageing as a risk factor for neurodegenerative disease. Nat. Rev. Neurol. **15**(10), 565–581 (2019)
10. Howell, B.R., et al.: The UNC/UMN baby connectome project (BCP): an overview of the study design and protocol development. Neuroimage **185**, 891–905 (2019)
11. Hu, J., Shen, L., Sun, G.: Squeeze-and-excitation networks. In: Proceedings of the IEEE Conference on Computer Vision and Pattern Recognition, pp. 7132–7141 (2018)
12. Jung, J., Cloutman, L.L., Binney, R.J., Ralph, M.A.L.: The structural connectivity of higher order association cortices reflects human functional brain networks. Cortex **97**, 221–239 (2017)
13. Kan, X., Dai, W., Cui, H., Zhang, Z., Guo, Y., Yang, C.: Brain network transformer. arXiv preprint arXiv:2210.06681 (2022)

14. Kawahara, J., et al.: Brainnetcnn: convolutional neural networks for brain networks; towards predicting neurodevelopment. Neuroimage **146**, 1038–1049 (2017)
15. Kim, B.H., Ye, J.C.: Understanding graph isomorphism network for rs-fMRI functional connectivity analysis. Front. Neurosci. **14**, 630 (2020)
16. Kipf, T.N., Welling, M.: Semi-supervised classification with graph convolutional networks. arXiv preprint arXiv:1609.02907 (2016)
17. Li, X., et al.: BrainGNN: interpretable brain graph neural network for fMRI analysis. Med. Image Anal. **74**, 102233 (2021)
18. Li, X., et al.: Pooling regularized graph neural network for fMRI biomarker analysis. In: Martel, A.L., et al. (eds.) MICCAI 2020. LNCS, vol. 12267, pp. 625–635. Springer, Cham (2020). https://doi.org/10.1007/978-3-030-59728-3_61
19. Li, Y., et al.: Brain connectivity based graph convolutional networks and its application to infant age prediction. IEEE Trans. Med. Imaging **41**(10), 2764–2776 (2022)
20. Nikolentzos, G., Dasoulas, G., Vazirgiannis, M.: K-hop graph neural networks. Neural Netw. **130**, 195–205 (2020)
21. Thiebaut de Schotten, M., Forkel, S.J.: The emergent properties of the connected brain. Science **378**(6619), 505–510 (2022)
22. Shi, G., Zhu, Y., Liu, W., Yao, Q., Li, X.: Heterogeneous graph-based multimodal brain network learning. arXiv e-prints pp. arXiv-2110 (2021)
23. Smith, S.M., et al.: Advances in functional and structural MR image analysis and implementation as FSL. Neuroimage **23**, S208–S219 (2004)
24. Van Den Heuvel, M.P., Pol, H.E.H.: Exploring the brain network: a review on resting-state fMRI functional connectivity. Eur. Neuropsychopharmacol. **20**(8), 519–534 (2010)
25. Velickovic, P., Cucurull, G., Casanova, A., Romero, A., Lio, P., Bengio, Y., et al.: Graph attention networks. Stat **1050**(20), 10–48550 (2017)
26. Wang, L., Li, K., Hu, X.P.: Graph convolutional network for fMRI analysis based on connectivity neighborhood. Netw. Neurosci. **5**(1), 83–95 (2021)
27. Weis, S., Patil, K.R., Hoffstaedter, F., Nostro, A., Yeo, B.T., Eickhoff, S.B.: Sex classification by resting state brain connectivity. Cereb. Cortex **30**(2), 824–835 (2020)
28. Wen, X., Wang, R., Yin, W., Lin, W., Zhang, H., Shen, D.: Development of dynamic functional architecture during early infancy. Cereb. Cortex **30**(11), 5626–5638 (2020)
29. Wu, K., Taki, Y., Sato, K., Hashizume, H., Sassa, Y., et al.: Topological organization of functional brain networks in healthy children: differences in relation to age, sex, and intelligence. PLoS ONE **8**(2), e55347 (2013)
30. Yao, D., Yang, E., Sun, L., Sui, J., Liu, M.: Integrating multimodal MRIs for adult ADHD identification with heterogeneous graph attention convolutional network. In: Rekik, I., Adeli, E., Park, S.H., Schnabel, J. (eds.) PRIME 2021. LNCS, vol. 12928, pp. 157–167. Springer, Cham (2021). https://doi.org/10.1007/978-3-030-87602-9_15
31. Yin, W., Li, L., Wu, F.X.: A graph attention neural network for diagnosing ASD with fMRI data. In: 2021 IEEE International Conference on Bioinformatics and Biomedicine (BIBM), pp. 1131–1136. IEEE (2021)
32. Zhang, H., et al.: Classification of brain disorders in rs-fMRI via local-to-global graph neural networks. IEEE Trans. Med. Imaging **42**(2), 444–455 (2023)

Dynamic Graph Neural Representation Based Multi-modal Fusion Model for Cognitive Outcome Prediction in Stroke Cases

Shuting Liu[1(✉)], Baochang Zhang[1], Rong Fang[2], Daniel Rueckert[1,3,4], and Veronika A. Zimmer[1,3]

[1] School of Computation, Information and Technology, Technical University of Munich, Munich, Germany
shuting.liu@tum.de

[2] Institute for Stroke and Dementia Research (ISD), LMU University Hospital, LMU Munich, Munich, Germany

[3] School of Medicine, Klinikum rechts der Isar, Technical University of Munich, Munich, Germany

[4] Department of Computing, Imperial College London, London, UK

Abstract. The number of stroke patients is growing worldwide and half of them will suffer from cognitive impairment. Therefore, the prediction of Post-Stroke Cognitive Impairment (PSCI) becomes more and more important. However, the determinants and mechanisms of PSCI are still insufficiently understood, making this task challenging. In this paper, we propose a multi-modal graph fusion model to solve this task. First, dynamic graph neural representation is proposed to integrate multi-modal information, such as clinical data and image data, which separates them into node-level and global-level properties rather than processing them uniformly. Second, considering the variability of brain anatomy, a subject-specific undirected graph is constructed based on the connections among 131 brain anatomical regions segmented from image data, while first-order statistical features are extracted from each brain region and internal stroke lesions as node features. Meanwhile, a novel missing information compensation module is proposed to reduce the impact of missing or incomplete clinical data. In the dynamic graph neural representation, two kinds of attention mechanisms are used to encourage the model to automatically localize brain anatomical regions that are highly relevant to PSCI prediction. One is node attention established between global tabular neural representation and nodes, the other is multi-head graph self-attention which changes the static undirected graph into multiple dynamic directed graphs and optimizes the broadcasting process of the graph. The proposed method studies 418 stroke patients and achieves the best overall performance with a balanced accuracy score of 79.6% on PSCI prediction, outperforming the competing models. The code is publicly available at github.com/fightingkitty/MHGSA.

Keywords: Post-stroke cognitive impairment · Graph neural representation · Fusion method

H. Greenspan et al. (Eds.): MICCAI 2023, LNCS 14227, pp. 338–347, 2023.
https://doi.org/10.1007/978-3-031-43993-3_33

1 Introduction

Post-Stroke Cognitive Impairment (PSCI) is common following stroke, and more than half of people will suffer from PSCI in the first year after stroke. Therefore, predicting cognitive impairment after stroke is an important task [2,5]. However, the determinants and mechanisms of PSCI are insufficiently understood, and doctors cannot make a clear risk prediction based on single factors, e.g., the locations of infarct and White Matter Hyperintensities (WMH) lesions, resulting in frequent underdiagnosis of PSCI [16,17]. Therefore, accurate prediction of PSCI becomes challenging and critical for devising appropriate post-stroke prevention strategies.

In recent years, Convolutional Neural Networks (CNN) have shown promising performance in brain disease diagnosis and prediction [4]. Compared with traditional machine learning methods [14], CNNs excel at extracting high-level information about neuroanatomy from Magnetic Resonance Imaging (MRI) [11]. However, brain MRIs provide only a partial view of the underlying changes that lead to cognitive decline, and the extracted features using CNNs are difficult to interpret by clinicians. Hence, some studies use a pre-defined set of image features, like cortical features, extracted from MRIs via Freesurfer [18]. However, the aforementioned approaches all focus on neuroimaging data, ignoring the importance of clinical data such as patient demographics, family history, or laboratory measurements. Some successful multi-modal fusion models have been proposed and achieved some improvements in predictive tasks, but image and clinical data are only integrated through simple concatenation or affine transformations [3,12], without considering the specificity of patient brain anatomy [10].

To overcome the aforementioned issues of existing methods, we propose a novel multi-modal fusion model to solve this challenging task. Our main contributions are as follows: (1) Both MRI data and clinical data are studied in this task, where subject-specific anatomy, infarcts localization and WMH distribution are explored from MRIs, while clinical (tabular) data provides personal information and clinical indicators such as stroke history and stroke severity. (2) Dynamic graphs neural representation is first proposed for PSCI prediction, which can not only integrate multi-modal information but also take subject-specific brain anatomy into account. An effective missing information compensation module is proposed to reduce the impact of incomplete clinical data. (3) A detailed ablation study is conducted to verify the effectiveness of each proposed module. The proposed method outperforms competing models by a large margin. Additionally, the Top-15 brain structural regions strongly associated with PSCI are uncovered by the proposed method, which further emphasizes the relevance of the proposed method.

2 Methodology

Our approach consists of three steps (illustrated in Fig. 1): (i) graph construction and node feature extraction, (ii) missing information compensation module, and (iii) dynamic graph neural representation.

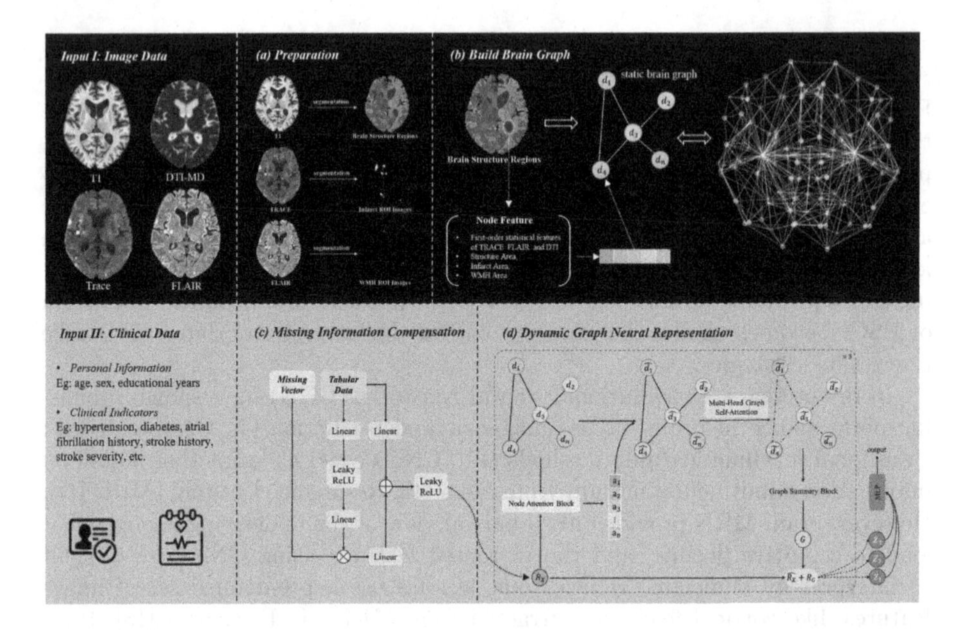

Fig. 1. The overview of our proposed method. (a) Data preparation, (b) graph building and node feature extraction, (c) missing information compensation module, (d) and dynamic graph neural representation.

2.1 Graph Construction and Node Feature Extraction

We represent the nodes of the graph with 131 structural brain regions obtained from the Hammers' brain atlas [7,9] (more details are given in Sect. 3.1). The edges of the graph are established based on the neighboring relationship of each brain structural area. Specifically, if two areas are adjacent, we create an edge between the corresponding nodes; otherwise, no edge is present. To each node, we assign a total of eighteen features, comprising fifteen first-order statistical features (maximum, minimum, median, mean, and variance) extracted from the corresponding brain structural regions in Trace Diffusion-Weighted MRI (DW-MRI) and Mean Diffusivity (MD) DW-MRI and FLAIR MRI, along with three stroke lesion features: structural volume, infarct volume, and WMH volume.

2.2 Missing Information Compensation Module

When dealing with clinical data, some patients may not have complete records, resulting in missing information. To mitigate the impact of missing data, we propose a module shown in Fig. 1(c). Rather than using a naive approach of filling in missing elements with an average value or zero, we suggest learning the offset value caused by missing elements based on the known information to compensate for the missing data. We denote a patient's 1D tabular record as $X = \{x_i\}$, and define a vector V with the same size as X. In V, bits for missing elements are represented as '1', and the remaining bits are '0'. We denote the set of indices of missing elements as $M = \{m|v_m = 1\}$. As a complete X is processed by a linear layer, each element $x_i^{'}$ can be expressed as

$$x_i^{'} = \sum_j w_{ij} x_j + b_i \tag{1}$$

When the value of some input values $x_m, m \in M$ is unknown due to missing tabular data, we can only compute $x_i^{''} = \sum_{j \notin M} w_{ij} x_j + b_i$ and the difference $x_i^{'} - x_i^{''}$ is the impact the missing tabular data can have. We propose a module to learn this difference to mitigate the effect of missing values by

$$\phi_2 \left(\phi_1(X) \bullet V\right) = \sum_m w_{im} x_m \qquad \forall m \in M \tag{2}$$

Here, ϕ_1 represents two normal linear layers, ϕ_2 is a linear layer without considering the bias, ϕ_3 represents a normal linear layers, $LReLU$ is a leaky rectified linear activation, and $\bullet$ is the element-wise multiplication operator. When some elements are missing, this module will estimate the offset; and when processing a complete clinical record, the output of Eq. (2) should always be zero. Hence, the neural representation of each patient's clinical tabular record will be calculated using Eq. (3).

$$R_X = LReLU \left(\phi_2 \left(\phi_1(X) \bullet V\right) + \phi_3(X)\right) \tag{3}$$

2.3 Dynamic Graph Neural Representation

We regard the tabular's neural representation R_X as global information and the node feature d_i in each subject-specific graph $G = \{D, E|d_i \in D\}$ as local information (here, E denotes the adjacency matrix): First, we set up an attention mechanism between R_X and d_i, which will capture the contribution of each node and enhance node feature. Then, a multi-head graph self-attention is introduced for this task, which pays attention to the weight of edges in the graphs. It transforms the static graph to several dynamically subgraphs, influencing the broadcasting process of the graph. Next, we use graph summary block which consists of a normal linear layer and a leaky rectified linear unit activation function (LReLU), to summarize the graph G and output a graph-based global feature R_G, i.e., graph's neural representation. Finally, we fuse the graph's neural representation R_G and the updated tabular's neural representation $\hat{R_X}$ via concatenation, which can be used as the input of a multilayer perceptron (MLP) for classification as shown in Fig. 1. Here, focal loss function is employed, which is formulated as,

$$FL\left(p_t\right) = -\alpha \left(1 - p\right)^{\gamma} log \left(p\right) \tag{4}$$

Here p is the predicted probability of the correct class, $\alpha = 0.2$ is the weighting factor for each sample, $\gamma = 0.05$ is a tunable focusing parameter.

Node Attention Block. This is a parameterized function that learns the mapping between query ($q_t \in \mathbb{R}^{1\times C}$) coming from the tabular's neural representation R_X, and the corresponding key ($k_g \in \mathbb{R}^{N\times C}$) and value ($v_g \in \mathbb{R}^{N\times C}$) representations in graph G. Here, N is the number of nodes, C is the channel of node feature. Hence, the node attention block fuses information from tabular (global level) and image data (local/node level) by measuring the correlation between q_t and k_g, and the attention weight ($A_1 \in \mathbb{R}^{1\times N}$) is computed as follow,

$$A_1 = softmax\left(\frac{q_t k_g^T}{\sqrt{C}}\right) \tag{5}$$

Using the computed attention weight, the output of the node attention block is computed to update graph G' as,

$$G^{'} = (1 + A_1^T)v_g \tag{6}$$

Multi-head Graph Self-attention Block. Inspired by [15] and in order to optimize the broadcasting of the graph, the multi-head self-attention is proposed to learn the edge weights during training. Following with the node attention, it is also a parameterized function that learns the mapping between the reused key ($k_g \in \mathbb{R}^{N\times C}$) and the query ($q_g \in \mathbb{R}^{N\times C}$) representation in graph G ($G \in \mathbb{R}^{N\times C}$). Then, the shapes of query q_g, key k_g and updated graph feature G' are reshaped as $[h, N, C/h]$. Here, h is the number of heads. The attention weight ($A_2 \in \mathbb{R}^{h\times N\times N}$) is computed by measuring the similarity between q_g and k_g according to,

$$A_2 = softmax\left(\frac{q_g k_g^T}{\sqrt{\frac{C}{h}}}\right) \tag{7}$$

Using the computed attention weight and considering the fixed adjacency matrix E, the output of graph multi-head self-attention block $G'' \in \mathbb{R}^{h\times N\times C/h}$ is computed and reshaped as $[N, C]$ for the preparation of starting the next graph neural representation unit.

$$G^{''} = E \cdot A_2 G^{'} \tag{8}$$

3 Materials and Experiments

3.1 Data and Preparation

Our study utilized clinical data from 418 stroke patients, collected within five days of the onset of their acute stroke symptoms, and consisted of both clinical data and image data. Each clinical tabular records contains 20 variables, including 3 variables of basic personal information (age, sex, education) and 17 variables of clinical indicators (i.e., smoking, drinking, hypertension, diabetes, atrial fibrillation, previous stroke, BMI, circulating low-density lipoprotein cholesterol levels, stroke severity, pre-stroke function, cognitive impairment in the acute post-stroke phase, stroke lesion volume, lacunes number, Fazekas score of WMH, Fazekas score of WMH in deep white matter, cerebral microbleeds number, perivascular spaces level). Patients were followed up, and clinical experts classified them as either PSCI negative or PSCI positive based on detailed neuropsychological tests conducted 12 months after stroke onset. Out of the 418 patients, there were 332 positive cases and 86 negative cases. All studies were approved by the local ethics committees. We split the patients into two groups with an equal proportion of PSCI candidates for five-fold cross validation, 80% for training and 20% for testing. In the preparation step, the MRIs are rigidly aligned using [1]. Based on each patient's T1 data, the whole brain was segmented into 131 structural regions using MALPEM [7,9], from which we constructed the subject-specific brain graphs. Meanwhile, based on each patient's Trace DW-MRI and FLAIR MRI, the infarct lesions and WMH lesions were semi-automatically segmented and manually corrected by a clinical expert.

3.2 Evaluation Measures

To validate the performance of our method, we employed five measures including Balanced Accuracy (BAcc), Classification Accuracy (Acc), Precision (Pre), Sensitivity (Sen), Specificity (Spe), and the area under the receiver operating characteristic curve (AUC). Among them, BAcc is a performance metric used to evaluate the effectiveness of a model on imbalanced datasets.

3.3 Experimental Design

To validate the performance of our method, we perform a series of experiments on our in-house dataset. Based on the multi-modal's information (i.e., clinical data and image data), we designed three experiments with different data inputs. (I) By only using the clinical data as input, the effectiveness of missing information compensation module is studied. (II) By only using the image-based features as input, the importance of graph-based analysis for PSCI prediction is investigated. (III) Using both as input, an ablation experiment is performed, where we learn the evolution process and the superiority of our proposed fusion model.

The proposed model is implemented on PyTorch library (version 1.13.0) with one NVIDIA GPU (Quadro RTX A6000), and trained for 400 epochs with a batch size of 32. Adam optimizer is employed with weight decay of 0.008, an initial learning rate of 0.0005; and CosineAnnealingLR scheduling technique is used to adjust the learning rate with the maximum number of iterations of 10.

4 Results

Table 1 summarizes the results of the five-fold cross-validation experiments. We used different models for the experiments, where $\boldsymbol{A}$ represents an MLP model, $\boldsymbol{B}$ represents an MLP model with the proposed missing information compensation module shown in Fig. 1(c), $\boldsymbol{C}$ represents another MLP model, $\boldsymbol{D}$ represents a

Graph Convolutional Network (GCN) [8], ***B+D*** represents a fusion model by concatenating the features from ***B*** and ***D***, ***B+E*** represents a fusion model by concatenating the features from ***B*** and ***E***, where ***E*** is a GCN with a multi-head graph self-attention block as explained in Sect. 2.3. Finally, ***F*** represents our proposed method as shown in Fig. 1.

Table 1. Experimental results of different methods. The input data is specified by *I* for clinical data only, *II* for imaging data only and *III* for both; **bold values** indicate best performance over all.

Exp.	Models	BAcc		Acc		AUC		Pre		Sen		Spe	
		mean	std	mean	std	mean	std	mean	std	mean	std	mean	std
I	***A***	0.688	0.066	0.695	0.033	0.732	0.071	0.367	0.052	0.675	0.134	0.700	0.032
	B	0.709	0.090	0.689	0.060	0.741	0.072	0.372	0.076	**0.745**	0.162	0.674	0.056
II	***C***	0.671	0.035	0.785	0.027	0.648	0.064	0.493	0.089	0.477	0.093	**0.865**	0.046
	D	0.700	0.054	0.738	0.108	0.711	0.080	0.461	0.117	0.634	0.129	0.765	0.159
III	***B+D***	0.742	0.035	0.796	0.028	0.746	0.057	0.508	0.053	0.649	0.058	0.835	0.028
	B+E	0.761	0.055	0.785	0.034	0.770	0.067	0.490	0.055	0.721	0.133	0.802	0.053
	F	**0.796**	0.049	**0.832**	0.038	**0.800**	0.083	**0.580**	0.083	0.734	0.086	0.858	0.041

4.1 Effectiveness of Missing Information Compensation

To assess the efficacy of compensating for missing information, a standard MLP model is trained solely on clinical data with missing elements filled in with zeros. Subsequently, the same MLP model is trained with a missing information compensation module, utilizing the identical training set. Comparing the performances of model ***A*** and ***B***, it is obvious that the proposed missing information compensation module plays an important role, achieving a BAcc of 0.709 ± 0.054 and AUC score of 0.741 ± 0.072, outperforming Method ***A***.

4.2 Importance of Graph-Based Analysis

To investigate the importance of graph-based analysis for this task, an MLP model embedded with GCNs is trained only on imaging-based features, and the comparison method is a standard MLP model. Comparing the performances of model ***C*** and ***D***, it is clear that the suggested graph-based model makes an notable improvement on both BAcc and AUC, with the rise of BAcc from 0.671 to 0.700 and of Auc from 0.648 to 0.711, which demonstrates that the consideration of subject-specific brain anatomy is very important for PSCI prediction.

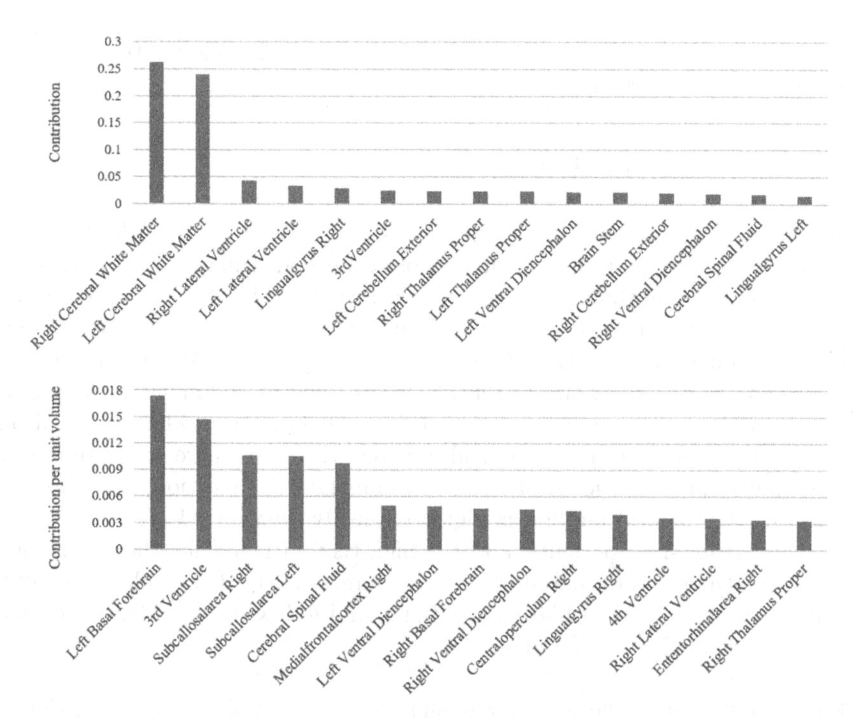

Fig. 2. The Top-15 brain structural regions ranking of contribution (top) and the Top-15 brain structural regions ranking of contribution per unit volume (bottom) that are strongly associated with PSCI discovered by our method.

4.3 Superiority of Proposed Fusion Model

Finally, we explore how to effectively fuse clinical data and image data for PSCI prediction. As a baseline approach, we directly concatenate the features learned from model $\boldsymbol{B}$ and $\boldsymbol{D}$. The first evolutionary process is to upgrade static brain map analysis to dynamic brain map analysis, thus the performance of model $\boldsymbol{B{+}E}$ is explored, where the proposed multi-head graph self-attention block is first embedded into the GCN layers. Then, a better fusion pattern is discovered, i.e., dynamic graph neural representation in our final proposed method. The experimental results demonstrate that $\boldsymbol{B{+}E}$ model results in a higher BAcc of 0.761 compared to baseline. Furthermore, our final proposed method outperforms all other methods, achieving a BAcc as high as 0.796 and an AUC of 0.800. To explore the contribution of each brain structural region in this task, we first average the attention map A_1 across all subjects and visualize it, as illustrated in Fig. 2(top). Additionally, to adjust for the influence of volume, we compute the contribution per unit volume for each region by dividing by its corresponding volume, as illustrated in Fig. 2(bottom). Our observations indicate that the network appears to concentrate on clinically significant regions that impact cognition, including the Cerebral White Matter, Basal Forebrain and 3rd Ventricle

[6,13]. Nevertheless, we have also identified some ambiguous relationships that necessitate further investigation.

5 Conclusion and Discussion

In this study, we tackle the challenge of predicting PSCI using both brain MRI and clinical non-imaging data. Our proposed method utilizes a dynamic graph neural representation that fully leverages the structural information of the brain. By incorporating node attention and self-attention, we effectively merge clinical tabular and image information. Our approach achieves a BAcc of 0.796 on the test dataset. Moreover, visual attention maps allow us to identify the contributions of different brain structures to this prediction task. The results of our study are consistent with prior medical research. However, there are certain brain regions that require further exploration by clinicians. In addition, there are several areas our proposed method is highly clinically relevant: (1) Risk prediction to identify (i) low-risk patients, and (ii) high-risk patients (for those early and targeted treatment and rehabilitation is recommended). (2) Detailed phenotyping and characterization of high-risk patients and follow-up. (3) Identification of novel biomarkers and risk factors.

Acknowledgements. The project was supported by the China Scholarship Council (File No. 202106210062) and ERC Grant Deep4MI (884622).

References

1. Avants, B.B., Tustison, N., Song, G., et al.: Advanced normalization tools (ANTS). Insight j **2**(365), 1–35 (2009)
2. Ball, E.L., et al.: Predicting post-stroke cognitive impairment using acute CT neuroimaging: a systematic review and meta-analysis. Int. J. Stroke **17**(6), 618–627 (2022)
3. Binzer, M., Hammernik, K., Rueckert, D., Zimmer, V.A.: Long-term cognitive outcome prediction in stroke patients using multi-task learning on imaging and tabular data. In: Rekik, I., Adeli, E., Park, S.H., Cintas, C. (eds.) PRIME 2022. LNCS, vol. 13564, pp. 137–148. Springer, Cham (2022). https://doi.org/10.1007/978-3-031-16919-9_13
4. Ebrahimighahnavieh, M.A., Luo, S., Chiong, R.: Deep learning to detect Alzheimer's disease from neuroimaging: a systematic literature review. Comput. Methods Programs Biomed. **187**, 105242 (2020)
5. Georgakis, M.K., et al.: Cerebral small vessel disease burden and cognitive and functional outcomes after stroke: a multicenter prospective cohort study. Alzheimer's Dementia **19**(4), 1152–1163 (2023)
6. Grothe, M., et al.: Reduction of basal forebrain cholinergic system parallels cognitive impairment in patients at high risk of developing Alzheimer's disease. Cereb. Cortex **20**(7), 1685–1695 (2010)
7. Heckemann, R.A., et al.: Brain extraction using label propagation and group agreement: pincram. PLoS ONE **10**(7), e0129211 (2015)

8. Kipf, T.N., Welling, M.: Semi-supervised classification with graph convolutional networks. arXiv preprint arXiv:1609.02907 (2016)
9. Ledig, C., et al.: Robust whole-brain segmentation: application to traumatic brain injury. Med. Image Anal. **21**(1), 40–58 (2015)
10. Lim, J.S., Lee, J.J., Woo, C.W.: Post-stroke cognitive impairment: pathophysiological insights into brain disconnectome from advanced neuroimaging analysis techniques. J. Stroke **23**(3), 297–311 (2021)
11. Liu, M., et al.: A multi-model deep convolutional neural network for automatic hippocampus segmentation and classification in Alzheimer's disease. Neuroimage **208**, 116459 (2020)
12. Pölsterl, S., Wolf, T.N., Wachinger, C.: Combining 3D image and tabular data via the dynamic affine feature map transform. In: de Bruijne, M., et al. (eds.) MICCAI 2021. LNCS, vol. 12905, pp. 688–698. Springer, Cham (2021). https://doi.org/10.1007/978-3-030-87240-3_66
13. Ray, N.J., et al.: Cholinergic basal forebrain structure influences the reconfiguration of white matter connections to support residual memory in mild cognitive impairment. J. Neurosci. **35**(2), 739–747 (2015)
14. Uysal, G., Ozturk, M.: Hippocampal atrophy based Alzheimer's disease diagnosis via machine learning methods. J. Neurosci. Methods **337**, 108669 (2020)
15. Vaswani, A., et al.: Attention is all you need. In: Advances in Neural Information Processing Systems, vol. 30 (2017)
16. Verdelho, A., et al.: Cognitive impairment in patients with cerebrovascular disease: a white paper from the links between stroke ESO dementia committee. Eur. Stroke J. **6**(1), 5–17 (2021)
17. Weaver, N.A., et al.: Strategic infarct locations for post-stroke cognitive impairment: a pooled analysis of individual patient data from 12 acute ischaemic stroke cohorts. Lancet Neurol. **20**(6), 448–459 (2021)
18. Zheng, G., et al.: A transformer-based multi-features fusion model for prediction of conversion in mild cognitive impairment. Methods **204**, 241–248 (2022)

Predicting Diverse Functional Connectivity from Structural Connectivity Based on Multi-contexts Discriminator GAN

Xiang Gao[1], Xin Zhang[1,3](✉), Lu Zhang[2], Xiangmin Xu[3,4], and Dajiang Zhu[2]

[1] School of Electronic and Information Engineering, South China University of Technology, Guangzhou, China
eexinzhang@scut.edu.cn
[2] Department of Computer Science and Engineering, University of Texas at Arlington, Arlington, TX, USA
[3] Pazhou Laboratory, Guangzhou, China
[4] School of Future Technology, South China University of Technology, Guangzhou, China

Abstract. Revealing structural-functional relationship is an important issue in neuroscience study since it helps to understand brain activities. Structural Connectivity (SC) represents the fibers connection between the brain regions, which is relatively static. Functional Connectivity (FC) represents the active signal correlations between the brain regions, which is relatively dynamic and diverse. Many works predict FC from SC and achieve unique FC prediction. However, FC is diverse since it represents brain activities. In this work, we propose the MCGAN, a **m**ulti-**c**ontexts discriminator based **g**enerative **a**dversarial **n**etwork for predicting diverse FC from SC. The proposed multi-contexts discriminator provides three kinds of supervisions to strengthen the generator, i.e. edge-level, node-level and graph-level. Since FC represents the connection of the brain regions, it can be regarded as edge-based graph. We adopt edge-based graph convolution method to model the context encoding. Moreover, to introduce the diversity of generated FC, we utilize monte-carlo mean samples to bring in more FC data for training. We validate our MCGAN on Human Connextome Project (HCP) dataset and Alzheimer's Disease Neuroimaging Initiative (ADNI) dataset. The results show that our method can generate diverse and meaningful FC from SC, revealing the one-to-many relationship between the individual SC and the multiple FC. The significance of this work is that once we have anatomical structure of brain represented by SC, we can predict diverse developments of brain activities represented by FC, which helps to reveal individual brain's static-dynamic structural-functional mode.

Keywords: Structural Connectivity · Functional Connectivity · Diverse Generation · GAN

H. Greenspan et al. (Eds.): MICCAI 2023, LNCS 14227, pp. 348–357, 2023.
https://doi.org/10.1007/978-3-031-43993-3_34

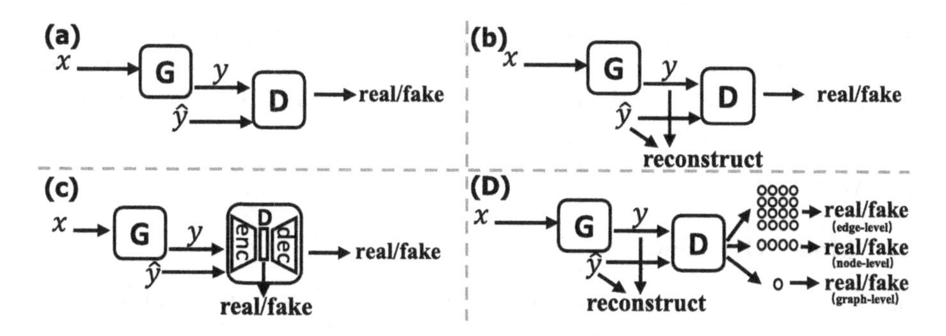

Fig. 1. Illustration of GAN [3] (a), GAN with reconstruct supervision in generator [9] (b), GAN with more supervisions in discriminator [14] (c), and our MCGAN (d).

1 Introduction

Recently, the study of exploring structural-functional relationship raises lots of attentions in neuroscience which helps to reveal individual behaviors of human brain [1]. Typically, Structural Connectivity (SC) [16] represents the fibers connection between the brain regions while Functional Connectivity (FC) [4] represents the Blood-Oxygen-Level-Dependent (BOLD) signal correlations between the brain regions. In comparing, SC is relatively static since it demonstrates the anatomical structure of brain, and FC is relatively dynamic and diverse since it demonstrates the development of the brain activities [2]. To explore the relationship and mapping between them, some works predict SC from FC [18,19] while some works map the structural to functional [21] and both of these works achieve unique prediction. It is accountable for predicting SC from FC since SC is relatively static. However, since the FC is relatively dynamic, the predicting FC from SC is quite challenging and it should be diverse predictions rather than deterministic prediction. In general, it is necessary and challenging to explore the one-to-many relationship between the one subjects's SC and the FC by predicting diverse FC from SC [2]. The significance of this work is that once we have anatomical structure of brain represented by SC, we can predict diverse developments of brain activities represented by FC, which helps to reveal individual brain's static-dynamic structural-functional mode.

In this work, we propose the diverse generations with the idea of conditional GAN [12], i.e. the input of the generator is random noise to realize diverse generations and is in condition of SC to generate FC. To improve the quality of generation, some works introduce the reconstruct supervision in generator while some works provide more supervisions in discriminator. In this work, we propose MCGAN, a multi-contexts discriminator based generative adversarial network to take both advantages. The comparison is shown in Fig. 1. Specifically, the multi-contexts discriminator provides three kinds of supervisions, i.e. edge-level, node-level and graph-level, to strengthen the generator. We adopt edge-based graph convolution method [11] to model the FC encoding. In addition, we utilize monte-carlo mean samples to enlarge the FC data for supervision. We validate our MCGAN on the HCP dataset [15] and ADNI dataset [10] and the results show that our method can generate diverse and meaningful FC. To the best of our knowledge, our method is the first to explore one-to-many SC-FC relationship.

2 Method

2.1 Model Overview

The framework of our proposed MCGAN is shown in Fig. 2 and detailed below.

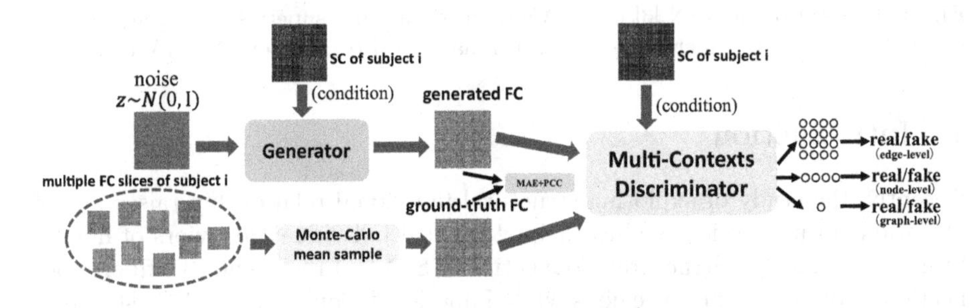

Fig. 2. Illustration of our proposed MCGAN.

The objective of this work is to predict diverse functional connectivity (FC) with the instruction of structural connectivity (SC). The proposed MCGAN is based on the generative adversarial network (GAN) framework [3], which consists of two parts. (a) Generator. For FC generation of subject i, the input of the generator is random noise which introduces the diverse generations and is in condition of SC of subject i. We utilize monte-carlo mean FC samples of subject i for supervision. (b) Multi-contexts Discriminator. The discriminator distinguishes the generated FC and real one. We introduce multi contexts supervision to discriminator, i.e. edge-level, node-level and graph-level, to strengthen the generator. We adopt edge-based graph convolution method [11] to model the FC encoding. In the prediction stage, for a specific subject, the generator predicts diverse FC with corresponding numbers of random noise input and in condition of the same SC of subject.

2.2 MCGAN

The adversarial objective of generating FC from SC with GAN model is defined as:

$$
\begin{aligned}
L_{adv} &= L^G + L^D, \\
L^G &= -E_{x,y}[logD(G(x,z))], \\
L^D &= -E_{x,y}[logD(x,y)] - E_{x,z}[log(1 - D(G(x,z)))].
\end{aligned} \tag{1}
$$

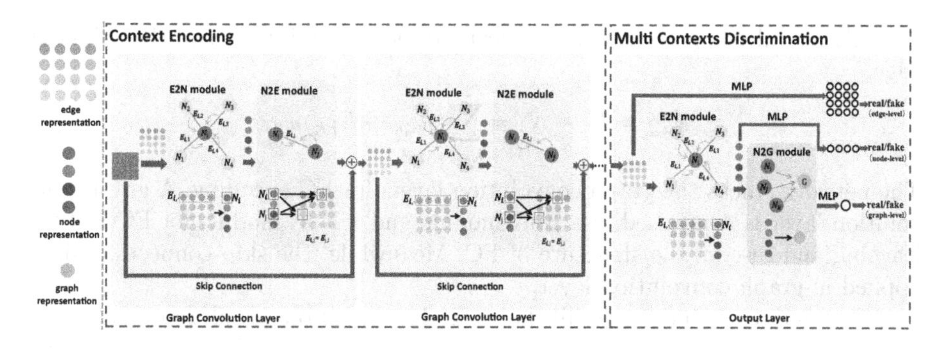

Fig. 3. Illustration of the proposed multi-contexts discriminator which consists of the context encoding and multi contexts discrimination.

where $z \in R^{n \times n}$ denotes the noise sample from $\mathcal{N}(0, I)$ to introduce the diversity of the generated FC. $x \in R^{n \times n}$ denotes the SC is the condition which instructs the generation of FC [12] and $y \in R^{n \times n}$ denotes the real FC.

Multi-contexts Discriminator

The original discriminator learns the global representation of synthetic or real data to distinguish them. However, it is insufficient to provide powerful feedback to encourage the generator to predict realistic data. Many works improve the discriminator and show that the stronger the discriminator is, the better the generator is [13,14]. Motivated by this insight, we propose a multi-contexts discriminator, which provides edge-level, node-level and graph-level supervisions to implicitly and explicitly strengthen the generator. The multi-contexts discriminator shown in Fig. 3 consists of two parts, context encoding and multi contexts discrimination.

Context Encoding. The context encoding is for feature extraction. For brain network, regions are represented as nodes and the links between regions are represented as edges. Since FC is defined as the correlation between the active brain regions, it represents the edge feature. However, most of the graph convolution networks (GCN) model the node feature extraction. To address the FC encoding, we adopt the edge-based graph convolution method for feature extraction [11]. There are three basic modules, i.e. E2N module (edge to node), N2E module (node to edge), N2G module (node to graph).

E2N module. Given a FC $E \in R^{n \times n}$, where $E_{i,j}$ denotes the edge between region i and region j, it aggregates the linking edges of region i into a node representation:

$$N_i = \sum_{k=1}^{n} E_{i,k} w_k, \quad i = 1, 2, ..., n \tag{2}$$

where w_i is trainable weight and N_i is the feature of $i^t h$ node.

N2E module. It propagates the feature of node i and node j to their linking edge:

$$E_{i,j} = N_i + N_j = \sum_{k=1}^{n} (E_{i,k} + E_{j,k}) w_k, \tag{3}$$

The network stacks the graph convolution layers for FC encoding. A graph convolution layer is comprised of a E2N module and a N2E module for FC feature learning and reserve the structure of FC. Meanwhile, the skip connection [5] is applied in graph convolution layer.

N2G module. It integrates all the nodes feature into a graph representation:

$$G = \sum_{i=1}^{n} N_i w_i. \tag{4}$$

The output layer of the network is comprised of a E2N module and a N2G module to encode the FC representation into the node representation and the graph representation in series.

Multi Contexts Discrimination. The original discriminator $D(\cdot)$ classifies the input data to be real or fake. In contrast, our proposed multi-contexts discriminator provides three kind of supervisions, i.e. edge-level, node-level and graph-level, to implicitly and explicitly strengthen the generator. Mathematically, our discriminator loss is comprised of three parts:

$$\begin{aligned}
&L^D = L^D_{edge} + L^D_{node} + L^D_{graph}, \\
&L^D_{edge} = -E_{x,y}[\sum_{i,j} log D_{edge}(x,y)_{i,j}] - E_{x,z}[\sum_{i,j} log(1 - D_{edge}(x, G(x,z))_{i,j})], \\
&L^D_{node} = -E_{x,y}[\sum_{n} log D_{node}(x,y)_n] - E_{x,z}[\sum_{n} log(1 - D_{node}(x, G(x,z))_n)], \\
&L^D_{graph} = -E_{x,y}[log D_{graph}(x,y)] - E_{x,z}[log(1 - D_{graph}(x, G(x,z)))].
\end{aligned} \tag{5}$$

In the output layer, we first use three MLP to transform the three kinds of context representation. Then the $D_{edge}(\cdot)$, $D_{node}(\cdot)$, $D_{graph}(\cdot)$ respectively classify the transformed edge representation, node representation and graph representation to be real or fake.

Correspondingly, the objective of the generator is:

$$\begin{aligned}
L^G = -E_{x,z}[&\sum_{i,j} log D_{edge}(x, G(x,z))_{i,j} + \\
&\sum_{n} log D_{node}(x, G(x,z))_n + log D_{graph}(x, G(x,z))].
\end{aligned} \tag{6}$$

The multi contexts supervision improves the discriminator, which implicitly strengthens the generator in adversarial training process. Meanwhile, it feedbacks the fine-grained information to the generator by backpropagation, which explicitly strengthens the generator and encourages it to predict realistic FC.

Monte-Carlo Mean Samples of FC
FC represents the development of brain activities so that it is relatively dynamic. While SC represents the fibers connection to indicate the anatomical structure of brain [2], which is relatively static. Since the limited of FC data and the supervision of the generated FC should not be limited in finite subspace, we augment the FC data via monte-carlo mean:

$$y = \frac{1}{m}\sum y_i. \tag{7}$$

where y_i denotes multiple real FC of each subject and m can be fixed or dynamic during training, which denotes the monte-carlo sampling, $m = 1, 2, ...M$, M is the maximum samples for each subject of dataset. The linear combination of FC enlarges the data space for supervision, making the generator predict diverse FC.

Reconstruction Loss of Generator
We adopt reconstruction loss in generator to make it predict realistic FC which includes the mean absolute error (MAE) and Pearson's correlation coefficient (PCC) between the generated FC and the training FC sample.

$$L_{rec} = E_{x,z,y}[||y - G(x,z)||] + E_{x,z,y}PCC(y, G(x,z)). \tag{8}$$

Full Objective
In summary, the objective function of the MCGAN is defined as:

$$L = L_{adv} + L_{rec}. \tag{9}$$

where L_{adv} optimizes the adversarial training of the generator and discriminator, L_{rec} optimizes the generator for realistic prediction.

3 Experiments

3.1 Setup

Datasets. We evaluate our MCGAN on the Human Connectome Project (HCP) dataset [15] and Alzheimer's Disease Neuroimaging Initiative (ADNI) dataset [10]. We apply the standard preprocessing procedures [20] for both datasets. We use diffusion magnetic resonance imaging (diffusion MRI) as the SC and resting state functional magnetic resonance imaging (rs-fMRI) as FC. We divide the BOLD signal into 20 slices for HCP and 10 slices for ADNI for each subject and obtain FC by calculating Pearson's correlation coefficient (PCC) between the segmented signal series. Both of the SC and FC are normalized.

Evaluation Metrics. (a) Quality. Instinctively, for each subject, the generated FC should be similar to one of the real FC. Therefore, for each generated FC of a subject, we calculate the minimum mean absolute error and the maximum Pearson's correlation coefficient to evaluate the quality of generation with the

Table 1. Performance comparison of the quality and diversity.

HCP/ADNI	Prediction	MAE ↓	PCC ↑	FID ↓
GAN [3]	Diverse	1.09/1.12	0.02/0.04	116.5/130.2
PixelGAN [9]	Diverse	0.28/0.30	0.52/0.49	63.5/70.2
UNet-GAN [14]	Diverse	0.89/0.96	0.09/0.05	40.5/49.2
MGCN-GAN [18]	Deterministic	0.12/0.12	0.80/0.76	76.6/79.2
MCGAN	Diverse	0.24/0.26	0.56/0.51	18.8/22.3

most similar real FC. (b) Diversity. We use Frechet Inception distance (FID) [6] to evaluate the diversity of the FC generation. FID compares the distribution of the generated and real data, to simultaneously evaluate the quality and diversity. The lower FID score indicates the better quality and diversity.

Implementation Details. Both of the generator and discriminator are in condition of SC and we set the almost equivalent parameters for them to ensure the efficient adversarial training. We respectively set the learning rate of 0.0001 and 0.0004 for generator and discriminator. At prediction stage, we generate corresponding number of FC samples with dataset for each subject (i.e. 20 for HCP and 10 for ADNI).

3.2 Results

Main Results. We compare our MCGAN with the original GAN [3], PixelGAN which is with reconstruct supervision in generator [9], UNet-GAN which is with more supervisions in discriminator [14]. In addition, many works perform the deterministic prediction on structural or functional prediction [8,17,18]. We compare our MCGAN with MGCN-GAN which is a deterministic prediction to map FC to SC [18]. The results are shown in Table 1. PixelGAN achieves good reconstruction but not well in diversity, which is demonstrated in original paper. UNet-GAN achieves ordinary diversity but fails to reconstruct meaningful FC. MGCN-GAN achieves the best reconstruction since it is a deterministic prediction method but lacks of diversity. In comparing, our MCGAN achieves quality-diversity trade-off.

Ablation Study. We conduct ablation study to investigate the contribution of different components and the results are shown in Table 2. The multi contexts discrimination improves the quality and diversity, indicating that it strengthens the generator for better prediction. The reconstruction provides the supervision to better quality and the monte-carlo mean samples provide the better diversity.

Visualization. We generate multiple FC for each subject and we visualize two FC of each subject. The visualization of the generated and real FC is shown in Fig. 4. Each column denotes the generated FC and the corresponding most similar real FC. It shows that our method achieves diverse predictions rather than only one deterministic prediction.

Table 2. Comparison of ablation study.

HCP/ADNI	MAE ↓	PCC ↑	FID ↓
MCGAN	0.24/0.26	0.56/0.51	18.8/22.3
w/o multi contexts discrimination	0.30/0.33	0.49/0.47	29.8/33.5
w/o mae&pcc	0.84/0.93	0.12/0.10	39.7/45.3
w/o monte-carlo samples	0.19/0.26	0.58/0.52	31.5/31.9

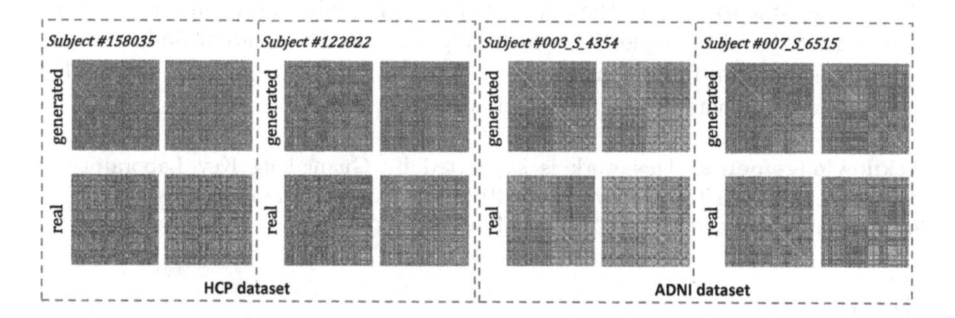

Fig. 4. Visualization of the generated and the real FC. Each column denotes the generated FC (first row) and the corresponding most similar real FC (second row) of multi-slices label of subject. It shows that our method can generate the diverse and realistic FC.

4 Discussion

Despite this work predicts diverse FC to explore one-to-many SC-FC relationship, the generated FC is not quite diverse enough to some extent. In our implementation, we attempt to use the diffusion model [7] to generate FC and it achieves better effect in diversity than this work. By visualization, we find that it generates diverse FC. However, since the diffusion model is trained with the only objective of noise prediction for its particularity and without the reconstruction task, it does not perform well in the evaluation metrics of MAE and PCC, indicating that it can not generate meaningful FC similar to this work. In general, the prediction of FC from SC should be diverse since the SC is relatively static and the FC is relatively dynamic and diverse. To the best of our knowledge, our method is the first to explore one-to-many SC-FC relationship and this task is challenging and significant. We lead to the future work for further exploration and there are many questions to be resolved. Moreover, to reveal individual brain's static-dynamic structural-functional mode, predicting original functional signals is more flexible than predicting FC since it directly explores the development of the brain activities.

5 Conclusion

In this work, for diverse generations, we propose a multi-contexts discriminator based GAN named MCGAN, which provides three kind of supervisions to strengthen the generator, including edge-level, node-level and graph-level. We adopt edge-based graph convolution method for FC encoding and we utilize monte-carlo mean samples to enlarge the FC data for supervision. The experiments show that our method can generate diverse and meaningful FC from SC. We are the first to explore the one-to-many relationship between one subject's individual SC and multiple FC, which helps to reveal individual brain's static-dynamic structural-functional mode. We lead to the future work for further exploration and there are many questions to be resolved.

Acknowledgements. This work is supported by Guangdong Key Laboratory of Human Digital Twin Technology (2022B1212010004) and Fundamental Research Funds for the Central Universities 2022ZYGXZR104.

References

1. Amiez, C., Petrides, M.: Neuroimaging evidence of the anatomo-functional organization of the human cingulate motor areas. Cereb. Cortex **24**(3), 563–578 (2014)
2. De Benedictis, A., et al.: Anatomo-functional study of the temporo-parieto-occipital region: dissection, tractographic and brain mapping evidence from a neurosurgical perspective. J. Anat. **225**(2), 132–151 (2014)
3. Goodfellow, I., et al.: Generative adversarial networks. Commun. ACM **63**(11), 139–144 (2020)
4. Gratton, C., et al.: Functional brain networks are dominated by stable group and individual factors, not cognitive or daily variation. Neuron **98**(2), 439–452 (2018)
5. He, K., Zhang, X., Ren, S., Sun, J.: Deep residual learning for image recognition. In: Proceedings of the IEEE Conference on Computer Vision and Pattern Recognition, pp. 770–778 (2016)
6. Heusel, M., Ramsauer, H., Unterthiner, T., Nessler, B., Hochreiter, S.: GANs trained by a two time-scale update rule converge to a local Nash equilibrium. In: NeurIPS, vol. 30 (2017)
7. Ho, J., Jain, A., Abbeel, P.: Denoising diffusion probabilistic models. In: NeurIPS, vol. 33, pp. 6840–6851 (2020)
8. Hu, D., et al.: Reference-relation guided autoencoder with deep CCA restriction for awake-to-sleep brain functional connectome prediction. In: de Bruijne, M., et al. (eds.) MICCAI 2021. LNCS, vol. 12903, pp. 231–240. Springer, Cham (2021). https://doi.org/10.1007/978-3-030-87199-4_22
9. Isola, P., Zhu, J.Y., Zhou, T., Efros, A.A.: Image-to-image translation with conditional adversarial networks. In: CVPR, pp. 1125–1134 (2017)
10. Jack, C.R., Jr., et al.: The Alzheimer's disease neuroimaging initiative (ADNI): MRI methods. J. Magn. Reson. Imaging: Official J. Int. Soc. Magn. Reson. Med. **27**(4), 685–691 (2008)
11. Li, Y., et al.: Brain connectivity based graph convolutional networks and its application to infant age prediction. IEEE Trans. Med. Imaging **41**(10), 2764–2776 (2022)

12. Mirza, M., Osindero, S.: Conditional generative adversarial nets. arXiv preprint arXiv:1411.1784 (2014)
13. Odena, A., Olah, C., Shlens, J.: Conditional image synthesis with auxiliary classifier GANs. In: ICML, pp. 2642–2651. PMLR (2017)
14. Schonfeld, E., Schiele, B., Khoreva, A.: A U-Net based discriminator for generative adversarial networks. In: CVPR, pp. 8207–8216 (2020)
15. Van Essen, D.C., et al.: The human connectome project: a data acquisition perspective. Neuroimage **62**(4), 2222–2231 (2012)
16. Yeh, F.C., et al.: Quantifying differences and similarities in whole-brain white matter architecture using local connectome fingerprints. PLoS Comput. Biol. **12**(11), e1005203 (2016)
17. Yu, X., et al.: Longitudinal infant functional connectivity prediction via conditional intensive triplet network. In: Wang, L., Dou, Q., Fletcher, P.T., Speidel, S., Li, S. (eds.) MICCAI 2022. LNCS, vol. 13438, pp. 255–264. Springer, Cham (2022). https://doi.org/10.1007/978-3-031-16452-1_25
18. Zhang, L., Wang, L., Zhu, D.: Recovering brain structural connectivity from functional connectivity via multi-GCN based generative adversarial network. In: Martel, A.L., et al. (eds.) MICCAI 2020. LNCS, vol. 12267, pp. 53–61. Springer, Cham (2020). https://doi.org/10.1007/978-3-030-59728-3_6
19. Zhang, L., Wang, L., Zhu, D., Initiative, A.D.N., et al.: Predicting brain structural network using functional connectivity. Med. Image Anal. **79**, 102463 (2022)
20. Zhu, D., et al.: Connectome-scale assessments of structural and functional connectivity in MCI. Hum. Brain Mapp. **35**(7), 2911–2923 (2014)
21. Zhu, Z., Huang, T., Zhen, Z., Wang, B., Wu, X., Li, S.: From sMRI to task-fMRI: a unified geometric deep learning framework for cross-modal brain anatomo-functional mapping. Med. Image Anal. **83**, 102681 (2023)

DeepGraphDMD: Interpretable Spatio-Temporal Decomposition of Non-linear Functional Brain Network Dynamics

Md Asadullah Turja[1(✉)], Martin Styner[1], and Guorong Wu[2]

[1] Department of Computer Science, University of North Carolina at Chapel Hill, Chapel Hill, USA
mturja@cs.unc.edu, styner@email.unc.edu

[2] Department of Psychiatry, University of North Carolina at Chapel Hill, Chapel Hill, USA
guorong_wu@med.unc.edu

Abstract. Functional brain dynamics is supported by parallel and overlapping functional network modes that are associated with specific neural circuits. Decomposing these network modes from fMRI data and finding their temporal characteristics is challenging due to their time-varying nature and the non-linearity of the functional dynamics. Dynamic Mode Decomposition (DMD) algorithms have been quite popular for solving this decomposition problem in recent years. In this work, we apply GraphDMD—an extension of the DMD for network data—to extract the dynamic network modes and their temporal characteristics from the fMRI time series in an interpretable manner. GraphDMD, however, regards the underlying system as a linear dynamical system that is sub-optimal for extracting the network modes from non-linear functional data. In this work, we develop a generalized version of the GraphDMD algorithm—DeepGraphDMD—applicable to arbitrary non-linear graph dynamical systems. DeepGraphDMD is an autoencoder-based deep learning model that learns Koopman eigenfunctions for graph data and embeds the non-linear graph dynamics into a latent linear space. We show the effectiveness of our method in both simulated data and the HCP resting-state fMRI data. In the HCP data, DeepGraphDMD provides novel insights into cognitive brain functions by discovering two major network modes related to fluid and crystallized intelligence.

Keywords: Component Analysis · Dynamic Mode Decomposition · Koopman Operator · Non-linear Graph Dynamical Systems · Functional Network Dynamics · State-space models · Graph Representation Learning

Supplementary Information The online version contains supplementary material available at https://doi.org/10.1007/978-3-031-43993-3_35.

H. Greenspan et al. (Eds.): MICCAI 2023, LNCS 14227, pp. 358–368, 2023.
https://doi.org/10.1007/978-3-031-43993-3_35

1 Introduction

The human brain has evolved to support a set of complementary and temporally varying brain network organizations enabling parallel and higher-order information processing [16,20,24]. Decoupling these networks from a non-linear mixture of signals (such as functional MRI) and extracting their temporal characteristics in an interpretable manner has been a long-standing challenge in the neuroscience community.

Conventional mode/component decomposition methods such as Principal Component Analysis (PCA) or Independent Component Analysis (ICA) assume the modes to be static [7,15,26] and thus sub-optimal for the functional networks generated by time-varying modes. Dynamic Mode Decomposition (DMD) can be treated as a dynamic extension of such component analysis methods since it allows its modes to oscillate over time with a fixed frequency [18]. This assumption is appropriate for the human brain as the functional brain organizations are supported by oscillatory network modes [2,8,11,23,27]. An extension of DMD for network data called GraphDMD [4] preserves the graph structure of the networks during the decomposition. In our work, we extend GraphDMD to a sequence of sliding window based dynamic functional connectivity (dNFC) networks to extract independent and oscillatory functional network modes.

Under the hood, GraphDMD regards the network sequence as a linear dynamical system (LDS) where a linear operator shifts the current network state one time-point in the future. The LDS assumption, however, is not optimal for modeling functional brain networks that exhibit complex non-linearity such as rapid synchronization and desynchronization as well as transient events [6]. Articles [3,13] propose switching linear dynamical system (SLDS) to tackle the nonlinearity of spatiotemporal data by a piecewise linear approximation. While these models offer interpretability, their shallow architecture limits their generalizability to arbitrary nonlinear systems. On the other hand, the methods in [5,10,21] model the non-linearity with a deep neural network. While these models have more representation capabilities compared to SLDS, the latent states are not interpretable. More importantly, all of these methods consider the node-level dynamics instead of the network dynamics.

Here, we propose a novel Deep Graph Dynamic Mode Decomposition (DeepGraphDMD) algorithm that applies to arbitrary non-linear network dynamics while maintaining interpretability in the latent space. Our method uses Koopman operator theory to lift a non-linear dynamical system into a linear space through a set of Koopman eigenfunctions (Fig. 1a). There has been a growing line of work that learns these measurement functions using deep autoencoder architectures [14,22]. Training these autoencoders for network data, however, has two unique challenges – 1. preserving the edge identity in the latent space so that the network modes are interpretable, 2. enforcing linearity in the latent space for the high dimensional network data. In DeepGraphDMD, we tackle the first challenge by indirectly computing the network embeddings by a novel node embedding scheme. For the second challenge, we introduce a sparse Koopman operator to reduce the complexity of the learning problem. We evaluate the

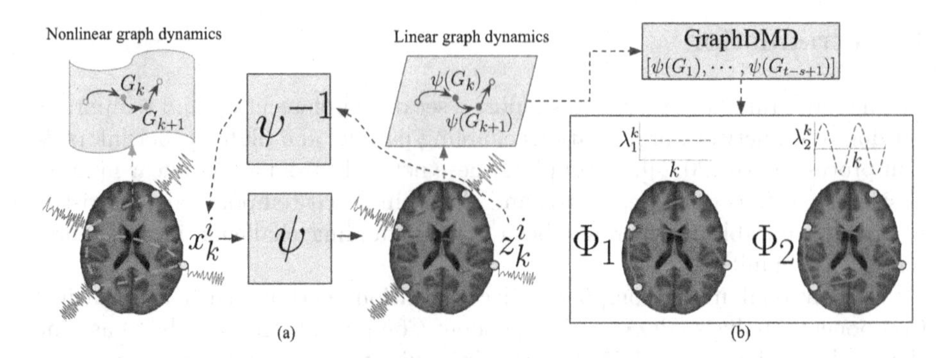

Fig. 1. (a) Illustration of the DeepGraphDMD model that embeds a nonlinear graph dynamical system into a linear space, and, (b) interpretable dynamic modes and their temporal characteristics after applying GraphDMD in the linear space.

effectiveness of our novel method in both simulated data and resting-state fMRI (rs-fMRI) data from Human Connectome Project.

2 Methodology

Let's assume $X \in \mathbb{R}^{n \times t}$ is a matrix containing the BOLD (blood-oxygen-level-dependent) signal of n brain regions (ROIs) in its rows at t time frames sampled at every $k\Delta t$ time points, where Δt is the temporal resolution. To compute the dynamic connectivity matrix at time point $k\Delta t$, a snapshot $X_k = X_{:,k:k+s}$ is taken in a sliding window of s time frames. A correlation matrix $G_k \in \mathbb{R}^{n \times n}$ is then computed from X_k by taking the pearson correlation between the rows of X_k, i.e., $G_k^{ij} = \texttt{pearson}(x_k^i, x_k^j)$ where x_k^i, x_k^j are the ith and j^{th} row of X_k respectively. This yields a sequence of graphs $G = [G_1, G_2, \cdots, G_{t-s+1}]$. Let's also assume that $g_k \in \mathcal{R}^{n^2}$ is a vectorized version of G_k, i.e. $g_k = vec(G_k)$ and $g \in \mathcal{R}^{n^2 \times (t-s+1)}$ is a matrix containing g_k in its columns. The goal is to decouple the overlapping spatiotemporal modes from the network sequence G using – 1. Graph Dynamic Mode Decomposition algorithm, and 2. a novel Deep Learning-based Graph Dynamic Mode Decomposition algorithm.

2.1 Graph Dynamic Mode Decomposition

GraphDMD [4] assumes that g_k follows an LDS:

$$g_{k+1} = Ag_k \tag{1}$$

where $A \in \mathbb{R}^{n^2 \times n^2}$ is a linear operator that shifts the current state g_k to the state at the next time frame g_{k+1}. To extract the low dimensional global network dynamics, GraphDMD projects A into a lower dimensional space $\hat{A}$ using tensor-train decomposition, applies eigendecomposition of $\hat{A}$, and projects the eigenvectors back to the original space which we refer to as dynamic modes (DMs). GraphDMD uses tensor-train decomposition to maintain the network structure of g_k and thus, the DMs from GraphDMD can be reshaped into $n \times n$ adjacency matrix forms. Let's assume these DMs are $\Phi_1, \Phi_2, \cdots, \Phi_r$ where $\Phi_p \in \mathbb{C}^{n \times n}$ and the corresponding eigenvalues are $\lambda_1, \lambda_2, \cdots, \lambda_r$ where $\lambda_p \in \mathbb{C}$ (Fig. 1b). Here, r is the total number of DMs. Φ_p corresponds to the coherent spatial mode and λ_p defines its temporal characteristics (growth/decay rate and frequencies). We can see this by unrolling Eq. 1 in time:

$$g_{k+1} = A^k g_1 = \sum_{p=1}^{r} \Phi_p \lambda_p^k b_p = \sum_{p=1}^{r} \Phi_p a_p^k \exp(\omega_p k \Delta t) b_p \tag{2}$$

where $\lambda_p = a_p \exp(\omega_p \Delta t)$, $\Phi^\dagger$ is the conjugate transpose of Φ, $b_p = vec(\Phi_p^\dagger) g_1$ is the projection of the initial value onto the DMD modes, $a_p = ||\lambda_p||$ is the growth/decay rate and $\omega_p = \text{Im}(\ln \lambda_p)/\Delta t$ is the angular frequency of Φ_p.

2.2 Adaptation of Graph-DMD for Nonlinear Graph Dynamics

Since the dynamics of the functional networks are often non-linear, the linearity assumption of Eq. 1 is sub-optimal. In this regard, we resort to Koopman operator theory to transform the non-linear system into an LDS using a set of Koopman eigenfunctions ψ, i.e., $\psi(g_{k+1}) = A\psi(g_k)$ [9]. We learn ψ using a deep autoencoder-based architecture—DeepGraphDMD—where the encoder and the decoder are trained to approximate ψ and ψ^{-1}, respectively. We enforce $\psi(g_k)$ to follow an LDS by applying Latent Koopman Invariant Loss [22] in the form:

$$\mathcal{L}_{lkis} = ||Y' - (Y'Y^\dagger)Y||_F^2 \tag{3}$$

where $Y = \begin{pmatrix} | & | & & | \\ \psi(g_1) & \psi(g_2) & \cdots & \psi(g_{t-s}) \\ | & | & & | \end{pmatrix}$, $Y' = \begin{pmatrix} | & | & & | \\ \psi(g_2) & \psi(g_3) & \cdots & \psi(g_{t-s+1}) \\ | & | & & | \end{pmatrix}$ are two matrices with columns stacked with $\psi(g_k)$ and $Y^\dagger$ is the right inverse of Y. After training, we reshape $\psi(g_k)$ into a $n \times n$ network $\psi(G_k)$ and generate the latent network sequence $\psi(G_1), \cdots, \psi(G_{t-s+1})$. We then apply GraphDMD (described in Sect. 2.1) on this latent and linearized network sequence to extract the DMs Φ_p and their corresponding λ_p.

However, there are two unique challenges of learning network embeddings using the DeepGraphDMD model: 1. the edge identity and, thus, the interpretability will be lost in the latent space if we directly embed g_k using ψ, and 2. $Y^\dagger$ doesn't exist, and thus $\mathcal{L}_{lkis}$ can't be computed because Y is low rank with the number of rows $\frac{n(n-1)}{2} >>$ the number of columns $t - s + 1$.

To solve the first problem, instead of learning $\psi(g_k)$ directly, we embed the BOLD signal x_k^i of each ROI independently using the encoder to learn the latent embeddings z_k^i (Fig. 1a). We then compute the pearson correlation between the latent embeddings of the ROIs to get the Koopman eigenfunctions of g_k i.e., $\psi(g_k^{ij}) = \texttt{pearson}(z_k^i, z_k^j)$. The weights of the encoder and decoder are shared across the ROIs.

The second problem arises because the Koopman operator A regresses the value of an edge at the next time-point as a linear combination of all the other edges at the current time-point, i.e., $g_{k+1}^{ij} = \sum_{p,q=1}^{N} w_{pq} g_k^{pq}$. This results in $\mathcal{O}(n^2)$ covariates with $t - s + 1 << \mathcal{O}(n^2)$ samples making the regression ill-posed. We propose a sparse Koopman operator where each edge g_k^{ij} is regressed using only the edges that share a common end-point with it, i.e., $g_{k+1}^{ij} = \sum_{p=1,p\neq i,j}^{n} w_{ip} g_k^{ip} + \sum_{q=1,q\neq i,j}^{n} w_{qj} g_k^{qj} + w_{ij} g_k^{ij}$ (Supplementary Fig. 1). Since there are only $\mathcal{O}(n)$ such edges, it solves the ill-posedness of the regression.

Other than $\mathcal{L}_{lkis}$, we also train the autoencoder with a reconstruction loss $\mathcal{L}_{recon}$ which is the mean-squared error (MSE) between x_k^i and the reconstructed output from the decoder $\hat{x}_k^i$. Moreover, a regularizer $\mathcal{L}_{reg}$ in the form of an MSE loss between g_k and the latent $\psi(g_k)$ is also added. The final loss is the following:

$$\mathcal{L} = \mathcal{L}_{recon} + \alpha \mathcal{L}_{lkis} + \beta \mathcal{L}_{reg} \tag{4}$$

where α and β are hyper-parameters. We choose α, β, and other hyperparameters using grid search on the validation set. The network architecture and the values of the hyper-parameters of DeepGraphDMD training are shown in Supplementary Fig. 1. The code is available in[1].

2.3 Window-Based GraphDMD

We apply GraphDMD in a short window of size 64 time frames with a step size of 4 time frames instead of the whole sequence G because, in real-world fMRI data, both the frequency and the structure of the DMs can change over time. We then combine the DMs across different sliding windows using the following post-processing steps:

Post-processing of the DMs: We first group the DMs within the frequency bins: 0–0.01 Hz, 0.01–0.04 Hz, 0.04–0.08 Hz, 0.08–0.12 Hz, and 0.12–0.16 Hz. We then cluster the DMs within each frequency bin using a clustering algorithm and select the cluster centroids as the representative DMs (except for the first bin where we average the DMs). We chose the optimal clustering algorithm to be Spherical KMeans [1] (among Gaussian Mixture Model, KMeans, Spherical KMeans, DBSCAN, and, KMedoids) and the optimal number of clusters to be 3 for every frequency bin based on silhouette analysis [17] (Supplementary Fig. 2). We use this frequency binning technique to allow for slight variations of ω of a DM over the scanning session. To align these representative DMs across subjects, we apply another round of spherical clustering on the DMs from all subjects and align them based on their cluster memberships.

3 Experiments

3.1 Dataset

We use rs-fMRI for 840 subjects from the HCP Dense Connectome dataset[2] [25]. Each fMRI image was acquired with a temporal resolution (Δt) of 0.72 s and a

[1] https://github.com/mturja-vf-ic-bd/DeepGraphDMD.git.
[2] https://www.humanconnectome.org/storage/app/media/documentation/s1200/HCP1200-DenseConnectome+PTN+Appendix-July2017.pdf.

2 mm isotropic spatial resolution using a 3T Siemens Skyra scanner. Individual subjects underwent four rs-fMRI runs of 14.4 min each (1200 frames per run). Group-ICA using FSL's MELODIC tool [7] was applied to parcellate the brain into 50 functional regions (ROIs). To find the correlation between cognition with the rs-fMRI data, we select two behavioral measures related to fluid intelligence: `CogFluidComp`, `PMAT24_A_CR` and one measure related to crystallized intelligence: `ReadEng`, and, the normalized scores of the fluid and crystallized cognition measures: `CogTotalComp`. We regress out the confounding factors: age, gender, and head motion from these behavioral measures using ordinary least squares [12].

3.2 Baseline Methods

We compare GraphDMD and DeepGraphDMD against three decomposition methods: Principal Component Analysis (PCA), Independent Component Analysis (ICA), and standard Dynamic Mode Decomposition (DMD) [18]. We use the `sklearn` decomposition library for PCA[3] and ICA[4] and the `pyDMD`[5] library for standard DMD. We apply PCA and ICA on g, and DMD directly on the bold signal X instead of g (for reasons described in Sect. 4.2). We choose the number of components (`n_components`) to be three for these decomposition methods, (Results for other `n_components` values are shown in Supplementary Table 1). The components are aligned across subjects using spherical clustering similar to the GraphDMD modes (Sect. 2.3). We also compare with static functional connectivity (sFC), which is the pairwise pearson correlation between brain regions across all time frames.

3.3 Simulation Study

We generate a sequence of dynamic adjacency matrices G using Eq. 2 from three time-varying modes Φ_1, Φ_2, Φ_3 with corresponding frequencies $\omega_1 \sim \mathcal{N}(0.1, 0.05)$, $\omega_2 \sim \mathcal{N}(1, 0.1)$, $\omega_3 \sim \mathcal{N}(2.5, 0.1)$ (Hz). Each Φ_p is a 32×32 block diagonal matrices with block sizes 16, 8, and 4. We choose $a_1 = 1.01$, $a_2 = 0.9$, $a_3 = 1.05$ and $b_1 = b_2 = b_3 = 1$. We simulate the process for $k = 1, \cdots, 29$ time-points yielding a sequence of 30 matrices of shape 32×32. We repeat the process ten times with different $\omega_1, \omega_2, \omega_3$ and generate ten matrix sequences. We apply PCA, ICA, and GraphDMD (Sect. 2.1) on G to extract three components and compare them against the ground truth modes using pearson correlation.

3.4 Application of GraphDMD and DeepGraphDMD in HCP Data

Comparison of DMs with sFC: The ground truth DMs are unknown for the HCP dataset; however, we can use the sFC as a substitute for the ground truth DM with $\omega = 0$ (static DM). sFC offsets the DMs with $\omega > 0$ as they have both

---
[3] sklearn.decomposition.PCA.
[4] sklearn.decomposition.FastICA.
[5] https://mathlab.github.io/PyDMD/dmd.html.

positive and negative cycles, and thus only retain the static DM. For comparison, we compute the pearson correlation between the DM within the frequency bin 0–0.01 Hz and sFC for both GraphDMD and DeepGraphDMD. For PCA and ICA, we take the maximum value of the correlation between sFC and the (PCA or ICA) components.

Regression Analysis of Behavioral Measures from HCP: In this experiment, we regress the behavioral measures with the DMs within each frequency bin (Sect. 2.3) using Elastic-net. As an input to the Elastic-net, we take the real part of the upper diagonal part of the DM and flatten it into a vector. We then train the Elastic-net in two ways—1. single-band: where we train the Elastic-net independently with the DMs in each frequency bin, and 2. multi-band: we concatenate two DMs in the frequency bins: 0–0.01 Hz and 0.08–0.12 Hz and regress using the concatenated vector. For evaluation, we compute the correlation coefficient r between the predicted and the true values of the measures.

4 Results

4.1 Simulation Study

In Fig. 2a, we show the results after applying PCA, ICA, and, GraphDMD on the simulated data described in Sect. 3.3. Since the DMs in this data are oscillating, the data generated from this process are more likely to be overlapping compared to when the modes are static. As a result, methods that assume static modes, such as PCA and ICA, struggle to decouple the DMs and discover modes in overlapping high-density regions. For example, in Mode 2 of ICA, we can see the remnants of Mode 3 in the blue boxes and Mode 1 (negative cycle) in the orange boxes. We observe similar scenarios in the Mode 3 of ICA and Mode 1, Mode 2, and Mode 3 of PCA (the red boxes). On the other hand, the DMs from GraphDMD have fewer remnants from other modes and closely resemble the ground truth. To empirically compare, the mean ($\pm$ std) pearson correlation for PCA, ICA, and, GraphDMD are 0.81($\pm$0.04), 0.88($\pm$0.03), and 0.98($\pm$0.01).

4.2 Application of GraphDMD and DeepGraphDMD in HCP Data

Comparison of DMs with sFC: The average pearson correlations with sFC across all the subjects are 0.6($\pm$0.09), 0.6($\pm$0.09), 0.84($\pm$0.09), and 0.86($\pm$0.05) for PCA, ICA, GraphDMD, and, DeepGraphDMD (Fig. 2b) respectively. This shows that the DMD-based methods can robustly decouple the static DM from time-varying DMs. In comparison, the corresponding PCA and ICA component has significantly lower correlation due to the overlap from the higher frequency components.

Regression Analysis of Behavioral Measures from HCP: We show the values of r across different methods in Table 1. We only show the results for two frequency bins 0–0.01 Hz and 0.08–0.12 Hz, as the DMs in the other bins are not significantly correlated ($r < 0.2$) with the behavioral measures (Supplementary Table 2). The table shows that multi-band training with the DMs from the DMD-based methods significantly improves the regression performance over the baseline methods. Compared to sFC, GraphDMD improves r by 22%, 6%, 0.7%, and, 3% for `CogFluidComp`, `PMAT24_A_CR`, `ReadEng`, `CogTotalComp`, respectively and DeepGraphDMD further improves the performance by 5%, 2.2%, 0.7%, and, 1.5%, respectively. Significant performance improvement for `CogFluidComp` can be explained by the DM within the bin 0.08–0.12 Hz. This DM provides additional information related to fluid intelligence ($r = 0.227$ for GraphDMD) to which the sFC doesn't have access. By considering non-linearity, DeepGraphDMD extracts more robust and less noisy DMs (Fig. 2b–c), and hence, it improves the regression performance by 8% compared to GraphDMD in this frequency bin. By contrast, the standard DMD algorithm yields unstable modes with $a_p << 1$ when applied to the network sequence G. These modes have no correspondence across subjects and thus can't be used for regression. We instead apply DMD on the BOLD signal X, but the DMD modes show little

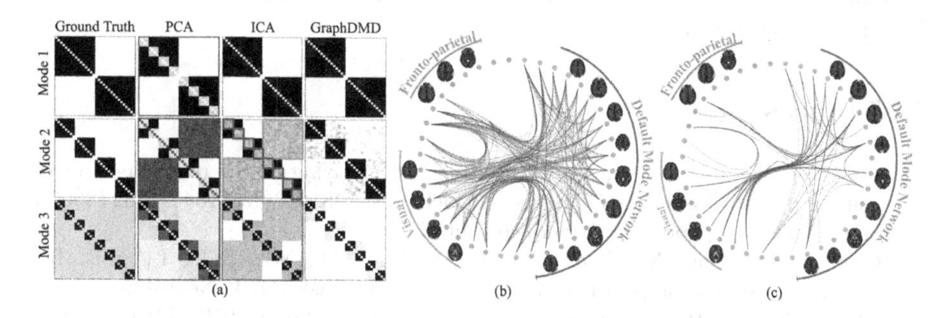

Fig. 2. (a) Ground truth network modes from simulated data (column 1) and extracted network modes from PCA (2nd column), ICA (3rd column), and, GraphDMD (4th column), (b) Circle plot of the average DMs with $\omega \approx 0$, (c) $\omega \in [0.08 - 0.12]$ from DeepGraphDMD organized based on common resting-state networks [19].

Table 1. Comparison of r for the behavioral measures across different methods.

	Frequency (Hz)	`CogFluidComp`	`PMAT24_A_CR`	`ReadEng`	`CogTotalComp`
sFC	N/A	**0.253 ± 0.003**	0.294 ± 0.004	0.407 ± 0.004	**0.440 ± 0.004**
PCA	N/A	0.109 ± 0.003	0.126 ± 0.003	0.224 ± 0.003	0.238 ± 0.003
ICA	N/A	0.148 ± 0.005	0.158 ± 0.004	0.239 ± 0.005	0.266 ± 0.006
DMD	0–0.01, 0.08–0.12	0.064 ± 0.002	0.169 ± 0.002	0.132 ± 0.003	0.138 ± 0.006
Graph DMD	0–0.01	0.254 ± 0.003	0.289 ± 0.004	0.402 ± 0.004	0.438 ± 0.003
	0.08–0.12	0.227 ± 0.004	0.193 ± 0.004	0.145 ± 0.004	0.248 ± 0.004
	0–0.01, 0.08–0.12	**0.308 ± 0.004**	0.312 ± 0.004	0.410 ± 0.003	**0.454 ± 0.004**
Deep Graph DMD	0–0.01	0.259 ± 0.003	0.290 ± 0.002	0.404 ± 0.002	0.439 ± 0.002
	0.08–0.12	0.245 ± 0.002	0.201 ± 0.004	0.144 ± 0.003	0.251 ± 0.004
	0–0.01, 0.08–0.12	**0.325 ± 0.003**	0.319 ± 0.003	0.413 ± 0.002	**0.461 ± 0.003**

correlation with the behavioral measures. PCA and ICA perform significantly worse than the baseline sFC method for all behavioral measures.

Traditional dynamical functional connectivity analysis methods (such as sliding window-based techniques) consider a sequence of network states. However, our results show that these states can be further decomposed into more atomic network modes. The importance of decoupling these network modes from non-linearly mixed fMRI signals using DeepGraphDMD has been shown in regressing behavioral measures from HCP data.

5 Conclusion

In this paper, we proposed a novel algorithm—DeepGraphDMD—to decouple spatiotemporal network modes in dynamic functional brain networks. Unlike other decomposition methods, DeepGraphDMD accounts for both the non-linear and the time-varying nature of the functional modes. As a result, these functional modes from DeepGraphDMD are more robust compared to their linear counterpart in GraphDMD and are shown to be correlated with fluid and crystallized intelligence measures.

References

1. Banerjee, A., Dhillon, I.S., Ghosh, J., Sra, S.: Clustering on the unit hypersphere using von Mises-Fisher distributions. J. Mach. Learn. Res. **6**(46), 1345–1382 (2005). http://jmlr.org/papers/v6/banerjee05a.html
2. Casorso, J., Kong, X., Chi, W., Van De Ville, D., Yeo, B.T., Liégeois, R.: Dynamic mode decomposition of resting-state and task fMRI. NeuroImage **194**, 42–54 (2019). https://doi.org/10.1016/j.neuroimage.2019.03.019. https://www.sciencedirect.com/science/article/pii/S1053811919301922
3. Fox, E., Sudderth, E., Jordan, M., Willsky, A.: Nonparametric Bayesian learning of switching linear dynamical systems. In: Koller, D., Schuurmans, D., Bengio, Y., Bottou, L. (eds.) Advances in Neural Information Processing Systems, vol. 21. Curran Associates, Inc. (2008). https://proceedings.neurips.cc/paper_files/paper/2008/file/950a4152c2b4aa3ad78bdd6b366cc179-Paper.pdf
4. Fujii, K., Takeishi, N., Hojo, M., Inaba, Y., Kawahara, Y.: Physically-interpretable classification of biological network dynamics for complex collective motions. Sci. Rep. **10** (2020). https://doi.org/10.1038/s41598-020-58064-w
5. Gao, Y., Archer, E.W., Paninski, L., Cunningham, J.P.: Linear dynamical neural population models through nonlinear embeddings. In: Lee, D., Sugiyama, M., Luxburg, U., Guyon, I., Garnett, R. (eds.) Advances in Neural Information Processing Systems, vol. 29. Curran Associates, Inc. (2016). https://proceedings.neurips.cc/paper_files/paper/2016/file/76dc611d6ebaafc66cc0879c71b5db5c-Paper.pdf
6. He, B.J.: Robust, transient neural dynamics during conscious perception. Trends Cogn. Sci. **22**(7), 563–565 (2018). https://doi.org/10.1016/J.TICS.2018.04.005. https://pubmed.ncbi.nlm.nih.gov/29764721/
7. Hyvarinen, A.: Fast and robust fixed-point algorithms for independent component analysis. IEEE Trans. Neural Netw. **10**(3), 626–634 (1999)

8. Ikeda, S., Kawano, K., Watanabe, S., Yamashita, O., Kawahara, Y.: Predicting behavior through dynamic modes in resting-state fMRI data. NeuroImage **247**, 118801 (2022). https://doi.org/10.1016/j.neuroimage.2021.118801. https://www.sciencedirect.com/science/article/pii/S1053811921010727
9. Koopman, B.O.: Hamiltonian systems and transformation in Hilbert space. Proc. Natl. Acad. Sci. U.S.A. **17**(5), 315 (1931). https://doi.org/10.1073/PNAS.17.5.315. https://www.ncbi.nlm.nih.gov/pmc/articles/PMC1076052/
10. Krishnan, R.G., Shalit, U., Sontag, D.A.: Structured inference networks for nonlinear state space models. In: AAAI Conference on Artificial Intelligence (2016)
11. Kunert-Graf, J.M., Eschenburg, K.M., Galas, D.J., Kutz, J.N., Rane, S.D., Brunton, B.W.: Extracting reproducible time-resolved resting state networks using dynamic mode decomposition. Front. Comput. Neurosci. **13** (2019). https://doi.org/10.3389/fncom.2019.00075. https://www.frontiersin.org/articles/10.3389/fncom.2019.00075
12. Liégeois, R., et al.: Resting brain dynamics at different timescales capture distinct aspects of human behavior. Nat. Commun. **10**(1), 2317 (2019)
13. Linderman, S., Johnson, M., Miller, A., Adams, R., Blei, D., Paninski, L.: Bayesian learning and inference in recurrent switching linear dynamical systems. In: Singh, A., Zhu, J. (eds.) Proceedings of the 20th International Conference on Artificial Intelligence and Statistics. Proceedings of Machine Learning Research, vol. 54, pp. 914–922. PMLR (2017). https://proceedings.mlr.press/v54/linderman17a.html
14. Lusch, B., Nathan Kutz, J., Brunton, S.L.: Deep learning for universal linear embeddings of nonlinear dynamics. Nat. Commun. **9**, 4950 (2018)
15. McKeown, M.J., et al.: Analysis of fMRI data by blind separation into independent spatial components. Hum. Brain Mapp. **6**(3), 160–188 (1998)
16. Osada, T., et al.: Parallel cognitive processing streams in human prefrontal cortex: parsing areal-level brain network for response inhibition. Cell Rep. **36**(12), 109732 (2021). https://doi.org/10.1016/j.celrep.2021.109732. https://www.sciencedirect.com/science/article/pii/S2211124721011815
17. Rousseeuw, P.J.: Silhouettes: a graphical aid to the interpretation and validation of cluster analysis. J. Comput. Appl. Math. **20**, 53–65 (1987). https://doi.org/10.1016/0377-0427(87)90125-7. https://www.sciencedirect.com/science/article/pii/0377042787901257
18. Schmid, P.J.: Dynamic mode decomposition of numerical and experimental data. J. Fluid Mech. **656**, 5–28 (2010)
19. Seitzman, B.A., et al.: A set of functionally-defined brain regions with improved representation of the subcortex and cerebellum. NeuroImage **206**, 116290 (2020). https://doi.org/10.1016/j.neuroimage.2019.116290. https://www.sciencedirect.com/science/article/pii/S105381191930881X
20. Sigman, M., Dehaene, S.: Brain mechanisms of serial and parallel processing during dual-task performance. J. Neurosci. **28**(30), 7585–7598 (2008)
21. Sussillo, D., Jozefowicz, R., Abbott, L., Pandarinath, C.: LFADS - latent factor analysis via dynamical systems (2016)
22. Takeishi, N., Kawahara, Y., Yairi, T.: Learning Koopman invariant subspaces for dynamic mode decomposition. In: Proceedings of the 31st International Conference on Neural Information Processing Systems, NIPS 2017, pp. 1130–1140. Curran Associates Inc., Red Hook (2017)

23. Turja, M.A., Wu, G., Yang, D., Styner, M.A.: Learning the latent heat diffusion process through structural brain network from longitudinal β-amyloid data. In: Heinrich, M., Dou, Q., de Bruijne, M., Lellmann, J., Schläfer, A., Ernst, F. (eds.) Proceedings of the Fourth Conference on Medical Imaging with Deep Learning. Proceedings of Machine Learning Research, vol. 143, pp. 761–773. PMLR (2021). https://proceedings.mlr.press/v143/turja21a.html
24. Turja, M.A., Zsembik, L.C.P., Wu, G., Styner, M.: Constructing consistent longitudinal brain networks by group-wise graph learning. In: Shen, D., et al. (eds.) MICCAI 2019. LNCS, vol. 11766, pp. 654–662. Springer, Cham (2019). https://doi.org/10.1007/978-3-030-32248-9_73
25. Van Essen, D.C., et al.: The human connectome project: a data acquisition perspective. Neuroimage **62**(4), 2222–2231 (2012)
26. Viviani, R., Grön, G., Spitzer, M.: Functional principal component analysis of fMRI data. Hum. Brain Mapp. **24**(2), 109–129 (2005)
27. Xiao, J., et al.: A spatio-temporal decomposition framework for dynamic functional connectivity in the human brain. NeuroImage **263**, 119618 (2022). https://doi.org/10.1016/j.neuroimage.2022.119618. https://www.sciencedirect.com/science/article/pii/S1053811922007339

Disentangling Site Effects with Cycle-Consistent Adversarial Autoencoder for Multi-site Cortical Data Harmonization

Fenqiang Zhao[1], Zhengwang Wu[1], Dajiang Zhu[2], Tianming Liu[3], John Gilmore[4], Weili Lin[1], Li Wang[1], and Gang Li[1(✉)]

[1] Department of Radiology and BRIC, University of North Carolina at Chapel Hill, Chapel Hill, NC, USA
gang_li@med.unc.edu
[2] Department of Computer Science and Engineering, University of Texas at Arlington, Arlington, TX, USA
[3] Department of Computer Science, University of Georgia, Athens, GA, USA
[4] Department of Psychiatry, University of North Carolina at Chapel Hill, Chapel Hill, NC, USA

Abstract. Modern multi-site neuroimaging studies are known to be biased by significant site effects observed in imaging data and their derived structural and functional features. Although many statistical models and deep learning methods have been proposed to eliminate the site effects while maintaining biological characteristics, they have two major drawbacks. *First*, statistical models are applicable for harmonizing regional-level data but are inherently not suitable to represent the complex non-linear mapping of vertex-wise cortical property maps. *Second*, existing deep learning methods can only harmonize data between two sites, which are practically less useful in multi-site data harmonization scenario and also ignore the rich information in the whole dataset. To address these issues, we develop a novel, flexible deep learning method to harmonize multi-site cortical surface property maps. Specifically, to detect and remove site effects, we employ a surface-based autoencoder and decompose the encoded cortical features into site-related and site-unrelated components and use an adversarial strategy to encourage the disentanglement. Then decoding the site-unrelated features with other site features can generate mappings across different sites. To learn more controllable and meaningful mappings, we enforce the cycle consistency between forward and backward mappings. Our method can thus efficiently learn rich information from the whole dataset and generate realistic harmonized surface maps at the target site. Experiments on harmonizing infant cortical thickness maps of 2,342 scans from four sites with different scanners and imaging protocols validate the superior performance of our method on both site effects removal and biological variability preservation compared to other methods. To the best of our knowledge, this is the largest validation of different methods on infant cortical data harmonization.

H. Greenspan et al. (Eds.): MICCAI 2023, LNCS 14227, pp. 369–379, 2023.
https://doi.org/10.1007/978-3-031-43993-3_36

1 Introduction

Modern large multi-site neuroimaging studies have shown increasing power to detect biological variability of interest and provided invaluable insights into the changes underlying neurodevelopmental and neurodegenerative disorders [8,12]. However, the aggregation of neuroimaging data across different sites and scanners typically introduces non-biological variability, also known as site effects [7]. Many harmonization methods are thus proposed to remove such unwanted site effects while preserving biological variability [14,20].

Inspired by image-to-image translation techniques in the computer vision field [30], many methods harmonized neuroimaging data in the image domain by synthesizing brain images among different sites [2,4,11]. However, image-based harmonization techniques cannot guarantee that the final derived structural or functional features are free of site effects, due to the huge complexity in neuroimaging data processing pipelines [23]. Alternatively, feature-based harmonization techniques have been proposed to directly mitigate the site effects in the final derived volumetric [19,21], structural [7,19,21,28], functional [26], or diffusion magnetic resonance imaging (dMRI) features [18,25]. Most of these methods are developed based on linear statistical models, e.g., Combat [7] and its variants [21], for harmonizing summarized cortical properties in each region of interest (ROI), e.g., cortical thickness [7] and surface area [20]. Although achieving promising results for long ROI-wise data (number of samples $n >$ number of features p), statistical models have inherent problems when applied to wide vertex-wise data ($n \ll p$) with spatially fine-grained cortical information [1,15]. This is because the growth in data dimension and possible associations in wide vertex-wise data makes the model more complex and consequently statistical inference becomes less tractable and precise [15]. Moreover, since the sources and underlying mechanisms of site effects are heterogeneous and not fully uncovered, linear models and their hypothesis on the model's parameter distribution may not sufficiently represent the complex non-linear mapping of vertex-wise data. Therefore, deep learning methods, that make minimal assumptions about the data-generating mechanisms and thus can automatically learn to fit the complex non-linear mappings, are more favored [15]. For example, Zhao et al. [28] proposed a surface-to-surface CycleGAN to harmonize cortical thickness maps between two sites, which, however, is inefficient and inconvenient in practice for harmonizing multi-site data, because a model needs to be re-trained between any two sites and ignores rich global information in the whole multi-site data.

To address these issues, we develop a novel, flexible deep learning-based method to harmonize multi-site cortical surface maps in a vertex-wise manner free of parcellation scheme, thus preserving the spatially detailed cortical measure information after harmonization and enabling comprehensive vertex-wise analysis in further studies. Our approach builds on a surface-based autoencoder and uses an adversarial strategy [5] to encourage the disentanglement of site-related and site-unrelated components. To learn a more controllable and meaningful generative model, we enforce the cycle consistency between forward and backward mappings, inspired by CycleGAN [30]. Our method also shares

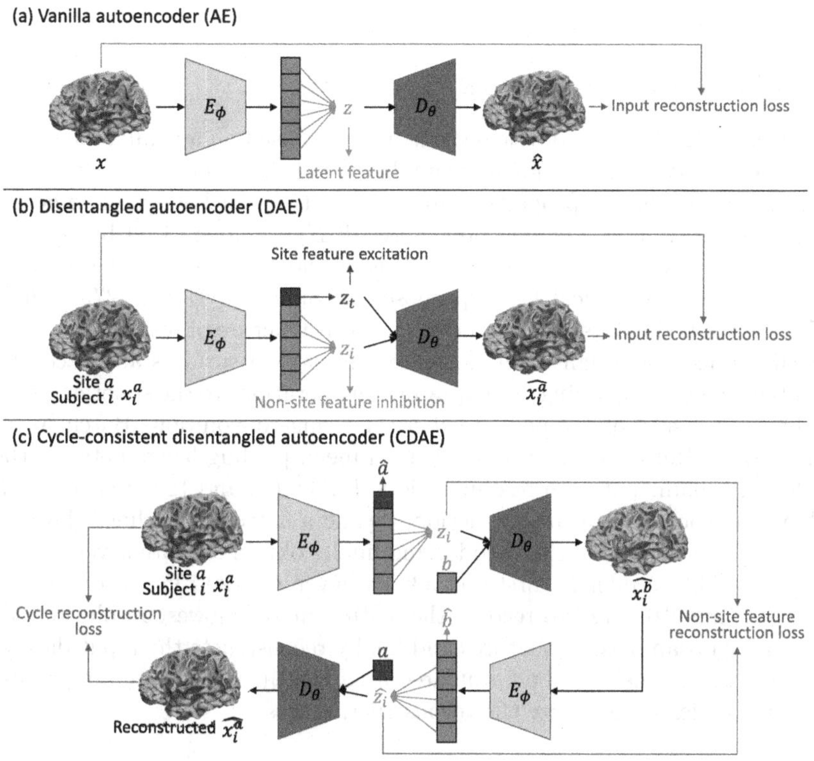

Fig. 1. Illustration of different autoencoder models for cortical surface data representation and harmonization.

certain similarity with the adversarial autoencoder [17], domain-adversarial neural networks [9], and guided variational autoencoder [5], but significantly differs in the operation space, network structure, and most importantly, the aim and application, where we focus on a more meaningful data harmonization task in the neuroimaging field. To sum up, the main contributions of this paper are:

1. We propose a novel method to fill the critical gap for multi-site vertex-wise cortical data harmonization, while there exist statistical models suitable for ROI-wise data harmonization and CycleGAN model for two-site data;
2. Taking advantage of disentanglement learning and adversarial strategies, we successfully learn a transparent, controllable, and meaningful generative model for efficiently mapping cortical surface data across different sites;
3. To the best of our knowledge, we performed the largest validation on infant cortical data harmonization with 2,342 scans from 4 sites and demonstrate the superior performance of our method compared to other methods.

2 Method

2.1 Vanilla Autoencoder (AE)

As shown in Fig. 1(a), we developed our method based on a vanilla autoencoder (AE) to enjoy its transparency and simplicity. Let $X = (x_1, ..., x_n)$ denote a set of input cortical surface property maps, e.g., cortical thickness map, where n is the number of samples. The encoder network E_ϕ parameterized by ϕ extracts the latent features $z : z = E_\phi(x)$, which is developed based on the fundamental operations of spherical CNN proposed in Spherical U-Net [27]. Spherical U-Net leverages the consistent and regular data structure of resampled spherical cortical surfaces to design the 1-ring filter on cortical surfaces and accordingly extends convolution, pooling, and upsampling operations to the spherical surface. Herein, E_ϕ consists of 5 repeated spherical 1-ring-convolution+Batch Normalization (BN)+ReLU layers with 4 spherical mean pooling layers between them. The feature channel at each resolution is 8, 16, 32, 64, and 128, respectively. To enable more compact feature extraction, we add a flatten and a linear layer with 512 neurons to the end of the encoder and finally obtain the latent vector z with size 1×512. The decoder D_θ first uses a linear layer with 20,736 neurons and then reshapes it into 162×128 to recover the feature map at lowest resolution. Then it gradually upsamples the features and finally reconstructs the input data with its original size, which can be formulated as: $\hat{x} = D_\theta(z)$. The training process of an AE thus tries to minimize the reconstruction loss:

$$\mathcal{L}_{recon}(\phi, \theta) = \sum_{i=1}^{n} \|x - D_\theta(E_\phi(x))\|^2 \tag{1}$$

2.2 Disentangled Autoencoder (DAE)

To detect and remove site effects from multi-site vertex-wise cortical measurements, we introduce a disentanglement learning strategy for its controllability and interpretability [5]. As shown in Fig. 1(b), suppose for training data $X = (x_1, \ldots, x_n)$, there are M sites in total and the corresponding ground-truth site labels are $T = (t_1, ..., t_n)$. Let $z = (z_t, z_i)$, where z_t defines a $1 \times M$ vector representing the site-wise classification probability and z_i represents remaining latent variables. We use the adversarial excitation and inhibition method [5] to encourage the disentanglement of the latent variables:

$$\mathcal{L}_{Excitation}(\phi, W_t) = -\sum_{i=1}^{n}[t_i \cdot log(W_t(z_t)) + (1 - t_i) \cdot log(1 - W_t(z_t))] \tag{2}$$

$$\mathcal{L}_{Inhibition}(\phi, W_i) = \sum_{i=1}^{n}[t_i \cdot log(W_i(z_i)) + (1 - t_i) \cdot log(1 - W_i(z_i))] \tag{3}$$

where W_t and W_i refer to the site classifiers using latent variables z_t and z_i, respectively, t_i is a one-hot vector encoding the ground-truth site label. The multi-class binary cross entropy loss in Eq. (2) thus encourages z_t to be the same as correct site label. Conversely, Eq. (3) is an inhibition process, as we want the remaining variables z_i to be site-unrelated. W_t is a simple sigmoid layer for outputting probability, while W_i consists of four linear layers with BN+ReLU+dropout layers between them and a sigmoid layer at the end. This is because z_t should be directly correlated with the site-wise predictions, while z_i with more features needs a stronger classifier to detect site-related information in it and adversarially train the encoder to extract site-unrelated features for z_i.

2.3 Cycle-Consistent Disentangled Autoencoder (CDAE)

Previously, a DAE, after successful training, will be directly used for image synthesis [5] or MR image harmonization [31] by combining site-unrelated variables with the target site label. Such image generation process may introduce new and even unseen style patterns in computer vision applications, which, however, are artifacts that are not meaningful and acceptable in medical imaging field [16].

To learn a more controllable and meaningful generative model, we propose to train the decoder with additional constraints on backward mapping. As shown in Fig. 1(c), after generating the surface map from source site a to target site b, denoted as $\widehat{x_i^b}$, we backward map it to the source site: $(\hat{a}, z_i) = E_\phi(x_i^a), \widehat{x_i^b} = D_\theta(b, z_i)$, $(\hat{b}, \hat{z}_i) = E_\phi(\widehat{x_i^b}), \widehat{x_i^a} = D_\theta(a, \hat{z}_i)$, and enforce the cycle-consistency loss to guarantee the generated surface map is meaningful to the original surface map:

$$\mathcal{L}_{cycle}(\phi, \theta) = \sum_{i=1}^{n} \|\widehat{x_i^a} - x_i^a\|^2. \quad (4)$$

We also add the cycle-consistency loss to the latent site-unrelated variable z_i to reinforce the correlation between the generated and original surface map:

$$\mathcal{L}_{latent}(\phi, \theta) = \sum_{i=1}^{n} \|\hat{z}_i - z_i\|^2, \quad (5)$$

and an explicit correlation loss to further reduce the ambiguity of indirect cycle-consistency losses for better preserving global structural information:

$$\mathcal{L}_{cc}(\phi, \theta) = \sum_{i=1}^{n} cov(\widehat{x_i^b}, x_i^a)/(\sigma_{\widehat{x_i^b}} \sigma_{x_i^a}) \quad (6)$$

where cov denotes the covariance, σ is the standard deviation. Besides, we also use the losses in AE and DAE to train the backward mapping. This means we reuse the site classifier W_t to adversarially train the decoder to generate fake surface maps at the target site that cannot be distinguished from the true maps, which is a standard GAN training process [10]. Finally, our model can automatically detect and disentangle the site-related feature from the input data using DAE losses, and generate more meaningful mappings across sites and better preserve individual variability using CDAE losses, thus fulfilling the requirement of data harmonization task.

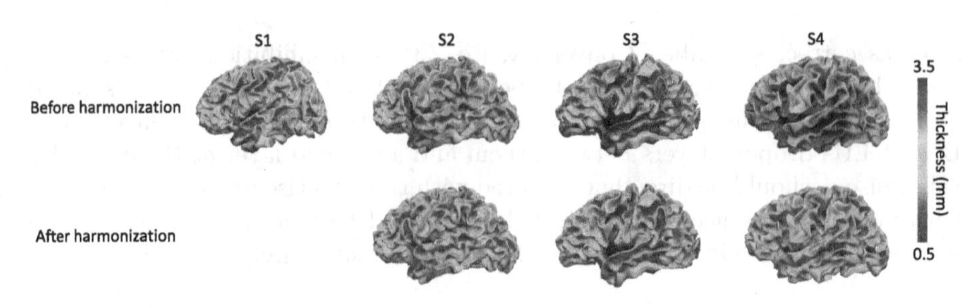

Fig. 2. Representative cortical thickness map harmonization results of four age-matched subjects from four different sites.

3 Experiments and Results

3.1 Experimental Setting

We evaluated our method using 4 large infant datasets (S1, S2, S3, and S4) acquired by different scanners and imaging protocols with resolutions ranging from isotropic 0.8 mm^3 (S1) to $1.25 \times 1.25 \times 1.95$ mm^3 (S3). S1 [13] and S4 [12] are two public datasets with 609 and 779 scans, respectively. S2 [22] and S3 [24] are two in-house datasets containing 335 and 619 scans, respectively. To the best of our knowledge, this is the largest collection of infant MRI datasets for harmonization and joint analysis purposes. All images were preprocessed using the infant-dedicated pipeline iBEAT V2.0 (http://www.ibeat.cloud/) [23]. Then the reconstructed cortical surfaces were mapped onto the sphere, nonlinearly aligned based on geometric features, and further resampled with 40,962 vertices [6]. Figure 2 shows typical reconstructed surfaces color-coded by cortical thickness in the first row. As can be seen, the site effects introduced by different acquisition methods largely dominate the data variance and will inevitably mislead the joint analysis of the four sites if without performing harmonization.

We implemented our method using PyTorch and public Spherical U-Net code [27,29]. We trained the models in an easy-to-hard manner by gradually adding losses from Eq. (1) to Eq. (6) using Adam optimizer with a fixed learning rate 5e-4. All surfaces were randomly split into training and testing sets with the proportion of 7:3. The weights of different loss terms are empirically set as 1.0, 1.0, 0.5, 0.5, 4.0, 5.0 for $\mathcal{L}_{recon}$, $\mathcal{L}_{Excitation}$, $\mathcal{L}_{Inhibition}$, $\mathcal{L}_{cycle}$, $\mathcal{L}_{latent}$, and $\mathcal{L}_{cc}$, respectively. All the experiments were run on a PC with an Nvidia RTX 3080-Ti GPU and an Intel Core i7-9700K CPU. We adopted the popular statistical model, Combat [7], as the baseline method for comparison. We used its public code with age and sex as biological covariates for harmonizing the four sites' cortical thickness in a vertex-wise manner and ROI-wise manner, referred to *Vertex-wise Combat* and *ROI-wise Combat*, respectively. Of note, Combat harmonizes multiple sites into one intermediate site, while our method maps less reliable sites (with low-quality images) to the more reliable site (with high-quality images), i.e., S1, in this work.

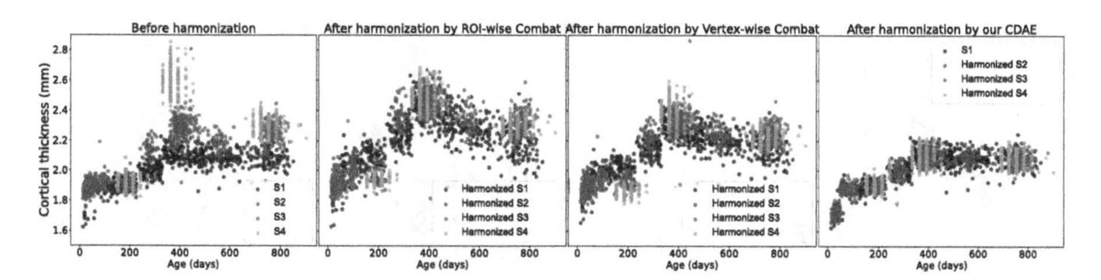

Fig. 3. Scatter plots of population-level developmental trajectories of average cortical thickness in four datasets after harmonization by different methods.

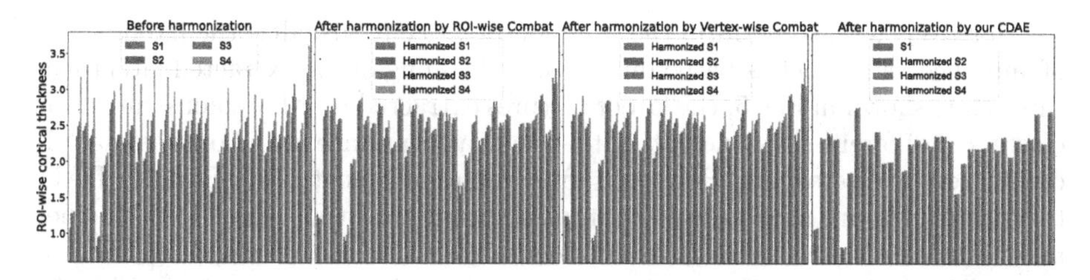

Fig. 4. ROI-wise average cortical thickness of 1-year-old infants at different sites. In each subplot, the x-axis represents the 36 ROIs of Desikan-Killiany parcellation [3].

3.2 Results

Validation on Removing Site Effects. To validate if site effects are successfully removed, we first show the population-level developmental trajectory of average cortical thickness in Fig. 3. To fairly compare with statistical models that use all data and deliver more reliable comparison with larger sample size, we draw the figures based on the whole dataset. As shown, the developmental trajectories from different sites are not consistent and comparable before harmonization and after harmonization using Combat, but are well harmonized to a common space and enable joint analysis using our method. We also compare the ROI-level thickness values before and after harmonization as performed in [7,28]. To be less biased by the data distribution of ages, we show the average thickness of 1-year-old infants for each ROI in Desikan-Killiany parcellation [3]. As shown in Fig. 4, the differences of ROI-wise thickness of different sites are significant before harmonization (all $p < 0.05$) and still remain after harmonization by Combat, but are not significant after harmonization by our method (all $p > 0.1$). Of note, although our method deals with vertex-wise data, compared to ROI-wise Combat, Fig. 3 and Fig. 4 still show better performance of our method on population-level and region-level site effects removal. These results indicate that Combat and other linear models indeed suffer from limited modeling ability and thus are not able to capture the complex non-linear mapping in heterogeneous multi-site data, while our method based on deep learning can effectively solve this problem with the proposed loss functions.

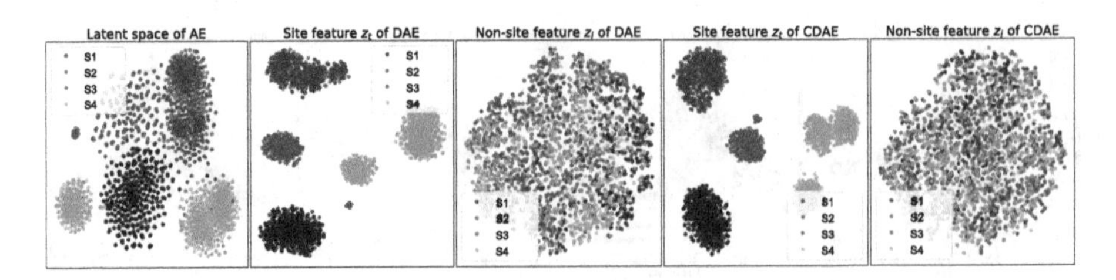

Fig. 5. t-SNE visualization of the latent variables of different autoencoders.

Moreover, we also perform the common practice to visualize the latent space of autoencoders, which is based on t-SNE embedding of the extracted latent features z. As shown in Fig. 5, the latent features extracted by AE demonstrate clear clusters of different sites, indicating that the data variance in the original data is dominated by site effects. However, after disentanglement learning using DAE, the site effects are successfully disentangled and removed, and CDAE further improves the results thanks to more constraints on backward mapping. Finally, to quantitatively evaluate site effects removal, we also attempted to predict the site from harmonized cortical thickness features [7]. Following the settings in [7], we used a support vector machine (SVM) model with radial basis kernel and the hyper-parameters of the model are selected using grid search based on 10-fold validation on the training set. After fitting on the training set, the SVM classification accuracies on the testing set are 74.3%, 92.2%, 51.4%, **40.2%** for ROI-wise Combat, Vertex-wise Combat, DAE, and CDAE, respectively. A lower accuracy of DAE and CDAE indicates that our method successfully removes site effects in the cortical thickness measurements, while Combat still preserves certain site-identifiable information.

Validation on Preserving Individual Variability. While it is important to validate if a harmonization method removes site effects, it is equally important to show that the method preserves biological variability after harmonization. Following [28], we computed the Euclidean distances between any two surface maps in the same site, thus forming a distance matrix, denoted as $H_{i,j}^{n \times n} = \|x_i - x_j\|^2$. We computed the Pearson correlation coefficient (PCC) of the distance matrices before and after harmonization to estimate if relative distances of different scans in the same site are preserved after harmonization. The average PCC values across sites are 0.853, 0.921, 0.842, **0.959** for ROI-wise Combat, Vertex-wise Combat, DAE, and CDAE, respectively, suggesting that our method preserves the most individual differences. From Fig. 2, we can also see that individual variability is well preserved while the site-related information is successfully removed with our method.

Validation on Downstream Task. We further investigate how different harmonization methods affect a downstream task. Same as in [7], we used a support vector regression (SVR) model with radial basis function to predict the scan age using cortical thickness features. After hyper-parameters selection and training, the mean absolute error on the testing set is 81.1, 59.6, 50.5, 56.3, and **45.1**

days for no harmonization, ROI-wise Combat, Vertex-wise Combat, DAE and CDAE, respectively. The coefficient of determination, or R^2, are 0.823, 0.865, 0.892, 0.884, **0.936** for no harmonization, ROI-wise Combat, Vertex-wise Combat, DAE, and CDAE, respectively. Note that Vertex-wise Combat improves the prediction accuracy of ROI-wise Combat possibly due to more spatial details preservation. All these results indicate that our CDAE model substantially increases the accuracy and trustworthiness of joint analysis of multi-site data compared to no harmonization and Combat-based harmonization methods.

4 Conclusion

In this paper, to address the important yet unsolved problem of multi-site vertex-wise cortical data harmonization, we proposed a novel, flexible deep learning-based method, named Cycle-consistent Disentangled Autoencoder (CDAE). Our CDAE takes advantage of the simplicity and transparency of autoencoder, the controllability and interpretability of disentanglement learning, and reinforced meaningfulness of cycle-consistency to successfully learn the complex non-linear mapping among heterogeneous multi-site data, which is inherently difficult and unsuitable for existing methods. Both visual and quantitative results on four datasets with 2,342 scans show the effectiveness of our method on both site effects removal and biological variability preservation. Our method will not only facilitate multi-site vertex-wise neuroimaging data analysis but also inspire novel directions in learning-based data harmonization.

Acknowledgements. This work was supported in part by the National Institutes of Health (NIH) under Grants MH116225, MH117943, MH123202, AG075582, and NS128534.

References

1. Bzdok, D.: Classical statistics and statistical learning in imaging neuroscience. Front. Neurosci. **11**, 543 (2017)
2. Cackowski, S., Barbier, E.L., Dojat, M., Christen, T.: ImUnity: a generalizable VAE-GAN solution for multicenter MR image harmonization. Med. Image Anal. **88**, 102799 (2023)
3. Desikan, R.S., et al.: An automated labeling system for subdividing the human cerebral cortex on MRI scans into gyral based regions of interest. Neuroimage **31**(3), 968–980 (2006)
4. Dewey, B.E., et al.: DeepHarmony: a deep learning approach to contrast harmonization across scanner changes. Magn. Reson. Imaging **64**, 160–170 (2019)
5. Ding, Z., et al.: Guided variational autoencoder for disentanglement learning. In: Proceedings of the IEEE/CVF Conference on Computer Vision and Pattern Recognition, pp. 7920–7929 (2020)
6. Fischl, B.: FreeSurfer. Neuroimage **62**(2), 774–781 (2012)
7. Fortin, J.P., et al.: Harmonization of cortical thickness measurements across scanners and sites. Neuroimage **167**, 104–120 (2018)

8. Frisoni, G.B., Fox, N.C., Jack, C.R., Jr., Scheltens, P., Thompson, P.M.: The clinical use of structural MRI in Alzheimer disease. Nat. Rev. Neurol. **6**(2), 67–77 (2010)
9. Ganin, Y., et al.: Domain-adversarial training of neural networks. J. Mach. Learn. Res. **17**(1), 1–35 (2016)
10. Goodfellow, I., et al.: Generative adversarial networks. Commun. ACM **63**(11), 139–144 (2020)
11. Guan, H., Liu, M.: DomainATM: domain adaptation toolbox for medical data analysis. NeuroImage **268**, 119863 (2023)
12. Hazlett, H.C., et al.: Early brain development in infants at high risk for autism spectrum disorder. Nature **542**(7641), 348–351 (2017)
13. Howell, B.R., et al.: The UNC/UMN baby connectome project (BCP): an overview of the study design and protocol development. Neuroimage **185**, 891–905 (2019)
14. Hu, F., et al.: Image harmonization: a review of statistical and deep learning methods for removing batch effects and evaluation metrics for effective harmonization. NeuroImage **274**, 120125 (2023)
15. Ij, H.: Statistics versus machine learning. Nat. Meth. **15**(4), 233 (2018)
16. Kazeminia, S., et al.: GANs for medical image analysis. Artif. Intell. Med. **109**, 101938 (2020)
17. Makhzani, A., Shlens, J., Jaitly, N., Goodfellow, I., Frey, B.: Adversarial autoencoders. arXiv preprint arXiv:1511.05644 (2015)
18. Moyer, D., Ver Steeg, G., Tax, C.M., Thompson, P.M.: Scanner invariant representations for diffusion MRI harmonization. Magn. Reson. Med. **84**(4), 2174–2189 (2020)
19. Pomponio, R.: Harmonization of large MRI datasets for the analysis of brain imaging patterns throughout the lifespan. Neuroimage **208**, 116450 (2020)
20. Solanes, A., et al.: Removing the effects of the site in brain imaging machine-learning-measurement and extendable benchmark. Neuroimage **265**, 119800 (2023)
21. Torbati, M.E., et al.: A multi-scanner neuroimaging data harmonization using ravel and combat. Neuroimage **245**, 118703 (2021)
22. Wang, F., et al.: Developmental topography of cortical thickness during infancy. Proc. Natl. Acad. Sci. **116**(32), 15855–15860 (2019)
23. Wang, L., Wu, Z., Chen, L., Sun, Y., Lin, W., Li, G.: iBEAT V2. 0: a multisite-applicable, deep learning-based pipeline for infant cerebral cortical surface reconstruction. Nat. Protoc. 18, 1488–1509 (2023)
24. Xia, K.: Genetic influences on longitudinal trajectories of cortical thickness and surface area during the first 2 years of life. Cereb. Cortex **32**(2), 367–379 (2022)
25. Xia, Y., Shi, Y.: Personalized DMRI harmonization on cortical surface. In: Wang, L., Dou, Q., Fletcher, P.T., Speidel, S., Li, S. (eds.) MICCAI 2022. LNCS, vol. 13436, pp. 717–725. Springer, Cham (2022). https://doi.org/10.1007/978-3-031-16446-0_68
26. Yamashita, A., et al.: Harmonization of resting-state functional MRI data across multiple imaging sites via the separation of site differences into sampling bias and measurement bias. PLoS Biol. **17**(4), e3000042 (2019)
27. Zhao, F., et al.: Spherical deformable U-Net: application to cortical surface parcellation and development prediction. IEEE Trans. Med. Imaging **40**(4), 1217–1228 (2021)
28. Zhao, F., et al.: Harmonization of infant cortical thickness using surface-to-surface cycle-consistent adversarial networks. In: Shen, D., et al. (eds.) MICCAI 2019. LNCS, vol. 11767, pp. 475–483. Springer, Cham (2019). https://doi.org/10.1007/978-3-030-32251-9_52

29. Zhao, F.: Spherical U-Net on cortical surfaces: methods and applications. In: Chung, A.C.S., Gee, J.C., Yushkevich, P.A., Bao, S. (eds.) IPMI 2019. LNCS, vol. 11492, pp. 855–866. Springer, Cham (2019). https://doi.org/10.1007/978-3-030-20351-1_67
30. Zhu, J.Y., Park, T., Isola, P., Efros, A.A.: Unpaired image-to-image translation using cycle-consistent adversarial networks. In: Proceedings of the IEEE International Conference on Computer Vision, pp. 2223–2232 (2017)
31. Zuo, L., et al.: Unsupervised MR harmonization by learning disentangled representations using information bottleneck theory. Neuroimage **243**, 118569 (2021)

SurfFlow: A Flow-Based Approach for Rapid and Accurate Cortical Surface Reconstruction from Infant Brain MRI

Xiaoyang Chen[1,2], Junjie Zhao[3], Siyuan Liu[4], Sahar Ahmad[1,2], and Pew-Thian Yap[1,2](✉)

[1] Department of Radiology, University of North Carolina, Chapel Hill, NC, USA
ptyap@med.unc.edu

[2] Biomedical Research Imaging Center, University of North Carolina, Chapel Hill, NC, USA

[3] Department of Computer Science, University of North Carolina, Chapel Hill, NC, USA

[4] College of Marine Engineering, Dalian Maritime University, Dalian, China

Abstract. The infant brain undergoes rapid changes in volume, shape, and structural organization during the first postnatal year. Accurate cortical surface reconstruction (CSR) is essential for understanding rapid changes in cortical morphometry during early brain development. However, existing CSR methods, designed for adult brain MRI, fall short in reconstructing cortical surfaces from infant MRI, owing to the poor tissue contrasts, partial volume effects, and rapid changes in cortical folding patterns. Here, we introduce an infant-centric CSR method in light of these challenges. Our method, *SurfFlow*, utilizes three seamlessly connected deformation blocks to sequentially deform an initial template mesh to target cortical surfaces. Remarkably, our method can rapidly reconstruct a high-resolution cortical surface mesh with 360k vertices in approximately one second. Performance evaluation based on an MRI dataset of infants 0 to 12 months of age indicates that SurfFlow significantly reduces geometric errors and substantially improves mesh regularity compared with state-of-the-art deep learning approaches.

Keywords: Infant Brain MRI · Cortical Surface Reconstruction · Topology Preservation

1 Introduction

The human brain undergoes dramatic changes in size, shape, and tissue architecture during the first postnatal year, driven by cellular processes [1] that lead to cortical folding and the formation of convoluted gyri and sulci on the cerebral

X. Chen and J. Zhao—Equal contribution.

This work was supported in part by the United States National Institutes of Health (NIH) through grants MH125479 and EB008374.

H. Greenspan et al. (Eds.): MICCAI 2023, LNCS 14227, pp. 380–388, 2023.
https://doi.org/10.1007/978-3-031-43993-3_37

surface. Understanding cortical development in these early postnatal years is crucial for comprehending later-stage functional development. Cortical surface reconstruction (CSR) is a step necessary for functional and anatomical visualization, brain morphology, and quantification of cortical growth. However, reconstructing the cortical surface from infant MRI poses significant challenges due to the low signal-to-noise ratio, pronounced motion artifacts, complex folding patterns, and rapid brain expansion [2].

CSR methods developed so far include FreeSurfer [5], DeepCSR [4], Voxel2Mesh [13], PialNN [8], and CorticalFlow [7]. However, these methods are optimized primarily for adult brain MRI, typically with clear cortical gyral and sulcal patterns and good tissue contrasts. Their performance deteriorates significantly when applied to infant MRI due to challenges like narrow sulcal spaces, low tissue contrast, and partial volume effects. To address these issues, specific methods like the dHCP pipeline [9] and Infant FreeSurfer [14] have been developed. However, the dHCP pipeline is only effective for neonatal data and Infant FreeSurfer often fails to produce accurate cortical surfaces, especially for the first postnatal year when the brain undergoes dynamic changes in tissue contrasts and morphology. Additionally, the computational inefficiency of Infant FreeSurfer, which takes approximately 10 h to reconstruct cortical surfaces for a single subject, limits its applicability to large-scale studies.

In this paper, we propose *SurfFlow*, a geometric deep learning model designed to reconstruct accurate cortical surfaces from infant brain MRI. Our model comprises three cascaded deformation blocks, each responsible for predicting a flow field and constructing a diffeomorphic mapping for each vertex through solving a flow ordinary differential equation (ODE). The flow fields deform template meshes progressively to finally produce accurate genus-zero cortical surfaces. Our work offers a threefold contribution. First, we propose an efficient dual-modal flow-based CSR method, enabling the creation of high-resolution and high-quality mesh representations for complex cortical surfaces. Second, we propose a novel loss function that effectively regularizes the lengths of mesh edges, leading to substantial improvements in mesh quality. Third, our method represents the first attempt in tackling the challenging task of directly reconstructing cortical surfaces from infant brain MRI. Our method outperforms state-of-the-art methods by a significant margin, judging based on multiple surface evaluation metrics.

2 Methods

2.1 Overview

As depicted in Fig. 1a, SurfFlow consists of three deformation blocks, each consisting of a 3D Unet [3] and a diffeomorphic mesh deformation (DMD) module. The final mesh, either the pial surface or the white surface, is achieved by composing the three diffeomorphic deformations generated by the DMD modules. Each Unet takes a T1w-T2w image pair as input and, except the first network, also receives flow fields predicted by previous deformation blocks to predict a new

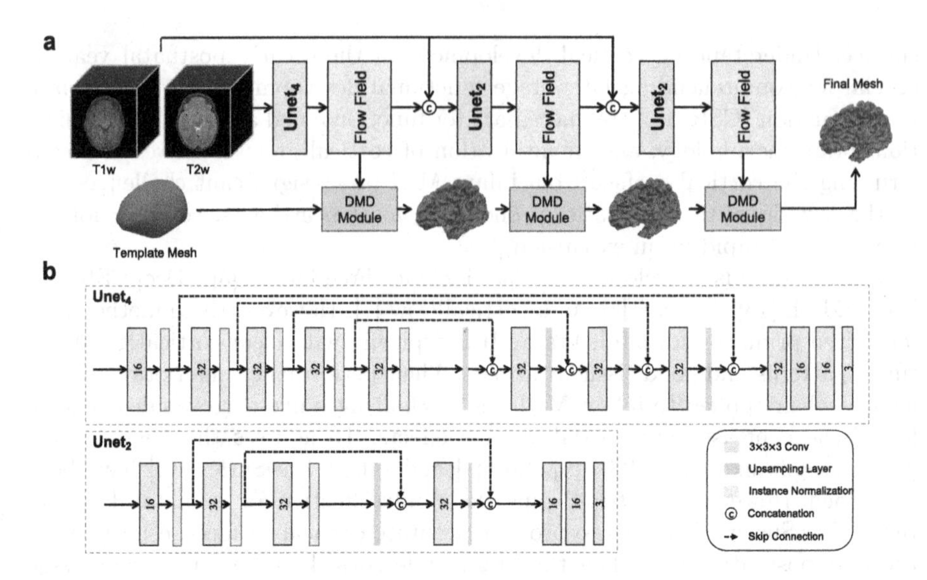

Fig. 1. (a) SurfFlow network structure. The network has three deformation blocks, each consisting of a 3D Unet and a diffeomorphic mesh deformation (DMD) module. (b) Unet architectures.

flow field. Figure 1b provides a detailed illustration of the design of the two different 3D Unets used. Each DMD module employs the flow field predicted within the same block and computes a diffeomorphic mapping $\phi_\theta(t; \mathbf{x})$ for each vertex $\mathbf{x}$ in the mesh generated by the previous deformation block through integration, assuming a stationary flow field. Formally, the dynamics of mesh deformation are formulated as a stationary flow ODE specified by a 3D Unet:

$$\frac{\mathrm{d}\phi_\theta(t; \mathbf{x})}{\mathrm{d}t} = \mathbf{V}(\phi_\theta(t; \mathbf{x})), \quad \phi_\theta(0; \mathbf{x}) = \mathbf{x}_0, \tag{1}$$

where θ denotes the parameters of the 3D Unet, $\mathbf{x}_0$ is the initial mesh before deformation in each stage. SurfFlow is trained stage-wise: a deformation block is trained with all previous deformation blocks frozen. Following CorticalFlow++ [10], the Runge-Kutta method is used for numerical integration.

2.2 Dual-Modal Input

Infant MRI exhibits three distinct phases during the first year of life. In the *infantile phase* ($\leq$5 months), gray matter (GM) shows higher signal intensity than white matter (WM) in T1w images. The *isointense phase* (5–8 months) corresponds to an increase in intensity of WM owing to myelination associated with brain maturation. This significantly lowers the contrast between GM and WM in T1w images and similarly T2w images. In the *early adult-like phase* ($\geq$ 8 months), the GM intensity is lower than WM in T1w images, similar to the tissue contrast in adult MRI. We propose to use both T1w and T2w images for complementary information needed for accurate surface reconstruction.

2.3 Loss Function

The loss function used for training *SurfFlow* is a weighted sum of Chamfer distance (CD; L_{cd}) and a piecewise edge length loss (PELL; L_{e}), balanced by parameter λ:

$$L_{\mathrm{total}} = L_{\mathrm{cd}} + \lambda L_{\mathrm{e}}. \tag{2}$$

Chamfer Distance Loss. The Chamfer distance is commonly used as the loss function for surface reconstruction. It measures the distance from a vertex in one mesh P to the closest vertex in another mesh Q bidirectionally:

$$L_{\mathrm{cd}} = \frac{1}{|P|} \sum_{p \in P} \min_{q \in Q} \|p - q\|_2^2 + \frac{1}{|Q|} \sum_{q \in Q} \min_{p \in P} \|p - q\|_2^2, \tag{3}$$

where p and q are vertices in P and Q, respectively.

Piecewise Edge Length Loss. We observed that vertices tend to be clustered in surfaces generated by CorticalFlow++ (Fig. 3). One possible cause is that the edge length loss used in CorticalFlow++ pushes edges to zero length. To address this problem, we propose a piecewise edge length loss that encourages edge lengths to lie within a suitable range. The proposed loss is formulated as

$$L_{\mathrm{e}} = \frac{1}{|P|} \sum_{p \in P} \frac{1}{|\mathcal{N}(p)|} \sum_{k \in \mathcal{N}(p)} \begin{cases} \epsilon - \|p - k\|_2^2 & \text{if } \|p - k\|_2^2 < \epsilon, \\ 0 & \text{if } \epsilon \leqslant \|p - k\|_2^2 \leqslant \gamma\epsilon, \\ \|p - k\|_2^2 - \gamma\epsilon & \text{otherwise,} \end{cases} \tag{4}$$

where P denotes the predicted mesh, p is a vertex in P, $\mathcal{N}(p)$ consists of the neighbors of p, and ϵ and γ are two tunable hyper-parameters that control the range position and width, respectively.

2.4 Deformation Computation in DMD Modules

Following CorticalFlow++, we use the fourth-order Runge-Kutta method to solve the stationary flow ODE over time interval $[0, 1]$ for both accuracy and stability. The solution $\mathbf{x}_1$ is obtained iteratively with

$$\mathbf{x}_{t_{n+1}} = \mathbf{x}_{t_n} + \frac{h}{6}\Big(\mathbf{k}_1 + 2\mathbf{k}_2 + 2\mathbf{k}_3 + \mathbf{k}_4\Big), \tag{5}$$

where $\mathbf{k}_1 = \mathbf{V}(\mathbf{x}_{t_n})$, $\mathbf{k}_2 = \mathbf{V}(\mathbf{x}_{t_n} + \mathbf{k}_1 \frac{h}{2})$, $\mathbf{k}_3 = \mathbf{V}(\mathbf{x}_{t_n} + \mathbf{k}_2 \frac{h}{2})$, and $\mathbf{k}_4 = \mathbf{V}(\mathbf{x}_{t_n} + \mathbf{k}_3 h)$. $\mathbf{V}(\mathbf{x})$ represents the trilinear interpolation of the flow field at position $\mathbf{x}$. h is the step size and is set to $1/30$. t_n is the time at n-th step, and $t_{n+1} = t_n$ + h.

2.5 Implementation Details

Our network was trained stage-wise. We froze the parameters of one deformation block once trained and then start the training of the next block. Each deformation block was trained for 27k iterations. Adam optimizer was used with an initial learning rate of 0.0001. The parameter λ of the loss function was set to 3.0 to balance the two loss terms. We set ϵ for PELL based on the average edge length determined from the training set. We set the γ to 4.0 and ϵ to 5×10^{-5}. Instance normalization (IN) [11] layers were added between convolutional and activation layers for faster convergence and improved performance.

3 Results

3.1 Data

The dataset includes aligned T1w and T2w image pairs from 121 subjects, 2 weeks to 12 months of age, from the Baby Connectome Project (BCP) [6]. Among them, 90 cases were used for training, 12 for validation, and 19 for testing. Ground truth cortical surfaces were generated with iBEAT v2.0 [12].

To obtain a smooth starting template, an average convex hull computed from the training dataset was re-meshed and triangularized. The Catmull-Clark subdivision algorithm was then applied to generate enough faces and vertices. Decimation was used to control the number of vertices. These steps were carried out in Blender[1].

3.2 Evaluation Metrics

For performance evaluation and comparison between different methods, we utilized the following metrics: Chamfer distance (CD), average symmetric surface distance (ASSD), 90% Hausdorff distance (HD), and normal consistency (NC). Their definitions are as follows:

$$\mathrm{CD} = \frac{1}{2}\Big(\sum_{p\in P} \min_{q\in Q} \frac{\|p-q\|^2}{|P|} + \sum_{q\in Q} \min_{p\in P} \frac{\|p-q\|^2}{|Q|}\Big),$$

$$\mathrm{ASSD} = \frac{1}{2}\Big(\sum_{p\in P} \min_{q\in Q} \frac{\|p-q\|}{|P|} + \sum_{q\in Q} \min_{p\in P} \frac{\|p-q\|}{|Q|}\Big),$$

$$\mathrm{HD} = \max\Big(\max_{p\in P} \min_{q\in Q} \|p-q\|, \max_{q\in Q} \min_{p\in P} \|p-q\|\}\Big),$$

$$\mathrm{NC} = \frac{1}{2}\Big(\frac{\sum_{p\in P}(\mathbf{n_p}\cdot\mathbf{n_{pq}})}{|P|} + \frac{\sum_{q\in Q}(\mathbf{n_q}\cdot\mathbf{n_{qp}})}{|Q|}\Big),$$

where P and Q are respectively the predicted and ground truth (GT) meshes, $\mathbf{n_p}$ and $\mathbf{n_q}$ are the normals at p and q, $\mathbf{n_{pq}}$ is the normal of the vertex in Q that is closest to p, and $\mathbf{n_{qp}}$ is the normal of the vertex in P that is closest to q.

[1] https://www.blender.org.

Table 1. Comparison of different CSR methods in reconstructing the white and pial surfaces of the left (L) and right (R) hemispheres. ↓ and ↑ indicate better performance with lower and higher metric values, respectively.

	Pial Surface							
	CD ↓		ASSD ↓		HD ↓		NC ↑	
	L	R	L	R	L	R	L	R
DeepCSR	4.69 (±2.10)	3.85 (±2.00)	1.26 (±0.25)	1.13 (±0.27)	5.10 (±1.13)	4.58 (±1.37)	0.75 (±0.02)	0.76 (±0.02)
CorticalFlow++	1.02 (±0.17)	0.67 (±0.10)	0.74 (±0.03)	0.63 (±0.04)	2.15 (±0.25)	1.55 (±0.12)	0.81 (±0.01)	0.84 (±0.01)
SurfFlow	0.39 (±0.14)	0.36 (±0.11)	0.48 (±0.03)	0.46 (±0.05)	0.98 (±0.08)	0.80 (±0.10)	0.90 (±0.01)	0.91 (±0.01)
	White Surface							
	CD ↓		ASSD ↓		HD ↓		NC ↑	
	L	R	L	R	L	R	L	R
DeepCSR	0.74 (±0.50)	0.50 (±0.90)	0.58 (±0.11)	0.60 (±0.16)	1.28 (±0.58)	1.27 (±0.68)	0.90 (±0.02)	0.90 (±0.02)
CorticalFlow++	1.13 (±0.43)	0.91 (±0.12)	0.86 (±0.04)	0.78 (±0.04)	1.96 (±0.11)	1.66 (±0.13)	0.72 (±0.13)	0.73 (±0.04)
SurfFlow	0.37 (±0.12)	0.41 (±0.15)	0.47 (±0.05)	0.49 (±0.06)	0.86 (±0.10)	0.90 (±0.12)	0.93 (±0.01)	0.92 (±0.01)

3.3 Results

We compared SurfFlow with CorticalFlow++ and DeepCSR. To ensure a fair comparison, we modified both CorticalFlow++ and DeepCSR to use dual-modal inputs. Specifically, for DeepCSR, we utilized the finest possible configuration, generating a 512^3 3D grid to predict the signed distance field for surface reconstruction using the marching cube algorithm. As depicted in Table 1, our evaluation demonstrates that SurfFlow outperforms the other two methods in all metrics. SurfFlow stands out with an average Chamfer distance of less than 0.5 mm, demonstrating significantly smaller errors compared with CorticalFlow++ and DeepCSR, which result in 22% to 1100% larger errors. Similar improvements were observed for evaluations with the ASSD, HD, and NC metrics. Furthermore, during our evaluation, we noticed that DeepCSR yielded higher errors for pial surface reconstruction due to numerous mesh topological artifacts. The utilization of implicit representation in DeepCSR does not guarantee a genus zero manifold. Visual comparisons (Figs. 2 and 3) further confirm that SurfFlow excels in reconstructing cortical gyri and sulci compared with CorticalFlow++ and DeepCSR. Moreover, SurfFlow demonstrates superior robustness with more consist results, as indicated lower standard deviations.

SurfFlow utilizes PELL to ensure that mesh edge lengths are within the desired range. We observed that this optimization leads to smoother, more uniform, and accurate meshes. In contrast, CorticalFlow++ shows limited accuracy in certain mesh areas, primarily due to its inability to ensure mesh uniformity.

This is evident in the zoomed-in areas depicted in Fig. 4, where the deficiency in mesh faces hinders accurate predictions. The superiority of SurfFlow in terms of mesh smoothness is further supported by the NC results.

Table 2. CSR performance with respect to modalities (T1w and T2w), instance normalization (IN), and piecewise edge length loss (PELL) for white and pial surfaces of the left (L) and right (R) hemispheres. ↓ and ↑ indicate better performance with lower and higher metric values, respectively.

		T1w	✓	✗	✓	✓	✓	✗	✓
		T2w	✗	✓	✓	✓	✗	✓	✓
		IN	✗	✗	✗	✓	✓	✓	✓
		PELL	✗	✗	✗	✗	✓	✓	✓
CD ↓	Pial	L	0.95 (±0.15)	0.89 (±0.16)	1.02 (±0.17)	0.66 (±0.21)	0.90 (±0.42)	0.98 (±0.23)	0.39 (±0.14)
		R	1.05 (±0.18)	1.08 (±0.24)	0.67 (±0.10)	0.66 (±0.17)	0.78 (±0.26)	0.88 (±0.31)	0.36 (±0.11)
	White	L	1.19 (±0.17)	1.31 (±0.27)	1.13 (±0.43)	0.85 (±0.17)	0.82 (±0.27)	0.93 (±0.27)	0.37 (±0.12)
		R	1.28 (±0.14)	1.32 (±0.24)	0.91 (±0.12)	0.90 (±0.18)	0.88 (±0.29)	0.99 (±0.36)	0.41 (±0.15)
ASSD ↓	Pial	L	0.74 (±0.04)	0.72 (±0.07)	0.74 (±0.03)	0.58 (±0.03)	0.64 (±0.11)	0.71 (±0.08)	0.48 (±0.03)
		R	0.77 (±0.05)	0.78 (±0.07)	0.63 (±0.04)	0.60 (±0.05)	0.60 (±0.05)	0.68 (±0.10)	0.46 (±0.05)
	White	L	0.86 (±0.04)	0.94 (±0.09)	0.86 (±0.04)	0.70 (±0.04)	0.68 (±0.10)	0.74 (±0.11)	0.47 (±0.05)
		R	0.92 (±0.05)	0.94 (±0.08)	0.78 (±0.04)	0.73 (±0.04)	0.70 (±0.10)	0.76 (±0.13)	0.49 (±0.06)
HD ↓	Pial	L	1.86 (±0.20)	1.79 (±0.17)	2.15 (±0.24)	1.36 (±0.16)	1.69 (±0.60)	1.89 (±0.28)	0.98 (±0.08)
		R	2.04 (±0.21)	2.09 (±0.19)	1.55 (±0.12)	1.42 (±0.18)	1.41 (±0.31)	1.51 (±0.33)	0.80 (±0.10)
	White	L	1.88 (±0.15)	1.92 (±0.22)	1.96 (±0.11)	1.57 (±0.12)	1.50 (±0.34)	1.62 (±0.26)	0.86 (±0.10)
		R	1.98 (±0.13)	1.94 (±0.23)	1.66 (±0.13)	1.70 (±0.18)	1.55 (±0.30)	1.67 (±0.35)	0.90 (±0.12)
NC ↑	Pial	L	0.80 (±0.02)	0.80 (±0.02)	0.81 (±0.01)	0.86 (±0.01)	0.85 (±0.04)	0.83 (±0.02)	0.90 (±0.01)
		R	0.80 (±0.01)	0.80 (±0.02)	0.84 (±0.01)	0.85 (±0.01)	0.82 (±0.05)	0.82 (±0.03)	0.91 (±0.01)
	White	L	0.66 (±0.06)	0.62 (±0.02)	0.72 (±0.04)	0.83 (±0.02)	0.81 (±0.08)	0.80 (±0.04)	0.93 (±0.01)
		R	0.65 (±0.06)	0.63 (±0.02)	0.73 (±0.04)	0.80 (±0.02)	0.79 (±0.08)	0.80 (±0.04)	0.92 (±0.01)

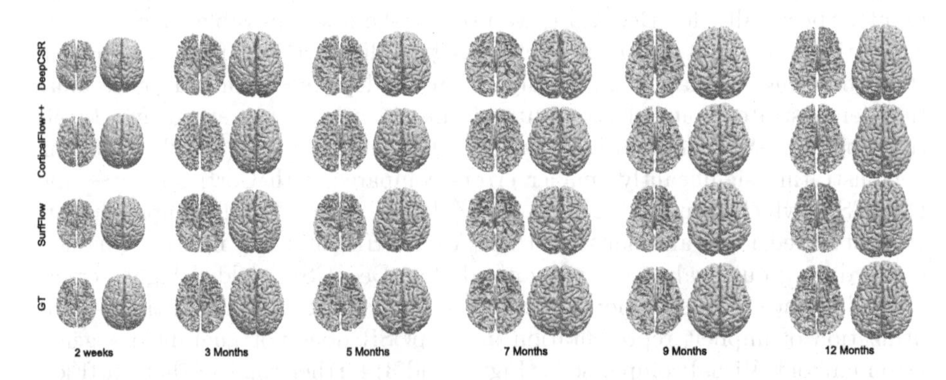

Fig. 2. Comparison of white and pial surfaces reconstructed via SurfFlow, CorticalFlow++, and DeepCSR for different time points.

3.4 Ablation Study

The results of an ablation study (Table 2) confirm that surface prediction performance improves (i) when both modalities are concurrently used; (ii) when

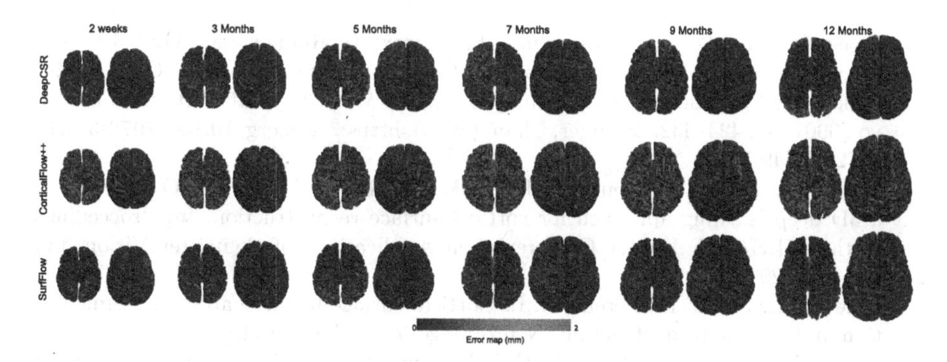

Fig. 3. Error map comparison between SurfFlow, CorticalFlow++, and DeepCSR.

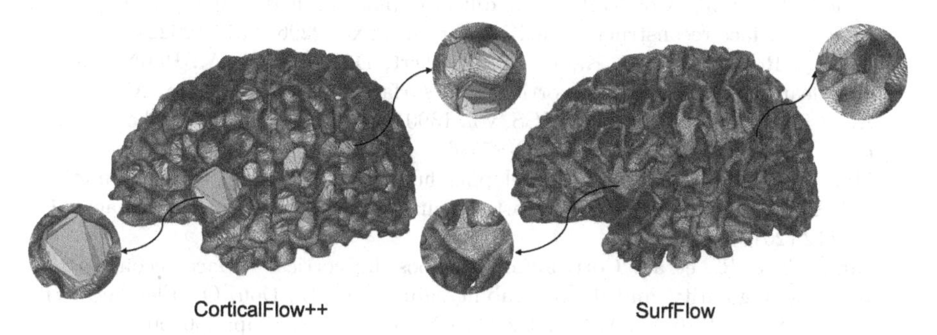

Fig. 4. Comparison of mesh uniformity.

instance normalization is used, and (iii) when PELL is used as opposed to the edge length loss in CorticalFlow++.

4 Conclusion

We presented SurfFlow—a flow-based deep-learning network to accurately reconstruct cortical surfaces. SurfFlow predicts a flow field to deform a template surface toward a target surface. It produces smooth and uniform surface meshes with sub-millimeter accuracy and outperforms CorticalFlow++ and DeepCSR in terms of surface accuracy and regularity.

References

1. Ahmad, S., et al.: Multifaceted atlases of the human brain in its infancy. Nat. Meth. **20**, 55–64 (2023)
2. Barkovich, M.J., Li, Y., Desikan, R.S., Barkovich, A.J., Xu, D.: Challenges in pediatric neuroimaging. Neuroimage **185**, 793–801 (2019)

3. Çiçek, Ö., Abdulkadir, A., Lienkamp, S.S., Brox, T., Ronneberger, O.: 3D U-Net: learning dense volumetric segmentation from sparse annotation. In: Ourselin, S., Joskowicz, L., Sabuncu, M.R., Unal, G., Wells, W. (eds.) MICCAI 2016. LNCS, vol. 9901, pp. 424–432. Springer, Cham (2016). https://doi.org/10.1007/978-3-319-46723-8_49
4. Cruz, R.S., Lebrat, L., Bourgeat, P., Fookes, C., Fripp, J., Salvado, O.: DeepCSR: a 3D deep learning approach for cortical surface reconstruction. In: Proceedings of the IEEE/CVF Winter Conference on Applications of Computer Vision, pp. 806–815 (2021)
5. Dale, A.M., Fischl, B., Sereno, M.I.: Cortical surface-based analysis: I. segmentation and surface reconstruction. NeuroImage **9**(2), 179–194 (1999)
6. Howell, B.R., et al.: The UNC/UMN baby connectome project (BCP): an overview of the study design and protocol development. Neuroimage **185**, 891–905 (2019)
7. Lebrat, L., et al.: CorticalFlow: a diffeomorphic mesh deformation module for cortical surface reconstruction. arXiv preprint arXiv:2206.02374 (2022)
8. Ma, Q., Robinson, E.C., Kainz, B., Rueckert, D., Alansary, A.: PialNN: a fast deep learning framework for cortical Pial surface reconstruction. In: Abdulkadir, A., et al. (eds.) MLCN 2021. LNCS, vol. 13001, pp. 73–81. Springer, Cham (2021). https://doi.org/10.1007/978-3-030-87586-2_8
9. Makropoulos, A., et al.: The developing human connectome project: a minimal processing pipeline for neonatal cortical surface reconstruction. Neuroimage **173**, 88–112 (2018)
10. Santa Cruz, R., et al.: CorticalFlow++: boosting cortical surface reconstruction accuracy, regularity, and interoperability. In: Wang, L., Dou, Q., Fletcher, P.T., Speidel, S., Li, S. (eds.) MICCAI 2022. LNCS, vol. 13435, pp. 496–505. Springer, Cham (2022). https://doi.org/10.1007/978-3-031-16443-9_48
11. Ulyanov, D., Vedaldi, A., Lempitsky, V.: Instance normalization: the missing ingredient for fast stylization. arXiv preprint arXiv:1607.08022 (2016)
12. Wang, L., Wu, Z., Chen, L., Sun, Y., Lin, W., Li, G.: iBEAT v2.0: a multisite-applicable, deep learning-based pipeline for infant cerebral cortical surface reconstruction. Nat. Protoc. **18**, 1488–1509 (2023)
13. Wickramasinghe, U., Remelli, E., Knott, G., Fua, P.: Voxel2Mesh: 3D mesh model generation from volumetric data. In: Martel, A.L., et al. (eds.) MICCAI 2020. LNCS, vol. 12264, pp. 299–308. Springer, Cham (2020). https://doi.org/10.1007/978-3-030-59719-1_30
14. Zöllei, L., Iglesias, J.E., Ou, Y., Grant, P.E., Fischl, B.: Infant FreeSurfer: an automated segmentation and surface extraction pipeline for T1-weighted neuroimaging data of infants 0–2 years. Neuroimage **218**, 116946 (2020)

Prior-Driven Dynamic Brain Networks for Multi-modal Emotion Recognition

Chuhang Zheng, Wei Shao, Daoqiang Zhang, and Qi Zhu(✉)

College of Computer Science and Technology, Key Laboratory of Brain-Machine Intelligence Technology, Ministry of Education, Nanjing University of Aeronautics and Astronautics, Nanjing 211106, China
zhuqinuaa@163.com

Abstract. Emotions are closely related to many mental and cognitive diseases, such as depression, mania, Parkinson's Disease, etc, and the recognition of emotion plays an important role in diagnosis of these diseases, which is mostly limited to the patient's self-description. Because emotion is always unstable, the objective quantitative methods are urgently needed for more accurate recognition of emotion, which can help improve the diagnosis performance for emotion related brain disease. Existing studies have shown that EEG and facial expressions are highly correlated, and combining EEG with facial expressions can better depict emotion-related information. However, most of the existing multi-modal emotion recognition studies cannot combine multiple modalities properly, and ignore the temporal variability of channel connectivity in EEG. In this paper, we propose a spatial-temporal feature extraction framework for multi-modal emotion recognition by constructing prior-driven Dynamic Functional Connectivity Networks (DFCNs). First, we consider each electrode as a node to construct the original dynamic brain networks. Second, we calculate the correlation between EEG and facial expression through cross attention, as a prior knowledge of dynamic brain networks, and embedded to obtain the final DFCNs representation with prior knowledge. Then, we design a spatial-temporal feature extraction network by stacking multiple residual blocks based on 3D convolutions, and non-local attention is introduced to capture the global information at the temporal level. Finally, we adopt the features from fully connected layer for classification. Experimental results on the DEAP dataset demonstrate the effectiveness of the proposed method.

Keywords: Multi-modal emotion recognition · Dynamic brain networks · Cross attention · Non-local attention · 3D convolutions

1 Introduction

In healthcare, affective computing can help measure the psychological state of patients automatically, especially for those with cognitive deficits. For example, the emotional state of hospitalized patients contributes to the early diagnosis

H. Greenspan et al. (Eds.): MICCAI 2023, LNCS 14227, pp. 389–398, 2023.
https://doi.org/10.1007/978-3-031-43993-3_38

of Parkinson's Disease (PD) [11]. In addition, for patients with neurological diseases, since neurological diseases are degenerative in nature, resulting in unstable cognitive function. The patients may not notice the symptoms of their disease, such as changes in their mood. Recent clinical diagnosis standards rely on the patients' self reports of their feelings to emotional disorders, but it may not be very accurate and stable. Therefore, we need to develop data-driven emotion identification method to improve the diagnosis of these disorders.

As EEG signals are directly related to high-level cognitive processes, EEG-based emotion recognition draws increasing attention in recent years [1]. Song et al. [15] proposed a dynamic graph convolutional network, which trained neural networks to dynamically learn the internal relationships between different EEG channels and extract more discriminative features. Zhang et al. [23] proposed a self-attention network to jointly model both local and global temporal information of EEG to reduce the effect of noise at the temporal level. These efforts do not take advantage of the complementary information between the modalities, which limits the performance of the model. Recently, a lot of works shown multi-modal data can provide complementary information to improve emotion recognition performance. Wang et al. [20] combined transformer encoders with attention based fusion to integrate EEG and eye movement data for emotion recognition. Ma et al. [10] designed a multi-modal residual long short-term memory network (MMResLSTM) to learn the correlation between EEG and peripheral physiological on multi-modal emotion recognition. However, the above work ignores correlations between EEG channels and fails to provide interpretable fusion model.

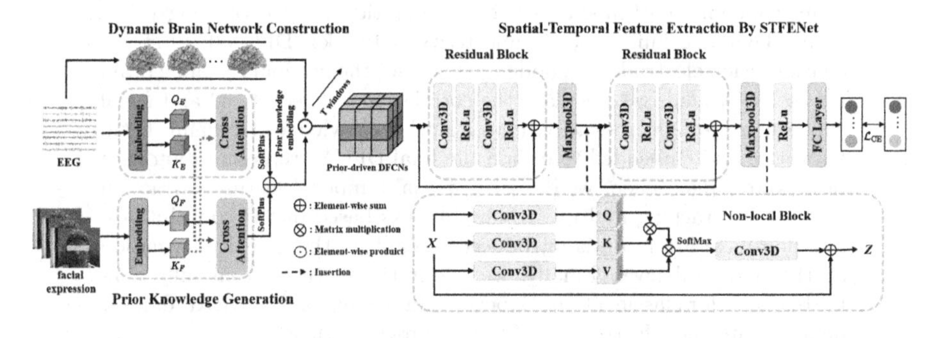

Fig. 1. Framework of our proposed multi-modal emotion recognition method.

Brain network analysis has been widely used in the field of disease diagnosis [8,22], which can describe the complex spatial relationships between brain regions of the brain. In recent years, researchers have migrated brain networks into emotion recognition. Wang et al. [21] implemented PageRank algorithm to rank the importance of brain network nodes, and screened important channels in emotion recognition according to the weight of channels. Huang et al. [6] proposed a novel neural decoding framework, which builds a bridge between emotions and brain regions, and captures their relationships by performing embedding propagation. However, the methods mentioned above regard the structure of brain network as static, ignoring that the variability of electrode channel connectivity over time. Since the multi-modal data is obtained from the synchronous stimulus in the same time period, this temporal level dynamic is particularly important in the multi-modal emotion recognition. In addition, integrating the heterogeneous data of EEG and facial expression also poses challenge to multi-modal emotion classification.

To overcome the above limitations, we design a spatial-temporal feature extraction framework based on prior-driven dynamic brain networks and apply it to emotion recognition. Specifically, we treat each electrode of EEG as a node of brain network, and then the dynamic functional connectivity networks (DFCNs) is constructed by Pearson correlation coefficient under non-overlapping time window. Besides, we calculate the correlation between EEG and facial expression across modal channels by cross attention mechanism, as the prior knowledge of DFCNs, and then embed it to above model obtain the final DFCNs representation. Finally, we implemented residual blocks and non-local attention to construct STFENet, so as to extract complex spatial-temporal feature and preserve the long-range dependencies in the time series.

2 Method

Figure 1 shows the framework of our approach, including four parts, i.e., the construction of dynamic brain networks, the representation and learning of cross-modal correlation between EEG and facial expression, the embedding of correlation into DFCNs as prior knowledge, and the extraction of spatial-temporal features of DFCNs for emotion recognition based on 3D convolutions.

Dynamic Brain Networks Construction: Functional Connectivity Networks (FCNs) ignore the temporal changes of brain connectivity. In this paper, we construct Dynamic Functional Connectivity Networks (DFCNs) to solve the above problem. First, each subject's EEG data can be represented as $X^E \in \mathbb{R}^{P\times T\times D}$, where P represents the number of channels, T is the number of time windows, and D represents the feature dimension. The t-th subsequence feature of P channels can be represented as a matrix $x(t) = [x_1(t), x_2(t), \cdots, x_p(t)] \in R^{P\times D}$, where $x_i(t) \in R^D$ represents the t-th subsequence feature extracted from the EEG time series of the i-th channel. According to the divided non-overlapping sliding time window, we build a functional connectivity network (i.e., matrix) by computing Pearson correlation coefficient between EEG from a pair of channels within the t-th time window:

$$C_{ij}(t) = \frac{\mathrm{cov}\left(x_i(t), x_j(t)\right)}{\sigma_{x_i(t)}\sigma_{x_j(t)}} \tag{1}$$

where cov denotes the covariance between two vectors, $\sigma_{x_i(t)}$ denotes the standard deviation of vector $x_i(t)$, $x_i(t)$ and $x_j(t)$ represent the EEG of a pair of channels i and j within the t-th time window, respectively. Thus, the original DFCNs of each subject $DFCNs_{\text{original}} = [C(1), C(2), \cdots, C(T)]^T \in R^{T\times P\times P}$ consists of T transient FCNs.

Prior Knowledge Embedding: Most of the existing multi-modal emotion recognition works aim to extract the features of different modalities respectively for fusion, which always lost the correlation between modalities. Existing studies have found there is high correlation between EEG and facial expression [5,12–14], but it is still challenging to find an appropriate way to fuse them. Therefore, we calculate the correlation of different modality data as prior knowledge to embed the previously constructed DFCN. Specifically, for each subject, $x^E \in R^{T\times P_E\times D}$, $x^F \in R^{T\times P_F\times D}$ represents the EEG and facial expression modality respectively, where T represents the number of time Windows, P_E and P_F represents the number of channels, and D represents the feature dimension. We perform different transformations by linear mapping pairs x^E and x^F:

$$Q_E = X^E W_Q^E, K_E = X^E W_K^E \tag{2}$$

$$Q_F = X^F W_Q^F, K_F = X^F W_K^F \tag{3}$$

where W_Q and W_K are the parameter matrices used to generate query and key, which are updated through network back-propagation during model training. We determined correlation scores across channel-dimension between modalities based on cross attention, by treating one modality as query and the other as key:

$$\text{Cor}(E, F) = \frac{Q_E K_F^T}{\sqrt{d_1}}, \text{Cor}(F, E) = \frac{Q_F K_E^T}{\sqrt{d_2}} \tag{4}$$

where, $\text{Cor}(E, F)$ and $\text{Cor}(F, E)$ represents the correlation score between the cross-modality channels, d_1, d_2 are normalized parameters equal to the dimension of K. It is worth noting that softmax is applied to the scoring weight of the equation Eq. 4. However, softmax proved to be overconfident in its results, which would result in the correlation scores of certain time windows being too high or too low, affecting the reliability of the prior knowledge. Therefore, we improve softmax to softplus to solve this problem while ensuring that the correlation matrix is non-negative. The calculated correlation matrix is as follows:

$$\text{Cor} = \text{softplus}(\text{Cor}(E, F) + \text{Cor}(F, E)) \tag{5}$$

At this point, we obtain the correlation between the cross-modal channels, and use it as the prior knowledge of DFCNs construction. We embed the modified prior knowledge into the previously constructed DFCNs by element-wise product:

$$DFCNs = DFCNs_{\text{original}} \odot \text{Cor} \tag{6}$$

where $\odot$ represents element-wise product. By the embedding of prior knowledge, we obtain the discriminative DFCNs representations with prior knowledge.

Spatial-Temporal Feature Extraction: Different from static brain networks, DFCNs can not only describe brain connectivity, but also contain the temporal-level volatility of brain connectivity. Most of the existing methods focus on extracting the temporal and spatial features of EEG separately, and concat them for feature fusion, which ignores the dynamic variations of electrode connectivity in the temporal dimension. 2D convolution has been widely used in the field of computer vision, but it is challenging to capture information at the temporal level. Previous studies has shown that 3D convolution operations can better model spatial information in continuous sequences [3,19]. So, we introduce 3D convolution to extract spatial-temporal feature of DFCNs simultaneously. Considering the DFCNs representation $X \in R^{C \times T \times P \times P}$ of each subject, where C is the number of channels, T represents the number of time windows, and P represents the number of electrode channels, then the m-th feature representation of the location (T, P, P) calculated by 3D convolution in space can be represented as:

$$v_m^{T,P,P} = \partial \left(b_m + \sum_{\sigma} \sum_{\varepsilon=0}^{P-1} \sum_{\rho=0}^{P-1} \sum_{\varphi=0}^{T-1} w_{c',m}^{T',P',P'} v_{c'}^{T+T',P+P',P+P'} \right) \tag{7}$$

where σ is the assigned activation function, b_m is the deviation, $w_{c',m}^{T',P',P'}$ represents the weight of the convolution kernel connected by the c'-th stacking channel to the feature representation of the position (T, P, P), and $v_{c'}^{T+T',P+P',P+P'}$ represents the characteristic value of the c'-th stacking channel at the position (T, P, P). To better capture the spatial-temporal topological structure in DFCNs, inspired by ResNet's remarkable success [4], we build a deeper network by stacking multiple residual blocks. A spatial-temporal feature extraction network (STFENet) is designed to extract spatial-temporal features of the DFCNs. The construction of STFENet is shown in the second half of Fig. 1. A residual block is used as the basic block, which includes two 3D convolutions, two activation functions and a residual connection. 3D Maxpooling is adopted between the multiple stacked residual block.

Since the operation of convolution will eventually focus on local areas, long-range dependencies which describe luxuriant emotion-related information will be lost to some extent. To solve this problem, we further introduce non-local block [18] to preserve information after the maxpooling layer. For a given input, non-local attention performs two different transformations:

$$\theta(x_i) = W_\theta x_i, \phi(x_i) = W_\phi x_i \tag{8}$$

where W_θ and W_ϕ is the weight to be learned, which is realized by 3D convolution in this paper. Then, non-local attention uses the self-attention term [17] to calculate the final features with the help of softmax:

$$y = \operatorname{softmax}\left(x^T W_\theta^T W_\phi x\right) g(x) \tag{9}$$

where g is implemented by $1 \times 1 \times 1$ convolution in this paper. Then, the non-local block can be defined as:

$$z = W_z y + x \tag{10}$$

where "+x" denotes the residual connection, and W_z represents the weight matrix. By the STFENet, we finally effectively extract the spatial-temporal emotion-related information in prior-driven DFCNs for the identification of emotions.

Table 1. Comparisons of different methods on DEAP dataset (Mean/Std%).

	method	Valence		Arousal	
		ACC	F1	ACC	F1
EEG based methods	SVM	58.88/11.47	67.81/13.28	58.05/15.26	66.71/20.54
	GSN	63.77/07.25	66.19/12.20	63.51/11.36	68.03/15.40
	DGCNN	63.28/08.15	65.17/10.22	62.61/12.93	69.22/15.17
Multi-modal based methods	MKL	59.02/12.94	67.37/16.97	58.96/13.89	64.78/18.69
	DCCA	61.04/11.73	65.46/12.29	59.58/11.89	65.78/16.98
	MMResLSTM	64.67/10.57	68.36/11.50	63.25/12.38	67.32/15.92
	ETF	64.63/09.35	66.35/13.41	65.64/10.29	66.63/16.97
	Ours	**67.36/05.58**	**69.17/09.01**	**68.47/08.45**	**74.68/13.36**

3 Experiment Results

Emotional Database: The DEAP dataset [9] collected EEG data from 32 healthy participants. The volunteers were asked to watch 40 one-minute videos and collect EEG signals from the subjects. The facial states of the first 22 subjects were recorded simultaneously. All participants rated each video on a 1–9 scale with the indicators, i.e. arousal, valence, dominance. We choose 18 subjects with both EEG signals and a complete facial video for the experiment. Same as many state-of-the-art studies [2], we turn the identification task into binary classification problem by setting the evaluation threshold of 5.

Data Pre-processing: For the EEG data, The 32-channel EEG signal with a duration of 63 s is down-sampled to 128HZ, and remove the first 3 s pre-trial baseline. Power spectral density (PSD) features is extracted from 3 s time windows with non-overlap through the Welch method in EEG, and 5 frequency bands are adopted, i.e. theta (4–8 Hz), slow alpha (8–10 Hz), alpha (8–12 Hz), beta (12–30 Hz), and gamma (30+ Hz) [9]. For the facial video data, referring to [12], we utilize OPENFACE to extract expression features from facial videos, including 3 face positions relative to the camera, 3 head position, 6 eye gaze directions and 17 facial action units. Similar to EEG, the face sequences are divided according to the 3-second non-overlapping sliding time window and the average value of each feature is taken.

Experiment Settings: In our experiment, we adopt the leave one subject out (LOSO) cross-validation strategy to verify the effectiveness of our method. Specifically, the samples are divided into 18 non-overlapping parts according to the subjects. The samples of one subject are selected as the test set while the remaining subjects are selected as the training set for each cross-validation. This process is repeated 18 times, and the average performance of the cross-validation is taken as the final result. Identification performance is measured by accuracy (ACC) and F1-Score. The proposed method is based on the Pytorch implementation, and the model mentioned in this study is trained on a single GPU (NVIDIA GeForce RTX3080). Adam algorithm is used to optimize this method, and the learning rate and batch size are set to 0.001 and 40, respectively.

Results and Discussion: We evaluate the performance of our method by calculating ACC and F1 on both valence and arousal. We also compare our method with many comparison methods, which can be divided into two categories: EEG based methods and multi-modal based methods. More specifically, EEG based methods are Support vector machine (SVM), GraphSleepNet (GSN) [7], Dynamical Graph-CNN (DGCNN) [15]. Multi-modal based methods include Multi-kernel learning (MKL), Deep-CCA (DCCA), MMResLSTM [10]. Emotion transformer fusion (ETF) [20]. For quantitative results in Table 1, firstly, most of the multi-modal based methods achieve higher performance than EEG-based methods, which shows the advantage of complementary information from multiple modalities. Secondly, our proposed method achieve the best emotion recognition performance. On valence and arousal, the average ACC and F1 of our method reached 67.36%, 69.17% and 68.47% and 74.68% respectively. The main reason for the superiority of our method is that we can not only use multi-modal data as prior knowledge to guide the construction of DFCNs, but also extract discriminate spatial-temporal features.

Table 2. Ablation study of our proposed method (Mean/Std%).

method	Valence		Arousal	
	ACC	F1	ACC	F1
Baseline	61.11/11.35	64.81/13.40	60.91/12.45	65.10/18.99
w/o PKE	64.15/09.73	67.19/12.48	65.58/10.18	68.12/16.10
w/o NL block	66.12/08.60	66.92/10.36	66.24/09.28	72.31/14.63
w/o STFENet	65.51/08.19	67.18/11.29	67.10/10.15	70.83/15.84
Ours	**67.36/05.58**	**69.17/09.01**	**68.47/08.45**	**74.68/13.36**

To evaluate the effectiveness of the different modules of our framework, we conduct several ablation experiments on DEAP dataset. Our method mainly contains two modules, Prior knowledge embedding (PKE) module and STFENet. Besides, we also evaluate the contribution of non-local block in STFENet. As can be seen in Table 2, every module used in our framework greatly improve the performance compared with baseline model, with an increase of 6.25%, 4.36% and 7.56%, 9.58% for ACC and F1 on valence and arousal, respectively. It can be seen that both non-local block and STFENet demonstrate the better performance of our proposed method. The reason lies in that STFENet is able to extract complex spatial-temporal feature, and non-local block of STFENet helps it preserve the long-range dependencies in the time series. Moreover, when we remove PKE module from our method, there comes a performance degradation. It suggested that the prior knowledge has vital guiding significance for the construction of DFCNs, so that it can better express emotion-related information.

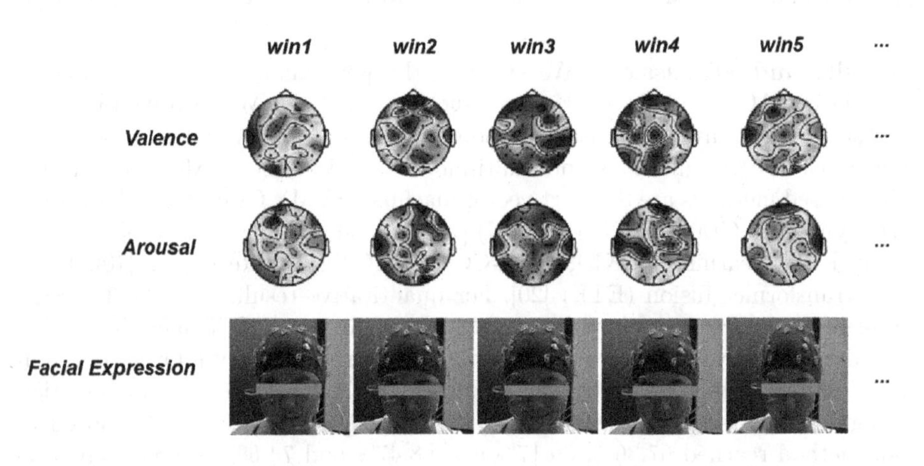

Fig. 2. Visualization of facial expression and its channel correlation with EEG in different time windows.

In addition, to further verify the feasibility of the prior knowledge embedded in our method, we visualize the facial expression of subject over several time windows and its channel correlation with EEG, as shown in Fig. 2. Firstly, from the channel correlation topographic map, the deeper the red of the brain area, the higher the correlation between EEG and facial expression. Conversely, the deeper the blue, the lower the correlation. On valence and arousal, high correlation areas focus on the binaural and prefrontal regions, which is in line with existing medical cognition [16]. As the stimulation method adopted by DEAP dataset is musical stimulation, the binaural region is highly activated. The prefrontal lobe plays a crucial role in emotional mobilization [2]. The experimental results show that our method can mine electrode channels that are highly correlated with emotion to provide prior knowledge to guide the construction of dynamic brain networks. Combined with the experimental results after removing PKE of our method in Table 2, it can be seen that embedding prior knowledge can achieve better emotion recognition performance. Therefore, the prior knowledge can better describe emotion-related information.

4 Conclusion

In this paper, we develop a spatial-temporal feature extraction framework based on prior-driven DFCNs for multi-modal emotion recognition. In our approach, not only the connectivity between EEG channels but also the dynamics of connectivity over time are jointly learned. Besides, we also calculate the correlation across modalities via cross attention to guide the construction of DFCNs. In addition, we build STFENet based on 3D convolution to model the spatial-temporal features contained in DFCNs to extract emotion-related spatial-temporal information and preserve the long-range dependencies in the time series. Experimental results show that our method outperforms the state-of-the-art methods.

Acknowledgement. This work was supported by the National Natural Science Foundation of China (Nos. 62076129, 62136004, and 62276130), the Key Research and Development Plan of Jiangsu Province (No. BE2022842), and the Fundamental Research Funds for the Central Universities (No. NS2023051).

References

1. Cai, Q., Cui, G.C., Wang, H.X.: EEG-based emotion recognition using multiple kernel learning. Mach. Intell. Res. **19**(5), 472–484 (2022)
2. Du, X., et al.: An efficient LSTM network for emotion recognition from multichannel EEG signals. IEEE Trans. Affect. Comput. **13**(3), 1528–1540 (2020)
3. Guo, S., Lin, Y., Li, S., Chen, Z., Wan, H.: Deep spatial-temporal 3D convolutional neural networks for traffic data forecasting. IEEE Trans. Intell. Transp. Syst. **20**(10), 3913–3926 (2019)
4. He, F., Liu, T., Tao, D.: Why ResNet works? Residuals generalize. IEEE Trans. Neural Netw. Learn. Syst. **31**(12), 5349–5362 (2020)
5. Huang, X., et al.: Multi-modal emotion analysis from facial expressions and electroencephalogram. Comput. Vis. Image Underst. **147**, 114–124 (2016)
6. Huang, Z., Du, C., Wang, Y., He, H.: Graph emotion decoding from visually evoked neural responses. In: Wang, L., Dou, Q., Fletcher, P.T., Speidel, S., Li, S. (eds.) MICCAI 2022. LNCS, vol. 13438, pp. 396–405. Springer, Cham (2022). https://doi.org/10.1007/978-3-031-16452-1_38
7. Jia, Z., et al.: GraphSleepNet: adaptive spatial-temporal graph convolutional networks for sleep stage classification. In: IJCAI, pp. 1324–1330 (2020)
8. Jie, B., Shen, D., Zhang, D.: Brain connectivity hyper-network for MCI classification. In: Golland, P., Hata, N., Barillot, C., Hornegger, J., Howe, R. (eds.) MICCAI 2014. LNCS, vol. 8674, pp. 724–732. Springer, Cham (2014). https://doi.org/10.1007/978-3-319-10470-6_90
9. Koelstra, S., et al.: DEAP: a database for emotion analysis; using physiological signals. IEEE Trans. Affect. Comput. **3**(1), 18–31 (2011)
10. Ma, J., Tang, H., Zheng, W.L., Lu, B.L.: Emotion recognition using multimodal residual LSTM network. In: Proceedings of the 27th ACM International Conference on Multimedia, pp. 176–183 (2019)
11. Prell, T., et al.: Specialized staff for the care of people with Parkinson's disease in Germany: an overview. J. Clin. Med. **9**(8), 2581 (2020)

12. Rayatdoost, S., Rudrauf, D., Soleymani, M.: Multimodal gated information fusion for emotion recognition from EEG signals and facial behaviors. In: Proceedings of the 2020 International Conference on Multimodal Interaction, pp. 655–659 (2020)
13. Siddharth, S., Jung, T.P., Sejnowski, T.J.: Impact of affective multimedia content on the electroencephalogram and facial expressions. Sci. Rep. **9**(1), 16295 (2019)
14. Soleymani, M., Pantic, M., Pun, T.: Multimodal emotion recognition in response to videos. IEEE Trans. Affect. Comput. **3**(2), 211–223 (2011)
15. Song, T., Zheng, W., Song, P., Cui, Z.: EEG emotion recognition using dynamical graph convolutional neural networks. IEEE Trans. Affect. Comput. **11**(3), 532–541 (2018)
16. Sun, Y., Ayaz, H., Akansu, A.N.: Multimodal affective state assessment using fNIRS+ EEG and spontaneous facial expression. Brain Sci. **10**(2), 85 (2020)
17. Vaswani, A., et al.: Attention is all you need. In: Advances in Neural Information Processing Systems, vol. 30 (2017)
18. Wang, X., Girshick, R., Gupta, A., He, K.: Non-local neural networks. In: Proceedings of the IEEE Conference on Computer Vision and Pattern Recognition, pp. 7794–7803 (2018)
19. Wang, Y., et al.: 3d auto-context-based locality adaptive multi-modality GANs for pet synthesis. IEEE Trans. Med. Imaging **38**(6), 1328–1339 (2018)
20. Wang, Y., Jiang, W.B., Li, R., Lu, B.L.: Emotion transformer fusion: complementary representation properties of EEG and eye movements on recognizing anger and surprise. In: 2021 IEEE International Conference on Bioinformatics and Biomedicine (BIBM), pp. 1575–1578. IEEE (2021)
21. Wang, Z.M., Zhang, J.W., He, Y., Zhang, J.: EEG emotion recognition using multichannel weighted multiscale permutation entropy. Appl. Intell. **52**(10), 12064–12076 (2022)
22. Yang, J., Zhu, Q., Zhang, R., Huang, J., Zhang, D.: Unified brain network with functional and structural data. In: Martel, A.L., et al. (eds.) MICCAI 2020. LNCS, vol. 12267, pp. 114–123. Springer, Cham (2020). https://doi.org/10.1007/978-3-030-59728-3_12
23. Zhang, Y., Liu, H., Zhang, D., Chen, X., Qin, T., Zheng, Q.: EEG-based emotion recognition with emotion localization via hierarchical self-attention. IEEE Trans. Affect. Comput. 1 (2022)

Unified Surface and Volumetric Inference on Functional Imaging Data

Thomas F. Kirk[1,2(✉)], Martin S. Craig[1,2], and Michael A. Chappell[1,2]

[1] Sir Peter Mansfield Imaging Center, School of Medicine, University of Nottingham, Nottingham, UK
tomfrankkirk@gmail.com
[2] Quantified Imaging, London, UK
https://physimals.github.io/physimals/

Abstract. Surface-based analysis methods for functional imaging data have been shown to offer substantial benefits for the study of the human cortex, namely in the localisation of functional areas and the establishment of inter-subject correspondence. A new approach for surface-based parameter estimation via non-linear model fitting on functional time-series data is presented. It treats the different anatomies within the brain in the manner that is most appropriate: surface-based for the cortex, volumetric for white matter, and using regions-of-interest for subcortical grey matter structures. The mapping between these different domains is incorporated using a novel algorithm that accounts for partial volume effects. A variational Bayesian framework is used to perform parameter inference in all anatomies simultaneously rather than separately. This approach, called hybrid inference, has been implemented using stochastic optimisation techniques. A comparison against a conventional volumetric workflow with post-projection on simulated perfusion data reveals improvements parameter recovery, preservation of spatial detail and consistency between spatial resolutions. At 4 mm isotropic resolution, the following improvements were obtained: 2.7% in SSD error of perfusion, 16% in SSD error of Z-score perfusion, and 27% in Bhattacharyya distance of perfusion distribution.

Keywords: surface analysis · neuroimaging · partial volume effect · variational Bayes

1 Introduction

In recent years, there has been increasing interest in performing surface-based analysis of magnetic resonance neuroimaging (MRI) data in the human cortex. A basic assumption that underpins this approach is that cortical activity will correlate better according to geodesic distance (along the surface) than geometric distance (straight-line). The benefits include improvements in the localisation of functional areas and the establishment of inter-subject correspondence [5]. Due to the fact that MRI data is typically acquired on a 3D voxel grid that does not directly correspond with a 2D surface, a necessary pre-processing step is to project the volumetric data onto the cortical surface, a complex operation for

H. Greenspan et al. (Eds.): MICCAI 2023, LNCS 14227, pp. 399–408, 2023.
https://doi.org/10.1007/978-3-031-43993-3_39

which there is no commonly-agreed upon solution [11,13]. Of particular relevance is the partial volume effect (PVE): because functional imaging voxels are of comparable size to the thickness of the human cortex, each is likely to contain a mixture of cortical grey matter (GM), white matter (WM) and cerebrospinal fluid (CSF). This mixing of signals from different tissues introduces confound into later analysis steps. Furthermore, because not all anatomies in the brain are amenable to surface analysis, data for the subcortex is usually processed in a volumetric manner. This leads to separate workflows for the different anatomies within the brain, as exemplified by the Human Connectome Project's (HCP) parallel *fMRISurface* and *fMRIVolume* pipelines [8].

The objective of this work was to develop a framework for parameter inference that is able to operate in a surface and volumetric manner *simultaneously.* This would remove the need for separate workflows whilst ensuring that each anatomy of interest is treated in an optimal manner, namely surface-based for the cortex, volumetric for WM, and region-of-interest (ROI) for subcortical GM structures. This principle corresponds directly with the HCP's concept of *grayordinates* [8]. In order to achieve this, a novel algorithm that maps data between different representations has been embedded within a modality-specific generative model. Via the use variational Bayesian techniques, it is then possible to perform non-linear model-fitting directly on volumetric timeseries data without pre-processing. The advantages of this approach are demonstrated on simulated perfusion MRI data.

2 Methods

Theory. The spatial locations within the brain at which parameter estimates are obtained are hereafter called *nodes.* A key step in performing inference is to define a mapping between nodes and the voxel data that has been acquired, which is then embedded within a generative model to yield a function that relates physiological parameters to the data they reconstruct. In the case of volumetric inference, this mapping is one-to-one: each node corresponds to exactly one voxel. The concept can be extended to include other types of node: alongside voxels for WM, surface vertices can represent the cortex and volumetric ROIs can represent subcortical GM structures. Collectively these are referred to as *hybrid* nodes, which correspond to greyordinates in the HCP terminology [8]. An illustration is given in Fig. 1.

In this work, the mapping for hybrid nodes previously introduced in [10] was used. It is constructed by calculating the volume of intersection between individual voxels and the geometric primitives that construct the cortical ribbon, WM tracts, or individual ROIs. Importantly, because each node corresponds by definition to exactly one tissue type, the mapping will be many-to-one in voxels that contain multiple tissues which is an explicit representation of PVE. The mapping takes the form of a non-square sparse matrix (number of nodes > voxels), an example of which is illustrated in Fig. 2.

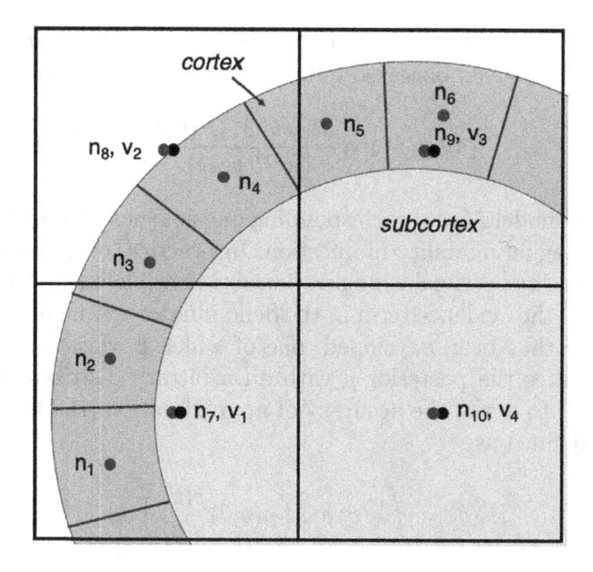

Fig. 1. Simplified representation of voxels and nodes around the cortical ribbon. The voxel grid is shown in black, and voxel centres are labelled v_1 to v_4. To represent cortical tissue, each volume primitive of the cortex (shaded brown) is assigned a node n_1 to n_6. For subcortical tissue (shaded blue), each volume primitive is assigned a node n_7 to n_{10}, which correspond exactly to the voxel centres v_1 to v_4. By definition, each node represents one tissue type. (Color figure online)

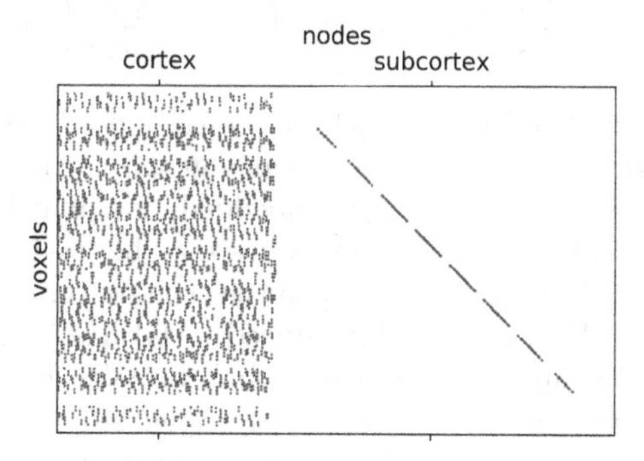

Fig. 2. Sparsity structure of the hybrid node to voxel mapping for a single cortical hemisphere at 32,000 vertex resolution intersecting a 3 mm isotropic voxel grid.

Having defined and embedded the mapping within a generative model, one can turn to the process of parameter inference. We adopt a Bayesian approach because this permits the incorporation of prior knowledge and the quantification of parameter uncertainty. Both attributes are valuable given the challenging signal-to-noise ratio (SNR) of functional imaging data. Under this approach, the objective is to obtain a posterior distribution p for the parameters θ of a generative model M given observed data y, which may be expressed as:

$$P(\theta|y, M) = \frac{P(y|\theta, M)P(\theta|M)}{P(y|M)} \tag{1}$$

The choice of model M and corresponding physiological parameters θ is determined by the imaging modality in question. In practice, evaluating this expression for all but the most trivial configurations is infeasible due to the integrations entailed (notably the evidence term in the denominator). A number of numerical approaches have thus been developed, one of which is variational Bayes (VB) which approximates the posterior p with an arbitrary distribution q and uses the free energy F to assess the accuracy of approximation [1]. Omitting M from notation, F is defined as:

$$F(\theta) = \int q(\theta) \log \left(p(y|\theta) \frac{p(\theta)}{q(\theta)} \right) d\theta \tag{2}$$

VB thus turns parameter inference into an optimisation problem: *what* q *best approximates* p *as measured by the free energy?* One means of implementing this is to use stochastic optimisation techniques. A Monte Carlo approximation to the objective function F may be obtained using an average over L randomly drawn samples θ^{*l} from $q(\theta)$:

$$F \approx \frac{1}{L} \sum_{l}^{L} \left[\log(p(y|\theta^{*l})) - \log \left(\frac{q(\theta^{*l})}{p(\theta^{*l})} \right) \right] \tag{3}$$

This strategy is referred to as stochastic variational Bayes (SVB). By constructing this expression as a computational graph, including the modality-specific generative model M with its embedded mapping from hybrid nodes to voxels, automatic differentiation techniques may be used to maximise F and thus derive the optimal approximation q to the true posterior p. The volumetric implementation of SVB previously introduced in [4] has been extended in this manner and the end result is referred to as hybrid SVB, hSVB. Note that this approach does not involve any learning or training: on each dataset, the inference starts afresh and operates on all data until approximate convergence of the optimisation problem has been reached.

A key advantage of the Bayesian approach is that it enables prior information to be incorporated into the inference. Such priors can either be distributional, for example an empirically-derived normal distribution; or spatial, which encode the belief that parameter values should correlate in adjacent regions of the brain [15]. A spatial prior is particularly useful as a means of mitigating the low SNR inherent to many functional imaging techniques because it applies regularisation in a principled manner that is determined solely by data quality as opposed to relying on user-set parameters. When applied to volumetric data, a drawback of the spatial prior is that it is unaware of underlying anatomy, and in particular PVE. This is because it is typically implemented using an isotropic Laplacian operator over the first-order neighbourhood of voxels, which is especially problematic at the cortical boundary where it will enforce a similarity constraint across the GM/WM boundary even though these tissues may have different parameter values. Under a hybrid approach, it is possible to restrict the spatial prior to operate only in anatomically contiguous regions. Specifically, one spatial prior is defined on the surface for the cortex, a second spatial prior is defined in the volume for non-cortical tissue, and hence regularisation no longer happens across tissue boundaries. In this work, the isotropic Laplacian on first-order neighbours was used for the volumetric prior, and the discrete cotangent Laplacian was used for the surface prior [6].

Evaluation. hSVB was evaluated using simulated mutli-delay arterial spin labelling (ASL) data, a perfusion modality which is sensitive to both cerebral blood flow (CBF) and arterial transit time (ATT). Inference on multi-delay ASL data requires fitting a non-linear model with two parameters which is more challenging than for single-delay ASL. A single subject's T1 MPRAGE anatomical image (1.5 mm isotropic resolution, TR 1.9 s, TE 3.74 ms, flip angle 8°) was processed to extract the left cortical hemisphere, which defined the ground truth anatomy from which ASL data was simulated. FreeSurfer was used to obtain mesh reconstructions of the white and pial cortical surfaces [7]; FSL FIRST was used to segment subcortcial GM structures [14]; FSL FAST was used to segment cerebrospinal fluid [16]; and finally Toblerone [9] was used to obtain WM PV estimates.

A single-compartment well-mixed ASL model was used to represent term M in Eq. 1 and simulate data [2]. Pseudo-continuous ASL with six post-label delays of [0.25, 0.5, 0.75, 1.0, 1.25, 1.5] s and 1.8 s label duration was used. For the cortex, a ground truth CBF map with a mean value of 60 units and sinusoidal variation of 40 units peak-to-peak was used, illustrated in Fig. 3, and a constant ATT of 1.1 s was used. In WM and subcortical GM structures, constant reference CBF and ATT values were assumed. Data was simulated at spatial resolutions of [2, 3, 4, 5] mm isotropic and SNR levels of $x \cdot 2^n, n \in [-1, -0.5, 0, 0.5, 1]$, i.e., from $x/2$ to $2x$, where x represents typical SNR estimated from a single subject's mutli-delay data acquired in a previous study [12]. In the results, the value x is omitted from notation, so SNR $= 2^0$ means 'typical SNR'. This variety of resolutions and SNR levels reflects the diversity of acquisitions seen in clinical practice. In all cases, zero-mean additive white noise was used.

hSVB has been implemented using TensorFlow version 2.9 running on Python 3.9.12. Optimisation was performed using RMSProp with a learning rate of 0.1, a decay factor of 0.97, and a sample size of 10. During training, a reversion to previous best state ('mean reversion') was performed when cost did not improve for 50 consecutive epochs; for all experiments, training was run until 20 such reversions had taken place which served as a proxy for convergence. Typically this implied training for around 2000 epochs. A folded normal distribution was used

Fig. 3. The ground truth cortical CBF map, illustrated on both the inflated sphere (left) and the native pial surface (right). The signal varies between 40 and 80 units with a mean of 60.

on all model parameters to restrict inference to positive values only. Runtime on a 6 core CPU was around 5 mins per inference, using around 4 GB of RAM.

The comparator method was based on BASIL, a conventional volumetric ASL processing workflow [3]. Specifically, the *oxford_asl* pipeline with partial volume effect correction was run using the same PV estimates from which the ASL data was simulated, and then the GM-specific parameter maps were projected onto the cortical surface using the same method as embedded within hSVB [10]. This method is referred to as BASIL-projected, BP. Runtime (single-threaded) was around 5 min per inference, using around 1 GB of RAM.

Performance was assessed by calculating the following metrics with respect to the ground truth cortical CBF map: sum of squared differences (SSD) of CBF; SSD of Z-transformed CBF, and Bhattacharyya distance of CBF distribution. The second and third metrics are included because they are sensitive to relative perfusion, whereas the first is sensitive to absolute CBF. A receiver-operator characteristic (ROC) analysis was performed using a binary classifier of varying threshold value t. Recalling that the ground truth CBF map had a mean value of 60 with extrema of ± 20, t was set at values of 5 to 15 with an increment of 1. At each t, areas of hypoperfusion were classified with a threshold value of $60 - t$ and hyperperfusion with $60 + t$. The area-under ROC (AUROC) for hypo and hyperperfusion was then calculated and the mean of the two taken. This yielded an AUROC score for each method at varying levels of threshold t.

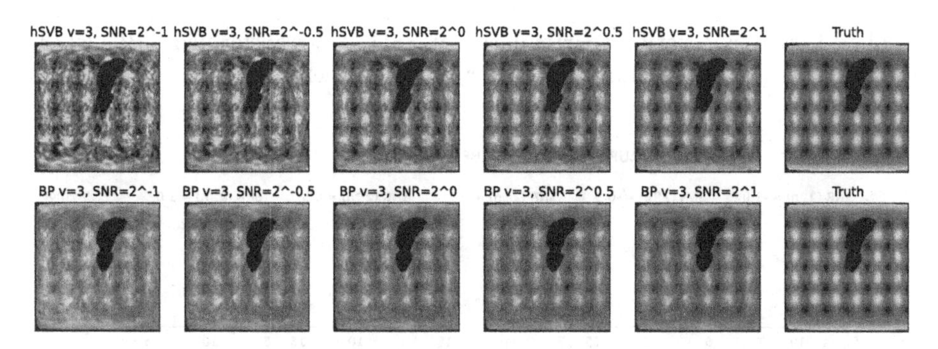

Fig. 4. Cortical CBF maps estimated from 3 mm data. Each map has been flattened onto a 2D plane for visual inspection. The blank spot corresponds to the corpus callosum which has no defined cortical thickness and therefore does not support parameter estimation.

3 Results

Figure 4 shows cortical CBF maps estimated by both methods on 3 mm data, flattened down onto a 2D plane for ease of inspection. For all SNR, hSVB's map retained more contrast than BP's: the bright spots were brighter, and vice-versa, particularly so at low SNR. At SNR $= 2^{-1}$, BP's result displayed a positive bias in CBF which was readily observed by comparison to other SNR levels.

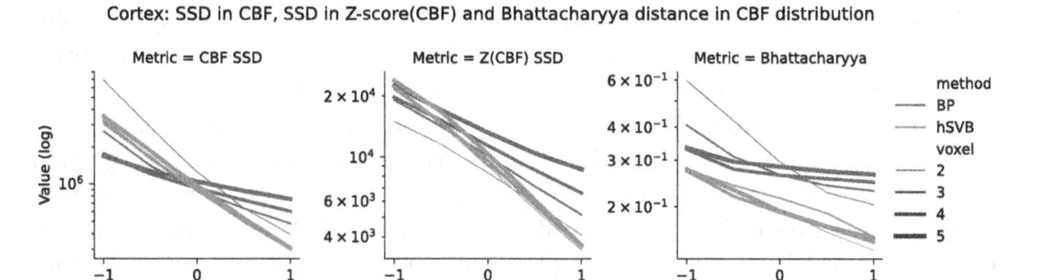

Fig. 5. SSD in CBF, SSD in Z-transformed CBF, and Bhattacharyya distance in CBF perfusion distribution, all evaluated for the cortex. Note in all panels the greater consistency in hSVB's scores across voxel sizes compared to BP.

Figure 5 shows SSD in CBF, SSD in Z-transformed CBF, and Bhattacharyya distance of CBF perfusion distribution, all for the cortex. For SSD, hSVB performed worse than BP at low SNR, and better for SNR $\geq 2^0$. By contrast, for SSD of Z-transformed CBF, hSVB performed better at almost all SNR. In Bhattacharyya distance, hSVB performed better at all voxel sizes and SNR. Of note in all panels was the consistency of hSVB's results across voxel sizes, whereas BP displayed greater variation (particularly for SSD of Z-transformed CBF, where the 2 mm result was substantially better than all others).

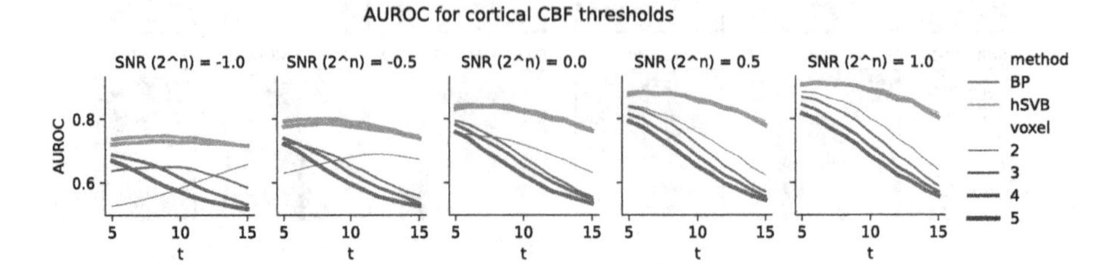

Fig. 6. AUROC scores for binary classification of cortical hypo and hyper-perfusion at varying threshold values t. For all comparisons, hSVB obtained a higher score than BP.

Figure 6 shows AUROC scores for binary classification of the estimated CBF maps returned by both methods. For all voxel sizes and SNR, the AUROC score for hSVB was higher than that of BP, particularly at high threshold values.

4 Discussion

The results presented here demonstrate that a hybrid approach to parameter inference using SVB (hSVB) offers a compelling alternative for the surface-based analysis of functional neuroimaging data. hSVB can operate directly on volumetric data without pre-processing and is able to able to apply the spatial prior in an anatomically-informed manner that respects tissue boundaries.

Applied to simulated ASL data, hSVB demonstrated a number of positive attributes in relation to a conventional volumetric workflow with post-projection (BP). Firstly, greater consistency in performance across voxel sizes was observed. As spatial resolution directly determines the extent of PVE within data, this implied that hSVB is more robust to PVE which is an important source of confound. Secondly, at higher levels of SNR, hSVB was able to deliver estimates that scored substantially better across a variety of metrics, which suggests it is well-placed to exploit future increases in SNR that result from advances in hardware and acquisition. Finally, hSVB was better able to discern relative perfusion differences, i.e., areas of abnormality. The trade-off for this was higher SSD errors in absolute perfusion values at low SNR.

A key point of divergence between hSVB and BP observed in this work was in the effect of the spatial prior at low SNR levels. Bayesian inference can be understood as an updating process whereby the prior distribution is modified to the extent that the observed data supports this. For low SNR data, the data will support less deviation from the priors. Referring to Fig. 4, BP tended towards globally homogenous solutions with high smoothing and low detail, whereas hSVB did the opposite: increasingly heterogenous ('textured', in the language of image analysis) solutions with low smoothing and high detail. In this scenario, the SSD of CBF metric is asymmetric: as smoothing increases towards a perfectly homogenous solution, SSD will approach a finite asymptotic value, whereas in the opposite outcome of increasingly extreme minima and maxima that preserve overall detail, SSD will increase towards infinity. It is believed that this mechanism explains the high SSD errors obtained by hSVB at low SNR, and also explains why hSVB performed much better on a relative basis (SSD of Z-score CBF), where the ability to distinguish perfusion variation matters more than the absolute values.

Two limitations of this work are that the ground truth cortical CBF map used to simulate data (Fig. 3) is clearly highly contrived, and the simulation did not include any of the imperfections normally seen in acquisition data, such as motion, intensity or geometric distortion artefacts. Though such problems are normally dealt with prior to performing inference via separate pre-processing operations, they can rarely be fully corrected, and thus some residual artefact may remain during inference which may degrade performance. The next step in the development of this work will be to try hSVB on human acquisition data to verify similar performance to the results presented here, which is currently in progress.

References

1. Attias, H.: A variational Bayesian framework for graphical models. In: Advances in Neural Information Processing Systems, pp. 209–215 (2000). ISBN 0262194503
2. Buxton, R.B., Frank, L.R., Wong, E.C., Siewert, B., Warach, S., Edelman, R.R.: A general kinetic model for quantitative perfusion imaging with arterial spin labeling. Magn. Reson. Med. **40**(3), 383–396 (1998). https://doi.org/10.1002/mrm.1910400308. ISBN 1522-2594
3. Chappell, M., Groves, A., Whitcher, B., Woolrich, M.: Variational Bayesian inference for a nonlinear forward model. IEEE Trans. Signal Process. **57**(1), 223–236 (2009). https://doi.org/10.1109/TSP.2008.2005752. ISBN 1053-587X
4. Chappell, M.A., Craig, M.S., Woolrich, M.W.: Stochastic variational Bayesian inference for a nonlinear forward model. arXiv e-prints arXiv:2007.01675 (2020). arXiv, eess.SP/2007.01675
5. Coalson, T.S., Van Essen, D.C., Glasser, M.F.: Lost in Space: The Impact of Traditional Neuroimaging Methods on the Spatial Localization of Cortical Areas. bioRxiv, p. 255620 (2018). Publisher: Cold Spring Harbor Laboratory
6. Desbrun, M., Meyer, M., Schröder, P., Barr, A.H.: Implicit fairing of irregular meshes using diffusion and curvature flow. In: Proceedings of the 26th Annual Conference on Computer Graphics and Interactive Techniques, SIGGRAPH 1999, pp. 317–324. ACM Press (1999). https://doi.org/10.1145/311535.311576. http://portal.acm.org/citation.cfm?doid=311535.311576
7. Fischl, B.: FreeSurfer. Neuroimage **62**(2), 774–781 (2012). https://doi.org/10.1016/j.neuroimage.2012.01.021.FreeSurfer. arXiv:0905.26710. ISBN 1095-9572 (Electronic) r053-8119 (Linking)

8. Glasser, M.F., et al.: The minimal preprocessing pipelines for the Human Connectome Project. NeuroImage **80**, 105–124 (2013). https://doi.org/10.1016/j.neuroimage.2013.04.127. arXiv, NIHMS150003. Publisher: Elsevier Inc. ISBN 1053-8119
9. Kirk, T.F., Coalson, T.S., Craig, M.S., Chappell, M.A.: Toblerone: surface-based partial volume estimation. IEEE Trans. Med. Imaging **39**(5), 1501–1510 (2020). https://doi.org/10.1109/TMI.2019.2951080
10. Kirk, T.F., Craig, M.S., Chappell, M.A.: Unified surface and volumetric projection of physiological imaging data (2022). https://doi.org/10.1101/2022.01.28.477071. https://www.biorxiv.org/content/10.1101/2022.01.28.477071v1, p. 2022.01.28.477071 Section: New Results
11. Lonjaret, L.T., Bakhous, C., Boutelier, T., Takerkart, S., Coulon, O.: ISA - an inverse surface-based approach for cortical fMRI data projection. In: 14th IEEE International Symposium on Biomedical Imaging, ISBI 2017, Melbourne, Australia, 18–21 April 2017, pp. 1104–1107. IEEE (2017). https://doi.org/10.1109/ISBI.2017.7950709
12. Mezue, M., Segerdahl, A.R., Okell, T.W., Chappell, M.A., Kelly, M.E., Tracey, I.: Optimization and reliability of multiple postlabeling delay pseudo-continuous arterial spin labeling during rest and stimulus-induced functional task activation. J. Cerebral Blood Flow Metab. **34**(12), 1919–1927 (2014). https://doi.org/10.1038/jcbfm.2014.163. https://www.ncbi.nlm.nih.gov/pubmed/25269517. Publisher: Nature Publishing Group
13. Operto, G., Bulot, R., Anton, J.L., Coulon, O.: Projection of fMRI data onto the cortical surface using anatomically-informed convolution kernels. NeuroImage **39**(1), 127–135 (2008). https://doi.org/10.1016/j.neuroimage.2007.08.039. http://linkinghub.elsevier.com/retrieve/pii/S1053811907007586
14. Patenaude, B., Smith, S.M., Kennedy, D.N., Jenkinson, M.: A Bayesian model of shape and appearance for subcortical brain segmentation. NeuroImage **56**(3), 907–922 (2011). https://doi.org/10.1016/J.NEUROIMAGE.2011.02.046. https://www.sciencedirect.com/science/article/pii/S1053811911002023. Publisher: Academic Press
15. Penny, W.D., Trujillo-Barreto, N.J., Friston, K.J.: Bayesian fMRI time series analysis with spatial priors. Neuroimage **24**(2), 350–362 (2005). https://doi.org/10.1016/j.neuroimage.2004.08.034
16. Zhang, Y., Brady, M., Smith, S.: Segmentation of brain MR images through a hidden Markov random field model and the expectation-maximization algorithm. IEEE Trans. Med. Imaging **20**(1), 45–57 (2001). https://doi.org/10.1109/42.906424. ISBN 0278-0062 (Print) 0278-0062 (Linking)

TractCloud: Registration-Free Tractography Parcellation with a Novel Local-Global Streamline Point Cloud Representation

Tengfei Xue[1,2], Yuqian Chen[1,2], Chaoyi Zhang[2], Alexandra J. Golby[1], Nikos Makris[1], Yogesh Rathi[1], Weidong Cai[2], Fan Zhang[1(✉)], and Lauren J. O'Donnell[1(✉)]

[1] Harvard Medical School, Boston, MA, USA
zhangfanmark@gmail.com, odonnell@bwh.harvard.edu
[2] The University of Sydney, Sydney, NSW, Australia

Abstract. Diffusion MRI tractography parcellation classifies streamlines into anatomical fiber tracts to enable quantification and visualization for clinical and scientific applications. Current tractography parcellation methods rely heavily on registration, but registration inaccuracies can affect parcellation and the computational cost of registration is high for large-scale datasets. Recently, deep-learning-based methods have been proposed for tractography parcellation using various types of representations for streamlines. However, these methods only focus on the information from a single streamline, ignoring geometric relationships between the streamlines in the brain. We propose *TractCloud*, a registration-free framework that performs whole-brain tractography parcellation directly in individual subject space. We propose a novel, learnable, local-global streamline representation that leverages information from neighboring and whole-brain streamlines to describe the local anatomy and global pose of the brain. We train our framework on a large-scale labeled tractography dataset, which we augment by applying synthetic transforms including rotation, scaling, and translations. We test our framework on five independently acquired datasets across populations and health conditions. TractCloud significantly outperforms several state-of-the-art methods on all testing datasets. TractCloud achieves efficient and consistent whole-brain white matter parcellation across the lifespan (from neonates to elderly subjects, including brain tumor patients) without the need for registration. The robustness and high inference speed of TractCloud make it suitable for large-scale tractography data analysis. Our project page is available at https://tractcloud.github.io/.

This work was supported by the following NIH grants: R01MH125860, R01MH119222, R01MH132610, R01MH074794, and R01NS125781.

Supplementary Information The online version contains supplementary material available at https://doi.org/10.1007/978-3-031-43993-3_40.

H. Greenspan et al. (Eds.): MICCAI 2023, LNCS 14227, pp. 409–419, 2023.
https://doi.org/10.1007/978-3-031-43993-3_40

Keywords: Diffusion MRI · Tractography · Registration-free white matter parcellation · Deep learning · Point cloud

1 Introduction

Diffusion MRI (dMRI) tractography is the only non-invasive method capable of mapping the complex white matter (WM) connections within the brain [2]. Tractography parcellation [12,28,42] classifies the vast numbers of streamlines resulting from whole-brain tractography to enable visualization and quantification of the brain's WM connections. (Here a streamline is defined as a set of ordered points in 3D space resulting from tractography [45]). In recent years, deep-learning-based methods have been proposed for tractography parcellation [5,7,14,15,17,20,33,34,36,37,39,44], of which many methods are designed to classify streamlines [7,15–17,37,39,44]. However, multiple challenges exist when using streamline data as deep network input. One well-known challenge is that streamlines can be equivalently represented in forward or reverse order [11,39], complicating their direct representation as vectors [7] or images [44]. Another challenge is that the geometric relationships between the streamlines in the brain have previously been ignored: existing parcellation methods [7,15–17,37,39,44] train and classify each streamline independently. Finally, computational cost can pose a challenge for the parcellation of large tractography datasets that can include thousands of subjects with millions of streamlines per subject.

In this work, we propose a novel point-cloud-based strategy that leverages neighboring and whole-brain streamline information to learn local-global streamline representations. Point clouds have been shown to be efficient and effective representations for streamlines [1,4,6,15,18,39] in applications such as tractography filtering [1], clustering [7], and parcellation [15,18,38,39]. One benefit of using point clouds is that streamlines with equivalent forward and reverse point orders (e.g., from cortex to brainstem or vice versa) can be represented equally. However, these existing methods focus on a single streamline (one point cloud) and ignore other streamlines (other point clouds) in the same brain that may provide important complementary information useful for tractography parcellation. In computer vision, point clouds are commonly used to describe scenes and objects (e.g., cars, tables, airplanes, etc.). However, point cloud segmentation methods from computer vision, which assign labels to points, cannot translate directly to the tractography field, where the task of interest is to label entire streamlines. Computer vision studies [21,26,32,35,40,41,46] have shown that point interactions *within one point cloud* can yield more effective features for downstream tasks. However, in tractography parcellation we are interested in the relationship *between multiple point clouds (streamlines)* in the brain. These other streamlines can provide detailed information about the local WM geometry surrounding the streamline to be classified, as well as global information about the location and pose of the brain that can reduce the need for image registration.

Affine or even nonrigid registration is needed for current tractography parcellation methods [13,28,42]. Recently, registration-free techniques have been proposed for tractography parcellation to handle computational challenges resulting

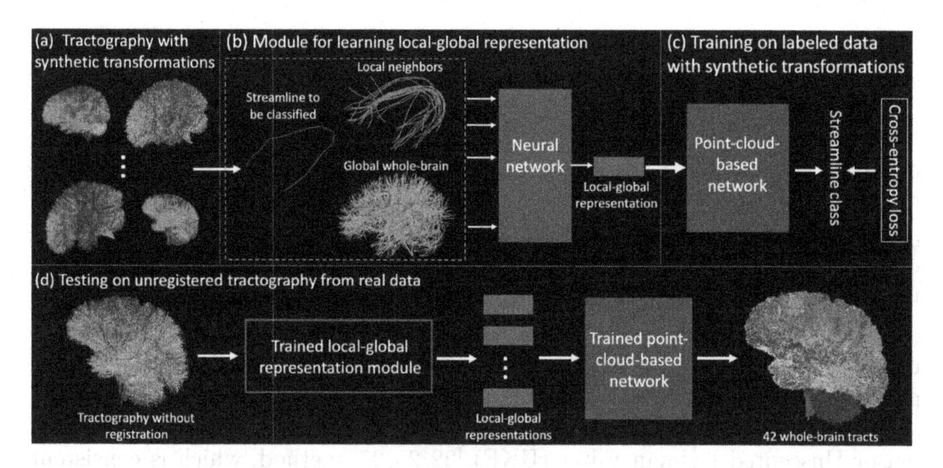

Fig. 1. TractCloud framework overview: (a) tractography with synthetic transformations, (b) module for learning local-global representation, (c) training on labeled data with synthetic transformations, (d) testing on unregistered tractography from real data.

from large inter-subject variability and to increase robustness to image registration inaccuracies [19,29]. Avoiding image registration can also reduce computational time and cost when processing very large tractography datasets with thousands of subjects. While other registration-free tractography parcellation techniques require Freesurfer input [29] or work with rigidly MNI-aligned Human Connectome Project data [19], our method can directly parcellate tractography in individual subject space.

In this study, we propose *TractCloud*, a registration-free tractography parcellation framework, as illustrated in Fig. 1. This paper has three main contributions. First, we propose a novel, learnable, local-global streamline representation that leverages information from neighboring and whole-brain streamlines to describe the local anatomy and global pose of the brain. Second, we leverage a training strategy using synthetic transformations of labeled tractography data to enable registration-free parcellation at the inference stage. Third, we implement our framework using two compared point cloud networks and demonstrate fast, registration-free, whole-brain tractography parcellation across the lifespan.

2 Methods

2.1 Training and Testing Datasets

We utilized a high-quality and large-scale dataset of 1 million labeled streamlines for model training and validation. The dataset was obtained from a WM tractography atlas [42] that was curated and annotated by a neuroanatomist. The atlas was derived from 100 registered tractography of young healthy adults in the Human Connectome Project (HCP) [30]. The training data includes 43 tract classes: 42 anatomically meaningful tracts from the whole brain and one tract category of "other streamlines," including, most importantly, anatomically implausible outlier streamlines. On average, the 42 anatomical tracts have 2539 streamlines with a standard deviation of 2693 streamlines.

For evaluation, we used a total of 120 subjects from four public datasets and one private dataset. These five datasets were independently acquired with different imaging protocols across ages and health conditions. (1) developing HCP (dHCP) [9]: 20 neonates (1 to 27 days); (2) Adolescent Brain Cognitive Development (ABCD) dataset [31]: 25 adolescents (9 to 11 years); (3) HCP dataset [30]: 25 young healthy adults (22 to 35 years, subjects not part of the training atlas); (4) Parkinson's Progression Markers Initiative (PPMI) dataset [23]: 25 older adults (51 to 75 years), including Parkinson's disease (PD) patients and healthy individuals; (5) Brain Tumor Patient (BTP) dataset: dMRI data from 25 brain tumor patients (28 to 70 years) were acquired at Brigham and Women's Hospital. dMRI acquisition parameters of datasets are shown in Supplementary Table S1. The two-tensor Unscented Kalman Filter (UKF) [22,25,27] method, which is consistent across ages, health conditions, and image acquisitions [42], was utilized to create whole-brain tractography for all subjects across the datasets mentioned above.

2.2 TractCloud Framework

Synthetic Transform Data Augmentation. To enable tractography parcellation without registration, we augmented the training data by applying synthetic transform-based augmentation (STA) including rotation, scaling, and translations. These transformations have been used in voxel-based WM segmentation [34], but no study has applied these transformations to study tractography, to our knowledge. In detail, we applied 30 random transformations to each subject tractography in the training dataset to obtain 3000 transformed subjects and 30 million streamlines. Transformations included: rotation from −45 to 45 °C along the left-right axis, from −10 to 10 °C along the anterior-posterior axis, and from −10 to 10 °C along the superior-inferior axis; translation from −50 to 50 mm along all three axes; scaling from −45% to 5% along all three axes. These transformations were selected based on typical differences between subjects due to variability in brain anatomy and volume, head position, and image acquisition protocol. Many methods are capable of tractography parcellation after affine registration [12,42]; therefore, with STA applied to the training dataset, our framework has the potential for registration-free parcellation.

Module for Local-Global Streamline Representation Learning. We propose a module (Fig. 2) to learn the proposed local-global representation, which benefits from information about the anatomy of the neighboring WM and the overall pose of the brain. We construct the input for the learning module by concatenating the coordinates of the original streamline (the one to be classified), its local neighbor streamlines, and global whole-brain streamlines. In detail, assume a brain has n streamlines, denoted by $S = \{s_1, s_2, \ldots, s_n\}$, $s_i \in \mathcal{R}^{m \times 3}$, where 3 is the dimensionality of the point coordinates and m is the number of points for each streamline ($m = 15$ as in [42,44]). For streamline s_i, we obtain a set of

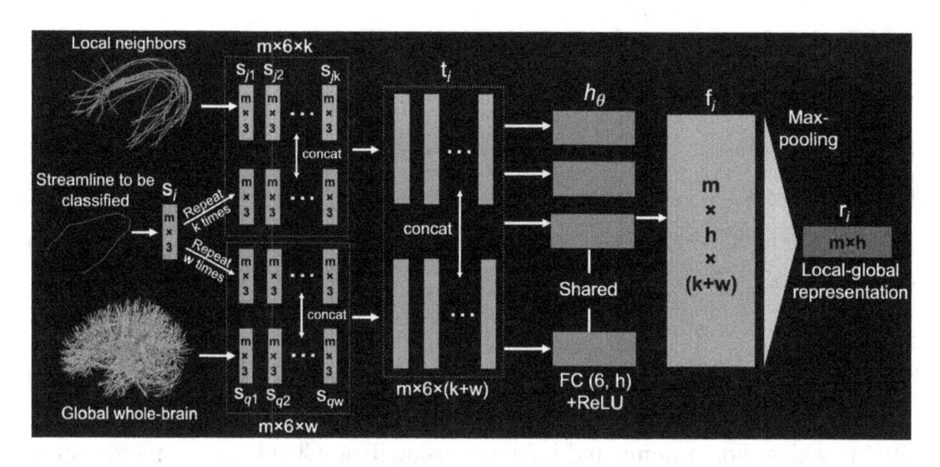

Fig. 2. The proposed module for learning local-global representation.

k nearest streamlines, *local*$(s_i) = \{s_{j1}, s_{j2}, \ldots, s_{jk}\}$, using a pairwise streamline distance [11]. From the whole brain, we also randomly select a set of w streamlines, *global*$(s_i) = \{s_{q1}, s_{q2}, \ldots, s_{qw}\}$. Then s_i, *local*(s_i), and *global*(s_i) are concatenated as shown in Fig. 2 to obtain the input of the module, $t_i \in \mathcal{R}^{m \times 6 \times (k+w)}$. The proposed module begins with a shared fully connected (FC) layer with ReLU activation function (h_Θ): $f_i = h_\Theta(t_i)$, $f_i \in \mathcal{R}^{m \times h \times (k+w)}$, where h is the output dimension of h_Θ (h=64 [3,26,32]). Finally, the local-global representation r_i is obtained through max-pooling $r_i = \text{pool}(f_i) \in \mathcal{R}^{m \times h}$. The network is trained in an end-to-end fashion where the local-global representation r_i is learned during training of the overall point-cloud-based classification network.

Network Structure for Streamline Classification. The local-global representation learning module can replace the first layer or module of typical point-cloud-based networks [3,26,32,46]. Here, we explore two widely used networks: PointNet [3] and Dynamic Graph Convolutional Neural Network (DGCNN) [32]. PointNet (see Fig. S1 for network details) encodes point-wise features individually, but DGCNN (see Fig. S2 for network details) encodes point-wise features by interacting with other points on a streamline. Both PointNet and DGCNN then aggregate features of all points through pooling to get a single streamline descriptor, which is input into fully connected layers for classification.

2.3 Implementation Details

To learn r_i, we used 20 local streamlines (selected from 10, 20, 50, 100) and 500 global streamlines (selected from 100, 300, 500, 1000). Our framework was trained with the Adam optimizer with a learning rate of 0.001 using cross-entropy loss. The epoch was 20, and the batch size was 1024. Training of our registration-free framework (TractCloud$_{\textit{reg-free}}$) with the large STA dataset took about 22 h and 10.9 GB GPU memory with Pytorch (v1.13) on an NVIDIA RTX A5000

Table 1. Results on the labeled training dataset with and without synthetic transformations. Bold and italic text indicates the best-performing method and the second-best-performing method, respectively. Abbreviations: Orig - Original, Acc - Accuracy.

Feature		Single Streamline				Local		Local + Global	
Data & Metric		SOTA Methods		Point Cloud Networks		TractCloud Effectiveness Study			
		Deep WMA	DCNN++	PointNet	DGCNN	PointNet $_{+loc}$	DGCNN $_{+loc}$	PointNet $_{+loc+glo}$	DGCNN $_{+loc+glo}$
Orig data	Acc	90.29	91.26	91.36	91.85	91.51	91.91	**92.28**	*91.99*
	F1	88.12	89.14	89.12	89.78	89.25	90.03	**90.36**	*90.10*
STA data	Acc	82.35	84.14	81.83	83.70	86.56	87.14	*91.57*	**91.69**
	F1	76.55	79.16	75.89	78.55	82.08	82.95	*89.40*	**89.65**

GPU machine. For training and inference using TractCloud$_{reg\text{-}free}$, tractography was centered at the mass center of the training atlas. TractCloud$_{reg\text{-}free}$ directly annotates tract labels at the inference stage without requiring the registration of an atlas.

3 Experiments and Results

3.1 Performance on the Labeled Atlas Dataset

We evaluated our method on the original labeled training dataset (registered and aligned) and its synthetic transform augmented (STA) data (unregistered and unaligned). We divided both the original and STA data into train/validation/test sets with the distribution of 70%/10%/20% by subjects (such that all streamlines from an individual subject were placed into only one set, either train or validation or test). For experimental comparison, we included two deep-learning-based state-of-the-art (SOTA) tractography parcellation methods: DCNN++ [37] and DeepWMA [44]. They were both designed to perform deep WM parcellation using CNNs, with streamline spatial coordinate features as input. We trained the networks based on the recommended settings in their papers and code. Two widely used point-cloud-based networks (PointNet [3] and DGCNN [32]), with a single streamline as input, were included as baseline methods. To evaluate the effectiveness of the local-global representation in TractCloud, we performed experiments using only local neighbor features (PointNet$_{+loc}$ and DGCNN$_{+loc}$) and both local neighbor and whole-brain global features (PointNet$_{+loc+glo}$ and DGCNN$_{+loc+glo}$). For all methods, we report two metrics (accuracy and macro F1) that are widely used for tractography parcellation [19,24,37,39,44]. The accuracy is reported as the overall accuracy of streamline classification, and the macro F1 score is reported as the mean across 43 tract classes (Table 1).

Table 1 shows that the TractCloud framework achieves the best performance on data with and without synthetic transformations (STA). Especially on STA data, TractCloud yields a large improvement in accuracy (up to 9.9%) and F1 (up to 13.8%), compared to PointNet and DGCNN baselines as well as SOTA methods. In addition, including local (PointNet$_{+loc}$ and DGCNN$_{+loc}$) and global

(PointNet$_{+loc+glo}$ and DGCNN$_{+loc+glo}$) features both improve the performance compared to baselines (PointNet and DGCNN) with a single streamline as input. This demonstrates the effectiveness of our local-global representation.

3.2 Performance on the Independently Acquired Testing Datasets

We performed experiments on five independently acquired, unlabeled testing datasets (dHCP, ABCD, HCP, PPMI, BTP) to evaluate the robustness and generalization ability of our TractCloud$_{reg\text{-}free}$ framework on unseen and unregistered data. All compared SOTA methods (DeepWMA, DCNN++) and TractCloud$_{regist}$ were tested on registered tractography, and only TractCloud$_{reg\text{-}free}$ was tested on unregistered tractography. Tractography was registered to the space of the training atlas using an affine transform produced by registering the baseline ($b = 0$) image of each subject to the atlas population mean T2 image using 3D Slicer [10]. For each method, we quantified the tract identification rate (TIR) and calculated the tract-to-atlas distance (TAD), and statistical significance tests were performed for results of TIR and TAD (Table 2). TIR measures if the tract is identified successfully when labels are not available [7,42,44]. Here, we chose 50 as the minimum number of streamlines for a tract to be considered as identified (The threshold of 50 is more strict than 10 or 20 in [7,42,44]). As a complementary metric for TIR, TDA measures the geometric similarity between identified tracts and corresponding tracts from the training atlas. For each testing subject's tract, we calculated the streamline-specific minimum average direct-flip distance [7,11,42] to the atlas tract and then computed the average across subjects and tracts to obtain TDA. We also recorded the computation time for tractography parcellation for every method (Table 2). The computation time was tested on a Linux workstation with an NVIDIA RTX A4000 GPU using tractography (0.28 million streamlines) from a randomly selected subject. To evaluate if differences in result values between our registration-free method (TractCloud$_{reg\text{-}free}$) and other methods are significant, we implemented a repeated measure ANOVA test for all methods across subjects, and then we performed multiple paired Student's t-tests between TractCloud$_{reg\text{-}free}$ method and each compared method. In addition, in order to evaluate how well our framework can perform without registration, we converted identified tracts into volume space and calculated the spatial overlap (weighted Dice) [8,43] between results of TractCloud$_{regist}$ and TractCloud$_{reg\text{-}free}$ (Table 3). Furthermore, we also provide a visualization of identified tracts in an example individual subject for every dataset across methods (Fig. 2).

As shown in Table 2, all methods achieve high TIRs on all datasets, and the TIR metric does not have significant differences across methods. This demonstrates that most tracts can be identified by all methods robustly. However, our registration-free framework (TractCloud$_{reg\text{-}free}$) obtains significantly lower TDA values (better quality of identified tracts) than all compared methods on ABCD, HCP, and PPMI datasets, where ages of test subjects are from 9 to 75 years old. On the very challenging dHCP (baby brain) dataset, TractCloud$_{reg\text{-}free}$ still significantly outperforms two SOTA methods. Note that TractCloud$_{reg\text{-}free}$ directly

Table 2. Results of tract identification rate (TIR) and tract distance to atlas (TDA) on five independently acquired testing datasets as well as computation time on a randomly selected subject. TIR results show no significant differences across methods (ANOVA $p > 0.05$), while TDA results do (ANOVA $p < 1 \times 10^{-10}$). Asterisks show that the difference between TractCloud$_{reg\text{-}free}$ and other methods is significant using a paired Student's t-test. ($*p < 0.05$, $**\,p < 0.001$). Abbreviations: TC - TractCloud.

Method	TIR (%) ↑					TDA (mm) ↓					Run Time
	dHCP	ABCD	HCP	PPMI	BTP	dHCP	ABCD	HCP	PPMI	BTP	
DeepWMA	98.8 ±1.4	100	100	100	99.9 ±0.5	6.81* ±1.1	5.66** ±0.9	5.09** ±0.7	5.94** ±1.0	6.24* ±1.2	113 s
DCNN++	98.7 ±1.9	100	100	100	99.8 ±0.9	6.90* ±1.4	5.69** ±0.9	5.08** ±0.7	5.95** ±0.9	6.43** ±1.8	102 s
TC$_{regist}$	99.2 ±1.6	100	100	100	99.9 ±0.5	6.53* ±1.1	5.60** ±0.9	5.06** ±0.7	5.87* ±1.0	6.09 ±1.1	97 s
TC$_{reg\text{-}free}$	97.7 ±3.1	100	100	100	99.9 ±0.5	6.71 ±1.3	5.52 ±0.8	5.02 ±0.6	5.85 ±0.9	6.11 ±1.0	58 s

Table 3. Tract spatial overlap (wDice) between TractCloud$_{regist}$ and TractCloud$_{reg\text{-}free}$

	dHCP	ABCD	HCP	PPMI	BTP
TSO	0.932 ± 0.14	0.965 ± 0.04	0.980 ± 0.02	0.977 ± 0.03	0.970 ± 0.05

works on unregistered tractography from neonate brains (much smaller than adult brains). In the challenging BTP (tumor patients) dataset, TractCloud$_{reg\text{-}free}$ obtains significantly lower TDA values than SOTA methods and comparable performance to TractCloud$_{regist}$. As shown in Table 2, our registration-free framework is much faster than other compared methods.

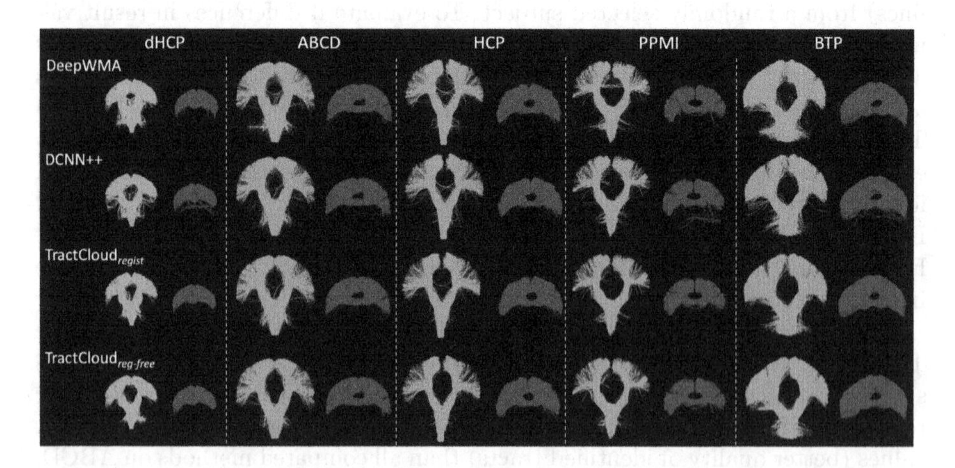

Fig. 3. Visualization of example tracts (corticospinal tract and corpus callosum IV) from each method, in the subject with the median TDA for each testing dataset.

The tract spatial overlap (wDice) is over 0.965 on all datasets, except for the challenging dHCP (wDice is 0.932) (Table 3). Overall, our registration-free framework is comparable to (or better than) our framework with registration.

Figure 3 shows visualization results of example tracts. All methods can successfully identify these tracts across datasets. It is visually apparent that the TractCloud$_{reg\text{-}free}$ framework obtains results with fewer outlier streamlines, especially on the challenging dHCP dataset.

4 Discussion and Conclusion

We have demonstrated TractCloud, a registration-free tractography parcellation framework with a novel, learnable, local-global representation of streamlines. Experimental results show that TractCloud can achieve efficient and consistent tractography parcellation results across populations and dMRI acquisitions, with and without registration. The fast inference speed and robust ability to parcellate data in original subject space will allow TractCloud to be useful for analysis of large-scale tractography datasets. Future work can investigate additional data augmentation using local deformations to potentially increase robustness to pathology. Overall, TractCloud demonstrates the feasibility of registration-free tractography parcellation across the lifespan.

References

1. Astolfi, P., et al.: Tractogram filtering of anatomically non-plausible fibers with geometric deep learning. In: Martel, A.L., et al. (eds.) MICCAI 2020. LNCS, vol. 12267, pp. 291–301. Springer, Cham (2020). https://doi.org/10.1007/978-3-030-59728-3_29
2. Basser, P.J., Pajevic, S., Pierpaoli, C., Duda, J., Aldroubi, A.: In vivo fiber tractography using DT-MRI data. Magn. Reson. Med. **44**(4), 625–632 (2000)
3. Charles, R.Q., Su, H., Kaichun, M., Guibas, L.J.: PointNet: Deep learning on point sets for 3D classification and segmentation. In: Proceedings of the IEEE Conference on Computer Vision and Pattern Recognition (CVPR), pp. 77–85 (2017)
4. Chen, Y., et al.: White matter tracts are point clouds: neuropsychological score prediction and critical region localization via geometric deep learning. In: Wang, L., Dou, Q., Fletcher, P.T., Speidel, S., Li, S. (eds.) Medical Image Computing and Computer Assisted Intervention-MICCAI 2022. MICCAI 2022. Lecture Notes in Computer Science. vol. 13431. Springer, Cham (2022). https://doi.org/10.1007/978-3-031-16431-6_17
5. Chen, Y., et al.: Deep fiber clustering: anatomically informed fiber clustering with self-supervised deep learning for fast and effective tractography parcellation. Neuroimage **273**, 120086 (2023)
6. Chen, Y., et al.: TractGeoNet: A geometric deep learning framework for pointwise analysis of tract microstructure to predict language assessment performance. arXiv 2307.0398 (2023)
7. Chen, Y., et al.: Deep fiber clustering: anatomically informed unsupervised deep learning for fast and effective white matter parcellation. In: de Bruijne, M., et al. (eds.) MICCAI 2021. LNCS, vol. 12907, pp. 497–507. Springer, Cham (2021). https://doi.org/10.1007/978-3-030-87234-2_47

8. Cousineau, M., et al.: A test-retest study on parkinson's PPMI dataset yields statistically significant white matter fascicles. Neuroimage Clin. **16**, 222–233 (2017)
9. Edwards, A.D., et al.: The developing human connectome project neonatal data release. Front. Neurosci. **16**, 886772 (2022)
10. Fedorov, A., et al.: 3D slicer as an image computing platform for the quantitative imaging network. Magn. Reson. Imaging **30**(9), 1323–1341 (2012)
11. Garyfallidis, E., et al.: QuickBundles, a method for tractography simplification. Front. Neurosci. **6**, 175 (2012)
12. Garyfallidis, E., et al.: Recognition of white matter bundles using local and global streamline-based registration and clustering. Neuroimage **170**, 283–295 (2018)
13. Garyfallidis, E., Ocegueda, O., Wassermann, D., Descoteaux, M.: Robust and efficient linear registration of white-matter fascicles in the space of streamlines. Neuroimage **117**, 124–140 (2015)
14. Gupta, V., Thomopoulos, S.I., Rashid, F.M., Thompson, P.M.: FiberNET: an ensemble deep learning framework for clustering white matter fibers. In: Descoteaux, M., Maier-Hein, L., Franz, A., Jannin, P., Collins, D.L., Duchesne, S. (eds.) MICCAI 2017. LNCS, vol. 10433, pp. 548–555. Springer, Cham (2017). https://doi.org/10.1007/978-3-319-66182-7_63
15. Kumaralingam, L., Thanikasalam, K., Sotheeswaran, S., Mahadevan, J., Ratnarajah, N.: Segmentation of whole-brain tractography: a deep learning algorithm based on 3D raw curve points. In: Wang, L., Dou, Q., Fletcher, P.T., Speidel, S., Li, S. (eds.) Medical Image Computing and Computer Assisted Intervention-MICCAI 2022. MICCAI 2022. Lecture Notes in Computer Science. vol. 13431. Springer, Cham (2022). https://doi.org/10.1007/978-3-031-16431-6_18
16. Legarreta, J.H., et al.: Filtering in tractography using autoencoders (FINTA). Med. Image Anal. **72**, 102126 (2021)
17. Legarreta, J.H., et al.: Clustering in tractography using autoencoders (CINTA). In: Computational Diffusion MRI, pp. 125–136 (2022)
18. Li, S., et al.: DeepRGVP: A novel Microstructure-Informed supervised contrastive learning framework for automated identification of the retinogeniculate pathway using dMRI tractography. In: ISBI (2023)
19. Liu, F., et al.: DeepBundle: fiber bundle parcellation with graph convolution neural networks. In: Graph Learning in Medical Imaging, pp. 88–95 (2019)
20. Liu, W., Lu, Q., Zhuo, Z., Liu, Y., Ye, C.: One-shot segmentation of novel white matter tracts via extensive data augmentation. In: Wang, L., Dou, Q., Fletcher, P.T., Speidel, S., Li, S. (eds.) Medical Image Computing and Computer Assisted Intervention-MICCAI 2022. MICCAI 2022. Lecture Notes in Computer Science. vol. 13431. Springer, Cham (2022). https://doi.org/10.1007/978-3-031-16431-6_13
21. Ma, X., Qin, C., You, H., Ran, H., Fu, Y.: Rethinking network design and local geometry in point cloud: a simple residual MLP framework. In: International Conference on Learning Representations (ICLR) (2022)
22. Malcolm, J.G., et al.: Filtered multitensor tractography. IEEE Trans. Med. Imaging **29**(9), 1664–1675 (2010)
23. Marek, K., et al.: The parkinson progression marker initiative (PPMI). Prog. Neurobiol. **95**(4), 629–635 (2011)
24. Ngattai Lam, P.D., et al.: TRAFIC: Fiber tract classification using deep learning. Proc. SPIE Int. Soc. Opt. Eng. **10574**, 1057412 (2018)
25. Norton, I., et al.: SlicerDMRI: open source diffusion MRI software for brain cancer research. Cancer Res. **77**(21), e101–e103 (2017)
26. Qi, C.R., Yi, L., Su, H., Guibas, L.J.: PointNet++: deep hierarchical feature learning on point sets in a metric space. In: NeurIPS, pp. 5105–5114 (2017)

27. Reddy, C.P., Rathi, Y.: Joint Multi-Fiber NODDI parameter estimation and tractography using the unscented information filter. Front. Neurosci. **10**, 166 (2016)
28. Román, C., et al.: Superficial white matter bundle atlas based on hierarchical fiber clustering over probabilistic tractography data. Neuroimage **262**, 119550 (2022)
29. Siless, V., et al.: Registration-free analysis of diffusion MRI tractography data across subjects through the human lifespan. Neuroimage **214**, 116703 (2020)
30. Van Essen, D.C., et al.: The WU-Minn human connectome project: an overview. Neuroimage **80**, 62–79 (2013)
31. Volkow, N.D., et al.: The conception of the ABCD study: from substance use to a broad NIH collaboration. Dev. Cogn. Neurosci. **32**, 4–7 (2018)
32. Wang, Y., Sun, Y., Liu, Z., Sarma, S.E., Bronstein, M.M., Solomon, J.M.: Dynamic graph CNN for learning on point clouds. ACM Trans. Graph. **38**(5), 1–12 (2019)
33. Wang, Z., et al.: Accurate corresponding fiber tract segmentation via FiberGeoMap learner. In: Medical Image Computing and Computer Assisted Intervention (MICCAI), pp. 143–152 (2022)
34. Wasserthal, J., Neher, P., Maier-Hein, K.H.: TractSeg - fast and accurate white matter tract segmentation. Neuroimage **183**, 239–253 (2018)
35. Xiang, T., Zhang, C., Song, Y., Yu, J., Cai, W.: Walk in the cloud: learning curves for point clouds shape analysis. In: Proceedings of the IEEE/CVF International Conference on Computer Vision (ICCV) (2021)
36. Xu, H., et al.: A registration- and uncertainty-based framework for white matter tract segmentation with only one annotated subject. In: ISBI (2023)
37. Xu, H., et al.: Objective detection of eloquent axonal pathways to minimize postoperative deficits in pediatric epilepsy surgery using diffusion tractography and convolutional neural networks. IEEE Trans. Med. Imaging **38**(8), 1910–1922 (2019)
38. Xue, T., et al.: SupWMA: Consistent and efficient tractography parcellation of superficial white matter with deep learning. In: ISBI (2022)
39. Xue, T., et al.: Superficial white matter analysis: an efficient point-cloud-based deep learning framework with supervised contrastive learning for consistent tractography parcellation across populations and dMRI acquisitions. Med. Image Anal. **85**, 102759 (2023)
40. Yan, X., et al.: PointASNL: robust point clouds processing using nonlocal neural networks with adaptive sampling. In: 2020 IEEE/CVF Conference on Computer Vision and Pattern Recognition (CVPR) (2020)
41. Yu, J., et al.: 3D medical point transformer: Introducing convolution to attention networks for medical point cloud analysis. arXiv 2112.04863 (2021)
42. Zhang, F., et al.: An anatomically curated fiber clustering white matter atlas for consistent white matter tract parcellation across the lifespan. Neuroimage **179**, 429–447 (2018)
43. Zhang, F., et al.: Test-retest reproducibility of white matter parcellation using diffusion MRI tractography fiber clustering. Hum. Brain Mapp. **40**(10), 3041–3057 (2019)
44. Zhang, F., et al.: Deep white matter analysis (DeepWMA): fast and consistent tractography segmentation. Med. Image Anal. **65**, 101761 (2020)
45. Zhang, F., et al.: Quantitative mapping of the brain's structural connectivity using diffusion MRI tractography: a review. Neuroimage **249**, 118870 (2022)
46. Zhao, H., et al.: Point transformer. In: 2021 IEEE/CVF International Conference on Computer Vision (ICCV) (2021)

Robust and Generalisable Segmentation of Subtle Epilepsy-Causing Lesions: A Graph Convolutional Approach

Hannah Spitzer[1,2], Mathilde Ripart[3], Abdulah Fawaz[4], Logan Z. J. Williams[4], MELD Project[3], Emma C. Robinson[4], Juan Eugenio Iglesias[5,6,7], Sophie Adler[3], and Konrad Wagstyl[8(✉)]

[1] Institute of Computational Biology, Helmholtz Center Munich, 85764 Munich, Germany
[2] Institute for Stroke and Dementia Research (ISD), LMU University Hospital, LMU Munich, Munich, Germany
[3] Department of Developmental Neuroscience, UCL Great Ormond Street Institute for Child Health, London WC1N 1EH, UK
[4] Department of Biomedical Engineering, School of Biomedical Engineering and Imaging Sciences, King's College London, London, UK
[5] Martinos Center for Biomedical Imaging, MGH and Harvard Medical School, Boston, USA
[6] Computer Science and Artificial Intelligence Laboratory, Massachusetts Institute of Technology, Boston 02139, USA
[7] Centre for Medical Image Computing, University College London, London WC1V 6LJ, UK
[8] Wellcome Centre for Human Neuroimaging, University College London, London WC1N 3AR, UK
k.wagstyl@ucl.ac.uk

Abstract. Focal cortical dysplasia (FCD) is a leading cause of drug-resistant focal epilepsy, which can be cured by surgery. These lesions are extremely subtle and often missed even by expert neuroradiologists. "Ground truth" manual lesion masks are therefore expensive, limited and have large inter-rater variability. Existing FCD detection methods are limited by high numbers of false positive predictions, primarily due to vertex- or patch-based approaches that lack whole-brain context. Here, we propose to approach the problem as semantic segmentation using graph convolutional networks (GCN), which allows our model to learn spatial relationships between brain regions. To address the specific challenges of FCD identification, our proposed model includes an auxiliary loss to predict distance from the lesion to reduce false positives and a weak supervision classification loss to facilitate learning from uncertain lesion masks. On a multi-centre dataset of 1015 participants with surface-based features and manual lesion masks from structural MRI data, the proposed GCN achieved an AUC of 0.74, a significant improvement against a previously used vertex-wise multi-layer perceptron (MLP)

Supplementary Information The online version contains supplementary material available at https://doi.org/10.1007/978-3-031-43993-3_41.

H. Greenspan et al. (Eds.): MICCAI 2023, LNCS 14227, pp. 420–428, 2023.
https://doi.org/10.1007/978-3-031-43993-3_41

classifier (AUC 0.64). With sensitivity thresholded at 67%, the GCN had a specificity of 71% in comparison to 49% when using the MLP. This improvement in specificity is vital for clinical integration of lesion-detection tools into the radiological workflow, through increasing clinical confidence in the use of AI radiological adjuncts and reducing the number of areas requiring expert review.

Keywords: Graph Convolutional Network · lesion segmentation · structural MRI

1 Introduction

Structural cerebral abnormalities commonly cause drug-resistant focal epilepsy, which may be cured with surgery. Focal cortical dysplasias (FCDs) are the most common pathology in children and the third most common pathology in adults undergoing epilepsy surgery [1]. However, 16–43% of FCDs are not identified on routine visual inspection of MRI data by radiologists [10]. Identification of these lesions on MRI is integral for presurgical planning. Furthermore, accurate identification of lesions assists with complete resection of the structural abnormality, which is associated with improved post-surgical seizure freedom rates [11].

There has been significant work seeking to automate the detection of FCDs, with the aim of identifying subtle structural abnormalities in patients with lesions not identified by visual inspection, termed "MRI negative" [12]. These algorithms are increasingly being evaluated prospectively on patients who are "MRI negative" with suspected FCD, where radiologists review algorithm outputs and evaluate all putative lesions. However, previous methods operate locally or semi-locally: using multilayer perceptrons (MLPs) which consider voxels or points on the cortical surface (vertices) individually [2,10], or convolutional neural networks which have to date typically been trained on patches of cortex [6]. One widely-available algorithm using such an approach was able to detect 63% of MRI negative examples, with an AUC of 0.64 [10]. Overall, although these algorithms show significant promise in finding subtle and previously unidentified lesions, they are commonly associated with high false positive rates which hampers clinical utility [12]. Detecting FCDs is particularly challenging due to small dataset sizes, high inter-annotator variability in manual lesion masks, and the large class imbalance, as FCDs typically only cover around 1% of the total cortex. Nevertheless the urgent clinical need to identify more of these subtle lesions motivates the development of methods to address these challenges.

Contributions. We propose a robust surface-based semantic segmentation approach to address the particular challenges of identifying FCDs (Fig. 1). Our three main contributions to address these challenges are: 1) Adapting nnU-Net [8], a state-of-the-art U-Net architecture, to a Graph Convolutional Network (GCN) for segmenting cortical surfaces. This creates a novel method for cortical segmentation in general and for FCD segmentation in particular, in which the model is

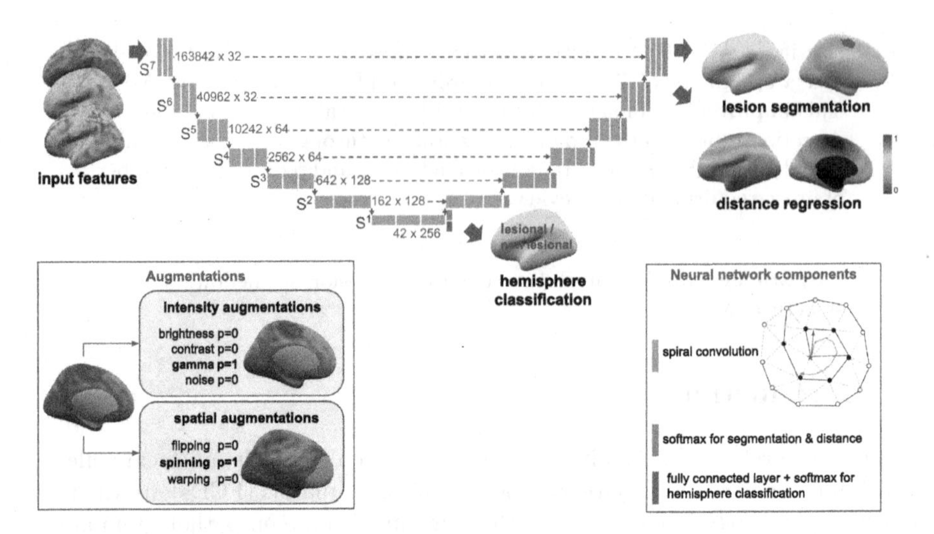

Fig. 1. Proposed GC-nnU-Net+dc model for lesion segmentation, with auxiliary distance regression and hemisphere classification tasks. Lower left box: Types of data augmentation employed. Examples show the result of gamma intensity augmentation (top) and spinning (bottom).

able to learn spatial relationships between brain regions. 2) Inclusion of a distance loss to help reduce false positives, and 3) Inclusion of a hemisphere classification loss to act as form of weak supervision, mitigating the impact of imperfect lesion masks. We directly evaluate the added value of each contribution on performance in comparison to a previously published MLP [10]. We hypothesised that the proposed GCN to segment FCDs would improve overall performance (AUC), in particular reducing the number of false positives (improved specificity) while retaining sensitivity. This improvement in classifier performance would facilitate clinical translation of automated FCD detection into clinical practice. All code to reproduce these results can be found at github.com/MELDProject/meld_graph.

2 Methods

2.1 Graph Convolutional Network (GCN) for Surface-Based Lesion Segmentation

We consider the lesion detection problem as a surface-based segmentation task. For this purpose, cortical surface-based features (intensity, curvature, etc.; see Sect. 3.1) are extracted from each brain hemisphere and registered using FreeSurfer [4] to a symmetrical template. This template was generated by successively upsampling an icosahedral icosphere, S^1, with 42 vertices and 80 triangular faces. Icospheres S^i, with i the resolution level of the icosphere, are triangulated spherical meshes, where S^{i+1} is generated from S^i by adding vertices at every edge. As input to our model we use icosphere S^7 (163842 vertices).

U-Net Architecture. To segment lesions on the icosphere, we created a graph-based re-implementation of nnU-Net [8,9]. Unlike typical imaging data represented on rectangular grids, surface-based data require customised convolutions, downsampling and upsampling steps. Here, we used spiral convolutions [7] which translates standard 2D convolutions to irregular meshes by defining the filter by an outward spiral. This ends up capturing a ring of information around the current node, similar to how a 2d filter captures a ring of information around the current pixel. We use a spiral length of 7, representing the central and adjacent 6 neighbours on a hexagonal mesh, roughly equivalent to a 3×3 2D kernel. For downsampling from S^{i+1} to S^i in the U-Net encoder, a similar translation of 2D max pooling is carried out by aggregating over all neighbours of the vertex at the higher-resolution S^{i+1}. Upsampling from S^i to S^{i+1} in the decoder is implemented via assigning the mean of each vertex in S^i to all neighbours at level S^{i+1}. In total, the U-Net contains seven levels (mirroring the seven icospheres S^7-S^1), and every level consists of three convolutional layers using spiral convolutions and leaky Relu as activation function (Fig. 1).

Loss Functions. Following best practices for U-Net segmentation models [8], we use both cross-entropy and dice as loss functions for the segmentation, where y is true labels, $\hat{y}$ is predicted, n is the number of vertices:

$$L_{ce} = -\sum_{i=1}^{n} y_i \log(\hat{y}_i) + (1 - y_i) \log(1 - \hat{y}_i) \tag{1}$$

$$L_{dice} = 1 - \frac{2 \sum_{i=1}^{n} y_i \hat{y}_i}{\sum_{i=1}^{n} y_i^2 + \sum_{i=1}^{n} \hat{y}_i^2 + \epsilon} \tag{2}$$

Distance Loss. To encourage the network to learn whole-brain context thereby reducing the number of false positives, we added an additional distance regression task. We train the model to predict the normalised geodesic distance d to the lesion boundary for every vertex, by applying an additional loss L_{dist} to the non-lesional prediction, $\hat{y}_{i,0}$ of the segmentation output for vertex i. We use the mean absolute error loss, weighted by the distance so as not to overly penalise small errors in predicting large distances from the lesion:

$$L_{dist} = \frac{1}{n} \sum_{i=1}^{n} \frac{|d_i - \hat{y}_{i,0}|}{d_i + 1} \tag{3}$$

Classification Loss. To mitigate uncertainty in the correspondence between lesion masks and lesions, we used a weakly-supervised classification loss L_{class}. For the ground truth c, examples were labelled as positive, if any of their vertices were annotated as positive. To predict this sample-level classification, we added a classification head to the deepest level (level 1) of the U-Net. The classification head contained a fully connected layer aggregating over all filters, followed by a fully-connected layer aggregating over all vertices, resulting in the classification output $\hat{c}$. This output was trained using cross-entropy:

$$L_{class} = -\sum_{i=1}^{n} c_i \log(\hat{c}_i) + (1 - c_i) \log(1 - \hat{c}_i) \tag{4}$$

Deep Supervision. To encourage the flow of gradients through the entire U-Net, we use deep supervision at levels $I_{ds} = [6, 5, 4, 3, 2, 1]$. Let L^i_{ce}, L^i_{dice}, L^i_{dist} be the cross-entropy, dice and distance losses applied to outputs at level i, respectively. The model is trained on a weighted sum of all the losses, with w^i_{ds} the loss weight at level i:

$$L = L_{ce} + L_{dice} + L_{dist} + L_{class} + \sum_{i \in I_{ds}} w^i_{ds}(L^i_{ce} + L^i_{dice} + L^i_{dist}) \tag{5}$$

2.2 Data Augmentation

Data augmentations consisted of spatial augmentations and intensity augmentations (Fig. 1), following recommendations outlined in nnU-Net. Spatial augmentation included rotation, inversion and non-linear deformations of the surface-based data [3]. Intensity-based augmentations included adding a Gaussian noise to the features intensity, adjusting the contrast, scaling the brightness by a uniform factor, and adding a gamma intensity transform.

3 Experiments and Results

3.1 Dataset and Implementation Details

Dataset. For the following experiments, we used a dataset of post-processed surface-based features and manual lesion masks from 618 patients with FCD and 397 controls [10]. This is a heterogeneous, clinically-acquired dataset, collated from 22 international epilepsy surgery centres, including paediatric and adult participants scanned on either 1.5T or 3T MRI scanners. Each centre received local ethical approval from their institutional review board (IRB) or ethics committee (EC) to retrieve and anonymise retrospective, routinely available clinical data. For each participant, MR images were processed using FreeSurfer [4] and 11 surface-based features (cortical thickness, grey-white matter intensity contrast, intrinsic curvature, sulcal depth, curvature and FLAIR intensity sampled as 6 intra- and sub-cortical depths) were extracted. FCDs were manually drawn by neuroradiologists to create 3D regions of interest (ROI) on T1 or fluid-attenuated inversion recovery (FLAIR) images. The ROIs were projected onto individual FreeSurfer surfaces and then the features and ROIs were registered to a bilaterally symmetrical template, fsaverage_sym, using folding-based registration. Post-processing included 10 mm full width at half-maximum surface-based smoothing of the per-vertex features, harmonisation of the data using Combat [5] (to account for scanners differences), inter- and intra-individual z-scoring to account for inter-regional differences and demographic differences, and computation of the asymmetry index of each feature. The final surface-based feature set consisted of the original, z-scored and asymmetry features, resulting in 33 input features.

In order to compare performance, the train/validation and test datasets were kept identical to those in the previously published vertex-wise classifier [10]. The train/validation cohort comprised 50% of the dataset and 5-fold cross validation was used to evaluate the models. The remaining 50% was withheld for final evaluation and comparison of models. Data from two independent sites (35 patients and 18 controls) were used to test the generalisability of the full model.

Implementation Details. The graph-based convolutional implementation of nnU-Net (GC-nnU-Net) had the following training parameters: batch size: 8, initial learning rate: 10^{-4}, learning rate decay: 0.9, momentum: 0.99, maximum epochs: 1000 (1 epoch is 1 complete view of training data), maximum patience: 400 epochs; and augmentation probabilities: inversion: 0.5, rotation & deformation: 0.2, Gaussian noise: 0.15, contrast: 0.15, brightness: 0.15, gamma: 0.15. Deep supervision weights were $w_{ds} = [0.5, 0.25, 0.125, 0.0625, 0.03125, 0.0150765]$ for levels $I_{ds} = [6, 5, 4, 3, 2, 1]$. Due to class imbalances, non-lesional hemispheres were undersampled during training to ensure 33% of training examples contained a lesion. The model from the epoch with the best validation loss is stored for evaluation (Fig. S1).

Hardware: High-performance cluster with Single NVIDIA A100 GPU, 1000 GiB RAM; Software: PyTorch 1.10.0+cu11.1, PyTorch Geometric 2.0.4, Python 3.9.13. Combined memory footprint of model and dataset while training is 49 GB.

Experiments. Using our graph-based adaptation of nnU-Net (GC-nnU-Net) and the previous MLP model as baseline, we ran an ablation study to measure the impact of the proposed auxiliary losses (Table 1). Each model was trained using the train/val cohort 5 times, withholding 20% of the cohort for validation and stopping criteria. Final test performance was computed by ensembling predictions across the 5-fold trained models, with uncertainty estimates calculated through bootstrapping. An additional experiment was carried out subsampling the training cohort at fixed fractions of 0.1, 0.2, 0.3, 0.4, 0.6 and 0.8 using the GC-nnU-Net+dc model (Fig. S2).

Evaluations. Model performances were compared according to their Area Under the Curve (AUC), which was calculated by computing the sensitivity and specificity at a range of prediction thresholds. For sensitivity calculations, due to uncertainty in the lesion masks, a lesion was considered detected if the prediction was within 20 mm of the original mask, as this corresponds with the inter-observer variability measured across annotators [10]. Specificity was defined by the presence of false positives in non-lesional examples. As an additional measure of model specificity, the number of false positive clusters in both patients

Table 1. Experiments

Experiment Name	Description
MLP [10]	vertex-wise multilayer perceptron
GC-nnU-Net	graph-based adaptation of nnU-Net
GC-nnU-Net+c	adding classification loss
GC-nnU-Net+d	adding distance loss
GC-nnU-Net+dc	adding distance loss and classification loss

Table 2. Comparison of models on the test dataset.

Experiment	AUC (±std)	Sensitivity	Specificity	Run time (min)
MLP [10]	0.64 (n.a.)	67%	49%	n.a.
GC-nnU-Net	$0.68^{\dagger\ddagger}$ (±0.004)	67%	64%	396.1
GC-nnU-Net+c	$0.74^{\dagger}$ (±0.008)	67%	66%	373.9
GC-nnU-Net+d	$0.69^{\dagger\ddagger}$ (±0.007)	67%	65%	426.9
GC-nnU-Net+dc	$0.74^{\dagger}$ (±0.005)	67%	71%	564.9

†Model performance significantly improved compared to MLP.
‡Model performance significantly worse compared to GC-nnU-Net+dc.

and controls were calculated. Model AUCs were statistically compared using t-tests, with correction for multiple comparisons using the Holm-Sidak method.

3.2 Results

Table 2 compares model performances on the withheld test set. GC-nnU-Net+dc, the graph-based implementation of nnU-Net with additional distance and classification losses, outperformed all other models. Examples of individual predictions using the MLP and GC-nnU-Net+dc model, as well as examples of the predicted geodesic distance from the lesion are presented in Fig. 2A,B. Figure 2C visualises the reduction in number of false positive clusters when using GC-nnU-Net+dc which is reflected in the significantly improved specificity. GC-nnU-Net+dc showed similarly improved specificity relative to the MLP on independent test sites, demonstrating good model generalisability (Table 3). In experiments varying the size of the training cohort, performance increased with sample size until around 220 subjects above which gains were negligible (Fig. S2).

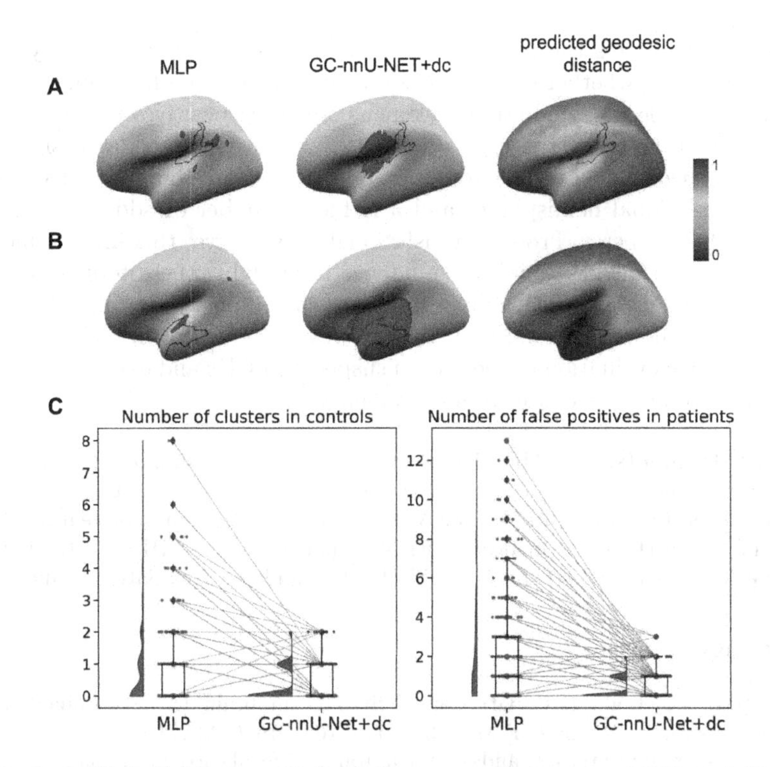

Fig. 2. A & B) Individual predictions for two patients using MLP and GC-nnU-Net+dc, as well as predicted geodesic distance from the lesion. Red: prediction. Black line: manual lesion mask. Patients A and B both have false positive predictions using the MLP, unlike the predictions from GC-nnU-Net+dc. C) Comparison of number of clusters in controls and patients between MLP and GC-nnU-Net+dc on the test dataset. Grey lines: change in number of clusters for individual participants. (Color figure online)

Table 3. Comparison of models on independent test sites.

Experiment	Sensitivity	Specificity	Median FP in patients [IQR]	Median clusters in patients [IQR]
MLP [10]	79%	17%	2.0 [1.0, 4.0]	1.0 [1.0, 2.75]
GC-nnU-Net+dc	79%	44%	1.0 [0.0, 1.0]	1.0 [0.0, 1.0]

FP: false positives, IQR: interquartile range.

4 Conclusions and Future Work

This paper presents a robust and generalisable graph convolutional approach for segmenting focal cortical dysplasias using surface-based cortical data. This approach outperforms specificity baselines by 22–27%, which is driven by three newly-proposed components. First, treating the hemispheric surface as a single connected graph allows the network to model spatial context. Second, a classification loss mitigates the impact of imprecise lesion masks by simplifying the task to predicting whether or not a lesion is present in every hemisphere. Third, a distance-from-lesion prediction task penalises false positives and encourages the network to consider the entire hemisphere. The results show a significant increase in specificity, both in terms of presence of any false positive predictions in non-lesional hemispheres and a reduced number of additional clusters in lesional hemispheres. From a translational perspective, this improvement in performance will increase clinical confidence in applying these tools to cases of suspected FCD, while additionally minimising the number of putative lesions an expert neuroradiologist would need to review. Future work will include systematic prospective evaluation of the tool in suspected FCDs and expansion of these approaches to multiple causes of focal epilepsy.

Acknowledgements. The MELD project, MR and SA are supported by the Rosetrees Trust (A2665) and Epilepsy Research UK (P2208). KSW is supported by the Wellcome Trust (215901/Z/19/Z). LZJW is supported by the Commonwealth Scholarship Commission (United Kingdom). JEI is supported by the NIH (1RF1MH123195, 1R01AG070988, 1R01EB031114, 1UM1MH130981) and the Jack Satter Foundation.

References

1. Blumcke, I., et al.: Histopathological findings in brain tissue obtained during epilepsy surgery. N. Engl. J. Med. **377**(17), 1648–1656 (2017)
2. David, B., et al.: External validation of automated focal cortical dysplasia detection using morphometric analysis. Epilepsia (2021)
3. Fawaz, A., et al.: Benchmarking geometric deep learning for cortical segmentation and neurodevelopmental phenotype prediction (2021)
4. Fischl, B.: FreeSurfer. Neuroimage **62**(2), 774–781 (2012)
5. Fortin, J.P., et al.: Harmonization of cortical thickness measurements across scanners and sites. Neuroimage **167**, 104–120 (2018)
6. Gill, R.S., et al.: Multicenter validation of a deep learning detection algorithm for focal cortical dysplasia. Neurology (2021)
7. Gong, S., Chen, L., Bronstein, M., Zafeiriou, S.: SpiralNet++: a fast and highly efficient mesh convolution operator. In: 2019 IEEE/CVF International Conference on Computer Vision Workshop (ICCVW). IEEE (2019)
8. Isensee, F., Jaeger, P.F., Kohl, S.A.A., Petersen, J., Maier-Hein, K.H.: nnU-Net: a self-configuring method for deep learning-based biomedical image segmentation. Nat. Methods **18**(2), 203–211 (2021)
9. Ronneberger, O., Fischer, P., Brox, T.: U-Net: convolutional networks for biomedical image segmentation. In: Navab, N., Hornegger, J., Wells, W.M., Frangi, A.F. (eds.) MICCAI 2015. LNCS, vol. 9351, pp. 234–241. Springer, Cham (2015). https://doi.org/10.1007/978-3-319-24574-4_28
10. Spitzer, H., et al.: Interpretable surface-based detection of focal cortical dysplasias: a Multi-centre Epilepsy Lesion Detection study. Brain **145**(11), 3859–3871 (2022)
11. Wagner, J., Urbach, H., Niehusmann, P., von Lehe, M., Elger, C.E., Wellmer, J.: Focal cortical dysplasia type IIb: completeness of cortical, not subcortical, resection is necessary for seizure freedom. Epilepsia **52**(8), 1418–1424 (2011)
12. Walger, L., et al.: Artificial intelligence for the detection of focal cortical dysplasia: challenges in translating algorithms into clinical practice. Epilepsia (2023)

Weakly Supervised Cerebellar Cortical Surface Parcellation with Self-Visual Representation Learning

Zhengwang Wu[1], Jiale Cheng[1], Fenqiang Zhao[1], Ya Wang[1], Yue Sun[1], Dajiang Zhu[2], Tianming Liu[3], Valerie Jewells[1], Weili Lin[1], Li Wang[1], and Gang Li[1](✉)

[1] Department of Radiology and Biomedical Research Imaging Center, University of North Carolina at Chapel Hill, Chapel Hill, NC, USA
gang_li@med.unc.edu
[2] Department of Computer Science and Engineering, University of Texas at Arlington, Arlington, TX, USA
[3] Department of Computer Science, University of Georgia, Athens, GA, USA

Abstract. The cerebellum (i.e., little brain) plays an important role in motion and balances control abilities, despite its much smaller size and deeper sulci compared to the cerebrum. Previous cerebellum studies mainly relied on and focused on conventional volumetric analysis, which ignores the extremely deep and highly convoluted nature of the cerebellar cortex. To better reveal localized functional and structural changes, we propose cortical surface-based analysis of the cerebellar cortex. Specifically, we first reconstruct the cerebellar cortical surfaces to represent and characterize the highly folded cerebellar cortex in a geometrically accurate and topologically correct manner. Then, we propose a novel method to automatically parcellate the cerebellar cortical surface into anatomically meaningful regions by a weakly supervised graph convolutional neural network. Instead of relying on registration or requiring mapping the cerebellar surface to a sphere, which are either inaccurate or have large geometric distortions due to the deep cerebellar sulci, our learning-based model directly deals with the original cerebellar cortical surface by decomposing this challenging task into two steps. First, we learn the effective representation of the cerebellar cortical surface patches with a contrastive self-learning framework. Then, we map the learned representations to parcellation labels. We have validated our method using data from the Baby Connectome Project and the experimental results demonstrate its superior effectiveness and accuracy, compared to existing methods.

Keywords: Cerebellar Cortex Parcellation · Representation Learning

1 Introduction

Parcellating the cerebellum (i.e., little brain) into neurobiologically meaningful regions of interest (ROIs) plays an important role in both structural and functional analysis [1, 2]. Therefore, automatic cerebellar parcellation methods are highly demanded to facilitate

H. Greenspan et al. (Eds.): MICCAI 2023, LNCS 14227, pp. 429–438, 2023.
https://doi.org/10.1007/978-3-031-43993-3_42

the analysis of increasingly larger cerebellum imaging datasets. Compared with the highly folded *cerebral* cortex, the *cerebellar* cortex has a more complex shape with even more tightly convoluted and much thinner cortical folds. Hence, few algorithms have been proposed for automatic cerebellar cortex parcellation.

Most existing methods mainly conduct the cerebellar cortex parcellation in the Euclidean space [3]. Generally, they treat the cerebellar cortex as a 3D volume, and then either use the traditional multi-atlas registration and label-fusion based methods or directly learn the cerebellar cortex parcellation model in the volumetric space, without reconstructing the cerebellar cortical surfaces and thus ignoring the topological and geometric properties of the cerebellar cortex. These methods are thus inappropriate for accurate and anatomically meaningful cerebellar cortex analysis, as the neighboring relationship in 3D space used in these methods cannot correctly reflect the actual cytoarchitectural neighborship. Specifically, two spatially neighboring points in the cerebellum volume do not necessarily mean they are neighboring in cytoarchitecture and geodesic. They could sit in completely different gyri or sulci, due to the extremely convoluted structure of the cerebellar cortex. This can cause severe consequences when requesting the parcellation consistency across "neighbors" and confuse the message aggregation during the parcellation model learning.

To address this issue, it is highly necessary to respect the cerebellar geometry, which provides macro-measurable imaging clues characterizing the cytoarchitecture of the cerebellar cortex. Therefore, in this paper, ***first,*** we reconstruct the cerebellar surfaces to restore its convoluted geometry. Specifically, we reconstruct the inner cerebellar surface (the interface between the cerebellar gray matter and white matter), and the pial/outer cerebellar cortical surface (the interface between the cerebellar gray matter and cerebrospinal fluid (CSF)). The reconstructed cerebellar surfaces are represented by triangular meshes, which can be regarded as an undirected graph. ***Second,*** we develop a graph convolutional neural network-based method to do the parcellation on the reconstructed cerebellar surfaces. Specifically, we first extract the informative surface patches from the cerebellar surfaces with the neighborship determined by their intrinsic geodesic distance. And then, we feed these surface patches into the graph convolutional neural network to train a parcellation model, which learns a highly nonlinear mapping from the geometric features of the training patches to the patch labels. One practical challenge for learning this mapping is the need for a large amount of manually labeled data, which is extremely expensive and time-consuming for vertex-level labeling. Therefore, we split the mapping learning into two steps, in the first step, we use massive patches to learn effective low-dimension representations in the latent space. Leveraging the concept of contrastive learning, we require similar representations for the patches from the same region, while distinct representations for patches from different regions. Then, in the second step, we learn the mapping from the low-dimensional latent representation to the parcellation labels. Of note, the representation learning only requires a tag to denote whether the patches are from the same region or not, which is clearly a weaker supervision compared to the accurate patch-level labels.

This paper makes three contributions: 1) we propose a practical pipeline to reconstruct the cerebellar cortical surface that respects its intrinsic geometry; 2) we conduct the parcellation on the original reconstructed cerebellar cortical surfaces, without the requirement to project them onto a simplified shape, like the sphere, which will inevitably introduce distortions during the projection; 3) we leverage the self-representation learning with weak supervision information to reduce the manual labeling cost for the parcellation network training. To the best of our knowledge, this is the first method that conducts the cerebellar cortex parcellation directly using the original reconstructed cerebellar surfaces.

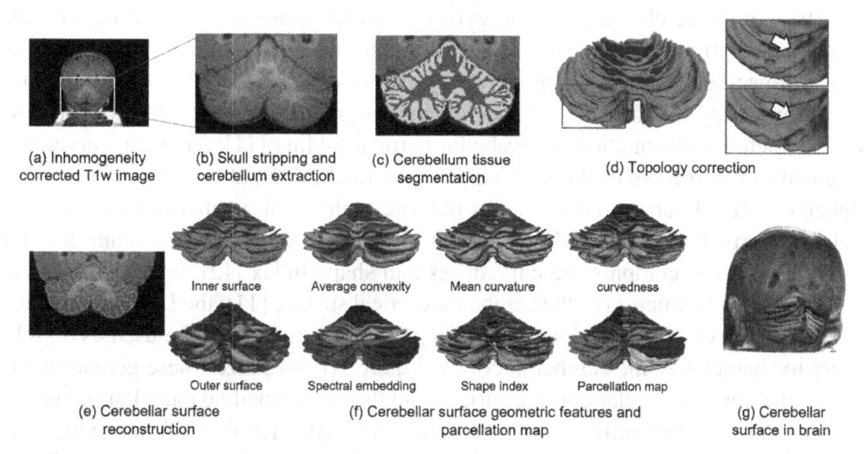

Fig. 1. Cerebellar surface reconstruction pipeline and the typical geometric features computation.

2 Method

Our cerebellar cortex parcellation method can be divided into 3 steps: 1) we reconstruct the cerebellar surfaces and compute the salient geometric features to characterize the geometry of the cerebellar surfaces; 2) we use weakly supervised information to learn effective latent representations from massive informative surface patches. These representations sit in low dimensional latent space, and they are similar for patches from the same region, while are distinct for patches from different regions; 3) we learn a straightforward mapping from the low dimensional latent space to the parcellation labels.

2.1 Cerebellar Surface Reconstruction and Geometric Feature Computation

Given the tissue map of the cerebellum [4], we can reconstruct the cerebellar cortical surfaces, including the inner cortical surface and the pial/outer cortical surface. The pipeline is illustrated in Fig. 1. Specifically, we first do the cerebellum tissue segmentation and then correct the topological errors. Then, we reconstruct the inner cerebellar surface and deform it to the outer cerebellar surface. This framework is similar to the typical cerebral cortical surface reconstruction pipelines. We adopt this typical strategy because, a) it can well preserve the geometry of the highly folded cerebellar cortex; b) it can establish the vertex-wise correspondence between the inner and outer cerebellar surfaces, which will substantially facilitate the subsequent analysis. More details about widely used cortical surface reconstruction pipelines can be referred to, such as FreeSurfer [5], dHCP [6], FastSurfer [7], HCP [8], and iBEAT V2.0 [9].

With the reconstructed cerebellar surfaces, we can then compute typical geometric features, including the average convexity, mean curvature, curvedness, shape index, and Laplacian spectral embedding [10]. These geometric features encode the local and global geometries of the cerebellar sulci and gyri, which can be used to identify the boundary of each parcellation region (typically appearing at the sulcal bottoms of the cerebellar surface). Specifically, the average convexity encodes the integrated normal movement of a vertex during inflating the inner cerebellar surface and mainly reflects the coarse-scale geometrical information of cerebellar cortical folding [11]; the mean curvature is computed as the average of the minimal and maximal principal curvatures of the inner cerebellar cortical surface and encodes the fine-scale local geometric information of cerebellar cortical folding [11]; also, based on the minimum and maximum principal curvatures, we can compute the curvedness and shape index [12], which characterize the local shape information of the cerebellar cortical surface [11]; the Laplacian spectral embedding reflects the vibration mode of a graph [10], which can be used as a global feature to characterize the cerebellar cortex. Figure 1(f) visualizes these geometric features on the inner cerebellar cortical surface and the corresponding parcellation map. In addition, since the cerebellar cortical surface is not separated into left and right hemispheres, we computed the vertex-wise distance to the geometric center of the surface as additional features, named centroid relation features, which can provide useful clues for the localization of symmetric regions.

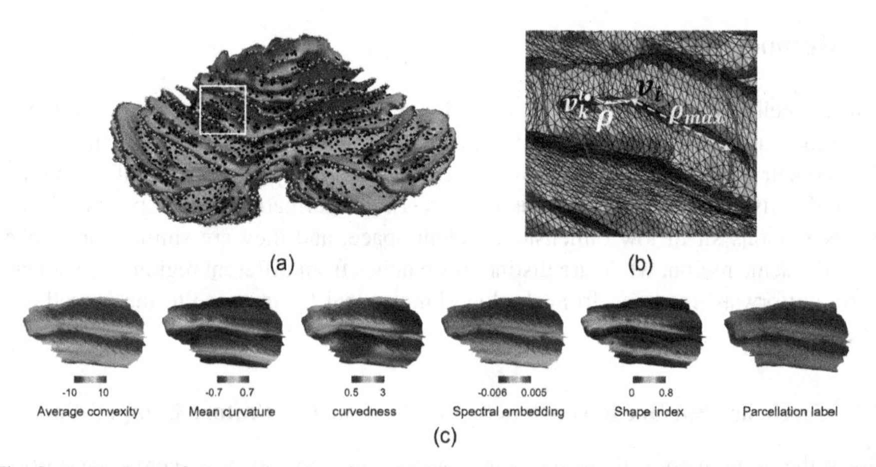

Fig. 2. Intrinsic surface patch extraction. (a) The patch sampling positions on the cerebellar surface. The white dot indicates the gyri samples, while the black dots indicate the sulci samples. (b) A local intrinsic patch bounded by the geodesic distance ρ_{max}. All vertices whose geodesic distance to the patch center lower than ρ_{max} will be included in the patch. (c) Typical geometric features on the extracted local intrinsic patch in (b).

2.2 Weakly Supervised Cerebellar Patch Representation Learning

Provided with the computed geometric features of the cerebellar surface, we can train a neural network to conduct the parcellation. However, treating the entire cerebellar surface as a single instance would require dense manual labeling for the network training, which is time-consuming and labor-intensive. Therefore, we choose the patch-wise strategy for the parcellation network training. The advantage is that it can significantly enlarge the training samples, while this comes at the cost of requiring the model to possess strong localization ability for the local patch. Directly learning a mapping from the geometric features of a local patch to the parcellation labels is very challenging for the cerebellar surface due to the shape complexity. The local sulci of the cerebellar surface are deeper, and the gyri appear more consistent shape patterns, compared to the cerebral cortical surfaces. This indicates that more training samples are required to enable the network to possess the localization ability from the local patches for a robust parcellation model.

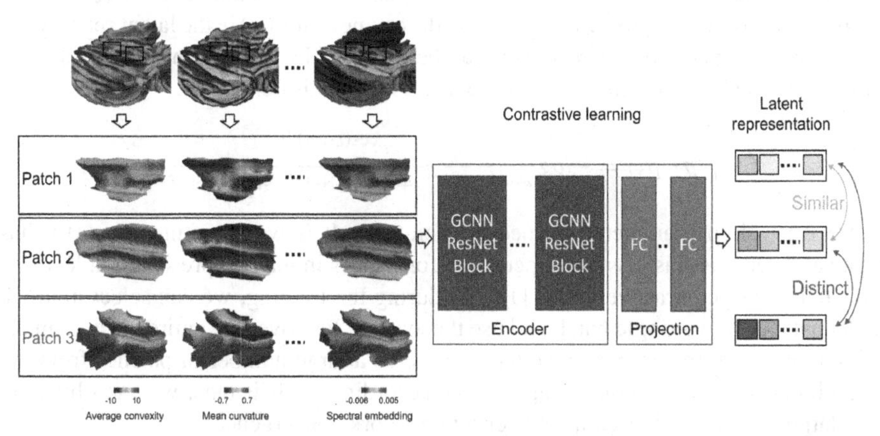

Fig. 3. Weakly supervised representation learning framework. After learning, we enforce similar latent representations between patches 1 and 2 (coming from the same region), but distinct representation between patches 1 (or 2) and patch 3.

To further reduce the cost of the expensive manual labeling, we split the parcellation network training into two steps. In the first step, we learn distinctive latent representations for patches from different regions in a weakly supervised manner. Specifically, we enforce similar representations for patches from the same region, while distinct representations for patches from different regions. This can be achieved with a contrastive learning framework [13]. Herein, we name it weakly supervised since in this step, we only require a tag to denote whether the patches are from the same region or not and we do not require the precise labels of which region it belongs to.

Following the above motivation, we can formulate the weakly supervised patch-based representation learning below. We first extract the local patches from the cerebellar surface, which can be regarded as an undirected graph (V, E), with V being the vertex set and E being the edge set. For any vertex, we can extract a local patch, which contains a set of vertices v_k^i whose geodesic distances to v_i are bounded by the predefined maximal geodesic distance ρ_{max}, as illustrated in Fig. 2. Considering the prior knowledge that parcellation boundaries of the cerebellar surface generally appear at the sulcal fundi or gyral crests, we sample more patches from those locations, and meanwhile reduce patch samples for vertices sitting on sulcal walls. This can be achieved by thresholding the mean curvature map of the cerebellar surface, since the sulcal fundi and gyral crests typically have larger curvature magnitude, as illustrated in Fig. 2(a), where white dots denote the sample points near gyri crests, and black dots denote the sample points near sulcal fundi. Figure 2(c) shows a typical surface patch with various geometric features and parcellation labels.

Once the local patches are extracted, we feed them into a graph convolution based neural network to learn their latent representation. Herein, we use the widely used residual network [14] as the basic block for constructing our latent representation learning framework. Figure 3 illustrates the learning framework. Specifically, given 3 local patches (p_1, p_2, p_3), where p_1 and p_2 are from the same region and p_3 is from a different region. After the encoding, we can denote their latent representations to be (z_1, z_2, z_3). Then, we enforce (z_1, z_2) to be similar (using the cosine criterion) in the latent space while (z_1, z_3) and (z_2, z_3) to be distinct. This can be achieved by minimizing the normalized temperature scaled cross entropy (NT-Xent) loss, which is defined as:

$$\mathcal{L}(p_1, p_2, p_3) = -log\frac{e^{\cos(z_1,z_2)/\tau}}{e^{\cos(z_1,z_2)/\tau} + e^{\cos(z_1,z_3)/\tau} + e^{\cos(z_2,z_3)/\tau}}$$

where τ is a hyperparameter (named temperature). It is worth noting that, a) it has been validated that this loss has superior performances in many representation learning tasks than other contrastive losses [13]; b) during the training, we can select multiple triplet patches into a single batch and use the average loss over the entire batch to make the training more robust. Since we have extracted a large number of patches from the cerebellar surface, after combining patches according to their tags, we can obtain an even larger set of training samples to ensure network convergence.

2.3 Mapping from Latent Space to Parcellation Labels

Through representation learning, we have encoded the patches with multiple channels of geometric features into the latent space. This step *not only* makes the extracted patches more distinct in the latent space from the regional perspective *but also* significantly reduces the potential feature dimension, which can greatly facilitate the parcellation network training, compared to training the parcellation network directly from the patch-wise geometric features.

Therefore, given the patch-wise latent representations, we further train a multilayer perceptron (MLP) to accomplish the parcellation task. Specifically, we use the parcellation labels from the manually labeled cerebellar surface to supervise the MLP training. Herein, the patch center label is used as the patch label to train the MLP and the popular cross-entropy loss is adopted.

However, since each patch is parcellated independently without considering the spatial consistency, it is possible to generate isolated parcellation labels and cause inconsistency in a geodesic neighborhood. To improve the parcellation, we further use the graph cuts method [15] to explicitly impose spatial consistency. Specifically, since most of the regions are separated at the sulci fundi of the cerebellar cortex according to the manual labeling protocol, we explicitly formulate parcellation as a cost minimization problem, i.e., $E = E_d + \lambda E_s$. Here, E_d is the data fitting term, which is defined as: $E_d = -\log p_v(l_v)$, where $p_v(l_v)$ is the probability of assigning vertex v as label l_v; E_s is the smoothness term, which is defined as: $E_s = \sum_{v^* \in \mathcal{N}_v} C_{v,v^*}(l_v, l_{v^*})$, where vertex v^* is the direct neighbor of v, and $C_{v,v^*}(l_v, l_{v^*})$ is the cost to label vertex v as l_v and also label vertex v^* as l_{v^*}. Herein, we used the formula from [16] to define $C_{v,v^*}(l_v, l_{v^*})$; finally, λ is a weight used to balance them.

Table 1. Performance comparison by different methods.

Comparison methods	Dice ratio
GCN [20]	76.03 ± 3.20
GIN [21]	75.20 ± 2.40
Graph U-Net [22]	73.17 ± 4.46
Graph SAGE [23]	73.52 ± 4.69
Proposed	81.26 ± 2.37

3 Experiments

3.1 Dataset and Implementation

To validate our method, we manually labeled 10 subjects from BCP dataset [17] in a vertex-wise manner using an inhouse developed toolkit. Specifically, for each cerebellum MRI, we first segmented it into white matter, gray matter, and CSF using a deep convolutional neural network [4]. Then, following the processing pipeline in Sect. 2.1, we reconstructed the geometrically accurate and topologically correct cerebellar surfaces [18, 19]. Finally, each inner cerebellar surface is labeled into 17 lobules following the SUIT parcellation strategy [3] by an expert. To ensure the labeling quality, each manually labeled region is further validated and cross-modified by another 2 experts to alleviate the subjective bias and ensure the labeling quality. Due to the limited subject number, we adopted these manually labeled cerebellums to validate our method in a 10-fold cross validation.

We implemented our method mainly based on the pytorch (https://pytorch.org/) and pytorch geometric (https://pyg.org/) packages. The graph convolution operator [20] implemented in the pytorch geometric package is adopted as the major building block to construct our network. For the contrastive learning framework, we used 18 ResNet blocks for the encoding and 3 fully connected layers for the projection. After the cerebellar cortical surface reconstruction, each scan typically has around 90k vertices. We extract 8k patches from each training scan. The ρ_{max} is set to 15mm. For the NT-Xent loss, the hyperparameter τ is set to 0.5. For the graph-cut cost, the weight balance λ is set to 1. For the testing subject, we perform the parcellation for each patch.

3.2 Comparison with the State-of-the-Art Methods

Since we conducted our parcellation on the original cerebellar surface, which can be regarded as a graph, we compared our method with state-of-the-art graph convolutional neural network based methods [20–23] for the cerebellum parcellation task. The Dice similarity coefficient (DSC) between the predicted label and the manual label is adopted as a quantitative evaluation of the parcellation performance. Table 1 reported the average DSC of the parcellation for all regions acquired by our proposed method and the comparison methods. It can be seen that our method achieves best performance, indicating the effectiveness of the latent patch feature embedding obtained by our proposed weakly supervised contrastive learning strategy.

Table 2. Different features' influence on parcellation performance.

Comparison methods	DSC
Raw geometric features	46.02 ± 4.13
Raw geometric features + spectral feature	72.72 ± 4.52
Raw geometric features + spectral feature + centroid relation feature	81.26 ± 2.37

3.3 Different Features' Influence Analysis

We also validated the influence of each geometric feature on the final parcellation performance. We conducted the ablation study by removing parts of the geometric features for the parcellation. Table 2 reported the average DSC using different feature combinations. It can be seen that the raw geometric features, i.e., the average convexity, mean curvature, curvedness, and shape index, are difficult to capture the localization ability. This is reasonable since the cerebellar cortex has much deeper sulci, and appear more similar geometric patterns, compared to the cerebral cortex. Therefore, the local patches have relatively low localization ability. However, after adding the spectral feature and the centroid relation feature, the parcellation performance is greatly improved.

4 Conclusion

In this paper, we propose an automated method for anatomically meaningful cerebellar cortical surface parcellation. We firstly reconstruct the geometric accurate and topologically correct cerebellar surfaces and then compute several widely used geometric features to comprehensively characterize the geometries of the cerebellar surface. Next, we extract local surface patches from the reconstructed cerebellar surfaces with the neighborship defined by the intrinsic geodesic metric. These extracted local surface patches are projected to a low dimensional latent space with a contrastive learning framework, i.e., patches from the same region are enforced to have similar representations, while patches from different regions have distinct representations. After that, we train a neural network to map the latent representations to the parcellation labels. Comparison to the state-of-the-art methods has validated the superior performance of our method. Currently, our work has two limitations, a) the quantitative evaluation is based on a small number of the subjects, due to the expensive manual labeling cost. In the future, we plan to involve and release more manually labeled cerebellar cortical surfaces to further improve the generalizability of the current framework, enhance the representation learning and validate our method on larger datasets; b) a graph cut post-processing is needed to remove potential inconsistent labelling. In the future, we plan to directly add the neighborhood smooth labeling constraint into the network cost function to obtain an end-to-end cerebellar cortical surface parcellation method.

Acknowledgements. This work was supported in part by the National Institutes of Health (NIH) under Grants MH116225, MH117943, MH123202, NS128534, and AG075582.

References

1. Davie, C.A., et al.: Persistent functional deficit in multiple sclerosis and autosomal dominant cerebellar ataxia is associated with axon loss. Brain **118**, 1583–1592 (1995)
2. Riva, D., Giorgi, C.: The cerebellum contributes to higher functions during development evidence from a series of children surgically treated for posterior fossa tumours. Brain **123**, 1051–1061 (2000)
3. Carass, A., et al.: Comparing fully automated state-of-the-art cerebellum parcellation from magnetic resonance images. Neuroimage **183**, 150–172 (2018)
4. Sun, Y., Gao, K., Niu, S., Lin, W., Li, G., Wang, L.: Semi-supervised transfer learning for infant cerebellum tissue segmentation. In: Liu, M., Yan, P., Lian, C., Cao, X. (eds.) MLMI 2020. LNCS, vol. 12436, pp. 663–673. Springer, Cham (2020). https://doi.org/10.1007/978-3-030-59861-7_67
5. Fischl, B.: FreeSurfer (2012)
6. Makropoulos, A., et al.: The developing human connectome project: a minimal processing pipeline for neonatal cortical surface reconstruction. Neuroimage **173**, 88–112 (2018)
7. Henschel, L., Conjeti, S., Estrada, S., Diers, K., Fischl, B., Reuter, M.: FastSurfer - a fast and accurate deep learning based neuroimaging pipeline. Neuroimage **219**, 117012 (2020)
8. Glasser, M.F., et al.: The minimal preprocessing pipelines for the human connectome project. Neuroimage **80**, 105–124 (2013)
9. Wang, L., Wu, Z., Chen, L., Sun, Y., Lin, W., Li, G.: iBEAT V2.0: a multisite-applicable, deep learning-based pipeline for infant cerebral cortical surface reconstruction. Nat. Protoc. 2023. 1–32 (2023)
10. Lombaert, H., Sporring, J., Siddiqi, K.: Diffeomorphic spectral matching of cortical surfaces. In: Gee, J.C., Joshi, S., Pohl, K.M., Wells, W.M., Zöllei, L. (eds.) IPMI 2013. LNCS, vol. 7917, pp. 376–389. Springer, Heidelberg (2013). https://doi.org/10.1007/978-3-642-38868-2_32
11. Fischl, B., Sereno, M.I., Dale, A.M.: Cortical surface-based analysis: II. Inflation, flattening, and a surface-based coordinate system. Neuroimage **9**, 195–207 (1999)
12. Goldman, R.: Curvature formulas for implicit curves and surfaces. In: Computer Aided Geometric Design, pp. 632–658. North-Holland (2005)
13. Chen, T., Kornblith, S., Norouzi, M., Hinton, G.: A simple framework for contrastive learning of visual representations. In: International Conference on Machine Learning, pp. 1575–1585. International Machine Learning Society (IMLS) (2020)

14. He, K., Zhang, X., Ren, S., Sun, J.: Deep residual learning for image recognition. In: IEEE Conference on Computer Vision and Pattern Recognition, pp. 770–778 (2016)
15. Boykov, Y., Kolmogorov, V.: An experimental comparison of min-cut/max-flow algorithms for energy minimization in vision. IEEE Trans. Pattern Anal. Mach. Intell. **26**, 1124–1137 (2004)
16. Wu, Z., et al.: Registration-free infant cortical surface parcellation using deep convolutional neural networks. In: International Conference on Medical Image Computing and Computer-Assisted Intervention, pp. 672–680 (2018)
17. Howell, B.R., et al.: The UNC/UMN baby connectome project (BCP): an overview of the study design and protocol development. Neuroimage **185**, 891–905 (2019)
18. Sun, L., et al.: Topological correction of infant white matter surfaces using anatomically constrained convolutional neural network. Neuroimage **198**, 114–124 (2019)
19. Li, G., et al.: Measuring the dynamic longitudinal cortex development in infants by reconstruction of temporally consistent cortical surfaces. Neuroimage **90**, 266–279 (2014)
20. Kipf, T.N., Welling, M.: Semi-supervised classification with graph convolutional networks. In: International Conference on Learning Representations. International Conference on Learning Representations. ICLR (2017)
21. Xu, K., Jegelka, S., Hu, W., Leskovec, J.: How powerful are graph neural networks? In: 7th International Conference on Learning Representations, ICLR 2019. International Conference on Learning Representations. ICLR (2019)
22. Gao, H., Ji, S.: Graph U-nets. IEEE Trans. Pattern Anal. Mach. Intell. **44**, 4948–4960 (2022)
23. Hamilton, W.L., Ying, R., Leskovec, J.: Inductive representation learning on large graphs. In: Advances in Neural Information Processing Systems, pp. 1025–1035. Neural Information Processing Systems Foundation (2017)

Maximum-Entropy Estimation of Joint Relaxation-Diffusion Distribution Using Multi-TE Diffusion MRI

Lipeng Ning[1,2](✉)

[1] Brigham and Women's Hospital, Boston, MA 02215, USA
[2] Harvard Medical School, Boston, MA 02215, USA
lning@bwh.harvard.edu

Abstract. Combined relaxation-diffusion MRI (rdMRI) is a technique to probe tissue microstructure using diffusion MRI data with multiple b-values and echo time. Joint analysis of rdMRI data can characterize the joint relaxation and diffusion distribution (RDD) function to examine heterogeneous tissue microstructure without using multi-component models. This paper shows that the problem of estimating RDD functions is equivalent to the multivariate Hausdorff moment problem by applying a change of variables. Three formulations of maximum entropy (ME) estimation problems are proposed to solve the inverse problem to derive ME-RDD functions in different parameter spaces. All three formulations can be solved by using convex optimization algorithms. The performance of the proposed algorithms is compared with the standard methods using basis functions based on simulations and *in vivo* rdMRI data. Results show that the proposed methods provide a more accurate estimation of RDD functions than basis-function methods.

Keywords: Diffusion MRI · T_2 relaxation · maximum entropy

1 Introduction

Diffusion magnetic resonance imaging (dMRI) is sensitive to water diffusion in biological tissue. Analytical models of dMRI signals have played an essential role in quantifying tissue microstructure in clinical studies. Several methods have been developed to use dMRI signals measured with a single or multiple b-values to estimate compartment-specific diffusivity or diffusion propagators [1,20,24]. But these methods are developed using dMRI data acquired with a fixed echo time (TE). Several studies have shown that joint modeling of dMRI with multiple TE can probe TE-dependent diffusivity and [23], tissue-specific T_2 relaxation rate [14] and the joint relaxation diffusion distribution (RDD) in each voxel [2,3,12,19]. More specifically, RDD functions describe the multidimensional distribution of T_2 relaxation rate and diffusivity in each voxel, providing a general framework to characterize heterogeneous tissue microstructure. The RDD functions were first applied to measure the structure of porous media

H. Greenspan et al. (Eds.): MICCAI 2023, LNCS 14227, pp. 439–448, 2023.
https://doi.org/10.1007/978-3-031-43993-3_43

[4,6,8,11]. It was generalized in [2,3,12] to probe the microstructure of biological tissue using rdMRI. A standard approach for estimating RDD functions is to represent the measurement signal using basis functions of different diffusivity and relaxation rates which may lead to biased estimation results because of the strong coupling between basis signals.

This work introduces the maximum entropy (ME) estimation method for more accurate estimation of RDD functions by adapting theories and techniques developed for the classical Hausdorff moment problems [10,13,15,18]. ME estimation is also a standard approach for high-resolution power spectral estimation of time series data which involves a similar trigonometric moment problem [7,21]. The ME power spectral estimation usually performs better than Fourier transform-based methods [21], which motivates this work to derive ME methods for estimating RDD functions and compare the results with basis function-based methods. To this end, we first show that the problem of estimation RDD function is equivalent to the multivariate Hausdorff moment problem by applying a change of variables. Three formulations of maximum entropy (ME) estimation problems are proposed to estimate ME-RDD functions in different parameter spaces. The performance of these methods is compared with results based on basis functions using simulations and *in vivo* data.

2 Method

2.1 On the Hausdorff Moment Problem

Let $s(b,t)$ denote the dMRI signal with b and t being the b-value and TE, respectively. It is a standard approach to represent $s(b,t)$ as

$$s(b,t) = \int_{I_D \times I_R} e^{-bD-tR} d\rho(D,R), \tag{1}$$

where $\rho(D,R)$ denotes the joint distribution of diffusion and relaxation coefficients and $I_D := [0, D_0]$ and $I_R = [0, R_0]$ denote the finite intervals for D and R. For simplicity, let $\boldsymbol{\theta} := [D, R]$, $\mathbf{x} := [b, t]$ and $\Theta := I_D \times I_R$. Then, (1) is equivalent to

$$s(\mathbf{x}) = \int_{\Theta} e^{-\mathbf{x}\cdot\boldsymbol{\theta}} p(\boldsymbol{\theta}) d\boldsymbol{\theta}, \tag{2}$$

where $\mathbf{x}\cdot\boldsymbol{\theta}$ denotes the inner product between $\mathbf{x}$ and $\boldsymbol{\theta}$. Assume that the signals are sampled at $b = 0, \delta_b, \ldots, n_b\delta_b$ and $t = t_{\min}, t_{\min} + \delta_t, \ldots, t_{\min} + n_t\delta_t$, where $t_{\min}$ denotes the shortest TE. Let $\mathbf{x}_{\min} = [0, t_{\min}]$. Then let $\mathbf{x}_{\mathbf{k}} := [k_1\delta_b, k_2\delta_t]$ with $\mathbf{k} = [k_1, k_2] \in \mathcal{K}$ where $\mathcal{K} = \{[k_1, k_2] \mid 0 \le k_1 \le n_b, 0 \le k_2 \le n_t\} \subset \mathcal{Z}_+^2$ represents the set of all feasible indices. Next, let $s_{\mathbf{k}} := s(\mathbf{x}_{\mathbf{k}})$ which satisfies that

$$s_{\mathbf{k}} = \int_{\Theta} e^{-\mathbf{x}_{\mathbf{k}}\cdot\boldsymbol{\theta} - \mathbf{x}_{\min}\cdot\boldsymbol{\theta}} p(\boldsymbol{\theta}) d\boldsymbol{\theta}, \tag{3}$$

$$= \int_{\Theta} e^{-\mathbf{x}_{\mathbf{k}}\cdot\boldsymbol{\theta}} \hat{p}(\boldsymbol{\theta}) d\boldsymbol{\theta}, \tag{4}$$

where $\hat{p}(\boldsymbol{\theta}) = e^{-\mathbf{x}_{\min}\cdot\boldsymbol{\theta}}p(\boldsymbol{\theta})$ is a scaled RDD function adjusted based on the non-zero minimum TE.

The Hausdorff moment problem focuses on the existence of distribution functions that satisfy a sequence of power moments [10,15]. To change $s_{\mathbf{k}}$ to power moments as in the Hausdorff moment problem, we define $\boldsymbol{\gamma} := [e^{-\delta_b\theta_1}, e^{-\delta_t\theta_2}]$, which takes value in the interval $\Gamma := [e^{-\delta_b D_0}, 1] \times [e^{-\delta_t R_0}, 1]$. For a vector $\mathbf{k}$, we define $\boldsymbol{\gamma}^{\mathbf{k}} = \prod_i \gamma_i^{k_i} = e^{-\mathbf{x}_{\mathbf{k}}\cdot\boldsymbol{\theta}}$ following the convention in [17]. Then Eq. (4) can be expressed using the new variables as

$$s_{\mathbf{k}} = \int_{\Gamma} \boldsymbol{\gamma}^{\mathbf{k}} f(\boldsymbol{\gamma}) d\boldsymbol{\gamma}, \tag{5}$$

with $f(\boldsymbol{\gamma}) = \frac{1}{\delta_b\delta_t\gamma_1\gamma_2}\hat{p}(\boldsymbol{\theta}(\boldsymbol{\gamma}))$. Thus, $s_{\mathbf{k}}$ can be considered as the power moments of the density function $f(\boldsymbol{\gamma})$ on the interval Γ. Therefore the problem of estimating RDD functions using finite rdMRI measurements is equivalent to a multivariate Hausdorff moment problem. We note that the unit interval is usually considered in Hausdorff moment problems. This can be obtained by changing the variable $\boldsymbol{\gamma}$ t $\hat{\boldsymbol{\gamma}} = [\frac{\gamma_1-\alpha_1}{1-\alpha_1}, \frac{\gamma_2-\alpha_2}{1-\alpha_2}]$ where $\boldsymbol{\alpha} = [\alpha_1, \alpha_2] = [e^{-\delta_b D_0}, e^{-\delta_t R_0}]$. Thus $\hat{\boldsymbol{\gamma}}$ takes the value on the unit interval I^2 whose moments can be computed using linear transforms of $s_{\mathbf{k}}$. To simplify notations, the analysis in the following subsection will be based on $s_{\mathbf{k}}$ and the distribution of $\boldsymbol{\gamma}$.

2.2 Maximum-Entropy Estimation

The RDD functions that satisfy the rdMRI data may not be unique. In moment problems, the maximum entropy (ME) method is a standard approach to estimate probability distributions and power spectral density functions [15]. Based on the three representations of rdMRI data in Eqs. (3), (4) and (5), three optimization problems are introduced to estimate ME-RDD functions below.

The first ME problem is developed based on Eq. (3) as below:

$$\begin{aligned} \max_{p(\boldsymbol{\theta})} \quad & \int_{\Theta} -p(\boldsymbol{\theta}) \ln p(\boldsymbol{\theta}) d\boldsymbol{\theta} \\ \text{s.t.} \quad & \int_{\Theta} e^{-\mathbf{x}_{\mathbf{k}}\cdot\boldsymbol{\theta} - \mathbf{x}_{\min}\cdot\boldsymbol{\theta}} p(\boldsymbol{\theta}) d\boldsymbol{\theta} = s_{\mathbf{k}}, \forall \mathbf{k} \in \mathcal{K}, \end{aligned} \tag{6}$$

where the objective function is the Shannon differential entropy of $p(\boldsymbol{\theta})$. The solutions to ME problems have been extensively investigated in moment problems [15]. Using the Lagrangian method, one can derive that the optimal solution has the following form

$$p_{\lambda}^{\text{me1}}(\boldsymbol{\theta}) = \exp\left(-\left(\sum_{\mathbf{k}\in\mathcal{K}} \lambda_{\mathbf{k}} e^{-\mathbf{x}_{\mathbf{k}}\cdot\boldsymbol{\theta} - \mathbf{x}_{\min}\cdot\boldsymbol{\theta}}\right) - 1\right) \tag{7}$$

for some coefficients $\lambda_{\mathbf{k}}$ with $\mathbf{k} \in \mathcal{K}$. The optimal parameters $\lambda_{\mathbf{k}}$ need to be solved to satisfy the constraints in Eq. (6).

Based on Eq. (4), the second ME problem is introduced as below:

$$\max_{\hat{p}} \int_{\Theta} -\hat{p}(\boldsymbol{\theta}) \ln \hat{p}(\boldsymbol{\theta}) d\boldsymbol{\theta} \tag{8}$$
$$\text{s.t.} \int_{\Theta} e^{-\mathbf{x_k}\cdot\boldsymbol{\theta}} \hat{p}(\boldsymbol{\theta}) d\boldsymbol{\theta} = s_{\mathbf{k}}, \forall \mathbf{k} \in \mathcal{K},$$

where $\hat{p}(\boldsymbol{\theta})$ is the pre-scaled RDD to adjust for the nonzero t_{min}. The optimal solution to Eq. (8) has the following form

$$\hat{p}_{\lambda}^{\text{me2}}(\boldsymbol{\theta}) = \exp\left(-\sum_{\mathbf{k}\in\mathcal{K}} \lambda_{\mathbf{k}} e^{-\mathbf{k}\cdot\boldsymbol{\theta}}\right). \tag{9}$$

It is noted that Eq. (9) does not include the constant -1 as in Eq. (7) since it can be absorbed by the variable λ_0. Then, $\hat{p}_{\text{ME2}}(\boldsymbol{\theta})$ can be scaled back to the original RDD function by $p_{\lambda}^{\text{me2}}(\boldsymbol{\theta}) = \exp\left(-\left(\sum_{\mathbf{k}\in\mathcal{K}} \lambda_{\mathbf{k}} \cdot e^{-\mathbf{x_k}\cdot\boldsymbol{\theta}}\right) + \mathbf{x}_{\text{min}} \cdot \boldsymbol{\theta}\right)$, which has a different form than the solution in Eq. (7).

The third ME-RDD is estimated based on the new variable $\boldsymbol{\gamma}$ as in Eq. (5) by solving the following problem:

$$\max_{f} \int_{\Gamma} -f(\boldsymbol{\gamma}) \ln f(\boldsymbol{\gamma}) d\boldsymbol{\gamma} \tag{10}$$
$$\text{s.t.} \int_{\Gamma} \boldsymbol{\gamma}^{\mathbf{k}} f(\boldsymbol{\gamma}) d\boldsymbol{\gamma} = s_{\mathbf{k}}, \forall \mathbf{k} \in \mathcal{K}. \tag{11}$$

By changing the variable $\boldsymbol{\gamma}$ back to $\boldsymbol{\theta}$, Eq. (10) is transformed to

$$\max_{\hat{p}} \int_{\Theta} -\hat{p}(\boldsymbol{\theta}) \left(\ln \hat{p}(\boldsymbol{\theta}) - \ln\left(\delta_b \delta_t e^{-\boldsymbol{\delta}\cdot\boldsymbol{\theta}}\right)\right) d\boldsymbol{\theta}$$
$$\text{s.t.} \int_{\Theta} e^{-\mathbf{x_k}\cdot\boldsymbol{\theta}} \hat{p}(\boldsymbol{\theta}) d\boldsymbol{\theta} = s_{\mathbf{k}}, \forall \mathbf{k} \in \mathcal{K},$$

where $\boldsymbol{\delta} = [\delta_b, \delta_t]$. It is interesting to note that the above objective function is equal to minimizing the *Kullback-Leibler* divergence, i.e., the relative entropy, between $\hat{p}(\boldsymbol{\theta})$ and $\delta_b \delta_t e^{-\boldsymbol{\delta}\cdot\boldsymbol{\theta}}$. The optimal solution has the following form

$$\hat{p}_{\lambda}^{\text{me3}}(\boldsymbol{\theta}) = \exp\left(-\left(\sum_{\mathbf{k}\in\mathcal{K}} \lambda_{\mathbf{k}} e^{-\mathbf{k}\cdot\boldsymbol{\theta}}\right) - \boldsymbol{\delta}\cdot\boldsymbol{\theta}\right). \tag{12}$$

Then, $\hat{p}_{\lambda}^{\text{me3}}(\boldsymbol{\theta})$ is scaled back to obtain

$$p_{\lambda}^{\text{me3}}(\boldsymbol{\theta}) = \exp\left(-\left(\sum_{\mathbf{k}\in\mathcal{K}} \lambda_{\mathbf{k}} \cdot e^{-\mathbf{x_k}\cdot\boldsymbol{\theta}}\right) + (\mathbf{x}_{\text{min}} - \boldsymbol{\delta})\cdot\boldsymbol{\theta}\right). \tag{13}$$

It is noted the difference between $p_{\lambda}^{\text{me1}}(\boldsymbol{\theta})$ and $p_{\lambda}^{\text{me2}}(\boldsymbol{\theta})$ is related to the non-zero offset x_{min}. The two solutions are equal if the shortest TE is zero, but it is impossible in practice. The difference between $p_{\lambda}^{\text{me2}}(\boldsymbol{\theta})$ and $p_{\lambda}^{\text{me3}}(\boldsymbol{\theta})$ is related to the sampling rate δ_b and δ_t. The difference is less significant, with a higher sampling rate in the b-value and TEs.

2.3 Dual Energy Minimization Problems

The optimal values of $\lambda_{\mathbf{k}}$ in Eqs. (7), (9) and (12) can be obtained by solving the dual formulation of the ME problems, which can be expressed as energy minimization problems based on the dual formulations.

For the solution in Eq. (7), the corresponding energy function, i.e., the dual objective function, is given by

$$\Delta_1(\boldsymbol{\lambda}) = \int_\Theta \exp\left(-\left(\sum_{\mathbf{k}\in\mathcal{K}} \lambda_{\mathbf{k}} e^{-\mathbf{x}_{\mathbf{k}}\cdot\boldsymbol{\theta}-\mathbf{x}_{\min}\cdot\boldsymbol{\theta}}\right)-1\right) d\boldsymbol{\theta} + \sum_{\mathbf{k}} \lambda_{\mathbf{k}} s_{\mathbf{k}}. \tag{14}$$

It can be shown that

$$\frac{\partial^2 \Delta_1}{\partial \boldsymbol{\lambda}_{\mathbf{k}} \partial \boldsymbol{\lambda}_{\ell}} = \int_\Theta e^{-\mathbf{x}_{\mathbf{k}}\cdot\boldsymbol{\theta}-\mathbf{x}_{\ell}\cdot\boldsymbol{\theta}-2\mathbf{x}_{\min}\cdot\boldsymbol{\theta}} p_\lambda^{\mathrm{me1}}(\boldsymbol{\theta}) d\boldsymbol{\theta}. \tag{15}$$

Thus, the Hessian matrix of $\Delta_1(\boldsymbol{\lambda})$ positive definite, indicating that $\Delta_1(\boldsymbol{\lambda})$ is a convex function. If $\Delta_1(\boldsymbol{\lambda})$ has a finite minimizer, then the minimizer satisfies that

$$\frac{\partial \Delta_1}{\partial \lambda_{\mathbf{k}}} = -\int_\Theta e^{-\mathbf{x}_{\mathbf{k}}\cdot\boldsymbol{\theta}-\mathbf{x}_{\min}\cdot\boldsymbol{\theta}} p_\lambda^{\mathrm{me1}}(\boldsymbol{\theta}) d\boldsymbol{\theta} + s_{\mathbf{k}} = 0. \tag{16}$$

Therefore, the optimal parameters for $p_\lambda^{\mathrm{me1}}(\boldsymbol{\theta})$ can be obtained from the minimizer of $\Delta_1(\boldsymbol{\lambda})$.

The optimal λ for Eq. (9) and Eq. (12) can be obtained by minimizing the following to convex energy functions:

$$\Delta_2(\lambda) = \int_\Theta \exp\left(-\left(\sum_{\mathbf{k}\in\mathcal{K}} \lambda_{\mathbf{k}} e^{-\mathbf{x}_{\mathbf{k}}\cdot\boldsymbol{\theta}}\right)\right) d\boldsymbol{\theta} + \sum_{\mathbf{k}} \lambda_{\mathbf{k}} s_{\mathbf{k}}, \tag{17}$$

$$\Delta_3(\lambda) = \int_\Theta \exp\left(-\left(\sum_{\mathbf{k}\in\mathcal{K}} \lambda_{\mathbf{k}} e^{-\mathbf{x}_{\mathbf{k}}\cdot\boldsymbol{\theta}}\right) - \boldsymbol{\delta}\cdot\boldsymbol{\theta}\right) d\boldsymbol{\theta} + \sum_{\mathbf{k}} \lambda_{\mathbf{k}} s_{\mathbf{k}}. \tag{18}$$

In this paper, the energy minimization problem was solved using a customized Newton algorithm with the Armijo line-search method [5]. The code and data used in this work are available at https://github.com/LipengNing/ME-RDD.

3 Examples

3.1 Synthetic Data

The proposed algorithms were examined using synthetic rdMRI data with an RDD function consisting of three Gaussian components with the mean at (1.5 $\mu\mathrm{m}^2$/ms, 10 ms^{-1}), (0.5 $\mu\mathrm{m}^2$/ms, 40 ms^{-1}), and (1.5 $\mu\mathrm{m}^2$/ms, 40 ms^{-1}) with the volume fraction being 0.2, 0.5 and 0.3, respectively. D and R were uncorrelated in each component, with the standard deviation being 0.01 $\mu m^2/ms$ and $5 ms^{-1}$. Simulated rdMRI signals had b-values at $b = 0, 0.5$ ms/$\mu m^2, \ldots, 5$ ms/μm^2 and TEs at $t = 50$ ms, 75 ms, $\ldots$, 200 ms, which can be achieved using an advanced MRI scanner for in vivo human brains such as the connectom scanner [9]. Then, independently and identically distributed Gaussian noise was added to simulated rdMRI signals with an average signal-to-noise ratio (SNR) from 100 to 600, similar to the range of the SNR of direction-averaged dMRI signals of in vivo human brains that scale according to the square root of the number of directions and voxel size.

3.2 Comparison Methods

For comparison, we applied the basis-function representation method, similar to the methods in [12], by representing the signals as

$$s(\mathbf{x}) = \sum_{n=1}^{N} c_n e^{-\mathbf{x}\cdot\boldsymbol{\theta}_n}, \tag{19}$$

where $\boldsymbol{\theta}_n$ are a set of predefined points on a discrete grid in Θ. We solved a constrained L_2 minimization problem to find the optimal non-negative coefficient c_n with minimum L_2 norm. To evaluate the performances, we computed the error of the center of mass (CE) of the estimated RDD functions in three regions defined using a watershed clustering approach; see the top left figure in Fig. 2. The CE in diffusivity and re Moreover, we also computed the volume-fraction error (VFE), i.e., VFE $= \sum_{k=1}^{K} |F_{\text{est}}(k) - F_{\text{true}}(k)|$, of the estimated RDD, where $F_{\text{est}}(k)$ and $F_{\text{true}}(k)$ denote the true and estimated volume fraction for each component.

3.3 In Vivo rdMRI

The proposed ME algorithms were applied to an in vivo rdMRI dataset acquired in our previous work [16]. The data was acquired from a healthy volunteer on a 3T Siemens Prisma scanner with the following parameters: voxel size = $2.5\times2.5\times2.5$mm^3, matrix size = $96\times96\times54$, TE = 71, 101, 131, 161 and 191 ms, TR=5.9 s, b = 700, 1400, 2100, 2800, 3500 s/mm^2 along 30 gradient directions together with 6 volumes at b = 0, simultaneous multi-slice (SMS) factor = 2, and iPAT = 2. The pulse width of the diffusion gradients and the diffusion time were fixed across scans. An additional pair of b=0 images with anterior-posterior (AP), and posterior-anterior (PA) phase encoding directions were acquired for distortion correction using FSL TOPUP/eddy. Then the data were further processed using the unring [22] tool.

4 Results

The first two figures in Fig. 1 show the error in the center of mass of D and R. The L2 approach based on basis functions had significantly overestimated D

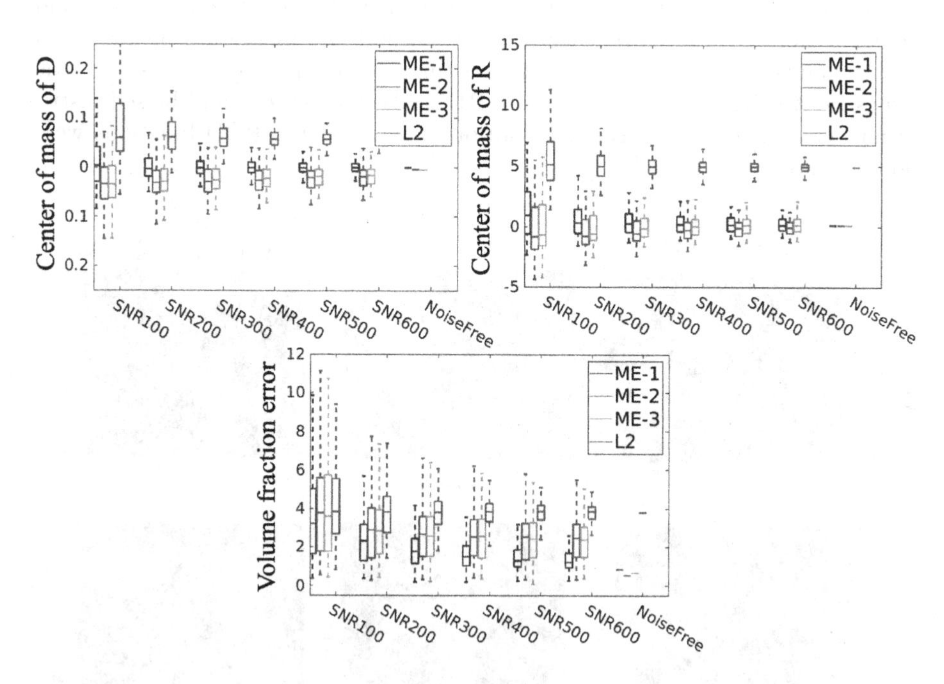

Fig. 1. Performance of ME-RDD and L2-based RDD functions using synthetic data for different simulated SNR levels.

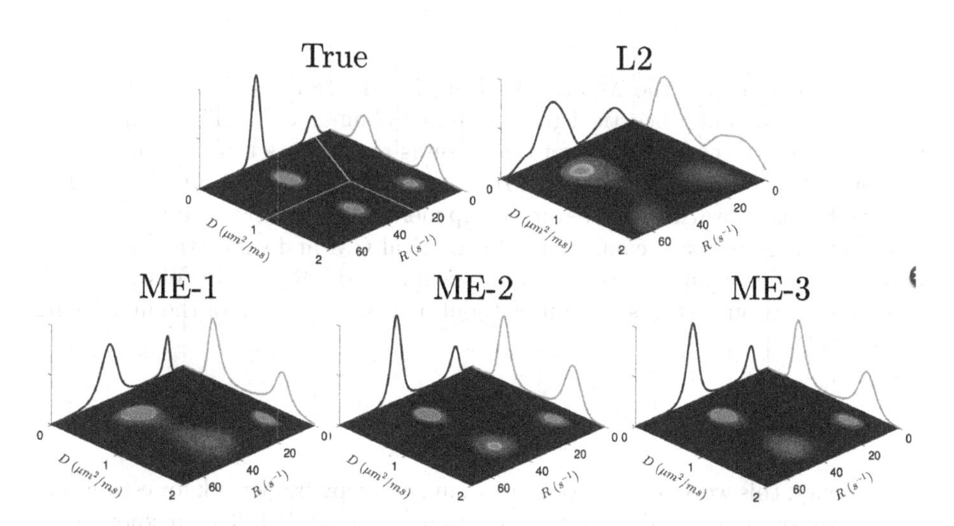

Fig. 2. Comparison of ME-RDD and L2-based functions with noise-free synthetic data.

and R, whereas the three ME methods had much lower estimation error and similar performance to each other. The L2 method also had a biased estimation of the volume fraction, as shown in the third figure. Figure 2 shows the true and estimated RDD functions. The L2-RDD has more spread and biased distributions compared to the true distribution. The three ME-RDD functions were more focal and less biased.

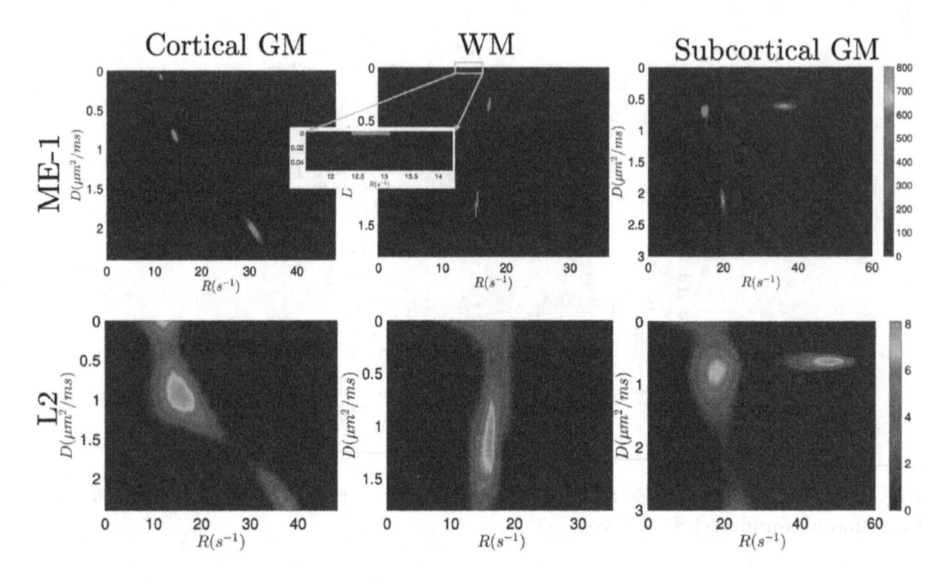

Fig. 3. Comparison of ME-RDD (ME-1)and L2-based functions using *in vivo* data.

Figure 3 compares the ME-RDD (using ME-1) and L2-RDD in the white matter, cortical, and subcortical gray matter of a human brain. The computation time to estimate the ME-RDD for one voxel is about 100 s using an Intel Xeon E5 cpu (2.20 GHz). The results for ME-2 and ME-3 were similar to ME-1. The ME-RDD has shown three separate components for each tissue type. But the L2-RDD has shown two components for cortical GM and subcortical GM with higher R values compared to the components in ME-RDD. For WM, the L2-RDD shows a continuous spectrum without a clear separation of the underlying components.

5 Summary

In summary, this work introduced a maximum-entropy framework for estimating the relaxation-diffusion distribution functions using rdMRI. To our knowledge, this is the first work showing that the estimation of multidimensional RDD functions is equivalent to the classical multivariate Hausdorff moment problem. Although this work focuses on the two dimensional RDD functions, the results generalize to the special cases for one dimensional relaxation or diffusion distribution functions. The contributions of this work also include the development of three algorithms to estimate RDD functions and the comparisons with the standard basis-function approach. The ME-RDD functions can be estimated using convex optimization algorithms. Experimental results have shown that the proposed methods provide more accurate parameters for each component and more accurate volume fractions compared to the standard basis function methods. Moreover, results based on in vivo data have shown that the proposed ME-RDD can resolve multiple components that cannot be distinguished by the basis function approach. The better performance ME-RDD functions compared to basis-function methods may relate to the superior performance of ME spectral estimation methods compared to Fourier transform-based methods [21]. We expect the improved spectral resolution will be useful in several clinical applications such as lesion and tumor detection. But further theoretical analysis on the performance of the ME methods is needed in future work. Moreover, further histological validations are needed to examine the biological basis of the RDD functions. Finally, we note a limitation of the proposed method is that the optimization algorithm may have a slow convergence speed because the Hessian matrices may not be well conditioned. Moreover, the results may be sensitive to measurement noise. Thus faster and more reliable computation algorithms will be developed in future work.

References

1. Basser, P.J., Mattiello, J., LeBihan, D.: MR diffusion tensor spectroscopy and imaging. Biophys. J. **66**, 259–267 (1994)
2. Benjamini, D., Basser, P.J.: Use of marginal distributions constrained optimization (MADCO) for accelerated 2D MRI relaxometry and diffusometry. J. Magn. Reson. **271**, 40–45 (2016)
3. Benjamini, D., Basser, P.J.: Towards clinically feasible relaxation-diffusion correlation MRI using MADCO. Microporous Mesoporous Mater. **269**, 93–96 (2018)
4. Bernin, D., Topgaard, D.: NMR diffusion and relaxation correlation methods: new insights in heterogeneous materials (2013)
5. Bertsekas, D.P.: Nonlinear Programming. 2nd edn (1999)
6. Callaghan, P.T., Arns, C.H., Galvosas, P., Hunter, M.W., Qiao, Y., Washburn, K.E.: Recent Fourier and Laplace perspectives for multidimensional NMR in porous media. Magn. Reson. Imaging **25**, 441–444 (2007)
7. Cover, T.M., Thomas, J.A.: Elements of Information Theory (1993)
8. De Almeida Martins, J.P., Topgaard, D.: Multidimensional correlation of nuclear relaxation rates and diffusion tensors for model-free investigations of heterogeneous anisotropic porous materials. Sci. Rep. **8**, 2488 (2018)
9. Fan, Q., et al.: Mapping the human connectome using diffusion MRI at 300 mT/m gradient strength: methodological advances and scientific impact. NeuroImage **254**, 118958 (2022)
10. Frontini, M., Tagliani, A.: Hausdorff moment problem and maximum entropy: on the existence conditions. Appl. Math. Comput. **218**(2), 430–433 (2011)

11. Hürlimann, M.D., Venkataramanan, L.: Quantitative measurement of two-dimensional distribution functions of diffusion and relaxation in grossly inhomogeneous fields. J. Magn. Reson. **157**, 31–42 (2002)
12. Kim, D., Doyle, E.K., Wisnowski, J.L., Kim, J.H., Haldar, J.P.: Diffusion-relaxation correlation spectroscopic imaging: a multidimensional approach for probing microstructure. Magn. Reson. Med. **78**, 2236–2249 (2017)
13. Landau, H.J.: Maximum entropy and the moment problem. Bull. Am. Math. Soc. **16**(1), 47–77 (1987)
14. McKinnon, E.T., Jensen, J.H.: Measuring intra-axonal T2 in white matter with direction-averaged diffusion MRI. Magn. Reson. Med. **81**, 2985–2994 (2019)
15. Mead, L.R., Papanicolaou, N.: Maximum entropy in the problem of moments. J. Math. Phys. **25**(8), 2404–2417 (1984)
16. Ning, L., Gagoski, B., Szczepankiewicz, F., Westin, C.F., Rathi, Y.: Joint relaxation-diffusion imaging moments to probe neurite microstructure. IEEE Trans. Med. Imaging **39**, 668–677 (2020)
17. Putinar, M., Scheiderer, C.: Multivariate moment problems: geometry and indeterminateness. Annali della Scuola Normale - Classe di Scienze **5**(2), 137–157 (2006)
18. Schmüdgen, K.: The Moment Problem. GTM, vol. 277. Springer, Cham (2017). https://doi.org/10.1007/978-3-319-64546-9
19. Slator, P.J., et al.: Combined diffusion-relaxometry microstructure imaging: current status and future prospects. Magn. Reson. Med. **86**(6), 2987–3011 (2021)
20. Tuch, D.S.: Q-ball imaging. Magn. Reson. Med. **52**, 1358–1372 (2004)
21. Ulrych, T.J., Bishop, T.N.: Maximum entropy spectral analysis and autoregressive decomposition. Rev. Geophys. **13**, 183–200 (1975)
22. Veraart, J., Fieremans, E., Jelescu, I.O., Knoll, F., Novikov, D.S.: Gibbs ringing in diffusion MRI. Magn. Reson. Med. **76**(1), 301–314 (2016)
23. Veraart, J., Novikov, D.S., Fieremans, E.: TE dependent diffusion Imaging (TEdDI) distinguishes between compartmental T2 relaxation times. NeuroImage **182**, 360–369 (2018)
24. Zhang, H., Schneider, T., Wheeler-Kingshott, C.A., Alexander, D.C.: NODDI: practical in vivo neurite orientation dispersion and density imaging of the human brain. NeuroImage **61**(4), 1000–1016 (2012)

Physics-Informed Conditional Autoencoder Approach for Robust Metabolic CEST MRI at 7T

Junaid R. Rajput[1,2(✉)], Tim A. Möhle[1], Moritz S. Fabian[1], Angelika Mennecke[1], Jochen A. Sembill[3], Joji B. Kuramatsu[3], Manuel Schmidt[1], Arnd Dörfler[1], Andreas Maier[2], and Moritz Zaiss[1]

[1] Institute of Neuroradiology, University Hospital Erlangen, Erlangen, Germany
[2] Pattern Recognition Lab Friedrich-Alexander-University Erlangen-Nürnberg, Erlangen, Germany
junaid.rajput@fau.de
[3] Department of Neurology, University Hospital Erlangen-Nürnberg, Erlangen, Germany

Abstract. Chemical exchange saturation transfer (CEST) is an MRI method that provides insights on the metabolic level. Several metabolite effects appear in the CEST spectrum. These effects are isolated by Lorentzian curve fitting. The separation of CEST effects suffers from the inhomogeneity of the saturation field B_1. This leads to inhomogeneities in the associated metabolic maps. Current B_1 correction methods require at least two sets of CEST-spectra. This at least doubles the acquisition time. In this study, we investigated the use of an unsupervised physics-informed conditional autoencoder (PICAE) to efficiently correct B_1 inhomogeneity and isolate metabolic maps while using a single CEST scan. The proposed approach integrates conventional Lorentzian model into the conditional autoencoder and performs voxel-wise B_1 correction and Lorentzian line fitting. The method provides clear interpretation of each step and is inherently generative. Thus, CEST-spectra and fitted metabolic maps can be created at arbitrary B_1 levels. This is important because the B_1 dispersion contains information about the exchange rates and concentration of metabolite protons, paving the way for their quantification. The isolated maps for tumor data showed a robust B_1 correction and more than 25% increase in structural similarity index (SSIM) with gadolinium reference image compared to the standard interpolation-based method and subsequent Lorentzian curve fitting. This efficient correction method directly results in at least 50% reduction in scan time.

Keywords: Physics-informed autoencoder · CEST · Bound loss · Lorentzian fit · Acquisition time · Interpretability

---

Supplementary Information The online version contains supplementary material available at https://doi.org/10.1007/978-3-031-43993-3_44.

H. Greenspan et al. (Eds.): MICCAI 2023, LNCS 14227, pp. 449–458, 2023.
https://doi.org/10.1007/978-3-031-43993-3_44

1 Introduction

Chemical exchange saturation transfer (CEST) is a novel metabolic magnetic resonance imaging (MRI) method that allows to detect molecules in tissue based on chemical exchange of their mobile protons with water protons [18]. CEST works by selectively saturating the magnetization of a specific pool of protons, such as those in metabolites or proteins, by applying narrow-band radiofrequency (RF) pulses at their respective Larmor frequency. Due to chemical exchange this saturation state is transferred to the water pool and a decrease in the detected water signal provides information about the concentration and exchange rate of the underlying molecules. This procedure is repeated for several RF frequencies to acquire the so-called CEST-spectrum in each voxel. CEST-MRI offers several promising contrasts that correlate with the diagnosis of diseases such as ischemic stroke [16], brain tumors [1], and neurodegenerative diseases [2,5]. The CEST-spectrum contains effects of proton pools of various chemical components in the tissue, typically isolated by a Lorentzian model [14] that is derived form the underlying physics of the Bloch-McConnell equations [8]. In this conventional method, several Lorentzian distributions are fitted to the CEST-spectrum using nonlinear least squares method [10], and the amplitude of each fitted distribution represents a particular metabolic map.

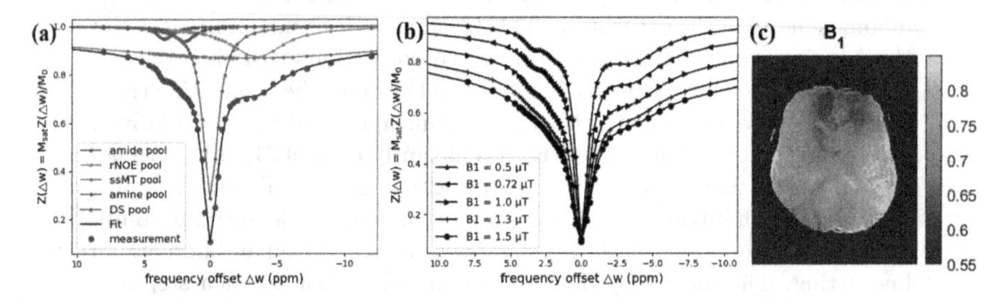

Fig. 1. (a) CEST-spectrum and the corresponding 5-pool Lorentzian fit. (b) CEST-spectrum dependence on B_1 saturation amplitude. (c) The B_1 inhomogeneity map at 7T.

The number of Lorentzian functions utilized in this process depends on the expected number of exchanging proton pools present in the spectrum. Figure 1a shows an example of an acquired CEST-spectrum and the corresponding 5-pool Lorentzian fit. Increasing the static magnetic field B_0 (e.g., with $B_0 = 7T$), enhances spectral resolution, but leads to significant variations in the B_1 amplitude of the saturating RF field across the field of view (cf. Fig. 1c). This B_1 inhomogeneity is corrected by acquiring CEST-spectra at various RF field strengths (cf. Fig. 1b) and then interpolating between them at fixed B_1 to produce the B_1-robust metabolic CEST contrast maps [14]. This B_1 correction increases the acquisition time at least twofold.

Hunger et al. shown that supervised learning can be used to generate B_1-robust CEST maps, coining the DeepCEST approach [3,4]. However, the previous work on generating the B_1-robust CEST contrasts rely on valid target data and the underlying assumptions to generate it, and can only create CEST maps at one particular B_1 level.

In this work, we developed a conditional autoencoder (CAE) [13] to generate B_1-homogeneous CEST-spectra at arbitrary B_1 levels, and a physics-informed autoencoder (PIAE) to fit the 5-pool Lorentzian model to the B_1 corrected CEST-spectra. This inclusion of physical knowledge in the form of known operators in neural nets (NN) is expected to reduce the absolute error margin of the model [6,7] and to increase its interpretability. Both CAE and PIAE are trained in an unsupervised end-to-end method that eliminates the shortcomings of conventional Lorentzian curve fitting and produces robust CEST contrast at arbitrary B_1 levels without the need for an additional acquisition scan. We called the proposed method physics-informed conditional autoencoder (PICAE).

2 Methods

Data Measurements. CEST imaging was performed in seven subjects, including two glioblastoma patients, after written informed consent was obtained to investigate the dependence of CEST effects on B_1 in brain tissue. The local ethics committee approved the study. All volunteers were measured at three B_1 field strengths 0.72 μT, 1.0 μT, and 1.5 μT. A method as described by Mennecke et al. [9] was used to acquire CEST data on a 7T whole-body MRI system (MAGNETOM Terra, Siemens Healthcare GmbH, Erlangen, Germany). Saturated images were obtained for 54 non-equidistant frequency offsets ranging from −100 ppm to +100 ppm. The acquisition time per B_1 level was 6:42 min. The acquisition of the B_1 map required an additional 1:06 min.

Conditional Autoencoder. We developed a conditional autoencoder (CAE) to solve the B_1 inhomogeneity problem, which is essential for the generation of metabolic CEST contrast maps at 7T. The left part of Fig. 2 describes the CAE. The encoding network of CAE took the raw CEST-spectrum and the corresponding effective B_1 value as input and generate a latent space that was concatenated once with the same B_1 input value and passed to the decoder that reconstruct the uncorrected B_1 CEST-spectrum, and another time the latent space was concatenated with the desired/specific effective B_1 value to reconstruct the CEST-spectrum at a specific B_1 saturation amplitude. Both decoders shared the weights (cf. Fig. 2). For the development of the CAE networks, we used the well-known fully concatenated (FC) layers with leaky ReLU activations except for the last layer of decoder, which had a linear activation. The encoder and decoder both consisted of 4 layers, where the layers of the encoder successively contain 128, 128, 64, 32 neurons, while the layers of the decoder successively contain 32, 64, 128, 128 neurons. The input, latent space and output layers had 55, 17 and 54 neurons respectively.

Physics-Informed Autoencoder. The Lorentzian model and its B_1-dispersion can be derived from the underlying spin physics described by

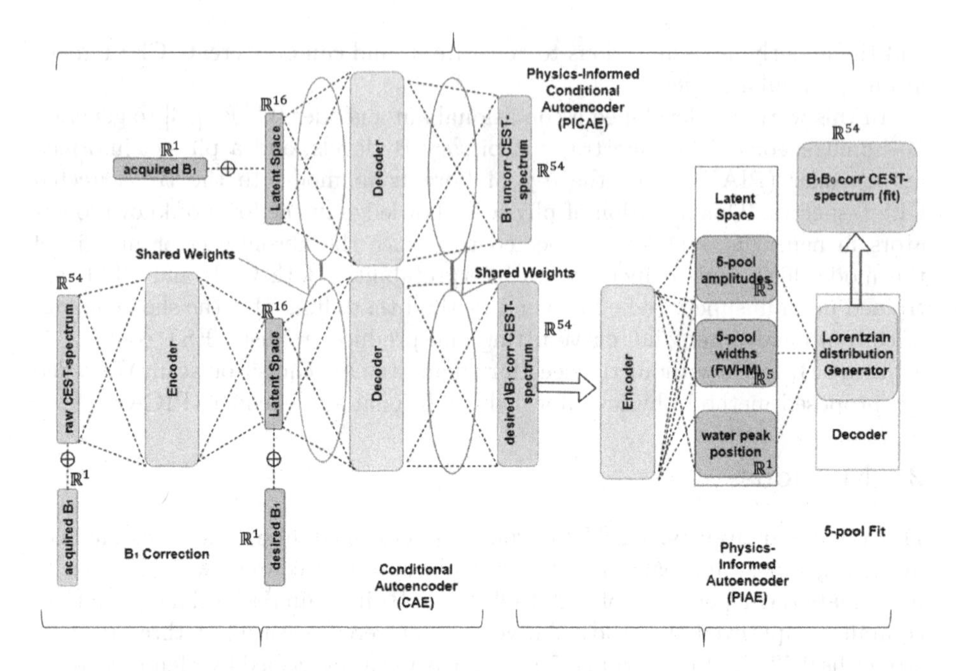

Fig. 2. Overview of the proposed physics-informed conditional autoencoder for B_1 inhomogeneity correction (left) and Lorentzian curve fitting (right).

the Bloch-McConnell equation system [8]. The physics-informed autoencoder (PIAE) utilized fully connected NN as encoder and Lorentzian distribution generator as a decoder to perform the pixel-wise 5-pool Lorentzian curve fit to the CEST-spectrum (water, amide, amine, NOE, MT) [14]. The 5-pool model was described as

$$Z(\Delta\omega) = 1 - L_{DS} - L_{ssMT} - L_{amine} - L_{rNOE} - L_{amide}, \tag{1}$$

where L denotes the Lorentz function. The direct saturation pool (water) was defined as

$$L_{DS} = \frac{A_{DS}}{1 + (\frac{\Delta\omega - \delta_{DS}}{\tau_{DS}/2})^2}. \tag{2}$$

The remaining other four pools were defined as

$$L_i = \frac{A_i}{1 + (\frac{\Delta\omega - \delta_{DS} - \delta_i}{\tau_i/2})^2}, \quad i \in amide, amine, rNOE, ssMT. \tag{3}$$

The right part of Fig. 2 describes the PIAE. The encoder of PIAE mapped the CEST-spectrum to the amplitudes A_i, the full width half maximum (FWHM) τ_i, and the water peak position δ_{DS} of the 5-pool Lorentzian model. Its encoder consisted of four FC layers, each with 128 neurons with leaky ReLU activations. It had three so-called FC latent space layers with linear activation for position and exponential activations for FWHM and amplitudes of 5-pool Lorentzian model. The positions of amide, rNOE, ssMT, and amine were fixed at 3.5 ppm, –3.5 ppm, –3 ppm, and 2 ppm, respectively, and shifted with respect to the predicted position of the water peak. The decoder of PIAE consisted of a Lorentzian distribution generator (cf. Fig. 2). It generated samples of the 5-pool distributions exactly at the offsets $\Delta\omega$ (i.e. between –100 ppm and 100 ppm) where the input CEST-spectrum was sampled, and combined them according to Eq. 1 to generate the input CEST spectrum with or without B_0 correction.

Bound Loss. The peak positions δ_i and widths τ_i of the pools had to be within certain bounds so that certain neurons in the latent space layer of PIAE would not be exchanged and provide the same pool parameters for all samples. We developed a simple cost function along the lines of the hinge loss [12], called the bound loss. Mathematically, it is defined as follows

$$bound\,loss = \left\{\begin{array}{l} abs(y-lb), \quad \text{if} \quad y < lb \\ abs(y-ub), \quad \text{if} \quad y > ub \\ 0, \quad \text{if} \quad y <= ub \quad \text{AND} \quad y >= lb \end{array}\right\}. \tag{4}$$

The bound loss increases linearly as the output of the latent space neurons of PIAE exceeds or recede from the boundaries. The lower and upper limits for positions and widths are given in Table 1 of the supplementary material.

Training and Evaluation. Four healthy volunteers formed the training and validation sets. The test set consisted of the two tumor patients and one healthy subject. To ensure that the outcomes were exclusively based on the CEST-spectrum and not influenced by spatial position, the training was carried out voxel-by-voxel. Consequently, there were approximately one million CEST-spectra for the training process. CAE was first trained with MSE loss. In this step, the CAE encoder was fed with the CEST-spectrum of a specific B_1 saturation amplitude, and it generated two CEST-spectra, one for the input B_1 saturation level and the other for the B_1 level injected into the latent space (cf. Fig. 2). Later, it was trained with a combination of MSE loss and perception loss (MSE loss between the latent space of the CEST-spectra at two different B_1 levels). To incorporate perception loss, we used two forward passes with two different B_1 CEST-spectra and used perception loss to generate a latent space that is independent of B_1 saturation amplitude. The following equation describes the loss of the second step.

$$CAE\,loss = MSE\,loss + 0.1 \cdot perceptionloss \tag{5}$$

PIAE, on the other hand, was trained with a combination of MSE loss and bound loss. The PIAE loss was described as follows

$$PIAE\,loss = MSE\,loss + bound\,loss, \tag{6}$$

for evaluation we input the uncorrected CEST-spectrum acquired at 1μT and generated corrected CEST-spectra at B_1 0.5, 0.72, 1.0, 1.3, 1.5 μT. PIAE encoder yielded the amplitudes of 5-pool for B_1 corrected CEST-spectrum. Its decoder reconstructed the B_1B_0 fitted CEST-spectrum. The B_0 correction simply refers to the shift of the position of the water peak to 0 ppm.

CEST Quantification. The multi-B_1 CEST-spectra allow quantification of CEST effects (amide, rNOE, amine) [14,15] down to the exchange rate and concentration. The amplitudes of the CEST contrasts were expressed according to the definition in [15] as follows

$$A_i = \underbrace{f_i k_i}_{Z_1} \underbrace{\frac{w_1^2}{\underbrace{k_i(k_i + r_{2i})}_{Z_2} + w_1^2}}_{\alpha} Z_{ref} \quad i \in amide, amine, rNOE, \tag{7}$$

where f_i, k_i, and r_{2i} express the concentrations, exchange rates, and relaxation rates of the pools. Z_{ref} defines the sum of all 5 distributions at the resonance frequency of the specific pool in B_1B_0 corrected CEST-spectrum and w_1 is the frequency of the oscillating field.

The amplitudes of CEST contrasts in the Lorentzian function have the B_1 dispersion function given by the labeling efficiency α (Eq. 7). The exchange rate occurs here separately from the concentration, which allows their quantification via the B_1 dispersion. Concentration and exchange rate were fitted as a product and denoted as Z_1 (quantified maps), and $k(k+r_2)$ was also fitted with the single term Z_2 using trust-region reflective least squares [10].

3 Results

The comparison of PICAE with the conventional method [9,14] is shown in columns 1 and 2 of Fig. 3. The top image in column 3 shows the T_1-weighted reference image enhanced with the exogenous contrast agent gadolinium (Gd-T_1w), and the bottom image shows the B_1-map. The tumor shows a typical so called gadolinium ring enhancement indicated by the arrow (a_{15}), which is also visible in the non-invasive and gadolinium-free CEST contrast maps (columns 1 and 2). The PICAE-CEST maps showed better visualization of this tumor feature compared to the conventional method. The proposed method yielded at least 25% increase in the structural similarity index (SSIM) with the Gd-T_1w image for the ring enhancement region. The contrast maps also appear less noisy and more homogeneous over the whole brain compared to the Lorentzian fit on the interpolated-corrected B_1 CEST-spectra [14]. To further evaluate the performance of PIAE and CAE, we b1-corrected the data using CAE and fitted it with the least squares method (CAE-Lorentzian fit). The comparison of the CEST maps produced by the conventional Lorentzian fit, the CAE-Lorentzian fit, and PICAE is shown in Table 1 using SSIM and gradient cross correlation (GCC)

[11] for the Tumor ring region. Both the CAE-Lorentzian fit and PICAE were better than the conventional method. CAE-Lorentzian fit even outperformed PICAE for rNOE metabolic map and has similar performance for amide, but it has much lower performance for amine.

Table 1. Camparision of PICAE with conventional and CAE-Lorentzian fit. SSIM and GCC are calculated with respected to the Gd image for tumor region.

Fit method	amide		rNOE		amine	
	SSIM	GCC	SSIM	GCC	SSIM	GCC
Conventional Lorentzian fit	0.09	0.13	0.10	0.18	0.09	0.07
CAE-Lorentzian fit	0.11	0.14	0.18	0.3	0.10	0.18
PICAE	0.12	0.13	0.14	0.27	0.16	0.27

The ability of PICAE to produce B_1-robust CEST maps at arbitrary levels is shown in Fig. 4, where different B_1 levels reveal different features of the heterogenous tumor. Quantification of chemical exchange rates and concentration, i.e., $Z_1 = f{\cdot}k$, is shown in column 4. Z_1 (quantified maps) further improve the visualization of the ring enhancement area. Column 5 shows the Z_2 maps, which are combination of the exchange rate k and the relaxation rate r_2. Quantified maps of amide, rNOE and amine for another tumor patient is shown in supplementary Fig. 1. The accuracy of the CAE to generate particular B_1 CEST-spectra is depicted using absolute error for acquisition at different B_1 levels (see supplementary Fig. 2). The performance was lower for B_1 0.72 μT, and 1.5 μT compared to 1 μT.

4 Discussion

In this work, we analyzed the use of an autoencoder approach to generate B_1-robust CEST contrast maps at arbitrary B_1 levels, which requires multiple acquisitions in conventional methods [14]. The proposed method reduces the acquisition time by at least half when only two acquisitions are performed for B_1 correction. Supervised learning (DeepCEST) can generate CEST maps that are not susceptible to B_1 inhomogeneity at a particular B_1, which already reduces acquisition time. However, DeepCEST was trained on data fitted using a conventional pipeline [9,14] which has suboptimal B_1 correction (cf. Fig. 3). Moreover, the different pools in the CEST spectrum are highlighted at different B_1 levels (cf. Fig. 4). An approach that can generate a B_1-robust CEST-spectrum at multiple B_1 levels allows quantifying the exchange rate and concentration of the CEST pools [15]. The optimal B_1 can often only be selected at post-processing during the analysis of clinical data, as some clinically important features appear better at certain B_1 levels (cf. Fig. 4). The proposed PICAE approach combines B_1 correction and Lorentzian curve fitting in a single step. The B_1 correction

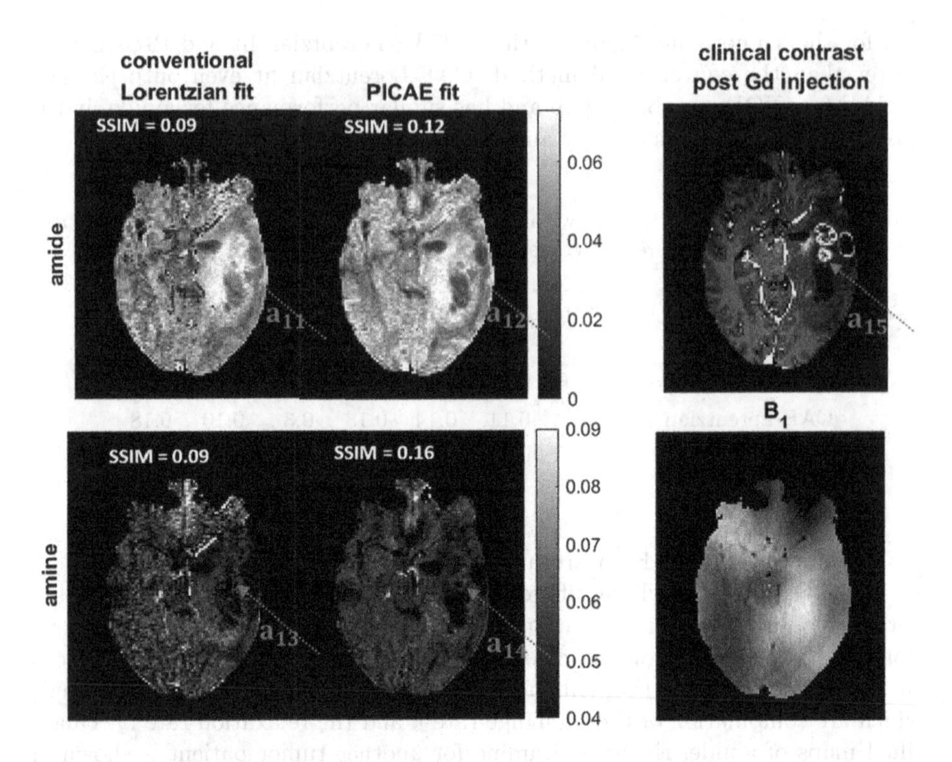

Fig. 3. Comparison of B_1 corrected CEST maps for conventional and PICAE methods (Columns 1 and 2). The top image in column 3 shows a reference Gd-T_1w image, while the bottom image shows the B_1 map. SSIM are calculated with respected to the Gd image for tumor region.

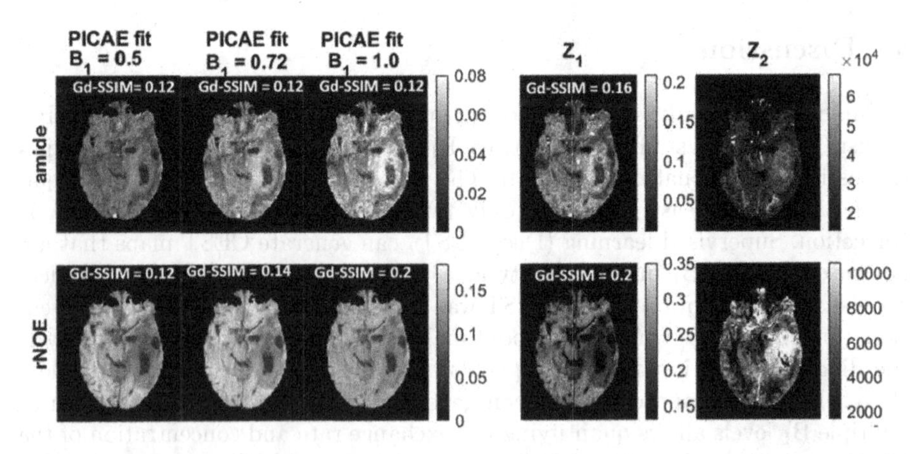

Fig. 4. B_1-robust CEST maps at 0.5, 0.72, and 1.0 μT (columns 1–3), quantified maps (column 4), and combination of exchange and relaxation rates (column 5). SSIM are calculated with respected to the Gd image for tumor region.

was performed with a CAE, while the Lorentzian line fitting was performed with a PIAE using NN as the encoder and Lorentzian distribution generator as the decoder. This allows interpretation of the model while overcoming the drawback of curve fitting, such as being prone to noise (cf. Fig. 3).

The bound loss ensured that the positions of the pools were not interchanged. Quantification was still performed using the nonlinear least squares fit according to Eq. 7. The main reason for this was that it does not affect the acquisition time and it is affected by the Z_{ref}. The training was performed voxel-wise to ensure that the results are based only on the CEST-spectrum. This also results in about 1 million CEST-spectra for the training. Figure 3 shows the superiority of PICAE over the standard method, and Fig. 4 shows that the results produced by PICAE are authentic because the quantification column Z_1 matches the amplitude images and follows Eq. 7. CAE-Lorentzian fitting showed comparable performance for amide and rNOE maps, but significantly lower performance for amine because it was still fitted using the least squares method, which is susceptible to noise in the input and takes up to 5 min to evaluate, compared to PICAE, which takes only a few seconds. The 1 μT acquisition performs better than 0.72 μT and 1.5 μT because it was trained for both lower and higher B_1 values compared to the other two acquisitions. The robustness of method for cyclic consistency [17] is displayed in supplementary Fig. 2, which also shows the interpretability of the method. The results of Fig. 3 and Fig. 4 also show the generalization capability of PICAE as it was trained without the tumor data.

5 Conclusion

In this work, we propose a PICAE method for evaluating 7T-CEST MRI that accounts for B_1 inhomogeneity in the input and predicts homogeneous metabolic CEST contrasts at arbitrary B_1 levels. The proposed generative and interpretable method enables (i) a reduction of scan time by at least 50%, (ii) the generation of reliable 7T-CEST contrast maps robust to B_1 inhomogeneity at multiple B_1 levels, (iii) a clear physical interpretation of the B_1 correction of the CEST-spectra and the fitting of the Lorentzian model to it, and (iv) the quantification of the CEST contrast maps.

References

1. Chen, L.Q., Pagel, M.D.: Evaluating pH in the extracellular tumor microenvironment using CEST MRI and other imaging methods. Adv. Radiol. **2015**, 25 (2015)
2. Dou, W., et al.: Chemical exchange saturation transfer magnetic resonance imaging and its main and potential applications in pre-clinical and clinical studies. Quant. Imaging Med. Surg. **9**(10), 1747–1766 (2019)
3. Glang, F., et al.: Linear projection-based CEST parameter estimation. NMR Biomed. (2022). https://doi.org/10.1002/nbm.4697
4. Hunger, L., et al.: Deepcest 7 t: fast and homogeneous mapping of 7 t CEST MRI parameters and their uncertainty quantification. Magn. Reson. Med. **89**, 1543–1556 (2023)

5. Lewerenz, J., Maher, P.: Chronic glutamate toxicity in neurodegenerative diseases-what is the evidence? Front. Neurosci. **9**, 469 (2015)
6. Maier, A.K., Schebesch, F., Syben, C., Würfl, T., Steidl, S., Choi, J.H., Fahrig, R.: Precision learning: Towards use of known operators in neural networks. In: 2018 24th International Conference on Pattern Recognition (ICPR), pp. 183–188 (2017)
7. Maier, A.K., et al.: Learning with known operators reduces maximum error bounds. Nat. Mach. Intell. **1**, 373–380 (2019)
8. Mcconnell, H.M.: Reaction rates by nuclear magnetic resonance. J. Chem. Phys. **28**, 430–431 (1958)
9. Mennecke, A., et al.: 7 tricks for 7 t CEST: improving reproducibility of multi-pool evaluation provides insights into effects of age and early stage parkinson's disease. NMR Biomed. **36**(6), e4717 (2022)
10. Moré, J.J., Sorensen, D.C.: Computing a trust region step. Siam J. Sci. Stat. Comput. **4**, 553–572 (1983)
11. Penney, G.P., Weese, J., Little, J.A., Desmedt, P., Hill, D.L.G., Hawkes, D.J.: A comparison of simularity measures for use in 2d–3d medical image registration. In: International Conference on Medical Image Computing and Computer-Assisted Intervention (1998)
12. Rosasco, L., de Vito, E., Caponnetto, A., Piana, M., Verri, A.: Are loss functions all the same? Neural Comput. **16**, 1063–1076 (2004)
13. Sohn, K., Lee, H., Yan, X.: Learning structured output representation using deep conditional generative models. In: NIPS (2015)
14. Windschuh, J., et al.: Correction of b1-inhomogeneities for relaxation-compensated CEST imaging at 7 t. NMR in Biomed. **28**, 529–537 (2015)
15. Zaiss, M., Jin, T., Kim, S.G., Gochberg, D.F.: Theory of chemical exchange saturation transfer MRI in the context of different magnetic fields. NMR Biomed. **35**, e4789 (2022)
16. Zhou, J., Payen, J.F., Wilson, D.A., Traystman, R.J., van Zijl, P.C.M.: Using the amide proton signals of intracellular proteins and peptides to detect PH effects in MRI. Nature Med. **9**, 1085–1090 (2003)
17. Zhu, J.Y., Park, T., Isola, P., Efros, A.A.: Unpaired image-to-image translation using cycle-consistent adversarial networks. In: 2017 IEEE International Conference on Computer Vision (ICCV), pp. 2242–2251 (2017)
18. van Zijl, P.C.M., Yadav, N.N.: Chemical exchange saturation transfer (CEST): what is in a name and what isn't? Magn. Reson. Med. **65**, 927–948 (2011)

A Coupled-Mechanisms Modelling Framework for Neurodegeneration

Tiantian He[1(✉)], Elinor Thompson[1], Anna Schroder[1], Neil P. Oxtoby[1], Ahmed Abdulaal[1], Frederik Barkhof[1,2,3], and Daniel C. Alexander[1]

[1] UCL Centre for Medical Image Computing, Department of Computer Science, University College London, London, UK
tiantian.he.20@ucl.ac.uk

[2] UCL Queen Square Institute of Neurology, University College London, London, UK

[3] Department of Radiology and Nuclear Medicine, Amsterdam University Medical Center, Vrije Universiteit, Amsterdam, The Netherlands

Abstract. Computational models of neurodegeneration aim to emulate the evolving pattern of pathology in the brain during neurodegenerative disease, such as Alzheimer's disease. Previous studies have made specific choices on the mechanisms of pathology production and diffusion, or assume that all the subjects lie on the same disease progression trajectory. However, the complexity and heterogeneity of neurodegenerative pathology suggests that multiple mechanisms may contribute synergistically with complex interactions, meanwhile the degree of contribution of each mechanism may vary among individuals. We thus put forward a coupled-mechanisms modelling framework which non-linearly combines the network-topology-informed pathology appearance with the process of pathology spreading within a dynamic modelling system. We account for the heterogeneity of disease by fitting the model at the individual level, allowing the epicenters and rate of progression to vary among subjects. We construct a Bayesian model selection framework to account for feature importance and parameter uncertainty. This provides a combination of mechanisms that best explains the observations for each individual. With the obtained distribution of mechanism importance for each subject, we are able to identify subgroups of patients sharing similar combinations of apparent mechanisms.

Keywords: Network spreading model · Disease progression · Bayesian model selection · Variational Inference

1 Introduction

Computational models of neurodegeneration aim to emulate the underlying physical process of how disease initiates and progresses over the brain from

E. Thompson and A. Schroder—Contributed equally to this work as the co-second authors.

Supplementary Information The online version contains supplementary material available at https://doi.org/10.1007/978-3-031-43993-3_45.

H. Greenspan et al. (Eds.): MICCAI 2023, LNCS 14227, pp. 459–469, 2023.
https://doi.org/10.1007/978-3-031-43993-3_45

a mechanistic point of view [6,10,25]. A better understanding of disease mechanisms and inter-individual variability through these models will aid in the development of new treatments and disease prevention strategies. Two types of components typically contribute to such models: models of pathology appearance - how and where pathology spontaneously appears in different brain regions; and models of pathology spreading - how pathology spreads from region to region. Both components are often linked to brain connectivity so that spontaneous appearance arises according to network topologies, and spreading is facilitated by network edges or brain-connectivity pathways.

Network metrics link spontaneous appearance of pathology to brain connectivity via various mechanistic hypotheses. For instance, Zhou et al. [27] relate disease patterns with several network topologies: i) centrality [3], indicating that regions with denser connections are more vulnerable to disease due to heavier nodal stress; ii) segregation [1], the converse of centrality, stating that regions with sparse connections are more susceptible due to lack of trophic sources; iii) shared vulnerability, expounding that connected regions have evenly distributed disease because they have common characteristics such as gene expressions [10].

For the spreading component, dynamical systems models emulate the spatio-temporal propagation process along the brain connectivity architecture [9,11,15,16,23]. One popular example is the network diffusion model [15] (NDM), which assumes that the pathology purely diffuses from epicenters. Weickenmeier et al. [25] combine the disease diffusion process with a local production term in a single model, which emulates the full process of how the protein diffuses from epicenters and replicates locally, gradually reaching a plateau. Thus, this model is able to reconstruct the process from disease onset to later stages.

However, such models make specific choices on the underlying mechanism in the particular physical process. The complexity and heterogeneity of neurodegenerative conditions suggests that multiple processes may contribute and vary among individuals. To avoid making such assumptions, Garbarino et al. [7] use a data-driven, linear combination of several network topological descriptors extracted from the structural connectome to match the disease patterns fitted by a Gaussian Process progression model. They find that the combination, which they refer to as a "mechanistic profile" better matches observed pathology patterns than any single characteristic. However, this considers appearance and spreading as interchangeable rather than interacting mechanisms. Secondly, although [7] does produce individual-level as well as group-level mechanistic profiles, the individual-level profiles assume that all the subjects lie on the same disease progression trajectory. This does not fully capture the heterogeneity (such as different epicenters [24], diffusion rate) within groups and underestimates the variability in composition of the mechanistic profile among subjects.

In this work, we introduce an alternative model framework that non-linearly couples the effects of spontaneous appearance and spreading. We construct a Bayesian framework with an appropriate sparsity structure to estimate the mechanistic profile, in a similar way to [7] but including interaction of model components and quantification of uncertainty. We account for the heterogeneity of neurodegenerative diseases in a more complex way than [7], by allowing factors like epicenters, rates of diffusion and production, and the weights of network metrics to vary among individuals. The resulting mechanistic profiles highlight distinct subgroups of individuals within an Alzheimer's disease (AD) cohort, in which each subgroup has similar combinations of network metrics.

2 Methodology

2.1 Model Definition

Baseline Model. One model of disease spreading [14,25] based on the Fisher-Kolmogorov equation, assumes that the concentration of toxic protein can be emulated by an ordinary differential equation (ODE) system, which involves the combination of two physical processes: 1) the diffusion of toxic proteins along the structural network from an epicentre(s), as described by the NDM [15]; 2) its local production and aggregation at each node. The diffusion component includes the graph Laplacian matrix $\mathbf{L}$ calculated from the structural connectome [15], as a substrate for disease diffusion with the rate k. The production and aggregation part provides a monotonically increasing trend converging to a plateau level v with a common speed α shared by all regions. The model evolves the pathology concentration $\mathbf{c}$ at time t according to:

$$\frac{d\mathbf{c}}{dt} = -k[\mathbf{L}\mathbf{c}(t)] + \alpha\mathbf{c}(t) \odot [v - \mathbf{c}(t)], \tag{1}$$

where $\odot$ denotes the element-wise product. It constructs the disease progression process from onset to late stage based on the single mechanism of network proximity. It assumes a constant local production rate across regions, thus the growth of concentration only depends on the level of biomarker propagating from the epicenter along the structural connectome to each node. This does not take into account the synergistic effect of other mechanisms.

Coupled Model. Following [25], we retain the network proximity mechanism for the diffusion component of our model, but we weight the local production process with the combination of P network metrics $\mathbf{M} = [\mathbf{m_1}, ..., \mathbf{m_P}]$. We use $\mathbf{w} = [w_1, ..., w_P]^T$ to represent the weighting, or extent of contribution of each characteristic. In contrast to the baseline model, by weighting the rate of production with the combination of network metrics, we further incorporate pathology appearance proportional to various brain network topologies to the disease spreading process. The new model expresses this coupling as follows:

$$\frac{d\mathbf{c}}{dt} = -k[\mathbf{L}\mathbf{c}(t)] + \alpha\mathbf{M}\mathbf{w} \odot \mathbf{c}(t) \odot [v - \mathbf{c}(t)]. \tag{2}$$

Network Metrics. The network metrics considered by the model are listed in Table 1 and a visualization is shown in Fig. 1. **StructureC**, **InvGeodist**, **FunctionC** represent the structural connectome, the matrix of the inverse of geodesic distance along the cortical surface and the functional connectome.

Table 1. Network metrics used as the interaction component with the spreading model

Metrics	formula	weights	mechanism
$\mathbf{m}_1$	$1 -$ **BetweennessCentrality(StructureC)**	w_1	Structural Segregation
$\mathbf{m}_2$	$1 -$ **ClusterCoefficient(StructureC)**	w_2	Structural Dispersion
$\mathbf{m}_3$	**ClosenessCentrality(StructureC)**	w_3	Structural Centrality
$\mathbf{m}_4$	**WeightedDegree(InvGeodist)**	w_4	Geodesic proximity
$\mathbf{m}_5$	**BetweennessCentrality(FunctionC)**	w_5	Functional centrality
$\mathbf{m}_6$	1	w_6	Even distribution

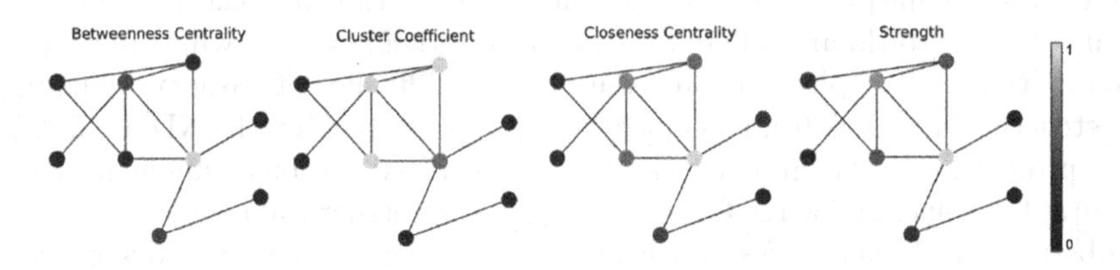

Fig. 1. Visualization of network metrics of the toy brain connectivity. BetweennessCentrality is the fraction of all shortest paths that contain a specific node. **ClusterCoefficient** quantifies the extent to which one node is clustered with its neighbours. **ClosenessCentrality** is proportional to the reciprocal of the sum of the shortest path length between the node and others. **WeightedDegree** is the sum of the weights of the connections. Calculation of metrics has been done using the Brain Connectivity Toolbox [18]. Metrics $\mathbf{m}_1$ to $\mathbf{m}_5$ have been normalized between [0,1] to maintain the same scale. Visualization is done using NetworkX (https://networkx.org/).

2.2 Bayesian Framework

In order to quantify the uncertainty of the estimation, we construct a Bayesian inference framework for our dynamic system, thus we are able to obtain distributions of parameters rather than deterministic values.

Parameter Distributions. In this work we focus on modelling the dynamics of tau protein, which is widely hypothesized to be a key causative agent in AD. Its concentration can be measured in vivo by positron emission tomography (PET). We assume for subject i at jth scan time t_{ij}, the measurement of tau concentration $\hat{\mathbf{c}}(t_{ij})$ follows a normal distribution

$$p(\hat{\mathbf{c}}(t_{ij}) \mid k_i, \alpha_i) \sim Normal\left(\mathbf{c}(t_{ij}, k_i, \alpha_i), \sigma^2\right), \tag{3}$$

where the mean is the model prediction with best-fit parameters and the error is quantified by the standard deviation σ. The time gap δ_{ij} (in years) between the baseline scan and the jth follow-up scan is available in the dataset. However the time from the disease onset to the baseline scan is unknown. Thus we need to estimate such time t_i^{onset} such that $t_{ij} = t_i^{onset} + \delta_{ij}$. This time parameterization enforces the relevant locations among all scans fixed by given δ_{ij}.

Descriptions of the key model parameters are displayed in Table 2. We enforce the rate of spreading and production to be positive by selecting a Half-Normal prior. The hyper-parameters of the rates are decided according to the research findings that the annual change rate of tau-PET signal is quite slight [12,20]. Explorations of the proper choice of the hyper-parameters has also been done by simulating different parameter levels and comparing the generated trajectories to the measured data distribution.

Table 2. Model parameters and their prior distribution[a]

Baseline Model	Coupled model	Interpretation	Prior distribution
k	k	Spreading rate	HalfNormal (std = 0.1)
α	α	Local production rate	HalfNormal (std = 0.1)
t_{onset}	t_{onset}	Pseudo onset time	Uniform (10,100)
σ	σ	Overall uncertainty	HalfNormal (std = 1)
	$\mathbf{w}$	Feature weights	Dirichlet ($\beta_1 \ldots \beta_P$)

[a] For ease of notation and clarity in the subsequent discussions, we drop the subject index i for the individual-level model parameters.

Feature Selection with Sparsity. We account for the feature importance by estimating the weights of network metrics

$$\mathbf{w} = [w_1, ..., w_P]^T \sim Dirichlet(\beta_1, \ldots, \beta_P).$$

We seek the minimal set of network components that explain the data to define the mechanistic profile. Thus, we apply sparsity to the weight in a Bayesian way by introducing the Horseshoe prior [4] to the hyper-parameters of the Dirichlet distribution:

$$\beta_l \mid \lambda_l, \tau \sim HalfNormal\left(0, \lambda_l^2 \tau^2\right), l = 1 \ldots P.$$

This horseshoe structure includes a global shrinkage parameter τ and local shrinkage parameters λ_l, each following a half Cauchy distribution:

$$\lambda_l, \tau \sim HalfCauchy(0, 1).$$

The flat tail from Cauchy Distribution allows the features with strong contribution to remain with a heavy tail of the density, while the sharp rise in density near 0 shrinks the weight of the features with weak signal.

2.3 Variational Inference

Suppose $\mathbf{x}$, $\mathbf{z}$ and θ represent the collections of observations, hidden variables and parameters respectively. Due to the complexity of the model structure, the posterior $p_\theta(\mathbf{z} \mid \mathbf{x})$ we are interested in is often intractable and hard to obtain analytically. Thus we use the variational distribution $q_\phi(\mathbf{z})$ with ϕ as the variational parameters to approximate the posterior. The evidence lower bound, $\text{ELBO} \equiv \text{E}_{q_\phi(\mathbf{z})}[\log p_\theta(\mathbf{x}, \mathbf{z}) - \log q_\phi(\mathbf{z})]$, can be used to approach the log likelihood, since the gap between them is the Kullback-Leibler divergence between the variational distribution and the posterior, which is larger or equal to 0:

$$\log p_\theta(\mathbf{x}) - \text{ELBO} = \text{KL}\left(\mathrm{q}_\phi(\mathbf{z}) \| \mathrm{p}_\theta(\mathbf{z} \mid \mathbf{x})\right) \geq 0. \tag{4}$$

Thus the objective of the optimization is to maximize the ELBO. We use a normal distribution with a diagonal covariance matrix as the variational distribution to sample the hidden variables in the latent space, and then use proper parameter transformation to obtain the constrained hidden variables. The process is accomplished with the use of Pyro [2], a probabilistic programming framework.

3 Experiments and Results

3.1 Data Processing

Brain Networks. Three types of connectivity were used to extract features for our coupled models: 1) the structural connectome, which contains the number of white matter fibre trajectories; 2) the matrix of the geodesic distance along the cortical surface; 3) the functional connectome, which reflects the synchrony of neural functioning among regions. The structural connectome is an average of 18 young and healthy subjects' connectomes from the Human Connectome Project [22]. We generated streamlines using probabilistic, anatomically constrained tractography, processed using MRtrix3 [21], and filtered the streamlines using the SIFT algorithm [19]. The geodesic distance matrix and the functional connectome are obtained from the Microstructure-Informed Connectomics Database [17], and is an average of 50 healthy subjects' matrices. We define the brain regions according to the Desikan-Killiany Atlas [5].

Tau-PET Data. We model the dynamics of tau protein measured by PET scans. We use the tau-PET standardized uptake value ratios (SUVRs) downloaded from the Alzheimer's Disease Neuroimaging Initiative (ADNI)[1] (adni.loni.usc.edu) [13]. We exclude subcortical regions, which are impacted by off-target binding of the radiotracer [8]. Two-component Gaussian mixture modelling is applied to the SUVR signals of all subjects. We treat the distribution with the lower mean as the distribution of negative signals, and define the mean plus one standard deviation of this distribution as the threshold for tau-positivity.

[1] Data used in preparation of this article were obtained from the Alzheimer's Disease Neuroimaging Initiative (ADNI) database (adni.loni.usc.edu). As such, the investigators within the ADNI contributed to the design and implementation of ADNI and/or provided data but did not participate in analysis or writing of this report. A complete listing of ADNI investigators can be found at: http://adni.loni.usc.edu/wp-content/uploads/how_to_apply/ADNI_Acknowledgement_List.pdf

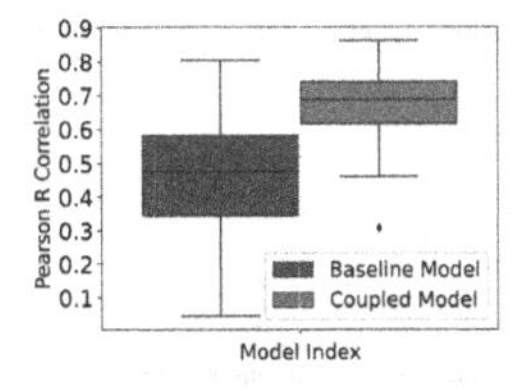

Fig. 2. Comparison of the overall fitting among all subjects. This boxplot visualizes the distribution of Pearson R correlations (between model fitting and the measured value) of all subjects fitted by the baseline and our coupled model respectively.

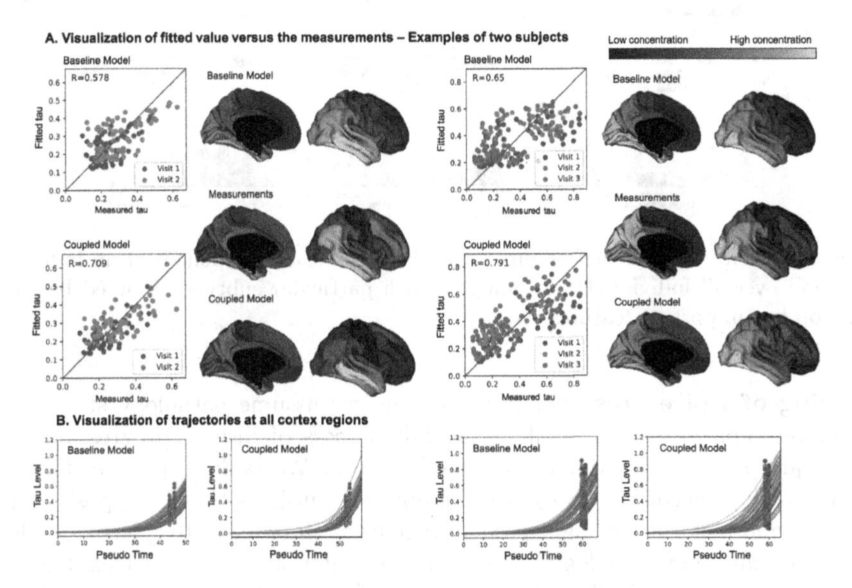

Fig. 3. Model comparison at individual level. We display examples from two subjects by comparing their fitting from the baseline model and the coupled model. A: scatter plots which show the model-fitted signals versus the measured signals and the visualization of signals on the brain. Each point in the scatter plots represents the signal in one region. B: visualizations of the individual-level trajectory. Each line represents the modelled trajectory for a region while each point represents the actual signal in that region at each visit. Improvement can be observed by the coupled model.

Selection of Subjects. We include N = 110 subjects with at least two tau-PET scans, amyloid beta positive status and at least one region with positive tau signal, including healthy, cognitively impaired and AD subjects, since we aim to focus on the people with a potential to accumulate abnormal pathological tangles. We normalise the data of all the subjects (i = 1,..,N) between 0 and 1 by $(tau_i - tau_{min})/(tau_{max} - tau_{min})$ where tau_{min} and tau_{max} are calculated across all the subjects and regions, thus the differences in data scales among subjects and regions are maintained.

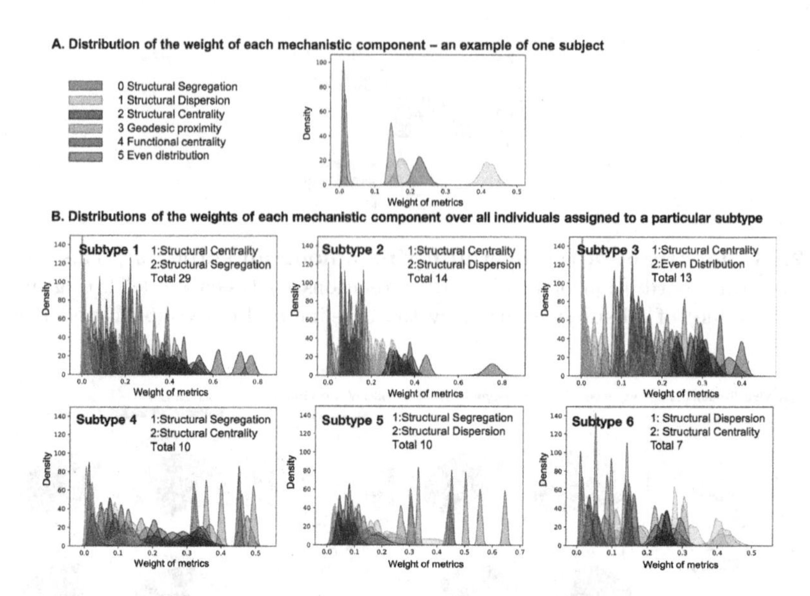

Fig. 4. Distributions of the weights for each mechanistic component of an individual (A) and over all individuals belonging to each particular subtype obtained based on the top two important features (B).

Setting of Epicentres. For initialization, we assume pathology starts from candidate epicentres, to simulate the full process of disease progression from very early stages, i.e. prior to the baseline scan. We rank 34 pairs of bilateral cortex regions according to the total number of subjects that have positive tau signals, and pick the top eight pairs of regions as the candidate epicentres where the propagation of pathology is likely to start: inferior temporal cortex, banks of the superior temporal sulcus, fusiform gyrus, lateral orbitofrontal cortex, middle temporal gyrus, entorhinal cortex, parahippocampal gyrus and temporal pole. The four epicentres identified by Vogel et al. [24] are all included.

3.2 Results

We fix the initial tau level at each candidate epicenter and the end level of the plateau of all the subjects to be 0.01 and 1.5 and fit each subject using the baseline model and our coupled model respectively. For evaluation we use Pearson R correlation between the measured and model fitted values.

Figure 2 compares the distribution of R correlations which reflect the performance of the baseline model and our coupled model on each individual. According to the one-side t test, the R correlations from the coupled model are significantly larger than those from the baseline model (p-value = 3.5732e−22). Especially, fitting of subjects with particularly low performance in the baseline model are noticeably improved.

Figure 3 visualizes the model-fitted tau signals versus measured tau signals in all 68 cortex regions. The distribution of tau fitted by our coupled model is closer to the measured pattern. Finally, we subtype the individuals according to the top two most dominant mechanisms of pathology appearance interacting with the network spreading model. Specifically, we encode each subject with a vector containing the rank of each metric according to the obtained weights, and assign the subjects having the same rank of the top two metrics to the same group. We consider a group containing at least 6 people (5% of all) as one subtype. As a result, 83 out of 110 subjects have been assigned to six subtypes. Figure 4A displays the feature importance distribution for an individual, while Fig. 4B places the feature distributions of the subjects belonging to the same subtype within each of the six plots. It can be observed that structural centrality appears most frequently as a dominant feature, followed by structural segregation.

4 Conclusions

We introduce a new Bayesian modelling framework that couples pathology appearance and spreading, by embedding the mechanistic profiles that consist of combinations of network metrics into the dynamic system of disease spreading. This improves the fitting of the observed pathology pattern, and provides a potential way to subtype subjects according to mechanistic profiles. For future work, we will validate the cohort-level mechanistic profiles derived from the identified subtypes using external datasets, and also verify the subtypes using other algorithms such as the SuStaIn [26]. Furthermore, we will incorporate uncertainty from connectomes. We will also perform further comparisons with other state-of-the-art models, such as the topological profiles by [7], which is currently hard to compare directly due to various differences in the model design.

Acknowledgment. TH, AS and AA are supported by the EPSRC funded UCL Centre for Doctoral Training in Intelligent, Integrated Imaging in Healthcare[EP/S021930/1]; TH, ET and DCA are supported by the Wellcome Trust(221915); DCA and FB are supported by the NIHR Biomedical Research Centre at UCLH and UCL; NPO acknowledges funding from a UKRI Future Leaders Fellowship(MR/S03546X/1).

References

1. Appel, S.H.: A unifying hypothesis for the cause of amyotrophic lateral sclerosis, parkinsonism, and Alzheimer disease. Ann. Neurol. J. Am. Neurol. Assoc. Child Neurol. Soc. **10**(6), 499–505 (1981)
2. Bingham, E., et al.: Pyro: deep universal probabilistic programming. J. Mach. Learn. Res. **20**(1), 973–978 (2019)
3. Buckner, R.L., et al.: Molecular, structural, and functional characterization of Alzheimer's disease: evidence for a relationship between default activity, amyloid, and memory. J. Neurosci. **25**(34), 7709 (2005). https://doi.org/10.1523/JNEUROSCI.2177-05.2005
4. Carvalho, C.M., Polson, N.G., Scott, J.G.: Handling sparsity via the horseshoe. In: Artificial Intelligence and Statistics, pp. 73–80. PMLR (2009)

5. Desikan, R.S., et al.: An automated labeling system for subdividing the human cerebral cortex on MRI scans into gyral based regions of interest. NeuroImage **31**(3), 968–980 (2006).https://doi.org/10.1016/J.NEUROIMAGE.2006.01.021
6. Garbarino, S., Lorenzi, M.: Investigating hypotheses of neurodegeneration by learning dynamical systems of protein propagation in the brain. NeuroImage **235**, 117980 (2021). https://doi.org/10.1016/j.neuroimage.2021.117980
7. Garbarino, S., et al.: Differences in topological progression profile among neurodegenerative diseases from imaging data. eLife **8**, e49298 (2019). https://doi.org/10.7554/eLife.49298
8. Groot, C., Villeneuve, S., Smith, R., Hansson, O., Ossenkoppele, R.: Tau PET imaging in neurodegenerative disorders. J. Nucl. Med. **63**(Supplement 1), 20S-26S (2022). https://doi.org/10.2967/jnumed.121.263196
9. Iturria-Medina, Y., Carbonell, F.M., Evans, A.C., Initiative, A.D.N.: Multimodal imaging-based therapeutic fingerprints for optimizing personalized interventions: application to neurodegeneration. Neuroimage **179**, 40–50 (2018)
10. Iturria-Medina, Y., Carbonell, F.M., Sotero, R.C., Chouinard-Decorte, F., Evans, A.C., Initiative, A.D.N.: Multifactorial causal model of brain (DIS) organization and therapeutic intervention: application to Alzheimer's disease. Neuroimage **152**, 60–77 (2017)
11. Iturria-Medina, Y., Sotero, R.C., Toussaint, P.J., Evans, A.C., Initiative, A.D.N.: Epidemic spreading model to characterize misfolded proteins propagation in aging and associated neurodegenerative disorders. PLoS Comput. Biol. **10**(11), e1003956 (2014)
12. Jack, C.R., Jr., et al.: Predicting future rates of tau accumulation on PET. Brain **143**(10), 3136–3150 (2020)
13. Landau, S., Jagust, W.: Flortaucipir (AV-1451) processing methods. Alzheimer's Disease Neuroimaging Initiative (2016)
14. Meisl, G., et al.: In vivo rate-determining steps of tau seed accumulation in Alzheimer's disease. Sci. Adv. **7**, eabh1448 (2021)
15. Raj, A., Kuceyeski, A., Weiner, M.: A Network diffusion model of disease progression in dementia. Neuron **73**(6), 1204–1215 (2012). https://doi.org/10.1016/j.neuron.2011.12.040
16. Raj, A., LoCastro, E., Kuceyeski, A., Tosun, D., Relkin, N., Weiner, M.: Network diffusion model of progression predicts longitudinal patterns of atrophy and metabolism in Alzheimer's disease. Cell Rep. **10**(3), 359–369 (2015)
17. Royer, J., et al.: An open MRI dataset for multiscale neuroscience. Sci. Data 9(1), 569 (2022). https://doi.org/10.1038/s41597-022-01682-y
18. Rubinov, M., Sporns, O.: Complex network measures of brain connectivity: uses and interpretations. Neuroimage **52**(3), 1059–1069 (2010)
19. Smith, R.E., Tournier, J.D., Calamante, F., Connelly, A.: SIFT2: enabling dense quantitative assessment of brain white matter connectivity using streamlines tractography. Neuroimage **119**, 338–351 (2015). https://doi.org/10.1016/j.neuroimage.2015.06.092
20. Smith, R., et al.: The accumulation rate of tau aggregates is higher in females and younger amyloid-positive subjects. Brain **143**(12), 3805–3815 (2020)
21. Tournier, J.D., et al.: MRtrix3: a fast, flexible and open software framework for medical image processing and visualisation. Neuroimage **202**, 116137 (2019). https://doi.org/10.1016/j.neuroimage.2019.116137

22. Van Essen, D.C., Smith, S.M., Barch, D.M., Behrens, T.E.J., Yacoub, E., Ugurbil, K.: WU-Minn HCP consortium: the WU-Minn human connectome project: an overview. Neuroimage **80**, 62–79 (2013). https://doi.org/10.1016/j.neuroimage.2013.05.041
23. Vogel, J.W., et al.: Spread of pathological tau proteins through communicating neurons in human Alzheimer's disease. Nat. Commun. **11**(1), 2612 (2020). https://doi.org/10.1038/s41467-020-15701-2
24. Vogel, J.W., et al.: Four distinct trajectories of tau deposition identified in Alzheimer's disease. Nat. Med. **27**(5), 871–881 (2021). https://doi.org/10.1038/s41591-021-01309-6
25. Weickenmeier, J., Kuhl, E., Goriely, A.: Multiphysics of Prionlike diseases: progression and atrophy. Phys. Rev. Lett. **121**(15), 158101 (2018). https://doi.org/10.1103/PhysRevLett.121.158101
26. Young, A.L., et al.: Uncovering the heterogeneity and temporal complexity of neurodegenerative diseases with subtype and stage inference. Nat. Commun. **9**(1), 4273 (2018)
27. Zhou, J., Gennatas, E.D., Kramer, J.H., Miller, B.L., Seeley, W.W.: Predicting regional neurodegeneration from the healthy brain functional connectome. Neuron **73**(6), 1216–1227 (2012). https://doi.org/10.1016/j.neuron.2012.03.004

A Texture Neural Network to Predict the Abnormal Brachial Plexus from Routine Magnetic Resonance Imaging

Weiguo Cao, Benjamin Howe, Nicholas Rhodes, Sumana Ramanathan, Panagiotis Korfiatis, Kimberly Amrami, Robert Spinner, and Timothy Kline(✉)

Mayo Clinic, Rochester, Minnesota 55905, USA
kline.timothy@mayo.edu

Abstract. Brachial plexopathy is a form of peripheral neuropathy, which occurs when there is damage to the brachial plexus (BP). However, the diagnosis of breast cancer related BP from radiological imaging is still a great challenge. This paper proposes a texture pattern based convolutional neural network, called TPPNet, to carry out abnormal prediction of BP from multiple routine magnetic resonance image (MRI) pulse sequences, i.e. T2, T1, and T1 post-gadolinium contrast administration. Different from classic CNNs, the input of the proposed TPPNet is multiple texture patterns instead of images. This allows for direct integration of radiomic (i.e. texture) features into the classification models. Beyond conventional radiomic features, we also developed a new family of texture patterns, called triple point patterns (TPPs), to extract huge number of texture patterns as representations of BP' heterogeneity from its MRIs. These texture patterns share the same size and show very stable properties under several geometric transformations. Then, the TPPNet is proposed to carry out the differentiation task of abnormal BP for our study. It has several special characteristics including 1) avoidance of image augmentation, 2) huge number of channels, 3) simple end-to-end architecture, 4) free from the interference of multi-texture-pattern arrangements. Ablation study and comparisons demonstrate that the proposed TPPNet yields outstanding performances with the accuracies of 96.1%, 93.5% and 93.6% over T2, T1 and post-gadolinium sequences which exceed at least 1.3%, 5.3% and 3.4% over state-of-the-art methods for classification of normal vs. abnormal brachial plexus.

Keywords: Brachial Plexus · Magnetic Resonance Imaging · Heterogeneity · Radiomics · Deep learning · Convolutional Neural Networks · Texture pattern

1 Introduction

Brachial plexopathy is a form of peripheral neuropathy [1]. It occurs when there is damage to the brachial plexus (BP) which is a complex nerve network under the skin of the shoulder. There is a wide range of disease that may cause a brachial plexopathy.

Supplementary Information The online version contains supplementary material available at https://doi.org/10.1007/978-3-031-43993-3_46.

H. Greenspan et al. (Eds.): MICCAI 2023, LNCS 14227, pp. 470–480, 2023.
https://doi.org/10.1007/978-3-031-43993-3_46

Radiation fibrosis, primary and metastatic lung cancer, and metastatic breast cancer account for almost three-fourths of causes [2]. Brachial plexus syndrome occurs not infrequently in patients with malignant disease. It is due to compression or direct invasion of the nerves by tumor which will bring many serious symptoms [3]. Our research focuses on the brachial plexopathy caused by metastatic breast cancers.

Magnetic resonance imaging (MRI) and ultrasound of the brachial plexus have become two reliable diagnostic tools for brachial plexopathy [4]. Automatic identification of the BP in MRI and ultrasound images has become a hot topic. Currently, most of relevant research in this field are focusing on Ultrasound modality [5–8]. Compared with ultrasound, MRI has become the primary imaging technique in the evaluation of brachial plexus pathology [9]. However, to our knowledge, radiomics related BP studies utilizing MRI have not been reported previously.

Many radiomics studies have experimentally demonstrated that image texture has great potential for differentiation of different tissue types and pathologies [10]. In the past several decades, many state-of-the-art methods have been proposed to extract texture patterns [11, 12]. However, how to most effectively combine texture features with deep learning, called deep texture, is still an open area of research. One prior approach, termed GLCM-CNN, was proposed to carry out a polyp differentiation task [13]. However, how to arrange these GLCMs to form the 3D volume to optimize the performance is a major challenge.

With the goal of classifying normal from abnormal BP, we explored the approach of deep texture learning. This paper constructed a BP dataset with the most commonly used BP MRIs in our clinical practice. Considering the shortcoming of traditional patterns, triple point pattern (TPP) is proposed for the quantitative representation of the heterogeneity of abnormal BP's. In contrast to GLCM-CNN, TPPNet is designed to train models by feeding TPP matrices as the input with a huge number of channels. Finally, we analyze the model's performance in the experimental section. The major contributions of this study include 1) directed triangle construction idea for TPP, 2) huge number of TPP matrices as the heterogeneity representations of BP, 3) TPPNet with 15 layers and huge number of channels, 4) the BP dataset containing MR images and their corresponding ROI masks.

2 Materials and Method

2.1 Dataset Preparation and Preprocessing

Following IRB approval for this study, we search for patients with metastatic breast cancer who had a breast cancer MRI performed between 2010 and 2020 and had morphologically positive BP on the MRI report from our electronic medical records (EMR) in * hospital. Totally, 189 patients including 141 normal patients and 41 abnormal ones are obtained. The range of the age are varying from 15 to 85 years old. The female patient number and male patient number are almost even. Their weights by kg are in the range of [40.8 kg, 145 kg]. All patient experiences three kinds of MRI sequences including T2, T1 and post-gadolinium. Approximately 25% sequences are conducted with GE SIGNA HDI 1.5T and the rest are produced by GE SIGNA™ Hero 3.0T. Patients are required to maintain a decubitus position while scanning. All series are scanned in sagittal view.

Slice thicknesses are varying in [1.2 mm, 6.0 mm] and approximate 96% of them share the slice thickness of 4mm. Their slice number is in the range of [38, 72]. Most of slice resolutions are 512 * 512 while less than 2% of them are 256 * 256.

Totally, we collect approximate 807 series which include 274 T2, 254 T1 and 279 Post-gadolinium. Since some scans are seriously degraded due to motion artifacts. Therefore, each case underwent several essential image adjustments such as multi-series splitting, two-series merging, slice swapping, artifact checking and boundary corrections.

Table 1. The final BP dataset including MRIs and their corresponding masks of T2, T1, Post-gadolinium (Pg).

Image	Total	Normal			Abnormal		
		T2	T1	Pg	T2	T1	Pg
MRI	462	123	123	123	31	31	31
Mask	462	123	123	123	31	31	31

To yield the ROI, firstly, we randomly sampled −40% of the sequences including both normal and abnormal ones that were manually segmented with ITK- snap by two skilled trainees [14, 15]. Then, the manual segmentations were utilized to train a 3D nnUNet model which was utilized to train the model which was used to predict ROIs for the rest series [16]. The predicted segmentations were manually divided into three groups, i.e. good, fair and poor. Good cases were added to the training set. This process was repeated until no improvements in the predictions for the remaining sequences was seen. The final dataset for radiomic analysis was constructed by merging the datasets for each sequence type. Only patients that had all three sequences segmented (T2, T1 and Post-gadolinium) were included in the dataset. Table 1 shows a breakdown of the final dataset.

2.2 Triple Point Pattern (TPP)

Theoretically, some texture pattern methods such as LBP, LTP, and GLDM, are based on single-variance pixel functions [17–19]. Therefore, they extract local texture features coded by the difference or difference counts between the concerned pixel and its neighboring ones. One obvious shortcoming is the absence of global properties which need other statistical methods as the aid to yield, such as histogram and invariants [20, 21]. Meanwhile, some other texture methods are generally defined by two-variance functions that only focus on two-variance patterns, such as (pixel, pixel), (pixel, neighbor count) [22–24]. In general, image textures extracted by these methods contain both local texture properties and global texture information. Their shortcomings might come from pattern shapes which might lead to the overfitting risk while combining with deep learning since the yielded texture matrix might have slim shapes or adaptive columns. In summary, as the requirement of the image texture and deep learning, ***an excellent image texture pattern should have some essential features including*** 1) local properties to characterize the micro-unit of the image texture, 2) global properties to represent the macro-structure of the image texture, 3) uniform shapes under nonuniform-shape images, 4) invariant or robustness under some common geometric transforms such as rotation, scaling and so on.

According above requirements, we developed a method to produce a serial of novel texture patterns by introducing a directed triangle idea with an adjacent triple pixel as a ternary group, called triple point pattern (TPP), to extract the local texture information. Then, a statistical method like histogram is employed to count the number of the same type of pixel-triplets within the ROI or throughout the whole image. Finally, a three-dimensional (3D) TPP matrix is formed to characterize the image texture globally as the following:

Two-dimensional image:

$$TPP_{(p_i,p_c,p_j)}(x,y,z) = \sum_{m=0}^{M-1}\sum_{j=0}^{N-1}\begin{cases} & I((m,n)+p_i) = x\& \\ 1 & I((m,n)+p_c) = y\& \\ & I\big((m,n)+p_j\big) = z \\ 0 & others \end{cases} \tag{1}$$

where I is a MxN image, x, y, and z is the pixel triplet, x,y,z $\in [0,L)$, L is its gray level, $p_c = (0,0)$ denotes the concerned pixel such as p_0 in Fig. 1, p_i and p_j are p_c's two adjacent pixels, $i,j \in [1,H]$, H is the number of p_c's adjacent points.

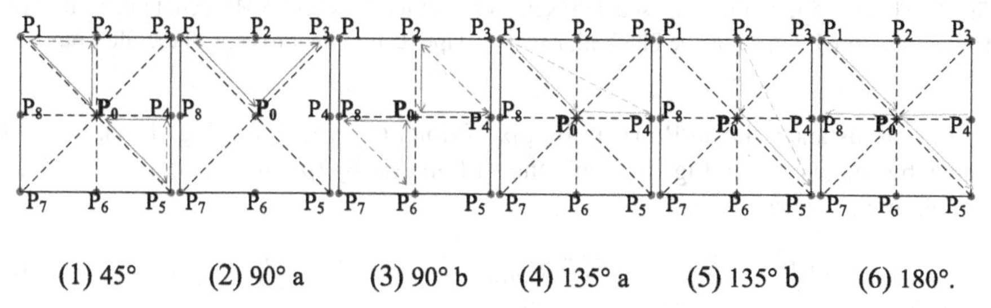

Fig. 1. Directed triangle idea for TPP construction in 2D images where p0 is the concerned pixel, p1 …p8 are its adjacent pixels. Totally, they are four modes according to four concerned angles, i.e. 45°, 90°, 135° and 180°. (1) is the 1st mode. Both (2) and (3) are the 2nd mode.

Three-dimensional image:

$$TPP_{(p_i,p_c,p_j)}(x,y,z) = \sum_{m=0}^{M-1}\sum_{n=0}^{N-1}\sum_{k=0}^{K-1}\begin{cases} & I((m,n,k)+p_i) = x\& \\ 1 & I((m,n,k)+p_c) = y\& \\ & I\big((m,n,k)+p_j\big) = z \\ 0 & others \end{cases} \tag{2}$$

where I is a three dimensional image with the shape of MxNxK, $p_c = (0,0,0)$, other parameters are similar to the Two-dimensional image.

In statistics, a TPP matrix should a 3D distribution of directed triangle. As shown in Fig. 1, the TPP is formed by the concerned pixel and its two adjacent pixels in two-dimensional(2D) images. Similarly, the TPP in 3D images is constructed by one concerned voxel and its two neighboring voxels. More details could be found in the supplementary material. As the construction idea of TPP, there are four independent modes categorized by the concerned angle, i.e. 45°, 90°, 135° and 180° in 2D images, which produce 8 TPPs, 8 TPPs, 8 TPPs and 8 TPPs respectively. Analogously, the 3D image has twelve independent angle modes, i.e. 35.26°, 45°, 54.74°, 60°, 70.53°, 90°, 109.47°, 120°, 125.26°, 135°, 144.74°, and 180°. These angle modes could generate 24 TPPs, 24 TPPs, 24 TPPs, 24 TPPs, 12 TPPs, 96 TPPs, 12 TPPs, 24 TPPs, 24 TPPs, 24 TPPs, 24 TPPs and 13 TPPs respectively. Totally there are 32 TPPs in 2D images and 325 TPPs in 3D images and every TPP could produce one corresponding TPP matrix.

By further analysis, we could find some TPP pairs have an isomorphism relationship since its TPP matrix could be generated by transposing or flipping another TPP matrix on some certain conditions when two triangles formed by the pixel triplet have the relevance of shifting or scaling.

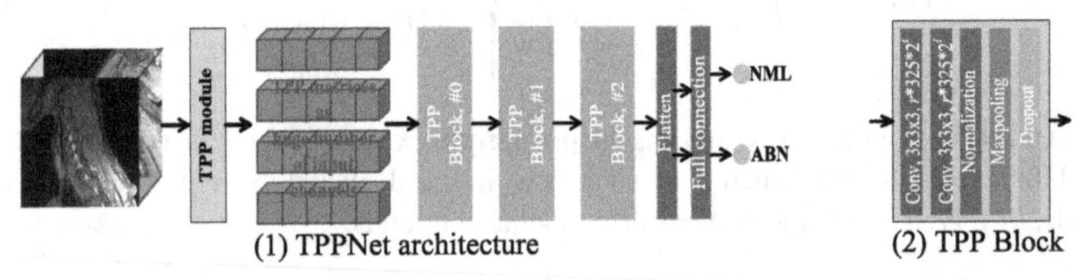

(1) TPPNet architecture (2) TPP Block

Fig. 2. The pipeline of the proposed TPPNet over 3D images where NML denote normal, ABN denotes abnormal, i in (2) is the block id, r is the adaptive argument to control the filter number.

For an instance, the TPP matrix by pixel-triplet (p_1, p_0, p_2) in Fig. 1 (1), the TPP matrix by (p_4, p_0, p_5) in Fig. 1 (1) and the TPP matrix by (p_6, p_0, p_8) in Fig. 1(3) have the following relevance:

$$TPP_{(p_1,p_0,p_2)} = T{*}F\left(TPP_{(p_4,p_0,p_5)}\right) = T{*}F\left(TPP_{(p_6,p_0,p_8)}\right) \tag{3}$$

where T denotes matrix transposing, F represents matrix flipping, * is the product operator. Since both T and F are continuous bijective mappings, these three TPPs are isomorphic.

In our study, these isomorphic TPP matrices are not dropped from the TPP matrix set because they are equivalent to image rotations and re-scaling. Image scaling can result in the image pixels increasing. The normalization could almost remove the effect of pixel increase caused by scaling transformations. Moreover, TPPs generated by 135° could also be treated as affine transformations. Therefore, data augmentation could be omitted when we combine TPP with deep learning for this study. As its definition, the TPP matrix should be a cubic array with the shape of $LxLxL$ where L is the gray level of the image.

2.3 TPPNet

The pipeline of the proposed method TPPNet is illustrated in **Fig. 2**. The TPP matrix calculation is the preprocessing module which feeds MRI and its ROI and yield TPP matrix set. The following step is the TPPNet architecture to yield training models. Based on the construction idea of TPP, the size of the TPP matrix depends on the gray level of the image. For the same image or ROI, the larger the gray level, the sparser the matrix will be. The sparse matrix would lead to overfitting while training the model. Therefore, the image requires a re-scaling step to lower its gray level to avoid the sparsity of the TPP matrix. Consequently, our proposed TPPNet only contains three convolution blocks consisting of 15 layers. Each block has two convolution, one normalization, one max-pooling and one dropout layer. ***It has four particular features as follows:***

1) Avoidance of image augmentation. Due to the stability of TPP matrix under rotation, scale and affine transformations, image augmentation could be omitted in the preprocessing step which can lead to image deformation.
2) Huge number of channels. TPPNet treats each TPP as an independent channel. For 2D images, there are at least 32 channels if more displacements of TPP is considered. Similarly, we could generate no less than 325 TPPs in 3D images.
3) Simple end-to-end architecture. We integrate the k-fold cross-validation, TPP generation and model training into one framework. Since the TPP matrix is always small, there are only 15 layers in TPP which could reduce the risk of overfitting issue met in deeper neural networks.
4) Free from the interference of multi-texture-pattern arrangements. Since each channel is corresponding with one TPP, it can solve the pattern arrangement issue occurred in GLCM-CNN.

Table 2. Test performances for different gray levels over T2, T1 and Post-gadolinium where Acc denotes accuracy and Pg denotes Post-gadolinium.

Gray level	T2		T1		Pg	
	Acc	Loss	Acc	Loss	Acc	Loss
8	0.942±0.032	0.319±0.028	0.902±0.022	0.395±0.207	0.929±0.023	0.292±0.087
12	**0.961±0.025**	**0.277±0.107**	**0.935±0.021**	0.307±0.164	0.936±0.027	**0.279±0.046**
16	0.948±0.034	0.350±0.333	0.922±0.028	**0.306±0.107**	0.921±0.041	0.313±0.053
20	0.948±0.026	0.342±0.126	0.922±0.035	0.363±0.098	**0.942±0.034**	0.302±0.124
24	0.948±0.034	0.348±0.162	0.929±0.023	0.309±0.067	0.922±0.046	0.322±0.176

3 Experiments

3.1 Preparations

Some important specificities of our computing platform contain: one AMD EPYC 7352 24-Core Processor, 1 TB memory and four Nivida A100-SXM GPUs with 320 GB GPU memory. The whole dataset is divided into three subsets according to the MR sequence, i.e. T2, T1 and post-gad. For each subset, both normal cases and abnormal cases were randomly and evenly split into five subgroups. A five-fold cross-validation scheme was employed to generate five cohorts. Totally, 15 cohorts were produced. Each cohort consists of training set, validation set and testing set by the ratio of 6:2:2.

3.2 Ablation Studies

Since all images in our dataset are 3D images, therefore, the initial channel is set 325 which is equal to the TPP number. The loss functions in the following experiments shared categorical_crossentropy. Nadam is adopted as the optimizer with the learning rate of 0.0001 and batch size of 8 for 200 epochs. All performances listed in this section are the average of performances with 5-fold cross-validation.

3.2.1 Impact of Gray Level

The image gray level determines the shape of the TPP matrix. To avoid its sparsity, we compute the TPP matrix set via Eq. (2) with gray levels of 8, 12, 16, 20, 24. While rescaling the image intensity, an arc tangent approach is utilized to yield the new image. The models are trained and tested over 15 cohorts. Their performances evaluated by accuracies are listed in Table2 which tells us that T2 sequence yields the highest accuracy of 96.1% when the gray level is 12. T1 and post-gadolinium also get acceptable results with the accuracies of 93.5% and 93.6% respectively. Other performances could be read in supplementary materials.

Table 3. Test performances for three intensity rescaling approaches over T2, T1 and Post-gadolinium where the gray level is set 12, Acc denotes accuracy, Pg denotes Post -gadolinium, and Artan is arc tangent function.

Rescaling approach	T2		T1		Pg	
	Acc	Loss	Acc	Loss	Acc	Loss
Min-max	0.947±0.027	0.303±0.082	0.928±0.026	0.312±0.138	0.920±0.028	0.475±0.327
Adaptive	0.921±0.035	0.385±0.158	0.902±0.022	0.462±0.092	0.856±0.020	0.517±0.092
Artan	**0.961±0.025**	**0.277±0.107**	**0.935±0.021**	**0.307±0.164**	**0.936±0.027**	**0.279±0.046**

Table 4. Test performances comparison between multi-channel and solo-channel over T2, T1 and Post-gadolinium where the intensity rescaling function is arc tangent, the gray level is set 20, Acc denotes accuracy and Pg denotes Post -gadolinium.

Input channel	T2		T1		Pg	
	Acc	Loss	Acc	Loss	Acc	Loss
325	**0.948±0.026**	0.342±0.126	**0.922±0.035**	**0.363±0.098**	**0.934±0.034**	**0.302±0.124**
1	0.863±0.027	**0.274±0.091**	0.810±0.145	0.458±0.105	0.811±0.047	0.590±0.107

3.2.2 Impact of Intensity Rescaling Approaches

Rescaling approaches of image intensity could also bring impacts on the BP's differentiation while producing the TPP matrix. The commonly used methods include min-max-linear approach [25], arc tangent approach [26], and adaptive rescaling approach [27]. To test the performances fairly, we test above rescaling methods at the same gray level 12. The yielded performances evaluated by accuracies are shown in Table 3 where arc tangent method achieves the highest accuracy of 96.1% over the T2 sequence. Other performances are shown in supplementary materials.

3.2.3 Multi-Channels vs Solo-Channel

We carry out experiments to train the TPPNet model and make tests with arc tangent rescaling approach under gray level 16. As a comparison, we test single channel mode as well. By sharing every TPP matrix's label with the original case label, our TPPNet works well by assigning one channel for the initial input. Once the trained model with solo channel is generated, all TPP matrices of the testing set could be tested. Hereafter, we adopted a voting method to determine if the prediction is normal, if the predicted probability is less than 0.5, otherwise, it is considered abnormal. Performances with accuracies and loss are listed in Table 4. Other performances are listed in supplementary materials.

3.3 Comparisons

We evaluated our proposed TPPNet by comparing it to the recent state-of-the-art approaches over our BP dataset including VGG16 [28], InceptionNet [29], MobileNet [30], GLCM-CNN [13], ViT [31]. The GLCM size of 32 × 32 is used in

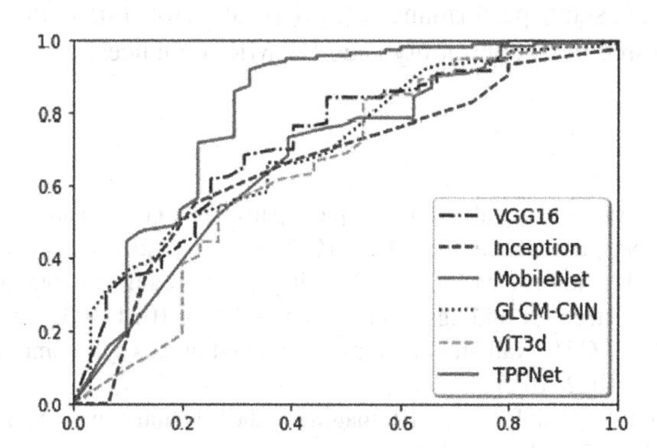

Fig. 3. ROC curves of TPPNet and five state-of-the-art approaches.

GLCM_CNN. All approaches shared the same image shape of 128 × 128 × 64 with 1 channel. The patch shape of ViT is 8 × 8 × 8, projection dim is 64, attention head number is set 4 with 8 transformer layers. The intensity rescaling step adopts the arc tangent approach. Other parameters are similar to the ablation study. Their performances are

Table 5. Performance comparison between TPPNet T1 sequence where Acc is accuracy, AUC denotes Spe denotes specificity. The grey level is set 12.

Approach	Acc	AUC	Sen	1-Spe	F1-Score
VGG16	0.837±0.106	0.739±0.110	0.910±0.125	0.548±0.124	0.749±0.114
Inception	0.805±0.014	0.465±0.102	1.000±0.000	0.033±0.067	0.474±0.061
Mobilenet	0.869±0.033	0.616±0.170	0.985±0.030	0.405±0.258	0.719±0.107
GLCM-CNN	0.851±0.019	0.704±0.113	0.975±0.050	0.362±0.249	0.707±0.073
ViT3D	0.882±0.020	0.675±0.163	0.969±0.043	0.533±0.245	0.782±0.058
TPPNet	**0.935±0.021**	**0.832±0.101**	**1.000±0.000**	**0.676±0.107**	**0.881±0.045**

demonstrated in Table 5 which contains the accuracy, AUC score, specificity, sensitivity, and F1 score. Their ROC curves over T1 sequence are plotted in Fig. 3. T2 and Post-gadolinium's performances could be found in supplementary materials.

4 Conclusions

In this paper, we develop an approach to carry out the pioneer study of differentiating abnormal BP from normal ones relevant to breast cancer. In particular, TPP is proposed to extract texture features as the representation of BP's heterogeneity from MRIs. Moreover, a TPPNet with huge number of initial channels is designed to train the model. To testify our proposed TPPNet, a BP dataset is constructed with 452 series including three most commonly used MR sequences in clinical practice, i.e. T2, T1 and Post-gadolinium. The best result is yielded when the gray level is 12, intensity rescaling method adopts arc tangent approach. Experimental outcomes also demonstrate that the proposed TPPNet not only exhibit more stable performances but also outperform six famous state-of-the-art approaches over three most commonly used BP MR sequences.

References

1. Shabeeb, D., Musa, A.E., et al.: Brachial plexopathy as a complication of radio-therapy: a systematic review. Curr. Cancer Ther. Rev. **15**(2), 110–120 (2020)
2. Wittenberg, K.H., Adkins, M.C.: MR imaging of nontraumatic brachial plex-opathies: frequency and spectrum of findings. Radiographics **20**(4), 1024–1032 (2004)
3. Nisce, L.Z., Chu, F.C.H.: Radiation therapy of brachial plexus syndrome from breast cancer. Radiology **91**(5), 1022–1025 (1968)
4. Lutz, A.M., Gold, G., Beaulieu, C.: MR imaging of the brachial plexus. Magn. Reson. Imaging Clin. N. Am. **20**(4), 791–826 (2012)
5. Wang, R., Shen, H., Zhou, M.: Ultrasound nerve segmentation of brachial plexus based on optimized resu-net. In: 2019 IEEE International Conference on Imaging Systems and Techniques (IST). pp. 1–6 (2019)
6. Pisda, K., Jain, P., Sisodia, D.S.: Deep networks for brachial plexus nerves segmentation and detection using ultrasound images. In: Garg, L., Kesswani, N., Vella, J.G., Xuereb, P.A., Lo, M.F., Diaz, R., Misra, S., Gupta, V., Randhawa, P. (eds.) ISMS 2020. LNNS, vol. 303, pp. 132–146. Springer, Cham (2022). https://doi.org/10.1007/978-3-030-86223-7_13

7. Wang, Y., Geng, J., Zhou, C., Zhang, Y.: Segmentation of ultrasound brachial plexus based on u-net. In: 2021 International Conference on Communications, Information System and Computer Engineering (CISCE). pp. 482–485 (2021)
8. Tian, D., Wang, Q., et al.: Brachial plexus nerve trunk recognition from ultra-sound images: a comparative study of deep learning models. IEEE Access **10**, 82003–82014 (2022)
9. Sureka, J., Cherian, R.A., Alexander, M., Thomas, B.P.: MRI of brachial plex-opathies. Clin. Radiol. **64**(2), 208–218 (2009)
10. Lambin, P., Leijenaar, R.T.H., et al.: Radiomics: the bridge between medical imaging and personalized medicine. Nat. Rev. Clin. Oncol. **14**(12), 749–762 (2017)
11. van Griethuysen, J.J.M., Fedorov, A., et al.: Computational radiomics system to decode the radiographic phenotype. Can. Res. **77**(21), e104–e107 (2017)
12. Ramel, A.A.: Analysis of membrane process model from black box to machine learning. J. Mach. Comput. **2**(1), 2788–7669 (2022)
13. Tan, J., Lei, B., et al.: 3D-GLCM CNN: a 3-dimensional gray-level cosoccur-rence matrix-based CNN model for polyp classification via CT colonography. IEEE Trans. Med. Imaging **39**(6), 2013–2024 (2020)
14. Yoo, T.S., Ackerman, M.J., et al.: Engineering and algorithm design for an image processing API: a technical report on ITK – the insight toolkit. In: Westwood, J. (ed.) Proceeding of Medicine Meets Virtual Reality, pp. 586–592 (2002)
15. McCormick, M., Liu, X., et al.: Enabling reproducible research and open science. Front. Neuroinform. **8**, 13 (2014)
16. Isensee, F., Jaeger, P.F., et al.: nnU-Net: a self-configuring method for deep learning-based biomedical image segmentation. Nat. Methods **18**(2), 203–211 (2021)
17. Ojala, T., Pietikainen, M., Maenpaa, T.: Multiresolution gray-scale and rotation invariant texture classification with local binary patterns. IEEE Trans. Pattern Anal. Mach. Intell. **24**(7), 971–987 (2002)
18. Tan, X., Triggs, B.: Enhanced local texture feature sets for face recognition under difficult lighting conditions. IEEE Trans. Image Process. **19**(6), 1635–1650 (2010)
19. Vallières, M., Zwanenburg, A., et al.: Responsible radiomics research for faster clinical translation. J. Nucl. Med. **59**(2), 189–193 (2018)
20. Güner, A., Alçin, Ö.F., Şengür, A.: Automatic digital modulation classification using extreme learning machine with local binary pattern histogram features. Measurement **145**, 214–225 (2019)
21. Bian, M., Liu, J. K., et al.: Verifiable privacy-enhanced rotation invariant LBP feature extraction in fog computing. In: IEEE Transactions on Industrial Informatics (2023)
22. Cao, W., Liang, Z., Gao, Y., et al.: A dynamic lesion model for differentiation of malignant and benign pathologies. Sci. Rep. **11**, 3485 (2021)
23. Galavis, P.E., Hollensen, C., et al.: Variability of textural features in FDG PET images due to different acquisition modes and reconstruction parameters. Acta Oncol. **49**(7), 1012–1016 (2010)
24. Doumou, G., Siddique, M., Tsoumpas, C., et al.: The precision of textural analysis in 18F-FDG-PET scans of oesophageal cancer. Eur. Radiol. **25**(9), 2805–2812 (2015)
25. Wahid, K.A., He, R., et al.: Intensity standardization methods in magnetic reso-nance imaging of head and neck cancer. Phy. Imaging Radiat. Oncol. **20**, 88–93 (2021)
26. Cao, W., Pomeroy, M.J., et al.: Lesion classification by model-based feature extraction: a differential affine invariant model of soft tissue elasticity. arXiv pre-print arXiv:2205.14029 (2022)
27. Pomeroy, M.J., Pickhardt, P., Liang, J., Lu, H.: Histogram-based adaptive gray level scaling for texture feature classification of colorectal polyps. In: Medical Imaging 2018: Computer-Aided Diagnosis, vol. 10575, pp. 507–513. SPIE (2018)

28. Simonyan, K., Zisserman, A.: Very deep convolution newtworks for large-scale image recognition. arXiv preprint arXiv:1409.1556 (2014)
29. Szegedy, C., Liu, W., et al.: Going deeper with convolutions. In: 2015 IEEE Conference on Computer Vision and Pattern Recognition (CVPR), Boston, MA, USA, pp. 1–9 (2015)
30. Howard A.G., Zhu, M., et al.: MobileNets: efficient convolutional neural net-works for mobile vision applications. arXiv:1704.04861 (2017)
31. Wu, B., et al.: Visual transformers: where do transformers really belong in vi-sion models?. In: 2021 IEEE/CVF International Conference on Computer Vision (ICCV), Montreal, QC, Canada, pp. 579–589 (2021)

Microscopy

Mitosis Detection from Partial Annotation by Dataset Generation via Frame-Order Flipping

Kazuya Nishimura[1(✉)], Ami Katanaya[2], Shinichiro Chuma[2], and Ryoma Bise[1]

[1] Kyushu University, Fukuoka, Japan
kazuya.nishimura@humna.ait.kyushu-u.ac.jp
[2] Kyoto University, Kyoto, Japan

Abstract. Detection of mitosis events plays an important role in biomedical research. Deep-learning-based mitosis detection methods have achieved outstanding performance with a certain amount of labeled data. However, these methods require annotations for each imaging condition. Collecting labeled data involves time-consuming human labor. In this paper, we propose a mitosis detection method that can be trained with partially annotated sequences. The base idea is to generate a fully labeled dataset from the partial labels and train a mitosis detection model with the generated dataset. First, we generate an image pair not containing mitosis events by frame-order flipping. Then, we paste mitosis events to the image pair by alpha-blending pasting and generate a fully labeled dataset. We demonstrate the performance of our method on four datasets, and we confirm that our method outperforms other comparisons which use partially labeled sequences. Code is available at https://github.com/naivete5656/MDPAFOF.

Keywords: Partial annotation · Mitosis detection · Fluorescent image

1 Introduction

Fluorescent microscopy is widely used to capture cell nuclei behavior. Mitosis detection is the task of detecting the moment of cell division from time-lapse images (the dotted circles in Fig. 1). Mitosis detection from fluorescent sequences is important in biological research, medical diagnosis, and drug development.

Conventionally tracking-based methods [1,19,19,21] and tracking-free methods [3,5–7] have been proposed for mitosis detection. Recently, deep-learning-based mitosis-detection methods have achieved outstanding performance [8–12,18]. However, training deep-learning methods require a certain amount of annotation for each imaging condition, such as types of cells and microscopy

---

Supplementary Information The online version contains supplementary material available at https://doi.org/10.1007/978-3-031-43993-3_47.

H. Greenspan et al. (Eds.): MICCAI 2023, LNCS 14227, pp. 483–492, 2023.
https://doi.org/10.1007/978-3-031-43993-3_47

and the density of cells. Collecting a sufficient number of labeled data covering the variability of cell type and cell density is time-consuming and labor-intensive.

Unlike cell detection and segmentation, which aims to recognize objects from a single image, mitosis detection aims to identify events from time series of images. Thus, it is necessary to observe differences between multiple frames to make mitosis events annotation. Comprehensively annotating mitosis events is time-consuming, and annotators may be missed mitosis events. Thus, we must carefully review the annotations to ensure that they are comprehensive.

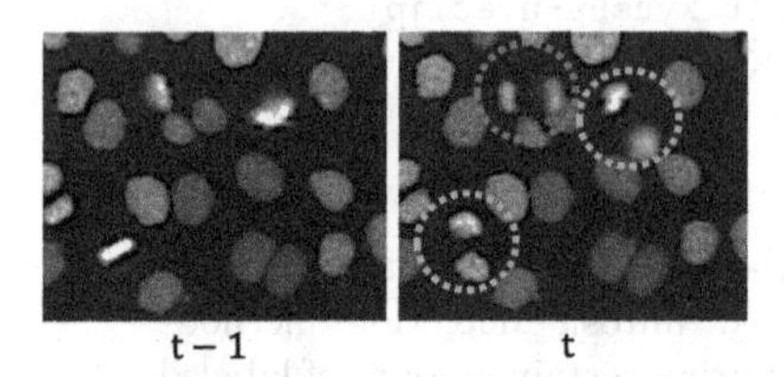

Fig. 1. Example of partially labeled frames. The mitosis shown in the red-dotted circle is annotated, and the mitoses shown in the light-blue-dotted circles are not annotated. (Color figure online)

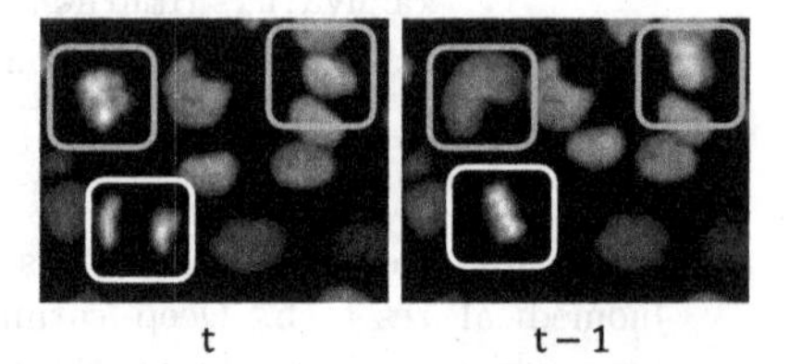

Fig. 2. Example of frame-order flipped images. A mitosis event is converted into a merge event (white rectangles). Non-mitotic cells still have non-mitotic cell movement (yellow rectangles). (Color figure online)

Partial annotation has been used as a way to reduce the annotation costs of cell and object detection [2,15,22]. Figure 1 shows an example of partially annotated frames. Some mitosis events are annotated (a red-dotted circle), and others are not (light-blue-dotted circles). The annotation costs are low because the annotator only needs to plot a few mitotic positions. In addition, this style of annotation allows for missing annotations. Therefore, it would be effective for mitosis detection.

Unlike supervised annotation, partial annotation can not treat unannotated areas as regions not containing mitosis events since the regions may contain mitosis events (Fig. 1). The regions naturally affect the training in the partial annotation setting. To avoid the effect of unlabeled objects in unlabeled regions, Qu *et al.* [15] proposed to use a Gaussian masked mean squared loss, which calculates the loss around the annotated regions. The loss function works in tasks in which foreground and background features have clearly different appearances, such as in cell detection. However, it does not work on mitosis detection since the appearance of several non-mitotic cells appears similar to mitosis cells; it produces many false positives.

In this paper, we propose a cell-mitosis detection method for fluorescent time-lapse images by generating a fully labeled dataset from partially annotated sequences. We achieve mitosis detection training in a mitosis detection model with the generated dataset. To generate the fully labeled dataset, we should consider two problems: (1) no label indicating regions not containing mitosis cells and (2) few mitosis annotations.

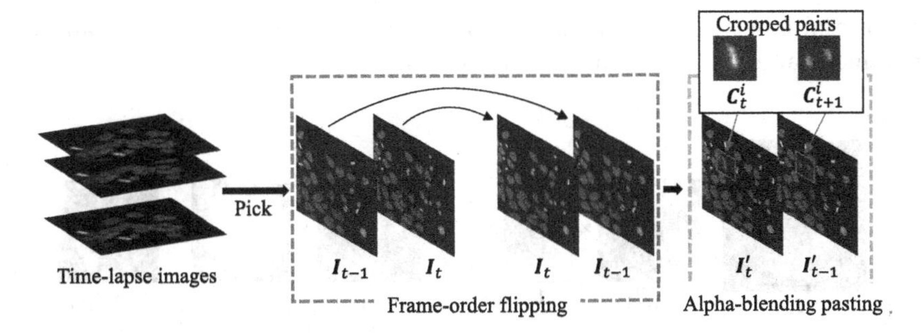

Fig. 3. Overview of dataset generation.

We can easily generate the regions not containing mitotic cells by using one image twice. However, such regions do not contribute to identifying mitotic cells and non-mitotic cells since the data do not show natural cell motions. For the training to be effective, the regions not containing mitotic cells should show the natural movements of cells. To generate such regions, we propose frame-order flipping which simply flips the frame order of a consecutive frame pair. As shown in the white rectangles in Fig. 2, we can convert a mitosis event to a cell fusion by flipping operation. Hence, the flipped pair is the region not containing mitosis cells. Even though we flipped the frame order, the non-mitotic cells still have natural time-series motion, as shown in the yellow rectangles in Fig. 2.

In addition, we can make the most of a few partial annotations by using copy-and-paste-based techniques. Unlike regular copy-and-paste augmentation [4] for supervised augmentation of instance segmentations which have object mask annotations, we only have point-level annotations. Thus, we propose to use alpha-blending pasting techniques which naturally blend two images.

Experiments conducted on four types of fluorescent sequences demonstrate that the proposed method outperforms other methods which use partial labels.

Related Work. Some methods used partially labeled data to train model [2,15,22]. Qu [15] proposed a Gaussian masked mean squared loss, which calculates the loss around the annotated areas. To more accurately identify negative and positive samples, positive unlabeled learning has been used for object detection [2,22]. These methods have used positive unlabeled learning on candidates detected by using partial annotation to identify whether the candidates are labeled objects or backgrounds. However, since the candidates detected by partial annotation include many false positives, the positive unlabeled learning does not work on mitosis detection. the appearance of the mitosis event and backgrounds in the mitosis detection task, it is difficult to estimate positive prior. These methods could not work on mitosis detection. The positive unlabeled learning requires a positive prior.

2 Method: Mitosis Detection with Partial Labels

Our method aims to detect coordinates and timing (t, x, y) of mitosis events from fluorescent sequences. For training, we use time-lapse images $\mathcal{I} = \{\boldsymbol{I}_t\}_{t=1}^{T}$

and partial labels (a set of annotated mitosis cells). Here, $\boldsymbol{I}_t$ denotes an image at frame t, and T is the total number of frames.

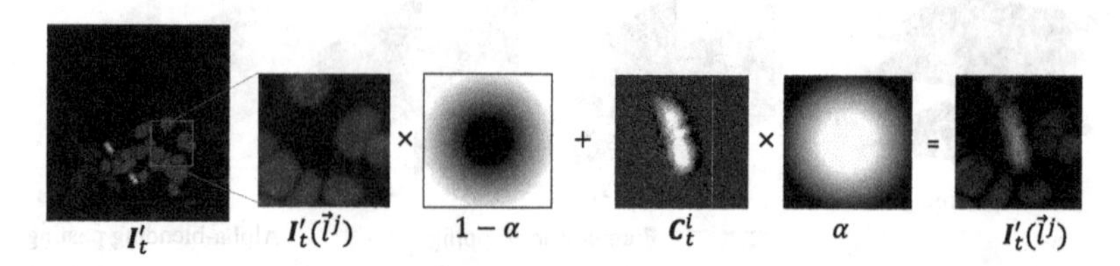

Fig. 4. Illustration of alpha-blending pasting.

Our method generates a fully labeled dataset $\mathcal{D}_p = \{(\boldsymbol{I}'_{t-1}, \boldsymbol{I}'_t), \mathcal{P}'_t\}_{t=1}^{T-1}$ from time-lapse images $\mathcal{I}$ and partial labels and then trains a mitosis detection model f_θ with the generated dataset. Here, $\boldsymbol{I}'_t$ is a generated image, and $\mathcal{P}'_t$ is a set of mitotic coordinates contained in $(\boldsymbol{I}'_{t-1}, \boldsymbol{I}'_t)$. Since our method trains the network with partial labels, it can eliminate the costs of checking for missed annotations.

2.1 Labeled Dataset Generation

Figure 3 shows an overview of our dataset generation. We randomly pick a pair of consecutive frames $(\boldsymbol{I}_{t-1}, \boldsymbol{I}_t)$ from time-lapse images $\mathcal{I}$. Since the pair may contain unannotated mitosis events, we forcibly convert the pair into a negative pair (*i.e.,* a pair which does not contain mitosis events) by using frame-order flipping. Next, we paste mitosis events to a generated pair using alpha-blending pasting and obtain a generated pair $(\boldsymbol{I}'_{t-1}, \boldsymbol{I}'_t)$. Since we know the pasted location, we can obtain the mitosis locations $\mathcal{P}'_t$ of the generated pair.

Negative Pair Generation with Frame-Order Flipping: In this step, we generate a pair not containing mitotic cells by using a simple augmentation-based frame-order flipping. Figure 3 shows an example of the pair images $(\boldsymbol{I}_{t-1}, \boldsymbol{I}_t)$. The pair may contain mitosis events. If we assume that the pair does not contain mitotic cells, it affects the training of the mitosis detection model f_θ. To prevent the pair from containing mitosis events, we flip the frame order and treat the flipped pair $(\boldsymbol{I}_t, \boldsymbol{I}_{t-1})$ as a pair of negative.

Since mitosis is the event that a cell divides into two daughter cells, the mitosis event is transformed into an event in which two cells fuse into one by flipping the order (Fig. 2). The flipped event can treat as a non-mitotic event. Note that the motivation behind using frame flipping is to be able to utilize pixels showing the motions of non-mitotic cells negatives by transforming mitosis into other events. Even if the order is flipped, the movements of the non-mitotic cell are still a non-mitotic cell feature, and we consider that these cells are effective for the training of the negative label.

Mitosis Label Utilization with Alpha-Blending Pasting: Next, we paste mitosis events to the flipped pair by using copy-and-paste techniques in order to utilize the positive labels effectively. Copy and paste augmentation has been

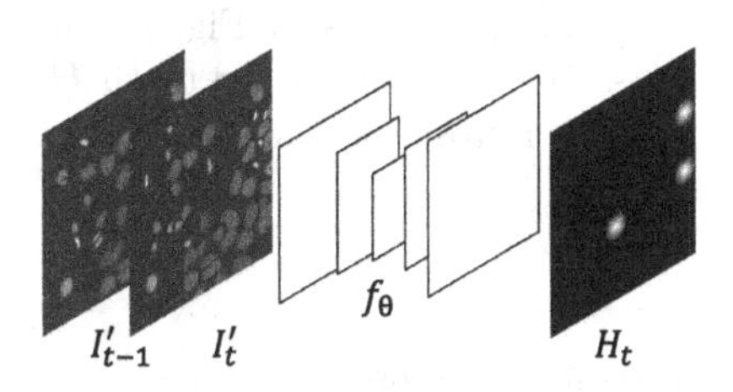

Fig. 5. Cell mitosis detection model.

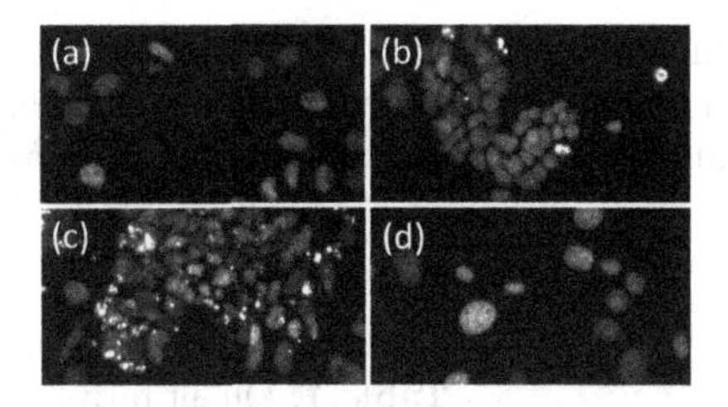

Fig. 6. Example of images in datasets. (a) HeLa, (b) ES, (c) ES-D, (d) Fib.

used for supervised augmentation of instance segmentation [4]. Unlike instance segmentation with object masks, we only have locations (t, x, y). A simple solution is to crop images around the mitosis position and copy and paste them to the target image, like in CutMix [23]. However, the cropped image naturally contains surrounding objects, and the generated image appears unnatural. Unnatural images cause the detection network to make biased predictions and reduce generalization performance. To avoid this problem, we propose alpha-blending pasting with a Gaussian blending mask. We blend two images by leaving the pixel value in the center and blurring the vicinity of the edge of the image.

First, we crop the image around the positive annotations and obtain a set of cropped pair $\mathcal{C} = \{(\boldsymbol{C}^i_{t-1}, \boldsymbol{C}^i_t)\}^N_{i=0}$ and initialize $(\boldsymbol{I}'_{t-1}, \boldsymbol{I}'_t) = (\boldsymbol{I}_t, \boldsymbol{I}_{t-1})$ and $\mathcal{P}'_t = \{\}$. Here, N is the total number of partial annotations, while $\boldsymbol{C}^i_{t-1}$ and $\boldsymbol{C}^i_t$ are images before and after the mitosis of the i-th annotation (Fig. 3). Define $\boldsymbol{I}'_t(\boldsymbol{l}^j)$, $\boldsymbol{I}'_{t-1}(\boldsymbol{l}^j)$ as a cropped pair image at the j-th random spatial location $\boldsymbol{l}^j$. We crop each image centered at $\boldsymbol{l}^j$ to a size that is the same as that of $\boldsymbol{C}^i_t$. We update the randomly selected patch $\boldsymbol{I}'_t(\boldsymbol{l}^j)$, $\boldsymbol{I}'_{t-1}(\boldsymbol{l}^j)$ by blending a randomly selected cropped pair $(\boldsymbol{C}^i_{t-1}, \boldsymbol{C}^i_t)$ with the following formula: $\boldsymbol{I}'_t(\boldsymbol{l}^j) = (1-\alpha) \odot \boldsymbol{I}'_t(\boldsymbol{l}^j) + \alpha \odot \boldsymbol{C}^i_t$, where α is a Gaussian blending mask (Fig. 4). We generate the blending mask by blurring a binary mask around the annotation with a Gaussian filter. We use a random sigma value for the Gaussian filter. Then, we add the paste location $\boldsymbol{l}^j$ to the set $\mathcal{P}'_t$. We repeat this process random k times.

2.2 Mitosis Detection with Generated Dataset

We modified a heatmap-based cell detection method [13] to work as a mitosis detection method. Figure 5 is an illustration of our mitosis detection model. Given two consecutive frames $(\boldsymbol{I}'_{t-1}, \boldsymbol{I}'_t)$, the network output heatmap $\hat{\boldsymbol{H}}_t$. We treat the channel axis as the time axis for the input. The first channel is $\boldsymbol{I}'_{t-1}$, and the second is $\boldsymbol{I}'_t$.

First, we generate individual heatmaps $\boldsymbol{H}^j_t$ for each pasted coordinate $\boldsymbol{l}^j = (l^j_x, l^j_y)$. $\boldsymbol{H}^j_t$ is defined as $\boldsymbol{H}^j_t(p_x, p_y) = \exp\left(-\frac{(l^j_x - p_x)^2 + (l^j_y - p_y)^2}{\sigma^2}\right)$, where p_x and p_y are the coordinates of $\boldsymbol{H}^j_t$ and σ is a hyper parameter that controls the spread of the peak. The ground truth of the heatmap at t is generated by taking the maximum through the individual heatmaps, $\boldsymbol{H}_t = \max_j(\boldsymbol{H}_t^j)$ ($\boldsymbol{H}_t$ in Fig. 5). The network is trained with the mean square error loss between the ground truth $\boldsymbol{H}_t$ and the output of the network $\hat{\boldsymbol{H}}_t$. We can find the mitosis position by finding a local maximum of the heatmap.

Table 1. Quantitative evaluation results (F1-score).

Method	1-shot					5-shot				
	HeLa	ES	ES-D	Fib	Ave	HeLa	ES	ES-D	Fib	Ave
Baseline [13]	0.356	0.581	0.06	0.235	0.308	0.591	0.756	0.277	0.210	0.459
GM [15]	0.315	0.303	0.057	0.119	0.199	0.501	0.523	0.123	0.230	0.344
PU [22]	0.030	0.095	0.012	0.053	0.048	0.463	0.538	0.375	0.224	0.400
PU-I [2]	0.499	0.177	0.035	0.115	0.207	0.474	0.420	0.037	0.141	0.268
Ours	**0.593**	**0.740**	**0.439**	**0.440**	**0.553**	**0.795**	**0.843**	**0.628**	**0.451**	**0.610**

3 Experiments

Dataset: We evaluated our method on four datasets. The first set is **HeLa** [20], in which live cell images of HeLa cells expressing H2B-GFP were captured with 1100 × 700 resolution [20][1]. Each sequence contains 92 fluorescent images with 141 mitosis events on average. The second set is **ES**, in which live cell images of mouse embryonic stem cells expressing H2B-mCherry were captured with 1024 × 1024 resolution. Each sequence contains 41 fluorescent images with 33 mitosis events on average. The third set is **ES-D** in which mouse embryonic stem cells expressing H2B-mCherry were induced to differentiate and used to capture live cell images. Each sequence contains 61 fluorescent images with 18 on average events on average. The fourth set is **Fib**, in which live cell images of mouse fibroblast cells expressing H2B-mCherry were captured with 1024 × 1024 resolution. Each sequence contains 42 fluorescent images with 11 mitosis events on average. Each dataset consists of four sequences of images. We performed four-fold cross-validation in which two sequences were used as training data, one as validation data, and one as test data. As shown in Fig. 6, the appearance and density are different depending on the dataset.

Implementation Details: We implemented our method within the Pytorch framework [14] and used a UNet-based architecture [16] for the mitosis-detection network. The model was trained with the Adam optimizer with a learning rate of 1e-3. σ, which controls the spread of the heatmap, was 6. The cropping size of the positive annotations was 40 pixels. We randomly change the number of pasting operations k between 1 and 10. We used random flipping, random cropping, and brightness change for the augmentation.

[1] We used the publicly available CTC data-set http://celltrackingchallenge.net/. We only use HeLa since the number of mitosis events in other cells is small.

Table 2. Ablation study. FOF: frame-order flipping, ABP: alpha-blending pasting.

Method	F1
w/o FOF	0.570
w/o ABP [23]	0.670
Ours	**0.795**

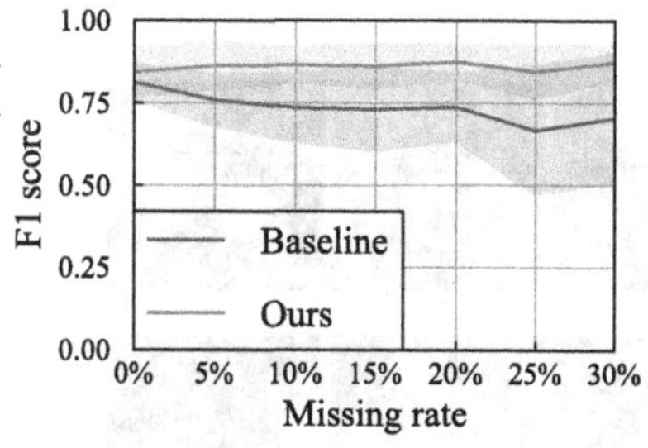

Fig. 7. Robustness against missing annotations.

Evaluation Metrics: We evaluated our method using the F1 score [18], which is widely used in mitosis detection. Given ground-truth coordinates and detected coordinates, we performed one-by-one matching. If the distance of the matched pair was within spatially 15 pixels and temporally 6, we associated the closest coordinate pairs. We treated the matched pair as true positives (TP), unassociated coordinates as false positives (FP), and unassociated ground-truth coordinates as false negatives (FN).

Comparisons: We conducted four comparisons that involved training the model with partially labeled data. For the first method, we trained the model by treating unlabeled pixels as non-mitosis ones (Baseline [13]). The second method used the Gaussian masked loss (GM [15]). The masked loss was calculated on the masked pixels around the positive-label pixels. Thus, the method ignored unlabeled pixels. The third method used positive unlabeled learning to identify mitosis from candidates obtained by the detection model trained with the masked loss (PU [22]). The fourth method generated pseudo-labels from the results of positive unlabeled learning and retrained the detection model with the pseudo-label (PU-I [2]).

In Table 1, we compared our method with previous methods in one and five-shot settings. We used N samples per sequence in the N-shot settings. For a robust comparison, we sampled one or five mitosis annotations under five seed conditions and took the average. Overall, our method outperformed all compared methods in F1 metric. GM [15], PU [22], and PU-I [2] are designed for detecting objects against simple backgrounds. Therefore, these methods are not suited to a mitosis detection task and are inferior to the baseline.

The baseline [13] treats unlabeled pixels as non-mitosis cell pixels. In the partially labeled setting, unlabeled pixels contain unannotated mitosis events, and unannotated mitosis affects performance. Unlike cell detection, mitosis detection requires identifying mitosis events from various non-mitotic cell motions, including motions that appear mitotic appearances. Although GM [15] can ignore unlabeled mitosis pixels with the masked loss, it is difficult to identify such non-mitosis motions. Therefore, GM estimates produce many false positives. PU [22] uses positive unlabeled learning to eliminate false positives from candidates obtained from the detection results with partial labels. However, positive unla-

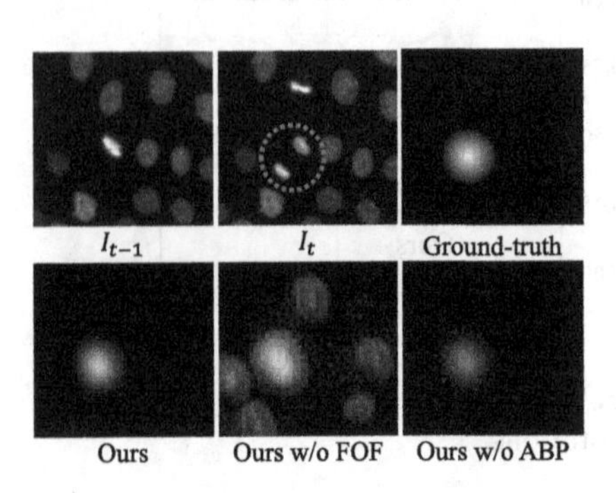

Fig. 8. Example of estimation results. Red dotted circles indicate mitosis events. (Color figure online)

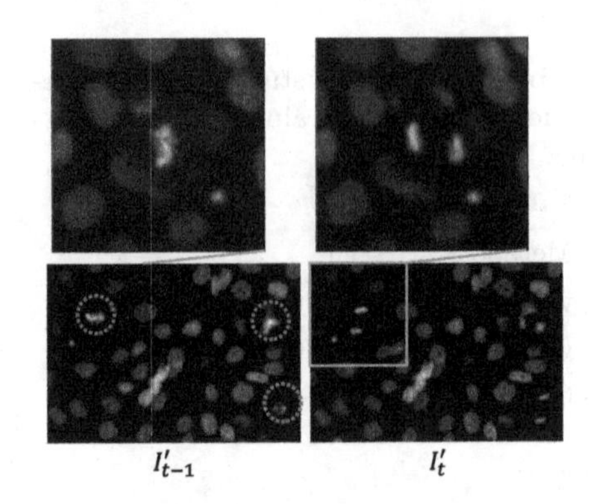

Fig. 9. Example of generated images. (Color figure online)

beled learning requires a positive prior in the candidates and a certain amount of randomly sampled positive samples. Since the candidates contain many false positives, the positive prior is difficult to estimate. In addition, there is no guarantee that positive unlabeled learning can work correctly with the selected N-shot annotations. Moreover, since positive unlabeled learning does not work in the mitosis detection task, PU-I [2] can not select accurate pseudo labels. Unlike these methods, our method can estimate mitosis events accurately. Since our method generates a fully labeled dataset from a partial label, it effectively uses a few partial annotations.

Effectiveness of Each Module: We performed an ablation study on the HeLa dataset to investigate the effectiveness of the proposed module. We used random augmentation (*i.e.,* random elastic transformation [17], brightness change, and gaussian noise) instead of using frame-order flipping (FOF). We generated $\boldsymbol{I}_t^{aug}$ by augmenting $\boldsymbol{I}_t$ and input the pair $(\boldsymbol{I}_t, \boldsymbol{I}_t^{aug})$ to the network. In the w/o ABP setting, we directly pasted cropped images on the target image as in CutMix [23]. Table 2 demonstrates that the proposed modules improve mitosis detection performance. Figure 8 shows examples of the estimation results for each condition. Without the FOF setting, the detection model estimates a high value for all moving cells, leading to over-detection. Without the ABP setting, the detection model overfits the directly pasted image. The directly pasted image tends to include unnatural boundaries on the edge, leading to missed detections in real images.

Robustness Against Missing Annotations: We confirmed the robustness of the proposed method against missing annotations on the ES dataset. We changed the missing annotation rate from 0% to 30%. A comparison with the supervised method in terms of F1-score is shown in Fig. 7. The performance of the supervised method deteriorates as the percentage of missing labels increases, whereas the performance of the proposed method remains steady. Since our method flips the frame order, we can avoid the effects of missing annotations.

Appearance of Generated Dataset: Fig. 9 shows an example of the generated image pair. The cropped mitosis image pairs were pasted on the red-dotted circle. It can be seen that the borders of the original image and the pasted image have been synthesized very naturally.

4 Conclusion

We proposed a mitosis detection method using partially labeled sequences with frame-order flipping and alpha-blending pasting. Our frame-order flipping transforms unlabeled data into non-mitosis labeled data through a simple flipping operation. Moreover, we generate various positive labels with a few positive labels by using alpha-blending pasting. Unlike directly using copy-and-paste, our method generates a natural image. Experiments demonstrated that our method outperforms other methods that use partially annotated sequences on four fluorescent microscopy images.

Acknowledgements. This work was supported by JSPS KAKENHI Grant Number JP21J21810 and JST ACT-X Grant Number JPMJAX21AK, Japan.

References

1. Debeir, O., Van Ham, P., Kiss, R., Decaestecker, C.: Tracking of migrating cells under phase-contrast video microscopy with combined mean-shift processes. IEEE Trans. Med. Imaging **24**(6), 697–711 (2005)
2. Fujii, K., Suehiro, D., Nishimura, K., Bise, R.: Cell detection from imperfect annotation by pseudo label selection using p-classification. In: MICCAI, pp. 425–434 (2021)
3. Gallardo, G.M., Yang, F., Ianzini, F., Mackey, M., Sonka, M.: Mitotic cell recognition with hidden markov models. In: Medical Imaging 2004: Visualization, Image-Guided Procedures, and Display, vol. 5367, pp. 661–668. SPIE (2004)
4. Ghiasi, G., et al.: Simple copy-paste is a strong data augmentation method for instance segmentation. In: CVPR, pp. 2918–2928 (2021)
5. Gilad, T., Reyes, J., Chen, J.Y., Lahav, G., Riklin Raviv, T.: Fully unsupervised symmetry-based mitosis detection in time-lapse cell microscopy. Bioinformatics **35**(15), 2644–2653 (2019)
6. Huh, S., Bise, R., Chen, M., Kanade, T., et al.: Automated mitosis detection of stem cell populations in phase-contrast microscopy images. IEEE Trans. Med. Imaging **30**(3), 586–596 (2010)
7. Liu, A.A., Tang, J., Nie, W., Su, Y.: Multi-grained random fields for mitosis identification in time-lapse phase contrast microscopy image sequences. IEEE Trans. Med. Imaging **36**(8), 1699–1710 (2017)
8. Lu, Y., Liu, A.A., Chen, M., Nie, W.Z., Su, Y.T.: Sequential saliency guided deep neural network for joint mitosis identification and localization in time-lapse phase contrast microscopy images. IEEE J. Biomed. Health Inf. **24**(5), 1367–1378 (2019)

9. Mao, Y., Yin, Z.: A hierarchical convolutional neural network for mitosis detection in phase-contrast microscopy images. In: MICCAI, pp. 685–692 (2016)
10. Mao, Y., Yin, Z.: Two-stream bidirectional long short-term memory for mitosis event detection and stage localization in phase-contrast microscopy images. In: MICCAI, pp. 56–64 (2017)
11. Nie, W.Z., Li, W.H., Liu, A.A., Hao, T., Su, Y.T.: 3d convolutional networks-based mitotic event detection in time-lapse phase contrast microscopy image sequences of stem cell populations. In: CVPR, pp. 55–62
12. Nishimura, K., Bise, R.: Spatial-temporal mitosis detection in phase-contrast microscopy via likelihood map estimation by 3dcnn. In: EMBC, pp. 1811–1815 (2020)
13. Nishimura, K., Wang, C., Watanabe, K., Bise, R., et al.: Weakly supervised cell instance segmentation under various conditions. Med. Image Anal. **73**, 102182 (2021)
14. Paszke, A., et al.: PyTorch: an imperative style, high-performance deep learning library. In: NeurIPS, vol. 32 (2019)
15. Qu, H., et al.: Weakly supervised deep nuclei segmentation using partial points annotation in histopathology images. IEEE Trans. Med. Imaging **39**(11), 3655–3666 (2020)
16. Ronneberger, O., Fischer, P., Brox, T.: U-net: convolutional networks for biomedical image segmentation. In: MICCAI, pp. 234–241 (2015)
17. Simard, P.Y., Steinkraus, D., Platt, J.C., et al.: Best practices for convolutional neural networks applied to visual document analysis. In: ICDAR, vol. 3 (2003)
18. Su, Y.T., Lu, Y., Liu, J., Chen, M., Liu, A.A.: Spatio-temporal mitosis detection in time-lapse phase-contrast microscopy image sequences: a benchmark. IEEE Trans. Med. Imaging **40**(5), 1319–1328 (2021)
19. Thirusittampalam, K., Hossain, M.J., Ghita, O., Whelan, P.F.: A novel framework for cellular tracking and mitosis detection in dense phase contrast microscopy images. IEEE J. Biomed. Health Inf. **17**(3), 642–653 (2013)
20. Ulman, V., Maška, M., Magnusson, K.E., Ronneberger, O., Haubold, C., Harder, N., Matula, P., Matula, P., Svoboda, D., Radojevic, M., et al.: An objective comparison of cell-tracking algorithms. Nat. Methods **14**(12), 1141–1152 (2017)
21. Yang, F., Mackey, M.A., Ianzini, F., Gallardo, G., Sonka, M.: Cell segmentation, tracking, and mitosis detection using temporal context. In: MICCAI, pp. 302–309 (2005)
22. Yang, Y., Liang, K.J., Carin, L.: Object detection as a positive-unlabeled problem. In: BMVC (2020)
23. Yun, S., Han, D., Oh, S.J., Chun, S., Choe, J., Yoo, Y.: CutMix: regularization strategy to train strong classifiers with localizable features. In: ICCV, pp. 6023–6032 (2019)

CircleFormer: Circular Nuclei Detection in Whole Slide Images with Circle Queries and Attention

Hengxu Zhang[1,2], Pengpeng Liang[3], Zhiyong Sun[1], Bo Song[1], and Erkang Cheng[1(✉)]

[1] Institute of Intelligent Machines, HFIPS, Chinese Academy of Sciences, Hefei, China
ekcheng@iim.ac.cn

[2] Institutes of Physical Science and Information Technology, Anhui University, Hefei, China

[3] School of Computer and Artificial Intelligence, Zhengzhou University, Zhengzhou, China

Abstract. Both CNN-based and Transformer-based object detection with bounding box representation have been extensively studied in computer vision and medical image analysis, but circular object detection in medical images is still underexplored. Inspired by the recent anchor free CNN-based circular object detection method (CircleNet) for ball-shape glomeruli detection in renal pathology, in this paper, we present CircleFormer, a Transformer-based circular medical object detection with dynamic anchor circles. Specifically, queries with circle representation in Transformer decoder iteratively refine the circular object detection results, and a circle cross attention module is introduced to compute the similarity between circular queries and image features. A generalized circle IoU (gCIoU) is proposed to serve as a new regression loss of circular object detection as well. Moreover, our approach is easy to generalize to the segmentation task by adding a simple segmentation branch to CircleFormer. We evaluate our method in circular nuclei detection and segmentation on the public MoNuSeg dataset, and the experimental results show that our method achieves promising performance compared with the state-of-the-art approaches. The effectiveness of each component is validated via ablation studies as well. Our code is released at: https://github.com/zhanghx-iim-ahu/CircleFormer.

Keywords: Circular Object Analysis · Circular Queries · Transformer

1 Introduction

Nuclei detection is a highly challenging task and plays an important role in many biological applications such as cancer diagnosis and drug discovery. Rectangle object detection approaches that use CNN have made great progress in the

H. Greenspan et al. (Eds.): MICCAI 2023, LNCS 14227, pp. 493–502, 2023.
https://doi.org/10.1007/978-3-031-43993-3_48

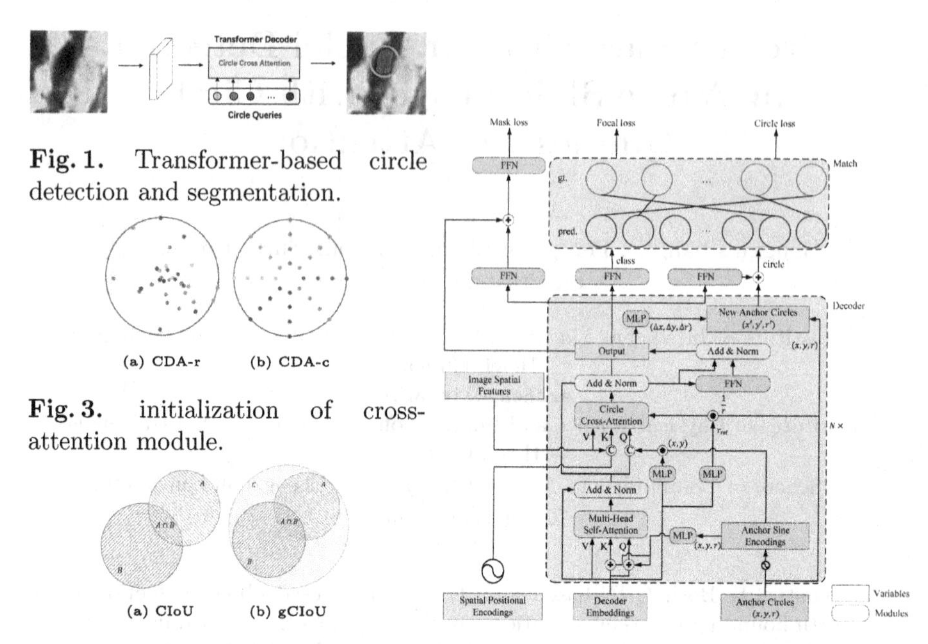

Fig. 1. Transformer-based circle detection and segmentation.

Fig. 3. initialization of cross-attention module.

Fig. 4. Different IoU evaluation metrics.

Fig. 2. Overview of the proposed method.

last decade [4,7,12,14,18]. These popular CNN models use boxes to represent objects that are not optimized for circular medical objects, such as detection of glomeruli in renal pathology. To address the problem, an anchor-free CNN-based circular object detection method CircleNet [16] is proposed for glomeruli detection. Different from CenterNet [18], CircleNet estimates the radius rather than the box size for circular objects. But it also suffers poor detection accuracy for overlapping objects and requires additional post-processing steps to obtain the final detection results.

Recently, DETR [1], a Transformer-based object detection method reformulates object detection as a set-to-set prediction problem, and it removes both the hand-crafted anchors and the non-maximum suppression (NMS) post-processing. Its variants ([3,10,11,15,19]) demonstrate promising results compared with CNN-based methods and DETR by improving the design of queries for faster training convergence. Built upon Conditional-DETR, DAB-DETR [10] introduces an analytic study of how query design affects rectangle object detection. Specifically, it models object query as 4D dynamic anchor boxes (x, y, w, h) and iteratively refine them by a sequence of Transformer decoders. However, recent studies on Transformer-based detection methods are designed for rectangle object detection in computer vision, which are not specifically designed for circular objects in medical images.

In this paper, we introduce CircleFormer, a Transformer-based circular object detection for medical image analysis. Inspired by DAB-DETR, we propose to use an anchor circle (x, y, r) as the query for circular object detection, where (x, y) is the center of the circle and r is the radius. We propose a novel circle cross attention module which enables us to apply circle center (x, y) to extract image features around a circle and make use of circle radius to modulate the cross attention map. In addition, a circle matching loss is adopted in the set-to-set prediction part to process circular predictions. In this way, our design of CircleFormer lends itself to circular object detection. We evaluate our CircleFormer on the public MoNuSeg dataset for nuclei detection in whole slide images. Experimental results show that our method outperforms both CNN-based methods for box detection and circular object detection. It also achieves superior results compared with recently Transformer-based box detection approaches. Meanwhile, we carry out ablation studies to demonstrate the effectiveness of each proposed component. To further study the generalization ability of our approach, we add a simple segmentation branch to CircleFormer following the recent query based instance segmentation models [2,17] and verify its performance on MoNuSeg as well.

2 Method

2.1 Overview

Our CircleFormer (Fig. 1) consists of a CNN backbone, a Transformer encoder module, a Transformer decoder and a prediction head to generate circular object results. The detail of the Transformer decoder is illustrated in Fig. 2.

2.2 Representing Query with Anchor Circle

Inspired by DAB-DETR, we represent queries in Transformer-based circular object detection with anchor circles. We denote $C_i = (x_i, y_i, r_i)$ as the i-th anchor, $x_i, y_i, r_i \in \mathbb{R}$. Its corresponding content part and positional part are $Z_i \in \mathbb{R}^D$ and $P_i \in \mathbb{R}^D$, respectively. The positional query P_i is calculated by:

$$P_i = \text{MLP}(\text{PE}(C_i)), \text{PE}(C_i) = \text{PE}(x_i, y_i, r_i) = \text{Concat}(\text{PE}(x_i), \text{PE}(y_i), \text{PE}(r_i)), \tag{1}$$

where positional encoding (PE) generates embeddings from floating point numbers, and the parameters of the MLP are shared among all layers.

In Transformer decoder, the self-attention and cross-attention are written as:

$$\text{Self-Attn}: Q_i = Z_i + P_i, K_i = Z_i + P_i, V_i = Z_i \tag{2}$$

$$\begin{aligned} \text{Cross-attn}: Q_i &= \text{Concat}(Z_i, \text{PE}(x_i, y_i) \cdot \text{MLP}^{(\text{csq})}(Z_i)) \\ K_{x,y} &= \text{Concat}(F_{x,y}, \text{PE}(x, y)),\ V_{x,y} = F_{x.y}, \end{aligned} \tag{3}$$

where $F_{x,y} \in \mathbb{R}^D$ denote the image feature at position (x, y) and an $\text{MLP}^{(\text{csq})} : \mathbb{R}^D \rightarrow \mathbb{R}^D$ is used to obtain a scaled vector conditioned on content information for a query.

By representing a circle query as (x, y, r), we can refine the circle query layer-by-layer in the Transformer decoder. Specifically, each Transformer decoder estimates relative circle information $(\Delta x, \Delta y, \Delta r)$. In this way, the circle query representation is suitable for circular object detection and is able to accelerate the learning convergence via layer-by-layer refinement scheme.

2.3 Circle Cross Attention

We propose circle-modulated attention and deformable circle cross attention to consider size information of circular object detection in cross attention module.

Circle-modulated Attention. The circle radius modulated positional attention map provides benefits to extract image features of objects with different scales.

$$\text{MA}((x, y), (x_{ref}, y_{ref})) = (\text{PE}(x) \cdot \text{PE}(x_{ref}) \frac{r_{i,ref}}{r_i} + \text{PE}(y) \cdot \text{PE}(y_{ref}) \frac{r_{i,ref}}{r_i}) / \sqrt{D}, \tag{4}$$

where r_i is the radius of the circle anchor A_i, and $r_{i,ref}$ is the reference radius calculated by $r_{i,ref} = \text{sigmoid}(\text{MLP}(C_i))$. sigmoid is used to normalize the prediction $r_{i,ref}$ to the range $[0, 1]$.

Deformable Circle Cross Attention. We modify standard deformable attention to deformable circle cross attention by applying radius information as constraint. Given an input feature map $F \in \mathbb{R}^{C \times H \times W}$, let i index a query element with content feature Z_i and a reference point P_i, the deformable circle cross attention feature is calculated by:

$$\begin{aligned} CDA(Z_i, P_i, F) = \sum_{m=1}^{M} W_m \sum_{k=1}^{K} Attn_{mik} \dot{W}'_m F \left((P_{ix} + \Delta r_{mik} \right. \\ \left. \times r_{i,ref} \times \cos \Delta\theta_{mik}, P_{iy} + \Delta r_{mik} \times r_{i,ref} \times \sin \Delta\theta_{mik}) \right), \end{aligned} \tag{5}$$

where m indexes the attention head, k indexes the sampled keys. M and K are the number of multi-heads and the total sampled key number. $W'_m \in \mathbb{R}^{D \times d}$, $W_m \in \mathbb{R}^{d \times D}$ are the learnable weights and $d = D/M$. $Attn_{mik}$ denotes attention weight of the kth sampling point in the mth attention head. Δr_{mik} and $\Delta\theta_{mik}$ are radius offset and angle offset, $r_{i,ref}$ is the reference radius. In circle deformable attention, we transform the offset in polar coordinates to Cartesian coordinates so that the reference point ends up in the circle anchor.

Rather than initialize the reference points by uniformly sampling within the rectangle as does Deformable DETR, we explore two ways to initialize the reference points within a circle, random sampling (CDA-r) and uniform sampling (CDA-c) (As in Fig. 3). Experiments show that CDA-c initialization of reference points outperforms others.

2.4 Circle Regression

A circle is predicted from a decoder embedding as $\hat{c}_i = \text{sigmoid}(\text{FFN}(f_i) + [A_i])$, where f is the decoder embedding. $\hat{c}_i = (\hat{x},\ \hat{y},\ \hat{r})$ consists of the circle center and circle radius. sigmoid is used to normalize the prediction $\hat{c}$ to the range $[0, 1]$. FFN aims to predict the unnormalized box, A_i is a circle anchor.

2.5 Circle Instance Segmentation

A mask is predicted from a decoder embedding by $\hat{m}_i = \text{FFN}(\text{FFN}(f_i) + f_i)$, where f is the decoder embedding. $\hat{m}_i \in \mathbb{R}^{28\times28}$ is the predicted mask. We use dice and BCE as the segmentation loss: $\mathcal{L}_{seg} = \lambda_{dice}\mathcal{L}_{dice}(m_i, \hat{m}_i) + \lambda_{bce}\mathcal{L}_{bce}(m_i, \hat{m}_i)$ between prediction $\hat{m}_i$ and the groundtruth m_i.

2.6 Generalized Circle IoU

CircleNet extends intersection over union (IoU) of bounding boxes to circle IoU (cIoU) and shows that the cIOU is a valid overlap metric for detection of circular objects in medical images. To address the difficulty optimizing non-overlapping bounding boxes, generalized IoU (GIoU) [13] is introduced as a loss for rectangle object detection tasks. We propose a generalized circle IoU (gCIoU) to compute the similarity between two circles: $gCIoU = \frac{C_A \cap C_B}{C_A \cup C_B} - \frac{|C_C - (C_A \cup C_B)|}{C_C}$, where C_A and C_B denotes two circles, and C_C is the smallest circle containing these two circles. We show that gCIoU can bring consistent improvement on circular object detection. Figure 4 shows the different measurements between two rectangles and circles. Different from CircleNet that only uses cIoU in the evaluation, we incorporate gCIoU in the training step. Then, we define the circle loss as: $\mathcal{L}_{circle}(c, \hat{c}) = \lambda_{gciou}\mathcal{L}_{gciou}(c, \hat{c}) + \lambda_c\|c - \hat{c}\|_1$, while $\mathcal{L}_{gciou}$ is generalized circle IoU loss, $\|\cdot\|_1$ is ℓ_1 loss, and $\lambda_{gciou}, \lambda_c \in \mathbb{R}$ are hyperparameters.

Circle Training Loss. Following DETR, i-th each element of the groundtruth set is $y_i = (l_i, c_i)$, where l_i is the target class label (which may be $\varnothing$) and $c_i = (x, y, r)$. We define the matching cost between the predictions and the groundtruth set as:

$$\mathcal{L}_{match}(y_i, \hat{y}_{\sigma(i)}) = \mathbb{I}_{\{l_i \neq \varnothing\}}\lambda_{focal}\mathcal{L}_{focal}(l_i, \hat{l}_{\sigma(i)}) + \mathbb{I}_{\{l_i \neq \varnothing\}}\mathcal{L}_{circle}(c_i, \hat{c}_{\sigma(i)}), \quad (6)$$

where $\sigma \in \mathfrak{S}_N$ is a permutation of all prediction elements, $\hat{y}_{\sigma(i)} = (\hat{l}_{\sigma(i)}, \hat{c}_{\sigma(i)})$ is the prediction, $\lambda_{focal} \in \mathbb{R}$ are hyperparameters, and $\mathcal{L}_{focal}$ is focal loss [8].

Finally, the overall loss is:

$$\mathcal{L}_{loss}(y_i, \hat{y}_{\hat{\sigma}(i)}) = \lambda_{focal}\mathcal{L}_{focal}(l_i, \hat{l}_{\hat{\sigma}(i)}) + \mathbb{I}_{\{l_i \neq \varnothing\}}\mathcal{L}_{circle}(c_i, \hat{c}_{\hat{\sigma}(i)}) + \mathcal{L}_{seg}, \quad (7)$$

where $\hat{\sigma}(i)$ is the index of prediction $\hat{y}$ corresponding to the i-th ground truth y after completing the match. m_i is the ground truth obtained by RoI Align [5] corresponding to $\hat{m}_i$.

3 Experiment

3.1 Dataset and Evaluation

MoNuSeg Dataset. MoNuSeg dataset is a public dataset from the 2018 Multi-Organ Nuclei Segmentation Challenge [6]. It contains 30 training/validataion tissue images sampled from a separate whole slide image of H&E stained tissue and 14 testing images of lung and brain tissue images. Following [16], we randomly sample 10 patches with size 512 ×512 from each image and create 200 training images, 100 validation images and 140 testing images.

Evaluation Metrics. We use AP for nuclei detection evaluation metrics as in CircleNet [16], and AP^m for the instance segmentation evaluation metrics. S and M are used to measure the performance of small scale with area less than 32^2 and median scale with area between 32^2 and 96^2.

3.2 Implementation Details

Two variants of our proposed method for nuclei detection, CircleFormer and CircleFormer-D are built with a circle cross attention module and a deformable circle cross attention module, respectively. CircleFormer-D-Joint (Ours) extends CircleFormer-D to include instance segmentation as additional output. All the models are with ResNet50 as backbone and the number of Transformer encoders and decoders is set to 6. The MLPs of the prediction heads share the same parameters. Since the maximum number of objects per image in the dataset is close to 1000, we set the number of queries to 1000. The parameter of focal loss for classification is set to $\alpha = 0.25$, $\gamma = 0.1$. λ_{focal} is set to 2.0 in the matching step and $\lambda_{focal} = 1.0$ in the final circle loss. We use $\lambda_{iou} = 2.0$, $\lambda_c = 5.0$, $\lambda_{dice} = 8.0$ and $\lambda_{bce} = 2.0$ in the experiments. All the models are initialized with the COCO pre-trained model [9].

3.3 Main Results

In Table 1, for nuclei detection, we compare our CircleFormer with CNN-based box detection, CNN-based circle detection and Transformer-based box detection. Compared to Faster-RCNN [12] with ResNet50 as backbone, CircleFormer and CircleFormer-D significantly improve box AP by 8.1% and 11.3%, respectively. CircleFormer and CircleFormer-D also surpass CircleNet [16] by 1.0% and 4.2% box AP. In summary, our CircleFormer designed for circular object detection achieves superior performance compared to both CNN-based box detection and CNN-based circle detection approaches.

Our CircleFormer with detection head also yields better performance than Transformer-based methods. DETR can not produce satisfied results due to its low convergence. CircleFormer imporoves box AP by 4.0% compared to DAB-DETR, and CircleFormer-D improves box AP by 13.4% and 3.3% compared to Deformable-DETR and DAB-Deformable-DETR. CircleFormer-D-Joint which

Table 1. Results of nuclei detection on Monuseg Dataset. Best and second-best results are colored red and blue, respectively. MS: Multi-Scale.

Methods	Output	MS	Backbone	AP↑	$AP_{(50)}$ ↑	$AP_{(75)}$ ↑	$AP_{(S)}$ ↑	$AP_{(M)}$ ↑
Faster-RCNN [12]	Box		ResNet-50	41.6	75.0	42.1	41.6	38.3
Faster-RCNN [12]	Box		ResNet-101	40.9	77.5	37.2	41.0	33.9
CornerNet [7]	Box		Hourglass-104	24.4	52.3	18.1	32.8	6.4
CenterNet-HG [18]	Box		Hourglass-104	44.7	84.6	42.7	45.1	39.5
CenterNet-DLA [18]	Box		DLA	39.9	82.6	31.5	40.3	33.8
CircleNet-HG [16]	Circle		Hourglass-104	48.7	85.6	50.9	49.9	33.7
CircleNet-DLA [16]	Circle		DLA	48.6	85.5	51.6	49.9	30.5
DETR [1]	Box		ResNet50	22.6	52.3	14.6	23.9	18.2
Deformable-DETR [19]	Box	✓	ResNet50	39.5	81.0	32.7	40.2	18.5
DAB-DETR [10]	Box		ResNet50	45.7	88.9	41.9	46.3	34.8
DAB-D-DETR [10]	Box	✓	ResNet50	49.6	**89.5**	51.5	50.1	31.9
CircleFormer	Circle		ResNet50	49.7	88.8	50.9	51.1	35.4
CircleFormer-D	Circle	✓	ResNet50	52.9	89.6	58.7	54.1	31.7
CircleFormer-D-Joint	Circle	✓	ResNet50	53.0	90.0	59.0	53.9	32.8

Table 2. Results of nuclei joint detection and segmentation on Monuseg Dataset. All the methods are with ResNet50 as backbone.

	Detection					Segmentation				
Methods	AP	$AP_{(50)}$	$AP_{(75)}$	$AP_{(S)}$	$AP_{(M)}$	AP^m	$AP^m_{(50)}$	$AP^m_{(75)}$	$AP^m_{(S)}$	$AP^m_{(M)}$
QueryInst [2]	40.2	77.7	36.7	40.8	21.4	38.0	76.2	33.6	38.0	39.4
SOIT [17]	44.8	82.6	45.2	45.5	27.5	41.3	80.6	38.8	41.3	40.7
Deformable-DETR-Joint	45.7	86.7	43.6	46.2	28.2	43.5	**84.8**	40.5	43.5	42.3
CircleFormer-D-Joint	**53.0**	**90.0**	**59.0**	**53.8**	**32.8**	**44.4**	84.5	**43.5**	**44.4**	**45.3**

jointly outputs detection and segmentation results additionally boosts the detection results of CircleFormer-D.

Experiments of joint nuclei detection and segmentation are listed in Table 2. Our method outperforms QueryInst [2], a CNN-based instance segmentation method and SOIT [17], an Transformer-based instance segmentation approach. We extend Transformer-based box detection method to provide additional segmentation output inside the detection region, denoted as Deformable-DETR-Joint. Our method with circular query representation largely improves both detection and segmentation results.

To summarize, our method with only detection head outperforms both CNN-based methods and Transformer based approaches in most evaluation metrics for circular nuclei detection task. Our CircleFormer-D-Joint provides superior results compared to CNN-based and Transformer-based instance segmentation methods. Also, our method with joint detection and segmentation outputs also improves the detection-only setting. We have provided additional visual analysis in the open source code repository.

3.4 Ablation Studies

We conduct ablation studies with CircleFormer on the nuclei detection task (Table 5).

Table 3. Results of the ablation study analyzing the effects of proposed components in CircleFormer on Monuseg Dataset. *wh*-MA: wh-Modulated Attention; *c*-MA: circle-Modulated Attention; SDA: standard deformable attention; CDA-r: Circle Deformable Attention with random initialization; CDA-c: Circle Deformable Attention with cirle initialization. † denotes CircleFormer. ‡ denotes CircleFormer-D.

Method	IoU			Cross Attention					AP				
	gIoU	CIoU	gCIoU	*wh*-MA	c-MA	SDA	CDA-r	CDA-c	AP ↑	$AP_{(50)}$ ↑	$AP_{(75)}$ ↑	$AP_{(S)}$ ↑	$AP_{(M)}$ ↑
	✓			✓		-	-	-	45.7	88.9	41.9	46.3	34.8
		✓			✓	-	-	-	48.6	88.3	48.8	49.6	30.7
†			✓		✓	-	-	-	49.7	88.8	50.9	51.1	35.4
	✓			-	-	✓			49.6	89.5	51.5	50.1	31.9
		✓		-	-	✓			50.8	88.2	54.4	51.7	26.6
		✓		-	-		✓		51.1	87.9	55.8	52.5	23.2
		✓		-	-			✓	51.1	86.8	56.6	52.7	30.1
			✓	-	-	✓			51.1	87.6	55.6	52.7	29.1
			✓	-	-		✓		51.8	88.4	57.2	53.1	30.2
‡			✓	-	-			✓	**52.9**	**89.6**	**58.7**	**54.1**	31.7

Table 4. Ablation Study of number of Multi-Head.

#	AP↑	$AP_{(50)}$ ↑	$AP_{(75)}$ ↑	$AP_{(S)}$ ↑	$AP_{(M)}$ ↑
1	50.3	86.9	54.4	51.6	30.3
2	50.4	87.5	53.9	51.6	**33.3**
4	51.3	88.6	54.9	52.2	31.1
8	**52.9**	**89.6**	**58.7**	**54.1**	31.7
16	50.4	88.2	53.6	51.6	28.3

Table 5. Ablation Study of number of reference points.

#	AP↑	$AP_{(50)}$ ↑	$AP_{(75)}$ ↑	$AP_{(S)}$ ↑	$AP_{(M)}$ ↑
1	49.3	86.4	52.5	50.6	25.6
2	51.2	88.4	55.7	52.4	29.6
4	**52.9**	**89.6**	**58.7**	**54.1**	**31.7**
8	50.8	88.1	54.7	52.0	29.2

Effects of the Proposed Components. For simplicity, we denote the two parts of Table 3 as P1 and P2.

In CircleFormer, the proposed circle-Modulated attention (c-MA) improves the performance of box AP from 45.7% to 48.6% box AP (Row 1 and Row 2 in P1). We replaced circle IoU (CIoU) loss with generalized circle IoU (gCIoU) loss, the performance is further boosted by 2.2% (Row 2 and Row 3 in P1).

We obtain similar observations of CircleFormer-D. When using standard deformable attention (SDA), learning cIoU loss gives a 1.2% improvement on box AP compared to using box IoU (Row 1 and Row 2 in P2). Replacing CIoU with gCIoU, the performances of SDA (Row 2 and Row 5 in P2), CDA-r (Row 3 and Row 6 in P2) and CDA-c (Row 4 and Row 7 in P2) are boostd by 0.3% box AP, 0.7% box AP and 1.8% box AP, respectively. Results show that the proposed gCIoU is a favorable loss for circular object detection.

Two multi-head initialization methods, random sampling (CDA-r) and uniform sampling (CDA-c), achieve similar results (Row 3 and Row 4 in P2) and both surpass SDA by 0.3% box AP (Row 2 and Row 3 in P2). By using gCIoU, CDA-r and CDA-c initialization methods surpasses SDA 1.1% box AP (Row 5 and Row 6 in P2), and 1.8% box AP (Row 5 and Row 7 in P2), respectively.

Numbers of Multi-head and Reference Points. We discuss how the number of Multi-heads in the Decoder affects the CircleFormer-D-DETR model. We vary the number of heads for multi-head attention and the performance of the model is shown in the Table 4. We find that the performance increases gradually as the number of heads increases up to 8. However, the performance drops when the number of head is 16. We assume increasing the number of heads brings too many parameters and makes the model difficult to converge. Similarly, we study the impact of the number of reference points in the cross attention module. We find that 4 reference points give the best performance. Therefore, we choose to use 8 attention heads of decoder and use 4 reference points in the cross attention module through all the experiments.

4 Conclusion

In this paper, we introduce CircleFormer, a Transformer-based circular medical object detection method. It formulates object queries as anchor circles and refines them layer-by-layer in Transformer decoders. In addition, we also present a circle cross attention module to compute the key-to-image similarity which can not only pool image features at the circle center but also leverage scale information of a circle object. We also extend CircleFormer to achieve instance segmentation with circle detection results. To this end, our CircleFormer is specifically designed for circular object analysis with DETR scheme.

Acknowledgments. This work is supported in part by NSFC (61973294), Anhui Provincial Key R&D Program (2022i01020020), and the University Synergy Innovation Program of Anhui Province, China (GXXT-2021-030).

References

1. Carion, N., Massa, F., Synnaeve, G., Usunier, N., Kirillov, A., Zagoruyko, S.: End-to-end object detection with transformers. In: Vedaldi, A., Bischof, H., Brox, T., Frahm, J.-M. (eds.) ECCV 2020. LNCS, vol. 12346, pp. 213–229. Springer, Cham (2020). https://doi.org/10.1007/978-3-030-58452-8_13
2. Fang, Y., et al.: Instances as queries. In: Proceedings of the IEEE/CVF International Conference on Computer Vision, pp. 6910–6919 (2021)
3. Gao, P., Zheng, M., Wang, X., Dai, J., Li, H.: Fast convergence of detr with spatially modulated co-attention. In: Proceedings of the IEEE/CVF International Conference on Computer Vision, pp. 3621–3630 (2021)
4. Girshick, R., Donahue, J., Darrell, T., Malik, J.: Rich feature hierarchies for accurate object detection and semantic segmentation. In: Proceedings of the IEEE Conference on Computer Vision and Pattern Recognition, pp. 580–587 (2014)
5. He, K., Gkioxari, G., Dollár, P., Girshick, R.: Mask r-cnn. In: Proceedings of the IEEE International Conference on Computer Vision, pp. 2961–2969 (2017)
6. Kumar, N., et al.: A multi-organ nucleus segmentation challenge. IEEE Trans. Med. Imaging **39**(5), 1380–1391 (2019)
7. Law, H., Deng, J.: CornerNet: detecting objects as paired keypoints. In: Proceedings of the European Conference on Computer Vision (ECCV), pp. 734–750 (2018)

8. Lin, T.Y., Goyal, P., Girshick, R., He, K., Dollár, P.: Focal loss for dense object detection. In: Proceedings of the IEEE International Conference on Computer Vision, pp. 2980–2988 (2017)
9. Lin, T.-Y., et al.: Microsoft COCO: common objects in context. In: Fleet, D., Pajdla, T., Schiele, B., Tuytelaars, T. (eds.) ECCV 2014. LNCS, vol. 8693, pp. 740–755. Springer, Cham (2014). https://doi.org/10.1007/978-3-319-10602-1_48
10. Liu, S., et al.: DAB-DETR: dynamic anchor boxes are better queries for DETR. In: International Conference on Learning Representations (2022). https://openreview.net/forum?id=oMI9PjOb9Jl
11. Meng, D., et al.: Conditional detr for fast training convergence. In: Proceedings of the IEEE/CVF International Conference on Computer Vision, pp. 3651–3660 (2021)
12. Ren, S., He, K., Girshick, R., Sun, J.: Faster r-cnn: Towards real-time object detection with region proposal networks. In: Advances in Neural Information Processing Systems, vol. 28 (2015)
13. Rezatofighi, H., Tsoi, N., Gwak, J., Sadeghian, A., Reid, I., Savarese, S.: Generalized intersection over union: A metric and a loss for bounding box regression. In: Proceedings of the IEEE/CVF Conference on Computer Vision and Pattern Recognition, pp. 658–666 (2019)
14. Tian, Z., Shen, C., Chen, H., He, T.: Fcos: Fully convolutional one-stage object detection. In: Proceedings of the IEEE/CVF International Conference on Computer Vision, pp. 9627–9636 (2019)
15. Wang, Y., Zhang, X., Yang, T., Sun, J.: Anchor detr: query design for transformer-based detector. In: Proceedings of the AAAI Conference on Artificial Intelligence, vol. 36, pp. 2567–2575 (2022)
16. Yang, H., et al.: CircleNet: anchor-free glomerulus detection with circle representation. In: Martel, A.L., et al. (eds.) MICCAI 2020. LNCS, vol. 12264, pp. 35–44. Springer, Cham (2020). https://doi.org/10.1007/978-3-030-59719-1_4
17. Yu, X., Shi, D., Wei, X., Ren, Y., Ye, T., Tan, W.: Soit: segmenting objects with instance-aware transformers. In: Proceedings of the AAAI Conference on Artificial Intelligence, vol. 36, pp. 3188–3196 (2022)
18. Zhou, X., Wang, D., Krähenbühl, P.: Objects as points. arXiv preprint arXiv:1904.07850 (2019)
19. Zhu, X., Su, W., Lu, L., Li, B., Wang, X., Dai, J.: Deformable DETR: deformable transformers for end-to-end object detection. In: International Conference on Learning Representations (2021)

A Motion Transformer for Single Particle Tracking in Fluorescence Microscopy Images

Yudong Zhang[1,2] and Ge Yang[1,2(✉)]

[1] School of Artificial Intelligence, University of Chinese Academy of Sciences, Beijing, China
{zhangyudong2020,ge.yang}@ia.ac.cn
[2] State Key Laboratory of Multimodal Artificial Intelligence Systems, Institute of Automation, Chinese Academy of Sciences, Beijing, China

Abstract. Single particle tracking is an important image analysis technique widely used in biomedical sciences to follow the movement of subcellular structures, which typically appear as individual particles in fluorescence microscopy images. In practice, the low signal-to-noise ratio (SNR) of fluorescence microscopy images as well as the high density and complex movement of subcellular structures pose substantial technical challenges for accurate and robust tracking. In this paper, we propose a novel Transformer-based single particle tracking method called Motion Transformer Tracker (MoTT). By using its attention mechanism to learn complex particle behaviors from past and hypothetical future tracklets (i.e., fragments of trajectories), MoTT estimates the matching probabilities between each live/established tracklet and its multiple hypothesis tracklets simultaneously, as well as the existence probability and position of each live tracklet. Global optimization is then used to find the overall best matching for all live tracklets. For those tracklets with high existence probabilities but missing detections due to e.g., low SNRs, MoTT utilizes its estimated particle positions to substitute for the missed detections, a strategy we refer to as relinking in this study. Experiments have confirmed that this strategy substantially alleviates the impact of missed detections and enhances the robustness of our tracking method. Overall, our method substantially outperforms competing state-of-the-art methods on the ISBI Particle Tracking Challenge datasets. It provides a powerful tool for studying the complex spatiotemporal behavior of subcellular structures. The source code is publicly available at https://github.com/imzhangyd/MoTT.git.

Keywords: Single particle tracking · Transformer · Multi-object tracking

Supplementary Information The online version contains supplementary material available at https://doi.org/10.1007/978-3-031-43993-3_49.

H. Greenspan et al. (Eds.): MICCAI 2023, LNCS 14227, pp. 503–513, 2023.
https://doi.org/10.1007/978-3-031-43993-3_49

1 Introduction

A commonly used method to observe the dynamics of subcellular structures, such as microtubule tips, receptors, and vesicles, is to label them with fluorescent probes and then collect their videos using a fluorescence microscope. Since these subcellular structures are often smaller than the diffraction limit of visible light, they often appear as individual particles with Airy disk-like patterns in fluorescence microscopy images, as shown e.g., in Fig. 1. To quantitatively study the dynamic behavior of these structures in live cells, these trajectories need to be recovered using single particle tracking techniques [14].

Most single particle tracking methods follow a two-step paradigm: particle detection and particle linking. Specifically, particles are detected first in each frame of the image sequence. The detected particles are then linked between consecutive frames to recover their complete trajectories. The contributions of this paper focus on particle linking. Classical particle linking methods [5,9,14] are usually based on joint probability data association (JPDA) [10,20], multiple hypothesis tracking (MHT) [16,19], etc. Many classical methods have been developed and evaluated in the 2012 International Symposium on Biomedical Imaging (ISBI) Particle Tracking Challenge [6]. However, classical methods require manual tuning of many model parameters and are usually designed for a specific type of dynamics, making it difficult to apply to complex dynamics. In addition, the performance of these methods tends to degrade when tracking dense particles.

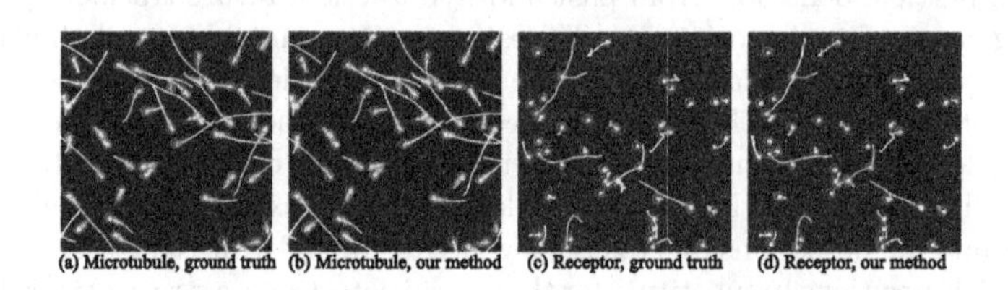

Fig. 1. Tracking performance of our method. (a–b) ground truth trajectories of microtubule tips in (a) versus trajectories recovered by our method in (b). (c–d) ground truth trajectories of receptors in (c) versus trajectories recovered by our method in (d). (a–d) colors are chosen randomly to differentiate between individual trajectories.

Deep learning provides a technique for automatically learning feature patterns and has been bringing performance improvements to many tasks. Recently, many deep learning-based single particle tracking methods have been developed. Many methods [21,25,26,30] use long short-term memory (LSTM) [13] modules to learn particle behavior. However, in [30], the matching probabilities between each tracklet and its multiple candidates are calculated independently, and there is no information exchange between multiple candidates. In [26,30], only detections in the next frame are used as candidates, which contain fewer motion features compared to hypothetical future tracklets. In [21,25], the number of their subnetworks grows exponentially with the depth of the hypothesis tree, making the network huge. And the trajectories will be disconnected due to missing detections. In addition, the source codes of most deep learning-based single particle tracking methods are not available, making them difficult to use for non-experts.

Cell tracking is closely related to particle tracking. There are different classes of cell tracking methods. An important category is tracking-by-evolution [7], which assumes spatiotemporal overlap between corresponding cells. It is not suitable for tracking particles because they generally do not overlap between frames. Another important category is tracking-by-detection. Some methods [18, 29] in this category assume coherence in motion of adjacent cells, which is not suitable for tracking particles that move independently from each other. There are also cell tracking methods [2] that rely on appearance features, which are not suitable for tracking particles because they lack appearance features.

Transformer [27] was originally proposed for modeling word sequences in machine translation tasks and has been used in various applications [3,4]. Recently, there have been many Transformer-based methods for motion forecasting [11,17,23], which improve the performance of motion forecasting in natural scenes (e.g., pedestrians, cars.). Compared to LSTM, Transformer shows advantages in sequence modeling by using the attention mechanism instead of sequence memory. However, to the best of our knowledge, Transformer has not been used for single particle tracking in fluorescence microscopy images.

In this paper, we propose a Transformer-based single particle tracking method MoTT, which is effective for different motion modes and different density levels of subcellular structures. The main contributions of our work are as follows: (1) We have developed a novel Transformer-based single particle tracking method MoTT. The attention mechanism of the Transformer is used to model complex particle behaviors from past and hypothetical future tracklets. To the best of our knowledge, we are the first to introduce Transformer-based networks to single particle tracking in fluorescence microscopy images; (2) We have designed an effective relinking strategy for those disconnected trajectories due to missed detections. Experiments have confirmed that the relinking strategy substantially alleviates the impact of missed detections and enhances the robustness of our tracking method; (3) Our method substantially outperforms competing state-of-the-art methods on the ISBI Particle Tracking Challenge dataset [6]. It provides a powerful tool for studying the complex spatiotemporal behavior of subcellular structures.

2 Method

Our particle tracking approach follows the two-step paradigm: particle detection and particle linking. We first use the detector DeepBlink [8] to detect particles at each frame. The detections of the first frame are initialized as the live tracklets. On each subsequent frame, we execute our particle linking method in four steps as follows. First (Sect. 2.1), for each live tracklet, we construct a hypothesis tree to generate its multiple hypothesis tracklets. Second (Sect. 2.2), all tracklets are

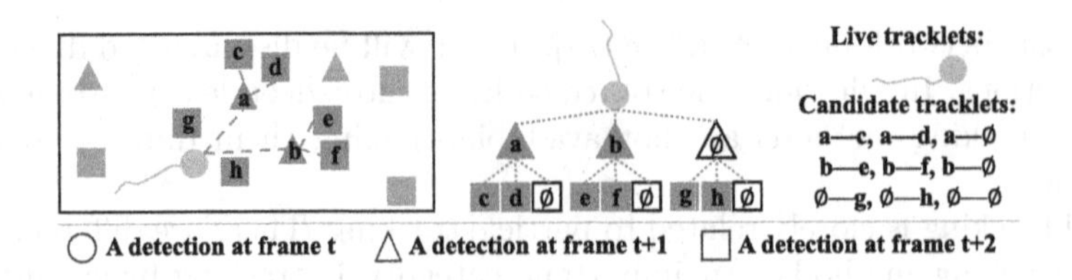

Fig. 2. An example of hypothesis tree construction with $m = 2$ and $d = 2$.

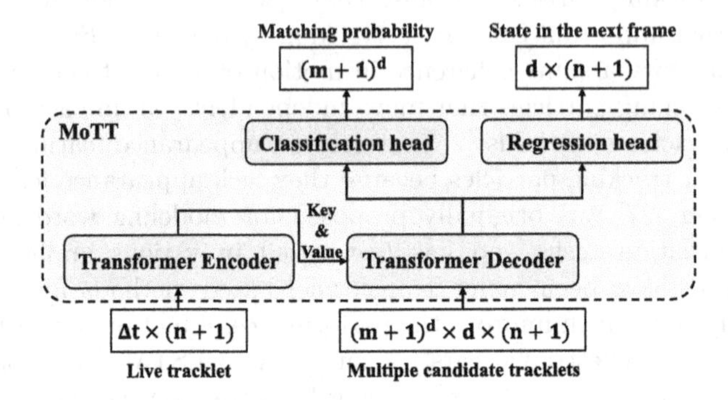

Fig. 3. MoTT network structure. Δt is the constant length of live tracklets, $n+1$ is the dimension number with the existence flag, d is the extended depth of hypothesis trees, $m+1$ is the number of hypothesis tracklets. See supplementary material for the details of the MoTT structure.

preprocessed and then fed into the proposed MoTT network to predict matching probabilities between each live tracklet and its multiple hypothesis tracklets, as well as the existence probability and position of each live tracklet in the next frame. Third (Sect. 2.3), we formulate a discrete optimization model to find the overall best matching for all live tracklets by maximizing the sum of the matching probabilities. Finally (Sect. 2.4), we design a track management scheme for trajectory initialization, updating, termination, and relinking.

2.1 Hypothesis Tree Construction

Assuming that the particle linking has been processed up to frame t. In order to find correspondence between the current live tracklets and the detections of frame $t+1$, we will build a hypothesis tree of depth d for each live tracklet, with its detection at frame t as the root node. To build the tree beyond the root node, we select m (real) detections of the next frame nearest to the current node as well as another null detection that represents a missing detection as children of the current node. If the current node is null, m (real) detections of the next frame nearest to the parent of the current node are selected. From the hypothesis tree, $(m+1)^d$ hypothesis tracklets will be obtained. Figure 2 shows an example of the hypothesis tree construction with $m = 2$ and $d = 2$.

2.2 MoTT Network

As shown in Fig. 3, We have designed a Transformer-based network, which contains a Transformer and two prediction head modules: classification head and regression head. Compared to the original Transformer, both the query masking and the positional encoding on the decoder are removed, since the input of the decoder is an unordered tracklet set. The classification head and regression head are constructed by fully connected layers.

For the generated tracklets from the previous step, the preprocessing is performed to make the length of all live tracklets equal to Δt, to convert position sequence to velocity sequence, and to add the existence flag making the coordinate dimension $n+1$. See supplementary material for the details of preprocessing. Then the preprocessed live tracklet is fed into the Transformer encoder, while the $(m+1)^d$ preprocessed hypothesis tracklets are fed into the Transformer decoder. The self-attention modules in the encoder and decoder are used to extract features of live tracklets and hypothesis tracklets, respectively. The cross-attention module is used to calculate the affinity between the live tracklet and its multiple candidate tracklets. The classification head outputs the predicted matching probabilities between the live tracklet and $(m+1)^d$ hypothesis tracklets. The regression head outputs the predicted existence probability and velocity of each live tracklet in the next frame. The existence probability represents the probability of the live tracklet existence in the next frame. The predicted velocity can be easily converted to the predicted position.

Training. We train the MoTT network in a supervised way, using the cross-entropy (CE) loss to supervise the output of the classification head and the mean square error (MSE) loss to supervise the output of the regression head. The target of classification head output is a class index in the range $[0, (m+1)^d)$ where $(m+1)^d$ is the number of hypothesis tracklets. The target of regression head output is the ground truth of the concatenation of normalized velocity and the existence flag.

Inference. In inference, we add a 1D max-pooling layer following the classification head to select the highest probability of the hypothesis tracklets with the same detection at frame $t+1$ as the matching probabilities between the live tracklet and the candidate detection at frame $t+1$. Then the $(m+1)$ predicted matching probabilities are normalized by softmax. The matching probabilities between the live tracklet and other detections besides the $m+1$ candidate detections are set to zero.

2.3 Modeling Discrete Optimization Problem

To find a one-to-one correspondence solution, we construct a discrete optimization formulation as (1), where p_{ij} is the predicted match probabilities between the live tracklet i and the detection j, and $a_{ij} \in \{0, 1\}$ is the indicator variable. In particular, $j = 0$ represents the null detection.

$$\max_{a} \sum_{i=1}^{M} \sum_{j=0}^{N} p_{ij} a_{ij}$$
$$s.t. \quad \sum_{j=0}^{N} a_{ij} = 1, i = 1, 2, ..., M \tag{1}$$
$$\sum_{i=1}^{M} a_{ij} \leqslant 1, j = 1, 2, ...N$$

The objective function aims at maximizing the sum of matching probabilities under the constraints that each live tracklet is matched to only one detection (real or null), and each real detection is matched by at most one tracklet. This optimization problem is solved by using Gurobi (a solver for mathematical programming) [12] to obtain a one-to-one correspondence solution.

2.4 Track Management

The one-to-one correspondence solution generally includes three situations. For each tracklet matched to a real detection, we add the matched real detection to the end of the live tracklet for updating. For each tracklet matched to a null detection, if the predicted existence probability is greater than a threshold p the predicted position is used to substitute for the null detection, else the live tracklet is terminated. In this way, the disconnected tracklets due to missing detections will be kept and be relinked when their detections emerge. For each detection that is not matched to any of the tracklets, a new live tracklet is initialized with this detection. After finishing particle linking on a whole movie, we remove the trajectories of length one, because they are considered false positive detections. See supplementary material for the details of track management.

3 Experimental Results

Datasets. The performance of our method is evaluated on ISBI Particle Tracking Challenge datasets (ISBI PTC, http://bioimageanalysis.org/track/) [6], which consist of movies of biological particles of four subcellular structures: microtubule tips, vesicles, receptors, and viruses. These movies cover three different particle motion modes, four different SNR levels, three different particle density levels, and two different coordinate dimensions. For each movie in the training set, we use the first 70% frames for training and the last 30% frames for validation.

Metrics. Metrics α, β, JSC_θ, JSC are used to evaluate the method performance [6]. Metric $\alpha \in [0, 1]$ quantifies the matching degree of ground truth and estimated tracks, while $\beta \in [0, \alpha]$ is penalized by false positive tracks additionally compared to α. $JSC_\theta \in [0, 1]$ and $JSC \in [0, 1]$ are the Jaccard similarity coefficients for entire tracks and track points, respectively. Higher values of the four metrics indicate better performance.

Table 1. Comparison with SOTA methods on microtubule movies of ISBI PTC datasets. Method 5, Method 1, and Method 2 are the overall top-three approaches in the 2012 ISBI Particle Tracking Challenge. See [6] for details of these three methods. "–" denotes that results are not reported in the papers. Bold represents the best performance. Trackpy [1], SORT [28], Bytetrack [31] and Ours use the same detections.

Density	Method	SNR = 4				SNR = 7			
		α	β	JSC_θ	JSC	α	β	JSC_θ	JSC
Low	Method5	0.750	0.728	0.917	0.874	0.803	0.787	0.939	0.894
	Method1	0.541	0.495	0.874	0.792	0.657	0.621	0.902	0.837
	Method2	0.562	0.259	0.356	0.369	0.694	0.686	0.959	**0.954**
	PMMS [22]	–	–	–	–	–	–	–	–
	DPT [26]	–	–	–	–	–	–	–	–
	SEF-GF-DPHT [25]	0.803	0.776	0.928	**0.890**	0.861	0.848	**0.970**	0.936
	DetNet-DPHT [21]	0.811	0.788	0.915	0.884	0.870	0.852	0.945	0.936
	Trackpy [1]	0.762	0.657	0.749	0.694	0.853	0.789	0.854	0.808
	SORT [28]	0.661	0.612	0.844	0.658	0.708	0.664	0.851	0.692
	Bytetrack [31]	0.800	**0.793**	**0.955**	0.840	0.801	0.792	0.955	0.813
	Ours	**0.835**	0.772	0.823	0.839	**0.904**	**0.870**	0.932	0.896
Med	Method5	0.460	0.402	0.696	0.523	0.511	0.450	0.739	0.558
	Method1	0.353	0.264	0.550	0.373	0.400	0.326	0.646	0.448
	Method2	0.465	0.225	0.363	0.341	0.564	0.535	**0.847**	0.763
	PMMS [22]	0.440	0.390	0.700	0.580	–	–	–	–
	DPT [26]	0.488	0.373	0.556	0.449	–	–	–	–
	SEF-GF-DPHT [25]	0.655	0.618	**0.839**	0.723	–	–	–	–
	DetNet-DPHT [21]	–	–	–	–	–	–	–	–
	Trackpy [1]	0.535	0.432	0.667	0.459	0.563	0.469	0.713	0.486
	SORT [28]	0.544	0.478	0.733	0.528	0.583	0.523	0.757	0.558
	Bytetrack [31]	0.555	0.495	0.717	0.552	0.582	0.528	0.721	0.567
	Ours	**0.814**	**0.719**	0.760	**0.769**	**0.869**	**0.792**	0.829	**0.823**
High	Method5	0.314	0.264	0.602	0.371	0.343	0.279	0.613	0.378
	Method1	0.272	0.210	0.544	0.299	0.293	0.231	0.582	0.322
	Method2	0.396	0.194	0.361	0.306	0.465	0.427	0.754	0.627
	PMMS [22]	0.350	0.300	0.630	0.460	–	–	–	–
	DPT [26]	0.414	0.313	0.524	0.389	–	–	–	–
	SEF-GF-DPHT [25]	0.548	0.501	**0.758**	0.605	–	–	–	–
	DetNet-DPHT [21]	–	–	–	–	–	–	–	–
	Trackpy [1]	0.410	0.311	0.603	0.340	0.410	0.315	0.622	0.335
	SORT [28]	0.432	0.354	0.645	0.407	0.465	0.390	0.664	0.436
	Bytetrack [31]	0.385	0.313	0.558	0.377	0.425	0.354	0.593	0.407
	Ours	**0.732**	**0.611**	0.660	**0.659**	**0.814**	**0.718**	**0.759**	**0.753**

Table 2. Comparison using the same ground truth detections on the microtubule, vesicle, and receptor scenarios.

Density	Method	Microtubule			Vesicle			Receptor		
		α	β	JSC_θ	α	β	JSC_θ	α	β	JSC_θ
Low	LAP [14]	0.850	0.852	0.923	**0.953**	**0.947**	**0.979**	0.940	0.931	0.962
	KF [15]	0.972	0.962	0.971	0.937	0.924	0.959	**0.964**	**0.955**	**0.972**
	Ours	**0.988**	**0.985**	**0.993**	0.926	0.891	0.925	0.949	0.921	0.943
Med	LAP [14]	0.486	0.394	0.662	0.753	0.703	0.704	0.742	0.686	0.826
	KF [15]	0.827	0.798	0.859	0.673	0.609	0.787	0.824	0.794	0.867
	Ours	**0.992**	**0.987**	**0.992**	**0.800**	**0.733**	**0.874**	**0.930**	**0.894**	**0.935**
High	LAP [14]	0.305	0.215	0.486	0.568	0.490	0.515	0.557	0.471	0.666
	KF [15]	0.679	0.616	0.735	0.477	0.389	0.643	0.658	0.591	0.724
	Ours	**0.987**	**0.980**	**0.988**	**0.652**	**0.544**	**0.748**	**0.903**	**0.851**	**0.910**

Implementation Details. In the following experiments, we set the length of live tracklets $\Delta t + 1 = 7$, the extension number $m = 4$, the depth of hypothesis tree $d = 2$, and the existence probability threshold p equals the mean of predicted existence probabilities of all live tracklets of current frame. See supplementary material for the ablation study on hyperparameters. We retrained the deepBlink network using simulated data generated by ISBI Challenge Track Generator. The MoTT model is implemented using PyTorch 1.8 and is trained on 1 NVIDIA GEFORCE RTX 2080 Ti with a batch size of 64 and an optimizer of Adam with an initial learning rate $lr = 10^{-3}$, as well as $betas = (0.9, 0.98)$ and $eps = 10^{-9}$.

3.1 Quantitative Performance

Comparison with the SOTA Methods. We compared our single particle tracking method with other SOTA methods, and the quantitative results on the microtubule scenario are shown in Table 1. Generally, our method outperforms other methods. Example visualization of tracking results can be found in Fig. 1.

Comparison Under the Same Ground Truth Detections. Under the ground truth detections, we compare our particle linking method with LAP [14] and KF (Kalman filter) [15]. The results in Table 2 show that our method generally outperforms other methods in both medium-density and high-density cases.

Effectiveness for All Scenarios. We perform our particle linking method using ground truth detections on the four scenarios with three density levels in the ISBI PTC dataset. The results (see the supplementary material) demonstrate the effectiveness of our method for both 2D and 3D single particle tracking.

3.2 Robustness Analysis

There are false positives (FPs) and false negatives (FNs) in actual detection results. Early study shows that FNs affect performance more than FPs [24]. We

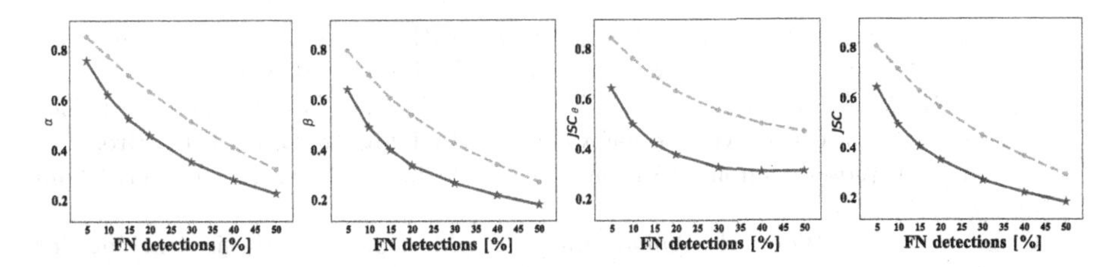

Fig. 4. Robustness analysis under different levels of FN detection. The performance with the relinking strategy (orange) is better than that without the relinking strategy (green) under different FN levels. (Color figure online)

evaluated the robustness of our method under different FN levels. The receptor particle with medium density is used in this experiment. We randomly drop 5%, 10%, 15%, 20%, 30%, 40%, 50% detections from ground truth detections. As Fig. 4 shows, the tracking performance with the relinking strategy is better than that without the relinking strategy under different FN levels. Therefore, the proposed relinking strategy alleviates the impact of missed detections and enhances the robustness of our tracking method.

4 Conclusion

In this paper, we proposed a novel Transformer-based method for single particle tracking in fluorescence microscopy images. We exploited the attention mechanism to model complex particle behaviors from past and hypothetical future tracklets. We designed a relinking strategy to alleviate the impact of missed detections due to e.g., low SNRs, and to enhance the robustness of our tracking method. Our experimental results show that our method is effective for all subcellular structures of ISBI Particle Tracking Challenge datasets, which cover different motion modes and different density levels. And our method achieves state-of-the-art performance on the microtubule movies of ISBI PTC datasets. In the future, we will test our method on other live cell fluorescence microscopy image sequences.

Acknowledgements. This work was supported in part by the Natural Science Foundation of China (grants 31971289, 91954201) and the Strategic Priority Research Program of the Chinese Academy of Sciences (grant XDB37040402).

References

1. Allan, D.B., Caswell, T., Keim, N.C., van der Wel, C.M., Verweij, R.W.: soft-matter/trackpy: v0.6.1, February 2023. https://doi.org/10.5281/zenodo.7670439
2. Ben-Haim, T., Raviv, T.R.: Graph neural network for cell tracking in microscopy videos. In: Avidan, S., Brostow, G., Cissé, M., Farinella, G.M., Hassner, T. (eds.) Computer Vision - ECCV 2022, pp. 610–626. Springer, Cham (2022). https://doi.org/10.1007/978-3-031-19803-8_36

3. Cai, J., et al.: MeMOT: multi-object tracking with memory. In: IEEE Computer Vision and Pattern Recognition Conference (CVPR), pp. 8090–8100 (2022)
4. Carion, N., Massa, F., Synnaeve, G., Usunier, N., Kirillov, A., Zagoruyko, S.: End-to-end object detection with transformers. In: Vedaldi, A., Bischof, H., Brox, T., Frahm, J.M. (eds.) Computer Vision - ECCV 2020, pp. 213–229. Springer, Cham (2020)
5. Chenouard, N., Bloch, I., Olivo-Marin, J.: Multiple hypothesis tracking for cluttered biological image sequences. IEEE Trans. Pattern Anal. Mach. Intell. (TPAMI) **35**(11), 2736–3750 (2013)
6. Chenouard, N., et al.: Objective comparison of particle tracking methods. Nat. Methods **11**(3), 281–289 (2014)
7. Dufour, A., Thibeaux, R., Labruyere, E., Guillen, N., Olivo-Marin, J.C.: 3-D active meshes: fast discrete deformable models for cell tracking in 3-D time-lapse microscopy. IEEE Trans. Image Process. **20**(7), 1925–1937 (2010)
8. Eichenberger, B.T., Zhan, Y., Rempfler, M., Giorgetti, L., Chao, J.A.: deepBlink: threshold-independent detection and localization of diffraction-limited spots. Nucleic Acids Res. **49**(13), 7292–7297 (2021)
9. Feng, L., Xu, Y., Yang, Y., Zheng, X.: Multiple dense particle tracking in fluorescence microscopy images based on multidimensional assignment. J. Struct. Biol. **173**(2), 219–228 (2011)
10. Fortmann, T., Bar-Shalom, Y., Scheffe, M.: Sonar tracking of multiple targets using joint probabilistic data association. IEEE J. Oceanic Eng. **8**(3), 173–184 (1983)
11. Giuliari, F., Hasan, I., Cristani, M., Galasso, F.: Transformer networks for trajectory forecasting. In: International Conference on Pattern Recognition (ICPR), pp. 10335–10342 (2021)
12. Gurobi Optimization, LLC: Gurobi Optimizer Reference Manual (2022). https://www.gurobi.com
13. Hochreiter, S., Schmidhuber, J.: Long short-term memory. Neural Comput. **9**(8), 1735–1780 (1997)
14. Jaqaman, K., et al.: Robust single-particle tracking in live-cell time-lapse sequences. Nat. Methods **5**(8), 695–702 (2008)
15. Kalman, R.E.: A new approach to linear filtering and prediction problems. J. Basic Eng. **82**(1), 35–45 (1960)
16. Kim, C., Li, F., Ciptadi, A., Rehg, J.M.: Multiple hypothesis tracking revisited. In: International Conference on Computer Vision (ICCV), pp. 4696–4704 (2015)
17. Liu, Y., Zhang, J., Fang, L., Jiang, Q., Zhou, B.: Multimodal motion prediction with stacked transformers. In: IEEE Computer Vision and Pattern Recognition Conference (CVPR), pp. 7577–7586 (2021)
18. Nguyen, J.P., Linder, A.N., Plummer, G.S., Shaevitz, J.W., Leifer, A.M.: Automatically tracking neurons in a moving and deforming brain. PLoS Comput. Biol. **13**(5), 1–19 (2017)
19. Reid, D.: An algorithm for tracking multiple targets. IEEE Trans. Autom. Control **24**(6), 843–854 (1979)
20. Rezatofighi, S.H., Milan, A., Zhang, Z., Shi, Q., Dick, A.R., Reid, I.D.: Joint probabilistic data association revisited. In: International Conference on Computer Vision (ICCV), pp. 3047–3055 (2015)
21. Ritter, C., Spilger, R., Lee, J.Y., Bartenschlager, R., Rohr, K.: Deep learning for particle detection and tracking in fluorescence microscopy images. In: IEEE International Symposium on Biomedical Imaging (ISBI), pp. 873–876 (2021)

22. Roudot, P., Ding, L., Jaqaman, K., Kervrann, C., Danuser, G.: Piecewise-stationary motion modeling and iterative smoothing to track heterogeneous particle motions in dense environments. IEEE Trans. Image Process. (TIP) **26**(11), 5395–5410 (2017)
23. Shi, S., Jiang, L., Dai, D., Schiele, B.: Motion transformer with global intention localization and local movement refinement. arXiv preprint arXiv:2209.13508 (2022)
24. Smal, I., Meijering, E.: Quantitative comparison of multiframe data association techniques for particle tracking in time-lapse fluorescence microscopy. Med. Image Anal. **24**(1), 163–189 (2015)
25. Spilger, R., et al.: A recurrent neural network for particle tracking in microscopy images using future information, track hypotheses, and multiple detections. IEEE Trans. Image Process. (TIP) **29**, 3681–3694 (2020)
26. Spilger, R., et al.: Deep particle tracker: automatic tracking of particles in fluorescence microscopy images using deep learning. In: Stoyanov, D., Taylor, Z., Carneiro, G., et al. (eds.) DLMIA ML-CDS 2018. LNCS, vol. 11045, pp. 128–136. Springer, Cham (2018). https://doi.org/10.1007/978-3-030-00889-5_15
27. Vaswani, A., et al.: Attention is all you need. In: Conference on Neural Information Processing Systems (NeurIPS), pp. 5998–6008 (2017)
28. Wojke, N., Bewley, A., Paulus, D.: Simple online and realtime tracking with a deep association metric. In: IEEE International Conference on Image Processing (ICIP), pp. 3645–3649 (2017)
29. Wu, Y., et al.: Rapid detection and recognition of whole brain activity in a freely behaving Caenorhabditis elegans. PLoS Comput. Biol. **18**(10), 1–27 (2022)
30. Yao, Y., Smal, I., Meijering, E.: Deep neural networks for data association in particle tracking. In: IEEE International Symposium on Biomedical Imaging (ISBI), pp. 458–461 (2018)
31. Zhang, Y., et al.: ByteTrack: multi-object tracking by associating every detection box. In: Avidan, S., Brostow, G., Cissé, M., Farinella, G.M., Hassner, T. (eds.) Computer Vision - ECCV 2022, pp. 1–21. Springer, Cham (2022). https://doi.org/10.1007/978-3-031-20047-2_1

B-Cos Aligned Transformers Learn Human-Interpretable Features

Manuel Tran[1,2,3], Amal Lahiani[1], Yashin Dicente Cid[1], Melanie Boxberg[3,5], Peter Lienemann[2,4], Christian Matek[2,6], Sophia J. Wagner[2,3], Fabian J. Theis[2,3], Eldad Klaiman[1], and Tingying Peng[2](✉)

[1] Roche Diagnostics, Penzberg, Germany
[2] Helmholtz AI, Helmholtz Munich, Neuherberg, Germany
tying84ster@gmail.com
[3] Technical University of Munich, Munich, Germany
[4] Ludwig Maximilian University of Munich, Munich, Germany
[5] Pathology Munich-North, Munich, Germany
[6] University Hospital Erlangen, Erlangen, Germany

Abstract. Vision Transformers (ViTs) and Swin Transformers (Swin) are currently state-of-the-art in computational pathology. However, domain experts are still reluctant to use these models due to their lack of interpretability. This is not surprising, as critical decisions need to be transparent and understandable. The most common approach to understanding transformers is to visualize their attention. However, attention maps of ViTs are often fragmented, leading to unsatisfactory explanations. Here, we introduce a novel architecture called the B-cos Vision Transformer (BvT) that is designed to be more interpretable. It replaces all linear transformations with the B-cos transform to promote weight-input alignment. In a blinded study, medical experts clearly ranked BvTs above ViTs, suggesting that our network is better at capturing biomedically relevant structures. This is also true for the B-cos Swin Transformer (Bwin). Compared to the Swin Transformer, it even improves the F1-score by up to 4.7% on two public datasets.

Keywords: transformer · self-attention · explainability · interpretability

1 Introduction

Making artificial neural networks more interpretable, transparent, and trustworthy remains one of the biggest challenges in deep learning. They are often still considered black boxes, limiting their application in safety-critical domains such

E. Klaiman—Equal contribution.

Supplementary Information The online version contains supplementary material available at https://doi.org/10.1007/978-3-031-43993-3_50.

H. Greenspan et al. (Eds.): MICCAI 2023, LNCS 14227, pp. 514–524, 2023.
https://doi.org/10.1007/978-3-031-43993-3_50

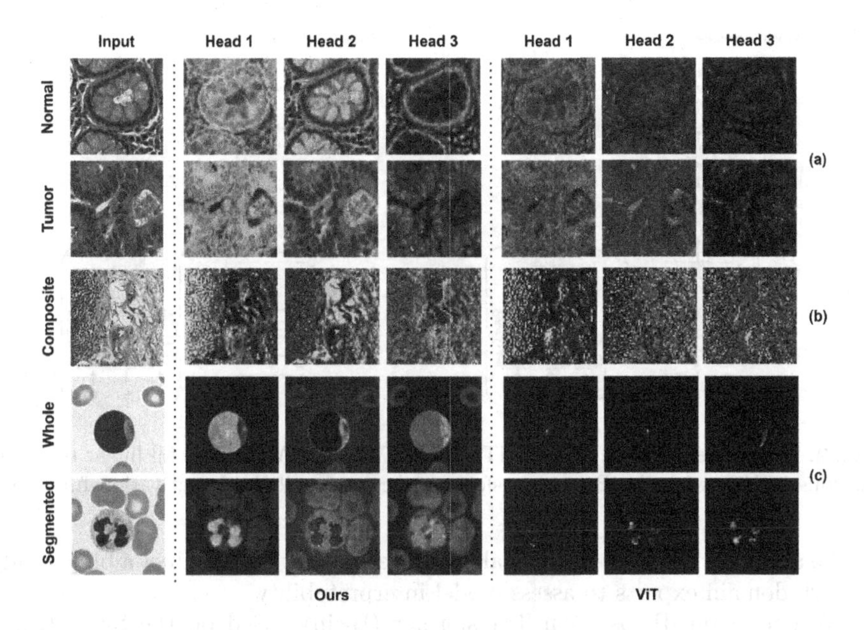

Fig. 1. Attention maps of ViT and BvT (ours) on the test set of (a) NCT-CRC-HE-100K, (b) TCGA-COAD-20X, and (c) Munich-AML-Morphology. BvT attends to various diagnostically relevant features such as cancer tissue, cells, and nuclei.

as healthcare. Histopathology is a prime example of this. For years, the number of pathologists has been decreasing while their workload has been increasing [23]. Consequently, the need for explainable computer-aided diagnostic tools has become more urgent.

As a result, research in explainable artificial intelligence is thriving [20]. Much of it focuses on convolutional neural networks (CNNs) [13]. However, with the rise of transformers [31] in computational pathology, and their increasing application to cancer classification, segmentation, survival prediction, and mutation detection tasks [26,32,33], the old tools need to be reconsidered. Visualizing filter maps does not work for transformers, and Grad-CAM [30] has known limitations for both CNNs and transformers.

The usual way to interpret transformer-based models is to plot their multi-head self-attention scores [8]. But these often lead to fragmented and unsatisfactory explanations [10]. In addition, there is an ongoing controversy about their trustworthiness [5]. To address these issues, we propose a novel family of transformer architectures based on the B-cos transform originally developed for CNNs [7]. By aligning the inputs and weights during training, the models are implicitly forced to learn more biomedically relevant and meaningful features (Fig. 1). Overall, our contributions are as follows:

- We propose the B-cos Vision Transformer (BvT) as a more explainable alternative to the Vision Transformer (ViT) [12].
- We extensively evaluate both models on three public datasets: NCT-CRC-HE-100K [18], TCGA-COAD-20X [19], Munich-AML-Morphology [25].

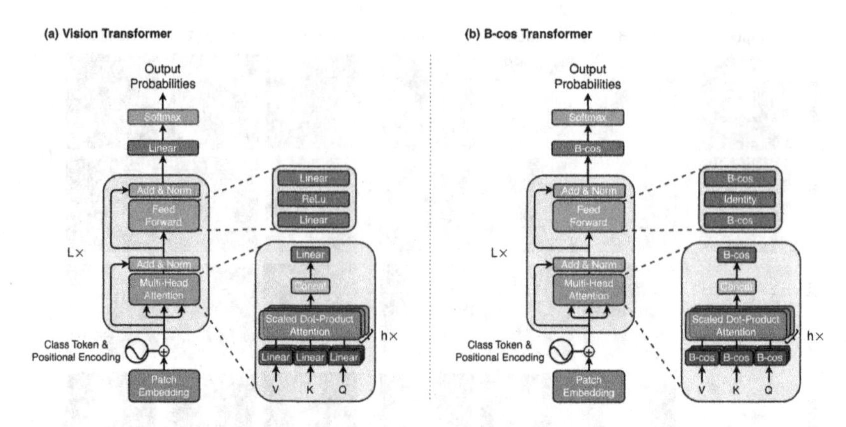

Fig. 2. The model architecture of ViT and BvT (ours). We replace all linear transformations in ViT with the B-cos transform and remove all ReLU activation functions.

- We apply various post-hoc visualization techniques and conduct a blind study with domain experts to assess model interpretability.
- We derive the B-cos Swin Transformer (Bwin) based on the Swin Transformer [21] (Swin) in a generalization study.

2 Related Work

Explainability, interpretability, and relevancy are terms used to describe the ability of machine learning models to provide insight into their decision-making process. Although these terms have subtle differences, they are often used interchangeably in the literature [15].

Recent research on understanding vision models has mostly focused on attribution methods [13,20], which aim to identify important parts of an image and highlight them in a saliency map. Gradient-based approaches like Grad-CAM [30] or attribution propagation strategies such as Deep Taylor Decomposition [27] and LRP [6] are commonly used methods. Perturbation-based techniques, such as SHAP [22], are another way to extract salient features from images. Besides saliency maps, one can also visualize the activations of the model using Activation Maximization [14].

However, it is still controversial whether the above methods can correctly reflect the behavior of the model and accurately explain the learned function (model-faithfulness [17]). For example, it has been shown that some saliency maps are independent of both the data on which the model was trained and the model parameters [2]. In addition, they are often considered unreliable for medical applications [4]. As a result, inherently interpretable models have been proposed as a more reliable and transparent solution. The most recent contribution are B-cos CNNs [7], which use a novel nonlinear transformation (the B-cos transformation) instead of the traditional linear transformation.

Compared to CNNs, there is limited research on understanding transformers beyond attention visualization [10]. Post-hoc methods such as Grad-CAM [30]

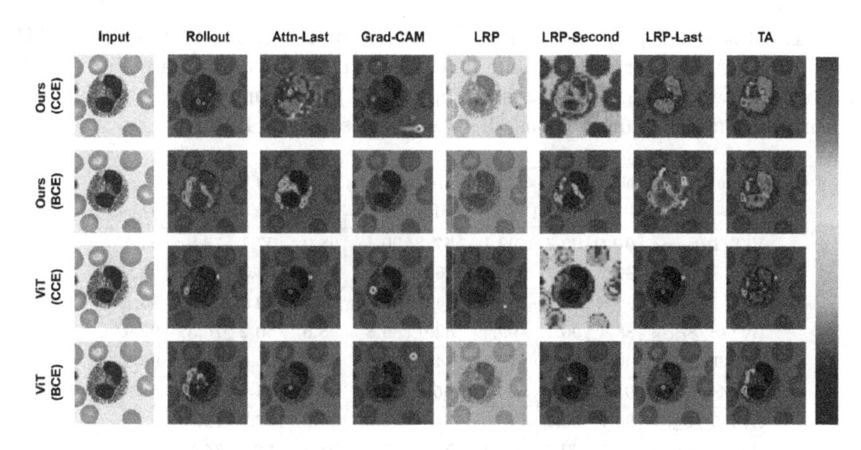

Fig. 3. Rollout, Attention-Last (Attn-Last), Grad-CAM, LRP, LRP of the second layer (LRP-Second), LRP of the last layer (LRP-Last), and Transformer Attribution (TA) applied on the test set of Munich-AML-Morphology. The image shows an eosinophil, which is characterized by its split, but connected nucleus, large specific granules (pink structures in the cytoplasm), and dense chromatin (dark spots inside the nuclei) [29]. Across all visualization techniques, BvT focuses on these exact features unlike ViT.

and Activation Maximization [14] used for CNNs can also be applied to transformers. But in practice, the focus is on visualizing the raw attention values (see Attention-Last [16], Integrated Attention Maps [12], Rollout [1], or Attention Flow [1]). More recent approaches such as Generic Attention [9], Transformer Attribution [10], and Conservative Propagation [3] go a step further and introduce novel visualization techniques that better integrate the attention modules with contributions from different parts of the network. Note that these methods are all post-hoc methods applied after training to visualize the model's reasoning.

On the other hand, the ConceptTransformer [28], achieves better explainability by cross-attending user-defined concept tokens in the classifier head during training. More recently, HIPT [11] combines multi-scale images and DINO [8] pre-training to learn hierarchical visual concepts in a self-supervised fashion. Unlike all of these methods, interpretability is already an integral part of our architecture. Therefore, these methods can be easily applied to our models. In Fig. 3 and Fig. 6, we show that the B-Cos Transformer produces superior feature maps over various post-hoc approaches – suggesting that our architecture does indeed learn human-plausible features that are independent of the specific visualization technique used.

3 Methods

We focus on the original Vision Transformer [12]: The input image is divided into non-overlapping patches, flattened, and projected into a latent space of dimension d. Class tokens [*cls*] are then prepended to these patch embeddings. In addition, positional encodings [*pos*] are added to preserve topological information. In the scaled dot-product attention [31], the model learns different features

Table 1. F1-score, top-1, and top-3 accuracy from the test set of NCT-CRC-HE-100K, Munich-AML-Morphology, and TCGA-COAD-20X. We compare ViT and BvT (ours) trained with categorical cross-entropy (CCE) and binary cross-entropy (BCE) loss using two model configurations: T/8 and S/8 (see Sect. 4).

Models	NCT			Munich			TCGA		
	F1	Top-1	Top-3	F1	Top-1	Top-3	F1	Top-1	Top-3
ViT-T/8_{CCE}	**90.9**	92.7	99.1	**57.3**	90.1	98.9	57.1	78.8	94.4
ViT-S/8_{CCE}	89.2	91.1	99.5	56.3	93.1	99.0	56.3	78.6	92.9
BvT-T/8_{CCE}	88.8	91.1	99.3	54.0	87.1	98.6	**61.0**	77.4	93.1
BvT-S/8_{CCE}	88.4	90.1	99.4	52.9	89.8	98.6	60.2	76.3	92.9
ViT-T/8_{BCE}	90.0	91.4	98.4	54.8	90.0	99.0	53.6	79.6	93.9
ViT-S/8_{BCE}	**90.2**	92.2	99.3	**55.4**	92.8	99.0	54.1	77.0	88.9
BvT-T/8_{BCE}	86.7	90.1	98.5	51.1	83.5	97.9	57.7	79.8	93.4
BvT-S/8_{BCE}	87.5	90.4	99.4	52.4	85.0	98.3	**59.0**	74.5	88.9

(query Q, key K, and value V) from the input vectors through a linear transformation. Both query and key are then correlated with a scaled dot-product and normalized with a softmax. These self-attention scores are then used to weight the value by importance:

$$\text{Attention}(Q, K, V) = \text{softmax}\left(\frac{QK^T}{\sqrt{d}}\right)V. \tag{1}$$

To extract more information, this process is repeated h times in parallel (multi-headed self-attention). Each self-attention layer is followed by a fully-connected layer consisting of two linear transformations and a ReLU activation.

We propose to replace all linear transforms in the original ViT (Fig. 2)

$$\text{Linear}(x, w) = w^T x = \|w\| \|x\| c(x, w), \tag{2}$$

$$c(x, w) = \cos(\angle(x, w)), \quad \angle\text{...angle between vectors} \tag{3}$$

with the B-cos* transform [7]

$$\text{B-cos}^*(x; w) = \underbrace{\|\hat{w}\|}_{=1} \|x\| |c(x, \hat{w})|^B \times \text{sgn}(c(x, \hat{w})), \tag{4}$$

where $B \in \mathbb{N}$. Similar to [7], an additional nonlinearity is applied after each B-cos* transform. Specifically, each input is processed by two B-cos* transforms, and the subsequent MaxOut activation passes only the larger output. This ensures that only weight vectors with higher cosine similarity to the inputs are selected, which further increases the alignment pressure during optimization. Thus, the final B-cos transform is given by

$$\text{B-cos}(x; w) = \max_{i \in \{1,2\}} \text{B-cos}^*(x; w_i). \tag{5}$$

To see the significance of these changes, we look at Eq. 4 and derive

$$\|\hat{w}\| = 1 \Rightarrow \text{B-cos}^*(x; w) \leq \|x\|. \tag{6}$$

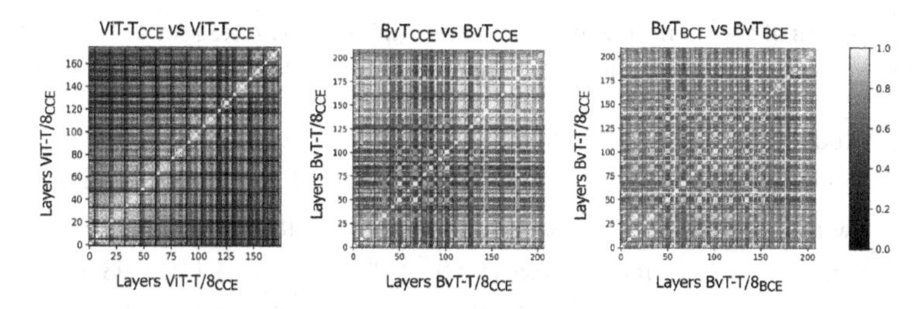

Fig. 4. We compute the central kernel alignment (CKA), which measures the representation similarity between each hidden layer. Since the B-cos transform aligns the weights with the inputs, BvT (ours) achieves a more uniform representation structure compared to ViT (values closer to 1). When trained with the binary cross-entropy loss (BCE) instead of the categorical cross-entropy loss (CCE), the alignment is higher.

Since $|c(x, \hat{w})| \leq 1$, equality is only achieved if x and w are collinear, i.e., if they are aligned. Intuitively, this forces the weight vector to be more similar to the input. Query, key, and value thus capture more patterns in an image – which the attention mechanism can then attend to. This can be shown visually by plotting the centered kernel alignment (CKA). It measures the similarity between layers by comparing their internal representation structure. Compared to ViTs, BvTs achieve a highly uniform representation across all layers (Fig. 4).

4 Implementation and Evaluation Details

Task-Based Evaluation: Cancer classification and segmentation is an important first step for many downstream tasks such as grading or staging. Therefore, we choose this problem as our target. We classify image patches from the public colorectal cancer dataset NCT-CRC-HE-100K [18]. We then apply our method to TCGA-COAD-20X [19], which consists of 38 annotated slides from the TCGA colorectal cancer cohort, to evaluate the effectiveness of transfer learning. This dataset is highly unbalanced and not color normalized compared

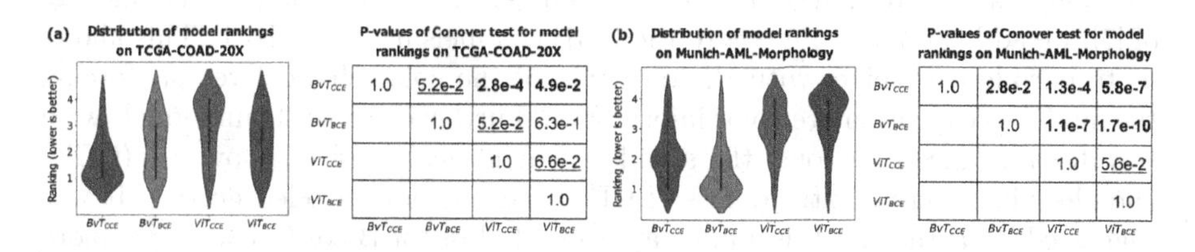

P-values of Conover test for model rankings on TCGA-COAD-20X

	BvT$_{CCE}$	BvT$_{BCE}$	ViT$_{CCE}$	ViT$_{BCE}$
BvT$_{CCE}$	1.0	5.2e-2	**2.8e-4**	**4.9e-2**
BvT$_{BCE}$		1.0	5.2e-2	6.3e-1
ViT$_{CCE}$			1.0	6.6e-2
ViT$_{BCE}$				1.0

P-values of Conover test for model rankings on Munich-AML-Morphology

	BvT$_{CCE}$	BvT$_{BCE}$	ViT$_{CCE}$	ViT$_{BCE}$
BvT$_{CCE}$	1.0	**2.8e-2**	**1.3e-4**	**5.8e-7**
BvT$_{BCE}$		1.0	**1.1e-7**	**1.7e-10**
ViT$_{CCE}$			1.0	5.6e-2
ViT$_{BCE}$				1.0

Fig. 5. In a blinded study, domain experts ranked models (lower is better) based on whether the models focus on biomedically relevant features that are known in the literature to be important for diagnosis. We then performed the Conover post-hoc test after Friedman with adjusted p-values according to the two-stage Benjamini-Hochberg procedure. BvT ranks above ViT with $p < 0.1$ (underlined) and $p < 0.05$ (bold).

Table 2. Results of Swin and Bwin (ours) experiments on the test set of NCT-CRC-HE-100K and Munich-AML-Morphology. We report F1-score, top-1, and top-3 accuracy.

Models	NCT			AML		
	F1	Top-1	Top-3	F1	Top-1	Top-3
Swin-T_{CCE}	89.1	92.1	99.0	48.2	94.1	98.8
Swin-T_{CCE} (modified)	89.8	92.0	99.6	49.1	94.2	98.9
Bwin-T_{CCE}	91.5	93.5	99.5	53.0	93.9	98.6
Bwin-T_{CCE} (modified)	**92.5**	94.3	99.6	**53.3**	93.8	98.7

to the first dataset. Additionally, we demonstrate that the B-cos Vision Transformer is adaptable to domains beyond histopathology by training the model on the single white blood cell dataset Munich-AML-Morphology [25], which is also highly unbalanced and also publicly available.

Domain-Expert Evaluation: Our primary objective is to develop an extension of the Vision Transformer that is more transparent and trusted by medical professionals. To assess this, we propose a blinded study with four steps: (i) randomly selecting images from the test set of TCGA-COAD-20X (32 samples) and Munich-AML-Morphology (56 samples), (ii) plotting the last-layer attention and transformer attributions for each image, (iii) anonymizing and randomly shuffling the outputs, (iv) submitting them to two domain experts in histology and cytology for evaluation. Most importantly, we show them all the available saliency maps without pre-selecting them to get their unbiased opinion.

Implementation Details: In our experiments, we compare different variants of the B-cos Vision Transformer and the Vision Transformer. Specifically, we implement two versions of ViT: ViT-T/8 and ViT-S/8. They only differ in parameter size (5M for T models and 22M for S models) and use the same patch size of 8. All BvT models (BvT-T/8 and BvT-S/8) are derivatives of the corresponding ViT models. The B-cos transform used in the BvT models has an exponent of $B = 2$. We use AdamW with a cosine learning rate scheduler for optimization and a separate validation set for hyperparameter selection. Following the findings of [7], we add $[1-r, 1-g, 1-b]$ to the RGB channels $[r, g, b]$ of BvT. This allows us to encode each pixel with the direction of the color channel vector, forcing the model to capture more color information. Furthermore, we train models with two different loss functions: the standard categorical cross-entropy loss (CCE) and the binary cross-entropy loss (BCE) with one-hot encoded entries. It was suggested in [7] that BCE is a more appropriate loss for B-cos CNNs. We explore whether this is also true for transformers in our experiments. Additional details on training, optimization, and datasets can be found in the Appendix.

5 Results and Discussion

Task-Based Evaluation: When trained from scratch, all BvT models underperform their ViT counterparts by about 2% on NCT-CRC-HE-100K and

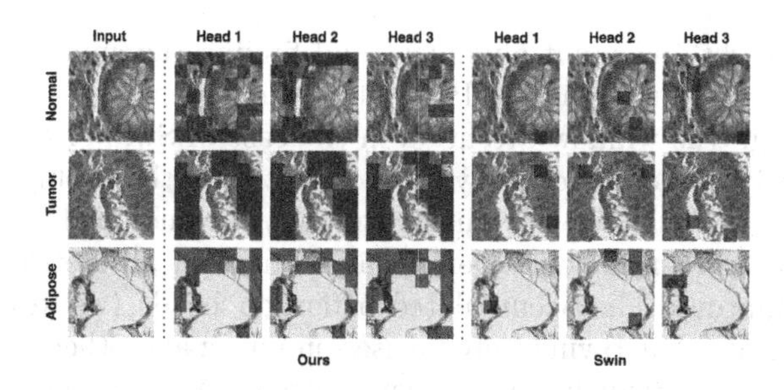

Fig. 6. Attention maps of the last layer of the modified Swin and Bwin (ours). Bwin focuses on cells and nuclei, while Swin mostly focuses on a few spots.

3% on Munich AML-Morphology (Table 1). However, when we use the pre-trained weights from NCT-CRC-HE-100K and transfer them to TCGA-COAD-20X for fine-tuning, BvT outperforms ViT by up to 5% (Table 1). We believe this is due to the simultaneous optimization of two objectives: classification loss and weight-input alignment. With a pre-trained model, BvT is likely to focus more on the former. In addition, we observe that models trained with BCE tend to perform worse than those trained with CCE. However, their saliency maps seem to be more interpretable (see Fig. 3).

Domain-Expert Evaluation: The results show that BvTs are significantly more trustworthy than ViTs ($p < 0.05$). This indicates that BvT consistently attends to biomedically relevant features such as cancer cells, nuclei, cytoplasm, or membrane [24] (Fig. 5). In many visualization techniques, we see that BvT, unlike ViT, focuses exclusively on these structures (Fig. 3). In contrast, ViT attributes high attention to seemingly irrelevant features, such as the edges of the cells. A third expert points out that ViT might overfit certain patterns in this dataset, which could aid the model in improving its performance.

6 Generalization to Other Architectures

We aim to explore whether the B-cos transform can enhance the interpretability of other transformer-based architectures. The Swin Transformer (Swin) [21] is a popular alternative to ViT (e.g., it is currently the SOTA feature extractor for histopathological images [33]). Swin utilizes window attention and feed-forward layers. In this study, we replace all its linear transforms with the B-cos transform, resulting in the B-cos Swin Transformer (Bwin). However, unlike BvT and ViT, it is not obvious how to visualize the window attention. Therefore, we introduce a modified variant here that has a regular ViT/BvT block in the last layer.

In our experiments (Table 2), we observe that Bwin outperforms Swin by up to 2.7% and 4.8% in F1-score on NCT-CRC-HE-100K and Munich-AML-Morphology, respectively. This is consistent with the observations made in Sect. 5: When BvT is trained from scratch, the model faces a trade-off between learning the weight and input alignment and finding the appropriate inductive bias to solve the classification task. By reintroducing many of the inductive biases of CNNs through the window attention in the case of Swin or transfer learning in the case of BvT, the model likely overcomes this initial problem.

Moreover, we would like to emphasize that the modified models have no negative impact on the model's performance. In fact, all metrics remain similar or even improve. The accumulated attention heads (we keep 50% of the mass) demonstrate that Bwin solely focuses on nuclei and other cellular features (Fig. 6). Conversely, Swin has very sparse attention heads, pointing to a few spots. Consistent with the BvT vs ViT blind study, our pathologists also agree that Bwin is more plausible than Swin ($p < 0.05$).

7 Conclusion

We have introduced the B-cos Vision Transformer (BvT) and the B-cos Swin Transformer (Bwin) as two alternatives to the Vision Transformer (ViT) and the Swin Transformer (Swin) that are more interpretable and explainable. These models use the B-cos transform to enforce similarity between weights and inputs. In a blinded study, domain experts clearly preferred both BvT and Bwin over ViT and Swin. We have also shown that BvT is competitive with ViT in terms of quantitative performance. Moreover, using Bwin or transfer learning for BvT, we can even outperform the original models.

Acknowledgements. M.T. and S.J.W. are supported by the Helmholtz Association under the joint research school "Munich School for Data Science".

References

1. Abnar, S., Zuidema, W.: Quantifying attention flow in transformers. In: 58th ACL, pp. 4190–4197. ACL (2020)
2. Adebayo, J., Gilmer, J., Muelly, M., Goodfellow, I., Hardt, M., Kim, B.: Sanity checks for saliency map. In: 32th NeurIPS, pp. 1–11. Curran Associates, Inc. (2018)
3. Ali, A., Schnake, T., Eberle, O., Montavon, G., Müller, K.R., Wolf, L.: XAI for transformers: better explanations through conservative propagation. In: 39th ICML, pp. 435–451. PMLR (2022)
4. Arun, N., et al.: Assessing the trustworthiness of saliency maps for localizing abnormalities in medical imaging. Radiol. Artif. Intell. **3**(6), 1–12 (2021)
5. Bibal, A., et al.: Is attention explanation? An introduction to the debate. In: 60th ACL, pp. 3889–3900. ACL (2022)
6. Binder, A., Montavon, G., Lapuschkin, S., Müller, K.-R., Samek, W.: Layer-wise relevance propagation for neural networks with local renormalization layers. In: Villa, A.E.P., Masulli, P., Pons Rivero, A.J. (eds.) ICANN 2016. LNCS, vol. 9887, pp. 63–71. Springer, Cham (2016). https://doi.org/10.1007/978-3-319-44781-0_8
7. Böhle, M., Fritz, M., Schiele, B.: B-Cos networks: alignment is all we need for interpretability. In: 2022 IEEE/CVF CVPR, pp. 10329–10338. IEEE (2022)
8. Caron, M., et al.: Emerging properties in self-supervised vision transformers. In: 2021 IEEE/CVF ICCV, pp. 9650–9660. IEEE (2021)

9. Chefer, H., Gur, S., Wolf, L.: Generic attention-model explainability for interpreting bi-modal and encoder-decoder transformers. In: 2021 IEEE/CVF CVPR, pp. 397–406. IEEE (2021)
10. Chefer, H., Gur, S., Wolf, L.: Transformer interpretability beyond attention visualization. In: 2021 IEEE/CVF CVPR, pp. 782–791. IEEE (2021)
11. Chen, R.J., et al.: Scaling vision transformers to gigapixel images via hierarchical self-supervised learning. In: 2022 IEEE/CVF CVPR, pp. 16144–16155. IEEE (2022)
12. Dosovitskiy, A., et al.: An image is worth 16 × 16 words: transformers for image recognition at scale. In: 9th ICLR, pp. 1–21. ICLR (2021)
13. Du, M., Liu, N., Hu, X.: Techniques for interpretable machine learning. Commun. ACM **63**(1), 68–77 (2020)
14. Erhan, D., Bengio, Y., Courville, A., Vincent, P.: Visualizing higher-layer features of a deep network. Technical reports University of Montreal, vol. 1341, no. (3), p. 1 (2009)
15. Gilpin, L.H., Bau, D., Yuan, B.Z., Bajwa, A., Specter, M., Kagal, L.: Explaining explanations: an overview of interpretability of machine learning. In: 2018 IEEE 5th DSAA, pp. 80–89. IEEE (2018)
16. Hollenstein, N., Beinborn, L.: Relative importance in sentence processing. In: 59th ACL and the 11th IJCNLP, pp. 141–150. ACL (2021)
17. Jacovi, A., Goldberg, Y.: Towards faithfully interpretable NLP systems: how should we define and evaluate faithfulness? In: Proceedings of the 58th ACL, pp. 4198–4205. ACL (2020)
18. Kather, J.N., et al.: Predicting survival from colorectal cancer histology slides using deep learning: a retrospective multicenter study. PLOS Med. **16**(1), 1–22 (2019)
19. Kirk, S., et al.: Radiology data from the Cancer Genome Atlas Colon Adenocarcinoma [TCGA-COAD] collection. The cancer imaging archive. Technical report, University of North Carolina, Brigham & Women's Hospital Boston, Roswell Park Cancer Institute (2016)
20. Linardatos, P., Papastefanopoulos, V., Kotsiantis, S.: Explainable AI: a review of machine learning interpretability methods. Entropy **23**(1), 1–45 (2020)
21. Liu, Z., et al.: Swin transformer: hierarchical vision transformer using shifted windows. In: 2021 IEEE/CVF ICCV, pp. 9992–10002. IEEE (2021)
22. Lundberg, S.M., Lee, S.I.: A unified approach to interpreting model predictions. In: 31th NeurIPS, pp. 4765–4774. Curran Associates, Inc. (2017)
23. Märkl, B., Füzesi, L., Huss, R., Bauer, S., Schaller, T.: Number of pathologists in Germany: comparison with European countries, USA, and Canada. Virchows Arch. **478**, 335–341 (2021)
24. Matek, C., Krappe, S., Münzenmayer, C., Haferlach, T., Marr, C.: Highly accurate differentiation of bone marrow cell morphologies using deep neural networks on a large image data set. Blood **138**(20), 1917–1927 (2021)
25. Matek, C., Schwarz, S., Spiekermann, K., Marr, C.: Human-level recognition of blast cells in acute myeloid leukemia with convolutional neural networks. Nat. Mach. Intell. **1**(11), 538–544 (2019)
26. Matsoukas, C., Haslum, J.F., Sorkhei, M., Söderberg, M., Smith, K.: What makes transfer learning work for medical images: feature reuse & other factors. In: 2022 IEEE/CVF CVPR, pp. 9225–9234. IEEE (2022)
27. Montavon, G., Lapuschkin, S., Binder, A., Samek, W., Müller, K.R.: Explaining nonlinear classification decisions with deep Taylor decomposition. Pattern Recogn. **65**, 211–222 (2017)

28. Rigotti, M., Miksovic, C., Giurgiu, I., Gschwind, T., Scotton, P.: Attention-based interpretability with concept transformers. In: 10th ICLR, pp. 1–16. ICLR (2022)
29. Rosenberg, H.F., Dyer, K.D., Foster, P.S.: Eosinophils: changing perspectives in health and disease. Nat. Rev. Immunol. **13**(1), 9–22 (2013)
30. Selvaraju, R.R., Cogswell, M., Das, A., Vedantam, R., Parikh, D., Batra, D.: Grad-CAM: visual explanations from deep networks via gradient-based localization. In: 2017 IEEE/CVF CVPR, pp. 618–626. IEEE (2017)
31. Vaswani, A., et al.: Attention is all you need. In: 31st NIPS, pp. 6000–6010. Curran Associates, Inc. (2017)
32. Wagner, S.J., et al.: Fully transformer-based biomarker prediction from colorectal cancer histology: a large-scale multicentric study. arXiv preprint arXiv:2301.09617 (2023)
33. Wang, X., et al.: Transformer-based unsupervised contrastive learning for histopathological image classification. Med. Image Anal. **81**(102559), 1–21 (2022)

PMC-CLIP: Contrastive Language-Image Pre-training Using Biomedical Documents

Weixiong Lin[1], Ziheng Zhao[1], Xiaoman Zhang[1,2], Chaoyi Wu[1,2], Ya Zhang[1,2], Yanfeng Wang[1,2], and Weidi Xie[1,2(✉)]

[1] Cooperative Medianet Innovation Center, Shanghai Jiao Tong University, Shanghai, China
[2] Shanghai AI Laboratory, Shanghai, China
{wx_lin,Zhao_Ziheng,xm99sjtu,wtzxxxwcy02,ya_zhang, wangyanfeng,weidi}@sjtu.edu.cn

Abstract. Foundation models trained on large-scale dataset gain a recent surge in CV and NLP. In contrast, development in biomedical domain lags far behind due to data scarcity. To address this issue, we build and release PMC-OA, a biomedical dataset with 1.6M image-caption pairs collected from PubMedCentral's OpenAccess subset, which is 8 times larger than before, PMC-OA covers diverse modalities or diseases, with majority of the image-caption samples aligned at finer-grained level, *i.e.*, subfigure and subcaption. While pretraining a CLIP-style model on PMC-OA, our model named PMC-CLIP outperform previous state-of-the-art models on various downstream tasks, including image-text retrieval on ROCO, MedMNIST image classification, Medical VQA, for example, +8.1% R@10 on image-text retrieval, +3.9% accuracy on image classification.

Keywords: Multimodal Dataset · Vision-Language Pretraining

1 Introduction

In the recent literature, development of foundational models has been the main driving force in artificial intelligence, for example, large language models [2,8,27,30] trained with either autoregressive prediction or masked token inpainting, and computer vision models [19,29] trained by contrasting visual-language features. In contrast, development in the biomedical domain lags far behind due to limitations of data availability from two aspects, (i) the expertise required for annotation, (ii) privacy concerns. This paper presents our preliminary study for constructing a **large-scale, high-quality, image-text** biomedical dataset using publicly available scientific papers, with **minimal manual efforts** involved.

Supplementary Information The online version contains supplementary material available at https://doi.org/10.1007/978-3-031-43993-3_51.

H. Greenspan et al. (Eds.): MICCAI 2023, LNCS 14227, pp. 525–536, 2023.
https://doi.org/10.1007/978-3-031-43993-3_51

In particular, we crawl figures and corresponding captions from scientific documents on PubMed Central, which is a free full-text archive of biomedical and life sciences journal literature at the U.S. National Institutes of Health's National Library of Medicine (NIH/NLM) [31]. This brings two benefits: (i) the contents in publications are generally well-annotated and examined by experts, (ii) the figures have been well-anonymized and de-identified. In the literature, we are clearly not the first to construct biomedical datasets in such manner, however, existing datasets [28,34,38] suffer from certain limitations in diversity or scale from today's standard. For example, as a pioneering work, ROCO [28] was constructed long time ago with only 81k radiology images. MedICAT [34] contains 217k images, but are mostly consisted of compound figures.

In this work, we tackle the above-mentioned limitations by introducing an automatic pipeline to generate dataset with subfigure-subcaption correspondence from scientific documents, including three major stages: medical figure collection, subfigure separation, subcaption separation & alignment. The final dataset, PMC-OA, consisting of 1.65M image-text pairs (not including samples from ROCO), covers a wide scope of diagnostic procedures and diseases, as shown in Fig. 1 and Fig. 3.

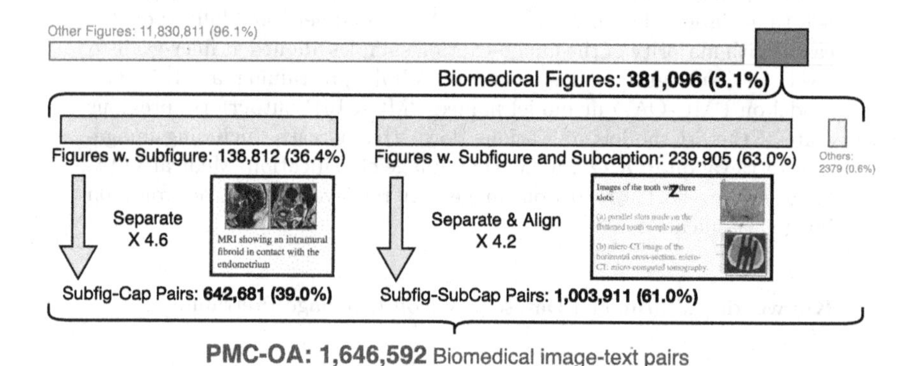

Fig. 1. Statistics over the pipeline and the collected PMC-OA.

Along with the constructed dataset, we train a CLIP-style vision-language model for the biomedical domain, termed as PMC-CLIP. To achieve such a goal, the model is trained on PMC-OA with standard image-text contrastive (ITC) loss, and to encourage the joint interaction of image and text, masked language modeling (MLM) is also applied. We evaluate the pre-trained model on several downstream tasks, including medical image-text retrieval, medical image classification, and medical visual question answering (VQA). PMC-CLIP achieves state-of-the-art performance on various downstream tasks, surpassing previous methods significantly.

Overall, in this paper, we make the following contributions: **First**, we propose an automatic pipeline to construct high-quality image-text biomedical datasets from scientific papers, and construct an image-caption dataset via the proposed pipeline, named PMC-OA, which is 8× larger than before. With the proposed pipeline, the dataset can be continuously updated. **Second**, we pre-train a vision-language model on the constructed image-caption dataset, termed as PMC-CLIP, to serve as a foundation model for biomedical domain. **Third**, we conduct thorough experiments on various downstream tasks (retrieval, classification, and VQA), and demonstrate state-of-the-art performance. The dataset and pre-trained model will be made available to the community.

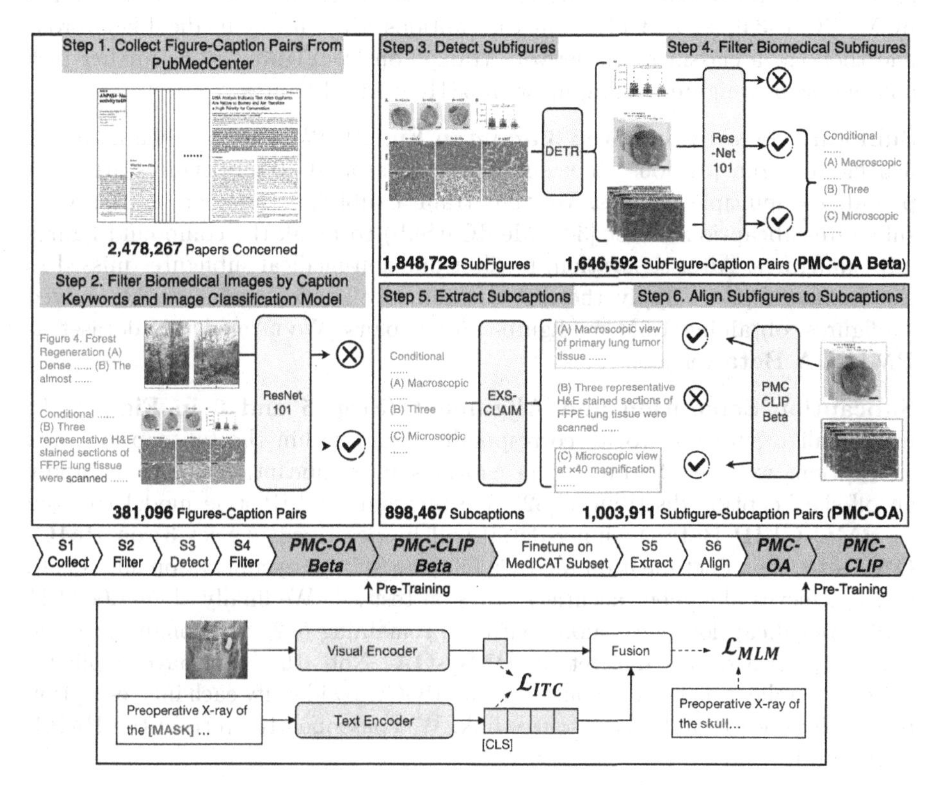

Fig. 2. The proposed pipeline to collect PMC-OA (upper) and the architecture of PMC-CLIP (bottom).

2 The PMC-OA Dataset

In this section, we start by describing the dataset collection procedure in Sect. 2.1, followed by a brief overview of PMC-OA in Sect. 2.2.

2.1 Dataset Collection

In this section, we detail the proposed pipeline to create PMC-OA, a large-scale dataset that contains 1.65M image-text pairs. The whole procedure consists of three major stages: (i) medical figure collection, (ii) subfigure separation, (iii) subcaption separation & alignment, as summarised in Fig. 2.

Medical Figure Collection (Step 1 and 2 in Fig. 2)**.** We first extract figures and captions from PubMedCentral (till 2022-09-16) [31]. 2.4M papers are covered and 12M figure-caption pairs are extracted. To derive medical figures, inspired by MedICat [34], we first filter out the captions without any medical keywords[1] and then use a classification network trained on DocFigure [14] to further pick out the medical figure, ending up with 381K medical figures.

Subfigure Separation (Step 3 and 4 in Fig. 2)**.** We randomly check around 300 figures from previous step, and find that around 80% of figures are compound, *i.e.* multiple pannels. We thus train a subfigure detector on MedICaT subfigure-subcaption subset [34] (MedICatSub) to break the compound figures into subfigures. After separation, to filter out non-medical subfigures missed in the former step, we apply the aforementioned classifier again on the derived subfigures, obtaining 1.6M subfigure-caption pairs. We termed this dataset as **PMC-OA Beta** version.

Subcaption Separation and Alignment (Step 5 and 6 in Fig. 2)**.** To further align subfigure to its corresponding part within the full caption, *i.e.*, subcaption, we need to break the captions into subcaptions first, we apply an off-shelf caption distributor [32]. We pretrain a CLIP-style model (termed as **PMC-CLIP-Beta**, training detail will be described in Sect. 3) on **PMC-OA-Beta**, then finetune it on MedICaTSub for subfigure-subcaption alignment, which achieves alignment accuracy=73% on test set. We finally align 1,003,911 subfigure-subcaption pairs, along with the remaining 642,681 subfigure-caption pairs, we termed this dataset as **PMC-OA**. Note that, we have explicitly removed duplication between our data and ROCO by identify each image-caption pair with paperID and image source link. We consequently pretrain the **PMC-CLIP** on it.

2.2 Dataset Overview

In this section, we provide a brief statistical overview of the collected dataset PMC-OA with UMLS parser [1] from three different perspectives, *i.e.*, diagnostic procedure, diseases and findings, and fairness. *First*, PMC-OA covers a wide range of diagnostic procedures, spanning from common (*CT, MRI*) to rare ones (*mitotic figure*), which is more diverse than before (Fig. 3(a)). *Second*, PMC-OA contains various diseases and findings, and is more up-to-date, covering new emergent diseases like COVID-19 (Fig. 3(b)). And the wide disease coverage in

[1] Follow Class TUI060 "*Diagnostic Procedure*" defined in UMLS [1].

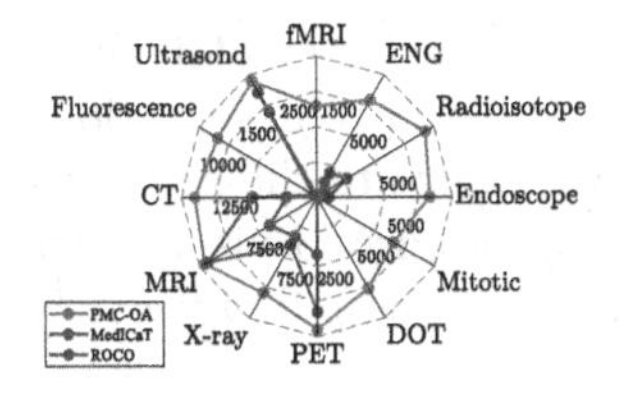

(a) Diagnostic procedure.

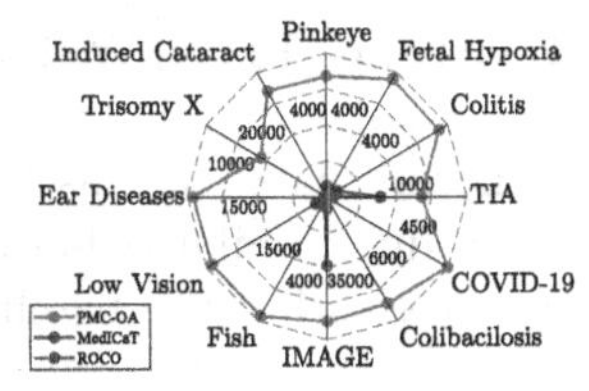

(b) Disease and findings.

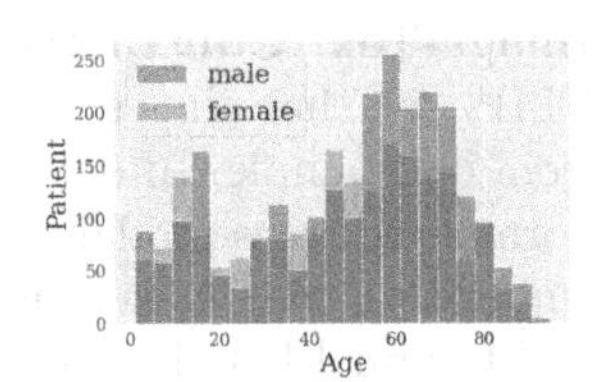

(c) Patients' age & gender.

Fig. 3. Statistical overview of PMC-OA.

our dataset supports learning the shared patterns of diseases, promoting accurate auto-diagnosis. *Third*, we also provide the sex-ratio across ages in Fig. 3(c), as we can see PMC-OA is approximately gender-balanced, with 54% males. The fairness on population ensures our dataset sightly suffers from patient characteristic bias, thus providing greater cross-center generalize ability.

Discussion. Compared to pioneering works [28,34] for constructing dataset based on PubMedCentral, our proposed PMC-OA is of larger scale, diversity, and has more accurate alignment: *First*, PMC-OA covers a wider range of papers (2.4M) than ROCO [28](1.8M) and MedICaT [34](131K), and thus enlarge our dataset(1.6M). *Second*, unlike ROCO [28], we maintain the non-radiology images, which makes PMC-OA a more diverse biomedical dataset as shown in Fig. 3. *Third*, to the best of our knowledge, we are the first to integrate subfigures separation, subcaptions separation and the alignment into the data collection pipeline, which explicitly enlarges our dataset (8 times of MedICaT and 20 times of ROCO), while reducing the noise as much as possible.

3 Visual-language Pre-training

With our constructed image-caption dataset, we further train a visual-language model, termed as PMC-CLIP as shown in Fig. 2 (bottom). We describe the architecture first and then introduce the two training objectives separately.

Architecture. Given N image-caption training pairs, *i.e.*, $\mathcal{D} = \{(\mathcal{I}_i, \mathcal{T}_i)|_{i=1}^{N}\}$, where $\mathcal{I}_i \in \mathbb{R}^{H \times W \times C}$ represents images, H, W, C are height, width, channel, and $\mathcal{T}_i$ represents the paired text. We aim to train a CLIP-style visual-language model with an image encoder Φ_{visual} and a text encoder Φ_{text}.

In detail, given a specific image-caption pair $(\mathcal{I}, \mathcal{T})$, we encode it separately with a ResNet-based Φ_{visual} and a BERT-based Φ_{text}, the embedding dimension is denoted as d and the text token length as l:

$$\boldsymbol{v} = \Phi_{\text{visual}}(\mathcal{I}) \in \mathbb{R}^{d}, \tag{1}$$

$$\boldsymbol{T} = \Phi_{\text{text}}(\mathcal{T}) \in \mathbb{R}^{l \times d}, \boldsymbol{t} = \boldsymbol{T}_0 \in \mathbb{R}^{d}, \tag{2}$$

where $\boldsymbol{v}$ represents the embedding for the whole image, $\boldsymbol{T}$ refers to the sentence embedding, and $\boldsymbol{t}$ denotes the embedding for [CLS] token.

Image-Text Contrastive Learning (ITC). We implement ITC loss following CLIP [29], that aims to match the corresponding visual and text representations from one sample. In detail, denoting batch size as b, we calculate the softmax-normalized cross-modality dot product similarity between the current visual/text embedding ($\boldsymbol{v}$ / $\boldsymbol{t}$) and all samples within the batch, termed as $p^{\mathrm{i2t}}, p^{\mathrm{t2i}} \in \mathbb{R}^b$, and the final ITC loss is:

$$L_{\mathrm{ITC}} = \mathbb{E}_{(\mathcal{I},\mathcal{T})\sim\mathcal{D}}\big[\mathrm{CE}(y^{\mathrm{i2t}}, p^{\mathrm{i2t}}) + \mathrm{CE}(y^{\mathrm{t2i}}, p^{\mathrm{t2i}})\big], \quad (3)$$

where $y^{\mathrm{i2t}}, y^{\mathrm{t2i}}$ refer to one-hot matching labels, CE refers to InfoNCE loss [25].

Masked Language Modeling (MLM). We implement MLM loss following BERT [7]. The network is trained to reconstruct the masked tokens from context contents and visual cues. We randomly mask the word in texts with a probability of 15% and replace it with a special token '[MASK]'. We concatenate the image embedding $\boldsymbol{v}$ with the text token embeddings $\boldsymbol{T}$, input it into a self-attention transformer-based fusion module Φ_{fusion}, and get the prediction for the masked token at the corresponding position in the output sequence, termed as $p^{\mathrm{mask}} = \Phi_{\mathrm{fusion}}(\boldsymbol{v}, \boldsymbol{T})$. Let y^{mask} denote the ground truth, and the MLM loss is:

$$L_{\mathrm{MLM}} = \mathbb{E}_{(\mathcal{I},\mathcal{T})\sim\mathcal{D}}\big[\mathrm{CE}(y^{\mathrm{mask}}, p^{\mathrm{mask}})\big] \quad (4)$$

Total Training Loss. The final loss is defined as $L = L_{\mathrm{ITC}} + \lambda L_{\mathrm{MLM}}$, where λ is a hyper-parameter deciding the weight of L_{MLM}, set as 0.5 by default.

Disscussion. While we recognize a lot of progress in VLP methodology [13, 35,36], PMC-CLIP is trained in an essential way to demonstrate the potential of the collected PMC-OA, and thus should be orthogonal to these works.

4 Experiment Settings

4.1 Pre-training Datasets

ROCO [28] is a image-caption dataset collected from PubMed [31]. It filters out all the compound or non-radiological images, and consists of 81K samples.

MedICaT [34] extends ROCO to 217K samples (image-caption pairs), however, 75% of its figures are compound ones, *i.e.* one figure with multiple subfigures.

MIMIC-CXR [15] is the largest chest X-ray dataset, containing 377,110 samples (image-report pairs). Each image is paired with a clinical report describing findings from doctors.

PMC-OA contains 1.65M image-text pairs, which we have explicitly conducted deduplication between ROCO.

4.2 Downstream Tasks

Image-Text Retrieval (ITR). ITR contains both image-to-text(I2T) and text-to-image(T2I) retrieval. We train PMC-CLIP on different datasets, and sample 2,000 image-text pairs from ROCO's testset for evaluation, following previous works [4,5,34]. **Note that**, as we have explicitly conducted deduplication, the results thus resemble *zero-shot* evaluation.

Classification. We finetune the model for different downstream tasks that focus on image classification. Spcifically, MedMINIST [37] contains 12 tasks for 2D images, and it covers primary data modalities in biomedical images, including Colon Pathology, Dermatoscope, Retinal OCT, etc.

Visual Question Answering (VQA). We evaluate on the official dataset split of SLAKE [22], and follow previous work's split [24] on VQA-RAD [18], where SLAKE is composed of 642 images and 14,028 questions and VQA-RAD contains 315 images and 3,515 questions. The questions in VQA-RAD and Slake are categorized as close-ended if answer choices are limited, otherwise open-ended. The image and text encoders are initialized from PMC-CLIP and fine-tuned, we refer the reader for more details in supplementary.

4.3 Implementation Details

For the visual and text encoders, we adopt ResNet50 [12] and PubmedBERT [11]. And we use 4 transformer layers for the fusion module. For input data, we resize each image to 224 × 224. During pre-training, our text encoder is initialized from PubmedBERT, while the vision encoder and fusion module are trained from scratch. We use AdamW [23] optimizer with $lr = 1 \times 10^{-4}$. We train on GeForce RTX 3090 GPUs with batch size 128 for 100 epochs. The first 10 epochs are set for warming up.

5 Result

We conduct experiments to validate our proposed dataset, and the effectiveness of model trained on it. In Sec. 5.1, we first compare with existing large-scale biomedical datasets on the image-text retrieval task to demonstrate the superiority of PMC-OA. In Sect. 5.2, we finetune the model (pre-trained on PMC-OA) across three different downstream tasks, namely, retrieval, classification, and visual question answering. And we also perform a thorough empirical study of the pretraining objectives and the model architectures in Sect. 5.3. **Note that**, for all experiments, we use the default setting: ResNet50 for image encoder, and pre-train with both ITC and MLM objectives, unless specified otherwise.

5.1 PMC-OA surpasses SOTA large-scale biomedical dataset

As shown in Table 1, we pre-train PMC-CLIP on different datasets and evaluate retrieval on ROCO test set. The performance can be largely improved by simply switching to our dataset, confirming the significance of it.

Table 1. Ablation studies on pre-training dataset.

Methods	Pretrain Data	DataSize	I2T			T2I		
			R@1	R@5	R@10	R@1	R@5	R@10
PMC-CLIP	ROCO	81 K	12.30	35.28	46.52	13.36	35.84	47.38
PMC-CLIP	MedICaT	173 K	17.44	41.08	52.72	17.14	40.42	51.71
PMC-CLIP	PMC-OA Beta	1.6 M	30.42	59.11	70.16	27.92	55.99	66.35
PMC-CLIP	PMC-OA	1.6 M	31.41	61.15	71.88	28.02	58.33	69.69

5.2 PMC-CLIP achieves SOTA across downstream tasks

To evaluate the learnt representation in PMC-CLIP, we compare it with several state-of-the-art approaches across various downstream tasks, including image-text retrieval, image classification, and visual question answering.

Image-Text Retrieval. As shown in Table 2, we report a state-of-the-art result on image-text retrieval. On I2T Rank@10, PMC-CLIP outperforms previous state-of-the-art by 8.1%. It is worth mentioning that, the training set of ROCO has been used during pretraining in M3AE [4], ARL [5]. While our dataset does not contain data from ROCO.

Table 2. Zero-shot Image-Text Retrieval on ROCO.

Methods	Pretrain Data	DataSize	I2T			T2I		
			R@1	R@5	R@10	R@1	R@5	R@10
ViLT [16]	COCO [20], VG [17], SBU, GCC	4.1M	11.90	31.90	43.20	9.75	28.95	41.40
METER [9]	COCO, VG, SBU [26], GCC [33]	4.1M	14.45	33.30	45.10	11.30	27.25	39.60
M3AE [4]	ROCO, MedICaT	233 K	19.10	45.60	61.20	19.05	47.75	61.35
ARL [5]	ROCO, MedICaT, CXR	233 K	23.45	50.60	62.05	23.50	49.05	63.00
PMC-CLIP	PMC-OA	1.6 M	31.41	61.15	71.88	28.02	58.33	69.69

Image Classification. To demonstrate the excellent transferability of PMC-CLIP, we validate it on MedMNIST and compare it with SOTA methods *i.e.*, DWT-CV [6] and SADAE [10]. We present the results of 3 of 12 sub-tests here, and the full results can be found in the supplementary material. As shown in Table 3, PMC-CLIP obtains consistently higher results, and it is notable that finetuning from PMC-CLIP achieves significant performance gains compared with training from scratch with ResNet.

Visual Question Answering. VQA requires model to learn finer grain visual and language representations. As Table 4 shows, we surpass SOTA method M3AE in 5 out of 6 results.

Table 3. Classification results on MedMNIST.

Methods	PneumoniaMNIST		BreastMNIST		DermaMNIST	
	AUC↑	ACC↑	AUC↑	ACC↑	AUC↑	ACC↑
ResNet50 [12]	96.20	88.40	86.60	84.20	91.20	73.10
DWT-CV [6]	95.69	88.67	89.77	85.68	91.67	74.75
SADAE [10]	98.30	91.80	91.50	87.80	92.70	75.90
PMC-CLIP	99.02	95.35	94.56	91.35	93.41	79.80

Table 4. VQA results on VQA-RAD and Slake.

Methods	VQA-RAD			Slake		
	Open	Closed	Overall	Open	Closed	Overall
MEVF-BAN [24]	49.20	77.20	66.10	77.80	79.80	78.60
CPRD-BAN [21]	52.50	77.90	67.80	79.50	83.40	81.10
M3AE [4]	67.23	83.46	77.01	80.31	87.82	83.25
PMC-CLIP	67.00	84.00	77.60	81.90	88.00	84.30

5.3 Ablation Study

Training Objectives. We pre-train PMC-CLIP with different objectives (*ITC, MLM*) for ablation studies, and summarize the results in the supplementary material (Table 5 in the supplementary). Here, we present a summary of the observations: *First*, ITC objective is essential for pretraining, and contributes most of the performance. *Second*, MLM using only text context works as a regularization term. *Third*, With incorporation of visual features, the model learns finer grain correlation between image-caption pairs, and achieve the best results.

Data Collection Pipeline. To demonstrate the effectiveness of subfigure-subcaption alignment, we compare PMC-CLIP with the model pretrained on dataset w/o alignment (Table 6(1–3) in the supplementary). The result verify that subfigure-subcaption alignment reduces dataset's noise thus enhance the pretrained model.

Visual Backbone. We have also explored different visual backbones, using the same setting as CLIP [29] (Table 6(4–7) in the supplementary). We observe that all ResNet variants have close performance with RN50, outperforming ViT-B/32, potentially due to the large patch size.

6 Conclusion

In this paper, we present a large-scale dataset in biomedical domain, named PMC-OA, by collecting image-caption pairs from abundant scientific documents. We train a CLIP-style model on PMC-OA, termed as PMC-CLIP, it achieves SOTA performance across various downstream biomedical tasks, including image-text retrieval, image classification, visual question answering. With the automatic collection pipeline, the dataset can be further expanded, which can be beneficial to the research community, fostering development of foundation models in biomedical domain.

Acknowledgement. This work is supported by the National Key R&D Program of China (No. 2022ZD0160702), STCSM (No. 22511106101, No. 18DZ2270700, No. 21DZ1100100), 111 plan (No. BP0719010), and State Key Laboratory of UHD Video and Audio Production and Presentation.

References

1. Bodenreider, Olivier: The unified medical language system (umls): integrating biomedical terminology. Nucleic Acids Research **32**, D267–D270 (2004)
2. Brown, Tom, et al.: Language models are few-shot learners. Advances in Neural Information Processing Systems **33**, 1877–1901 (2020)
3. Nicolas Carion, Francisco Massa, Gabriel Synnaeve, Nicolas Usunier, Alexander Kirillov, and Sergey Zagoruyko. End-to-end object detection with transformers. In Computer Vision-ECCV 2020: 16th European Conference, Glasgow, UK, August 23–28, 2020, Proceedings, Part I 16, pages 213–229. Springer, 2020
4. Zhihong Chen, Yuhao Du, Jinpeng Hu, Yang Liu, Guanbin Li, Xiang Wan, and Tsung-Hui Chang. Multi-modal masked autoencoders for medical vision-and-language pre-training. In Medical Image Computing and Computer Assisted Intervention-MICCAI 2022: 25th International Conference, Singapore, September 18–22, 2022, Proceedings, Part V, pages 679–689. Springer, 2022
5. Zhihong Chen, Guanbin Li, and Xiang Wan. Align, reason and learn: Enhancing medical vision-and-language pre-training with knowledge. In Proceedings of the 30th ACM International Conference on Multimedia, pages 5152–5161, 2022
6. Cheng, Jianhong, Kuang, Hulin, Zhao, Qichang, Wang, Yahui, Lei, Xu., Liu, Jin, Wang, Jianxin: Dwt-cv: Dense weight transfer-based cross validation strategy for model selection in biomedical data analysis. Future Generation Computer Systems **135**, 20–29 (2022)
7. Jacob Devlin et al. Bert: Pre-training of deep bidirectional transformers for language understanding. ArXiv preprint ArXiv:1810.04805, 2018
8. Ming Ding et al. Cogview2: Faster and better text-to-image generation via hierarchical transformers. ArXiv preprint ArXiv:2204.14217, 2022
9. Zi-Yi Dou et al. An empirical study of training end-to-end vision-and-language transformers. In Proceedings of the IEEE/CVF Conference on Computer Vision and Pattern Recognition, pages 18166–18176, 2022
10. Ge, Xiaolong, et al.: A self-adaptive discriminative autoencoder for medical applications. IEEE Transactions on Circuits and Systems for Video Technology **32**(12), 8875–8886 (2022)
11. Yu, Gu., Tinn, Robert, Cheng, Hao, Lucas, Michael, Usuyama, Naoto, Liu, Xiaodong, Naumann, Tristan, Gao, Jianfeng, Poon, Hoifung: Domain-specific language model pretraining for biomedical natural language processing. ACM Transactions on Computing for Healthcare (HEALTH) **3**(1), 1–23 (2021)

12. Kaiming He, Xiangyu Zhang, Shaoqing Ren, and Jian Sun. Deep residual learning for image recognition. In Proceedings of the IEEE Conference on Computer Vision and Pattern Recognition, pages 770–778, 2016
13. Shih-Cheng Huang, Liyue Shen, Matthew P Lungren, and Serena Yeung. Gloria: A multimodal global-local representation learning framework for label-efficient medical image recognition. In Proceedings of the IEEE/CVF International Conference on Computer Vision, pages 3942–3951, 2021
14. KV Jobin, Ajoy Mondal, and CV Jawahar. Docfigure: A dataset for scientific document figure classification. In 2019 International Conference on Document Analysis and Recognition Workshops, volume 1, pages 74–79. IEEE, 2019
15. Johnson, Alistair EW., et al.: Mimic-cxr, a de-identified publicly available database of chest radiographs with free-text reports. Scientific Data **6**(1), 317 (2019)
16. Wonjae Kim et al. Vilt: Vision-and-language transformer without convolution or region supervision. In International Conference on Machine Learning, pages 5583–5594. PMLR, 2021
17. Krishna, Ranjay, et al.: Visual genome: Connecting language and vision using crowdsourced dense image annotations. International Journal of Computer Vision **123**, 32–73 (2017)
18. Jason J Lau et al. A dataset of clinically generated visual questions and answers about radiology images. Scientific Data, 5(1), 1–10, 2018
19. Li, Junnan, Selvaraju, Ramprasaath, Gotmare, Akhilesh, Joty, Shafiq, Xiong, Caiming, Hoi, Steven Chu Hong.: Align before fuse: Vision and language representation learning with momentum distillation. Advances in Neural Information Processing Systems **34**, 9694–9705 (2021)
20. Tsung-Yi Lin et al. Microsoft coco: Common objects in context. In Computer Vision-ECCV 2014: 13th European Conference, Zurich, Switzerland, September 6–12, 2014, Proceedings, Part V 13, pages 740–755. Springer, 2014
21. Bo Liu et al. Contrastive pre-training and representation distillation for medical visual question answering based on radiology images. In Medical Image Computing and Computer Assisted Intervention-MICCAI 2021: 24th International Conference, Strasbourg, France, September 27-October 1, 2021, Proceedings, Part II 24, pages 210–220. Springer, 2021
22. Bo Liu et al. Slake: A semantically-labeled knowledge-enhanced dataset for medical visual question answering. In 2021 IEEE 18th International Symposium on Biomedical Imaging (ISBI), pages 1650–1654. IEEE, 2021
23. Ilya Loshchilov and Frank Hutter. Decoupled weight decay regularization. In International Conference on Learning Representations, 2017
24. Binh D Nguyen et al. Overcoming data limitation in medical visual question answering. In Medical Image Computing and Computer Assisted Intervention-MICCAI 2019, pages 522–530. Springer, 2019
25. Aaron van den Oord et al. Representation learning with contrastive predictive coding. ArXiv preprint ArXiv:1807.03748, 2018
26. Vicente Ordonez et al. Im2text: Describing images using 1 million captioned photographs. Advances in Neural Information Processing Systems, 24, 2011
27. Long Ouyang et al. Training language models to follow instructions with human feedback. ArXiv preprint ArXiv:2203.02155, 2022
28. Obioma Pelka, Sven Koitka, Johannes Rückert, Felix Nensa, and Christoph M Friedrich. Radiology objects in context (roco): a multimodal image dataset. In MICCAI Workshop on Large-scale Annotation of Biomedical Data and Expert Label Synthesis (LABELS) 2018, pages 180–189. Springer, 2018

29. Alec Radford et al. Learning transferable visual models from natural language supervision. In International Conference on Machine Learning, pages 8748–8763. PMLR, 2021
30. Aditya Ramesh et al. Hierarchical text-conditional image generation with clip latents. ArXiv preprint ArXiv:2204.06125, 2022
31. Richard J Roberts. Pubmed central: The genbank of the published literature, 2001
32. Eric Schwenker et al. Exsclaim!-an automated pipeline for the construction of labeled materials imaging datasets from literature. ArXiv preprint ArXiv:2103.10631, 2021
33. Piyush Sharma et al. Conceptual captions: A cleaned, hypernymed, image alt-text dataset for automatic image captioning. In Proceedings of the 56th Annual Meeting of the Association for Computational Linguistics (Volume 1: Long Papers), pages 2556–2565, 2018
34. Sanjay Subramanian et al. Medicat: A dataset of medical images, captions, and textual references. In Findings of EMNLP, 2020
35. Zifeng Wang, Zhenbang Wu, Dinesh Agarwal, and Jimeng Sun. Medclip: Contrastive learning from unpaired medical images and text. arXiv preprint arXiv:2210.10163, 2022
36. Chaoyi Wu, Xiaoman Zhang, Ya Zhang, Yanfeng Wang, and Weidi Xie. Medklip: Medical knowledge enhanced language-image pre-training. MedRxiv, pages 2023–01, 2023
37. Yang, Jiancheng, et al.: Medmnist v2-a large-scale lightweight benchmark for 2d and 3d biomedical image classification. Scientific Data **10**(1), 41 (2023)
38. Zheng Yuan, Qiao Jin, Chuanqi Tan, Zhengyun Zhao, Hongyi Yuan, Fei Huang, and Songfang Huang. Ramm: Retrieval-augmented biomedical visual question answering with multi-modal pre-training. arXiv preprint arXiv:2303.00534, 2023

Self-supervised Dense Representation Learning for Live-Cell Microscopy with Time Arrow Prediction

Benjamin Gallusser, Max Stieber, and Martin Weigert(✉)

École Polytechnique Fédérale de Lausanne (EPFL), Lausanne, Switzerland
{benjamin.gallusser,max.stieber,martin.weigert}@epfl.ch

Abstract. State-of-the-art object detection and segmentation methods for microscopy images rely on supervised machine learning, which requires laborious manual annotation of training data. Here we present a self-supervised method based on *time arrow prediction pre-training* that learns dense image representations from raw, unlabeled live-cell microscopy videos. Our method builds upon the task of predicting the correct order of time-flipped image regions via a single-image feature extractor followed by a time arrow prediction head that operates on the fused features. We show that the resulting dense representations capture inherently time-asymmetric biological processes such as cell divisions on a pixel-level. We furthermore demonstrate the utility of these representations on several live-cell microscopy datasets for detection and segmentation of dividing cells, as well as for cell state classification. Our method outperforms supervised methods, particularly when only limited ground truth annotations are available as is commonly the case in practice. We provide code at https://github.com/weigertlab/tarrow.

Keywords: Self-supervised learning · Live-cell microscopy

1 Introduction

Live-cell microscopy is a fundamental tool to study the spatio-temporal dynamics of biological systems [4,24,26]. The resulting datasets can consist of terabytes of raw videos that require automatic methods for downstream tasks such as classification, segmentation, and tracking of objects (*e.g.* cells or nuclei). Current state-of-the-art methods rely on supervised learning using deep neural networks that are trained on large amounts of ground truth annotations [6,25,31]. The manual creation of these annotations, however, is laborious and often constitutes a practical bottleneck in the analysis of microscopy experiments [6]. Recently, self-supervised representation learning (SSL) has emerged as a promising approach to alleviate this problem [1,3]. In SSL one first defines a *pretext task*

Supplementary Information The online version contains supplementary material available at https://doi.org/10.1007/978-3-031-43993-3_52.

H. Greenspan et al. (Eds.): MICCAI 2023, LNCS 14227, pp. 537–547, 2023.
https://doi.org/10.1007/978-3-031-43993-3_52

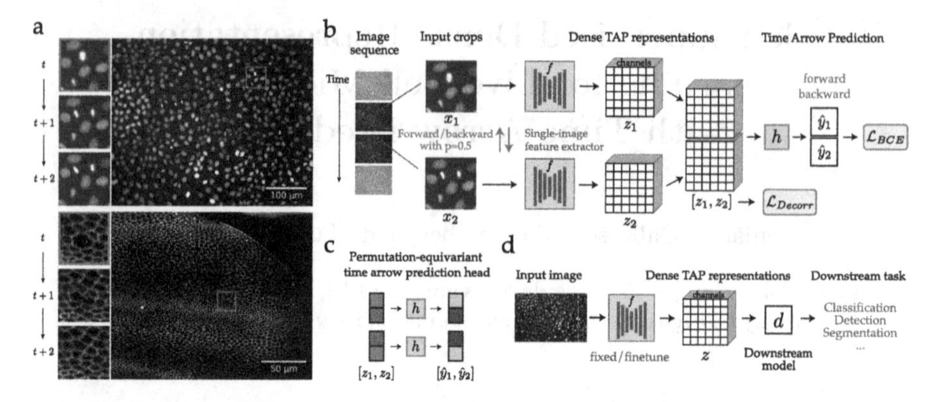

Fig. 1. a) Example frames from two live-cell microscopy videos. Top: *MDCK* cells with labeled nuclei [28], Bottom: *Drosophila* wing with labeled membrane [4]. Insets show three consecutive time points containing cell divisions. **b)** Overview of TAP: We create crops (x_1, x_2) from consecutive time points of a given video. After randomly flipping the input order (forward/backward), each crop is passed through a dense feature extractor f creating pixel-wise TAP representations (z_1, z_2). These are stacked and fed to the time arrow prediction head h. **c)** We design h to be permutation-equivariant ensuring consistent classification of temporally flipped inputs. **d)** The learned TAP representations z are used as input to a downstream model d.

which can be formulated solely based on *unlabeled* images (*e.g.* inpainting [8], or rotation prediction [5]) and tasks a neural network to solve it, with the aim of generating latent representations that capture high-level image semantics. In a second step, these representations can then be either *finetuned* or used directly (*e.g.* via *linear probing*) for a *downstream task* (*e.g.* image classification) with available ground truth [7,10,18]. Importantly, a proper choice of the pretext task is crucial for the resulting representations to be beneficial for a specific downstream task.

In this paper we investigate whether *time arrow prediction*, *i.e.* the prediction of the correct order of temporally shuffled image frames extracted from live-cell microscopy videos, can serve as a suitable pretext task to generate meaningful representations of microscopy images. We are motivated by the observation that for most biological systems the temporal dynamics of local image features are closely related to their semantic content: whereas static background regions are time-symmetric, processes such as cell divisions or cell death are inherently time-asymmetric (*cf.* Fig. 1a). Importantly, we are interested in *dense* representations of individual images as they are useful for both image-level (*e.g.* classification) or pixel-level (*e.g.* segmentation) downstream tasks. To that end, we propose a time arrow prediction pre-training scheme, which we call TAP, that uses a feature extractor operating on single images followed by a time arrow prediction head operating on the fused representations of consecutive time points. The use of time arrow prediction as a pretext task for natural (*e.g.* youtube) videos was introduced by Pickup *et al.* [19] and has since then seen numerous applications for image-level tasks, such as action recognition, video retrieval, and motion classification [2,11,14,15,22,30]. However, to the best of our knowledge, SSL via time arrow prediction has not yet been studied in the context of live-cell microscopy. Concretely our contributions are: *i)* We introduce the time arrow prediction pretext task to the domain of live-cell microscopy and propose the TAP pre-training scheme, which learns dense representations (in contrast to only image-level representations) from raw, unlabeled live-cell microscopy videos, *ii)* we propose a custom (permutation-equivariant) time arrow prediction head that enables robust training, *iii)* we show via attribution maps that the representations learned by TAP capture biologically relevant processes such as cell divisions, and finally *iv)* we demonstrate that TAP representations are beneficial for common image-level and pixel-level downstream tasks in live-cell microscopy, especially in the low training data regime.

2 Method

Our proposed TAP pre-training takes as input a set $\{I\}$ of live-cell microscopy image sequences $I \in \mathbb{R}^{T\times H\times W}$ with the goal to produce a feature extractor f that generates c-dimensional dense representations $z = f(x) \in \mathbb{R}^{c\times H\times W}$ from single images $x \in R^{H\times W}$ (*cf.* Fig. 1b for an overview of TAP). To that end, we randomly sample from each sequence I pairs of smaller patches $x_1, x_2 \in \mathbb{R}^{h\times w}$ from the same spatial location but consecutive time points $x_1 \subset I_t, x_2 \subset I_{t+1}$. We next flip the order of each pair with equal probability $p = 0.5$, assign it the corresponding label y (*forward* or *backward*) and compute dense representations $z_1 = f(x_1)$ and $z_2 = f(x_2)$ with $z_1, z_2 \in \mathbb{R}^{c\times h\times w}$ via a fully convolutional feature extractor f. The stacked representations $z = [z_1, z_2] \in \mathbb{R}^{2\times c\times h\times w}$ are fed to a *time arrow prediction head* h, which produces the classification logits $\hat{y} = [\hat{y}_1, \hat{y}_2] = h([z_1, z_2]) = h([f(x_1), f(x_2)]) \in \mathbb{R}^2$. Both f and h are trained jointly to minimize the loss

$$\mathcal{L} = \mathcal{L}_{BCE}(y, \hat{y}) + \lambda \mathcal{L}_{Decorr}(z) \;, \tag{1}$$

where $\mathcal{L}_{BCE}$ denotes the standard softmax + binary cross-entropy loss between the ground truth label y and the logits $\hat{y} = h(z)$, and $\mathcal{L}_{Decorr}$ is a loss term that promotes z to be decorrelated across feature channels [12,33] via maximizing the diagonal of the softmax-normalized correlation matrix A_{ij}:

$$\mathcal{L}_{Decorr}(\tilde{z}) = -\frac{1}{c}\log \sum_{i=1}^{c} A_{ii} \;, \quad A_{ij} = \mathrm{softmax}(\tilde{z}_i^T \cdot \tilde{z}_j/\tau) = \frac{e^{\tilde{z}_i^T \cdot \tilde{z}_j/\tau}}{\sum_{j=1}^{c} e^{\tilde{z}_i^T \cdot \tilde{z}_j/\tau}} \tag{2}$$

Here $\tilde{z} \in \mathbb{R}^{c\times 2hw}$ denotes the stacked features z flattened across the non-channel dimensions, and τ is a temperature parameter. Throughout the experiments we use $\lambda = 0.01$ and $\tau = 0.2$. Note that instead of creating image pairs from consecutive video frames we can as well choose a custom time step $\Delta t \in \mathbb{N}$ and sample $x_1 \subset I_t$ and $x_2 \subset I_{t+\Delta t}$, which we empirically found to work better for datasets with high frame rate.

Permutation-equivariant Time Arrow Prediction Head: The time arrow prediction task has an inherent symmetry: flipping the input $[z_1, z_2] \rightarrow [z_2, z_1]$ should flip the logits $[\hat{y}_1, \hat{y}_2] \rightarrow [\hat{y}_2, \hat{y}_1]$. In other words, h should be *equivariant* wrt. to permutations of the input. In contrast to common models (*e.g.* ResNet [9]) that lack this symmetry, we here directly incorporate this inductive bias via a *permutation-equivariant head* h that is a generalization of the set permutation-equivariant layer proposed in [32] to dense inputs. Specifically, we choose $h = h_1 \circ \ldots \circ h_L$ as a chain of permutation-equivariant layers h_l:

$$h_l : \mathbb{R}^{2 \times c \times h \times w} \rightarrow \mathbb{R}^{2 \times \tilde{c} \times h \times w}$$

$$h_l(z)_{tmij} = \sigma\Big(\sum_n L_{mn} z_{t,n,i,j} + \sum_{s,n} G_{mn} z_{s,n,i,j}\Big) , \tag{3}$$

with weight matrices $L, G \in \mathbb{R}^{\tilde{c} \times c}$ and a non-linear activation function σ. Note that L operates independently on each temporal axis and thus is trivially permutation equivariant, while G operates on the temporal sum and thus is permutation invariant. The last layer h_L includes an additional global average pooling along the spatial dimensions to yield the final logits $\hat{y} \in \mathbb{R}^2$.

Augmentations: To avoid overfitting on artificial image cues that could be discriminative of the temporal order (such as a globally consistent cell drift, or decay of image intensity due to photo-bleaching) we apply the following augmentations (with probability 0.5) to each image patch pair x_1, x_2: flips, arbitrary rotations and elastic transformations (jointly for x_1 and x_2), translations for x_1 and x_2 (independently), spatial scaling, additive Gaussian noise, and intensity shifting and scaling (jointly+independently).

3 Experiments

3.1 Datasets

To demonstrate the utility of TAP for a diverse set of specimen and microscopy modalities we use the following four different datasets:

HeLa. Human cervical cancer cells expressing histone 2B-GFP imaged by fluorescence microscopy every 30 min [29] . The dataset consists of four videos with overall 368 frames of size 1100×700. We use $\Delta t = 1$ for TAP training.
Mdck. Madin-Darby canine kidney epithelial cells expressing histone 2B-GFP (*cf.* Fig. 3b), imaged by fluorescence microscopy every 4 min [27,28]. The dataset consists of a single video with 1200 frames of size 1600×1200. We use $\Delta t \in \{4, 8\}$.
Flywing. *Drosphila melanogaster* pupal wing expressing Ecad::GFP (*cf.* Fig. 3a), imaged by spinning disk confocal microscopy every 5 min [4,20]. The dataset consists of three videos with overall 410 frames of size 3900×1900. We use $\Delta t = 1$.

YEAST. *S. cerevisiae* cells (*cf.* Fig. 3c) imaged by phase-contrast microscopy every 3 min [16,17]. The dataset consists of five videos with overall 600 frames of size 1024 × 1024. We use $\Delta t \in \{1, 2, 3\}$.

For each dataset we heuristically choose Δt to roughly correspond to the time scale of observable biological processes (*i.e.* larger Δt for higher frame rates).

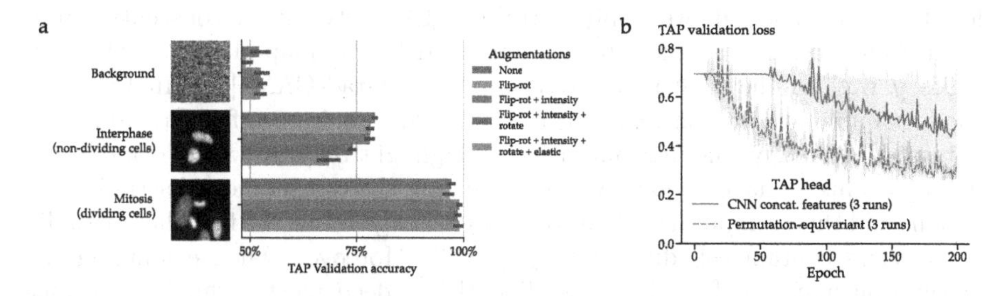

Fig. 2. a) TAP validation accuracy for different image augmentations on crops of background, interphase (non-dividing), and mitotic (dividing) cells (from HELA dataset). **b)** TAP validation loss during training on FLYWING for a regular CNN time arrow prediction head (green) and the proposed permutation-equivariant head (orange). We show results of three runs per model.

3.2 Implementation Details:

For the feature extractor f we use a 2D U-NET [21] with depth 3 and $c = 32$ output features, batch normalization and leaky ReLU activation (approx. 2M params). The time arrow prediction head h consists of two permutation-equivariant layers with batch normalization and leaky ReLU activation, followed by global average pooling and a final permutation-equivariant layer (approx. 5k params). We train all TAP models for 200 epochs and 10^5 samples per epoch, using the Adam optimizer [13] with a learning rate of 4×10^{-4} with cyclic schedule, and batch size 256. Total training time for a single TAP model is roughly 8h on a single GPU. TAP is implemented in PyTorch.

3.3 Time Arrow Prediction Pretraining

We first study how well the time arrow prediction pretext task can be solved depending on different image structures and used data augmentations. To that end, we train TAP networks with an increasing number of augmentations on HELA and compute the TAP classification accuracy for consecutive image patches x_1, x_2 that contain either background, interphase (non-dividing) cells, or mitotic (dividing) cells. As shown in Fig. 2a, the accuracy on background regions is approx. 50% irrespective of the used augmentations, suggesting the absence of predictive cues in the background for this dataset. In contrast, on regions with cell divisions the accuracy reaches almost 100%, confirming that

TAP is able to pick up on strong time-asymmetric image features. Interestingly, the accuracy for regions with non-dividing cells ranges from 68% to 80%, indicating the presence of weak visual cues such as global drift or cell growth. When using more data augmentations the accuracy decreases by roughly 12% points, suggesting that data augmentation is key to avoid overfitting on confounding cues.

Next we investigate which regions in full-sized videos are most discriminative for TAP. To that end, we apply a trained TAP network on consecutive full-sized frames x_1, x_2 and compute the dense attribution map of the classification logits y wrt. to the TAP representations z via Grad-CAM [23]. In Fig. 3 we show example attribution maps on top of single raw frames for three different datasets. Strikingly, the attribution maps highlight only a few distributed, yet highly localized image regions. When inspecting the top six most discriminative regions and their temporal context for a single image frame, we find that virtually all of them contain cell divisions (*cf.* Fig. 3). Moreover, when examining the attribution maps for full videos, we find that indeed most highlighted regions correspond to mitotic cells, underlining the strong potential of TAP to reveal time-asymmetric biological phenomena from raw microscopy videos alone (*cf.* Supplementary Video 1).

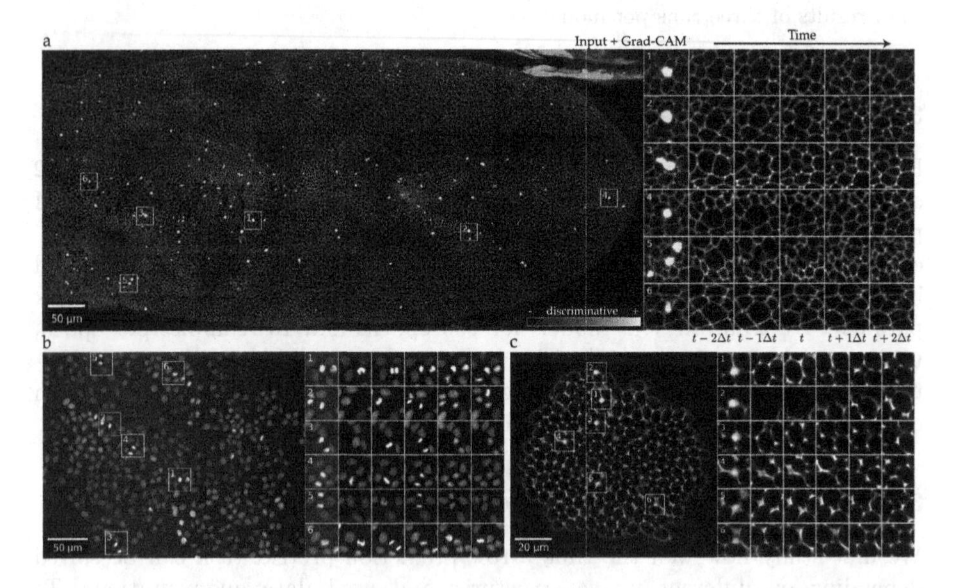

Fig. 3. A single image frame overlayed with TAP attribution maps (computed with Grad-CAM [23]) for **a)** FLYWING, **b)** MDCK, and **c)** YEAST. Insets show the top six most discriminative regions and their temporal context (± 2 timepoints). Note that across all datasets almost all regions contain cell divisions. Best viewed on screen.

Finally, we emphasize the positive effect of the permutation-equivariant time arrow prediction head on the training process. When we originally used a regular

CNN-based head, we consistently observed that the TAP loss stagnated during the initial training epochs and decreased only slowly thereafter (*cf.* Fig. 2b). Using the permutation-equivariant head alleviated this problem and enabled a consistent loss decrease already from the beginning of training.

3.4 Downstream Tasks

We next investigate whether the learned TAP representations are useful for common supervised downstream tasks, where we especially focus on their utility in the low training data regime. First we test the learned representations on two image-level classification tasks, and later on two dense segmentation tasks.

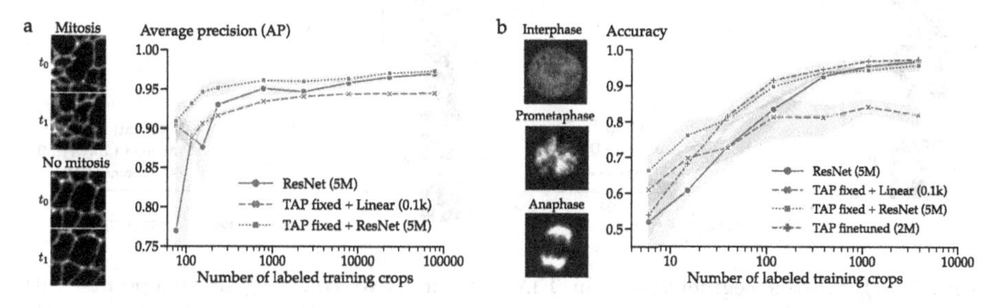

Fig. 4. a) Mitosis classification on FLYWING for two consecutive timepoints with TAP representations *vs.* a supervised ResNet baseline (green). **b)** Cell state classification in MDCK with fixed/fine-tuned TAP representations *vs.* a supervised ResNet baseline (green). We show results of three runs per model, # of params in parenthesis.

Mitosis Classification on Flywing: Since TAP attribution maps strongly highlight cell divisions, we consider predicting mitotic events an appropriate first downstream task to evaluate TAP. To that end, we generate a dataset of 97k crops of size $2 \times 96 \times 96$ from FLYWING and label them as mitotic/non-mitotic (16k/81k) based on available tracking data [20]. We train TAP networks on FLYWING and use a small ResNet architecture ($\approx$ 5M params) that is trained from scratch as a supervised baseline. In Fig. 4a we show average precision (AP) on a held-out test set while varying the amount of available training data. As expected, the performance of the supervised baseline drops substantially for low amounts of training data and surprisingly is already outperformed by a linear classifier (100 params) on top of TAP representations (*e.g.* 0.90 *vs.* 0.77 for 76 labeled crops). Training a small ResNet on fixed TAP representations consistently outperforms the supervised baseline even if hundreds of annotated cell divisions are available for training (*e.g.* 0.96 *vs.* 0.95 for 2328 labeled crops with $\sim$ 400 cell divisions), confirming the value of TAP representations to detect mitotic events.

Cell State Classification on Mdck: Next we turn to the more challenging task of distinguishing between cells in interphase, prometaphase and anaphase from MDCK. This dataset consists of 4800 crops of size 80×80 that are labeled with one of the three classes (1600 crops/class). Again we use a ResNet as supervised baseline and report in Fig. 4b test classification accuracy for varying amount of training data. As before, both a linear classifier as well as a ResNet trained on fixed TAP representations outperform the baseline especially in the low data regime, with the latter showing better or comparable results across the whole data regime (*e.g.* 0.90 vs. 0.83 for 117 annotated cells). Additionally, we finetune the pretrained TAP feature extractor for this downstream task, which slightly improves the results given enough training data. Notably, already at 30% training data it reaches the same performance (0.97) as the baseline model trained on the full training set.

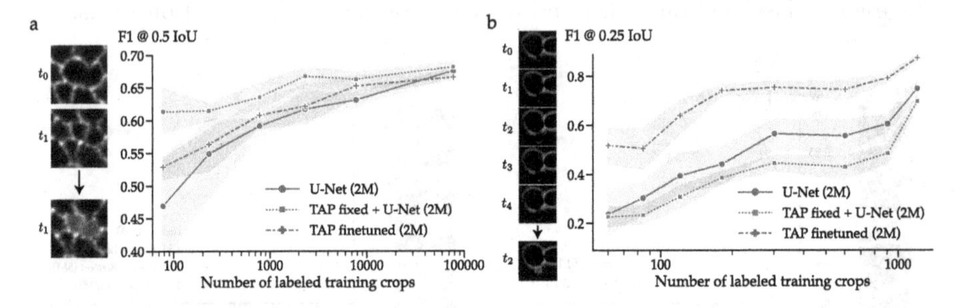

Fig. 5. a) Mitosis segmentation in FLYWING for two consecutive timepoints with fixed/finetuned TAP representations *vs.* a supervised U-NET baseline (green). We report F1 @ 0.5 IoU after removing objects smaller than 64 pixels. **b)** Emerging bud detection in YEAST from five consecutive timepoints with fixed/finetuned TAP representations versus a supervised U-NET baseline (green). We report F1 @ 0.25 IoU on 2D+time objects. We show results of three runs per model, # of params in parenthesis.

Mitosis Segmentation on Flywing: We now apply TAP on a pixel-level downstream task to fully exploit that the learned TAP representations are dense. We use the same dataset as for FLYWING mitosis classification, but now densely label post-mitotic cells. We predict a pixel-wise probability map, threshold it at 0.5 and extract connected components as objects. To evaluate performance, we match a predicted/ground truth object if their intersection over union (IoU) is greater than 0.5, and report the F1 score after matching. The baseline model is a U-NET trained from scratch. Training a U-NET on fixed TAP representations always outperforms the baseline, and when only using 3% of the training data it reaches similar performance as the baseline trained on all available labels (0.67 vs. 0.68, Fig. 5a). Interestingly, fine-tuning TAP only slightly outperforms the supervised baseline for this task even for moderate amounts of training data, suggesting that fixed TAP representations generalize better for limited-size datasets.

Emerging Bud Detection on Yeast: Finally, we test TAP on the challenging task of segmenting emerging buds in phase contrast images of yeast colonies. We train TAP networks on YEAST and generate a dataset of 1205 crops of size $5 \times 192 \times 192$ where we densely label yeast buds in the central frame (defined as buds that appeared less than 13 frames ago) based on available segmentation data [17]. We evaluate all methods on held out test videos by interpreting the resulting 2D+time segmentations as 3D objects and computing the F1 score using an IoU threshold of 0.25. The baseline model is again a U-NET trained from scratch. Surprisingly, training with fixed TAP representations performs slightly worse than the baseline for this dataset (Fig. 5b), possibly due to cell density differences between TAP training and test videos. However, fine-tuning TAP features outperforms the baseline by a large margin (*e.g.* 0.64 *vs*. 0.39 for 120 frames) across the full training data regime, yielding already with 15% labels the same F1 score as the baseline using all labels.

4 Discussion

We have presented TAP, a self-supervised pretraining scheme that learns biologically meaningful representations from live-cell microscopy videos. We show that TAP uncovers sparse time-asymmetric biological processes and events in raw unlabeled recordings without any human supervision. Furthermore, we demonstrate on a variety of datasets that the learned features can substantially reduce the required amount of annotations for downstream tasks. Although in this work we focus on 2D+t image sequences, the principle of TAP should generalize to 3D+t datasets, for which dense ground truth creation is often prohibitively expensive and therefore the benefits of modern deep learning are not fully tapped into. We leave this to future work, together with the application of TAP to cell tracking algorithms, in which accurate mitosis detection is a crucial component.

Acknowledgements. We thank Albert Dominguez (EPFL) and Uwe Schmidt for helpful comments, Natalie Dye (PoL Dresden) and Franz Gruber for providing the FLYWING dataset, Benedikt Mairhörmann and Kurt Schmoller (Helmholtz Munich) for providing additional YEAST training data, and Alan Lowe (UCL) for providing the MDCK dataset. M.W. and B.G. are supported by the EPFL School of Life Sciences ELISIR program and CARIGEST SA.

References

1. Chen, T., Kornblith, S., Norouzi, M., Hinton, G.: A simple framework for contrastive learning of visual representations. In: ICML, pp. 1597–1607 (2020)
2. Dorkenwald, M., Xiao, F., Brattoli, B., Tighe, J., Modolo, D.: SCVRL: shuffled contrastive video representation learning. In: CVPR, pp. 4132–4141 (2022)
3. Ericsson, L., Gouk, H., Loy, C.C., Hospedales, T.M.: Self-supervised representation learning: introduction, advances, and challenges. IEEE Sig. Process. Mag. **39**(3), 42–62 (2022)
4. Etournay, R., Popović, M., Merkel, M., Nandi, A., Blasse, C., Aigouy, B., et al.: Interplay of cell dynamics and epithelial tension during morphogenesis of the Drosophila pupal wing. eLife **4**, e07090 (2015)
5. Gidaris, S., Singh, P., Komodakis, N.: Unsupervised representation learning by predicting image rotations. In: ICLR, OpenReview.net (2018)

6. Greenwald, N.F., Miller, G., Moen, E., Kong, A., Kagel, A., et al.: Whole-cell segmentation of tissue images with human-level performance using large-scale data annotation and deep learning. Nat. Biotechnol. **40**(4), 555–565 (2021)
7. Han, H., Dmitrieva, M., Sauer, A., Tam, K.H., Rittscher, J.: Self-supervised voxel-level representation rediscovers subcellular structures in volume electron microscopy. In: CVPRW, pp. 1874–1883 (2022)
8. He, K., Chen, X., Xie, S., Li, Y., Dollár, P., Girshick, R.: Masked autoencoders are scalable vision learners. In: CVPR, pp. 16000–16009 (2022)
9. He, K., Zhang, X., Ren, S., Sun, J.: Deep residual learning for image recognition. In: CVPR, pp. 770–778 (2016)
10. Hsu, J., Gu, J., Wu, G., Chiu, W., Yeung, S.: Capturing implicit hierarchical structure in 3d biomedical images with self-supervised hyperbolic representations. In: NeurIPS, vol. 34, pp. 5112–5123 (2021)
11. Hu, K., Shao, J., Liu, Y., Raj, B., Savvides, M., Shen, Z.: Contrast and order representations for video self-supervised learning. In: ICCV, pp. 7939–7949 (2021)
12. Hua, T., Wang, W., Xue, Z., Ren, S., Wang, Y., Zhao, H.: On feature decorrelation in self-supervised learning. In: CVPR, pp. 9598–9608 (2021)
13. Kingma, D.P., Ba, J.: Adam: a method for stochastic optimization. In: ICLR (2015)
14. Lee, H.Y., Huang, J.B., Singh, M., Yang, M.H.: Unsupervised representation learning by sorting sequences. In: ICCV, pp. 667–676 (2017)
15. Misra, I., Zitnick, C.L., Hebert, M.: Shuffle and learn: unsupervised learning using temporal order verification. In: ECCV, pp. 527–544 (2016)
16. Padovani, F., Mairhörmann, B., Falter-Braun, P., Lengefeld, J., Schmoller, K.M.: Segmentation, tracking and cell cycle analysis of live-cell imaging data with Cell-ACDC. BMC Biol. **20**, 174 (2022)
17. Padovani, F., Mairhörmann, B., Lengefeld, J., Falter-Braun, P., Schmoller, K.: Cell-ACDC: segmentation, tracking, annotation and quantification of microscopy imaging data (dataset). https://zenodo.org/record/6795124 (2022)
18. Pathak, D., Krahenbuhl, P., Donahue, J., Darrell, T., Efros, A.A.: Context encoders: feature learning by inpainting. In: CVPR, pp. 2536–2544 (2016)
19. Pickup, L.C., et al.: Seeing the arrow of time. In: CVPR, pp. 2043–2050 (2014)
20. Piscitello-Gómez, R., Gruber, F.S., Krishna, A., Duclut, C., Modes, C.D., et al.: Core PCP mutations affect short time mechanical properties but not tissue morphogenesis in the Drosophila pupal wing. bioRxiv (2022)
21. Ronneberger, O., Fischer, P., Brox, T.: U-net: convolutional networks for biomedical image segmentation. In: Navab, N., Hornegger, J., Wells, W.M., Frangi, A.F. (eds.) MICCAI 2015. LNCS, vol. 9351, pp. 234–241. Springer, Cham (2015). https://doi.org/10.1007/978-3-319-24574-4_28
22. Schiappa, M.C., Rawat, Y.S., Shah, M.: Self-supervised learning for videos: a survey. ACM Comput. Surv. **55**(13s), 1–37 (2022)
23. Selvaraju, R.R., Cogswell, M., Das, A., Vedantam, R., Parikh, D., Batra, D.: Grad-CAM: visual explanations from deep networks via gradient-based localization. In: ICCV, pp. 618–626 (2017)
24. Stelzer, E.H.K., et al.: Light sheet fluorescence microscopy. Nat. Rev. Methods Primers **1**(1), 1–25 (2021)
25. Stringer, C., Wang, T., Michaelos, M., Pachitariu, M.: Cellpose: a generalist algorithm for cellular segmentation. Nat. Methods **18**(1), 100–106 (2021)
26. Tomer, R., Khairy, K., Keller, P.J.: Shedding light on the system: studying embryonic development with light sheet microscopy. Curr. Opin. Genet. Dev. **21**(5), 558–565 (2011)

27. Ulicna, K., Vallardi, G., Charras, G., Lowe, A.: Mdck cell tracking reference dataset. https://rdr.ucl.ac.uk/articles/dataset/Cell_tracking_reference_dataset/16595978
28. Ulicna, K., Vallardi, G., Charras, G., Lowe, A.R.: Automated deep lineage tree analysis using a bayesian single cell tracking approach. Front. Comput. Sci. **3**, 734559 (2021)
29. Ulman, V., Maška, M., Magnusson, K.E.G., Ronneberger, O., Haubold, C., et al.: An objective comparison of cell-tracking algorithms. Nat. Methods **14**(12), 1141–1152 (2017). https://doi.org/10.1038/nmeth.4473
30. Wei, D., Lim, J., Zisserman, A., Freeman, W.T.: Learning and using the arrow of time. In: CVPR, pp. 8052–8060 (2018)
31. Weigert, M., Schmidt, U., Haase, R., Sugawara, K., Myers, G.: Star-convex polyhedra for 3d object detection and segmentation in microscopy. In: WACV, pp. 3666–3673 (2020)
32. Zaheer, M., Kottur, S., Ravanbakhsh, S., Poczos, B., Salakhutdinov, R.R., Smola, A.J.: Deep sets. In: NeurIPS (2017)
33. Zbontar, J., Jing, L., Misra, I., LeCun, Y., Deny, S.: Barlow twins: self-supervised learning via redundancy reduction. In: ICML, pp. 12310–12320 (2021)

Learning Large Margin Sparse Embeddings for Open Set Medical Diagnosis

Mingyuan Liu[1], Lu Xu[1], and Jicong Zhang[1,2](✉)

[1] School of Biological Science and Medical Engineering, Beihang University, Beijing, China
{liumingyuan95,xulu181221,jicongzhang}@buaa.edu.cn
[2] Hefei Innovation Research Institute, Beihang University, Hefei, Anhui, China

Abstract. Fueled by deep learning, computer-aided diagnosis achieves huge advances. However, out of controlled lab environments, algorithms could face multiple challenges. Open set recognition (OSR), as an important one, states that categories unseen in training could appear in testing. In medical fields, it could derive from incompletely collected training datasets and the constantly emerging new or rare diseases. OSR requires an algorithm to not only correctly classify known classes, but also recognize unknown classes and forward them to experts for further diagnosis. To tackle OSR, we assume that known classes could densely occupy small parts of the embedding space and the remaining sparse regions could be recognized as unknowns. Following it, we propose Open Margin Cosine Loss (OMCL) unifying two mechanisms. The former, called Margin Loss with Adaptive Scale (MLAS), introduces angular margin for reinforcing intra-class compactness and inter-class separability, together with an adaptive scaling factor to strengthen the generalization capacity. The latter, called Open-Space Suppression (OSS), opens the classifier by recognizing sparse embedding space as unknowns using proposed feature space descriptors. Besides, since medical OSR is still a nascent field, two publicly available benchmark datasets are proposed for comparison. Extensive ablation studies and feature visualization demonstrate the effectiveness of each design. Compared with state-of-the-art methods, MLAS achieves superior performances, measured by ACC, AUROC, and OSCR.

Keywords: Open set recognition · Computer aided diagnosis · Image classification

1 Introduction and Related Work

Deep learning achieves great success in image-based disease classification. However, the computer-aided diagnosis is far from being solved when considering

---

Supplementary Information The online version contains supplementary material available at https://doi.org/10.1007/978-3-031-43993-3_53.

H. Greenspan et al. (Eds.): MICCAI 2023, LNCS 14227, pp. 548–558, 2023.
https://doi.org/10.1007/978-3-031-43993-3_53

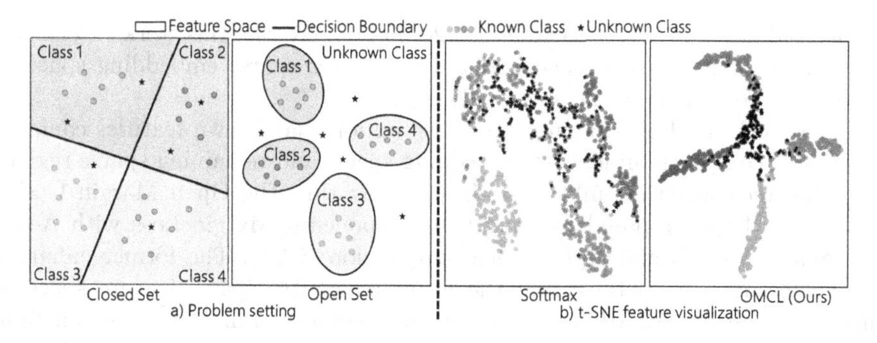

Fig. 1. a) Diagrams of the open set recognition problem, and b) the feature visualization of original closed set classifier and our proposed OMCL.

various requirements in real-world applications. As an important one, open set recognition (OSR) specifies that diseases unseen in training could appear in testing [23]. It is practical in the medical field, caused by the difficulties of collecting a training dataset exhausting all diseases, and by the unpredictably appearing new or rare diseases. As a result, an OSR-informed model should not only accurately recognize known diseases but also detect unknowns and report them. Clinically, these models help construct trustworthy computer-aided systems. By forwarding unseen diseases to experts, not only the misdiagnosis of rare diseases could be avoided, but an early warning of a new disease outbreak could be raised.

There are many fields related to OSR but are essentially different. In classification with reject options [7,9], samples with low confidence are rejected to avoid misclassification. However, since its closed set nature, unknown classes could still be misclassified confidently [8,23]. Anomaly detection, novelty detection, and one-class classification [21] aim at recognizing unknowns but ignore categorizing the known classes. In outlier detection or one-/few-show learning [27], samples of novel classes appear in training. In zero-shot learning [29], semantic information from novel classes could be accessed. Such as zebra, an unknown class, could be identified given the idea that they are stripped horses, and abundant samples of horse and stripe patterns. Differently, OSR knows nothing about novel classes and should have high classification accuracy of the known meanwhile recognize unknowns, as illustrated in Fig. 1a). Due to limited space, some reviews [8,22,34] are recommended for more comprehensive conceptual distinctions.

Most OSR researches focus on natural images, while medical OSR is still in its infancy. In medical fields, representative work like T3PO [6] introduces an extra task to predict the input image augmentation, and samples with low probabilities are regarded as unknowns. CSL [32] uses generative adversarial neural networks (GAN) to generate proxy images and unknown anchors. As for natural images, a line of work tries to simulate unknowns using generated adversarial or counterfactual samples using GAN [14,20,28,33]. However, whether unknown patterns could be generated by learning from the known is unclear. Some works learn descriptive feature representations. They enhance better feature separation between unknowns and knowns or assume the known features following certain distributions so that samples away from distributional centers could be recognized as unknowns [3,5,10,17,18,24,35]. Differently, this work categorizes densely distributed known features and recognizes sparse embedding space as unknowns, regardless of the specific distribution.

This work tackles OSR under the assumption that known features could be assembled compactly in feature embedding space, and remaining sparse regions could be recognized as unknowns. Inspired by this, the Open Margin Cosine Loss (OMCL) is proposed merging two components, Margin Loss with Adaptive Scale (MLAS) and Open-Space Suppression (OSS). The former enhances known feature compactness and the latter recognizes sparse feature space as unknown. Specifically, MLAS introduces the angular margin to the loss function, which reinforces the intra-class compactness and inter-class separability. Besides, a learnable scaling factor is proposed to enhance the generalization capacity. OSS generates feature space descriptors that scatter across a bounded feature space. By categorizing them as unknowns, it opens a classifier by recognizing sparse feature space as unknowns and suppressing the overconfidence of the known. An embedding space example is demonstrated in Fig. 1b), showing OMCL learns more descriptive features and more distinguishing known-unknown separation.

Considering medical OSR is still a nascent field, besides OMCL, we also proposed two publicly available benchmark datasets. One is microscopic images of blood cells, and the other is optical coherence tomography (OCT) of the eye fundus. OMCL shows good adaptability to different image modalities.

Our contributions are summarized as follows. **Firstly**, we propose a novel approach, OMCL for OSR in medical diagnosis. It reinforces intra-class compactness and inter-class separability, and meanwhile recognizes sparse feature space as unknowns. **Secondly**, an adaptive scaling factor is proposed to enhance the generalization capacity of OMCL. **Thirdly**, two benchmark datasets are proposed for OSR. Extensive ablation experiments and feature visualization demonstrate the effectiveness of each design. The superiority over state-of-the-art methods indicates the effectiveness of our method and the adaptability of OMCL on different image modalities.

2 Method

In Sect. 2.1, the open set problem and the formation of cosine Softmax are introduced. The two mechanisms MLAS and OSS are sequentially elaborated in Sect. 2.2 and 2.3, followed by the overall formation of OMCL in Sect. 2.4.

2.1 Preliminaries

Problem Setting: Both closed set and open set classifiers learn from the training set $\mathcal{D}_{train} = \{(\boldsymbol{x}_i, y_i)\}_{i=1}^{N}$ with N image-label pairs $(\boldsymbol{x}_i, y_i)$, where $y_i \in \mathcal{Y} = \{1, 2, ..., C\}$ is a class label. In testing, closed set testing data $\mathcal{D}_{test}$ shares the same label space $\mathcal{Y}$ with the training data. However, in the open set problem, unseen class $y_i = C + 1$ could appear in testing *i.e.* $y_i \in \mathcal{Y}_{open} = \{1, 2, ..., C, C + 1\}$.

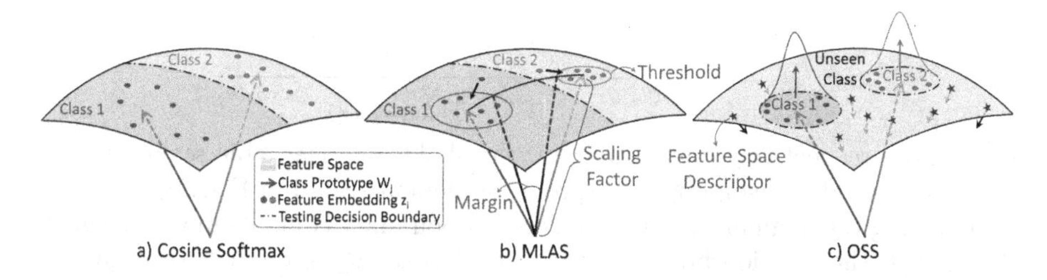

Fig. 2. Geometric interpretation of the cosine Softmax, MLAS, and OSS. MLAS introduces angular margin m and threshold t to reinforce the intra-class compactness and the inter-class separability, together with an adaptive scaling factor to enhance the adaptability. OSS opens a classifier by recognizing the sparse feature space as unknowns using the proposed feature space descriptors.

Cosine Loss: The cosine Softmax is used as the basis of the OMCL. It transfers feature embeddings from the Euclidian space to a hyperspherical one, where feature differences depend merely on their angular separation rather than spatial distance. Given an image $\boldsymbol{x}_i$, its vectorized feature embedding $\boldsymbol{z}_i$, and its label y_i, the derivation progress of the cosine Softmax is

$$S_{cos} = \underbrace{\frac{e^{\boldsymbol{W}_{y_i}^T \boldsymbol{z}_i}}{\sum_{j=1}^{C} e^{\boldsymbol{W}_j^T \boldsymbol{z}_i}}}_{ConventioanlForm} = \frac{e^{\|\boldsymbol{W}_{y_i}\|\|\boldsymbol{z}_i\| cos(\theta_{y_i,i})}}{\sum_{j=1}^{C} e^{\|\boldsymbol{W}_j\|\|\boldsymbol{z}_i\| cos(\theta_{j,i})}} = \underbrace{\frac{e^{s\cdot cos(\theta_{y_i,i})}}{\sum_{j=1}^{C} e^{s\cdot cos(\theta_{j,i})}}}_{CosineForm}, \tag{1}$$

where $\boldsymbol{W}_j$ denotes the weights of the last fully-connected layer (bias is set to 0 for simplicity). $\| \boldsymbol{W}_j \| = 1$ and $\| \boldsymbol{z}_i \| = s$ are manually fixed to constant numbers 1 and s by L2 normalization. s is named the scaling factor. $cos(\theta_{j,i})$ denotes the angle between $\boldsymbol{W}_j$ and $\boldsymbol{z}_i$. By doing so, the direction of W_j could be regarded as the prototypical direction of class j as shown in Fig. 2a). Samples with large angular differences from their corresponding prototype will be punished and meanwhile class-wise prototypes will be pushed apart in the angular space.

Compared with Softmax, the cosine form has a more explicit geometric interpretation, promotes more stabilized weights updating, and learns more discriminative embeddings [15,16,26]. Moreover, the L2 normalization constrains features to a bounded feature space, which allows us to generate feature space descriptors for opening a classifier (will be further discussed in Sect. 2.3).

2.2 Margin Loss with Adaptive Scale (MLAS)

MLAS serves three purposes. 1) By applying angular margin, the intra-class compactness and the inter-class separability are strengthened. 2) The threshold could represent the potential probability of the unknowns, which not only prepares for the open set but also learns more confident probabilities of the knowns. 3) A trainable scaling factor is designed to strengthen the generalization capacity. MLAS is:

$$S_{MLAS} = \frac{e^{s\cdot(cos(\theta_{y_i,i})-m)}}{e^{s\cdot(cos(\theta_{y_i,i})-m)} + e^{s\cdot t} + \sum_{j=1,j\neq y_i}^{C} e^{s\cdot cos(\theta_{j,i})}} \quad (2)$$

m, t, and s respectively denote margin, threshold, and learnable scaling factor, with corresponding geometric interpretation demonstrated in Fig. 2b).

By using the angular margin, the decision boundary could be more stringent. Without it, the decision boundary is $cos(\theta_{1,i}) > cos(\theta_{2,i})$ for the i-th sample of class 1. It becomes $cos(\theta_{1,i}) > cos(\theta_{2,i}) + m$ when using the margin, which leads to stronger intra-class compactness. Moreover, the angular similarities with other classes are punished in the denominator to increase inter-class separability.

The threshold t could be regarded as an extra dimension that prepares for unknown classes. Given the conventional input of Softmax as $[q_i^1, q_i^2, ..., q_i^C] \in \mathbb{R}^C$, ours could be understood as $[q_i^1, q_i^2, ..., q_i^C, t] \in \mathbb{R}^{C+1}$. Since t is added, the class-wise output q_i^c before Softmax is forced to have a higher value to avoid misclassification (at least larger than t). It reinforces more stringent learning and hence increases the feature compactness in the hyperspherical space.

A large s makes the distribution more uniform, and a small s makes it collapses to a point mass. In this work, it is learnable, with a learning rate $0.1\times$ the learning rate of the model. It theoretically offers stronger generalization capacity to various datasets and is experimentally observed to converge to different values in different data trails and could boost performances.

LMCL [26] and NMCL [15] are the most similar arts to ours. Differently, from the task perspective, these designs are proposed for closed-world problems. From the method perspective, an OSS mechanism is designed to tackle OSR leveraging generate pseudo-unknown features for discriminative learning. Moreover, an adaptive scaling factor is introduced for increasing generalization.

2.3 Open-Space Suppression (OSS)

OSS generates feature space descriptors of bounded feature space. By categorizing them into an extra $C+1$ class, samples in sparse feature space could be recognized as unknown and the overconfidence of the known is suppressed.

OSS selects points scattered over the entire feature space, named descriptors, representing pseudo-unknown samples. Different from existing arts that generate pseudo-unknowns by learning from known samples, the OSS selects points scattered over the feature space. It guarantees all space could be possibly considered for simulating the potential unknowns. By competing with the known features, feature space with densely distributed samples is classified as the known, and the sparse space, represented by the descriptors, will be recognized as unknown.

In this work, the corresponding descriptor set, with M samples, is $\mathcal{D}_{desc} = \{(\boldsymbol{z}_i, C+1)\}_{i=1}^{M}$, where $\boldsymbol{z}_i \in \mathbb{U}[-s, s]^d$ subject to $\| \boldsymbol{z}_i \| = s$. $\mathbb{U}[-s, s]$ denotes random continuous uniform distribution ranges between $-s$ to s, and d is the dimension of feature embeddings. s is trainable and the descriptors are dynamically generated with the training. Figure 2c) demonstrates the geometric interpretation. During training, descriptors are concatenated with the training samples at the input of the last fully-connected layer, to equip the last layer with the discrimination capacity of known and unknown samples. The OSS is

$$S_{OSS} = \frac{e^{s \cdot t}}{e^{s \cdot t} + \sum_{j=1}^{C} e^{s \cdot cos(\theta_{j,i})}} \tag{3}$$

where t and s follow the same definition in MLAS.

Most similar arts like AL [25] attempts to reduce misclassification by abandoning ambiguous training images. Differently, we focus on OSR and exploit a novel discriminative loss with feature-level descriptors for OSR.

2.4 Open Margin Cosine Loss (OMCL)

OMCL unifies MLAS and OSS into one formula, which is

$$L_{OMCL} = -\frac{1}{N+M} \sum_{i=1}^{N+M} \mathbb{I}_i log(S_{cos}) + \lambda \mathbb{I}_i log(S_{MLAS}) + \lambda \mathbb{I}_i log(S_{OSS}) \tag{4}$$

$\mathbb{I}_i$ equals 1 if the i-th sample is training data, and equals 0 if it belongs to the feature space descriptors. λ is a weight factor. Since the output of the channel $C+1$ is fixed as t, no extra weights $\boldsymbol{W}_{C+1}$ are trained in the last fully-connected layer. As a result, OMCL does not increase the number of trainable weights in a neural network. During testing, just as in other works [2,3], the maximum probability of known classes is taken as the index of unknowns, where a lower known probability indicates a high possibility of unknowns.

3 Result

3.1 Datasets, Evaluation Metrics, and Implementation Details

Two datasets are adapted as new benchmarks for evaluating the OSR problem. Following protocols in natural images [2,4], half of the classes are selected as known and reminders as unknowns. Since the grouping affects the results, it is randomly repeated K times, leading to K independent data trials. The average results of K trials are used for evaluation. The specific groupings are listed in the supplementary material, so that future works could follow it for fair comparisons.
BloodMnist contains 8 kinds of individual normal cells with 17,092 images [1]. Our setting is based on the closed set split and prepossessing from [31]. Classes are selected 5 rounds (K=5). In each trial, images belonging to 4 chosen classes are selected for training and closed-set evaluation. Images belonging to the other 4 classes in testing data are used for open set evaluation.
OCTMnist has 109,309 optical coherence tomography (OCT) images [13], preprocessed following [31]. Among the 4 classes, 1 is healthy and the other 3 are retinal diseases. In data trail splitting, the healthy class is always in the known set, which is consistent with real circumstances, and trails equal to 3 (K=3).

Table 1. Comparison with state-of-the-art methods. The average of multiple trials is reported.

Method (Pub'Year)	BloodMnist K=5			OCTMnist K=3		
	Acc$_c$%	AUROC$_o$%	OSCR$_o$%	Acc$_c$%	AUROC$_o$%	OSCR$_o$%
Baseline [12] (ICLR'17)	98.0	84.3	84.4	94.3	64.1	62.8
GCPL [30] (CVPR'18)	98.1	85.5	85.0	94.8	65.5	64.2
RPL [3] (ECCV'20)	98.0	86.8	86.3	93.7	65.9	64.2
ARPL+CS [2] (TPAMI'21)	**98.5**	87.6	87.1	95.9	77.7	75.8
DIAS [19] (ECCV'22)	98.4	86.3	85.7	96.0	74.1	72.5
OMCL (Ours)	98.3	**88.6**	**88.0**	**96.8**	**78.9**	**77.8**

Table 2. Ablation studies on the effectiveness of MLAS and OSS in OMCL. Each result is the average of 5 trials on BloodMnist dataset.

MLAS	OSS	Acc$_c$%	AUROC$_o$%	OSCR$_o$%
		98.3	84.7	84.2
✓		98.3	85.4	84.9
	✓	98.3	86.8	86.3
✓	✓	98.3	**88.6**	**88.0**

Metrics: Following previous arts [2,19], accuracy (ACC$_c$) validates closed set classification. Area Under the Receiver Operating Characteristic (AUROC$_o$), a threshold-independent value, measures the open set performances. Open Set Classification Rate (OSCR$_o$) [4], considers both open set recognition and closed set accuracy, where a larger OSCR indicates better performance.

Implementation Details: The classification network is ResNet18 [11], optimized by Adam with an initial learning rate of 1e−3 and a batch size 64. The number of training epochs is 200 and 100 for BloodMnist and OCTMnist respectively because the number of training samples in BloodMnist is smaller. Margin m, threshold t, λ are experimentally set to −0.1, 0.1, and 0.5 respectively. Images are augmented by random crop, random horizontal flip, and normalization.

3.2 Comparison with State-of-the-Art Methods

As demonstrated in Table 1, the proposed OMCL surpasses state-of-the-art models, including typical discriminative methods, baseline [12], GCPL [30], and RPL [3]; latest generative model DIAS [19]; and ARPL+CS [2] that hybrids both. All methods are implemented based on their official codes. Their best results after hyperparameter finetunes are reported. Results show the OMCL maintains the accuracy, meanwhile could effectively recognize unknowns.

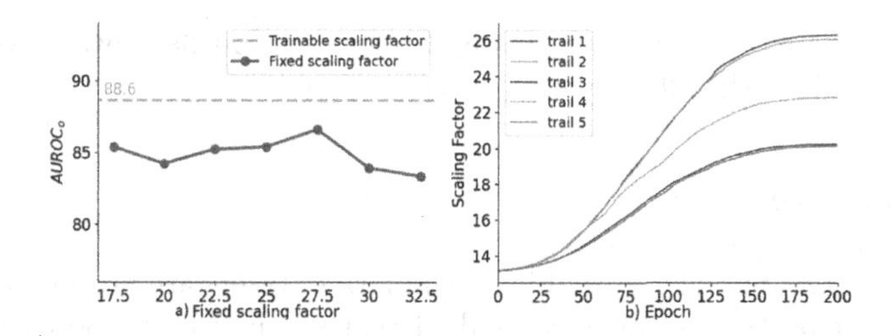

Fig. 3. Ablation studies of adaptive scaling factor on BloodMnist dataset.

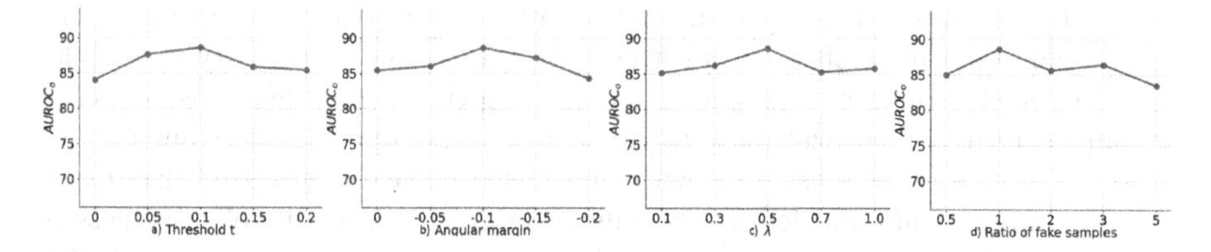

Fig. 4. Ablation studies of hyperparameters. Each result is the average of 5 trails on the BloodMnist dataset.

3.3 Ablation Studies

Effectiveness of MLAS and OSS: Table 2 demonstrates the respective contributions of MLAS and OSS in OMCL. Each of them enhances the performances and they could work complementarily to further improve performances.

Ablation Study of Adaptive Scaling Factor: Fig. 3a) demonstrates the effectiveness of the adaptive scaling factor. Quantitatively, the adaptive design surpasses a fixed one. Moreover, Fig. 3b) displays the scaling factor will converge to different values in different training trials. Both results demonstrate the effectiveness and the generalization capacity of the adaptive design.

Ablation Study of Hyperparameters t, m, ***and*** λ: Fig. 4a), b), and c) respectively show the influence on results when using different hyperparameters. t and m are the threshold and angular margin, presented in Eq. 2, and λ is the trade-off parameter in Eq. 4 .

Ablation Study of M: Fig. 4d) illustrates the effect of the number of feature space descriptors upon results. The ratio 1:1 is experimentally validated as a proper ratio. Because a randomly generated descriptor could be extremely close to a known feature point, but classified as a novel category, which may disturb the training. If the number of descriptors is far more than that of the training samples (the 5 times shown in Fig. 4), the performance gets lower.

Feature Visualization: Fig. 1b) visualizes the t-SNE results of features $\boldsymbol{z}$ of both known and unknown classes after dimension reduction. For each class, 200 samples are visualized and the perplexity of the t-SNE is set to 30. It shows that OMCL could learn better intra-class compactness and inter-class separability.

Moreover, samples of unknown classes tend to be pushed away from known classes, incidcating the effectiveness of our designs.

4 Conclusion

In this paper, two publicly available benchmark datasets are proposed for evaluating the OSR problem in medical fields. Besides, a novel method called OMCL is proposed, under the assumption that known features could be assembled compactly in feature space and the sparse regions could be recognized as unknowns. The OMCL unifies two mechanisms, MLAS and OSS, into a unified formula. The former reinforces intra-class compactness and inter-class separability of samples in the hyperspherical feature space, and an adaptive scaling factor is proposed to empower the generalization capability. The latter opens a classifier by categorizing sparse regions as unknown using feature space descriptors. Extensive ablation experiments and feature visualization demonstrate the effectiveness of each design. Compared to recent state-of-the-art methods, the proposed OMCL performs superior, measured by ACC, AUROC, and OSCR.

Acknowledgement. This work is supported by Beijing Natural Science Foundation (Grant Number: Z200024), the Industrialization Research Project of Hefei Innovation Research Institute, Beihang University (Grant Number: BHKX-20-01), and the University Synergy Innovation Program of Anhui Province (Grant Number: GXXT-2019-044).

References

1. Acevedo, A., Merino, A., Alférez, S., Molina, Á., Boldú, L., Rodellar, J.: A dataset of microscopic peripheral blood cell images for development of automatic recognition systems. Data Brief **30**, 105474 (2020)
2. Chen, G., Peng, P., Wang, X., Tian, Y.: Adversarial reciprocal points learning for open set recognition. IEEE T-PAMI **44**(11), 8065–8081 (2021)
3. Chen, G., et al.: Learning open set network with discriminative reciprocal points. In: Vedaldi, A., Bischof, H., Brox, T., Frahm, J.-M. (eds.) ECCV 2020. LNCS, vol. 12348, pp. 507–522. Springer, Cham (2020). https://doi.org/10.1007/978-3-030-58580-8_30
4. Dhamija, A.R., Günther, M., Boult, T.: Reducing network agnostophobia. In: NeurIPS, vol. 31 (2018)
5. Fontanel, D., Cermelli, F., Mancini, M., Bulo, S.R., Ricci, E., Caputo, B.: Boosting deep open world recognition by clustering. IEEE Robot. Autom. Lett. **5**(4), 5985–5992 (2020)
6. Galdran, A., Hewitt, K.J., Ghaffari Laleh, N., Kather, J.N., Carneiro, G., González Ballester, M.A.: Test time transform prediction for open set histopathological image recognition. In: Wang, L., Dou, Q., Fletcher, P.T., Speidel, S., Li, S. (eds.) MICCAI 2022. LNCS, vol. 13432, pp. 263–272. Springer, Cham (2022). https://doi.org/10.1007/978-3-031-16434-7_26
7. Geifman, Y., El-Yaniv, R.: Selective classification for deep neural networks. In: NeurIPS, vol. 30 (2017)

8. Geng, C., Huang, S., Chen, S.: Recent advances in open set recognition: a survey. IEEE T-PAMI **43**(10), 3614–3631 (2020)
9. Grandvalet, Y., Rakotomamonjy, A., Keshet, J., Canu, S.: Support vector machines with a reject option. In: NeurIPS, vol. 21 (2008)
10. Hassen, M., Chan, P.K.: Learning a neural-network-based representation for open set recognition. In: Proceedings of the 2020 SIAM International Conference on Data Mining, pp. 154–162 (2020)
11. He, K., Zhang, X., Ren, S., Sun, J.: Deep residual learning for image recognition. In: CVPR, pp. 770–778 (2016)
12. Hendrycks, D., Gimpel, K.: A baseline for detecting misclassified and out-of-distribution examples in neural networks. In: ICLR (2017)
13. Kermany, D.S., et al.: Identifying medical diagnoses and treatable diseases by image-based deep learning. Cell **172**(5), 1122–1131 (2018)
14. Kong, S., Ramanan, D.: OpenGAN: open-set recognition via open data generation. In: ICCV, pp. 813–822 (2021)
15. Liu, B., et al.: Negative margin matters: understanding margin in few-shot classification. In: Vedaldi, A., Bischof, H., Brox, T., Frahm, J.-M. (eds.) ECCV 2020. LNCS, vol. 12349, pp. 438–455. Springer, Cham (2020). https://doi.org/10.1007/978-3-030-58548-8_26
16. Liu, W., Wen, Y., Yu, Z., Li, M., Raj, B., Song, L.: SphereFace: deep hypersphere embedding for face recognition. In: CVPR, pp. 212–220 (2017)
17. Liu, Z., Miao, Z., Zhan, X., Wang, J., Gong, B., Yu, S.X.: Large-scale long-tailed recognition in an open world. In: CVPR, pp. 2537–2546 (2019)
18. Lu, J., Xu, Y., Li, H., Cheng, Z., Niu, Y.: PMAL: open set recognition via robust prototype mining. In: AAAI, vol. 36, pp. 1872–1880 (2022)
19. Moon, W., Park, J., Seong, H.S., Cho, C.H., Heo, J.P.: Difficulty-aware simulator for open set recognition. In: Avidan, S., Brostow, G., Cissé, M., Farinella, G.M., Hassner, T. (eds.) ECCV 2022. LNCS, vol. 13685, pp. 365–381. Springer, Cham (2022). https://doi.org/10.1007/978-3-031-19806-9_21
20. Neal, L., Olson, M., Fern, X., Wong, W.-K., Li, F.: Open set learning with counterfactual images. In: Ferrari, V., Hebert, M., Sminchisescu, C., Weiss, Y. (eds.) ECCV 2018. LNCS, vol. 11210, pp. 620–635. Springer, Cham (2018). https://doi.org/10.1007/978-3-030-01231-1_38
21. Pang, G., Shen, C., Cao, L., Hengel, A.V.D.: Deep learning for anomaly detection: a review. ACM Comput. Surv. (CSUR) **54**(2), 1–38 (2021)
22. Salehi, M., Mirzaei, H., Hendrycks, D., Li, Y., Rohban, M.H., Sabokrou, M.: A unified survey on anomaly, novelty, open-set, and out-of-distribution detection: solutions and future challenges. arXiv preprint arXiv:2110.14051 (2021)
23. Scheirer, W.J., de Rezende Rocha, A., Sapkota, A., Boult, T.E.: Toward open set recognition. IEEE T-PAMI **35**(7), 1757–1772 (2012)
24. Shu, Y., Shi, Y., Wang, Y., Huang, T., Tian, Y.: P-ODN: prototype-based open deep network for open set recognition. Sci. Rep. **10**(1), 7146 (2020)
25. Thulasidasan, S., Bhattacharya, T., Bilmes, J., Chennupati, G., Mohd-Yusof, J.: Combating label noise in deep learning using abstention. arXiv preprint arXiv:1905.10964 (2019)
26. Wang, H., et al.: CosFace: large margin cosine loss for deep face recognition. In: CVPR, pp. 5265–5274 (2018)
27. Wang, Y., Yao, Q., Kwok, J.T., Ni, L.M.: Generalizing from a few examples: a survey on few-shot learning. ACM Comput. Surv. (CSUR) **53**(3), 1–34 (2020)
28. Wang, Y., Li, B., Che, T., Zhou, K., Liu, Z., Li, D.: Energy-based open-world uncertainty modeling for confidence calibration. In: ICCV, pp. 9302–9311 (2021)

29. Xian, Y., Schiele, B., Akata, Z.: Zero-shot learning-the good, the bad and the ugly. In: CVPR, pp. 4582–4591 (2017)
30. Yang, H.M., Zhang, X.Y., Yin, F., Liu, C.L.: Robust classification with convolutional prototype learning. In: CVPR, pp. 3474–3482 (2018)
31. Yang, J., et al.: MedMNIST v2-a large-scale lightweight benchmark for 2D and 3D biomedical image classification. Sci. Data **10**(1), 41 (2023)
32. Yu, Z., Shi, Y.: Centralized space learning for open-set computer-aided diagnosis. Sci. Rep. **13**(1), 1630 (2023)
33. Yue, Z., Wang, T., Sun, Q., Hua, X.S., Zhang, H.: Counterfactual zero-shot and open-set visual recognition. In: CVPR, pp. 15404–15414 (2021)
34. Zhang, X.Y., Liu, C.L., Suen, C.Y.: Towards robust pattern recognition: a review. Proc. IEEE **108**(6), 894–922 (2020)
35. Zhou, D.W., Ye, H.J., Zhan, D.C.: Learning placeholders for open-set recognition. In: CVPR, pp. 4401–4410 (2021)

Exploring Unsupervised Cell Recognition with Prior Self-activation Maps

Pingyi Chen[1,2,3], Chenglu Zhu[2,3], Zhongyi Shui[1,2,3], Jiatong Cai[2,3], Sunyi Zheng[2,3], Shichuan Zhang[1,2,3], and Lin Yang[2,3](✉)

[1] College of Computer Science and Technology, Zhejiang University, Hangzhou, China

[2] Research Center for Industries of the Future, Westlake University, Hangzhou, China

{chenpingyi,yanglin}@westlake.edu.cn

[3] School of Engineering, Westlake University, Hangzhou, China

Abstract. The success of supervised deep learning models on cell recognition tasks relies on detailed annotations. Many previous works have managed to reduce the dependency on labels. However, considering the large number of cells contained in a patch, costly and inefficient labeling is still inevitable. To this end, we explored label-free methods for cell recognition. Prior self-activation maps (PSM) are proposed to generate pseudo masks as training targets. To be specific, an activation network is trained with self-supervised learning. The gradient information in the shallow layers of the network is aggregated to generate prior self-activation maps. Afterward, a semantic clustering module is then introduced as a pipeline to transform PSMs to pixel-level semantic pseudo masks for downstream tasks. We evaluated our method on two histological datasets: MoNuSeg (cell segmentation) and BCData (multi-class cell detection). Compared with other fully-supervised and weakly-supervised methods, our method can achieve competitive performance without any manual annotations. Our simple but effective framework can also achieve multi-class cell detection which can not be done by existing unsupervised methods. The results show the potential of PSMs that might inspire other research to deal with the hunger for labels in medical area.

Keywords: Unsupervised method · Self-supervised learning · Cell recognition

1 Introduction

Cell recognition serves a key role in exploiting pathological images for disease diagnosis. Clear and accurate cell shapes provide rich details: nucleus structure,

---

Supplementary Information The online version contains supplementary material available at https://doi.org/10.1007/978-3-031-43993-3_54.

H. Greenspan et al. (Eds.): MICCAI 2023, LNCS 14227, pp. 559–568, 2023.
https://doi.org/10.1007/978-3-031-43993-3_54

cell counts, and cell density of distribution. Hence, pathologists are able to conduct a reliable diagnosis according to the information from the segmented cell, which also improves their experience of routine pathology workflow [5,14].

In recent years, the advancement of deep learning has facilitated significant success in medical images [17,18,20]. However, the supervised training requires massive manual labels, especially when labeling cells in histopathology images. A large number of cells are required to be labeled, which results in inefficient and expensive annotating processes. It is also difficult to achieve accurate labeling because of the large variations among different cells and the variability of reading experiences among pathologists.

Work has been devoted to reducing dependency on manual annotations recently. Qu et al. use points as supervision [19]. It is still a labor-intensive task due to the large number of objects contained in a pathological image. With regard to unsupervised cell recognition, traditional methods can segment the nuclei by clustering or morphological processing. But these methods suffer from worse performance than deep learning methods. Among AI-based methods, some works use domain adaptation to realize unsupervised instance segmentation [2,9,12], which transfers the source domain containing annotations to the unlabeled target. However, their satisfactory performance depends on the appropriate annotated source dataset. Hou et al. [10] proposed to synthesize training samples with GAN. It relies on predefined nuclei texture and color. Feng et al. [6] achieved unsupervised detection and segmentation by a mutual-complementing network. It combines the advantage of correlation filters and deep learning but needs iterative training and finetuning.

CNNs with inductive biases have priority over local features of the nuclei with dense distribution and semi-regular shape. In this paper, we proposed a simple but effective framework for unsupervised cell recognition. Inspired by the strong representation capability of self-supervised learning, we devised the prior self-activation maps (PSM) as the supervision for downstream cell recognition tasks. Firstly, the activation network is initially trained with self-supervised learning like predicting instance-level contrastiveness. Gradient information accumulated in the shallow layers of the activation network is then calculated and aggregated with the raw input information. These features extracted from the activation network are then clustered to generate pseudo masks which are used for downstream cell recognition tasks. In the inferring stage, the networks which are supervised by pseudo masks are directly applied for cell detection or segmentation. To evaluate the effectiveness of PSM, we evaluated our method on two datasets. Our framework achieved comparable performance on cell detection and segmentation on par with supervised methods. Code is available at https://github.com/cpystan/PSM.

2 Method

The structure of our proposed method is demonstrated in Fig. 1. Firstly, an activation network U_{ss} is trained with self-supervised learning. After the back-propagation of gradients, gradient-weighted features are exploited to generate the self-activation maps (PSM). Next is semantic clustering where the PSM is

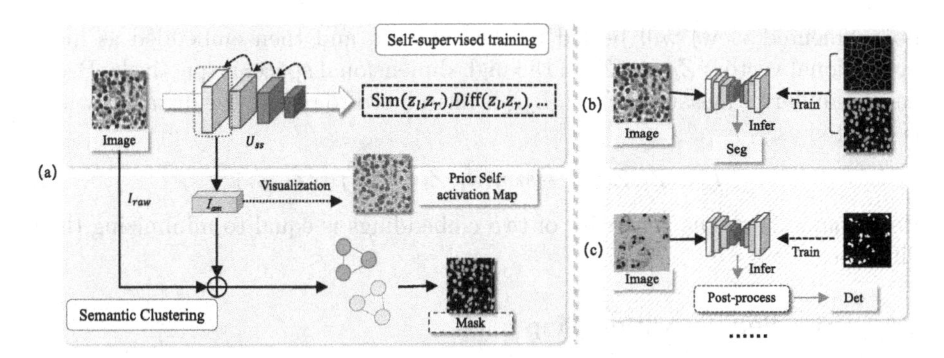

Fig. 1. The framework of our proposed model. (a) The top block shows the activation network which aggregates gradient information to generate self-activation maps. The bottom block presents the work process of semantic clustering. (b–c) The right part is the cell detection and the cell segmentation network which are supervised by the pseudo masks.

combined with the raw input to generate pseudo masks. These pseudo masks can be used as supervision for downstream tasks. Related details are discussed in the following.

2.1 Proxy Task

We introduce self-supervised learning to encourage the network to focus on the local features in the image. And our experiments show that neural networks are capable of adaptively recognizing nuclei with dense distribution and semi-regular shape. Here, we have experimented with several basic proxy tasks below.

ImageNet Pre-training: It is straightforward to exploit the models pre-trained on natural images. In this strategy, we directly extract the gradient-weighted feature map in the ImageNet pre-trained network and generate prior self-activation maps.

Contrastiveness: Following the contrastive learning [3] methods, the network is encouraged to distinguish between different patches. For each image, its augmented view will be regarded as the positive sample, and the other image sampled in the training set is defined as the negative sample. The network is trained to minimize the distance between positive samples. It also maximizes the distance between the negative sample and the input image. The optimization goal can be denoted as:

$$L_{dis}(Z_l, Z_r, Z_n) = diff(Z_l, Z_r) - diff(Z_l, Z_n), \tag{1}$$

where L_{dis} is the loss function. Z_l, Z_r, and Z_n are representations of the input sample, the positive sample, and the negative sample, respectively. In addition, $diff(\cdot)$ is a function that measures the difference of embeddings.

Similarity: LeCun et al. [4] proposed a Siamese network to train the model with a similarity metric. We also adopted a weight-shared network to learn the similarity discrimination task. In specific, the pair of samples (each input and its augmented view) will be fed to the network, and then embedded as high-dimensional vectors Z_l and Z_r in the high-dimensional space, respectively. Based on the similarity measure $sim(\cdot)$, L_{dis} is introduced to reduce the distance, which is denoted as,

$$L_{dis}(Z_l, Z_r) = -sim(Z_l, Z_r) = diff(Z_l, Z_r). \tag{2}$$

Here, maximizing the similarity of two embeddings is equal to minimizing their difference.

2.2 Prior Self-activation Map

The self-supervised model U_{ss} is constructed by sequential blocks which contain several convolutional layers, batch normalization layers, and activation layers. The self-activation map of a certain block can be obtained by nonlinearly mapping the weighted feature maps A^k:

$$I_{am} = ReLU(\sum_k \alpha_k A^k), \tag{3}$$

where I_{am} is the prior self-activation map. A^k indicates the k-th feature map in the selected layer. α_k is the weight of each feature map, which is defined by global-average-pooling the gradients of output z with regard to A^k:

$$\alpha_k = \frac{1}{N}\sum_i \sum_j \frac{\partial z}{\partial A^k_{ij}}, \tag{4}$$

where i, j denote the height and width of output, respectively, and N indicates the input size. The obtained features are visualized in the format of the heat map which is later transformed to pseudo masks by clustering.

Semantic Clustering. We construct a semantic clustering module (SCM) which converts prior self-activation maps to pseudo masks. In SCM, the original information is included to strengthen the detailed features. It is defined as:

$$I_f = I_{am} + \beta \cdot I_{raw}, \tag{5}$$

where I_f denotes the fused semantic map, I_{raw} is the raw input, β is the weight of I_{raw}.

To generate semantic labels, an unsupervised clustering method K-Means is selected to directly split all pixels into several clusters and obtain foreground and background pixels. Given the semantic map I_f and its N pixel features $F = \{f_i | i \in \{1, 2, ...N\}\}$, N features are partitioned into K clusters $S = \{S_i | i \in \{1, 2, ...K\}\}$, $K < N$. The goal is to find S to reach the minimization of within-class variances as follows:

$$min \sum_{i=1}^{K} \sum_{f_j \in S_i} ||f_j - c_i||^2, \tag{6}$$

where c_i denotes the centroid of each cluster S_i. After clustering, the pseudo mask I_{sg} can be obtained.

2.3 Downstream Tasks

In this section, we introduce the training and inferring of cell recognition models.

Cell Detection. For the task of cell detection, a detection network is trained under the supervision of pseudo mask I_{sg}. In the inferring stage, the output of the detection network is a score map. Then, it is post-processed to obtain the detection result.

The coordinates of cells can be got by searching local extremums in the score map, which is described below:

$$T_{(m,n)} = \begin{cases} 1, & p_{(m,n)} > p_{(i,j)}, \quad \forall (i,j) \in D_{(m,n)}, \\ 0, & otherwise, \end{cases} \tag{7}$$

where $T_{(m,n)}$ denotes the predicted label at location of (m,n), p is the value of the score map and $D_{(m,n)}$ indicates the neighborhood of point (m,n). $T_{(m,n)}$ is exactly the detection result.

Cell Segmentation. Due to the lack of instance-level supervision, the model does not perform well in distinguishing adjacent objects in the segmentation. To further reduce errors and uncertainties, the Voronoi map I_{vor} which can be transformed from I_{sg} is utilized to encourage the model to focus on instance-wise features. In the Voronoi map, the edges are labeled as background and the seed points are denoted as foreground. Other pixels are ignored.

We train the segmentation model with these two types of labels. The training loss function can be formulated as below,

$$\begin{aligned} L = \lambda[y \log I_{vor} + (1-y)\log(1-I_{vor})] \\ + (1-y)\log(1-I_{sg}), \end{aligned} \tag{8}$$

where λ is the partition enhancement coefficient. In our experiment, we discovered that false positives hamper the effectiveness of segmentation due to the ambiguity of cell boundaries. Since that, only the background of I_{sg} will be concerned to eliminate the influence of false positives in instance identification.

3 Experiments

3.1 Implementation Details

Dataset. We validated the proposed method on the public dataset of Multi-Organ Nuclei Segmentation (MoNuSeg) [13] and Breast tumor Cell Dataset (BCData) [11]. MoNuSeg consists of 44 images of size 1000 $\times$ 1000 with around 29,000 nuclei boundary annotations. BCData is a public large-scale breast tumor dataset containing 1338 immunohistochemically Ki-67 stained images of size 640 $\times$ 640.

Table 1. Results on MoNuSeg. According to the requirements of each method, various labels are used: localization (Loc) and contour (Cnt). * indicates the model is trained from scratch with the same hyperparameter as ours.

Methods	Loc	Cnt	Pixel-level		Object-level	
			IoU	F1 score	Dice	AJI
Unet* [20]	✔	✔	0.606	0.745	0.715	0.511
MedT [24]	✔	✔	0.662	0.795	–	–
CDNet [8]	✔	✔	–	–	0.832	0.633
Competition Winner [13]	✔	✔	–	–	–	0.691
Qu et al. [19]	✔	✗	0.579	0.732	0.702	0.496
Tian et al. [22]	✔	✗	0.624	0.764	0.713	0.493
CellProfiler [1]	✗	✗	–	0.404	0.597	0.123
Fiji [21]	✗	✗	–	0.665	0.649	0.273
CyCADA [9]	✗	✗	–	0.705	–	0.472
Hou et al. [10]	✗	✗	–	0.750	–	0.498
Ours	✗	✗	0.610	0.762	0.724	0.542

Evaluation Metrics. In our experiments on MoNuSeg, F1-score and IOU are employed to evaluate the segmentation performance. Denote TP, FP, and FN as the number of true positives, false positives, and false negatives. Then F1-score and IOU can be defined as: $F1 = 2TP/(2TP + FP + FN)$, $IOU = TP/(TP+FP+FN)$. In addition, common object-level indicators such as Dice coefficient and Aggregated Jaccard Index (AJI) [13] are also considered to assess the segmentation performance.

In the experiment on BCData, precision (P), recall (R), and F1-score are used to evaluate the detection performance. Predicted points will be matched to ground-truth points one by one. And those unmatched points are regarded as false positives. Precision and recall are: $P = TP/(TP + FP)$, and $R = TP/(TP + FN)$. In addition, we introduce MP and MN to evaluate the cell counting results. 'MP' and 'MN' denote the mean average error of positive and negative cell numbers.

Hyperparameters. Res2Net101 [7] is adopted as the activation network U_{ss} with random initialization of parameters. The positive sample is augmented by rotation. The weights β are set to 2.5 and 4 for training in MoNuSeg and BCData, respectively. The weight λ is 0.5. The analysis for β and λ is included in the supplementary. Pixels of the fused semantic map will be decoupled into three piles by K-Means. The following segmentation and detection are constructed with ResNet-34. They are optimized using CrossEntropy loss by the Adam optimizer for 100 epochs with the initial learning rate of $1e^{-4}$. The function $diff(\cdot)$ is instantiated as the measurement of Manhattan distance.

Table 2. Results on BCData. According to the requirements of each method, various labels are used: localization (Loc) and the number (Num) of cells.

Methods	Loc	Num	Backbone	P	R	F1 score	MP↓	MN↓
CSRNet [15]	✔	✔	ResNet50	0.824	0.834	0.829	9.24	24.90
SC-CNN [23]	✔	✔	ResNet50	0.770	0.828	0.798	9.18	20.60
U-CSRNet [11]	✔	✔	ResNet50	0.869	0.857	0.863	10.04	18.09
TransCrowd [16]	✗	✔	Swin-Transformer	–	–	–	13.08	33.10
Ours	✗	✗	ResNet34	0.855	0.771	0.811	8.89	28.02

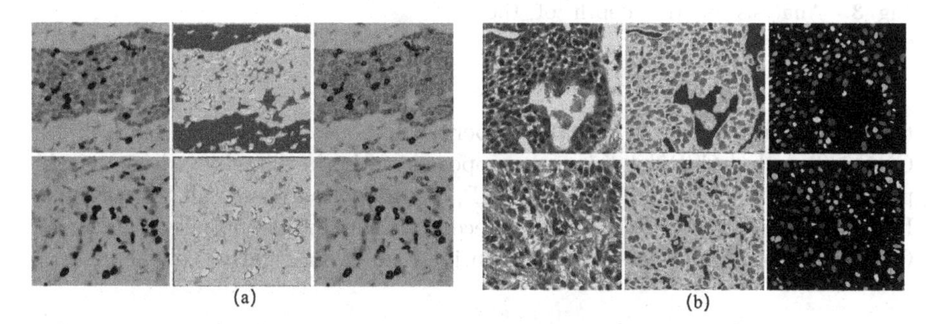

Fig. 2. Visualization. (a) Typical results of multi-class detection, the red dot and green dot represent positive and negative cells respectively. (b) Typical results of cell segmentation. (Color figure online)

3.2 Result

This section includes the discussion of results which are visualized in Fig. 2

Segmentation. In MoNuSeg Dataset, four fully-supervised methods Unet [20], MedT [24], CDNet [8], and the competition winner [13] are adopted to estimate the upper limit as shown in the first four rows of Table 1. Two weakly-supervised models trained with only point annotations are also adopted as the comparison. Compared with the method [22] fully exploiting localization information, ours can achieve better performance without any annotations in object-level metrics (AJI). In addition, two unsupervised methods using traditional image processing tools [1,21] and two unsupervised methods [9,10] with deep learning are compared. Our framework has achieved promising performance because robust low-level features are exploited to generate high-quality pseudo masks.

Detection. Following the benchmark of BCData, metrics of detection and counting are adopted to evaluate the performance as shown in Table 2. The first three methods are fully supervised methods which predict probability maps to achieve detection.

Furthermore, TransCrowd [16] with the backbone of Swin-Transformer is employed as the weaker supervision trained by cell counts regression. By con-

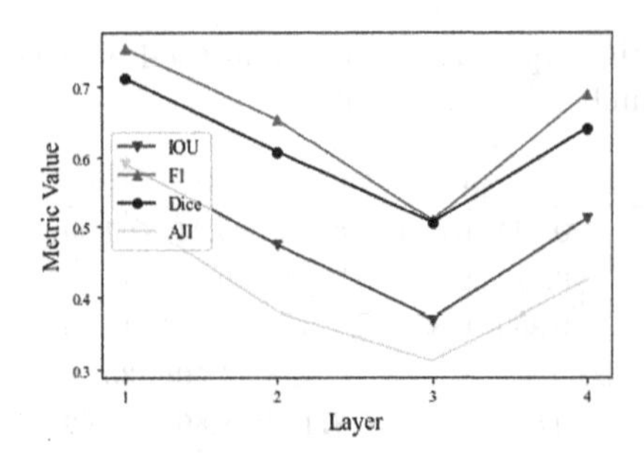

Fig. 3. Analysis of the depth of the extracted layer.

Table 3. The training strategy analysis

Training Strategy	F1	AJI
ImageNet Pretrained	0.658	0.443
Similarity	**0.750**	**0.542**
Contrastiveness	0.741	0.498

trast, even without any annotation supervision, compared to CSRNet [15], NP-CNN [23] and U-CSRNet [11], our proposed method still achieved comparable performance. Especially in terms of MP, our model surpasses all the baselines. It is challenging to realize multi-class recognition in an unsupervised framework. Our method still achieves not bad counting results on negative cells.

Ablation Study. Ablation experiments are built on MoNuSeg. In our pipeline, the activation network can be divided into four layers which consists of multiple basic units including ReLU, BacthNorm, and Convolution. We exploit the prior self-activation maps generated from different depths in the model after training with the same proxy tasks. As shown in Fig. 3, the performance goes down and up with we extracting features from deeper layers. Due to the relatively small receptive field, the shallowest layer in the activation network is the most capable to translate local descriptions

We have also experimented with different types of proxy tasks in a self-supervised manner, as shown in Table 3. We can see that relying on the pre-trained models with external data can not improve the results of subsequent segmentation. The model achieves similar pixel-level performance (F1) when learning similarity or contrastiveness. But similarity learning makes the model performs better in object-level metrics (AJI) than contrastive learning. The high intra-domain similarity hinders the comparison between constructed image pairs. Unlike natural image datasets containing diverse samples, the minor inter class differences in biomedical images may not fully exploit the superiority of contrastive learning.

4 Conclusion

In this paper, we proposed the prior self-activation map (PSM) based framework for unsupervised cell segmentation and multi-class detection. The framework is composed of an activation network, a semantic clustering module (SCM), and the networks for cell recognition. The proposed PSM has a strong capability of learning low-level representations to highlight the area of interest without the need for manual labels. SCM is designed to serve as a pipeline between representations from the activation network and the downstream task. And our segmentation and detection network are supervised by the pseudo masks. In the whole training process, no manual annotation is needed. Our unsupervised method was evaluated on two publicly available datasets and obtained competitive results compared to the methods with annotations. In the future, we will apply our PSM to other types of medical images to further release the dependency on annotations.

Acknowledgement. This study was partially supported by the National Natural Science Foundation of China (Grant no. 92270108), Zhejiang Provincial Natural Science Foundation of China (Grant no. XHD23F0201), and the Research Center for Industries of the Future (RCIF) at Westlake University.

References

1. Carpenter, A.E., et al.: CellProfiler: image analysis software for identifying and quantifying cell phenotypes. Genome Biol. **7**, 1–11 (2006). https://doi.org/10.1186/gb-2006-7-10-r100
2. Chen, C., Dou, Q., Chen, H., Qin, J., Heng, P.A.: Synergistic image and feature adaptation: towards cross-modality domain adaptation for medical image segmentation. In: Proceedings of the AAAI Conference on Artificial Intelligence, vol. 33, pp. 865–872 (2019)
3. Chen, T., Kornblith, S., Norouzi, M., Hinton, G.: A simple framework for contrastive learning of visual representations. In: International Conference on Machine Learning, pp. 1597–1607. PMLR (2020)
4. Chopra, S., Hadsell, R., LeCun, Y.: Learning a similarity metric discriminatively, with application to face verification. In: 2005 IEEE Computer Society Conference on Computer Vision and Pattern Recognition (CVPR 2005), vol. 1, pp. 539–546. IEEE (2005)
5. Elston, C.W., Ellis, I.O.: Pathological prognostic factors in breast cancer. I. The value of histological grade in breast cancer: experience from a large study with long-term follow-up. Histopathology **19**(5), 403–410 (1991)
6. Feng, Z., et al.: Mutual-complementing framework for nuclei detection and segmentation in pathology image. In: Proceedings of the IEEE/CVF International Conference on Computer Vision, pp. 4036–4045 (2021)
7. Gao, S.H., Cheng, M.M., Zhao, K., Zhang, X.Y., Yang, M.H., Torr, P.: Res2Net: a new multi-scale backbone architecture. IEEE Trans. Pattern Anal. Mach. Intell. **43**(2), 652–662 (2021)
8. He, H., et al.: CDNet: centripetal direction network for nuclear instance segmentation. In: Proceedings of the IEEE/CVF International Conference on Computer Vision, pp. 4026–4035 (2021)
9. Hoffman, J., et al.: CyCADA: cycle-consistent adversarial domain adaptation. In: Dy, J., Krause, A. (eds.) Proceedings of the 35th International Conference on Machine Learning. Proceedings of Machine Learning Research, vol. 80, pp. 1989–1998. PMLR (2018)

10. Hou, L., Agarwal, A., Samaras, D., Kurc, T.M., Gupta, R.R., Saltz, J.H.: Robust histopathology image analysis: to label or to synthesize? In: Proceedings of the IEEE/CVF Conference on Computer Vision and Pattern Recognition, pp. 8533–8542 (2019)
11. Huang, Z., et al.: BCData: a large-scale dataset and benchmark for cell detection and counting. In: Martel, A.L., et al. (eds.) MICCAI 2020. LNCS, vol. 12265, pp. 289–298. Springer, Cham (2020). https://doi.org/10.1007/978-3-030-59722-1_28
12. Kim, T., Jeong, M., Kim, S., Choi, S., Kim, C.: Diversify and match: a domain adaptive representation learning paradigm for object detection. IEEE (2019)
13. Kumar, N., Verma, R., Sharma, S., Bhargava, S., Vahadane, A., Sethi, A.: A dataset and a technique for generalized nuclear segmentation for computational pathology. IEEE Trans. Med. Imaging **36**(7), 1550–1560 (2017)
14. Le Doussal, V., Tubiana-Hulin, M., Friedman, S., Hacene, K., Spyratos, F., Brunet, M.: Prognostic value of histologic grade nuclear components of Scarff-Bloom-Richardson (SBR). An improved score modification based on a multivariate analysis of 1262 invasive ductal breast carcinomas. Cancer **64**(9), 1914–1921 (1989)
15. Li, Y., Zhang, X., Chen, D.: CSRNet: dilated convolutional neural networks for understanding the highly congested scenes. In: Proceedings of the IEEE Conference on Computer Vision and Pattern Recognition (CVPR) (2018)
16. Liang, D., Chen, X., Xu, W., Zhou, Y., Bai, X.: TransCrowd: weakly-supervised crowd counting with transformers. Sci. China Inf. Sci. **65**, 160104 (2021). https://doi.org/10.1007/s11432-021-3445-y
17. Liu, D., et al.: Nuclei segmentation via a deep panoptic model with semantic feature fusion. In: IJCAI, pp. 861–868 (2019)
18. Naylor, P., Laé, M., Reyal, F., Walter, T.: Segmentation of nuclei in histopathology images by deep regression of the distance map. IEEE Trans. Med. Imaging **38**(2), 448–459 (2018)
19. Qu, H., et al.: Weakly supervised deep nuclei segmentation using points annotation in histopathology images. In: International Conference on Medical Imaging with Deep Learning, pp. 390–400. PMLR (2019)
20. Ronneberger, O., Fischer, P., Brox, T.: U-Net: convolutional networks for biomedical image segmentation. In: Navab, N., Hornegger, J., Wells, W.M., Frangi, A.F. (eds.) MICCAI 2015. LNCS, vol. 9351, pp. 234–241. Springer, Cham (2015). https://doi.org/10.1007/978-3-319-24574-4_28
21. Schindelin, J., et al.: Fiji: an open-source platform for biological-image analysis. Nat. Methods **9**(7), 676–682 (2012)
22. Tian, K., et al.: Weakly-supervised nucleus segmentation based on point annotations: a coarse-to-fine self-stimulated learning strategy. In: Martel, A.L., et al. (eds.) MICCAI 2020. LNCS, vol. 12265, pp. 299–308. Springer, Cham (2020). https://doi.org/10.1007/978-3-030-59722-1_29
23. Tofighi, M., Guo, T., Vanamala, J.K.P., Monga, V.: Prior information guided regularized deep learning for cell nucleus detection. IEEE Trans. Med. Imaging **38**(9), 2047–2058 (2019)
24. Valanarasu, J.M.J., Oza, P., Hacihaliloglu, I., Patel, V.M.: Medical transformer: gated axial-attention for medical image segmentation. In: de Bruijne, M., et al. (eds.) MICCAI 2021. LNCS, vol. 12901, pp. 36–46. Springer, Cham (2021). https://doi.org/10.1007/978-3-030-87193-2_4

Prompt-Based Grouping Transformer for Nucleus Detection and Classification

Junjia Huang[1,2], Haofeng Li[2,3], Weijun Sun[4], Xiang Wan[2], and Guanbin Li[1(✉)]

[1] School of Computer Science and Engineering, Research Institute of Sun Yat-sen University in Shenzhen, Sun Yat-sen University, Guangzhou, China
liguanbin@mail.sysu.edu.cn
[2] Shenzhen Research Institute of Big Data, Shenzhen, China
[3] The Chinese University of Hong Kong, Shenzhen, China
[4] Guangdong University of Technology, Guangzhou, China

Abstract. Automatic nuclei detection and classification can produce effective information for disease diagnosis. Most existing methods classify nuclei independently or do not make full use of the semantic similarity between nuclei and their grouping features. In this paper, we propose a novel end-to-end nuclei detection and classification framework based on a grouping transformer-based classifier. The nuclei classifier learns and updates the representations of nuclei groups and categories via hierarchically grouping the nucleus embeddings. Then the cell types are predicted with the pairwise correlations between categorical embeddings and nucleus features. For the efficiency of the fully transformer-based framework, we take the nucleus group embeddings as the input prompts of backbone, which helps harvest grouping guided features by tuning only the prompts instead of the whole backbone. Experimental results show that the proposed method significantly outperforms the existing models on three datasets.

Keywords: Nuclei classification · Prompt tuning · Clustering · Transformer

1 Introduction

Nucleus classification is to identify the cell types from digital pathology image, assisting pathologists in cancer diagnosis and prognosis [3,30]. For example, the

This work was supported in part by the Chinese Key-Area Research and Development Program of Guangdong Province (2020B0101350001), in part by the National Natural Science Foundation of China (No. 62102267, NO. 61976250), in part by the Guangdong Basic and Applied Basic Research Foundation (2023A1515011464, 2020B1515020048), in part by the Shenzhen Science and Technology Program (JCYJ20220818103001002, JCYJ20220530141211024), and the Guangdong Provincial Key Laboratory of Big Data Computing, The Chinese University of Hong Kong, Shenzhen.

J. Huang and H. Li—Contribute equally to this work.

Supplementary Information The online version contains supplementary material available at https://doi.org/10.1007/978-3-031-43993-3_55.

H. Greenspan et al. (Eds.): MICCAI 2023, LNCS 14227, pp. 569–579, 2023.
https://doi.org/10.1007/978-3-031-43993-3_55

involvement of tumor-infiltrating lymphocytes (TILs) is a critical prognostic variable for the evaluation of breast/lung cancer [4,29]. It is a challenge to infer the nucleus types due to the diversity and unbalanced distribution of nuclei. Thus, we aim to automatically classify cell nuclei in pathological images.

A number of methods [7,10,14,23–25,33,34] have been proposed for automatic nuclei segmentation and classification. Most of them use a U-shape model [28] for training to produce dense predictions with expensive pixel-level labels. In this paper, we aim to obtain the location and category of cells, which only needs affordable labels of centroids or bounding boxes. The task can be solved by generic object detector [17,26,27], but they are usually built for everyday objects whose positions and combinations are quite random. Differently, in pathological images, experts often identify nuclear communities via their relationships and spatial distribution. Some recent methods resort to the spatial contexts among nuclei. Abousamra *et al.* [1] adopt a spatial statistical function to model the local density of cells. Hassan *et al.* [11] build a location-based graph for nuclei classification. However, the semantics similarity and dissimilarity between nucleus instances as well as the category representations have not been fully exploited.

Based on these observations, we develop a learnable Grouping Transformer based Classifier (GTC) that leverages the similarity between nuclei and their cluster representations to infer their types. Specifically, we define a number of nucleus clusters with learnable initial embeddings, and assign nucleus instances to their most correlated clusters by computing the correlations between clusters and nuclei. Next, the cluster embeddings are updated with their affiliated instances, and are further grouped into the categorical representations. Then, the cell types can be well estimated using the correlations between the nuclei and the categorical embeddings. We propose a novel fully transformer-based framework for nuclei detection and classification, by integrating a backbone, a centroid detector, and the grouping-based classifier. However, the transformer framework has a relatively large number of parameters, which could cause high costs in fine-tuning the whole model on large datasets. On the other hand, there exist domain gaps in the pathological images of different organs, staining, and institutions, which makes it necessary to fine-tune models to new applications. Thus, it is of great significance to tune our proposed transformer framework efficiently.

Inspired by the prompt tuning methods [13,16,20] which train continuous prompts with frozen pretrained models for natural language processing tasks, we propose a grouping prompt based learning strategy for efficient tuning. We prepend the embeddings of nucleus clusters to the input space and freeze the entire pre-trained transformer backbone so that these group embeddings act as prompt information to help the backbone extract grouping-aware features. Our contributions are: (1) a prompt-based grouping transformer framework for end-to-end detection and classification of nuclei; (2) a novel grouping prompt learning mechanism that exploits nucleus clusters to guide feature learning with low tuning costs; (3) Experimental results show that our method achieves the state-of-the-art on three public benchmarks.

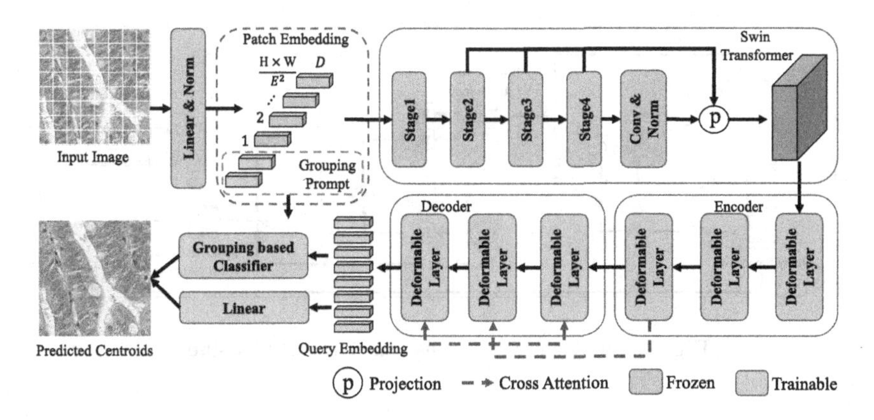

Fig. 1. The architecture of Prompt-based Grouping Transformer.

2 Methodology

As shown in Fig. 1, We propose a novel framework, Prompt-based Grouping Transformer (PGT), which directly outputs the coordinates of nuclei centroids and leverages *grouping prompts* for cell-type prediction. In the architecture, the detection and classification parts are interdependent and can be trained together. The proposed framework consists of a transformer-based nucleus detector, a grouping transformer-based classifier, and a grouping prompt learning strategy, which are presented in the following.

2.1 Transformer-Based Centroid Detector

Backbone. We adopt Swin Transformer [21] as the backbone to learn deep features. The pixel-level feature maps output from Stage 2 to Stage 4 of the backbone are extracted. Then the Stage-4 feature map is downsampled with a 3×3 convolution of stride 2 to yield another lower-resolution feature map. We obtain four feature maps in total. The channel number of each feature map is aligned via a 1×1 convolution layer and a group normalization operator.

Encoder and Decoder. The encoder and decoder have 3 deformable attention layers [35], respectively. The multi-scale feature maps output by the backbone are fed into the encoder in which the pixel-level feature vectors in all these feature maps are updated via deformable self-attention. After the attention layers, we send each feature vector into 2 fully connected (FC) layers separately to obtain the fine-grained categorical scores of each pixel. Only the Q feature vectors with the highest confidence are preserved as object embeddings and their position coordinates are recorded as reference points. Each decoder layer utilizes cross-attention to enhance the object embeddings by taking them as queries/values and the updated feature maps as keys. The enhanced query embeddings are fed into 2 FC layers to regress position offsets which are added to and refine the

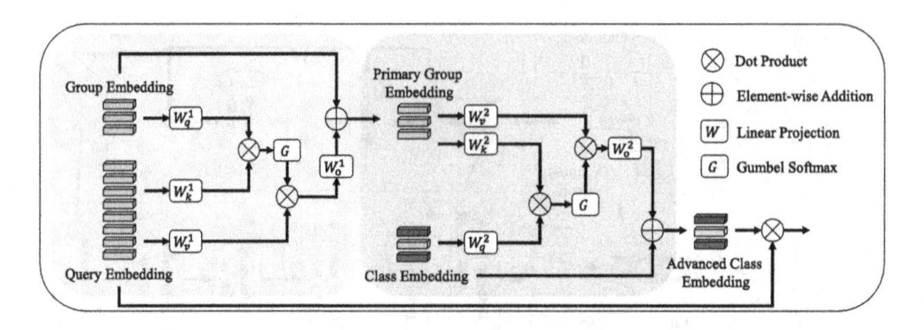

Fig. 2. The Grouping Transformer based Classifier.

reference points. The reference points output by the last decoder layer are the finally detected nucleus centroids. The last query embeddings from the decoder are sent to the proposed classifier for cell type prediction.

2.2 Grouping Transformer Based Classifier

In Fig. 2, we develop a Grouping Transformer based Classifier (GTC) that takes grouping prompts $g \in \mathbb{R}^{G \times D}$ and query embeddings $q \in \mathbb{R}^{Q \times D}$ as inputs, and yields categorical scores for each nucleus query. To divide the queries into primary groups, The similarity matrix $S \in \mathbb{R}^{G \times Q}$ between the query embeddings and the grouping prompts is built via inner product and Gumbel-Softmax [12] operation as Eq. (1):

$$S = softmax(W_q^1 g \cdot (W_k^1 q)^T + \gamma/\tau), \tag{1}$$

where W_q^1 and W_k^1 are the weights of learnable linear projections, $\gamma \in \mathbb{R}^{G \times Q}$ are i.i.d random samples drawn from the distribution $Gumbel(0, 1)$ and τ denotes the Softmax temperature. Then we utilize the hard assignment strategy [31,32] and assign the query embedding to different groups as Eq. (2):

$$\hat{S} = one\text{-}hot(argmax(S)) + S - sg(S), \tag{2}$$

where $argmax(S)$ returns a $1 \times Q$ vector, and $one\text{-}hot(\cdot)$ converts the vector to a binary $G \times Q$ matrix. sg is the stop gradient operator for better training of the one-hot function [31,32]. Then we merge the embeddings belonging to the same group into a primary group via Eq. (3):

$$g_p = g + W_o^1 \frac{\hat{S} \cdot W_v^1 q}{\sum_{i=1}^{G} \hat{S}_i} \tag{3}$$

where g_p denotes the embeddings of primary groups, W_v^1 and W_o^1 are learnable linear weights. To separate the primary groups into the cell categories, we measure the similar matrix between the primary groups g_p and learnable class embeddings $c_e \in \mathbb{R}^{C \times D}$ to yield advanced class embeddings $c_a \in \mathbb{R}^{C \times D}$, in the same way as Eq.(1)–(3). To classify each centroid query, we measure the similarity between each query embedding and the advanced class embeddings.

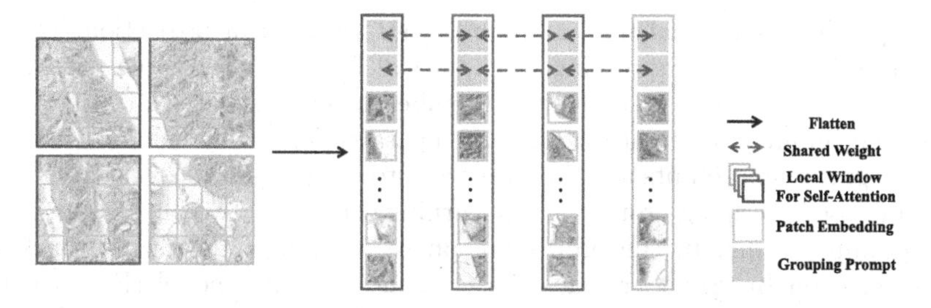

Fig. 3. The inputs with grouping prompts of the Shift-Window transformer backbone.

The category whose advanced embedding is most similar to a query, is assigned to the centroid query. The classification results $c \in \mathbb{R}^{C \times Q}$ are computed as: $c = c_a \cdot q^T$.

2.3 Loss Function

The proposed method outputs a set of centroid proposals $\{(x_q, y_q)|q \in \{1, \cdots, Q\}\}$ with a decoder layer, and their corresponding cell-type scores $\{c_q|q \in \{1, \cdots, Q\}\}$ with our proposed classifier. To compute the loss with detected centroids, we use the Hungarian algorithm [15] to assign K target centroids (ground truth) to proposal centroids and get P positive (matched) samples and $Q - P$ negative (unmatched) samples. The overall loss is defined as Eq. (4):

$$L(y, \hat{y}) = \frac{1}{P} \sum_{i=1}^{P} \left(\omega_1 ||(x_i, y_i) - (\hat{x}_i, \hat{y}_i)||_2^2 + \omega_2 FL(c_i, \hat{c}_i)\right) + \omega_3 \sum_{j=P+1}^{Q} FL(c_j, \hat{c}_j), \tag{4}$$

where $\omega_1, \omega_2, \omega_3$ are weight terms, (x_i, y_i) is the ith matched centroid coordinates, $(\hat{x}_i, \hat{y}_i)$ is the target coordinates. c_i and c_j denote the categorical scores of matched and unmatched samples, respectively. As the target of unmatched samples, $\hat{c}_j$ is set to an empty category. $FL(\cdot)$ is the Focal Loss [18] for training the proposed classifier. We adopt the deep supervision strategy [35]. In the training, each decoder layer produces the side outputs of centroids and query embeddings that are fed into a GTC for classifying nuclei. For the 3 decoder layers, they yield 3 sets of detection and classification results for the loss in Eq. (4).

2.4 Grouping Prompts Based Tuning

To avoid the inefficient fine-tuning of the backbone, we propose a new and simple learning strategy based on grouping prompts, as shown in Fig. 1. We inject a set of prompt embeddings as extra input of the Swin-Transformer [21], and only tune the prompts instead of the backbone. To learn group-aware representations, we further propose to share the embeddings of prompts with those of initial groups in the proposed GTC. Such prompt embeddings are define as *Grouping Prompts*.

For a typical Swin-Transformer backbone, an input pathological image $I \in \mathbb{R}^{H \times W \times 3}$ is divided into $\frac{HW}{E^2}$ image patches of size $E \times E$. We first embed each image patch into a D-dimensional latent space via a linear projection. Then we randomly initialize the grouping prompts $g \in \mathbb{R}^{G \times D}$ as learnable parameters, and concatenate them with the patch embeddings as input. Note that in the backbone, input patch embeddings are separated into different local windows and the grouping prompts are also inserted into each window, as shown in Fig. 3. Our proposed grouping prompt based learning consists of two phases, pre-tuning and prompt-tuning. In the pre-tuning phase, we adopt the Swin-b backbone pre-trained on ImageNet, replace the GTC head in our model (Fig. 1) with 2 FC layers, and train the overall framework without prompts and GTC. In the prompt-tuning phase, grouping prompts are added to the input of the backbone and GTC, while the backbone parameters are frozen.

3 Experiments and Results

3.1 Datasets and Implementation Details

CoNSeP[1] [10] is a colorectal nuclear dataset with three types, consisting of 41 H&E stained image tiles from 16 colorectal adenocarcinoma whole-slide images (WSIs). The WSIs are at 20× magnification and the size of the slides is 500 × 500. We split them following the official partition [1,10].

BRCA-M2C[2] [1] is a breast cancer dataset with three types and consists of 120 image tiles from 113 patients. The WSIs are at 20× magnification and the size of the slides ranges from 465 × 465 to 504 × 504. We follow the work [1] to apply the SLIC [2] algorithm to generate superpixels as instances and split them into 80/10/30 slides for training/validation/testing.

Lizard[3] [9] has 291 histology images of colon tissue from six datasets, containing nearly half a million labeled nuclei in H&E stained colon tissue. The WSIs are at 20× magnification with an average size of 1,016 × 917 pixels.

Our implementation and the setting of hyper-parameters are based on MMDetection [5]. The number of grouping prompts G is 64. Random crop, flipping, and scaling are used for data augmentation. Our method is trained with PyTorch on a 48 GB GPU (NVIDIA A100) for 12–24 h (depending on the dataset size). More details are listed in the supplementary material.

3.2 Comparison with the State-of-the-Art

The proposed method is compared with the state-of-the-art models: the existing methods for detecting and classifying cells in pathological images, i.e., Hover-Net [10], MCSpatNet [1], SONNET [7], and the sate-of-the-art methods for object detection in natural images, i.e., DDOD [6], TOOD [8], DAB-DETR [19] and Uper-Net with ConvNeXt backbone [22]. As shown in Table 1, our method exceeds all

[1] https://warwick.ac.uk/fac/cross_fac/tia/data/hovernet/.
[2] https://github.com/TopoXLab/Dataset-BRCA-M2C/.
[3] https://warwick.ac.uk/fac/cross_fac/tia/data/lizard/.

Table 1. Comparison with existing methods on CoNSeP, BRCA-M2C and Lizard. For each dataset, we report the F-score of each class (F_c^k), the mean F-score over all classes ($\overline{F_c}$) and the detection F-score (F_d). $F_c^{Infl.}$, $F_c^{Epi.}$, $F_c^{Stro.}$, $F_c^{Neu.}$, $F_c^{Lym.}$, $F_c^{Pla.}$, $F_c^{Eos.}$ and $F_c^{Con.}$ denote the F-score for the inflammatory, epithelial, stromal, neutrophils, lymphocytes, plasma, Eosinophil and connective tissue cells, respectively. For each row, the best result is in **bold** and the second best is underlined.

	F-score↑	Hovernet [10]	DDOD [6]	TOOD [8]	MCSpatNet [1]	SONNET [7]	DAB-DETR [19]	ConvNeXt [22]	(Ours)
		2019	2021	2021	2021	2022	2022	2022	-
CoNSeP	$F_c^{Infl.}$	0.514	0.516	0.622	0.583	0.563	0.531	0.618	**0.623**
	$F_c^{Epi.}$	0.604	0.436	0.616	0.608	0.502	0.440	0.625	**0.639**
	$F_c^{Stro.}$	0.391	0.429	0.382	0.527	0.366	0.443	0.542	**0.577**
	$\overline{F_c}$	0.503	0.494	0.540	0.573	0.477	0.471	0.595	**0.613**
	F_d	0.621	0.554	0.608	0.722	0.590	0.619	0.715	**0.738**
BRCA-M2C	$F_c^{Infl.}$	0.454	0.394	0.400	0.424	0.343	0.437	0.423	**0.473**
	$F_c^{Epi.}$	0.577	0.544	0.559	0.627	0.411	0.634	0.636	**0.686**
	$F_c^{Stro.}$	0.339	0.373	0.315	0.387	0.281	0.380	0.353	**0.409**
	$\overline{F_c}$	0.457	0.437	0.425	0.479	0.345	0.484	0.471	**0.523**
	F_d	0.74	0.659	0.662	0.794	0.653	0.705	0.785	**0.799**
Lizard	$F_c^{Neu.}$	0.210	0.025	0.029	0.105	0.09	0.142	0.205	**0.301**
	$F_c^{Epi.}$	0.665	0.584	0.615	0.601	0.599	0.653	0.714	**0.762**
	$F_c^{Lym.}$	0.472	0.342	0.404	0.457	0.538	0.544	0.611	**0.664**
	$F_c^{Pla.}$	0.376	0.130	0.152	0.228	0.370	0.356	0.333	**0.403**
	$F_c^{Eos.}$	0.367	0.124	0.157	0.220	0.365	0.295	0.403	**0.457**
	$F_c^{Con.}$	0.492	0.347	0.383	0.484	0.143	0.559	0.578	**0.644**
	$\overline{F_c}$	0.430	0.259	0.290	0.349	0.351	0.425	0.474	**0.538**
	F_d	0.729	0.561	0.606	0.713	0.682	0.656	0.764	**0.779**

the other methods on three benchmarks with both detection and classification metrics. Specifically, on the CoNSeP dataset, our approach achieves 1.6% higher F-score on the detection (F_d) and 1.8% higher F-score on the classification ($\overline{F_c}$) than the second best methods MCSpatNet [1] and UperNet [22]. On BRCA-M2C dataset, our method has 0.5% higher F_d and 3.9% higher $\overline{F_c}$, compared with the second best models MCSpatNet [1] and DAB-DETR [19]. Besides, on Lizard dataset, our method outperforms UperNet [22] by more than 1.5% and 6.4% on F_d and $\overline{F_c}$, respectively. Meanwhile, we conduct t-test on CoNSeP dataset for statistical significance test. The details are listed in the supplementary material. The visual comparisons are shown in Fig. 4. With the context information from surrounding nuclei, our method effectively reduces the misclassification rate of the lymphocytes and neutrophil categories (Blue and Red).

3.3 Ablation Analysis

The strengths of the grouping transformer based classifier and the grouping prompts are verified on CoNSeP dataset, as shown in Table 2. Prompt-based Grouping Transformer (PGT) is our proposed detection and classification architecture with grouping prompts and the GTC (in Fig. 1), while the

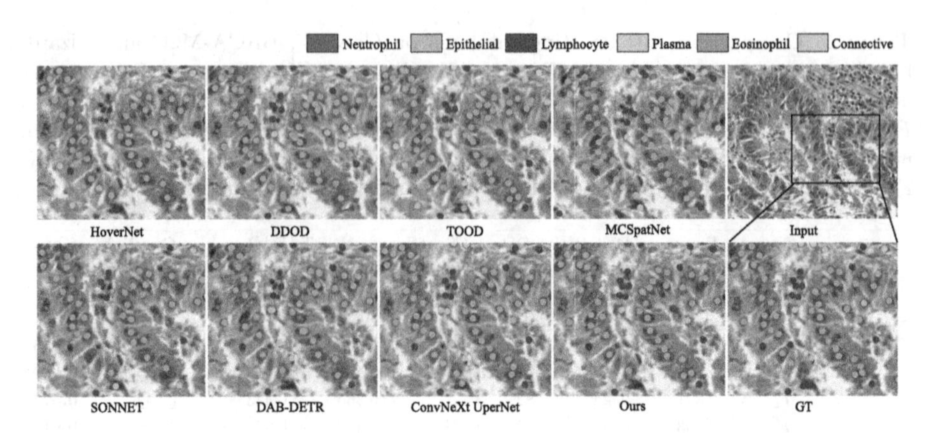

Fig. 4. The visualization results on CoNSeP dataset. (Color figure online)

Table 2. Ablation study on CoNSeP. PGT is the overall detection-classification framework. PT denotes training the network with Prompt Tuning. GTC means using the Grouping Transformer-based Classifier. * means freezing the weights of the backbone.

Methods	$F_c^{Infl.}$	$F_c^{Epi.}$	$F_c^{Stro.}$	$\overline{F_c}$	F_d	Tuned Params (M)
PGT (Full)	0.631	0.641	0.572	0.615	0.735	102.2
w/o GTC & PT (Baseline)	0.599	0.600	0.570	0.590	0.714	95.767
w/o PT*	0.602	0.604	0.558	0.588	0.713	15.321
w/o GTC*	0.615	0.604	0.564	0.594	0.724	8.895
w/ detached GTC & PT*	0.577	0.623	0.545	0.582	0.714	15.429
PGT* (Ours)	**0.623**	**0.639**	**0.577**	**0.613**	**0.738**	15.379

'Baseline' has no these two settings. PT means using naive prompt tuning. GTC means classifying nuclei with the grouping transformer. Our method achieves comparable results to the fully fine-tuning PGT with tuning only 15% parameters. Compared to the Baseline, our method yields 2.4% higher F_d and 2.3% higher $\overline{F_c}$, respectively, which shows the effective combination of the grouping classifier and prompts. 'detached GTC & PT' means that group features and prompts are independent. Our method surpasses the detached setting by 2.4% in F_d and 3.1% in $\overline{F_c}$, which suggests that sharing embeddings of groups and prompts is effective. With a frozen backbone, the performances of 'w/o PT' and 'w/o GTC' are both dropping, which verifies the strength of the prompt tuning and the GTC module, respectively.

Table 3 shows **the effect of different numbers of grouping prompts** on CoNSeP dataset. When the number of groups is small, the classification result is inferior. When the group number is large than 64, the groups may contain too few nuclei to capture their common patterns. It is suggested to set the group number to a moderate value such as 64.

Table 3. The effects of the number of grouping prompts G on CoNSeP.

F-score↑	8	16	32	64	128
F_d	0.727	0.724	0.726	**0.738**	0.723
$\overline{F_c}$	0.600	0.599	0.604	**0.613**	0.583

Table 4. $\overline{F_d}$ denotes the mean of detection F-scores of all testing images. * means p-value ≤0.05. ** means p-value ≤0.01.

F-score↑	Hovernet [10]	DDOD [6]	TOOD [8]	MCSpatNet [1]	SONNET [7]	DAT-DETR [19]	ConvNeXt -UperNet [22]	PGT* (Ours)
$\overline{F_d}$	0.615	0.545	0.625	0.706	0.582	0.615	0.698	**0.728**
p-value	0.001*	0.000**	0.000*	0.027*	0.000**	0.000*	0.012*	–

The Statistical Tests. As shown in Table 4, We calculate F_d of each testing image as sample data and conduct t-test to obtain p-values on the CoNSeP dataset. The p-values are computed between our method and the others.

4 Conclusion

We propose a new prompt-based grouping transformer framework that is fully transformer-based, and can achieve end-to-end nuclei detection and classification. In our framework, a grouping-based classifier groups nucleus features into cluster and category embeddings whose correlations with nuclei are used for identifying cell types. We further propose a novel learning scheme, which shares group embeddings with prompt tokens and extracts features guided by nuclei groups with less tuning costs. The results not only suggest that our method can obtain competitive performance on nuclei classification, but also indicate that the proposed prompt learning strategy can enhance the tuning efficiency.

References

1. Abousamra, S., et al.: Multi-class cell detection using spatial context representation. In: ICCV, pp. 4005–4014 (2021)
2. Achanta, R., Shaji, A., Smith, K., Lucchi, A., Fua, P., Süsstrunk, S.: SLIC superpixels compared to state-of-the-art superpixel methods. IEEE Trans. Pattern Anal. Mach. Intell. **34**(11), 2274–2282 (2012)
3. Aeffner, F., et al.: Introduction to digital image analysis in whole-slide imaging: a white paper from the digital pathology association. J. Pathol. Inform. **10**(1), 9 (2019)
4. Bremnes, R.M., et al.: The role of tumor-infiltrating lymphocytes in development, progression, and prognosis of non-small cell lung cancer. J. Thorac. Oncol. **11**(6), 789–800 (2016)
5. Chen, K., et al.: MMDetection: open MMLab detection toolbox and benchmark. arXiv preprint arXiv:1906.07155 (2019)

6. Chen, Z., Yang, C., Li, Q., Zhao, F., Zha, Z.J., Wu, F.: Disentangle your dense object detector. In: ACM MM, pp. 4939–4948 (2021)
7. Doan, T.N., Song, B., Vuong, T.T., Kim, K., Kwak, J.T.: SONNET: a self-guided ordinal regression neural network for segmentation and classification of nuclei in large-scale multi-tissue histology images. JBHI **26**(7), 3218–3228 (2022)
8. Feng, C., Zhong, Y., Gao, Y., Scott, M.R., Huang, W.: TOOD: task-aligned one-stage object detection. In: ICCV, pp. 3490–3499 (2021)
9. Graham, S., et al.: Lizard: a large-scale dataset for colonic nuclear instance segmentation and classification. In: ICCV Workshop, pp. 684–693 (2021)
10. Graham, S., et al.: HoVer-Net: simultaneous segmentation and classification of nuclei in multi-tissue histology images. Med. Image Anal. **58**, 101563 (2019)
11. Hassan, T., Javed, S., Mahmood, A., Qaiser, T., Werghi, N., Rajpoot, N.: Nucleus classification in histology images using message passing network. Med. Image Anal. **79**, 102480 (2022)
12. Jang, E., Gu, S., Poole, B.: Categorical reparameterization with gumbel-softmax. In: ICLR (2017)
13. Jia, M., et al.: Visual prompt tuning. In: Avidan, S., Brostow, G., Cissé, M., Farinella, G.M., Hassner, T. (eds.) ECCV 2022. LNCS, vol. 13693, pp. 709–727. Springer, Cham (2022). https://doi.org/10.1007/978-3-031-19827-4_41
14. Kiran, I., Raza, B., Ijaz, A., Khan, M.A.: DenseRes-Unet: segmentation of overlapped/clustered nuclei from multi organ histopathology images. Comput. Biol. Med. **143**, 105267 (2022)
15. Kuhn, H.W.: The Hungarian method for the assignment problem. Naval Res. Logist. Q. **2**(1–2), 83–97 (1955)
16. Lester, B., Al-Rfou, R., Constant, N.: The power of scale for parameter-efficient prompt tuning. In: EMNLP, Punta Cana, Dominican Republic, pp. 3045–3059. Association for Computational Linguistics (2021)
17. Li, X., Li, Q., et al.: Detection and classification of cervical exfoliated cells based on faster R-CNN. In: IEEE 11th International Conference on Advanced Infocomm Technology (ICAIT), pp. 52–57 (2019)
18. Lin, T.Y., Goyal, P., Girshick, R., He, K., Dollár, P.: Focal loss for dense object detection. In: ICCV, pp. 2980–2988 (2017)
19. Liu, S., et al.: DAB-DETR: dynamic anchor boxes are better queries for DETR. In: ICLR (2022)
20. Liu, X., et al.: P-tuning: prompt tuning can be comparable to fine-tuning across scales and tasks. In: Proceedings of the 60th Annual Meeting of the Association for Computational Linguistics (Volume 2: Short Papers), Dublin, Ireland, pp. 61–68. Association for Computational Linguistics (2022)
21. Liu, Z., et al.: Swin transformer: hierarchical vision transformer using shifted windows. In: ICCV, pp. 10012–10022 (2021)
22. Liu, Z., Mao, H., Wu, C.Y., Feichtenhofer, C., Darrell, T., Xie, S.: A convnet for the 2020s. In: CVPR, pp. 11976–11986 (2022)
23. Liu, Z., Wang, H., Zhang, S., Wang, G., Qi, J.: NAS-SCAM: neural architecture search-based spatial and channel joint attention module for nuclei semantic segmentation and classification. In: Martel, A.L., et al. (eds.) MICCAI 2020. LNCS, vol. 12261, pp. 263–272. Springer, Cham (2020). https://doi.org/10.1007/978-3-030-59710-8_26
24. Lou, W., Li, H., Li, G., Han, X., Wan, X.: Which pixel to annotate: a label-efficient nuclei segmentation framework. IEEE TMI **42**(4), 947–958 (2022)

25. Lou, W., et al.: Multi-stream cell segmentation with low-level cues for multi-modality images. In: Competitions in Neural Information Processing Systems, pp. 1–10. PMLR (2023)
26. Nair, L.S., Prabhu, R., Sugathan, G., Gireesh, K.V., Nair, A.S.: Mitotic nuclei detection in breast histopathology images using YOLOv4. In: 12th International Conference on Computing Communication and Networking Technologies, pp. 1–5 (2021)
27. Obeid, A., Mahbub, T., Javed, S., Dias, J., Werghi, N.: NucDETR: end-to-end transformer for nucleus detection in histopathology images. In: Qin, W., Zaki, N., Zhang, F., Wu, J., Yang, F. (eds.) CMMCA 2022. LNCS, vol. 13574, pp. 47–57. Springer, Cham (2022). https://doi.org/10.1007/978-3-031-17266-3_5
28. Ronneberger, O., Fischer, P., Brox, T.: U-Net: convolutional networks for biomedical image segmentation. In: Navab, N., Hornegger, J., Wells, W.M., Frangi, A.F. (eds.) MICCAI 2015. LNCS, vol. 9351, pp. 234–241. Springer, Cham (2015). https://doi.org/10.1007/978-3-319-24574-4_28
29. Salgado, R., et al.: The evaluation of tumor-infiltrating lymphocytes (TILs) in breast cancer: recommendations by an International TILs Working Group 2014. Ann. Oncol. **26**(2), 259–271 (2015)
30. Sirinukunwattana, K., Raza, S.E.A., Tsang, Y.W., Snead, D.R., Cree, I.A., Rajpoot, N.M.: Locality sensitive deep learning for detection and classification of nuclei in routine colon cancer histology images. IEEE TMI **35**(5), 1196–1206 (2016)
31. Van Den Oord, A., Vinyals, O., et al.: Neural discrete representation learning. In: NeurIPS, vol. 30 (2017)
32. Xu, J., et al.: GroupViT: semantic segmentation emerges from text supervision. In: CVPR, pp. 18134–18144 (2022)
33. Zeng, Z., Xie, W., Zhang, Y., Lu, Y.: RIC-Unet: an improved neural network based on Unet for nuclei segmentation in histology images. IEEE Access **7**, 21420–21428 (2019)
34. Zhou, H.Y., et al.: SSMD: semi-supervised medical image detection with adaptive consistency and heterogeneous perturbation. Med. Image Anal. **72**, 102117 (2021)
35. Zhu, X., Su, W., Lu, L., Li, B., Wang, X., Dai, J.: Deformable DETR: deformable transformers for end-to-end object detection. In: ICLR (2021)

PAS-Net: Rapid Prediction of Antibiotic Susceptibility from Fluorescence Images of Bacterial Cells Using Parallel Dual-Branch Network

Wei Xiong[1], Kaiwei Yu[2], Liang Yang[2], and Baiying Lei[1](✉)

[1] National-Regional Key Technology Engineering Laboratory for MedicalUltrasound, Guangdong Key Laboratory for Biomedical Measurements and Ultrasound Imaging, School of Biomedical Engineering, Health Science Center, Shenzhen University, Shenzhen 518060, China
leiby@szu.edu.cn

[2] School of Medicine, Southern University of Science and Technology, Shenzhen 518055, China

Abstract. In recent years, the emergence and rapid spread of multi-drug resistant bacteria has become a serious threat to global public health. Antibiotic susceptibility testing (AST) is used clinically to determine the susceptibility of bacteria to antibiotics, thereby guiding physicians in the rational use of drugs as well as slowing down the process of bacterial resistance. However, traditional phenotypic AST methods based on bacterial culture are time-consuming and laborious (usually 24–72 h). Because delayed identification of drug-resistant bacteria increases patient morbidity and mortality, there is an urgent clinical need for a rapid AST method that allows physicians to prescribe appropriate antibiotics promptly. In this paper, we present a parallel dual-branch network (i.e., PAS-Net) to predict bacterial antibiotic susceptibility from fluorescent images. Specifically, we use the feature interaction unit (FIU) as a connecting bridge to align and fuse the local features from the convolutional neural network (CNN) branch (C-branch) and the global representations from the Transformer branch (T-branch) interactively and effectively. Moreover, we propose a new hierarchical multi-head self-attention (HMSA) module that reduces the computational overhead while maintaining the global relationship modeling capability of the T-branch. PAS-Net is experimented on a fluorescent image dataset of clinically isolated Pseudomonas aeruginosa (PA) with promising prediction performance. Also, we verify the generalization performance of our algorithm in fluorescence image classification on two HEp-2 cell public datasets.

Keywords: Parallel dual-branch network · Feature interaction unit · Hierarchical multi-head self-attention · Antibiotic susceptibility prediction

1 Introduction

In recent years, the overuse and misuse of antibiotics have led to an increase in the rate of bacterial antibiotic resistance worldwide [1, 2]. The increasing number of multi-drug resistant strains not only poses a serious threat to human health, but also poses great

H. Greenspan et al. (Eds.): MICCAI 2023, LNCS 14227, pp. 580–591, 2023.
https://doi.org/10.1007/978-3-031-43993-3_56

difficulties in clinical anti-infection treatment [3]. To address this issue, clinicians rely on antibiotic susceptibility testing (AST) to determine bacterial susceptibility to antibiotics, thus guiding rational drug use. However, the traditional AST method requires overnight culture of the bacteria in the presence of antibiotics, which is time-consuming and laborious (usually 24–72 h). Such delays prevent physicians from determining effective antibiotic treatments promptly. Therefore, there is an urgent clinical need for a rapid AST method that allows physicians to prescribe appropriate antibiotics in an informed manner, which is essential to improve patient outcomes, shorten the treatment duration, and slow down the progression of bacterial resistance.

In this paper, we take *Pseudomonas aeruginosa* (PA) as the research object and observe the difference in shape and distribution of bacterial aggregates formed by sensitive and multi-drug resistant bacteria through fluorescent images, so we want to use image recognition technology to distinguish these two types of bacteria for the purpose of rapid prediction of antibiotic susceptibility. However, we recognize that this classification task presents several challenges (as shown in Fig. 1). Firstly, in images of sensitive PA and multi-drug resistant Pseudomonas aeruginosa (MDRPA), inter-class variation is low, but intra-class variation is high. Secondly, some images have exposure problems due to the high intensity of bacterial aggregation. Lastly, there are low signal-to-noise ratio of the images, coupled with possible image artifacts resulting from inhomogeneous staining or inappropriate manipulation.

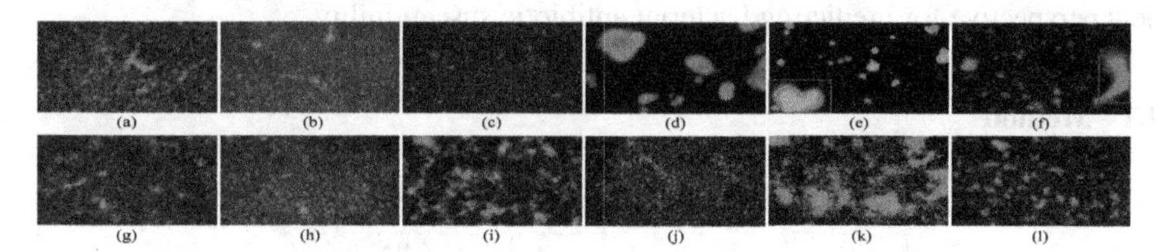

Fig. 1. Examples of fluorescent images of sensitive PA (first row) and MDRPA (second row).

In recent years, deep learning techniques have made a splash in the field of image recognition with their impressive performance and have provided powerful support for a wide range of applications in biomedical research and clinical practice. Notably, deep learning methods based on convolutional neural network (CNN, e.g., ResNet [4], ResNeXt [5], ResNeSt [6]) are widely used in microscopic image classification tasks. For instance, Waisman *et al.* [7] utilized transmission light microscopy images to train a CNN to distinguish pluripotent stem cells from early differentiated cells. Riasatian *et al.* [8] proposed a novel network based on DenseNet [9] and fine-tuned and trained it with various configurations of histopathology images. Recently, due to the successful application of ViT [10] to image classification tasks, many research efforts (e.g., DeiT [11], PVT [12], Swin Transformer [13]) have attempted to introduce the power of self-attention mechanism [14] into computer vision. For example, He *et al.* [15] applied a spatial pyramidal Transformer network to learn long-range contextual information for skin lesion analysis for skin disease classification.

The above studies show that two deep learning frameworks, CNN and Transformer, are effective in microscopy image classification tasks. CNN is good at extracting local features, but its receptive field is limited by the size of the convolution kernel and cannot effectively capture the global information in the image. Meanwhile, in visual Transformer, its self-attention module is good at capturing feature dependencies over long distances, but ignores local feature information. However, these two kinds of feature information are very important for the classification of microscope images with complex features. To tackle this issue, this paper builds a hybrid model that maximizes the advantages of CNN and Transformer, thus enhancing the feature representation of the network. To achieve the complementary advantages of these two techniques, we propose a parallel dual-branch network named PAS-Net, specifically designed to enable rapid prediction of bacterial antibiotic susceptibility. The main contributions of this study are as follows:

1) We develop a parallel dual-branch classification network to realize the interactive learning of features throughout the whole process through feature interaction unit (FIU), which can better integrate local features of CNN branch (C-branch) and global representations of Transformer branch (T-branch).
2) We propose a more efficient hierarchical multi-head self-attention (HMSA) module, which utilizes a local-to-global attention mechanism to simulate the global information of an image, while effectively reducing the computational costs and memory consumption.

To the best of our knowledge, this study represents the first attempt to use deep learning techniques to realize rapid AST based on PA fluorescence images, which provides a new perspective for predicting bacterial antibiotic susceptibility.

1.1 Method

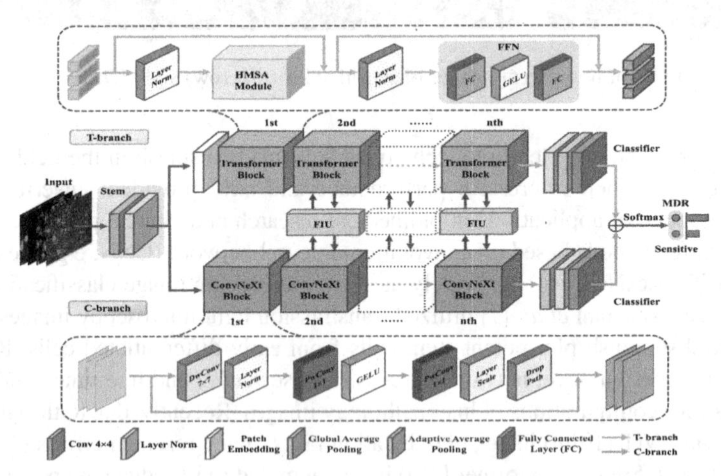

Fig. 2. Overall architecture of the proposed PAS-Net. DwConv: Depthwise convolution. PwConv: Pointwise convolution. FFN: Feed forward network.

Figure 2 shows the overview of our proposed PAS-Net. The model consists of four parts: the Stem module, the parallel C-branch and T-branch, and the FIU connecting the dual branches. The Stem module, which is a 4×4 convolution with stride 4 followed by a layer normalization for quadruple downsampling of the input image. The C-branch and T-branch are stacked with 12 ConvNeXt [16] blocks and Transformer blocks respectively. FIU is applied from the second feature extraction layer, because the initial features of the first feature extraction layer are the same and all come from Stem module, so there is no need for interaction.

1.2 Feature Interaction Unit

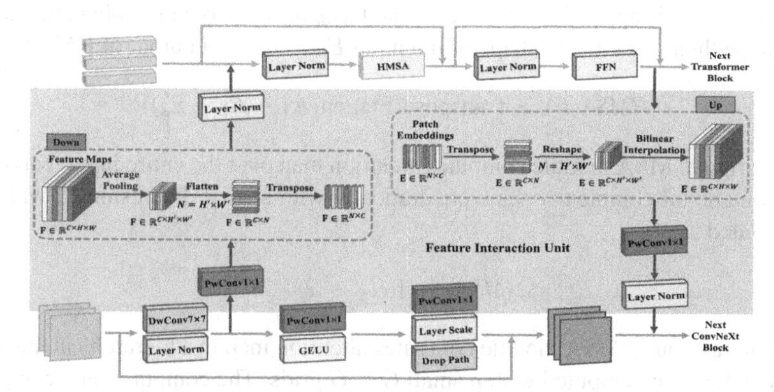

Fig. 3. FIU implementation details. The data stream into FIU is divided into two directions, denoted as CNN → Transformer and Transformer → CNN.

Feature dimension mismatch exists between feature map from C-branch and vector sequence from T-branch. Therefore, our network use FIU as a bridge to effectively combine the local features and the global representation in an interactive manner to eliminate the misalignment between the two features, as shown in Fig. 3.

CNN→Transformer: The feature map is first aligned with the dimensions of the patch embedding by 1×1 convolution. Then, the feature resolution is adjusted using the down-sampling module to complete the alignment of the spatial dimensions. Finally, the feature maps are summed with the patch embedding of the T-branch.

Transformer→CNN: After going through the HMSA module and FFN, the patch embedding is fed back from the T-branch to the C-branch. An up-sampling module needs to be used first for the patch embedding to align the spatial scales. The patch embedding is then aligned to the number of channels of the feature map by 1×1 convolution, and added to the feature map of the C-branch.

1.3 Hierarchical Multi-head Self-attention

Figure 4 shows the detailed structure of HMSA module. To be able to compute attention in a hierarchical manner, we reshape the input patch embedding $\boldsymbol{E}$ back to the patch map $\boldsymbol{E}_p$. Firstly, the patch map $\boldsymbol{E}_p$ is divided into small grids of size $G \times G$, *i.e.*, each grid contains $G \times G$ (set $G = 4$ in this paper) pixel points. Then, a 1×1 pointwise convolution is performed on $\boldsymbol{E}_{p'}$ to obtain three matrices $\boldsymbol{Q} = \boldsymbol{E}_{p'}\boldsymbol{W}^Q$ 、 $\boldsymbol{K} = \boldsymbol{E}_{p'}\boldsymbol{W}^K$ and $\boldsymbol{V} = \boldsymbol{E}_{p'}\boldsymbol{W}^V$, respectively, where $\boldsymbol{W}^Q$, $\boldsymbol{W}^K$ and $\boldsymbol{W}^V$ are three learnable weight matrices with shared parameters that are updated together with the model parameters during training. After that, we compute local attention $\boldsymbol{A}_0$ within each small grid using the self-attention mechanism, which can be defined as:

$$\text{Attention}(\boldsymbol{Q}, \boldsymbol{K}, \boldsymbol{V}) = \text{Softmax}\left(\frac{\boldsymbol{Q}\boldsymbol{K}^T}{\sqrt{d}}\right)\boldsymbol{V}. \quad (1)$$

Then Eq. (1) is applied once more on the basis of $\boldsymbol{A}_0$ to obtain global attention $\boldsymbol{A}_1$. We reshape them back to the shape of the input $\boldsymbol{E}_{p'}$. The final output of HMSA is

$$HMSA(E) = \text{Transpose}(\text{Flatten}(\mathrm{A}_1 + \mathrm{A}_0 + \mathrm{E}_\mathrm{p}')). \quad (2)$$

The original MSA module computes attention map over the entire input feature, and its computational complexity scale quadratically with spatial dimension N, which can be calculated as:

$$\Omega(MSA) = 4ND^2 + 2N^2D. \quad (3)$$

In contrast, our HMSA module computes attention map in a hierarchical manner so that $\boldsymbol{A}_0$ and $\boldsymbol{A}_1$ are computed within small $G \times G$ grids. The computational complexity of HMSA is

$$\Omega(HMSA) = 3ND^2 + 2NG^2D. \quad (4)$$

With this approach, only a limited number of image blocks need to be processed in each step, thus significantly reducing the computational effort of the module from $O(N^2)$ to $O(NG^2)$, where G^2 is much smaller than N. For example, the size of the input image is 224×224, if the patch is divided according to the size of 4×4, the division will get $(224 / 4)^2 = 3136$ patches, i.e., $N = 3136$. However, we set G to 4, so the computational complexity of the HMSA module is greatly reduced.

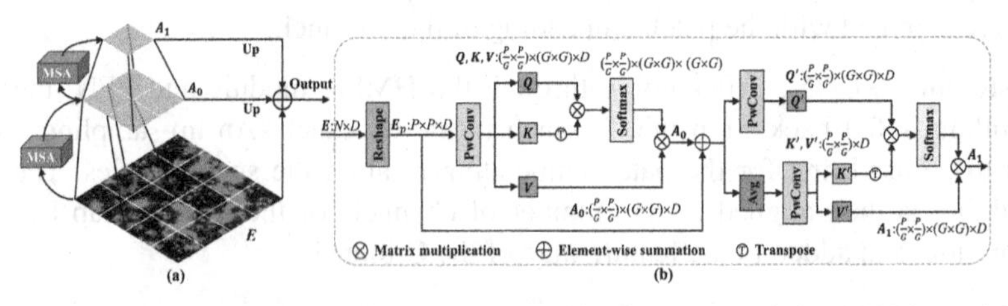

Fig. 4. Illustration of proposed HMSA. (a) Hierarchical structure of the HMSA module. (b) Implementation details of the HMSA module.

2 Experiments and Results

2.1 Experimental Setup

The fluorescent images of PA come from a local medical school. We screen out 12 multidrug resistant strains and 11 sensitive strains. Our dataset has 2625 fluorescent images of PA, 1233 images of sensitive PA and 1392 images of MDRPA. We randomly divide the data into a training set and a test set in a ratio of 9:1. To better train the network model and prevent overfitting, we perform five data enhancement operations on each image, including horizontal flip, vertical flip and rotation at different angles (90°, 180°, 270°). Finally, our data volume is expanded to 15,750 images, including 14,178 training images and 1,572 test images.

To achieve comprehensive and objective assessment of the classification performance of the proposed method, we select eight classification evaluation metrics, including accuracy (Acc), precision (Pre), recall, specificity (Spec), F1-score (F1), Kappa, area under the receiver operating characteristic (ROC) curve (AUC). All experiments are implemented by configuring the PyTorch framework on NVIDIA GTX 2080Ti GPU with 11 GB of memory.

2.2 Results

In this paper, we adopt Conformer [17] as the baseline of our network, and then make adjustments and improvements to optimize its performance on the PA fluorescence image dataset. Table 1 shows the results of the ablation experiments for different modules in the network. Among them, "Baseline + CB" indicates that the ResNet block in the original C-branch is replaced by the ConvNeXt block, reflecting the impact of the performance-enhanced C-branch on the classification performance. "Baseline + CB + Stem" replaces the convolutional module of the standard ResNet network in baseline with the Stem module on top of the modified C-branch. "Baseline + CB + Stem + HMSA" represents the replacement of the traditional MSA module in Baseline with the efficient HMSA module proposed in this paper on the basis of "Baseline + CB + Stem". The proposed HMSA module replaces the traditional MSA module in baseline, which achieves the improvement of network efficiency and classification performance.

In order to evaluate the classification performance of the proposed method, we choose ten state-of-the-art image classification methods for comparison, including 5 CNN networks: ResNet50 [4], ResNeXt50 [5], ResNeSt50 [6], ConvNeXt-T [16] and DenseNet121 [9], and 5 Transformer-related networks: ViT-B/16 [10], DeiT-S [11], PVT-M [12], Swin-T [13] and CeiT-S [18]. The results of the comparative experiment are illustrated in Table 2. We can observe that our dual-branch network achieves the best performance on our dataset, and outperforms the CeiT-S by 7.08%, 6.7%, and 6.66% in accuracy, recall and F1-score, respectively.

To further analyze and compare the computational complexity of different methods, we compare the number of model parameters (#Param) and the number of floating-point operations per second (FLOPs). In general, the higher the number of parameters and operations, the higher the performance of the model, but at the same time, the greater the computational and storage overhead. It can be seen that the accuracy of ViT is 5% lower

Table 1. Ablation experiments of different modules in PAS-Net (%).

Model	Acc	Pre	Recall	Spec	F1	Kappa	Youden	AUC
Baseline	91.72 ± 2.71	92.59 ± 2.98	91.83 ± 4.38	91.59 ± 3.75	92.14 ± 2.64	83.38 ± 5.41	83.42 ± 5.34	97.46 ± 1.30
Baseline + CB	94.08 ± 2.39	93.46 ± 3.23	95.63 ± 2.74	92.32 ± 4.14	94.49 ± 2.19	88.09 ± 4.82	87.95 ± 4.90	98.48 ± 1.04
Baseline + CB + Stem	94.19 ± 1.64	93.50 ± 1.84	95.73 ± 2.19	92.45 ± 2.25	94.59 ± 1.55	88.32 ± 3.30	88.18 ± 3.28	98.11 ± 0.45
Baseline + CB + Stem + HMSA	$\mathbf{96.04 \pm 1.35}$	$\mathbf{95.81 \pm 1.59}$	$\mathbf{96.80 \pm 2.36}$	$\mathbf{95.18 \pm 1.91}$	$\mathbf{96.28 \pm 1.30}$	$\mathbf{92.04 \pm 2.70}$	$\mathbf{91.97 \pm 2.64}$	$\mathbf{99.42 \pm 0.37}$

Table 2. Classification performance camparison of state-of-the-art methods on the test set (%).

Model	Acc	Pre	Recall	Spec	F1	Kappa	AUC	#Param	FLOPs
ResNet50	87.97 ± 1.74	88.85 ± 2.89	88.66 ± 5.15	87.19 ± 4.17	88.62 ± 1.94	75.86 ± 3.43	95.58 ± 1.23	23.5M	8.45G
ResNeXt50	89.62 ± 1.50	89.40 ± 1.73	91.31 ± 3.03	87.71 ± 2.41	90.31 ± 1.51	79.14 ± 2.98	95.50 ± 0.71	23.0M	8.78G
ResNeSt50	90.67 ± 1.90	89.88 ± 2.31	92.96 ± 3.76	88.07 ± 3.10	91.34 ± 1.90	81.23 ± 3.79	96.38 ± 0.75	25.4M	11.02G
ConvNeXt-T	87.91 ± 0.93	84.63 ± 1.43	94.41 ± 1.91	80.57 ± 2.38	89.23 ± 0.83	75.55 ± 1.88	95.85 ± 0.44	27.8M	8.90G
DenseNet121	89.92 ± 1.89	89.77 ± 2.53	91.65 ± 6.09	87.95 ± 3.90	90.53 ± 2.18	79.74 ± 3.68	96.31 ± 0.99	7.0M	5.94G
ViT-B/16	82.63 ± 2.47	83.20 ± 4.03	84.75 ± 4.72	80.25 ± 6.72	83.81 ± 2.18	65.08 ± 5.07	90.84 ± 1.23	86.0M	37.30G
DeiT-S	85.67 ± 1.57	88.44 ± 4.81	84.52 ± 4.93	86.98 ± 6.54	86.21 ± 1.48	71.31 ± 3.21	94.14 ± 1.12	21.7M	10.13G
PVT-M	86.72 ± 0.58	84.16 ± 1.68	92.46 ± 2.37	80.23 ± 2.94	88.07 ± 0.54	73.16 ± 1.19	93.52 ± 0.54	43.7M	14.09G
Swin-T	88.57 ± 1.68	90.07 ± 3.67	88.58 ± 5.92	88.58 ± 4.64	89.09 ± 2.11	77.09 ± 3.26	95.52 ± 0.49	29.8M	4.45G
CeiT-S	88.96 ± 1.29	89.24 ± 1.43	90.10 ± 3.21	87.67 ± 2.08	89.62 ± 1.40	77.82 ± 2.55	95.41 ± 0.55	23.9M	10.51G
Ours	**96.04 ± 1.35**	**95.81 ± 1.59**	**96.80 ± 2.36**	**95.18 ± 1.91**	**96.28 ± 1.30**	**92.04 ± 2.70**	**99.42 ± 0.37**	43.4M	23.37G

than that of ResNet50, but its model complexity is about three times higher. The number of model parameters of PVT-M is similar to that of our PAS-Net, but the accuracy is much worse. The number of parameters of our proposed PAS-Net is 43.4M and FLOPs is 23.37G, indicating that the network achieves a good balance between the number of parameters, FLOPs, accuracy and classification consistency.

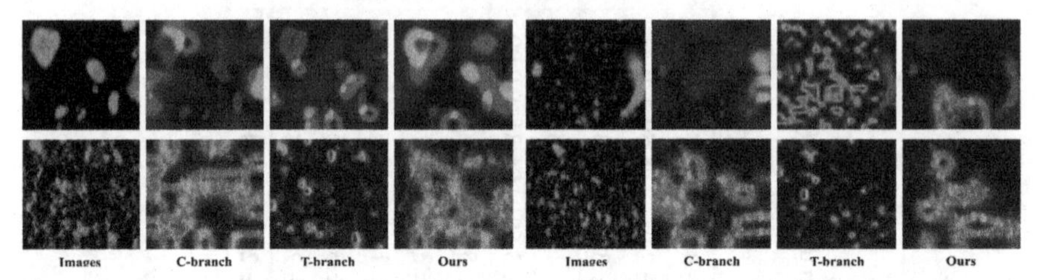

Fig. 5. Grad-cam is used to highlight the discriminant regions of interest for predicting sensitive and multi-drug resistant bacteria. The first column shows four images from the test set, the first two for sensitive PA and the last two for MDRPA.

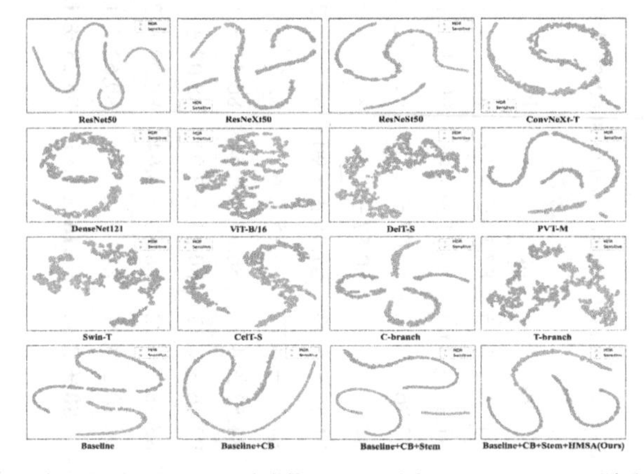

Fig. 6. Two-dimensional t-SNE maps of different models on our test set. Pink points represent MDRPA, blue points represent sensitive PA.

To verify the interpretability of the proposed PAS-Net and understand its classification effect more intuitively and effectively, we visualize the results using Grad-CAM, as shown in Fig. 5. From the second column vertically, we can see that the C-branch only focuses on local edge parts or incorrectly highlights some regions that are not relevant to the discrimination, as shown in the heat map in the first and second rows. From the third column, we can see that the T-branch can obtain the global attention map, but at the same time it produces some worthless and redundant features. A side-by-side comparison shows that the heat map in the fourth column can focus well on some discriminative regions with distinct features and reflect the correlation between local regions. For example, our network can effectively capture the bacterial aggregates with clear edges and largest area in the first image, and also establish the long-range feature dependencies among the three small bacterial aggregates near the lower right corner; for the second image, our dual-branch network corrects the error of focusing the C-branch to the exposure position because of the Transformer's ability to learn the global feature representation For the third and fourth images, the network nicely combines the discriminative regions focused on by the C-branch and T-branch, capturing both the local features of larger bacterial aggregates and learning the distributional dependencies among bacterial aggregates. This shows to some extent that our proposed model effectively exploits the advantages of CNN and Transformer and maximizes the retention of local features and global representation.

We also use the t-SNE dimensionality reduction algorithm to map the feature vectors learned from the last feature extraction layer of different networks onto a two-dimensional plane, as shown in Fig. 6. The visualization allows us to observe the clustering of the image features extracted by these networks. Compared with other models, the features extracted by our proposed PAS-Net can better distinguish the sensitive bacteria (blue) from the multi-drug resistant bacteria (pink).

2.3 Robustness to HEp-2 Dataset

To further verify the effectiveness of PAS-Net in fluorescent image classification tasks, we also apply our method to two HEp-2 cell public datasets, ICPR 2012 and I3A Task1. ICPR 2012 dataset uses average class accuracy (ACA) as the evaluation metric, which is the same concept as the accuracy mentioned above, while I3A Task1 uses mean class accuracy (MCA). We select four deep learning techniques for classification of HEp-2 cells for comparison, respectively, and the results are shown in Table 3. Without using pre-trained weights for migration learning and data augmentation, our network achieves 81.61% and 98.71% accuracy on ICPR 2012 dataset and I3A Task1 dataset, respectively, and the experimental results demonstrate the generalizability of the proposed PAS-Net for fluorescent image classification tasks.

Table 3: Algorithm comparison on ICPR2012 dataset and I3A Task1 dataset (%).

ICPR 2012	Method	ACA	I3A Task1	Method	MCA
Gao et al. [19]	Seven layers CNN	74.8	Gao et al. [19]	Seven layers CNN	96.76
Phan et al. [20]	VGG-16 + SVM	77.1	Jia et al. [21]	VGG-like network	98.26
Jia et al. [21]	VGG-like network	79.29	Li et al. [22]	Deep residual inception model	98.37
Liu et al. [23]	DACN	81.2	Lei et al. [24]	Cross-modal transfer learning	98.42
Ours	PAS-Net	81.61	Ours	PAS-Net	98.71

3 Conclusion

In this paper, we develop a PAS-Net framework for rapid prediction of antibiotic susceptibility from bacterial fluorescence images only. PAS-Net is a parallel dual-branch feature interaction network. FIU is a connecting bridge to align and fuse the local features from the C-branch and the global representation from the T-branch, which enhances the feature representation ability of the network. We design a HMSA module with less computational overhead to improve the computational efficiency of the model. The experimental results demonstrate that our method is feasible and effective in PA fluorescence image classification task, and can assist clinicians in determining bacterial antibiotic susceptibility.

Acknowledgement. This work was supported National Natural Science Foundation of China (Nos. 62101338, 61871274, 32270196 and U1902209), National Natural Science Foundation of Guangdong Province (2019A1515111205), Shenzhen Key Basic Research Project (KCXFZ20201221173213036, JCYJ20220818095809021, SGDX202011030958020–07, JCYJ201908081556188–06, and JCYJ20190808145011 -259), Shenzhen Peacock Plan Team Project (grants number KQTD20200909113758–004).

References

1. Holmes, A.H., et al.: Understanding the mechanisms and drivers of antimicrobial resistance. Lancet **387**, 176–187 (2016)
2. Dadgostar, P.: Antimicrobial resistance: implications and costs. Infect. Drug Resist. **12**, 3903–3910 (2019)
3. Ferri, M., Ranucci, E., Romagnoli, P., Giaccone, V.: Antimicrobial resistance: a global emerging threat to public health systems. Crit. Rev. Food Sci. Nutr. **57**, 2857–2876 (2017)
4. He, K., Zhang, X., Ren, S., Sun, J.: Deep residual learning for image recognition. In: Proceedings of the IEEE Conference on Computer Vision and Pattern Recognition, pp. 770–778 (2016)
5. Xie, S., Girshick, R., Dollár, P., Tu, Z., He, K.: Aggregated residual transformations for deep neural networks. In: Proceedings of the IEEE Conference on Computer Vision and Pattern Recognition, pp. 1492–1500 (2017)
6. Zhang, H., et al.: Resnest: split-attention networks. In: Proceedings of the IEEE/CVF Conference on Computer Vision and Pattern Recognition, pp. 2736–2746 (2022)
7. Waisman, A., et al.: Deep learning neural networks highly predict very early onset of pluripotent stem cell differentiation. Stem Cell Rep. **12**, 845–859 (2019)
8. Riasatian, A., et al.: Fine-Tuning and training of densenet for histopathology image representation using TCGA diagnostic slides. Med. Image Anal. **70**, 102032 (2021)
9. Huang, G., Liu, Z., Van Der Maaten, L., Weinberger, K.Q.: Densely connected convolutional networks. In: Proceedings of the IEEE Conference on Computer Vision and Pattern Recognition, pp. 4700–4708 (2017)
10. Dosovitskiy, A., et al.: An image is worth 16 $\times$ 16 words: transformers for image recognition at scale. arXiv preprint arXiv:.11929 (2020)
11. Touvron, H., Cord, M., Douze, M., Massa, F., Sablayrolles, A., Jégou, H.: Training data-efficient image transformers & distillation through attention. In: International Conference on Machine Learning, pp. 10347–10357. PMLR (2021)

12. Wang, W., et al.: Pyramid vision transformer: a versatile backbone for dense prediction without convolutions. In: Proceedings of the IEEE/CVF International Conference on Computer Vision, pp. 568–578 (2021)
13. Liu, Z., et al.: Swin transformer: hierarchical vision transformer using shifted windows. In: Proceedings of the IEEE/CVF International Conference on Computer Vision, pp. 10012–10022 (2021)
14. Vaswani, A., et al.: Attention is all you need. Adv. neural inf. Process. Syst. **30** (2017)
15. He, X., Tan, E.-L., Bi, H., Zhang, X., Zhao, S., Lei, B.: Fully transformer network for skin lesion analysis. Med. Image Anal. **77**, 102357 (2022)
16. Liu, Z., Mao, H., Wu, C.Y., Feichtenhofer, C., Darrell, T., Xie, S.: A convnet for the 2020s. In: Proceedings of the IEEE/CVF Conference on Computer Vision and Pattern Recognition, pp. 11976–11986 (2022)
17. Peng, Z., et al.: Conformer: local features coupling global representations for visual recognition. In: Proceedings of the IEEE/CVF International Conference on Computer Vision, pp. 367–376 (2021)
18. Yuan, K., Guo, S., Liu, Z., Zhou, A., Yu, F., Wu, W.: Incorporating convolution designs into visual transformers. In: Proceedings of the IEEE/CVF International Conference on Computer Vision, pp. 579–588 (2021)
19. Gao, Z., Wang, L., Zhou, L., Zhang, J.: HEp-2 cell image classification with deep convolutional neural networks. IEEE j. Biomed. Health Inform. **21**, 416–428 (2016)
20. Phan, H.T.H., Kumar, A., Kim, J., Feng, D.: Transfer learning of a convolutional neural network for HEp-2 cell image classification. In: 2016 IEEE 13th International Symposium on Biomedical Imaging (ISBI), pp. 1208–1211. IEEE (2016)
21. Jia, X., Shen, L., Zhou, X., Yu, S.: Deep convolutional neural network based HEp-2 cell classification. In: 2016 23rd International Conference on Pattern Recognition (ICPR), pp. 77–80. IEEE (2016)
22. Li, Y., Shen, L.: A deep residual inception network for HEp-2 cell classification. In: Cardoso, M.J., Arbel, T., Carneiro, G., Syeda-Mahmood, T., Tavares, J.M.R.S., Moradi, M., Bradley, A., Greenspan, H., Papa, J.P., Madabhushi, A., Nascimento, J.C., Cardoso, J.S., Belagiannis, V., Lu, Z. (eds.) DLMIA/ML-CDS -2017. LNCS, vol. 10553, pp. 12–20. Springer, Cham (2017). https://doi.org/10.1007/978-3-319-67558-9_2
23. Liu, J., Xu, B., Shen, L., Garibaldi, J., Qiu, G.: HEp-2 cell classification based on a deep autoencoding-classification convolutional neural network. In: 2017 IEEE 14th International Symposium on Biomedical Imaging (ISBI 2017), pp. 1019–1023. IEEE (2017)
24. Lei, H., et al.: A deeply supervised residual network for HEp-2 cell classification via cross-modal transfer learning. Pattern Recogn. **79**, 290–302 (2018)

Diffusion-Based Data Augmentation for Nuclei Image Segmentation

Xinyi Yu[1], Guanbin Li[2], Wei Lou[3], Siqi Liu[1], Xiang Wan[1], Yan Chen[4], and Haofeng Li[1](✉)

[1] Shenzhen Research Institute of Big Data, Shenzhen, China
lhaof@sribd.cn
[2] School of Computer Science and Engineering, Research Institute of Sun Yat-sen University in Shenzhen, Sun Yat-sen University, Guangzhou, China
[3] The Chinese University of Hong Kong, Shenzhen, China
[4] Shenzhen Health Development Research and Data Management Center, Shenzhen, China

Abstract. Nuclei segmentation is a fundamental but challenging task in the quantitative analysis of histopathology images. Although fully-supervised deep learning-based methods have made significant progress, a large number of labeled images are required to achieve great segmentation performance. Considering that manually labeling all nuclei instances for a dataset is inefficient, obtaining a large-scale human-annotated dataset is time-consuming and labor-intensive. Therefore, augmenting a dataset with only a few labeled images to improve the segmentation performance is of significant research and application value. In this paper, we introduce the first diffusion-based augmentation method for nuclei segmentation. The idea is to synthesize a large number of labeled images to facilitate training the segmentation model. To achieve this, we propose a two-step strategy. In the first step, we train an unconditional diffusion model to synthesize the *Nuclei Structure* that is defined as the representation of pixel-level semantic and distance transform. Each synthetic nuclei structure will serve as a constraint on histopathology image synthesis and is further post-processed to be an instance map. In the second step, we train a conditioned diffusion model to synthesize histopathology images based on nuclei structures. The synthetic histopathology images paired with synthetic instance maps will be added to the real dataset for

This work is supported by Chinese Key-Area Research and Development Program of Guangdong Province (2020B0101350001), and the Guangdong Basic and Applied Basic Research Foundation (2023A1515011464, 2020B1515020048), and the National Natural Science Foundation of China (No. 62102267, No. 61976250), and the Shenzhen Science and Technology Program (JCYJ20220818103001002, JCYJ20220530141211024), and the Guangdong Provincial Key Laboratory of Big Data Computing, The Chinese University of Hong Kong, Shenzhen.

Supplementary Information The online version contains supplementary material available at https://doi.org/10.1007/978-3-031-43993-3_57.

H. Greenspan et al. (Eds.): MICCAI 2023, LNCS 14227, pp. 592–602, 2023.
https://doi.org/10.1007/978-3-031-43993-3_57

training the segmentation model. The experimental results show that by augmenting 10% labeled real dataset with synthetic samples, one can achieve comparable segmentation results with the fully-supervised baseline.

Keywords: Data augmentation · Nuclei segmentation · Diffusion models

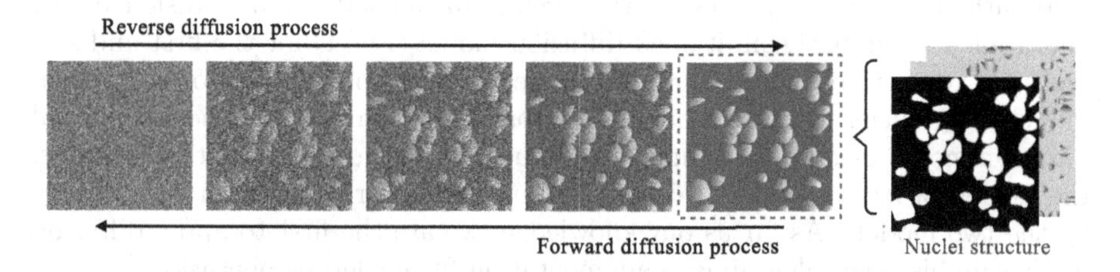

Fig. 1. The illustration of diffusion model in the context of nuclei structure.

1 Introduction

Nuclei segmentation is a fundamental step in medical image analysis. Accurately segmenting nuclei helps analyze histopathology images to facilitate clinical diagnosis and prognosis. In recent years, many deep learning based nuclei segmentation methods have been proposed [5,18,19,23]. Most of these methods are fully-supervised so the great segmentation performance usually relies on a large number of labeled images. However, manually labeling the pixels belonging to all nucleus boundaries in an image is time-consuming and requires domain knowledge. In practice, it is hard to obtain an amount of histopathology images with dense pixel-wise annotations but feasible to collect a few labeled images. A question is raised naturally: can we expand the training dataset with a small proportion of images labeled to reach or even exceed the segmentation performance of the fully-supervised baseline? Intuitively, since the labeled images are samples from the population of histopathology images, if the underlying distribution of histopathology images is learned, one can generate infinite images and their pixel-level labels to augment the original dataset. Therefore, it is demanded to develop a tool that is capable of learning distributions and generating new paired samples for segmentation.

Generative adversarial network (GANs) [2,4,12,16,20] have been widely used in data augmentation [11,22,27,31]. Specially, a newly proposed GAN-based method can synthesize labeled histopathology image for nuclei segmentation [21]. While GANs are able to generate high quality images, they are known for unstable training and lack of diversity in generation due to the adversarial training strategy. Recently, diffusion models represented by denoising diffusion probabilistic model (DDPM) [8] tend to overshadow GANs. Due to the theoretical basis and impressive performance of diffusion models, they were soon applied to a variety of vision tasks, such as inpainting, superresolution [30], text-to-image translation, anomaly detection and segmentation [1,9,24,26]. As likelihood-based models, diffusion models do not require adversarial training and outperform GANs on the diversity of generated images [3], which are naturally more suitable for data augmentation. In this paper, we propose a novel diffusion-based augmentation framework for nuclei segmentation. The proposed method consists of two steps: unconditional nuclei structure synthesis and conditional histopathology image synthesis. We develop an unconditional diffusion model and a nuclei-structure conditioned diffusion model (Fig. 1) for the first and second step, respectively. On the training stage, we train the unconditional diffusion model using nuclei structures calculated from instance maps and the conditional diffusion model using paired images and nuclei structures. On the testing stage, the nuclei structures and the corresponding images are generated successively by the two models. As far as our knowledge, we are the first to apply diffusion models on histopathology image augmentation for nuclei segmentation.

Our contributions are: (1) a diffusion-based data augmentation framework that can generate histopathology images and their segmentation labels from scratch; (2) an unconditional nuclei structure synthesis model and a conditional histopathology image synthesis model; (3) experiments show that with our method, by augmenting only 10% labeled training data, one can obtain segmentation results comparable to the fully-supervised baseline.

2 Method

Our goal is to augment a dataset containing a limited number of labeled images with more samples to improve the segmentation performance. To increase the diversity of labeled images, it is preferred to synthesize both images and their corresponding instance maps. We propose a two-step strategy for generating new labeled images. Both steps are based on diffusion models. The overview of the proposed framework is shown in Fig. 2. In this section, we introduce the two steps in detail.

2.1 Unconditional Nuclei Structure Synthesis

In the first step, we aim to synthesize more instance maps. Since it is not viable to directly generate an instance map, we instead choose to generate its surrogate – *nuclei structure*, which is defined as the concatenation of pixel-level semantic and distance transform. Pixel-level semantic is a binary map where 1 or 0 indicates whether a pixel belongs to a nucleus or not. The distance transform consists of the horizontal and the vertical distance transform, which are obtained by calculating the normalized distance of each pixel in a nucleus to the horizontal and the vertical line passing through the nucleus center [5]. Clearly, the nuclei structure is a 3-channel map with the same size as the image. As nuclei instances can be identified from the nuclei structure, we can easily construct the corresponding instance map by performance marker-controlled watershed algorithm on the nuclei structure [29]. Therefore, the problem of synthesizing instance map transfers to synthesizing nuclei structures. We deploy an unconditional diffusion model to learn the distribution of nuclei structures.

Denote a true nuclei structure as $\mathbf{y}_0$, which is sampled from real distribution $q(\mathbf{y})$. To maximize data likelihood, the diffusion model defines a forward and a reverse process. In the forward process, small amount of Gaussian noise are successively added to the sample $\mathbf{y}_0$ in T steps by:

$$\mathbf{y}_t = \sqrt{1-\beta_t}\mathbf{y}_{t-1} + \sqrt{\beta_t}\epsilon_{t-1}, t = 1, ..., T, \tag{1}$$

where $\epsilon_t \sim \mathcal{N}(0, \mathbf{I})$ and $\{\beta_t \in (0,1)\}_{t=1}^{T}$ is a variance schedule. The resulting sequence $\{\mathbf{y}_0, ..., \mathbf{y}_T\}$ forms a Markov chain. The conditional probability of $\mathbf{y}_t$ given $\mathbf{y}_{t-1}$ follows a Gaussian distribution:

$$q(\mathbf{y}_t|\mathbf{y}_{t-1}) = \mathcal{N}(\mathbf{y}_t; \sqrt{1-\beta_t}\mathbf{y}_{t-1}, \beta_t\mathbf{I}). \tag{2}$$

In the reverse process, since $q(\mathbf{y}_{t-1}|\mathbf{y}_t)$ cannot be easily estimated, a model $p_\theta(\mathbf{y}_{t-1}|\mathbf{y}_t)$ (typically a neural network) will be learned to approximate $q(\mathbf{y}_{t-1}|\mathbf{y}_t)$. Specifically, $p_\theta(\mathbf{y}_{t-1}|\mathbf{y}_t)$ is a also Gaussian distribution:

$$p_\theta(\mathbf{y}_{t-1}|\mathbf{y}_t) = \mathcal{N}(\mathbf{y}_{t-1}; \mu_\theta(\mathbf{y}_t, t), \mathbf{\Sigma}_\theta(\mathbf{y}_t, t)), \tag{3}$$

The objective function is the variational lower bound loss: $L = L_T + L_{T-1} + ... + L_0$, where every term except L_0 is a KL divergence between two Gaussian distributions. In practice, a simplified version of L_t is commonly used [8]:

$$L_t^{simple} = \mathbb{E}_{\mathbf{y}_0,\epsilon_t}\|\epsilon_t - \epsilon_\theta(\sqrt{\bar{\alpha}_t}\mathbf{y}_t + \sqrt{1-\bar{\alpha}_t}\epsilon_t, t)\|^2, \tag{4}$$

where $\alpha_t = 1 - \beta_t$ and $\bar{\alpha}_t = \prod_{i=1}^{t}\alpha_i$. Clearly, the optimization objective of the neural network parameterized by θ is to predict the Gaussian noise ϵ_t from the input $\mathbf{y}_t$ at time t.

After the network is trained, one can progressively denoise a random point from $\mathcal{N}(0, \mathbf{I})$ by T steps to produce a new sample:

$$\mathbf{y}_{t-1} = \frac{1}{\sqrt{\alpha_t}}(\mathbf{y}_t - \frac{1-\alpha_t}{\sqrt{1-\bar{\alpha}_t}}\epsilon_\theta(\mathbf{y}_t, t)) + \sigma_t\mathbf{z}, \mathbf{z} \sim \mathcal{N}(0, \mathbf{I}) \tag{5}$$

For synthesizing nuclei structures, we train an unconditional DDPM on nuclei structures calculated from real instance maps. Following [8], the network of this unconditional DDPM has a U-Net architecture.

2.2 Conditional Histopathology Image Synthesis

In the second step, we synthesize histopathology images conditioned on nuclei structures. Without any constraint, an unconditional diffusion model will generate diverse samples. There are usually two ways to synthesize images constrained

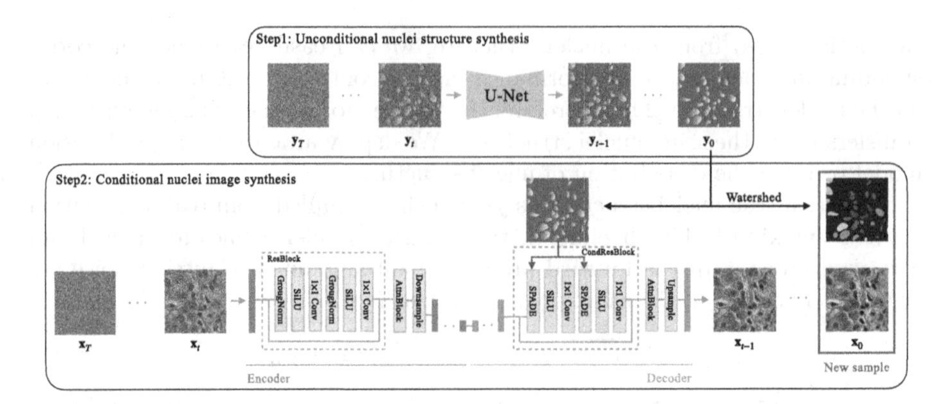

Fig. 2. The proposed diffusion-based data augmentation framework. We first generate a nuclei structure with an unconditional diffusion model, and then generate images conditioned on the nuclei structure. The instance map from the nuclei structure is paired with the synthetic image to forms a new sample.

by certain conditions: classifier-guided diffusion [3] and classifier-free guidance [10]. Since classifier-guided diffusion requires training a separate classifier which is an extra cost, we choose classifier-free guidance to control sampling process.

Let $\epsilon_\theta(\mathbf{x}_t, t)$ and $\epsilon_\theta(\mathbf{x}_t, t, \mathbf{y})$ be the noise predictor of unconditional diffusion model $p_\theta(\mathbf{x}|\mathbf{y})$ and conditional diffusion model $p_\theta(\mathbf{x})$, respectively. The two models can be learned with one neural network. Specifically, $p_\theta(\mathbf{x}|y)$ is trained on paired data $(\mathbf{x}_0, \mathbf{y}_0)$ and $p_\theta(\mathbf{x})$ can be trained by randomly discarding y (i.e. $\mathbf{y} = \emptyset$) with a certain *drop_rate* $\in (0, 1)$ so that the model learns unconditional and conditional generation simultaneously. The noise predictor $\epsilon'_\theta(\mathbf{x}_t, t, y)$ of classifier-free guidance is a combination of the above two predictors:

$$\epsilon'_\theta(\mathbf{x}_t, t, y) = (w + 1)\epsilon_\theta(\mathbf{x}_t, t, y) - w\epsilon_\theta(\mathbf{x}_t, t), \tag{6}$$

where $\epsilon_\theta(\mathbf{x}_t, t) = \epsilon_\theta(\mathbf{x}_t, t, \mathbf{y} = \emptyset)$, w is a scalar controlling the strength of classifier-free guidance.

Unlike the network of unconditional nuclei structure synthesis which inputs the noisy nuclei structure $\mathbf{y}_t$ and outputs the prediction of $\epsilon_t(\mathbf{y}_t, t)$, the network of conditional nuclei image synthesis takes the noisy nuclei image $\mathbf{x}_t$ and the corresponding nuclei structure $\mathbf{y}$ as inputs and the prediction of $\epsilon_t(\mathbf{x}_t, t, \mathbf{y})$ as output. Therefore, the conditional network should be equipped with the ability to well align the paired histopathology image and nuclei structure. Since nuclei structures and histopathology images have different feature spaces, simply concatenating or passing them through a cross-attention module [7,15,17] before entering the U-Net will degrade image fidelity and yield unclear correspondence between synthetic nuclei image and its nuclei structure. Inspired by [28], we embed information of the nuclei structure into feature maps of nuclei image by the spatially-adaptive normalization (SPADE) module [25]. In other words, the spatial and morphological information of nuclei modulates the normalized feature maps such that the nuclei are generated in the right places while the background is left to be created freely. We include the SPADE module in different levels of the network to utilize the multi-scale information of nuclei structure. The network of conditional nuclei image synthesis also applies a U-Net architecture. The encoder is a stack of Resblocks and attention blocks (AttnBlocks). Each Resblock consists of 2 GroupNorm-SiLU-Conv and each Attnblocks calculates the self-attention of the input feature map. The decoder is a stack of CondResBlocks and attention blocks. Each CondResBlock consists of SPADE-SiLU-Conv which takes both feature map and nuclei structure as inputs.

3 Experiments and Results

3.1 Implementation Details

Datasets. We conduct experiments on two datasets: MoNuSeg [13] and Kumar [14]. The MoNuSeg dataset has 44 labeled images of size 1000×1000, 30 for training and 14 for testing. The Kumar dataset consists of 30 1000×1000 labeled images from seven organs of The Cancer Genome Atlas (TCGA) database. The dataset is splited into 16 training images and 14 testing images.
Paired Sample Synthesis. To validate the effectiveness of the proposed augmentation method, we create 4 subsets of each training dataset with 10%, 20%, 50% and 100% nuclei instance labels. Precisely, we first crop all images of each dataset into 256×256 patches with stride 128, then obtain the features of all patches with pretrained ResNet50 [6] and cluster the patches into 6 classes by K-means. Patches close to the cluster centers are selected. The encoder and decoder of the two networks have 6 layers with channels 256, 256, 512, 512, 1024 and 1024. For the unconditional nuclei structure synthesis network, each layer of the encoder and decoder has 2 ResBlocks and last 3 layers contain AttnBlocks. The network is trained using the AdamW optimizer with a learning rate of 10^{-4} and a batch size of 4. For the conditional histopathology image synthesis network, each layer of the encoder and the decoder has 2 ResBlocks and 2 CondResBlocks respectively, and last 3 layers contain AttnBlocks. The network is first trained in a fully-conditional style ($drop\_rate = 0$) and then finetuned in a classifier free style ($drop\_rate = 0.2$). We use AdamW optimizer with learning rates of 10^{-4} and 2×10^{-5} for the two training stages, respectively. The batch size is set to be 1. For the diffusion process of both steps, we set the total diffusion timestep T to 1000 with a linear variance schedule $\{\beta_1, ..., \beta_T\}$ following [8].

For MoNuSeg dataset, we generate 512/512/512/1024 synthetic samples for 10%/20%/50%/100% labeled subsets; for Kumar dataset, 256/256/256/512 synthetic samples are generated for 10%/20%/50%/100% labeled subsets. The synthetic nuclei structures are generate by the nuclei structure synthesis network and the corresponding images are generated by the histopathology image synthesis network with the classifier-free guidance scale $w = 2$. Each follows the reverse diffusion process with 1000 timesteps [8]. We then obtain the augmented subsets by adding the synthetic paired images to the corresponding labeled subsets.

Nuclei segmentation. The effectiveness of the proposed augmentation method can be evaluated by comparing the segmentation performance of using the four labeled subsets and using the corresponding augmented subsets to train a segmentation model. We choose to train two nuclei segmentation models – Hover-Net [5] and PFF-Net [18]. To quantify the segmentation performance, we use two metrics: Dice coefficient and Aggregated Jaccard Index (AJI) [14].

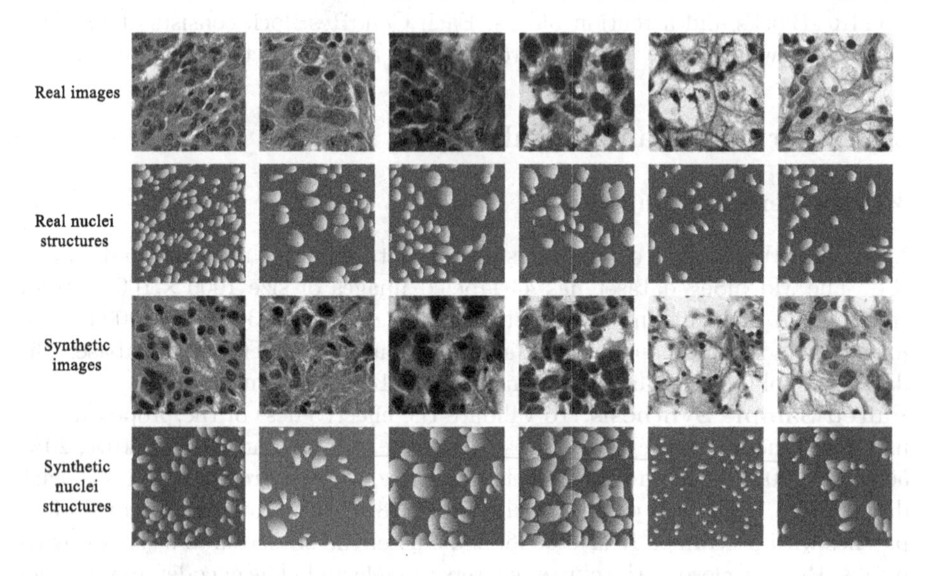

Fig. 3. Visualization of synthetic samples. The first and second row show selected patches and corresponding nuclei structures from the 10% labeled subset of MoNuSeg dataset. The third and fourth row show selected synthetic images and corresponding nuclei with similar style as the real one in the same column.

3.2 Effectiveness of the Proposed Data Augmentation Method

Fig. 3 shows the synthetic samples from the models trained on the subset with 10% labeled images. We have the following observations. First, the synthetic samples look realistic: the patterns of synthetic nuclei structures and textures of synthetic images are close to the real samples. Second, due to the conditional mechanism of the image synthesis network and the classifier-guidance sampling, the synthetic images are well aligned with the corresponding nuclei structures, which is the prerequisite to be additional segmentation training samples. Third, the synthetic nuclei structures and images show great diversity: the synthetic samples resemble different styles of the real ones but with apparent differences.

We then train segmentation models on the four labeled subsets of MoNuSeg and Kumar dataset and corresponding augmented subsets with both real and synthetic labeled images. With a specific labeling proportion, say 10%, we name the original subset as 10% labeled subset and the augmented on as 10% augmented subset. Specially, 100% labeled subset is the fully-supervised baseline. Table 1 show the segmentation performances with Hover-Net. For MoNuSeg dataset, it is clear that the segmentation metrics drop with fewer labeled images. For example, with only 10% labeled images, Dice and AJI reduce by 2.4% and 3.1%, respectively. However, by augmenting the 10% labeled subset, Dice and AJI exceed the fully-supervised baseline by 0.9% and 1.3%. For the 20% and 50% case, the two metrics obtained by augmented subset are of the same level as using all labeled images. Note that the metrics of 10% augmented subset are higher than those of 20% augmented subset, which might be attributed to the indetermination of the diffusion model training and sampling. Interestingly, augmenting the full dataset also helps: Dice increases by 1.3% and AJI increases by 1.6% compared with the original full dataset. Therefore, the proposed augmentation method consistently improves segmentation performance of different labeling proportion. For Kumar dataset, by augmenting 10% labeled subset, AJI increases to a level comparable with that using 100% labeled images; by augmenting 20% and 50% labeled subset, AJIs exceed the fully-supervised baseline. These results demonstrate the effectiveness of the proposed augmentation method that we can achieve the same or higher level segmentation performance of the fully-supervised baseline by augmenting a dataset with a small amount of labeled images.

Generalization of the Proposed Data Augmentation. Moreover, we have similar observations when using PFF-Net as the segmentation model. Table 2 shows the segmentation results with PFF-Net. For both MoNuSeg and Kumar datasets, all the four labeling proportions metrics notably improve with synthetic samples. This indicates the generalization of our proposed augmentation method.

Table 1. Effectiveness of the proposed data augmentation method with Hover-Net.

Training data	MoNuSeg		Kumar	
	Dice	AJI	Dice	AJI
10% labeled	0.7969	0.6344	0.8040	0.5939
10% augmented	0.8291	0.6785	0.8049	0.6161
20% labeled	0.8118	0.6501	0.8078	0.6098
20% augmented	0.8219	0.6657	0.8192	0.6255
50% labeled	0.8182	0.6603	0.8175	0.6201
50% augmented	0.8291	0.6764	0.8158	0.6307
100% labeled	0.8206	0.6652	0.8150	0.6183
100% augmented	0.8336	0.6810	0.8210	0.6301

Table 2. Generalization of the proposed data augmentation method with PFF-Net.

Training data	MoNuSeg		Kumar	
	Dice	AJI	Dice	AJI
10% labeled	0.7489	0.5290	0.7685	0.5965
10% augmented	0.7764	0.5618	0.8051	0.6458
20% labeled	0.7691	0.5629	0.7786	0.6087
20% augmented	0.7891	0.5927	0.8019	0.6400
50% labeled	0.7663	0.5661	0.7797	0.6175
50% augmented	0.7902	0.5998	0.8104	0.6524
100% labeled	0.7809	0.5708	0.8032	0.6461
100% augmented	0.7872	0.5860	0.8125	0.6550

4 Conclusion

In this paper, we propose a novel diffusion-based data augmentation method for nuclei segmentation in histopathology images. The proposed unconditional nuclei structure synthesis model can generate nuclei structures with realistic nuclei shapes and spatial distribution. The proposed conditional histopathology image synthesis model can generate images of close resemblance to real histopathology images and high diversity. Great alignments between synthetic images and corresponding nuclei structures are ensured by the special design of the conditional diffusion model and classifier-free guidance. By augmenting datasets with a small amount of labeled images, we achieved even better segmentation results than the fully-supervised baseline on some benchmarks. Our work points out the great potential of diffusion models in paired sample synthesis for histopathology images.

References

1. Amit, T., Shaharbany, T., Nachmani, E., Wolf, L.: SegDiff: image segmentation with diffusion probabilistic models. arXiv preprint arXiv:2112.00390 (2021)
2. Arjovsky, M., Chintala, S., Bottou, L.: Wasserstein generative adversarial networks. In: International Conference on Machine Learning, pp. 214–223. PMLR (2017)
3. Dhariwal, P., Nichol, A.: Diffusion models beat GANs on image synthesis. Adv. Neural Inf. Process. Syst. **34**, 8780–8794 (2021)
4. Goodfellow, I., et al.: Generative adversarial nets. In: Advances in Neural Information Processing Systems, vol. 27 (2014)
5. Graham, S., Vu, Q.D., Raza, S.E.A., Azam, A., Tsang, Y.W., Kwak, J.T., Rajpoot, N.: Hover-net: Simultaneous segmentation and classification of nuclei in multi-tissue histology images. Med. Image Anal. **58**, 101563 (2019)
6. He, K., Zhang, X., Ren, S., Sun, J.: Deep residual learning for image recognition. In: Proceedings of the IEEE Conference on Computer Vision and Pattern Recognition, pp. 770–778 (2016)

7. He, X., Yang, S., Li, G., Li, H., Chang, H., Yu, Y.: Non-local context encoder: robust biomedical image segmentation against adversarial attacks. In: Proceedings of the AAAI Conference on Artificial Intelligence, vol. 33, pp. 8417–8424 (2019)
8. Ho, J., Jain, A., Abbeel, P.: Denoising diffusion probabilistic models. Adv. Neural Inf. Process. Syst. **33**, 6840–6851 (2020)
9. Ho, J., Saharia, C., Chan, W., Fleet, D.J., Norouzi, M., Salimans, T.: Cascaded diffusion models for high fidelity image generation. J. Mach. Learn. Res. **23**(1), 2249–2281 (2022)
10. Ho, J., Salimans, T.: Classifier-free diffusion guidance. In: NeurIPS 2021 Workshop on Deep Generative Models and Downstream Applications (2021)
11. Isola, P., Zhu, J.Y., Zhou, T., Efros, A.A.: Image-to-image translation with conditional adversarial networks. In: Proceedings of the IEEE conference on Computer Vision and Pattern Recognition, pp. 1125–1134 (2017)
12. Karras, T., Laine, S., Aila, T.: A style-based generator architecture for generative adversarial networks. In: Proceedings of the IEEE/CVF Conference on Computer Vision and Pattern Recognition, pp. 4401–4410 (2019)
13. Kumar, N., et al.: A multi-organ nucleus segmentation challenge. IEEE Trans. Med. Imaging **39**(5), 1380–1391 (2019)
14. Kumar, N., Verma, R., Sharma, S., Bhargava, S., Vahadane, A., Sethi, A.: A dataset and a technique for generalized nuclear segmentation for computational pathology. IEEE Trans. Med. Imaging **36**(7), 1550–1560 (2017)
15. Li, H., Chen, G., Li, G., Yu, Y.: Motion guided attention for video salient object detection. In: Proceedings of the IEEE/CVF International Conference on Computer Vision, pp. 7274–7283 (2019)
16. Li, H., Li, G., Lin, L., Yu, H., Yu, Y.: Context-aware semantic inpainting. IEEE Trans. Cybern. **49**(12), 4398–4411 (2018)
17. Li, H., Li, G., Yang, B., Chen, G., Lin, L., Yu, Y.: Depthwise nonlocal module for fast salient object detection using a single thread. IEEE Trans. Cybern. **51**(12), 6188–6199 (2020)
18. Liu, D., Zhang, D., Song, Y., Huang, H., Cai, W.: Panoptic feature fusion net: a novel instance segmentation paradigm for biomedical and biological images. IEEE Trans. Image Process. **30**, 2045–2059 (2021)
19. Liu, D., et al.: Unsupervised instance segmentation in microscopy images via panoptic domain adaptation and task re-weighting. In: Proceedings of the IEEE/CVF conference on Computer Vision and Pattern Recognition, pp. 4243–4252 (2020)
20. Lou, W., Li, H., Li, G., Han, X., Wan, X.: Which pixel to annotate: a label-efficient nuclei segmentation framework. IEEE Trans. Medical Imaging **42**(4), 947–958 (2022)
21. Lou, W., et al.: Multi-stream cell segmentation with low-level cues for multi-modality images. In: Competitions in Neural Information Processing Systems, pp. 1–10. PMLR (2023)
22. Mirza, M., Osindero, S.: Conditional generative adversarial nets. arXiv preprint arXiv:1411.1784 (2014)
23. Naylor, P., Laé, M., Reyal, F., Walter, T.: Segmentation of nuclei in histopathology images by deep regression of the distance map. IEEE Trans. Med. Imaging **38**(2), 448–459 (2018)
24. Nichol, A.Q., et al.: Glide: Towards photorealistic image generation and editing with text-guided diffusion models. In: International Conference on Machine Learning, pp. 16784–16804. PMLR (2022)

25. Park, T., Liu, M.Y., Wang, T.C., Zhu, J.Y.: Semantic image synthesis with spatially-adaptive normalization. In: Proceedings of the IEEE/CVF conference on Computer Vision and Pattern Recognition, pp. 2337–2346 (2019)
26. Rombach, R., Blattmann, A., Lorenz, D., Esser, P., Ommer, B.: High-resolution image synthesis with latent diffusion models. In: Proceedings of the IEEE/CVF Conference on Computer Vision and Pattern Recognition, pp. 10684–10695 (2022)
27. Shaham, T.R., Dekel, T., Michaeli, T.: Singan: learning a generative model from a single natural image. In: Proceedings of the IEEE/CVF International Conference on Computer Vision, pp. 4570–4580 (2019)
28. Wang, W., et al.: Semantic image synthesis via diffusion models. arXiv preprint arXiv:2207.00050 (2022)
29. Yang, X., Li, H., Zhou, X.: Nuclei segmentation using marker-controlled watershed, tracking using mean-shift, and kalman filter in time-lapse microscopy. IEEE Trans. Circ. Syst. I: Regul. Papers **53**(11), 2405–2414 (2006)
30. Yue, J., Li, H., Wei, P., Li, G., Lin, L.: Robust real-world image super-resolution against adversarial attacks. In: Proceedings of the 29th ACM International Conference on Multimedia, pp. 5148–5157 (2021)
31. Zhu, J.Y., Park, T., Isola, P., Efros, A.A.: Unpaired image-to-image translation using cycle-consistent adversarial networks. In: Proceedings of the IEEE International Conference on Computer Vision, pp. 2223–2232 (2017)

Unsupervised Learning for Feature Extraction and Temporal Alignment of 3D+t Point Clouds of Zebrafish Embryos

Zhu Chen, Ina Laube, and Johannes Stegmaier(✉)

Institute of Imaging and Computer Vision, RWTH Aachen University, Aachen, Germany
{Zhu.Chen,Ina.Laube,Johannes.Stegmaier}@lfb.rwth-aachen.de

Abstract. Zebrafish are widely used in biomedical research and developmental stages of their embryos often need to be synchronized for further analysis. We present an unsupervised approach to extract descriptive features from 3D+t point clouds of zebrafish embryos and subsequently use those features to temporally align corresponding developmental stages. An autoencoder architecture is proposed to learn a descriptive representation of the point clouds and we designed a deep regression network for their temporal alignment. We achieve a high alignment accuracy with an average mismatch of only 3.83 min over an experimental duration of 5.3 h. As a fully-unsupervised approach, there is no manual labeling effort required and unlike manual analyses the method easily scales. Besides, the alignment without human annotation of the data also avoids any influence caused by subjective bias.

Keywords: Embryo Development · Point Clouds · Unsupervised Learning · Autoencoder

1 Introduction

Zebrafish are widely used model organisms in many experiments due to their fully-sequenced genome, easy genetic manipulation, high fecundity, external fertilization, rapid development and nearly transparent embryos [3]. The spatiotemporal resolution of modern light-sheet microscopes allows imaging the embryonic development at the single-cell level [4]. Fluorescently labeled nuclei can be detected, segmented and tracked in these data sets and the extracted 3D+t

Z. Chen and I. Laube—Equal contrib.; funded by the German Research Foundation DFG (STE2802/1-1).

Supplementary Information The online version contains supplementary material available at https://doi.org/10.1007/978-3-031-43993-3_58.

H. Greenspan et al. (Eds.): MICCAI 2023, LNCS 14227, pp. 603–612, 2023.
https://doi.org/10.1007/978-3-031-43993-3_58

point clouds can be used for analyzing the development of a single embryo in unprecedented detail [4]. In many experiments, an important task is to compare the level of growth between different individuals, especially the research on mutants and the growth under particular conditions like pollution and exposure to potentially harmful chemicals [5,6]. Thus, automatically obtaining an accurate temporal alignment that synchronizes the developmental stages of two or more individuals is an important component of such comparative analyses.

Existing approaches are mostly based on the automatic alignment of manually identified landmarks or operate in the image domain. However, to the best of our knowledge there is no approach available yet to obtain a spatiotemporal alignment of 3D+t point clouds of biological specimens in an automatic and unsupervised fashion. In [5], the authors generate sets of landmarks based on the deformation of embryos, and the landmarks are paired to generate the temporal registration between two sequences of embryos. Michelin *et al.* [13] formulate the temporal alignment problem as an image-to-sequence registration applicable to more complex organisms like the floral meristem. The method introduced in [14] pairs the 3D images of ascidian embryos by finding the symmetry plane and by computing the transformation that optimizes the cell-to-cell mapping. In [15], landmarks of multiple mouse embryos are manually identified and subsequently used to automatically obtain a spatiotemporal alignment with their custom-made Tardis method. In the past couple of years, deep neural networks emerged as a powerful approach to learn descriptive representations of images and point clouds that can be flexibly used for various tasks. The authors of [2] use an image-based convolutional neural network and a PointNet-based [8] architecture in a supervised fashion to obtain an automatic staging of zebrafish embryos.

In this work, we present a deep learning-based method for the temporal alignment of 3D+t point clouds of developing embryos. Firstly, an autoencoder is employed to extract descriptive features from the point clouds of each time frame. As an autoencoder designed explicitly for point clouds, FoldingNet [1] is used as the basic architecture and we propose several modifications to improve its applicability to point cloud data of developing organisms. As the next step, the extracted latent features of the autoencoder are used in a regression network to temporally align different embryos. The final output are pairwisely aligned time frames of two different 3D+t point clouds. We show that the autoencoder learns discriminative and descriptive feature vectors that allow to recover the temporal ordering of different time frames. In addition to quantitatively assessing the alignment accuracy of the regression network, we demonstrate the effectiveness of the latent features by visualizing the reconstructed 3D point clouds and low-dimensional representations obtained with t-SNE [7]. Being a fully-automatic and unsupervised approach, our method does not require any time-consuming human labeling effort and additionally avoids any subjective biases.

2 Methods

Our method is based on the FoldingNet [1] as the point cloud feature extractor. Several modifications are added to both the network and the loss function. The temporal alignment is realized with a regression network using the latent features and a subsequent consistency-check is used as a postprocessing to further improve the results. Finally, we introduce a new method to synthesize validation data sets with known ground truth from a set of unlabeled embryo point clouds.

2.1 Autoencoder

FoldingNet: FoldingNet is an autoencoder specifically designed for 3D point clouds [1]. In the encoder, a k-nearest-neighbor graph is built and the local information of each sub-region is extracted using the graph layers. A global max-pooling operation is applied to the local features to obtain the one-dimensional latent feature vector in the bottleneck layer. As a symmetric function, the pooling operation adds permutation invariance to cope with the disorderliness of point clouds. In the decoder, the feature vector is duplicated and concatenated with a fixed grid of points. The latent features lead to the deformation of the point grid in a 3-layer multi-layer perceptron (MLP) and a 3D structure is constructed. With a second folding operation, more details are rebuilt and the input is reconstructed. Unlike the original FoldingNet [1], we use a spherical template (M evenly distributed 3D points on a spherical surface) instead of a planar template, which provides a better initialization for reconstructing spherical objects from the learned representations. Since the shape of the embryos varies from a hemisphere to an ellipsoid during development, the spherical template simplifies the folding operation and improves the reconstruction quality in combination with the Modified Chamfer Distance loss (see next section).

Modified Chamfer Distance: The Chamfer Distance (CD) is one of the most widely used similarity measures for point clouds and is used as the loss function of FoldingNet. The discrepancy between two point clouds is calculated as the sum of the distances between the closest pairs of points as follows:

$$L_{CD}\left(S_{in}, S_{out}\right) = \frac{1}{|S_{in}|} \sum_{p \in S_{in}} \min_{q \in S_{out}} \|p - q\|_2 + \frac{1}{|S_{out}|} \sum_{q \in S_{out}} \min_{p \in S_{in}} \|q - p\|_2, \quad (1)$$

where S_{in}, S_{out} are the input and reconstructed point clouds, respectively. However, this approach does not consider the local density distribution since all points are treated independently. In this work, we introduce the Modified Chamfer Distance (MCD), in which the point-to-region distance replaces the point-to-point distance. The new loss is the summation of the k-nearest-neighbors of each point in the target point cloud and defined as:

$$L_{MCD}(S_{in}, S_{out}) = \frac{1}{|S_{in}|} \sum_{p \in S_{in}} \frac{1}{k} \sum_{i}^{k} \min_{q_i \in S_{out}} d(p, q_i) + \frac{1}{|S_{out}|} \sum_{q \in S_{out}} \frac{1}{k} \sum_{j}^{k} \min_{p_j \in S_{in}} d(q, p_j), \tag{2}$$

where d is the Euclidean distance between a point p and its respective nearest neighbor q_i. As the embryos grow, the density distribution changes significantly in different parts. The utilized data set [4] is spatially prealigned such that the animal and vegetal pole align with the y-axis and the prospective dorsal part with the positive x-axis. During epiboly, the embryo grows from a hemisphere to a complete sphere in the negative y-direction. As development progresses into the bud stage, density increases in the direction of the positive x-axis and thus the center of gravity moves to the right. We found that a FoldingNet trained with MCD consistently yielded more accurate reconstructions and alignment results compared to CD (Suppl. Fig. 1, Suppl. Fig. 6). In Fig. 1, the input and the reconstructed point clouds of an embryo from the hold-out test set are visualized using ParaView [9] to qualitatively illustrate the effectiveness of the FoldingNet that was trained with the MCD loss.

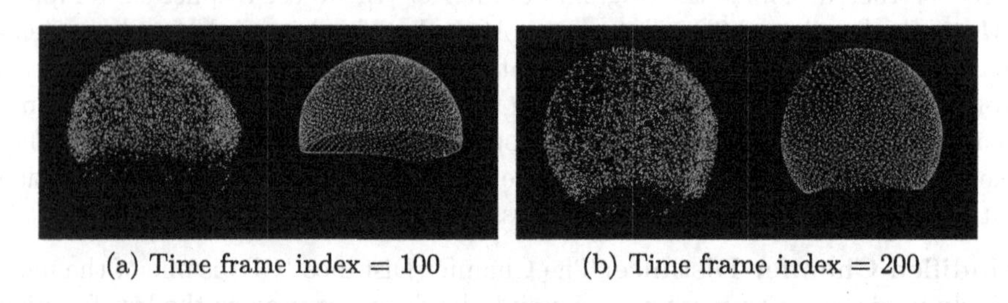

(a) Time frame index = 100 (b) Time frame index = 200

Fig. 1. Comparison of raw 3D point clouds (left sub-panels) and the reconstructions of our modified FoldingNet that was trained with MCD and the spherical point template (right sub-panels). Shape and density distribution of the reconstructions are nicely preserved, *i.e.*, the learned representation successfully condenses the properties of the input point clouds (see Suppl. Fig. 2 for additional examples).

2.2 Regression Network

The features extracted by the autoencoder are used to generate the temporal alignment of different embryos. We select one embryo as the reference and train an MLP regression network that maps autoencoder-generated feature vectors to frame numbers. The regression MLP consists of a sequence of fully-connected layers with ReLU activation. The input is a latent feature vector with length

256. In each dense layer, the size of the vector is reduced by a factor of two. The penultimate layer converts the 8-dimensional vector directly to the output node. The mean squared error is utilized as the loss function to compare the output to the ground truth: the sequence of time frame indices from 1 to 370. To temporally synchronize a new embryo with the reference, we present extracted features of all its frames to the trained regression network and generate a new index sequence. The desired alignment result is obtained by comparing the generated sequence with the reference embryo (Fig. 2).

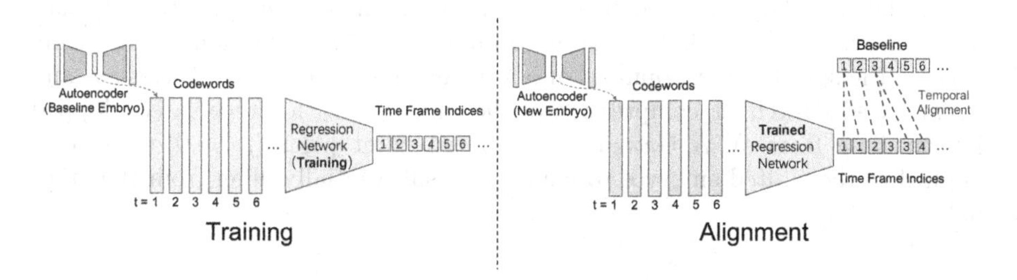

Fig. 2. The temporal alignment network is trained with a baseline embryo and learns to map each feature vector to the corresponding reference frame index (left). The trained network is then applied to feature vectors of a new embryo and the predicted frame index sequence indicates how the embryo should be aligned to the reference (right).

Postprocessing: By definition, the sequence of time frame indices must be monotonically increasing. However, since the alignment results are generated from the point cloud of each time frame individually, the relationship between the time points in the sequence is not considered. So the generated sequences are not guaranteed to be monotonically increasing. We use a simple postprocessing strategy to generate monotonically increasing and more accurate alignment results. For an aligned sequence with oscillations, we generate its monotonically increasing upper and lower boundaries, and the desired result is obtained by calculating the mean values of those boundaries (Suppl. Fig. 3).

3 Experimental Results and Discussions

3.1 Data Sets and Evaluation

The data set used in this work was published in [4] and consists of four wild-type zebrafish embryos that were imaged from 4.7 to 10.0 hpf (hours post fertilization) with one minute time intervals. Each embryo is represented as a 3D point cloud and has 370 time frames. The staging of these data sets was performed at a single time point (10 hpf) and the 370 preceding frames were selected irrespective of potential developmental differences. We thus only know that the temporal windows of the four embryos largely overlap but do not have frame-accurate annotations of the actual developmental time (see [2,4,12] for details). To simplify the training approach, a fixed number of points is randomly chosen from each point cloud using the PyTorch Geometric library [10] and we use the data loader implemented in the PyTorch 3D Points library [11]. We choose 4096 points since the size of the original point clouds ranges from 4160 to 19794. For improved generalizability and orientation invariance of the FoldingNet-based autoencoder, the point clouds are randomly rotated between 0 and 360° along each axis as data augmentation. For additional augmentation, we generated randomized synthetic variants of the four embryos as described in [2]. Since there are no labeled data to evaluate the temporal alignment results, we introduce a new method to artificially generate ground truth for validation by randomly varying the speed of development of a selected embryo (sin, cos, Gaussian, and linear-based stretching/compression of the time axis with interpolated/skipped intermediate frames). A Gaussian-distributed point jitter ($\mu = 0, \sigma^2 = 5$) is applied to the shifted embryos to make them substantially differ from the originals.

3.2 Experimental Settings

The FoldingNet-based autoencoder is implemented with PyTorch Lightning. The number of neighbors is set to 16 in the KNN-graph layers and to 20 in the MCD. The autoencoder is trained with an initial learning rate of 0.0001 for 250 epochs and a batch size of 1. At each iteration, the embryo from a single time frame with the size 4096×3 is given to the network and the encoder converts it to a 1D codeword of length 256, which is the latent feature vector. For network training with the four wild-type embryos, we use a 4-fold cross-validation scheme with three embryos for training and one for testing in each fold. The regression network is trained for 700 epochs with a learning rate of 0.00001. The hyperparameters of the regression network are determined empirically based on the convergence of the training loss, since there is no validation or test set available. The temporal alignment is validated with the embryo from the test set to make the result independent from training the autoencoder and the embryo is aligned to its shifted variants. To reduce the influence of randomness, we repeat the alignment of each test embryo three times and take the average value of all experiments of the four embryos.

3.3 Experimental Results

Temporal Alignment Results: The resulting alignment with different shifting methods is depicted in Fig. 3. In Fig. 3(a) and Fig. 3(b), a cosine- and sine-shifted embryo is aligned to the original one, where the cosine-based temporal shifting lets the embryo develop faster at the beginning and then slow down while the sine-based shifting is defined contrarily. In Fig. 3(c), a Gaussian-distributed random difference is added to the shifted embryo. Furthermore, Fig. 3(d) illustrates an embryo that develops approximately three times faster than the original one. The alignment error is defined as the average number of mismatched indices, which is the difference between the sequence of the aligned indices and the ground truth in the x-direction. As a result, an average mismatch of only 3.83 min in a total developmental period of 5.3 h is achieved (Table 1).

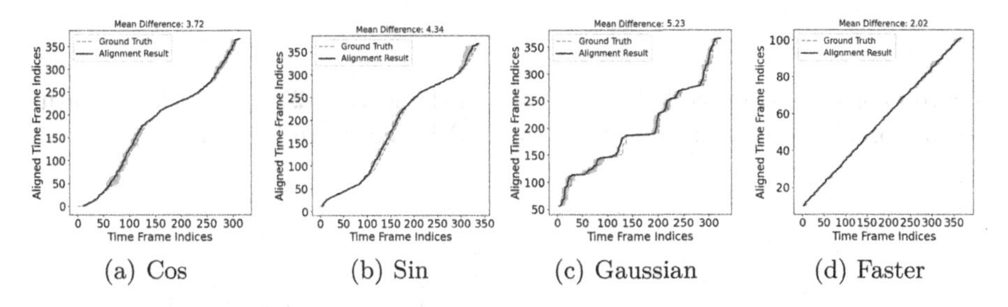

(a) Cos (b) Sin (c) Gaussian (d) Faster

Fig. 3. Alignment results of the embryos with different shifting methods. The black line indicates the average result of all experiments and the area shaded in gray represents the variance. The alignment error is calculated as the average number of mismatched time frame indices. (Color figure online)

Impact of Spatial Transformations: In the previous experiments, the embryos to be aligned were generated by changing the development speed and by scattering the points to make them different from the original ones. However, the embryos to be aligned in real applications could be located and oriented differently in space. Thus, we add some data augmentation for the alignment network to increase its rotation invariance. The point clouds are randomly rotated between $\pm 20^\circ$ along each axis. Moreover, the shifted embryos to be aligned are rotated between $\pm 15^\circ$. The average alignment error obtained for rotated embryos is 3.48 min (Suppl. Fig. 4). However, a larger variance can be observed, which indicates that rotation can have an impact on the alignment accuracy in some cases. Since the overall accuracy is still high, however, our approach proofs to be robust for aligning embryos with slightly varying orientations. In addition to the orientation, embryos may also be positioned differently in the sample chamber. To potentially improve the translation invariance, we tested if centering all point clouds at the origin of the coordinate axes before inputting them to the alignment network has a positive effect (Suppl. Fig. 5). Although centering makes the approach invariant to spatial translation, we find that the alignment accuracy is reduced (aligned sequences have a more significant variance, and the average error increases to 5.74 min). We hypothesize that the relative displacement of the point cloud's centroid to the origin of the coordinate system (which is removed by centering) may play an important role in determining the developmental stage and the level of completion of the epiboly phase. An overview of obtained alignment results is provided in Table 1.

Visualizing the Learned Representation: To confirm that the autoencoders actually learned a representation suitable for temporal alignment, we visualize

Table 1. Results of the average temporal alignment errors in minutes.

Shifting Method	Cosine	Sine	Gaussian	Faster	Average
Scattering	3.72	4.34	5.23	2.02	3.83
Scattering + Rot	3.90	2.89	5.00	2.13	3.48
Scattering + Rot. + Cent	7.16	5.64	7.38	2.78	5.74

a chronologically color-coded scatter plot of the learned 256-dimensional feature vectors of all 370 time frames using the t-SNE algorithm [7] (Fig. 4).

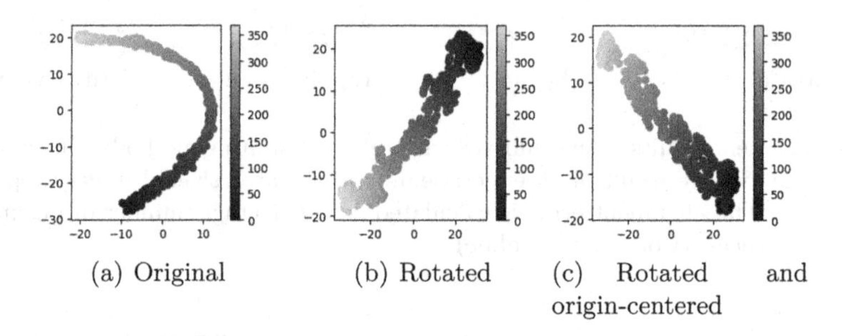

(a) Original (b) Rotated (c) Rotated and origin-centered

Fig. 4. Visualization of the feature vectors using the t-SNE algorithm. The color-code represents the time frame index and changes smoothly as the time increases.

The original features are clustered to a narrow band and the color changes smoothly as the index increases, which indicates that different time frames are well distinguishable using the representation learned by the autoencoder. When the point clouds are rotated and centered, the projections of the feature vectors become more dispersed as illustrated in Fig. 4(b) and 4(c). Nevertheless, the color changes smoothly and the features are suitable for distinguishing different time frames. This is in line with the observed increased variance for the temporal alignment results with maintained good average alignment accuracy.

4 Conclusion

In this work we present a fully-unsupervised approach to temporally align 3D+t point clouds of zebrafish embryos. A FoldingNet-based autoencoder is implemented to extract low-dimensional descriptive features from large-scaled 3D+t point clouds of embryonic development. Several modifications are made to the network and the loss function to improve their applicability for this application. The embryos are temporally aligned by applying a regression network to the features extracted by the autoencoder. A postprocessing method is designed to provide consistent and accurate alignments. As no frame-accurate ground truth is available yet, we assess the effectiveness of our method via a 4-fold cross validation and a synthetically generated ground truth. An average mismatch of only 3.83 min in a developmental period of 5.3 h is achieved. Finally, we performed several ablation studies to show the impact of rotation and spatial translation of the point clouds to the alignment results. By aligning embryos with different spatial locations and deflected central axis, a relatively small error rate of 5.74 min can still be achieved. According to feedback from a biological expert the achievable manual alignment accuracy is on the order of 30 min and potentially exhibits intra- and inter-rater variabilities. As the first unsupervised method designed for the automatic spatiotemporal alignment of 3D+t point clouds, our method achieves high accuracy and eradicates the need for any human interaction. This will particularly help to minimize human effort, to speedup experimental analysis and to avoid any subjective biases.

In future works, our approach could be applied to more data sets and other model organisms with different scales and development periods to further validate its applicability. We're currently conducting an extensive effort to obtain frame-accurate manual labels from multiple raters in a randomized study, to better assess the actual performance that we can expect under real-world conditions including intra- and inter-rater variability. In the long term, we envision an iteratively optimized spatiotemporal average model of multiple wild-type embryos to finally obtain a 3D+t reference atlas that can be used to precisely analyze developmental differences of corresponding anatomical regions across experiments.

References

1. Yang, Y., Feng, C., Shen, Y., Tian, D.: FoldingNet: point cloud auto-encoder via deep grid deformation. In: Proceedings of the IEEE Conference on Computer Vision and Pattern Recognition (CVPR), pp. 206–215 (2018)
2. Traub, M., Stegmaier, J.: Towards automatic embryo staging in 3D+t microscopy images using convolutional neural networks and PointNets. In: Simulation and Synthesis in Medical Imaging, pp. 153–163 (2020)
3. Teame, T., et al.: The use of zebrafish (Danio rerio) as biomedical models. Anim. Front. **9**(3), 68–77 (2019)
4. Kobitski, A.Y., et al.: An ensemble-averaged, cell density-based digital model of zebrafish embryo development derived from light-sheet microscopy data with single-cell resolution. Sci. Rep. **5**(1), 8601 (2015)
5. Guignard, L., Godin, C., Fiuza, U.M., Hufnagel, L., Lemaire, P., Malandain, G.: Spatio-temporal registration of embryo images. In: 2014 IEEE 11th International Symposium on Biomedical Imaging (ISBI), pp. 778–781 (2014)
6. Castro-González, C., et al.: A digital framework to build, visualize and analyze a gene expression atlas with cellular resolution in zebrafish early embryogenesis. PLoS Comput. Biol. **10**(6), 1–13 (2014)
7. Van der Maaten, L., Hinton, G.: Visualizing data using t-SNE. J. Mach. Learn. Res. **9**(11), 2579–2605 (2008)
8. Qi, C.R., Su, H., Mo, K., Guibas, L.J.: PointNet: deep learning on point sets for 3D classification and segmentation. In: Proceedings of the IEEE Conference on Computer Vision and Pattern Recognition (CVPR), pp. 652–660 (2017)

9. Ahrens, J., Geveci, B., Law, C.C.: ParaView: An End-User Tool for Large-Data Visualization. In: The Visualization Handbook (2005)
10. Fey, M., Lenssen, J.E.: Fast graph representation learning with PyTorch geometric. In: ICLR 2019 Workshop on Representation Learning on Graphs and Manifolds (2019)
11. Chaton, T., Chaulet, N., Horache, S., Landrieu, L.: Torch-Points3D: a modular multi-task framework for reproducible deep learning on 3D point clouds. In: 2020 International Conference on 3D Vision (3DV), pp. 1–10 (2020)
12. Schott, B.: EmbryoMiner: a new framework for interactive knowledge discovery in large-scale cell tracking data of developing embryos. PLoS Comput. Biol. **14**(4), 1–18 (2018)
13. Michelin, G., et al.: Spatio-temporal registration of 3D microscopy image sequences of arabidopsis floral meristems. In: 2016 IEEE 13th International Symposium on Biomedical Imaging (ISBI), pp. 1127–1130 (2016)
14. Michelin, G., Guignard, L., Fiuza, U.M., Lemaire, P., Godine, C., Malandain, G.: Cell pairings for ascidian embryo registration. In: 2015 IEEE 12th International Symposium on Biomedical Imaging (ISBI), pp. 298–301 (2015)
15. McDole, K., et al.: In Toto imaging and reconstruction of post-implantation mouse development at the single-cell level. Cell **175**(3), 859–876 (2018)

3D Mitochondria Instance Segmentation with Spatio-Temporal Transformers

Omkar Thawakar[1(✉)], Rao Muhammad Anwer[1,2], Jorma Laaksonen[2], Orly Reiner[3], Mubarak Shah[4], and Fahad Shahbaz Khan[1,5]

[1] MBZUAI, Masdar City, UAE
omkar.thawakar@mbzuai.ac.ae
[2] Aalto University, Espoo, Finland
[3] Weizmann Institute of Science, Rehovot, Israel
[4] University of Central Florida, Orlando, USA
[5] Linköping University, Linköping, Sweden

Abstract. Accurate 3D mitochondria instance segmentation in electron microscopy (EM) is a challenging problem and serves as a prerequisite to empirically analyze their distributions and morphology. Most existing approaches employ 3D convolutions to obtain representative features. However, these convolution-based approaches struggle to effectively capture long-range dependencies in the volume mitochondria data, due to their limited local receptive field. To address this, we propose a hybrid encoder-decoder framework based on a split spatio-temporal attention module that efficiently computes spatial and temporal self-attentions in parallel, which are later fused through a deformable convolution. Further, we introduce a semantic foreground-background adversarial loss during training that aids in delineating the region of mitochondria instances from the background clutter. Our extensive experiments on three benchmarks, Lucchi, MitoEM-R and MitoEM-H, reveal the benefits of the proposed contributions achieving state-of-the-art results on all three datasets. Our code and models are available at https://github.com/OmkarThawakar/STT-UNET.

Keywords: Electron Microscopy · Mitochondria instance segmentation · Spatio-Temporal Transformer · Hybrid CNN-Transformers

1 Introduction

Mitochondria are membrane-bound organelles that generate the primary energy required to power the cell activities, thereby crucial for metabolism. Mitochondrial dysfunction, which occurs when mitochondria are not functioning properly

Supplementary Information The online version contains supplementary material available at https://doi.org/10.1007/978-3-031-43993-3_59.

H. Greenspan et al. (Eds.): MICCAI 2023, LNCS 14227, pp. 613–623, 2023.
https://doi.org/10.1007/978-3-031-43993-3_59

has been witnessed as a major factor in numerous diseases, including noncommunicable chronic diseases (*e.g*, cardiovascular and cancer), metabolic (*e.g*, obesity) and neurodegenerative (*e.g*, Alzheimer and Parkinson) disorders [23,25]. Electron microscopy (EM) images are typically utilized to reveal the corresponding 3D geometry and size of mitochondria at a nanometer scale, thereby facilitating basic biological research at finer scales. Therefore, automatic instance segmentation of mitochondria is desired, since manually segmenting from a large amount of data is particularly laborious and demanding. However, automatic 3D mitochondria instance segmentation is a challenging task, since complete shape of mitochondria can be sophisticated and multiple instances can also experience entanglement with each other resulting in unclear boundaries. Here, we look into the problem of accurate 3D mitochondria instance segmentation.

Earlier works on mitochondria segmentation employ standard image processing and machine learning methods [20,21,33]. Recent approaches address [4, 15,26] this problem by leveraging either 2D or 3D deep convolutional neural network (CNNs) architectures. These existing CNN-based approaches can be roughly categorized [36] into bottom-up [3,4,14,15,28] and top-down [12]. In case of bottom-up mitochondria instance segmentation approaches, a binary segmentation mask, an affinity map or a binary mask with boundary instances is computed typically using a 3D U-Net [5], followed by a post-processing step to distinguish the different instances. On the other hand, top-down methods typically rely on techniques such as Mask R-CNN [7] for segmentation. However, Mask R-CNN based approaches struggle due to undefined bounding-box scale in EM data volume.

When designing a attention-based framework for 3D mitochondria instance segmentation, a straightforward way is to compute joint spatio-temporal self-attention where all pairwise interactions are modelled between all spatio-temporal tokens. However, such a joint spatio-temporal attention computation is computation and memory intensive as the number of tokens increases linearly with the number of input slices in the volume. In this work, we look into an alternative way to compute spatio-temporal attention that captures long-range global contextual relationships without significantly increasing the computational complexity. Our contributions are as follows:

- We propose a hybrid CNN-transformers based encoder-decoder framework, named STT-UNET. The focus of our design is the introduction of a split spatio-temporal attention (SST) module that captures long-range dependencies within the cubic volume of human and rat mitochondria samples. The SST module independently computes spatial and temporal self-attentions in parallel, which are then later fused through a deformable convolution.
- To accurately delineate the region of mitochondria instances from the cluttered background, we further introduce a semantic foreground-background (FG-BG) adversarial loss during the training that aids in learning improved instance-level features.
- We conduct experiments on three commonly used benchmarks: Lucchi [20], MitoEM-R [36] and MitoEM-H [36]. Our STT-UNET achieves state-of-the-art

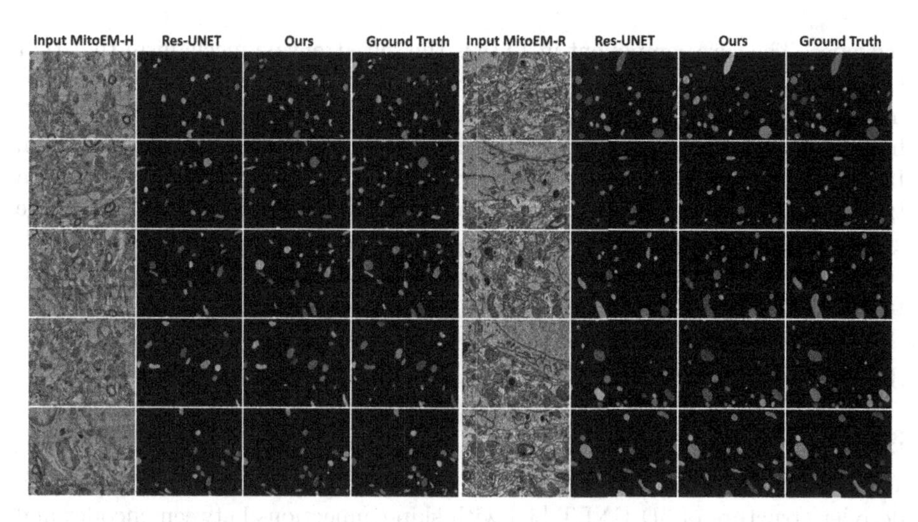

Fig. 1. Qualitative 3D instance segmentation comparison between the recent Res-UNET [16] and our proposed STT-UNET approach on the example input regions from MitoEM-H and MitoEM-R validation sets. Here, we present the corresponding segmentation predictions of the baseline and our approach along with the ground truth. Our STT-UNET approach achieves superior segmentation performance by accurately segmenting 16% more cell instances in these examples, compared to Res-UNET-R.

segmentation performance on all three datasets. On Lucchi test set, our STT-UNET outperforms the recent [4] with an absolute gain of 3.0% in terms of Jaccard-index coefficient. On MitoEM-H val. set, STT-UNET achieves AP-75 score of 0.842 and outperforms the recent 3D Res-UNET [16] by 3.0%. Figure 1 shows a qualitative comparison between our STT-UNET and 3D Res-UNET [16] on examples from MitoEM-R and MitoEM-H datasets.

2 Related Work

Most recent approaches for 3D mitochondria instance segmentation utilize convolution based designs within the "U-shaped" 3D encoder-decoder architecture. In such an architecture, the encoder aims to generate a low-dimensional representation of the 3D data by gradually performing the downsampling of the extracted features. On the other hand, the decoder performs upsampling of these extracted feature representations to the input resolution for segmentation prediction. Although such a CNN-based designs [11,16,34] has achieved promising segmentation results compared to traditional methods, they struggle to effectively capture long-range dependencies due to their limited local receptive field. Inspired from success in natural language processing [32], recently vision transformers (ViTs) [6,13,19,30,31] have been successfully utilized in different computer vision problems due to their capabilities at modelling long-range dependencies and enabling the model to attend to all the elements in the input sequence. The core component in ViTs is the self-attention mechanism that that learns the relationships between sequence elements by performing relevance estimation of one item to other items. The other attention such as [1,8,10,29,35] have demonstrated remarkable efficacy in effectively managing volumetric data. Inspired by ViTs [10,19] and based on the observation that attention-based vision transformers architectures are an intuitive design choice for modelling long-range global contextual relationships in volume data, we investigate designing a CNN-transformers based framework for the task of 3D mitochondria instance segmentation.

3 Method

3.1 Baseline Framework

We base our approach on the recent Res-UNET [16], which utilizes encoder-decoder structure of 3D UNET [34] with skip-connections between encoder and decoder. Here, 3D input patch of mitochondria volume ($32 \times 320 \times 320$) is taken from the entire volume of ($400 \times 4096 \times 4096$). The input volume is denoised using an interpolation network adapted for medical images [9]. The denoised volume is then processed utilizing an encoder-decoder structure containing residual anisotropic convolution blocks (ACB). The ACB contains three layers of 3D convolutions with kernels ($1 \times 3 \times 3$), ($3 \times 3 \times 3$), ($3 \times 3 \times 3$) having skip connections between first and third layers. The decoder outputs semantic mask and instance boundary, which are then post-processed using connected component labelling to generate final instance masks. We refer to [16] for more details.

Limitations: As discussed above, the recent Res-UNET approach utilizes 3D convolutions to handle the volumetric input data. However, 3D convolutions are designed to encode short-range spatio-temporal feature information and struggle to model global contextual dependencies that extend beyond the designated receptive field. In contrast, the self-attention mechanism within the vision transformers possesses the capabilities to effectively encode both local and global long-range dependencies by directly performing a comparison of feature activations at all the space-time locations. In this way, self-attention mechanism goes much beyond the receptive field of the conventional convolutional filters. While self-attention has been shown to be beneficial when combined with convolutional layers for different medical imaging tasks, to the best of our knowledge, no previous attempt to design spatio-temporal self-attention as an exclusive building block for the problem of 3D mitochondria instance segmentation exists in literature. Next, we present our approach that effectively utilizes an efficient spatio-temporal attention mechanism for 3D mitochondria instance segmentation.

3.2 Spatio-Temporal Transformer Res-UNET (STT-UNET)

Figure 2(a) presents the overall architecture of the proposed hybrid transformers-CNN based 3D mitochondria instance segmentation approach, named STT-UNET. It comprises a denoising module, transformer based encoder-decoder

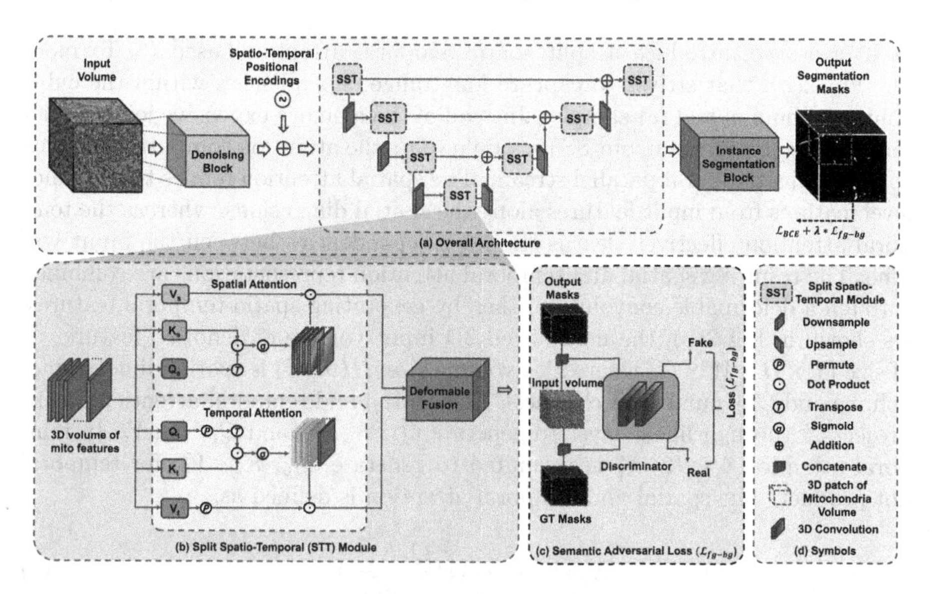

Fig. 2. (a) Overall architecture of our STT-UNET framework for 3D mitochondria instance segmentation. A 3D volume patch of mitochondria is first pre-processed using the interpolation network. The resulting reconstructed volume is then fed to our split spatio-temporal attention based encoder-decoder to generate the semantic-level mitochondria segmentation masks. The focus of our design is the introduction of split spatio-temporal attention (SST) module within the encoder-decoder. **(b)** The SST module first computes spatial and temporal attentions independently, which are later combined through a deformable convolution. Consequently, the semantic masks from the decoder are then input to the instance segmentation module to generate the final instance masks. The entire framework is trained using the standard BCE loss (L_{BCE}) and our semantic foreground-background (FG-BG) adversarial loss (L_{fg-bg}). **(c)** The L_{fg-bg} loss improves the instance-level features, thereby aiding in the better separability of the region of mitochondria instances from the cluttered background.

with split spatio-temporal attention and an instance segmentation block. The denoising module alleviates the segmentation faults caused by anomalies in the EM images, as in the baseline. The denoising is performed by convolving the current frame with two adjacent frames using predicted kernels, thereby generating the resultant frame by adding the convolution outputs. The resulting denoised output is then processed by our transformer based encoder-decoder with split spatio-temporal attention to generate the semantic masks. Consequently, these semantic masks are post-processed by an instance segmentation module using a connected component labelling scheme, thereby generating the final instance-level segmentation output prediction. To further enhance the semantic segmentation quality with cluttered background we introduced semantic adversarial loss which leads to improved semantic segmentation in noisy background.

Split Spatio-Temporal Attention based Encoder-Decoder: Our STT-UNET framework comprises four encoder and three decoder layers. Within each layer, we introduce a split spatio-temporal attention-based (SST) module, Fig. 2(b), that strives to capture long-range dependencies within the cubic volume of human and rat samples. Instead of the memory expensive joint spatio-temporal representation, our SST module splits the attention computation into a spatial and a temporal parallel stream. The spatial attention refines the instance level features from input features along the spatial dimensions, whereas the temporal attention effectively learns the inter-dependencies between the input volume. The resulting spatial and temporal attention representations are combined through a deformable convolution, thereby generating spatio-temporal features. As shown in Fig 2(b), the normalized 3D input volume of denoised features X of size $(T \times H \times W \times C)$ where T is volume size, $(H \times W)$ is spatial dimension of volume and C is number of channels. The spatial and temporal attention blocks project X through linear layer to generate Q_s, K_s, V_s and Q_t, K_t, V_t. In temporal attention Q_t, K_t, V_t is permuted to generate Q_{tp}, K_{tp}, V_{tp} for temporal dot product. The spatial and temporal attention is defined as,

$$X_s = softmax(\frac{Q_s K_s^T}{\sqrt{d_k}})V_s \tag{1}$$

$$X_t = softmax(\frac{Q_{tp} K_{tp}^T}{\sqrt{d_k}})V_{tp} \tag{2}$$

where, X_s is spatial attention map, X_t is temporal attention map and d_k is dimension of Q_s and K_s. To fuse spatial and temporal attention maps, X_s and X_t, we employ deformable convolution. The deformable convolution generates offsets according to temporal attention map X_t and by using these offsets the spatial attention map X_s is aligned. The deformable fusion is given as,

$$X = \int_{c=1}^{C} \sum_{k_n \in R} W(k_n) \cdot X_s(k_0 + k_n + \Delta K_n) \tag{3}$$

where, C is no of channels, X is spatially aligned attention map with respect to X_t. W is the weight matrix of kernels, X_s is spatial attention map, k_0 is starting position of kernel, k_n is enumerating along all the positions in kernel size of R and ΔK_n is the offset sampled from temporal attention map X_t. We empirically observe that fusing spatial and temporal features through a deformable convolution, instead of concatenation through a conv. layer or addition, leads to better performance. The resulting spatio-temporal features of decoder are then input to instance segmentation block to generate final instance masks, as in baseline.

Semantic FG-BG Adversarial Loss: As discussed earlier, a common challenge in mitochondria instance segmentation is to accurately delineate the region of mitochondria instances from the cluttered background. To address this, we introduce a semantic foreground-background (FG-BG) adversarial loss during the training to enhance the FG-BG separability. Here, we introduce the auxiliary discriminator network D with two layers of 3D convolutions with stride 2 during the training as shown in Fig. 2(c). The discriminator takes the input volume I

along with the corresponding mask as an input. Here, the mask M is obtained either from the ground truth or predictions, such that all mitochondria instances within a frame are marked as foreground. While the discriminator D attempts to distinguish between ground truth and predicted masks (M_{gt} and M_{pred}, respectively), the model Ψ learns to output semantic mask such that the predicted masks M_{pred} are close to ground truth M_{gt}. Let $\mathbf{F}_{gt} = \text{CONCAT}(\mathbf{I}, \mathbf{M}_{gt})$ and $\mathbf{F}_{pr} = \text{CONCAT}(\mathbf{I}, \mathbf{M}_{pred})$ denote the real and fake input, respectively, to the discriminator D. Similar to [11], the adversarial loss is then given by,

$$L_{fg-bg} = \min_{\Psi} \max_{D} \Psi[\log D(F_{gt})] + \Psi[\log(1 - D(F_{pr}))] + \lambda_1 \Psi[D(F_{gt}) - D(F_{pr})] \quad (4)$$

Consequently, the overall loss for training is: $L = L_{BCE} + \lambda \cdot L_{fg-bg}$, Where, L_{BCE} is BCE loss, $\lambda = 0.5$ and L_{fg-bg} is semantic adversarial loss.

Table 1. State-of-the-art comparison in terms of AP on Mit-EM-R and MitoEM-H validation sets. Best results are in bold.

Methods	MitoEM-R	MitoEM-H
Wei [36]	0.521	0.605
Nightingale [24]	0.715	0.625
Li [17]	0.890	0.787
Chen [16]	0.917	0.82
STT-UNET (Ours)	**0.958**	**0.849**

Table 2. State-of-the-art comparison in terms of Jaccard and DSC on Lucchi test set. Best results are in bold.

Methods	Jaccard	DSC
Yuan [37]	0.865	0.927
Casser [2]	0.890	0.942
Res-UNET-R [16]	0.895	0.945
Res-UNET-R + MRDA [4]	0.897	0.946
STT-UNET (Ours)	**0.913**	**0.962**

4 Experiments

Dataset: We evaluate our approach on three datasets: MitoEM-R [36], MitoEM-H [36] and Lucchi [22]. The MitoEM [36] is a dense mitochondria instance segmentation dataset from ISBI 2021 challenge. The dataset consists of 2 EM image volumes (30 μm^3) of resolution of $8 \times 8 \times 30$ nm, from rat tissues (MitoEM-R) and human tissue (MitoEM-H) samples, respectively. Each volume has 1000 grayscale images of resolution (4096×4096) of mitochondria, out of which train set has 400, validation set contains 100 and test set has 500 images. Lucchi [22] is a sparse mitochondria semantic segmentation dataset with training and test volume size of $165 \times 1024 \times 768$.

Implementation Details: We implement our approach using Pytorch1.9 [27] (rcom env) and models are trained using 2 AMD MI250X GPUs. During training of MitoEM, for the fair comparison, we adopt same data augmentation technique from [36]. The 3D patch of size ($32 \times 320 \times 320$) is input to the model and trained using batch size of 2. The model is optimized by Adam optimizer with learning rate of $1e^{-4}$. Unlike baseline [16], we do not follow multi-scale training and perform single stage training for 200k iterations. For Lucchi, we follow training details of [16,36] for semantic segmentation. For fair comparison with previous works, we use the same evaluation metrics as in the literature for both datasets. We use 3D AP-75 metric [36] for MitoEM-R and MitoEM-H datasets. For Lucchi, we use jaccard-index coefficient (Jaccard) and dice similarity coefficient (DSC).

4.1 Results

State-of-the-Art Comparison: Table 1 shows the comparison on MitoEm-R and MitoEM-H validation sets. Our STT-UNET achieves state-of-the-art performance on both sets. Compared to the recent [16], our STT-UNET achieves an absolute gains of 4.1% and 2.9% on MitoEM-R and MitoEM-H validation sets, respectively. Note that [16] employs two decoders for MitoEM-H. In contrast, we utilize only a single decoder for both MitoEM-H and MitoEM-R sets, while still achieving improved segmentation performance. Fig 3 presents the segmentation predictions of our approach on example input regions from the validation set. Our approach achieves promising segmentation results despite the noise in the input samples. Table 2 presents the comparison on Lucchi test set. Our method sets a new state-of-the-art on this dataset in terms of both Jaccard and DSC.

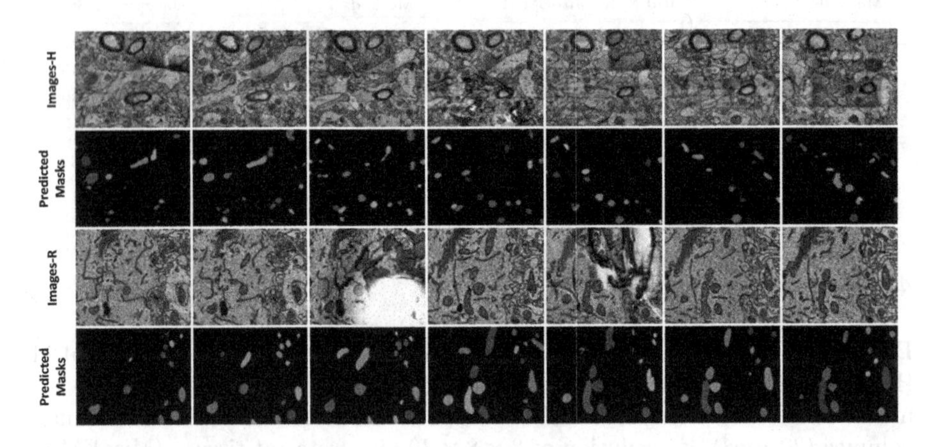

Fig. 3. Qualitative 3D instance segmentation results of our STT-UNET on the example input regions from MitoEM-H and MitoEM-R val sets. Our STT-UNET achieves promising results on these input examples containing noise.

Table 3. Baseline performance comparison.

Methods	MitoEM-R	MitoEM-H
Baseline	0.921	0.823
+ SST	0.948	0.839
+ L_{fg-bg}	**0.958**	**0.849**

Table 4. Ablation study on the impact of feature fusion.

Feature Fusion	MitoEM-R	MitoEM-H
addition	0.950	0.841
concat	0.952	0.842
def-conv	**0.958**	**0.849**

Table 5. Ablation study on the impact of design choice.

Deisgn choice	MitoEM-R	MitoEM-H
spatial	0.914	0.812
spatial-temporal	0.922	0.817
temporal-spatial	0.937	0.832
spatial|temporal	**0.958**	**0.849**

Ablation Study: Table 3 shows a baseline comparison when progressively integrating our contributions: SST module and semantic foreground-background adversarial loss. The introduction of SST module improves performance from 0.921 to 0.941 with a gain of 2.7%. The performance is further improved by 1%, when introducing our semantic foreground-background adversarial loss. Our final approach achieves absolute gains of 3.7% and 2.6% over the baseline on MitoEM-R and MitoEM-H, respectively. We also compare our approach with other attention mechanism in literature such as divided space-time attention [1] and axial attention [35] with our method achieving favorable results with gain of 0.9% and 1.1%, respectively likely due to computing spatial and temporal in parallel and later fusing them through a deformable convolution. Further, we compare our approach with [16] on MitoEM-v2 test set achieving a gain of 4% on MitoEM-R, where the postprocessing from [18] is used to differentiate the mitochondria instances for both methods. Table 4 shows ablation study with feature fusion strategies in our SST module: addition, concat and deformable-conv. The best results are obtained with deformable-conv on both datasets. For encoding spatial and temporal information, we analyze two design choices with SST module: cascaded and split, as shown in Table 5. The best results are obtained using our split design choice (row 3) with spatial and temporal information encoded in parallel and later combined. We also evaluate with different input volumes: 4,8,16,32. We observe best results are obtained when using 32 input volume.

5 Conclusion

We propose a hybrid CNN-transformers based encoder-decoder approach for 3D mitochorndia instance segmentation. We introduce a split spatio-temporal attention (SST) module to capture long-range dependencies within the cubic volume of human and rat mitochondria samples. The SST module computes spatial and temporal attention in parallel, which are later fused. Further, we introduce a semantic adversarial loss for better delineation of mitochondria instances from background. Experiments on three datasets demonstrate the effectiveness of our approach, leading to state-of-the-art segmentation performance.

References

1. Bertasius, G., Wang, H., Torresani, L.: Is space-time attention all you need for video understanding? In: ICML, vol. 2, p. 4 (2021)
2. Casser, V., Kang, K., Pfister, H., Haehn, D.: Fast mitochondria detection for connectomics. In: Medical Imaging with Deep Learning, pp. 111–120. PMLR (2020)
3. Chen, H., Qi, X., Yu, L., Heng, P.: DCAN: deep contour-aware networks for accurate gland segmentation. In: Proceedings of the IEEE Conference on Computer Vision and Pattern Recognition, pp. 2487–2496 (2016)
4. Chen, Q., Li, M., Li, J., Hu, B., Xiong, Z.: Mask rearranging data augmentation for 3D mitochondria segmentation. In: Wang, L., Dou, Q., Fletcher, P.T., Speidel, S., Li, S. (eds.) MICCAI 2022. LNCS, vol. 13434, pp. 36–46. Springer, Cham (2022). https://doi.org/10.1007/978-3-031-16440-8_4

5. Çiçek, Ö., Abdulkadir, A., Lienkamp, S.S., Brox, T., Ronneberger, O.: 3D U-Net: learning dense volumetric segmentation from sparse annotation. In: Ourselin, S., Joskowicz, L., Sabuncu, M.R., Unal, G., Wells, W. (eds.) MICCAI 2016. LNCS, vol. 9901, pp. 424–432. Springer, Cham (2016). https://doi.org/10.1007/978-3-319-46723-8_49
6. Dosovitskiy, A., et al.: An image is worth 16x16 words: transformers for image recognition at scale. In: ICLR (2021)
7. He, K., Gkioxari, G., Dollár, P., Girshick, R.: Mask R-CNN. In: Proceedings of the IEEE International Conference on Computer Vision, pp. 2961–2969 (2017)
8. Ho, J., Kalchbrenner, N., Weissenborn, D., Salimans, T.: Axial attention in multidimensional transformers. arXiv preprint arXiv:1912.12180 (2019)
9. Huang, W., Chen, C., Xiong, Z., Zhang, Y., Liu, D., Wu, F.: Learning to restore ssTEM images from deformation and corruption. In: Bartoli, A., Fusiello, A. (eds.) ECCV 2020. LNCS, vol. 12535, pp. 394–410. Springer, Cham (2020). https://doi.org/10.1007/978-3-030-66415-2_26
10. Huang, Z., Wang, X., Huang, L., Huang, C., Wei, Y., Liu, W.: CCNet: criss-cross attention for semantic segmentation. In: Proceedings of the IEEE/CVF International Conference on Computer Vision (ICCV) (2019)
11. Isola, P., Zhu, J.Y., Zhou, T., Efros, A.A.: Image-to-image translation with conditional adversarial networks. In: CVPR (2017)
12. Januszewski, M., et al.: High-precision automated reconstruction of neurons with flood-filling networks. Nat. Meth. **15**(8), 605–610 (2018)
13. Khan, S., Naseer, M., Hayat, M., Zamir, S.W., Khan, F.S., Shah, M.: Transformers in vision: a survey. ACM Comput. Surv. (CSUR) **54**(10s), 1–41 (2022)
14. Lee, K., Zung, J., Li, P., Jain, V., Seung, H.S.: Superhuman accuracy on the SNEMI3D connectomics challenge. arXiv preprint arXiv:1706.00120 (2017)
15. Li, M., Chen, C., Liu, X., Huang, W., Zhang, Y., Xiong, Z.: Advanced deep networks for 3D mitochondria instance segmentation. arXiv preprint arXiv:2104.07961 (2021)
16. Li, M., Chen, C., Liu, X., Huang, W., Zhang, Y., Xiong, Z.: Advanced deep networks for 3D mitochondria instance segmentation. In: 2022 IEEE 19th International Symposium on Biomedical Imaging (ISBI), pp. 1–5. IEEE (2022)
17. Li, Z., Chen, X., Zhao, J., Xiong, Z.: Contrastive learning for mitochondria segmentation. In: 2021 43rd Annual International Conference of the IEEE Engineering in Medicine & Biology Society (EMBC), pp. 3496–3500. IEEE (2021)
18. Lin, Z., Wei, D., Lichtman, J., Pfister, H.: PyTorch connectomics: a scalable and flexible segmentation framework for EM connectomics. arXiv preprint arXiv:2112.05754 (2021)
19. Liu, Z., et al.: Swin transformer: hierarchical vision transformer using shifted windows. In: CVPR (2021)
20. Lucchi, A., Li, Y., Smith, K., Fua, P.: Structured image segmentation using kernelized features. In: Fitzgibbon, A., Lazebnik, S., Perona, P., Sato, Y., Schmid, C. (eds.) ECCV 2012. LNCS, vol. 7573, pp. 400–413. Springer, Heidelberg (2012). https://doi.org/10.1007/978-3-642-33709-3_29
21. Lucchi, A., et al.: Learning structured models for segmentation of 2-D and 3-D imagery. IEEE Trans. Med. Imaging **34**(5), 1096–1110 (2014)
22. Lucchi, A., Smith, K., Achanta, R., Knott, G., Fua, P.: Supervoxel-based segmentation of mitochondria in EM image stacks with learned shape features. IEEE Trans. Med. Imaging **31**(2), 474–486 (2011)
23. McBride, H.M., Neuspiel, M., Wasiak, S.: Mitochondria: more than just a powerhouse. Curr. Biol. **16**(14), R551–R560 (2006)

24. Nightingale, L., de Folter, J., Spiers, H., Strange, A., Collinson, L.M., Jones, M.L.: Automatic instance segmentation of mitochondria in electron microscopy data. BioRxiv, pp. 2021–05 (2021)
25. Nunnari, J., Suomalainen, A.: Mitochondria: in sickness and in health. Cell **148**(6), 1145–1159 (2012)
26. Oztel, I., Yolcu, G., Ersoy, I., White, T., Bunyak, F.: Mitochondria segmentation in electron microscopy volumes using deep convolutional neural network. In: 2017 IEEE International Conference on Bioinformatics and Biomedicine (BIBM), pp. 1195–1200. IEEE (2017)
27. Paszke, A., et al.: PyTorch: an imperative style, high-performance deep learning library. In: NeurIPS, vol. 32 (2019)
28. Ronneberger, O., Fischer, P., Brox, T.: U-Net: convolutional networks for biomedical image segmentation. In: Navab, N., Hornegger, J., Wells, W.M., Frangi, A.F. (eds.) MICCAI 2015. LNCS, vol. 9351, pp. 234–241. Springer, Cham (2015). https://doi.org/10.1007/978-3-319-24574-4_28
29. Shaker, A., Maaz, M., Rasheed, H., Khan, S., Yang, M.H., Khan, F.S.: UNETR++: delving into efficient and accurate 3d medical image segmentation. arXiv preprint arXiv:2212.04497 (2022)
30. Shamshad, F., et al.: Transformers in medical imaging: a survey. arXiv preprint arXiv:2201.09873 (2022)
31. Touvron, H., Cord, M., Douze, M., Massa, F., Sablayrolles, A., Jégou, H.: Training data-efficient image transformers & distillation through attention. In: ICML (2021)
32. Vaswani, A., et al.: Attention is all you need. In: NeurIPS (2017)
33. Vazquez-Reina, A., Gelbart, M., Huang, D., Lichtman, J., Miller, E., Pfister, H.: Segmentation fusion for connectomics. In: 2011 International Conference on Computer Vision, pp. 177–184. IEEE (2011)
34. Wang, C., MacGillivray, T., Macnaught, G., Yang, G., Newby, D.: A two-stage 3D UNet framework for multi-class segmentation on full resolution image. arXiv preprint arXiv:1804.04341 (2018)
35. Wang, H., Zhu, Y., Green, B., Adam, H., Yuille, A., Chen, L.-C.: Axial-DeepLab: stand-alone axial-attention for panoptic segmentation. In: Vedaldi, A., Bischof, H., Brox, T., Frahm, J.-M. (eds.) ECCV 2020. LNCS, vol. 12349, pp. 108–126. Springer, Cham (2020). https://doi.org/10.1007/978-3-030-58548-8_7
36. Wei, D., et al.: MitoEM dataset: large-scale 3D mitochondria instance segmentation from EM images. In: Martel, A.L., et al. (eds.) MICCAI 2020. LNCS, vol. 12265, pp. 66–76. Springer, Cham (2020). https://doi.org/10.1007/978-3-030-59722-1_7
37. Yuan, Z., Yi, J., Luo, Z., Jia, Z., Peng, J.: EM-Net: centerline-aware mitochondria segmentation in EM images via hierarchical view-ensemble convolutional network. In: 2020 IEEE 17th International Symposium on Biomedical Imaging (ISBI), pp. 1219–1222. IEEE (2020)

Prompt-MIL: Boosting Multi-instance Learning Schemes via Task-Specific Prompt Tuning

Jingwei Zhang[1(✉)], Saarthak Kapse[1], Ke Ma[2], Prateek Prasanna[1], Joel Saltz[1], Maria Vakalopoulou[3], and Dimitris Samaras[1]

[1] Stony Brook University, Stony Brook, USA
{jingwezhang,samaras}@cs.stonybrook.edu,
{saarthak.kapse,prateek.prasanna}@stonybrook.edu,
Joel.Saltz@stonybrookmedicine.edu

[2] Snap Inc., New York, USA
kemma@cs.stonybrook.edu

[3] CentraleSupélec, University of Paris-Saclay, Paris, France
maria.vakalopoulou@centralesupelec.fr

Abstract. Whole slide image (WSI) classification is a critical task in computational pathology, requiring the processing of gigapixel-sized images, which is challenging for current deep-learning methods. Current state of the art methods are based on multi-instance learning schemes (MIL), which usually rely on pretrained features to represent the instances. Due to the lack of task-specific annotated data, these features are either obtained from well-established backbones on natural images, or, more recently from self-supervised models pretrained on histopathology. However, both approaches yield task-agnostic features, resulting in performance loss compared to the appropriate task-related supervision, if available. In this paper, we show that when task-specific annotations are limited, we can inject such supervision into downstream task training, to reduce the gap between fully task-tuned and task agnostic features. We propose Prompt-MIL, an MIL framework that integrates prompts into WSI classification. Prompt-MIL adopts a prompt tuning mechanism, where only a small fraction of parameters calibrates the pretrained features to encode task-specific information, rather than the conventional full fine-tuning approaches. Extensive experiments on three WSI datasets, TCGA-BRCA, TCGA-CRC, and BRIGHT, demonstrate the superiority of Prompt-MIL over conventional MIL methods, achieving a relative improvement of 1.49%–4.03% in accuracy and 0.25%–8.97% in AUROC while using fewer than 0.3% additional parameters. Compared to conventional full fine-tuning approaches, we fine-tune less than 1.3% of the parameters, yet achieve a relative improvement of 1.29%–13.61% in accuracy and 3.22%–27.18% in AUROC and reduce GPU memory consumption by 38%–45% while training 21%–27% faster.

Supplementary Information The online version contains supplementary material available at https://doi.org/10.1007/978-3-031-43993-3_60.

H. Greenspan et al. (Eds.): MICCAI 2023, LNCS 14227, pp. 624–634, 2023.
https://doi.org/10.1007/978-3-031-43993-3_60

Keywords: Whole slide image classification · Multiple instance learning · Prompt tuning

1 Introduction

Whole slide image (WSI) classification is a critical task in computational pathology enabling disease diagnosis and subtyping using automatic tools. Owing to the paucity of patch-level annotations, multiple instance learning (MIL) [9,18,24] techniques have become a staple in WSI classification. Under an MIL scheme, WSIs are divided into tissue patches or instances, and a feature extractor is used to generate features for each instance. These features are then aggregated using different pooling or attention-based operators to provide a WSI-level prediction. ImageNet pretrained networks have been widely used as MIL feature extractors. More recently, self-supervised learning (SSL), using a large amount of unlabeled histopathology data, has become quite popular for WSI classification [5,13] as it outperforms ImageNet feature encoders.

Most existing MIL methods do not fine-tune their feature extractor together with their classification task; this stems from the requirement for far larger GPU memory than is available currently due to the gigapixel nature of WSIs, e.g. training a WSI at 10x magnification may require more than 300 Gb of GPU memory. Recently, researchers have started to explore optimization methods to enable end-to-end training of the entire network and entire WSI within GPU memory [21,25,29]. These methods show better performance compared to conventional MIL; they suffer, however, from two limitations. First, they are ImageNet-pretrained and do not leverage the powerful learning capabilities of histology-trained SSL models. Second, these are mostly limited to convolutional architectures rather than more effective attention-based architectures such as vision transformers [7].

Motivation: To improve WSI-level analysis, we explore end-to-end training of the entire network using SSL pretrained ViTs. To achieve this, we use the patch batching and gradient retaining techniques in [25]. However, we find that conventional fine-tuning approaches, where the entire network is fine-tuned, achieve low performance. For example, on the BRIGHT dataset [2], the accuracy drops more than 5% compared to the conventional MIL approaches. The poor performance is probably caused by the large network over-fitted to the limited downstream training data, leading to suboptimal feature representation. Indeed, especially for weakly supervised WSI classification, where annotated data for downstream tasks is significantly less compared to natural image datasets, conventional fine-tuning schemes can prove to be quite challenging.

To address the subpar performance of SSL-pretrained vision transformers, we utilize the prompt tuning techniques. Initially proposed in natural language processing, a prompt is a trainable or a pre-defined natural language statement that is provided as additional input to a transformer to guide the neural network towards learning a specific task or objective [3,12]. Using prompt tuning we *fine-tune only the prompt and downstream network without re-training the large backbone* (e.g. GPT-3 with 17B parameters). This approach is parameter efficient [12,15] and has been shown to better inject task-specific information and reduce the overfitting in downstream tasks, particularly in limited data scenarios [8,23]. Recently, prompts have also been adopted in computer vision and demonstrated superior performance compared to conventional fine-tuning methods [10]. Prompt tuning performs well even when only limited labeled data is available for training, making it particularly attractive in computational pathology. The process of prompt tuning thus involves providing a form of limited guidance during the training of downstream tasks, with the goal of minimizing the discrepancy between feature representations that are fully tuned to the task and those that are not task-specific.

In this paper, we propose a novel framework, Prompt-MIL, which uses prompts for WSI-level classification tasks within an MIL paradigm. Our contributions are:

- **Fine-tuning:** Unlike existing works in histopathology image analysis, Prompt-MIL is fine-tuned using prompts rather than conventional full fine-tuning methods.
- **Task-specific representation learning:** Our framework employs an SSL pretrained ViT feature extractor with a trainable prompt that calibrates the representations making them task-specific. By doing so, only the prompt parameters together with the classifier, are optimized. This avoids potential overfitting while still injecting task-specific knowledge into the learned representations.

Extensive experiments on three public WSI datasets, TCGA-BRCA, TCGA-CRC, and BRIGHT demonstrate the superiority of Prompt-MIL over conventional MIL methods, achieving a relative improvement of 1.49%–4.03% in accuracy and 0.25%–8.97% in AUROC by using only less than 0.3% additional parameters. Compared to the conventional full fine-tuning approach, we fine-tune less than 1.3% of the parameters, yet achieve a relative improvement of 1.29%–13.61% in accuracy and 3.22%–27.18% in AUROC. Moreover, compared to the full fine-tuning approach, our method reduces GPU memory consumption by 38%–45% and trains 21%–27% faster. To the best of our knowledge, this is the first work where prompts are explored for WSI classification. While our method is quite simple, it is versatile as it is agnostic to the MIL scheme and can be easily applied to different MIL methods. Our code is available at https://github.com/cvlab-stonybrook/PromptMIL.

2 Method

Our Prompt-MIL framework consists of three components: a frozen feature model to extract features of tissue patches, a classifier that performs an MIL scheme of feature aggregation and classification of the WSIs, and a trainable prompt. Given a WSI and its label y, the image is tiled into n tissue patches/instances $\{x_1, x_2, \ldots, x_n\}$ at a predefined magnification. As shown in Fig. 1, the feature model $F(\cdot)$ computes n feature representations from the corresponding n patches:

$$\begin{aligned} h &= [h_1, h_2, \ldots, h_n] \\ &= [F(x_1, \mathbb{P}), F(x_2, \mathbb{P}), \ldots, F(x_n, \mathbb{P})], \end{aligned} \tag{1}$$

where h_i denotes the feature of the ith patch, h is the concatenation of all h_i, and $\mathbb{P} = \{p_i, i = 1, 2, \ldots, k\}$ is the trainable prompt consisting of k trainable tokens. The classifier $G(\cdot)$ applies an MIL scheme to predict the label $\hat{y}$ and calculate the loss $\mathcal{L}$ as:

$$\mathcal{L} = \mathcal{L}_{cls}(\hat{y}, y) = \mathcal{L}_{cls}(G(h), y), \tag{2}$$

where the $\mathcal{L}_{cls}$ is a classification loss.

2.1 Visual Prompt Tuning

The visual prompt tuning is the key component of our framework. As shown in Fig. 1(b), our feature model $F(\cdot)$ is a ViT based architecture. It consists of a patch embedding layer L_0 and l sequential encoding layers $\{L_1, L_2, \ldots, L_l\}$. The ViT first divides an input image x_i into w smaller patches $[z_1, z_2, \ldots, z_w]$ and embeds them into w tokens:

$$\mathbb{T}_z^0 = L_0([z_1, z_2, \ldots, z_w]) = \{t_1^0, t_2^0, \ldots, t_w^0\}, \tag{3}$$

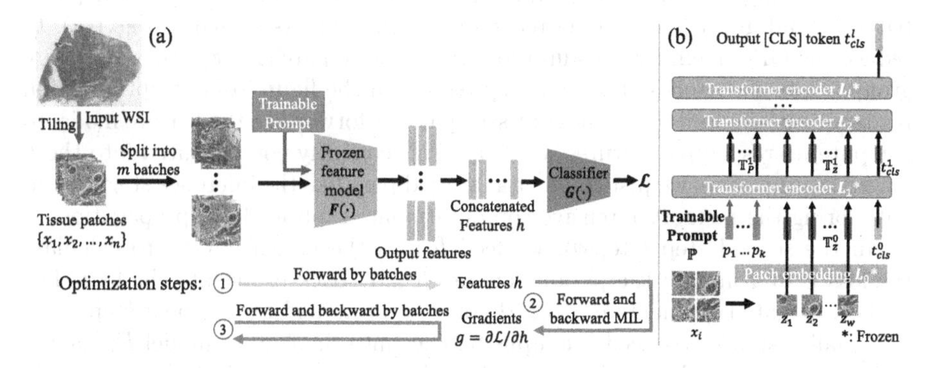

Fig. 1. Overview of the proposed method. (a) Overall structure of our training pipeline. Tissue patches tiled from the input WSI are grouped into separate batches, which are fed into a frozen feature model $F(\cdot)$ to compute their respective features. The features are subsequently concatenated into the feature h and a classifier $G(\cdot)$ applies an MIL scheme on h to predict the label and calculate the loss $\mathcal{L}$. (b) Structure of the feature model $F(\cdot)$ with the additional prompt. An input image x_i is cropped into w small patches $z_1, \ldots, z_w$. k trainable prompt tokens, together with the embedding of small patches and a class token t_{cls}^0, are fed into l layers of Transformer encoders. The output feature corresponding to x_i is the last class token t_{cls}^l. The feature model $F(\cdot)$ is frozen and only the prompt is trainable.

where t_i^0 is the embedding token of z_i and $\mathbb{T}_z^0$ is the collection of such tokens. These tokens $\mathbb{T}_z^0$ are concatenated with a class token t_{cls}^0 and a prompt $\mathbb{P}$: The class token is used to aggregate information from all other tokens. The prompt consists of k trainable tokens $\mathbb{P} = \{p_i | i = 1, 2, \ldots, k\}$. The concatenation is fed into l layers of the Transformer encoders:

$$[\mathbb{T}_z^1, \mathbb{T}_P^1, t_{cls}^1] = L_1([\mathbb{T}_z^0, \mathbb{P}, t_{cls}^0]) \quad (4)$$

$$[\mathbb{T}_z^i, \mathbb{T}_P^i, t_{cls}^i] = L_i([\mathbb{T}_z^{i-1}, \mathbb{T}_P^{i-1}, t_{cls}^{i-1}]), i = 2, 3, \ldots, l \quad (5)$$

$$\mathbb{T}_P^i = \{p_j^i | j = 1, 2, \ldots, k\}, \quad (6)$$

where p_j^i is the jth output prompt token of the ith Transformer encoder and $\mathbb{T}_P^i$ is the collection of all k such output prompt tokens, which are not trainable. The output feature of x_i is defined as the last class token: $h_i = t_{cls}^l$.

2.2 Optimization

Our overall loss function is defined as

$$\begin{aligned} \mathcal{L} &= \mathcal{L}_{cls}(G(H), y) \\ &= \mathcal{L}_{cls}(G([F(x_1, \mathbb{P}), F(x_2, \mathbb{P}), \ldots, F(x_n, \mathbb{P})]), y), \end{aligned} \quad (7)$$

where only the parameters of the $G(\cdot)$ and the prompt $\mathbb{P}$ are optimized, while the feature extractor model $F(\cdot)$ is frozen.

Training the entire pipeline in an end-to-end fashion on gigapixel images is infeasible using the current hardware. To address this issue, we utilize the patch batching and gradient retaining techniques from [25]. As shown in Fig. 1(a), to reduce the GPU memory consumption, the n tissue patches $\{x_1, x_2, \ldots, x_n\}$ are grouped into m batches. The first step (step① in the figure) of our optimization is to sequentially feed m batches of tissue patches forward to the feature model to compute its respective features which are subsequently concatenated into the h matrix. In this step, we just conduct a forward pass like the inference stage, without storing the memory-intensive computational graph for back-propagation.

In the second step (step②), we feed h into the classifier $G(\cdot)$ to calculate the loss $\mathcal{L}$ and update the parameters of $G(\cdot)$ by back-propagate the loss. The back-propagated gradients $g = \partial\mathcal{L}/\partial h$ on h are retained for the next step.

Finally (step③), we feed the input batches into the feature model $F(\cdot)$ again and use the output h and the retained gradients g from the last step to update the trainable prompt tokens. In particular, the gradients on the jth prompt token p_j are calculated as:

$$\begin{aligned} \frac{\partial\mathcal{L}}{\partial p_j} &= \frac{\partial\mathcal{L}}{\partial h}\frac{\partial h}{\partial p_j} \\ &= \sum_i \frac{\partial\mathcal{L}}{\partial h_i}\frac{\partial h_i}{\partial p_j} = \sum_i g_i \frac{\partial h_i}{\partial p_j}, \end{aligned} \quad (8)$$

where g_i is the gradient calculated with respect to h_i.

To sum up, in each step, we only update either F or G given the current batch, which avoid storing the gradients of the whole framework for all the input patches. This patch batching and gradient retaining techniques make the end-to-end training feasible.

In this study, we use DSMIL [13] as the classifier and binary cross entropy as the classification loss $\mathcal{L}_{cls}$ when the task is a tumor sub-type classification or cross entropy otherwise.

3 Experiments and Discussion

3.1 Datasets

We assessed Prompt-MIL using three histopathological WSI datasets: TCGA-BRCA [14], TCGA-CRC [19], and BRIGHT [2]. These datasets were utilized for both the self-supervised feature extractor pretraining and the end-to-end fine-tuning (with or without prompts), including the MIL component. Note that the testing data were not used in the SSL pretraining. **TCGA-BRCA** contains 1034 diagnostic digital slides of two breast cancer subtypes: invasive ductal carcinoma (IDC) and invasive lobular carcinoma (ILC). We used the same training, validation, and test split as that in the first fold cross validation in [5]. The cropped patches (790K training, 90K test) were extracted at 5$\times$ magnification. **TCGA-CRC** contains 430 diagnostic digital slides of colorectal cancer for a binary classification task: chromosomal instability (CIN) or genome stable (GS). Following the common 4-fold data split [1,16], we used the first three folds for training (236 GS, 89 CIN), and the fourth for testing (77 GS, 28 CIN). We further split 20% (65 slides) training data as a validation set. The cropped patches (1.07M training, 370K test) were extracted at 10$\times$ magnification. **BRIGHT** contains 503 diagnostic slides of breast tissues. We used the official training (423 WSIs) and test (80 WSIs) splits. The task involves classifying non-cancerous (196 training, 25 test) vs. pre-cancerous (66 training, 23 test) vs. cancerous (161 training, 32 test). We further used 20% (85 slides) training slides for validation. The cropped patches (1.24M training, 195K test) were extracted at 10$\times$ magnification.

Table 1. Comparison of accuracy and AUROC on three datasets. Reported metrics (in %age) are the average across 3 runs. "Num. of Parameters" represents the number of optimized parameters

Dataset	TCGA-BRCA		TCGA-CRC		BRIGHT		Num. of
Metric	Accuracy	AUROC	Accuracy	AUROC	Accuracy	AUROC	Parameters
Conventional MIL	92.10	96.65	73.02	69.24	62.08	80.96	70k
Full fine-tuning	88.14	93.78	74.53	56.63	56.13	75.87	5.6M
Prompt-MIL (ours)	**93.47**	**96.89**	**75.47**	**75.45**	**64.58**	**81.31**	70k + 192

3.2 Implementation Details

We cropped non-overlapping 224 $\times$ 224 sized patches in all our experiments and used ViT-Tiny (ViT-T/16) [7] for feature extraction. For SSL pretraining, we leveraged the DINO framework [4] with the default hyperparameters, but adjusted the batch size to 256 and employed the global average pooling for token aggregation. We pretrained separate ViT models on the TCGA-CRC datasets for 50 epochs, on the BRIGHT dataset for 50 epochs, and on the BRCA dataset for 30 epochs. For TCGA-BRCA, we used the AdamW [17] optimizer with a learning rate of $1e-4$, $1e-2$ weight decay, and trained for 40 epochs. For TCGA-CRC, we also used the AdamW optimizer with a learning rate of $5e-4$ and trained for 40 epochs. For Bright, we used the Adam [11] optimizer with a learning rate of $1e-4$, $5e-2$ weight decay and trained for 40 epochs. We applied a cosine annealing learning rate decay policy in all our experiments. For the MIL baselines, we employed the same hyperparameters as above. For all full fine-tuning experiments, we used the learning rate in the corresponding prompt experiment as the base learning rate. For parameters in the feature model $F(\cdot)$, which are SSL pretrained, we use 1/10 of the base learning rate. For parameters in the Classifier $G(\cdot)$, which are randomly initialized, we use the base learning rate. We train the full tuning model for 10 more epochs than our prompt training to allow full convergence. This training strategy is optimized using the validation datasets. All model implementations were in PyTorch [20] on a NVIDIA Tesla V100 or a Nvidia Quadro RTX 8000.

3.3 Results

We chose overall accuracy and Area Under Receiver Operating Characteristic curve (AUROC) as the evaluation metrics.

Evaluation of Prompt Tuning Performance: We compared the proposed Prompt-MIL with two baselines: 1) a conventional MIL model with a frozen feature extractor [13], 2) fine-tuning all parameters in the feature model (full fine-tuning). Table 1 highlights that our Prompt-MIL consistently outperformed both. Compared to the conventional MIL method, Prompt-MIL added negligible parameters (192, less than 0.3% of the total parameters), achieving a relative improvement of 1.49% in accuracy and 0.25% in AUROC on TCGA-BRCA, 3.36% in accuracy and 8.97% in AUROC on TCGA-CRC, and 4.03% in accuracy and 0.43% in AUROC on BRIGHT. The observed improvement can be attributed to a more optimal alignment between the feature representation learned during the SSL pretraining and the downstream task, i.e., the prompt explicitly calibrated the features toward the downstream task.

The computationally intensive full fine-tuning method under-performed conventional MIL and Prompt-MIL. Compared to the full fine-tuning method, our method achieved a relative improvement of 1.29% to 13.61% in accuracy and 3.22% to 27.18% in AUROC on the three datasets. Due to the relatively small amount of slide-level labels (few hundred to a few thousands) fully fine tuning

5M parameters in the feature model might suffer from overfitting. In contrast, our method contained less than 1.3% of parameters compared to full fine-tuning, leading to robust training.

Table 2. Comparison of GPU memory consumption and training speed per slide benchmarked on the BRIGHT dataset between the full fine-tuning and our prompt tuning on four slides with different sizes. Our prompt method requires far less memory and is significantly faster.

	WSI size	$44k \times 21k$	$26k \times 21k$	$22k \times 17k$	$11k \times 16k$
	#Tissue patches	9212	4765	2307	1108
GPU Mem.	Full fine-tuning	21.81G	18.22G	16.37G	12.71G
	Prompt (ours)	**12.04G**	**10.66G**	**10.00G**	**7.90G**
	Reduction percentage	44.79%	41.50%	38.92%	37.84%
Time per slide	Full fine-tuning	17.73 s	8.92 s	4.37 s	2.15 s
	Prompt (ours)	**13.92** s	**7.09** s	**3.35**s	**1.56** s
	Reduction percentage	21.49%	20.51%	23.32%	27.27%

Table 3. Comparison of accuracy and AUROC on three datasets for a pathological foundation model.

Dataset	TCGA-BRCA		BRIGHT	
Metric	Accuracy	AUROC	Accuracy	AUROC
ViT-small [27]	91.75	97.03	54.17	76.76
ViT-small w/ Prompt-MIL	**92.78**	**97.53**	**57.50**	**78.29**

Evaluation of Time and GPU Memory Efficiency: Prompt-MIL is an efficient method requiring less GPU memory to train and running much faster than full fine-tuning methods. We evaluated the training speed and memory consumption of our method and compared to the full fine-tuning baseline on four different sized WSIs in the BRIGHT dataset. As shown in Table 2, our method consumed around 38% to 45% less GPU memory compared to full fine-tuning and was 21% to 27% faster. As we scaled up the WSI size (i.e. WSIs with more number of patches), the memory cost difference between Prompt-MIL and full fine-tuning further widened.

Evaluation on the Pathological Foundation Models: We demonstrated our Prompt-MIL also had a better performance when used with the pathological foundation model. Foundational models refer to those trained on large-scale pathology datasets (e.g. the entire TCGA Pan-cancer dataset [28]). We utilized the publicly available [26,27] ViT-Small network pretrained using MoCo v3 [6] on all the slides from TCGA [28] and PAIP [22]. In Table 3, we showed that our method robustly boosted the performance on both TCGA (the same domain as the foundation model trained on) and BRIGHT (a different domain). The improvement is more prominent in BRIGHT, which further confirmed that Prompt-MIL aligns the feature extractor to be more task-specific.

Table 4. Performance with a different number of prompt tokens. For two different WSI classification tasks, one token was enough to boost the performance of the conventional MIL schemes.

Dataset	TCGA-BRCA		BRIGHT	
#prompt tokens k	Accuracy	AUROC	Accuracy	AUROC
$k = 1$	**93.47**	**96.89**	**64.58**	**81.31**
$k = 2$	93.13	**96.93**	60.41	79.74
$k = 3$	**93.47**	96.86	59.17	76.75

Ablation Study: An ablation was performed to study the effect of the number of trainable prompt tokens on downstream tasks. Table 4 shows the accuracy and AUROC of our Prompt-MIL model with 1, 2 and 3 trainable prompt tokens ($k = 1, 2, 3$) on the TCGA-BRCA and the BRIGHT datasets. On the TCGA-BRCA dataset, our Prompt-MIL model with 1 to 3 prompt tokens reported similar performance. On the BRIGHT dataset, the performance of our model dropped with the increased number of prompt tokens. Empirically, this ablation study shows that for classification tasks, one prompt token is sufficient to boost the performance of conventional MIL methods.

4 Conclusion

In this work, we introduced a new framework, Prompt-MIL, which combines the use of Multiple Instance Learning (MIL) with prompts to improve the performance of WSI classification. Prompt-MIL adopts a prompt tuning mechanism rather than a conventional full fine-tuning of the entire feature representation. In such a scheme, only a small fraction of parameters calibrates the pretrained representations to encode task-specific information, so the entire training can be performed in an end-to-end manner. We applied our proposed method to three publicly available datasets. Extensive experiments demonstrated the superiority of Prompt-MIL over the conventional MIL as well as the conventional fully fine-tuning methods. Moreover, by fine-tuning much fewer parameters compared to fully fine-tuning, our method is GPU memory efficient and fast. Our proposed approach also showed promising potentials in transferring foundation models. We will further explore the task-specific features that are captured by our prompt toward explainability of these models.

Acknowledgements. This work was partially supported by the ANR Hagnodice ANR-21-CE45-0007, the NSF IIS-2212046, the NSF IIS-2123920, the NIH 1R21CA258493-01A1, the NCI UH3CA225021 and Stony Brook University Provost Funds. The content is solely the responsibility of the authors and does not necessarily represent the official views of the National Institutes of Health.

References

1. Bilal, M., et al.: Development and validation of a weakly supervised deep learning framework to predict the status of molecular pathways and key mutations in colorectal cancer from routine histology images: a retrospective study. Lancet Digital Health **3**(12), e763–e772 (2021)
2. Brancati, N., et al.: BRACS: a dataset for breast carcinoma subtyping in H&E histology images. Database **2022**, baac093 (2022)
3. Brown, T., et al.: Language models are few-shot learners. Adv. Neural. Inf. Process. Syst. **33**, 1877–1901 (2020)
4. Caron, M., et al.: Emerging properties in self-supervised vision transformers. In: Proceedings of the IEEE/CVF International Conference on Computer Vision, pp. 9650–9660 (2021)
5. Chen, R.J., et al.: Scaling vision transformers to gigapixel images via hierarchical self-supervised learning. In: Proceedings of the IEEE/CVF Conference on Computer Vision and Pattern Recognition (CVPR), pp. 16144–16155, June 2022
6. Chen, X., Xie, S., He, K.: An empirical study of training self-supervised vision transformers. In: Proceedings of the IEEE/CVF International Conference on Computer Vision, pp. 9640–9649 (2021)
7. Dosovitskiy, A., et al.: An image is worth 16 × 16 words: transformers for image recognition at scale. In: International Conference on Learning Representations (2021)
8. Gu, Y., Han, X., Liu, Z., Huang, M.: PPT: pre-trained prompt tuning for few-shot learning. In: Proceedings of the 60th Annual Meeting of the Association for Computational Linguistics (vol. 1: Long Papers), pp. 8410–8423 (2022)
9. Hou, L., Samaras, D., Kurc, T.M., Gao, Y., Davis, J.E., Saltz, J.H.: Patch-based convolutional neural network for whole slide tissue image classification. In: Proceedings of the IEEE Conference on Computer Vision and Pattern Recognition, pp. 2424–2433 (2016)
10. Jia, M., et al.: Visual prompt tuning. In: Avidan, S., Brostow, G., Cissé, M., Farinella, G.M., Hassner, T. (eds.) Computer Vision-ECCV 2022: 17th European Conference, Tel Aviv, Israel, 23–27 October 2022, Proceedings, Part XXXIII, pp. 709–727. Springer, Cham (2022). https://doi.org/10.1007/978-3-031-19827-4_41
11. Kingma, D.P., Ba, J.: Adam: a method for stochastic optimization. In: Bengio, Y., LeCun, Y. (eds.) 3rd International Conference on Learning Representations, ICLR 2015, San Diego, CA, USA, 7–9 May 2015, Conference Track Proceedings (2015)
12. Lester, B., Al-Rfou, R., Constant, N.: The power of scale for parameter-efficient prompt tuning. In: Proceedings of the 2021 Conference on Empirical Methods in Natural Language Processing, pp. 3045–3059 (2021)
13. Li, B., Li, Y., Eliceiri, K.W.: Dual-stream multiple instance learning network for whole slide image classification with self-supervised contrastive learning. In: Proceedings of the IEEE/CVF Conference on Computer Vision and Pattern Recognition, pp. 14318–14328 (2021)

14. Lingle, W., et al.: Radiology data from the Cancer Genome Atlas Breast Invasive Carcinoma [TCGA-BRCA] collection. Cancer Imaging Arch. **10**, K9 (2016)
15. Liu, X., et al.: P-tuning: prompt tuning can be comparable to fine-tuning across scales and tasks. In: Proceedings of the 60th Annual Meeting of the Association for Computational Linguistics (vol. 2: Short Papers), pp. 61–68 (2022)
16. Liu, Y., et al.: Comparative molecular analysis of gastrointestinal adenocarcinomas. Cancer Cell **33**, 721–735.e8 (2018)
17. Loshchilov, I., Hutter, F.: Decoupled weight decay regularization. In: ICLR (2018)
18. Lu, M.Y., Williamson, D.F., Chen, T.Y., Chen, R.J., Barbieri, M., Mahmood, F.: Data-efficient and weakly supervised computational pathology on whole-slide images. Nat. Biomed. Eng. **5**(6), 555–570 (2021)
19. Network, C.G.A., et al.: Comprehensive molecular characterization of human colon and rectal cancer. Nature **487**, 330–337 (2012)
20. Paszke, A., et al.: PyTorch: an imperative style, high-performance deep learning library. In: Advances in Neural Information Processing Systems, vol. 32 (2019)
21. Pinckaers, H., Van Ginneken, B., Litjens, G.: Streaming convolutional neural networks for end-to-end learning with multi-megapixel images. IEEE Trans. Pattern Anal. Mach. Intell. **44**(3), 1581–1590 (2020)
22. Platform, P.A.: PAIP (2021). Data retrieved from PAIP, http://www.wisepaip.org/paip/
23. Schucher, N., Reddy, S., de Vries, H.: The power of prompt tuning for low-resource semantic parsing. In: Proceedings of the 60th Annual Meeting of the Association for Computational Linguistics (vol. 2: Short Papers), pp. 148–156 (2022)
24. Shao, Z., et al.: TransMIL: transformer based correlated multiple instance learning for whole slide image classification. Adv. Neural. Inf. Process. Syst. **34**, 2136–2147 (2021)
25. Takahama, S., et al.: Multi-stage pathological image classification using semantic segmentation. In: Proceedings of the IEEE/CVF International Conference on Computer Vision, pp. 10702–10711 (2019)
26. Wang, X., et al.: TransPath: transformer-based self-supervised learning for histopathological image classification. In: de Bruijne, M., et al. (eds.) MICCAI 2021, Part VIII. LNCS, vol. 12908, pp. 186–195. Springer, Cham (2021). https://doi.org/10.1007/978-3-030-87237-3_18
27. Wang, X., et al.: Transformer-based unsupervised contrastive learning for histopathological image classification. Med. Image Anal. **81**, 102559 (2022)
28. Weinstein, J.N., et al.: The cancer genome atlas pan-cancer analysis project. Nat. Genet. **45**(10), 1113–1120 (2013)
29. Zhang, J., et al.: Gigapixel whole-slide images classification using locally supervised learning. In: Wang, L., Dou, Q., Fletcher, P.T., Speidel, S., Li, S. (eds.) International Conference on Medical Image Computing and Computer-Assisted Intervention, pp. 192–201. Springer, Cham (2022). https://doi.org/10.1007/978-3-031-16434-7_19

Pick and Trace: Instance Segmentation for Filamentous Objects with a Recurrent Neural Network

Yi Liu(✉), Su Peng, Jeffrey Caplan, and Chandra Kambhamettu

University of Delaware, Newark, DE 19716, USA
yliu@udel.edu

Abstract. Filamentous objects are ubiquitous in biomedical images, and segmenting individual filaments is fundamental for biomedical research. Unlike common objects with well-defined boundaries and centers, filaments are thin, non-rigid, varying in shape, and often densely overlapping. These properties make it extremely challenging to extract individual filaments. This paper proposes a novel approach to extract filamentous objects by transforming an instance segmentation problem into a sequence modeling problem. Our approach first identifies filaments' tip points, and we segment each instance by tracing them from each tip with a sequential encoder-decoder framework. The proposed method simulates the process of humans extracting filaments: pick a tip and trace the filament. As few datasets contain instance labels of filaments, we first generate synthetic filament datasets for training and evaluation. Then, we collected a dataset of 15 microscopic images of microtubules with instance labels for evaluation. Our proposed method can alleviate the data shortage problem since our proposed model can be trained with synthetic data and achieve state-of-art results when directly evaluated on the microtubule dataset and P. rubescens dataset. We also demonstrate our approaches' capabilities in extracting short and thick elongated objects by evaluating on the C. elegans dataset. Our method achieves a comparable result compared to the state-of-art method with faster processing time. Our code is available at https://github.com/VimsLab/DRIFT.

Keywords: Instance Segmentation · Filament Tracing · Recurrent Neural Network

1 Introduction

Filamentous objects, such as microtubules, actin filaments, and blood vessels, play a fundamental role in biological systems. For example, microtubules (See Fig. 1) form part of cell cytoskeleton structures and are involved in various cellular activities such as movement, transportation, and key signaling events. To

---

Supplementary Information The online version contains supplementary material available at https://doi.org/10.1007/978-3-031-43993-3_61.

H. Greenspan et al. (Eds.): MICCAI 2023, LNCS 14227, pp. 635–645, 2023.
https://doi.org/10.1007/978-3-031-43993-3_61

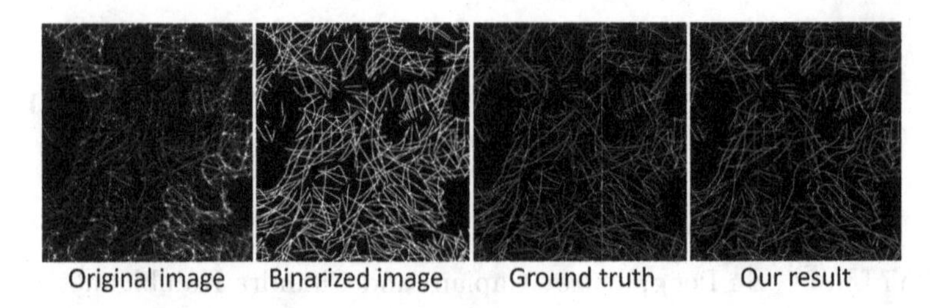

Fig. 1. Qualitative result of our method on a full resolution microscopic image of microtubules. The image has a size of 1376 × 1504 and contains 558 filaments.

understand the mechanism of filamentous objects, quantifying their properties, including quantity, length, curvature, and distribution, is fundamental for biological research. Instance segmentation is often the first step for quantitative analysis. However, as filaments are very thin, non-rigid, and usually span over the image intersecting each other, extracting individual filaments is very challenging.

Since the advent of deep learning, region-based instance segmentation methods [4,7,8,13,27], which segment instances within the detected bounding boxes, have shown remarkable performance on objects with well-defined centers and boundaries. However, filaments are non-rigid objects and span widely across the image. Each filament has a distinct shape varying in length and deformation, while segments of different filaments share a similar appearance. These properties make it extremely hard for region-based methods to detect the centers and bounding boxes for filaments and segment the target within the detected region.

Region-free methods utilize learned embedding [1,2,5,14] or affinity graph [6,19] to separate instances. These methods rely on pixel-level prediction to group instances and do not directly extract the complete shape of instances. Since filaments may be densely clustered and overlapping, these methods have difficulties in disentangling filaments in complicated scenes. Liu *et al.* [17] address the overlapping challenge by proposing an orientation-aware network to disentangle filaments at intersections, but it requires a heuristic post-processing to form instances. Hirsch *et al.* [9] predict dense patches representing instances' shape for each pixel and assemble them to form instances with affinity graph. Effectively extracting longer filaments requires predicting a larger shape patch for each pixel, leading to a longer computational time.

Lacking instance-level labels of filaments is another challenge. Most existing approaches for filaments extraction adopt traditional computer vision techniques such as morphological operations [29], template matching [28,29] and active contour [26]. These methods follow the segment-break-regroup strategy. They first obtain the binary segmentation, then break it at intersections and regroup the segments into filaments by geometric properties. The performance heavily relies on manual parameter tuning.

When a human tries to manually extract filaments in Fig. 1, directly pointing out each filament can be challenging. Instead, a human would first identify each

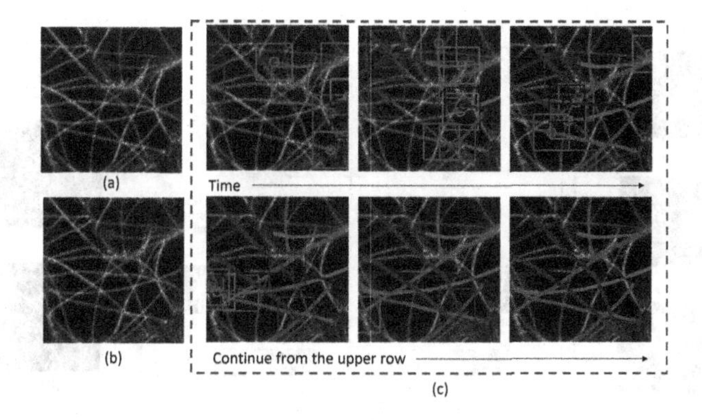

Fig. 2. Our methods mimic humans extracting filaments. (a). Image of microtubules. (b). Pick: tip points detection. (red circles are detected tips). (c). Trace: Successive stages in the process of tracing 10 instances. The boxes are the areas that the model is processing. The circles are patches' centers. (Color figure online)

filament's tip and then trace each filament. Inspired by this human behavior, we introduce Deep Recurrent Instance Filament Tracer (DRIFT) for instance segmentation on filamentous objects. Figure 2 shows an example of how DRIFT mimics a human and sequentially extracts filaments. As shown in Fig. 3, the pick module in DRIFT first detects all tip points as candidates. The recurrent neural network (RNN) based tracing module will 'look' at the patch around tip points, segment the object within the patch, and predict the next location. The trace module sequentially segments the object until a stop flag is predicted. The RNN learns where 'it' comes from, where 'it' is, and where 'it' goes next.

Our method is fundamentally different from [11,12,23,25]. Ren *et al.* [23] and Amaia [25] *et al.* use the RNN model to sequentially predict objects' bounding boxes, which are essentially region-based segmentation methods. Januszewski *et al.* [11,12] propose a flood-filling network to repeatedly perform segmentation within a set of manually defined patches to grow object masks. While the iterations in [11,12] are heuristic, our method learns to trace the filaments and sequentially segments the targets.

The major contributions of this work are as follows: (1). To our knowledge, our method is the first method that converts the instance segmentation into a sequence modeling problem. Our proposed method mimics human tracing and extracting individual filament, tackling the challenges of extracting filamentous objects. (2). We propose a synthetic filament dataset for training and evaluation. Our model trained on synthetic datasets can be applied to various real filament datasets and thus alleviate the data shortage problem. (3). We collected a dataset of 15 microscopic images of microtubules with instance labels for evaluation. (4). Our method is evaluated on four different datasets and compared with several competing approaches. Our method shows better or comparable results regarding accuracy and computational time.

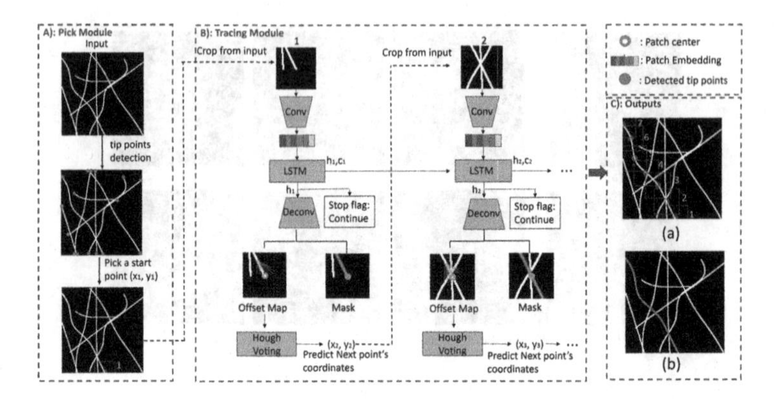

Fig. 3. Flowchart of DRIFT for instance segmentation on filaments. The example is shown with a synthetic filament image. A). Pick module for tip points detection. B). Tracing module for individual filament extraction. C-a). Red boxes are the sequence of patches processed. C-b). The final extracted filament. (Color figure online)

2 Method

Figure 3 shows the framework of our proposed method. The pick module detects tip points for all filaments. Then we crop the patches around tip points, and the tracing module will encode these patches into patch embeddings with a convolutional block. The tip point's patch embedding is used to initialize the hidden state of RNN. Then a decoder will decode the hidden state output and predict a stop flag and the object's mask within the current patch. The decoder also outputs an offset map, where each pixel predicts a vector pointing to the next center. We use the offset map to locate the next center via Hough voting. The model sequentially segment instances until the stop flag turns on.

2.1 Pick: Tip Points Detection Module

We adapt the U-shaped structure from [3,15] to regress the tip points' heatmap and use a maximum filter to acquire coordinates of tip points. Network details are included in supplementary materials.

2.2 Trace: A Recurrent Network for Filament Tracing

Network Description. After tip points are detected, the tracing module will trace and extract each instance. As shown in Fig. 4, we use patch size 64 as an example to describe our network design. Since we have converted the instance segmentation problem into a sequence learning problem, we use Long Short Term Memory (LSTM) [10] to encode the sequence of patches. We use one LSTM layer with a hidden size of 512 and an input size of 256.

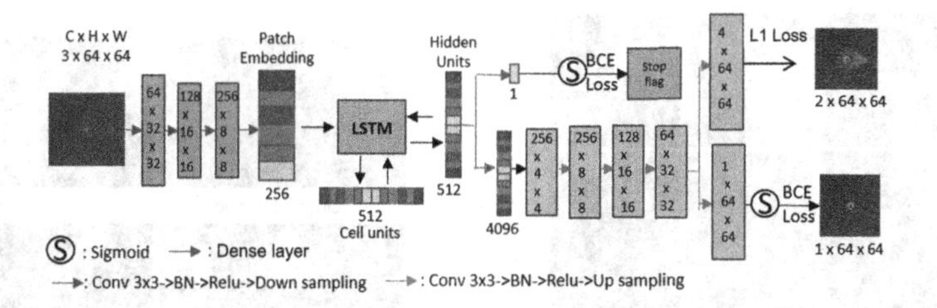

Fig. 4. Network structure for tracing module

The tracing module takes patches as input, and encode each patch into a 256 embedding vector by 3 downsampling blocks followed by a dense layer. The input of LSTM layer is the encoded embedding vector. At each step, the decoder outputs a stop flag, offset maps, and mask. We use a dense layer with a sigmoid activation function to predict the stop flag, which takes the hidden unit as input. As stop flag prediction is a classification problem, it takes an independent branch. The other branch uses a dense layer and decodes the hidden unit to a vector size of 4096, which is reshaped to $256 \times 4 \times 4$. The following layers include 3 conv3x3-bn-relu-upsampling blocks. The offset map prediction is a regression problem, and mask prediction is a binary classification problem. We split the current branch into two branches. The offset map prediction includes a conv3x3-bn-relu-upsampling block and outputs the offset map with a size of $2 \times 64 \times 64$. The offset map includes a horizontal offset channel and a vertical offset channel. The mask branch includes a conv3x3-bn-relu-upsampling-sigmoid block.

Predicting the Next Points. At each step, the decoder regresses offset maps where each pixel predicts a vector pointing to the center of the next patch. We use Hough Voting to decide the exact coordinates of the next center. Each pixel casts a vote to the next point, generating a heatmap of the number of votes for each pixel. The highest response point will be selected as the next center.

Loss Function. We use binary cross entropy (BCE) for the tip prediction. For the tracing module, we use BCE for stop flag and mask prediction and L_1 loss for offset prediction. The final loss for tracing module is

$$L_t = \sum_{t=1}^{t=T} (\lambda_1 L_{bce}(s_t, \hat{s_t}) + \lambda_2 L_{bce}(A, \hat{A}) + \lambda_3 Lreg(M, \hat{M}))$$

T is the number of steps for tracing, and s, A, M stand for stop flag, binary mask, and offset maps. $\lambda_1, \lambda_2, \lambda_3$ are the balance parameters and set as one.

Training and Inference. We use patches as input for training, and the labels are the corresponding offset map, binary mask, and stop flag. The offset map is generated by computing the distance vector between pixels in the current patch

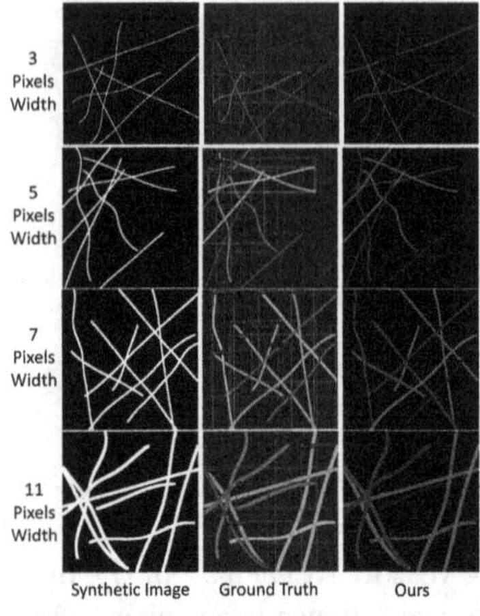

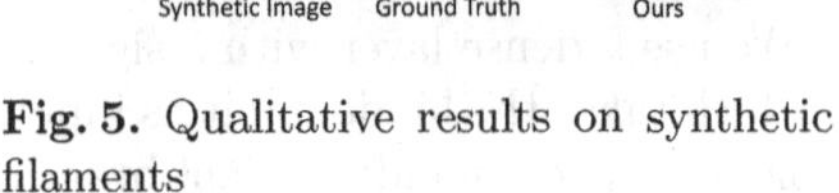

Fig. 5. Qualitative results on synthetic filaments

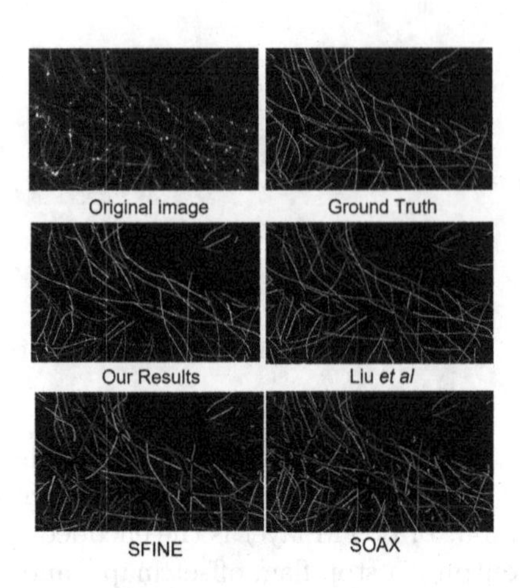

Fig. 6. Qualitative results on our microtubule dataset

to the center of the next patch. The tracing module only run Hough voting to predict the next point' coordinate and crop the next patch during inference.

3 Experiments

Our model is implemented with Pytorch [22] and trained on one RTX 2080 Ti GPU. We convert each instance into patch sequences with a step size of 30 pixels, and patch size of 64×64. We evaluate our approach on four datasets.

Table 1. Quantitative results on synthetic dataset. E, F, G, H are evaluated with our models trained on A, B, C, D respectively, which are highlighted with $\star$. AveInsNum: average number of instance per image. AveLength: average length of filaments in the dataset. AveInsNum and AveLength are reported as $Mean \pm std$.

Datasets					MRCNN [8]			Liu *et al.* [17]			Ours		
	Size	AveInsNum	Width	AveLength	AP	$AP_{0.5}$	$AP0.75$	AP	$AP_{0.5}$	$AP_{0.75}$	AP	$AP_{0.5}$	$AP_{0.75}$
A	256	6.94 ± 2.94	3	221.38 ± 40.89	0.00	0.00	0.00	0.48	0.67	0.56	**0.57**	**0.81**	**0.65**
B	256	6.61 ± 2.38	5	219.92 ± 39.95	0.01	0.02	0.01	0.45	0.71	0.51	**0.51**	**0.77**	**0.57**
C	256	6.86 ± 2.28	7	225.37 ± 38.44	0.07	0.19	0.04	0.39	0.66	**0.43**	**0.40**	**0.67**	0.42
D	256	6.76 ± 2.52	11	222.38 ± 39.54	0.19	0.46	0.14	0.30	0.41	0.31	**0.42**	**0.71**	**0.42**
E	512	9.28 ± 1.62	3	371.94 ± 106.86	0.00	0.00	0.00	0.45	0.69	0.52	**0.66**$\star$	**0.81**$\star$	**0.67**$\star$
F	512	9.16 ± 1.57	5	373.94 ± 104.29	0.01	0.02	0.01	0.47	0.68	0.54	**0.64**$\star$	**0.80**$\star$	**0.70**$\star$
G	512	9.44 ± 1.72	7	381.64 ± 103.56	0.03	0.25	0.13	0.48	0.69	0.55	**0.56**$\star$	**0.76**$\star$	**0.64**$\star$
H	512	9.75 ± 1.94	11	382.75 ± 106.81	0.17	0.44	0.10	0.42	0.60	0.44	**0.52**$\star$	**0.72**$\star$	**0.56**$\star$

Table 2. Quantitative Results

(a) Microtubule dataset

Method	AP	$AP_{0.5}$	$AP_{0.75}$	Time (min/image)
SOAX [26]	0.1860	0.3166	0.202	25
SFINE [29]	0.2222	0.4206	0.1894	5
Liu *et al.* [17]	0.3698	0.5774	0.3689	50
Ours	**0.3882**	**0.6040**	**0.3786**	**5**

(b) P. rubescens dataset

Image	Manual Count [28]	Zeder *et al.* [28]	Liu *et al.* [17]	ours
1	25.8 ± 1.5	27	25	25
2	18.5 ± 0.5	23	22	18
3	27.5 ± 1.5	27	26	27
4	44.0 ± 0	44	44	44
5	36.8 ± 0.4	40	37	36
6	21.8 ± 0.4	21	24	22
7	17.5 ± 1.1	18	18	17
Total	191.8 ± 3.6	200	196	189

(c) C. elegans dataset

Method	AP	$AP_{0.5}$	$AP_{0.75}$	Time (sec/image)
Semi-conv Ops [21]	0.569	0.885	0.661	–
M-RCNN [8]	0.559	0.865	0.641	–
Harmonic Emb. [14]	0.724	0.900	0.723	–
PatchPerPix [9]	**0.775**	**0.939**	**0.891**	13
Ours	0.745	0.935	0.828	**4**

3.1 Synthetic Dataset

We first create eight synthetic filament datasets. The statistics of generated datasets are shown in Table 1, and samples are shown in Fig. 5. Each dataset contains 1000 images with random filaments varying in widths. We split each dataset into 700, 200, 100 images for training, validating, and testing. We train our network on A–D (image size 256×256) for 20 epochs and evaluate our model on the their test set. We also directly evaluate the models trained with dataset A–D on dataset E–H (image size 512×512) respectively. We report the average precision (AP) in COCO evaluation criteria [16]. As shown in Table 1, our model trained with smaller size images achieves better results on the unseen datasets (E, F, G, H) with larger images and longer filaments. This is because our model learns how to trace filaments, and longer filaments do not affect the performance of our model. Also, the lower density of filament in E–H making it easier for our model to trace filaments. In addition, thicker filaments create larger overlapping areas and make it harder to separate the filaments. Therefore, the performance decreases from dataset A to D and E to H.

As shown in Table 1, MRCNN [8] achieves zero AP for filaments with a width below five, as segmentation is performed at strides of eight. Our approach outperforms Liu *et al.* [17] in all metrics except $AP_{0.75}$ of dataset C.

3.2 Microtubule Dataset

We annotated 15 microscopic images of microtubules with a size of 1376×1504. The number of filaments per image is 631 ± 167, and the length of filaments is 104 ± 96. We use a modified U-net [18,24] to obtain the binary segmentation. As the average width of microtubules is five, we directly evaluate the microtubule dataset with model trained on synthetic dataset B (see table 1). We compare our approach against SOAX [26], SFINE [29], and deep learning method in [17].

Figure 1 shows qualitative results in full size, and Fig. 6 presents qualitative comparison with detailed area. Our method and Liu *et al.* [17]'s approach can better extract long and crossing filaments. SOAX [26] and SFINE [29]'s break-regroup strategies struggle to regroup segments at intersections and create fragments. Table 2a presents the quantitative comparisons with AP. Our approach has shown a better performance than [17] regarding process time and accuracy.

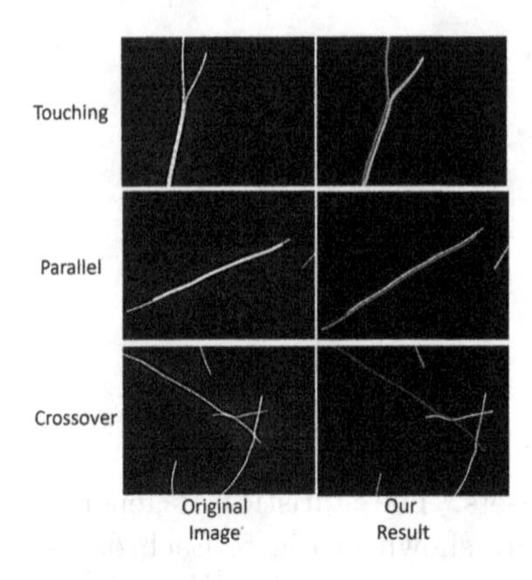

Fig. 7. Qualitative results on P. rubescens dataset.

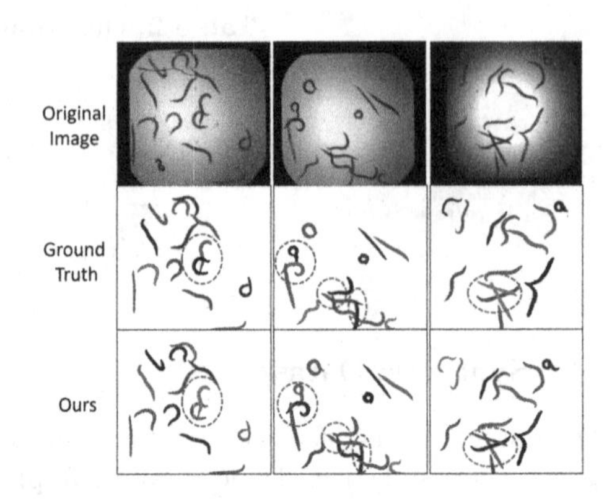

Fig. 8. Qualitative results on C.elegans dataset. Red circle highlight complicated overlapping area (Color figure online)

3.3 P. Rubescens Dataset

P. rubescens is a type of filamentous cyanobacteria, and Zeder *et al.* [28] provide a dataset (Fig. 7) of seven 5000×5000 microscopic images of P. rubescens. We apply our model trained with synthetic dataset A (see Table 1) to the binary predictions and follow the evaluation scheme in [28] by comparing the quantity per image. Table 2b shows our result is close to the manual count.

3.4 C. Elegans Dataset

We further investigate our model's performance on the C. elegans roundworm dataset (Fig. 8) from the Broad Bioimage Benchmark Collection [20]. The dataset contains 100 696×520 images with an average of 30 roundworms per image. Different from P. rubescens and microtubules, roundworms are much thicker and shorter. We convert each instance in the training set into a sequence of points with a step size of 30 and generate the corresponding 64×64 patches for training. The network is trained and evaluated following the set up in [14,21].

Table 2c shows the quantitative comparison between our approach and previous methods [8,9,14,21]. Our method achieves comparable $AP_{0.5}$ to SOTA, and our method runs 9 s faster than SOTA. Red circles in Fig. 8 show our approach handles complex crossover areas.

4 Conclusion

We present a novel method for filament extraction by transforming the instance segmentation problem to a sequence modeling problem. Our method comprises of a sequential encoder-decoder framework to imitate humans extracting filaments and address the challenges brought by filaments' properties, including crossover, spanning and non-rigidity. The experiments show that our method can achieve better or comparable results on filament datasets from different domains. Our method can alleviate the data shortage problem as our models trained on synthetic dataset achieve a better performance on microtubules and P. rubescens dataset. We also train and evaluate our model on C. elegans dataset, achieving comparable results with thicker and shorter filaments. Our method exhibits limitations in tracing "Y"-shaped junctions due to the limited directional information in 2D images. Future work will focus on extending the current method to 3D data.

Acknowledgement. This work was supported by the National Institute of General Medical Sciences grant (R01 GM097587) from the National Institutes of Health. Microscopy equipment was acquired with an NIH-shared instrumentation grant (S10 OD016361) and, access was supported by the NIH-NIGMS (P20 GM103446 and P20 GM139760) and the State of Delaware.

References

1. Chen, H., Qi, X., Yu, L., Heng, P.A.: DCAN: deep contour-aware networks for accurate gland segmentation. In: Proceedings of the IEEE Conference on Computer Vision and Pattern Recognition, pp. 2487–2496 (2016)
2. Chen, L., Strauch, M., Merhof, D.: Instance segmentation of biomedical images with an object-aware embedding learned with local constraints. In: Shen, D., et al. (eds.) MICCAI 2019. LNCS, vol. 11764, pp. 451–459. Springer, Cham (2019). https://doi.org/10.1007/978-3-030-32239-7_50
3. Chen, Y., Wang, Z., Peng, Y., Zhang, Z., Yu, G., Sun, J.: Cascaded pyramid network for multi-person pose estimation. In: Proceedings of the IEEE Conference on Computer Vision and Pattern Recognition, pp. 7103–7112 (2018)
4. Dai, J., He, K., Li, Y., Ren, S., Sun, J.: Instance-sensitive fully convolutional networks. In: Leibe, B., Matas, J., Sebe, N., Welling, M. (eds.) ECCV 2016. LNCS, vol. 9910, pp. 534–549. Springer, Cham (2016). https://doi.org/10.1007/978-3-319-46466-4_32
5. De Brabandere, B., Neven, D., Van Gool, L.: Semantic instance segmentation with a discriminative loss function. arXiv preprint arXiv:1708.02551 (2017)
6. Gao, N., et al.: SSAP: single-shot instance segmentation with affinity pyramid. In: Proceedings of the IEEE/CVF International Conference on Computer Vision, pp. 642–651 (2019)
7. Hariharan, B., Arbeláez, P., Girshick, R., Malik, J.: Simultaneous detection and segmentation. In: Fleet, D., Pajdla, T., Schiele, B., Tuytelaars, T. (eds.) ECCV 2014. LNCS, vol. 8695, pp. 297–312. Springer, Cham (2014). https://doi.org/10.1007/978-3-319-10584-0_20
8. He, K., Gkioxari, G., Dollár, P., Girshick, R.: Mask R-CNN. In: 2017 IEEE International Conference on Computer Vision (ICCV), pp. 2980–2988. IEEE (2017)
9. Hirsch, P., Mais, L., Kainmueller, D.: PatchPerPix for instance segmentation. arXiv preprint arXiv:2001.07626 (2020)
10. Hochreiter, S., Schmidhuber, J.: Long short-term memory. Neural Comput. **9**(8), 1735–1780 (1997)

11. Januszewski, M., et al.: High-precision automated reconstruction of neurons with flood-filling networks. Nat. Methods **15**(8), 605–610 (2018)
12. Januszewski, M., Maitin-Shepard, J., Li, P., Kornfeld, J., Denk, W., Jain, V.: Flood-filling networks. arXiv preprint arXiv:1611.00421 (2016)
13. Ke, L., Tai, Y.W., Tang, C.K.: Deep occlusion-aware instance segmentation with overlapping bilayers. In: Proceedings of the IEEE/CVF Conference on Computer Vision and Pattern Recognition, pp. 4019–4028 (2021)
14. Kulikov, V., Lempitsky, V.: Instance segmentation of biological images using harmonic embeddings. In: Proceedings of the IEEE/CVF Conference on Computer Vision and Pattern Recognition, pp. 3843–3851 (2020)
15. Lin, T.Y., Dollár, P., Girshick, R., He, K., Hariharan, B., Belongie, S.: Feature pyramid networks for object detection (2017)
16. Lin, T.-Y., et al.: Microsoft COCO: common objects in context. In: Fleet, D., Pajdla, T., Schiele, B., Tuytelaars, T. (eds.) ECCV 2014. LNCS, vol. 8693, pp. 740–755. Springer, Cham (2014). https://doi.org/10.1007/978-3-319-10602-1_48
17. Liu, Y., Kolagunda, A., Treible, W., Nedo, A., Caplan, J., Kambhamettu, C.: Intersection to overpass: instance segmentation on filamentous structures with an orientation-aware neural network and terminus pairing algorithm. In: Proceedings of the IEEE/CVF Conference on Computer Vision and Pattern Recognition Workshops, pp. 125–133 (2019)
18. Liu, Y., et al.: Densely connected stacked U-network for filament segmentation in microscopy images. In: Leal-Taixé, L., Roth, S. (eds.) ECCV 2018. LNCS, vol. 11134, pp. 403–411. Springer, Cham (2019). https://doi.org/10.1007/978-3-030-11024-6_30
19. Liu, Y., et al.: Affinity derivation and graph merge for instance segmentation. In: Ferrari, V., Hebert, M., Sminchisescu, C., Weiss, Y. (eds.) ECCV 2018. LNCS, vol. 11207, pp. 708–724. Springer, Cham (2018). https://doi.org/10.1007/978-3-030-01219-9_42
20. Ljosa, V., Sokolnicki, K.L., Carpenter, A.E.: Annotated high-throughput microscopy image sets for validation. Nat. Methods **9**(7), 637–637 (2012)
21. Novotny, D., Albanie, S., Larlus, D., Vedaldi, A.: Semi-convolutional operators for instance segmentation. In: Ferrari, V., Hebert, M., Sminchisescu, C., Weiss, Y. (eds.) ECCV 2018. LNCS, vol. 11205, pp. 89–105. Springer, Cham (2018). https://doi.org/10.1007/978-3-030-01246-5_6
22. Paszke, A., et al.: Automatic differentiation in PyTorch (2017)
23. Ren, M., Zemel, R.S.: End-to-end instance segmentation with recurrent attention. In: Proceedings of the 2017 IEEE Conference on Computer Vision and Pattern Recognition (CVPR), Honolulu, HI, USA, pp. 21–26 (2017)
24. Ronneberger, O., Fischer, P., Brox, T.: U-Net: convolutional networks for biomedical image segmentation. In: Navab, N., Hornegger, J., Wells, W.M., Frangi, A.F. (eds.) MICCAI 2015. LNCS, vol. 9351, pp. 234–241. Springer, Cham (2015). https://doi.org/10.1007/978-3-319-24574-4_28
25. Salvador, A., et al.: Recurrent neural networks for semantic instance segmentation. arXiv preprint arXiv:1712.00617 (2017)
26. Xu, T., et al.: SOAX: a software for quantification of 3D biopolymer networks. Sci. Rep. **5**, 9081 (2015)
27. Yi, J., et al.: Multi-scale cell instance segmentation with keypoint graph based bounding boxes. In: Shen, D., et al. (eds.) MICCAI 2019. LNCS, vol. 11764, pp. 369–377. Springer, Cham (2019). https://doi.org/10.1007/978-3-030-32239-7_41

28. Zeder, M., Van den Wyngaert, S., Köster, O., Felder, K.M., Pernthaler, J.: Automated quantification and sizing of unbranched filamentous cyanobacteria by model-based object-oriented image analysis. Appl. Environ. Microbiol. **76**(5), 1615–1622 (2010)
29. Zhang, Z., Nishimura, Y., Kanchanawong, P.: Extracting microtubule networks from superresolution single-molecule localization microscopy data. Mol. Biol. Cell **28**(2), 333–345 (2017)

BigFUSE: Global Context-Aware Image Fusion in Dual-View Light-Sheet Fluorescence Microscopy with Image Formation Prior

Yu Liu[1], Gesine Müller[2], Nassir Navab[1,3], Carsten Marr[4], Jan Huisken[2(✉)], and Tingying Peng[5(✉)]

[1] Technical University of Munich, Munich, Germany
[2] Georg-August-University Goettingen, Goettingen, Germany
[3] Johns Hopkins University, Baltimore, USA
[4] Institute of AI for Health, Helmholtz Munich - German Research Center for Environmental Health, Neuherberg, Germany
[5] Helmholtz AI, Helmholtz Munich - German Research Center for Environmental Health, Neuherberg, Germany
tingying.peng@helmholtz-muenchen.de

Abstract. Light-sheet fluorescence microscopy (LSFM), a planar illumination technique that enables high-resolution imaging of samples, experiences "defocused" image quality caused by light scattering when photons propagate through thick tissues. To circumvent this issue, dual-view imaging is helpful. It allows various sections of the specimen to be scanned ideally by viewing the sample from opposing orientations. Recent image fusion approaches can then be applied to determine in-focus pixels by comparing image qualities of two views locally and thus yield spatially inconsistent focus measures due to their limited field-of-view. Here, we propose BigFUSE, a global context-aware image fuser that stabilizes image fusion in LSFM by considering the global impact of photon propagation in the specimen while determining focus-defocus based on local image qualities. Inspired by the image formation prior in dual-view LSFM, image fusion is considered as estimating a focus-defocus boundary using Bayes' Theorem, where (*i*) the effect of light scattering onto focus measures is included within *Likelihood*; and (*ii*) the spatial consistency regarding focus-defocus is imposed in *Prior*. The expectation-maximum algorithm is then adopted to estimate the focus-defocus boundary. Competitive experimental results show that BigFUSE is the first dual-view LSFM fuser that is able to exclude structured artifacts when fusing information, highlighting its abilities of automatic image fusion.

Keywords: Light-sheet Fluorescence Microscopy (LSFM) · Multi-View Image Fusion · Bayesian

H. Greenspan et al. (Eds.): MICCAI 2023, LNCS 14227, pp. 646–655, 2023.
https://doi.org/10.1007/978-3-031-43993-3_62

1 Introduction

Light-sheet fluorescence microscopy (LSFM), characterized by orthogonal illumination with respect to detection, provides higher imaging speeds than other light microscopies, e.g., confocal microscopy, via gentle optical sectioning [5,14,15], which makes it well-suited for whole-organism studies [11]. At macroscopic scales, however, light scattering degrade image quality. It leads to images from deeper layers of the sample being of worse quality than from tissues close to the illumination source [6,17]. To overcome the negative effect of photon propagation, dual-view LSFM is introduced, in which the sample is sequentially illuminated from opposing directions, and thus portions of the specimen with inferior quality in one view will be better in the other [16] (Fig. 1a). Thus, image fusion methods that combine information from opposite views into one volume are needed.

To realize dual-view LSFM fusion, recent pipelines adapt image fusers for natural image fusion to weigh between views by comparing the local clarity of images [16,17]. For example, one line of research estimates focus measures in a transformed domain, e.g., wavelet [7] or contourlet [18], such that details with various scales can be considered independently [10]. However, the composite result often exhibits global artifacts [8]. Another line of studies conducts fusion in the image space, with pixel-level focus measures decided via local block-based representational engineering such as multi-scale weighted multi-scale weighted gradient [20] and SIFT [9]. Unfortunately, spatially inconsistent focus measures are commonly derived for LSFM, considering the sparse structures of the biological sample involved by the limited field-of-view (FOV).

Apart from limited FOV, another obstacle that hinders the adoption of natural image fusion methods into dual-view LSFM is the inability to distinguish sample structures from structural artifacts [1]. For example, ghost artifacts, surrounding the sample as a result of scattered illumination light [3], can be observed, as it only appears in regions far from the light source after light travels through scattering tissues. Yet, when ghosts appear in one view, the same region in the opposite view would be background, i.e., no signal. Thus, ghosts will be erroneously transferred to the result by conventional fusion studies, as they are considered as owning richer information than its counterpart in the other view.

Here, we propose BigFUSE to realize spatially consistent image fusion and exclude ghost artifacts. Main contributions are summarized as follows:

- BigFUSE is the first effort to think of dual-view LSFM fusion using Bayes, which maximizes the conditional probability of fused volume regarding image clarity, given the image formation prior of opposing illumination directions.
- The overall focus measure along illumination is modeled as a joint consideration of both global light scattering and local neighboring image qualities in the contourlet domain, which, together with the smoothness of focus-defocus, can be maximized as *Likelihood* and *Prior* in Bayesian.
- Aided by a reliable initialization, BigFUSE can be efficiently optimized by utilizing expectation-maximum (EM) algorithm.

2 Methods

An illustration of BigFUSE for dual-view LSFM fusion is given in Fig. 1. First, pixel-level focus measures are derived for two opposing views separately, using nonsubsampled contourlet transform (NSCT) (Fig. 1b). Pixel-wise photon propagation maps in tissue are then determined along light sheet via segmentation (Fig. 1c). The overall focus measures are thus modeled as the inverse of integrated photon scattering along illumination conditioned on the focus-defocus change, i.e., *Likelihood*, whereas the smoothness of focus-defocus is ensured via *Prior*. Finally, the focus-defocus boundary is optimized via EM (Fig. 1d).

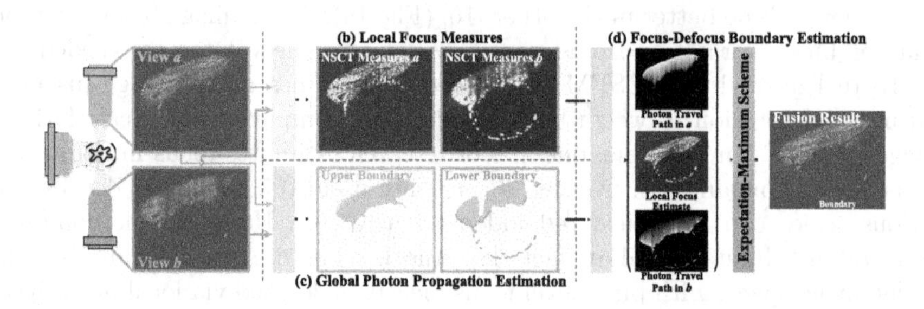

Fig. 1. An overview of BigFUSE (see text for explanation)

2.1 Revisiting Dual-View LSFM Fusion Using Bayes

BigFUSE first rethinks dual-view LSFM fusion from a Bayesian perspective, that is, the conditional probability of fused volume in terms of "in-focusness", is given on not only a pair of image inputs, but also prior knowledge that these two images are illiminated from opposing orientations respectively:

$$\mathrm{Y} = \underset{\mathrm{Y}}{\operatorname{argmax}}\, p(\mathrm{Y}|(\mathrm{X}^a, \mathrm{X}^b), \mathcal{P}) = \underset{\mathrm{Y}}{\operatorname{argmax}} \{p(\mathrm{Y}|\mathcal{P}) p((\mathrm{X}^a, \mathrm{X}^b)|\mathrm{Y}, \mathcal{P})\} \tag{1}$$

where $\mathrm{Y} \in \mathbb{R}^{M \times N}$ is our predicted fusion with minimal light scattering effect. X^a and X^b are two image views illuminated by light source a and b, respectively. We choose $\mathrm{Y} \in \{\mathrm{X}^a, \mathrm{X}^b\}$ depending on their competitive image clarity at each pixel. Priors $\mathcal{P}$ denote our empirical favor of X^a against X^b at each pixel if photons travel through fewer scattering tissues from source a than b, and vice versa. Due to the non-positive light scattering effect along illumination path, there is only one focus-defocus change per column for dual-view LSFM in Fig. 1a. Thus, fusion is equivalent to estimating a focus-defocus boundary ω defined as a function associating focus-defocus changes to column indexes:

$$\omega = \underset{\omega}{\operatorname{argmax}}\, p(\omega) p((\mathrm{X}^a, \mathrm{X}^b)|\omega) \tag{2}$$

which can be further reformulated by logarithm:

$$\omega = \underset{\omega}{\operatorname{argmax}}\, p(\omega)p((\mathrm{X}^a, \mathrm{X}^b)|\omega) = \underset{\omega \in \{1,2,\ldots,N\}}{\operatorname{argmax}} \log(p(\omega) \prod_{i=1}^{N} p((\mathrm{X}^a_{:,i}, \mathrm{X}^b_{:,i})|\omega_i))$$
$$= \underset{\omega \in \{1,2,\ldots,N\}}{\operatorname{argmax}} \left\{ \log(p(\omega)) + \sum_{i=1}^{N} \log(p((\mathrm{X}^a_{:,i}, \mathrm{X}^b_{:,i})|\omega_i)) \right\} \quad (3)$$

where ω_i denotes the focus-defocus changeover at i-th column, $\mathrm{X}_{:,i}$ is the i-th column of X. Next, estimating ω is decomposed into: (i) define the column-wise image clarity, i.e., *log-likelihood* $\log(p((\mathrm{X}^a_{:,i}, \mathrm{X}^b_{:,i})|\omega_i))$; ($ii$) consider the belief on a spatially smooth focus-defocus boundary, namely *log-prior* $\log(p(\omega))$.

2.2 Image Clarity Characterization with Image Formation Prior

In LSFM, log-likelihood $\log(p((\mathrm{X}^a_{:,i}, \mathrm{X}^b_{:,i})|\omega_i))$ can be interpreted as the probability of observing $(\mathrm{X}^a_{:,i}, \mathrm{X}^b_{:,i})$ given the hypothesis that the focus-defocus change is determined as ω_i in the i-th column:

$$\underset{\omega_i}{\operatorname{argmax}} \left\{ \log(p((\mathrm{X}^a_{:,i}, \mathrm{X}^b_{:,i})|\omega_i)) \right\} = \underset{\omega_i}{\operatorname{argmax}} \left\{ c(\mathrm{X}^a_{1:\omega_i,i} \oplus \mathrm{X}^b_{\omega_i+1:M,i}) \right\} \quad (4)$$

where $\oplus$ is a concatenation, $c(\bullet)$ is the column-wise image clarity to be defined.

Estimating Pixel-Level Image Clarity in NSCT. To define $c(\bullet)$, BigFUSE first uses NSCT, a shift-invariant image representation technique, to estimate pixel-level focus measures by characterizing salient image structures [18]. Specifically, NSCT coefficients S^a and S^b are derived for two opposing LSFM views, where $\mathrm{S} = \{\mathrm{S}_{i_0}, \mathrm{S}_{i,l} | (1 \leq i \leq i_0, 1 \leq l \leq 2^{l_i})\}$, S_{i_0} is the lowpass coefficient at the coarsest scale, $\mathrm{S}_{i,l}$ is the bandpass directional coefficient at i-th scale and l-th direction. Local image clarity is then projected from S^a and S^b [18]:

$$\mathrm{F}_{i,l} = \mathrm{R}_{i,l} \times \mathrm{D}\sigma_i = \frac{|\mathrm{S}_{i,l}|}{\bar{\mathrm{S}}_{i_0}} \times \sqrt{\frac{1}{2^{l_i}} \sum_{r=1}^{2^{l_i}} [|S_{i,r}| - \bar{S}_i]^2}, \ \ \bar{\mathrm{S}}_i = \frac{1}{2^{l_i}} \sum_{r=1}^{2^{l_i}} |\mathrm{S}_{i,r}| \quad (5)$$

where $\mathrm{R}_{i,l}$ is local directional band-limited image contrast and $\bar{\mathrm{S}}_{i_0}$ is the smoothed image baseline, whereas $\mathrm{D}\sigma_i$ highlights image features that are distributed only on a few directions, which is helpful to exclude noise [18]. As a result, pixel-level image clarity $\mathrm{F} = \sum_{j=1}^{j_0} \sum_{l=1}^{2^{\wedge} l_j} \mathrm{F}_{j,l}$ is quantified for respective LSFM view.

Reweighting Image Clarity Measures by Photon Traveling Path. Pixel-independent focus measures may be adversely sensitive to noise, due to the limited receptive field when characterizing local image clarities. Thus, BigFUSE proposes to integrate pixel-independent image clarity measures along columns by taking into consideration the photon propagation in depth. Specifically, given a pair of pixels $(\mathrm{X}^a_{m,n}, \mathrm{X}^b_{m,n})$, $\mathrm{X}^a_{m,n}$ is empirically more in-focus than $\mathrm{X}^a_{m,n}$, if photons travel through fewer light-scattering tissues from illumination objective a than from b to get to position (m, n), and vice versa. Therefore, BigFUSE defines column-level image clarity measures as:

$$c(\mathrm{X}^a_{1:\omega_i,i} \oplus \mathrm{X}^b_{\omega_i+1:M,i}) = \sum\nolimits_{j=1}^{\omega_i} \mathrm{A}_{j,i}\mathrm{F}^a_{j,i} + \sum\nolimits_{j=\omega_i+1}^{M} \mathrm{A}_{j,i}\mathrm{F}^b_{j,i} \tag{6}$$

where $\mathrm{A}_{:,i}$ is to model the image deterioration due to light scattering. To visualize photon traveling path, BigFUSE uses OTSU thresholding for foreground segmentation (followed by AlphaShape to generalize bounding polygons), and thus obtains sample boundary, i.e., incident points of light sheet, which we refer to as p^u and p^l for opposing views a and b respectively. Since the derivative of $\mathrm{A}_{:,i}$ implicitly refers to the spatially varying index of refraction within the sample, which is nearly impossible to accurately measure from the physics perspective, we model it using a piecewise linear model, without loss of generality:

$$\mathrm{A}_{j,i} = \begin{cases} 1 - 0.5 \times |j - p^u| / \omega_i, & j < \omega_i \\ 0.5 + 0.5 \times |j - p^l| / (M - \omega_i), & j \geq \omega_i \\ 0, & \text{otherwise} \end{cases} \tag{7}$$

As a result, $\log(p((\mathrm{X}^a_{:,i}, \mathrm{X}^b_{:,i})|\omega_i))$ is obtained as summed pixel-level image clarity measures with integral factors conditioned on photon propagation in depth.

2.3 Least Squares Smoothness of Focus-Defocus Boundary

With log-likelihood $\log(p((\mathrm{X}^a_{:,i}, \mathrm{X}^b_{:,i})|\omega_i))$ considering the focus-defocus consistency along illuminations using image formation prior in LSFM, BigFUSE then ensures consistency across columns in $p(\omega)$. Specifically, the smoothness of ω is characterized as a window-based polynomial fitness using linear least squares:

$$\log(p(\omega)) = \sum\nolimits_{i=1}^{N} \log(p(\omega_i)) = \sum\nolimits_{i=1}^{N} \|\omega_{i-s:i+s} - \hat{\upsilon}_i\|^2 \tag{8}$$

where $\hat{\upsilon}_i = c_i\Omega_i = [c_{i,0}, c_{i,1}, \dots, c_{i,Q}][\boldsymbol{\omega}^{\mathrm{T}}_{i-s}, \boldsymbol{\omega}^{\mathrm{T}}_{i-s+1}, \dots, \boldsymbol{\omega}^{\mathrm{T}}_{i+s}]$, $\boldsymbol{\omega}_i = [\omega_i^Q, \omega_i^{Q-1}, \dots, 1]$, c_i is the parameters to be estimated, the sliding window is with a size of $2s + 1$.

2.4 Focus-Defocus Boundary Inference via EM

Finally, in order to estimate the ω together with the fitting parameter c, BigFUSE reformulates the posterior distribution in Eq. (2) as follows:

$$\{\hat{\omega}, \hat{c}\} = \underset{\omega,c}{\operatorname{argmax}} \left\{ \begin{array}{l} \sum_{j=1}^{N} \left\{ \sum_{i=1}^{\omega_j} \mathrm{A}_{i,j}\mathrm{F}^a_{i,j} + \sum_{i=\omega_j+1}^{M} \mathrm{A}_{i,j}\mathrm{F}^b_{i,j} \right\} \\ +\lambda\{\sum_{i=1}^{N} \|\omega_{i-s:i+s} - \hat{\upsilon}_i\|^2\} \end{array} \right\} \tag{9}$$

where λ is the trade-off parameter. Here, BigFUSE alternates the estimations of ω, and c, and iterates until the method converges. Specifically, given $c^{(n)}$ for the n-th iteration, $\omega_i^{(n+1)}$ is estimated by maximizing (E-step):

$$\omega_i^{(n+1)} = \underset{\omega_i}{\text{argmax}} \left\{ \sum\nolimits_{j=1}^{\omega_i} \mathrm{A}_{j,i}\mathrm{F}_{j,i}^a + \sum\nolimits_{j=\omega_i+1}^{M} \mathrm{A}_{j,i}\mathrm{F}_{j,i}^b + \lambda \|\omega_{i-s:i+s} - \hat{\upsilon}_i\|^2 \right\} \tag{10}$$

which can be solved by iterating over $\{i|1 \leq i \leq M\}$. BigFUSE then updates $c_i^{(n+1)}$ based on least squares estimation:

$$c_i^{(n+1)} = (\Omega^{\mathrm{T}}\Omega)^{-1}\Omega^{\mathrm{T}}\omega_{i-s:i+s} \tag{11}$$

Table 1. BigFUSE achieves best quantitative results on synthetic blur.

	DWT [16]	NSCT [18]	dSIFT [9]	BF [19]	$\mathcal{S}(\bullet)$	$\mathcal{P}(\bullet)$	**BigFUSE**
EMSE ($\times 10^{-5}$)	6.81 ±0.72	4.66 ±0.56	5.17 ±0.72	1.55 ±0.25	1.43 ±0.34	0.96 ±0.08	**0.94 ±0.09**
SSIM	0.974 ±0.02	0.996 ±0.03	0.93 ±0.01	0.994 ±0.03	0.993 ±0.02	0.993 ±0.02	**0.998 ±0.01**

Additionally, A^{n+1} is updated based on Eq. (7) subject to $\omega^{(n+1)}$ (M-step). BigFUSE proposes to initialize ω based on F^a and F^b:

$$\omega_i^{(0)} = p^u + |\Phi|, \quad \Phi = \{j|\mathrm{F}_{j,i}^a > \mathrm{F}_{j,i}^b, p^u \leq j \leq p^l\} \tag{12}$$

where $|\Phi|$ denotes the total number of elements in Φ.

2.5 Competitive Methods

We compare BigFUSE to four baseline methods: (*i*) DWT [16]: a multi-resolution image fusion method using discrete wavelet transform (DWT); (*ii*) NSCT [18]: another multi-scale image fuser but in the NSCT domain; (*iii*) dSIFT [9]: a dense SIFT-based focus estimator in the image space; (*iv*) BF [19]: a focus-defocus boundary detection method that considers region consistency in focus; and two BigFUSE variations: (*v*) $\mathcal{S}(\bullet)$: built by disabling smooth constraint; (*vi*) $\mathcal{P}(\bullet)$: formulated by replacing the weighted summation of pixel-level clarity measures for overall characterization, by a simple average.

To access the blind image fusion performance, we adopt three fusion quality metrics, Q_{mi} [4], Q_g [12] and Q_s [13]. Specifically, Q_{mi}, Q_g and Q_s use mutual information, gradient or image quality index to quantify how well the information or features of the inputs are transferred to the result, respectively. In the simulation studies where ground truth is available, mean square error (EMSE) and structural similarity index (SSIM) are used for quantification.

3 Results and Discussion

3.1 Evaluation on LSFM Images with Synthetic Blur

We first evaluate BigFUSE in fusing dual-view LSFM blurred by simulation. Here, we blur a mouse brain sample collected in [2] with spatially varying Gaussian filter for thirty times, which is chemically well-cleared and thus provides an almost optimal ground truth for simulation, perform image fusion, and compare the results to the ground truth. BigFUSE achieves the best EMSE and SSIM, statistically surpassing other approaches (Table 1, $p < 0.05$ using Wilcoxon signed-rank test). Only BigFUSE and BF realize information fusing without damaging original images with top two EMSE. In comparison, DWT, NSCT and dSIFT could distort the original signal when fusing (green box in Fig. 2).

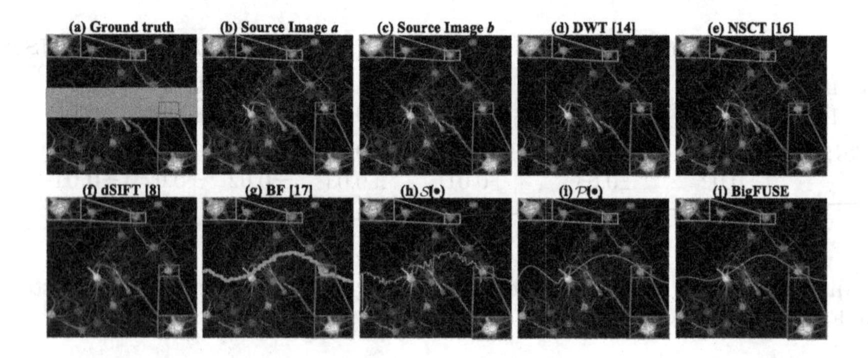

Fig. 2. Visualization of fusion quality with respect to ground-truth. Focus-defocus boundary is given in blue curve, if applicable. (Color figure online)

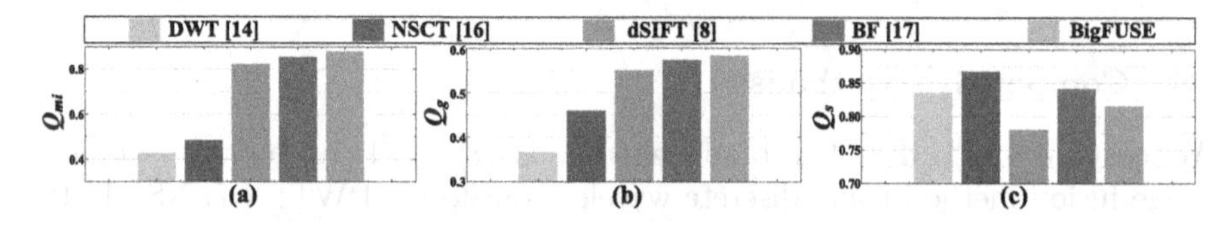

Fig. 3. Comparison between BigFUSE and baselines on Q_{mi}, Q_g and Q_s.

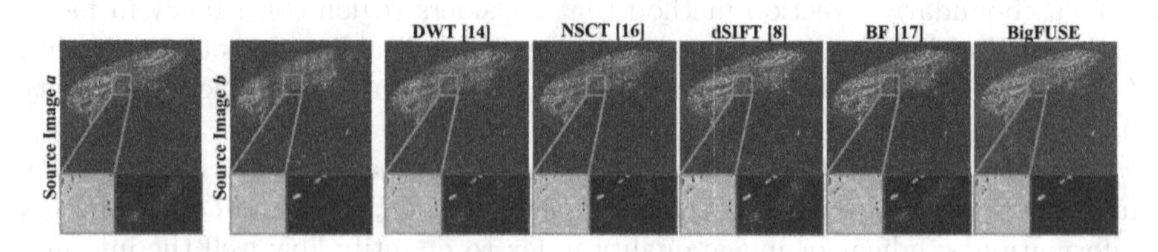

Fig. 4. Visualization of fusing a zebrafish embryo.

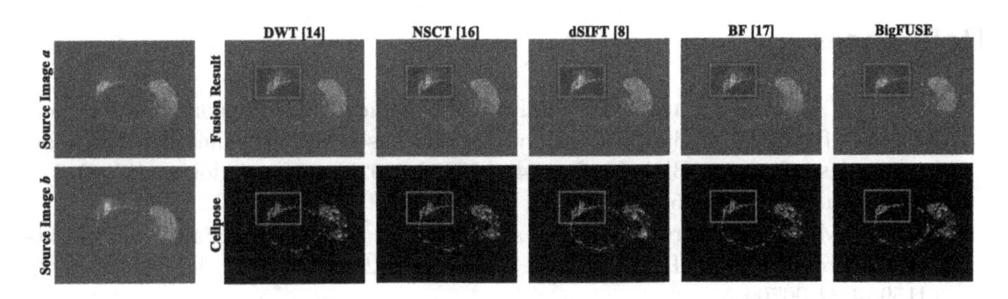

Fig. 5. Visualization of Cellpose result.

3.2 Evaluation on LSFM Images with Real Blur

BigFUSE is then evaluated against baseline methods on real dual-view LSFM. A large sample volume, zebrafish embryo ($282 \times 2048 \times 2048$ for each view), is imaged using a Flamingo Light Sheet Microscope. BigFUSE takes roughly nine minutes to process this zebrafish embryo, using a T4 GPU with 25 GB system RAM and 15 GB GPU RAM. In Fig. 4, inconsistent boundary is detected by BF, while methods like DWT and NSCT generate structures that do not exist in either input. Moreover, only BigFUSE can exclude ghost artifact from the result (red box), as BigFUSE is the only pipeline that considers image formation prior. Additionally, we demonstrate the impact of bigFUSE on a specific downstream task in Fig. 5, i.e., segmentation by Cellpose. Only the fusion result provided by bigFUSE allows reasonable predicted cell pose, given that ghosting artifacts are excluded dramatically. This explains why the Q_s in Fig. 3 is suboptimal, since BigFUSE do not allow for the transmission of structural ghosts to the output.

4 Conclusion

In this paper, we propose BigFUSE, a image fusion pipline with image formation prior. Specifically, image fusion in dual-view LSFM is revisited as inferring a focus-defocus boundary using Bayes, which is essential to exclude ghost artifacts. Furthermore, focus measures are determined based on not only pure image representational engineering in NSCT domain, but also the empirical effects of photon propagation in depth embeded in the opposite illumination directions in dual-view LSFM. BigFUSE can be efficiently optimized using EM. Both qualitative and quantitative evaluations show that BigFUSE surpasses other state-of-the-art LSFM fusers by a large margin. BigFUSE will be made accessible.

Acknowledgements. Y.L. is supported by the China Scholarship Council (No.20210 6020050).

References

1. Azam, M.A., et al.: A review on multimodal medical image fusion: compendious analysis of medical modalities, multimodal databases, fusion techniques and quality metrics. Comput. Biol. Med. **144**, 105253 (2022). https://doi.org/10.1016/j.compbiomed.2022.105253
2. Dean, K.M., et al.: Isotropic imaging across spatial scales with axially swept light-sheet microscopy. Nat. Protoc. **17**(9), 2025–2053 (2022). https://doi.org/10.1038/s41596-022-00706-6
3. Fahrbach, F.O., Simon, P., Rohrbach, A.: Microscopy with self-reconstructing beams. Nat. Photonics **4**(11), 780–785 (2010). https://doi.org/10.1038/nphoton.2010.204
4. Hossny, M., Nahavandi, S., Creighton, D.: Comments on 'information measure for performance of image fusion' (2008)
5. Keller, P.J., Stelzer, E.H.: Quantitative in vivo imaging of entire embryos with digital scanned laser light sheet fluorescence microscopy. Curr. Opin. Neurobiol. **18**(6), 624–632 (2008). https://doi.org/10.1016/j.conb.2009.03.008
6. Krzic, U., Gunther, S., Saunders, T.E., Streichan, S.J., Hufnagel, L.: Multiview light-sheet microscope for rapid in toto imaging. Nat. Meth. **9**(7), 730–733 (2012). https://doi.org/10.1038/nmeth.2064
7. Lewis, J.J., O'Callaghan, R.J., Nikolov, S.G., Bull, D.R., Canagarajah, N.: Pixel- and region-based image fusion with complex wavelets. Inf. Fusion **8**(2), 119–130 (2007). https://doi.org/10.1016/j.inffus.2005.09.006
8. Li, S., Kang, X., Hu, J.: Image fusion with guided filtering. IEEE Trans. Image Process. **22**(7), 2864–2875 (2013). https://doi.org/10.1109/TIP.2013.2244222
9. Liu, Y., Liu, S., Wang, Z.: Multi-focus image fusion with dense sift. Inf. Fusion **23**, 139–155 (2015). https://doi.org/10.1016/j.inffus.2014.05.004
10. Liu, Y., Wang, L., Cheng, J., Li, C., Chen, X.: Multi-focus image fusion: a survey of the state of the art. Inf. Fusion **64**, 71–91 (2020). https://doi.org/10.1016/j.inffus.2020.06.013
11. Medeiros, G., et al.: Confocal multiview light-sheet microscopy. Nat. Commun. **6**(1), 8881 (2015). https://doi.org/10.1038/ncomms9881
12. Petrovic, V., Xydeas, C.: Objective image fusion performance characterisation. In: Tenth IEEE International Conference on Computer Vision (ICCV 2005), Volume 1, vol. 2, pp. 1866–1871. IEEE (2005). https://doi.org/10.1109/ICCV.2005.175
13. Piella, G., Heijmans, H.: A new quality metric for image fusion. In: Proceedings 2003 International Conference on Image Processing (Cat. No. 03CH37429). vol. 3, pp. III-173. IEEE (2003). https://doi.org/10.1109/ICIP.2003.1247209
14. Power, R.M., Huisken, J.: A guide to light-sheet fluorescence microscopy for multiscale imaging. Nat. Meth. **14**(4), 360–373 (2017). https://doi.org/10.1038/nmeth.4224
15. Reynaud, E.G., Peychl, J., Huisken, J., Tomancak, P.: Guide to light-sheet microscopy for adventurous biologists. Nat. Meth. **12**(1), 30–34 (2015). https://doi.org/10.1038/nmeth.3222
16. Rubio-Guivernau, J.L., et al.: Wavelet-based image fusion in multi-view three-dimensional microscopy. Bioinformatics **28**(2), 238–245 (2012). https://doi.org/10.1093/bioinformatics/btr609
17. Verveer, P.J., et al.: Restoration of light sheet multi-view data with the Huygens fusion and deconvolution wizard. Microscopy today **26**(5), 12–19 (2018). https://doi.org/10.1017/S1551929518000846

18. Zhang, Q., Guo, B.: Multifocus image fusion using the nonsubsampled contourlet transform. Signal Process. **89**(7), 1334–1346 (2009). https://doi.org/10.1016/j.sigpro.2009.01.012
19. Zhang, Y., Bai, X., Wang, T.: Boundary finding based multi-focus image fusion through multi-scale morphological focus-measure. Inf. Fusion **35**, 81–101 (2017). https://doi.org/10.1016/j.inffus.2016.09.006
20. Zhou, Z., Li, S., Wang, B.: Multi-scale weighted gradient-based fusion for multi-focus images. Inf. Fusion **20**, 60–72 (2014). https://doi.org/10.1016/j.inffus.2013.11.005

LUCYD: A Feature-Driven Richardson-Lucy Deconvolution Network

Tomáš Chobola[1,4], Gesine Müller[2], Veit Dausmann[3], Anton Theileis[3], Jan Taucher[3], Jan Huisken[2], and Tingying Peng[4(✉)]

[1] Technical University of Munich, Munich, Germany
[2] Georg-August-University Göttingen, Göttingen, Germany
[3] GEOMAR Helmholtz Centre for Ocean Research Kiel, Kiel, Germany
[4] Helmholtz AI, Helmholtz Munich - German Research Center for Environmental Health, Neuherberg, Germany
tingying.peng@helmholtz-muenchen.de

Abstract. The process of acquiring microscopic images in life sciences often results in image degradation and corruption, characterised by the presence of noise and blur, which poses significant challenges in accurately analysing and interpreting the obtained data. This paper proposes LUCYD, a novel method for the restoration of volumetric microscopy images that combines the Richardson-Lucy deconvolution formula and the fusion of deep features obtained by a fully convolutional network. By integrating the image formation process into a feature-driven restoration model, the proposed approach aims to enhance the quality of the restored images whilst reducing computational costs and maintaining a high degree of interpretability. Our results demonstrate that LUCYD outperforms the state-of-the-art methods in both synthetic and real microscopy images, achieving superior performance in terms of image quality and generalisability. We show that the model can handle various microscopy modalities and different imaging conditions by evaluating it on two different microscopy datasets, including volumetric widefield and light-sheet microscopy. Our experiments indicate that LUCYD can significantly improve resolution, contrast, and overall quality of microscopy images. Therefore, it can be a valuable tool for microscopy image restoration and can facilitate further research in various microscopy applications. We made the source code for the model accessible under https://github.com/ctom2/lucyd-deconvolution/.

Keywords: Deconvolution · Deblurring · Denoising · Microscopy

1 Introduction

Microscopy is one of the most widely used imaging techniques that allows life scientists to analyse cells, tissues and subcellular structures with a high level of detail. However, microscopy images often suffer from degradation such as blur, noise and other artefacts, which may result in an inaccurate quantification and

H. Greenspan et al. (Eds.): MICCAI 2023, LNCS 14227, pp. 656–665, 2023.
https://doi.org/10.1007/978-3-031-43993-3_63

hinder downstream analysis. Therefore, deconvolution techniques are necessary to restore the images to improve their quality, thus increasing the accuracy of downstream tasks [7,12,16]. Image deconvolution is a well-studied task in computer vision and imaging sciences that aims to recover a sharp and clear object out of a degraded input. The mathematical representation of image corruption can be expressed as:

$$y = x * K + n, \tag{1}$$

where $*$ represents convolution, y denotes the resulting image of an object x, which has been blurred with a point spread function (PSF) K, and degraded by noise n.

Two classic image deconvolution methods widely used in microscopy and medical imaging are Wiener filter [18] and Richardson-Lucy algorithm (RL) [9,11]. The Wiener filter is a linear filter that is applied to the frequency domain representation of the blurred image. It assumes the Gaussian noise distribution and thus minimises the mean squared error between the restored image and the original one. The RL method, on the other hand, is an iterative algorithm that works in the spatial domain, usually leading to better reconstruction than Wiener filter. It assumes Poisson noise distribution and seeks to estimate the corresponding sharp image x in a fixed number of iterations or until a convergence criterion is met [5]. While being simple and effective, the main limitations of both methods are their susceptibility to noise amplification [3,13] and the assumption that the accurate PSF is known. In practice, however, PSF is challenging to obtain and is often unknown or varies across the image, which leads to inaccurate reconstructions of the sharp image. Moreover, as an iterative method, RL is computationally costly for three-dimensional (3D) data [4].

In the computer vision field, numerous deep learning models have been trained on large datasets with the objective to learn a direct mapping between input and output domains [1,6,10,14,15]. Some of these models have also been adapted for use in microscopy, such as the U-Net-based content-aware image restoration networks (CARE) [17]. These methods have exhibited exceptional performance in tasks such as super-resolution and denoising. However, the interpretability of these methods is limited, and given their data-driven nature, the quantity and quality of training data can be a restricting factor, particularly in biomedical applications where data pairs are often scarce or not available.

Inspired by the RL algorithm, the Richardson-Lucy Network (RLN) [8] was recently designed to overcome the problem of data-driven models by embedding the RL formula for iterative image restoration into a neural network and substituting convolutions of the measured PSF kernel with learnable convolutional layers. Although being more compact than U-Net, the low capacity of RLN makes it insufficiently robust to different blur intensities and noise levels, requiring the network to be re-trained whenever there is a shift in the input image domain. This reduces the efficacy of the method.

To address the limitations of existing methods, we propose a novel lightweight model called LUCYD, which integrates the RL deconvolution formula and a U-shaped network. The main contributions of this paper can be summarised as:

1. LUCYD is a lightweight deconvolution method that embeds the RL deconvolution formula into a deep convolutional network that leverages the features extracted by a U-shaped module while maintaining low computational costs for processing 3D microscopy images and a high level of interpretability.
2. The proposed method outperforms existing deconvolution methods on both real and synthetic datasets, based on qualitative assessment and quantitative evaluation metrics, respectively.
3. We show that LUCYD has strong resistance to noise and can generalise well to new and unseen data. This ability makes it a valuable tool for practical applications in microscopy imaging fields where image quality is critical for downstream tasks yet training data are often scarce or unavailable.

2 Method

The overall architecture of the proposed model is illustrated in Fig. 1 and it comprises three main components: a **correction module**, an **update module**, and a **bottleneck** that is shared between the two modules. The data flow in the model is based on the following iterative RL formula:

$$z_{(k)} = \underbrace{z_{(k-1)}}_{x \text{ estimate}} \cdot \underbrace{\left(\frac{y}{z_{(k-1)} * K} * K^{\top} \right)}_{\text{update term}}, \tag{2}$$

which aims to recover x in k steps. We bypass the requirement of $k-1$ preceding iterations with the correction module that generates a mask M to form an intermediate sharp image estimation through a single forward pass, allowing to rapidly process 3D data, as follows:

$$\tilde{z} = y + M. \tag{3}$$

Next, inspired by Li *et al.* [8], we adopt the three-step approach to decompose the RL update term from Eq. 2 in the update module:

$$\text{(a) FP} = y * f, \text{(b) DV} = y/\text{FP}, \text{(c) } u = \text{DV} * b. \tag{4}$$

Specifically, we replace convolutions with a known PSF in steps (a) and (c) with forward projector f and backward projector b, which consist of sets of learnable convolutional layers. The produced update term u allows us to recondition the estimate $\tilde{z}$ from the correction module into a sharp image through multiplication, i.e. the last step of image formation in the RL formula: $x' = \tilde{z} \cdot u$. The whole network can then be expressed as follows,

$$x' = \tilde{z} \cdot \left(\frac{y}{y * f} * b \right). \tag{5}$$

By adhering to the image formation steps as prescribed by the RL formula, we maintain a high degree of interpretability, critical for real-world scenarios, where the accuracy and reliability of the generated results are of utmost importance.

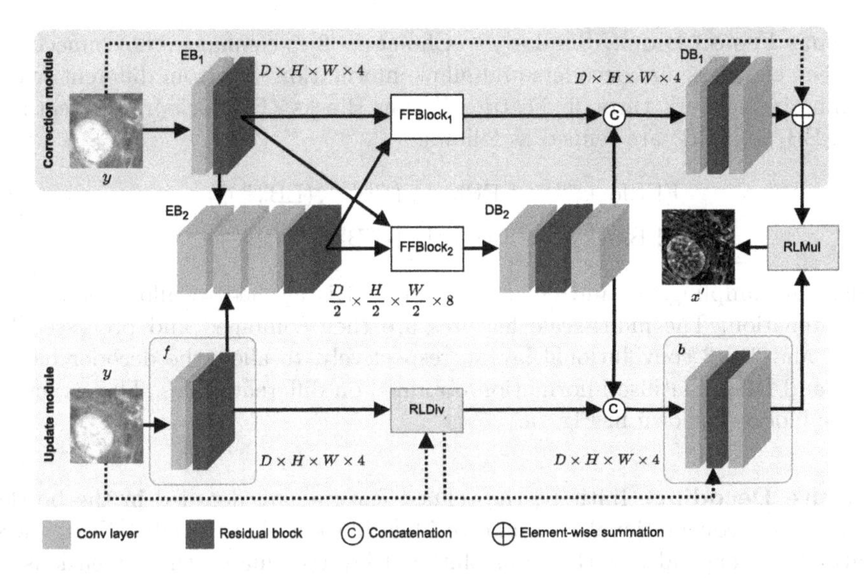

Fig. 1. The architecture of LUCYD consists of a correction module, an update module and a bottleneck that is shared between the two modules.

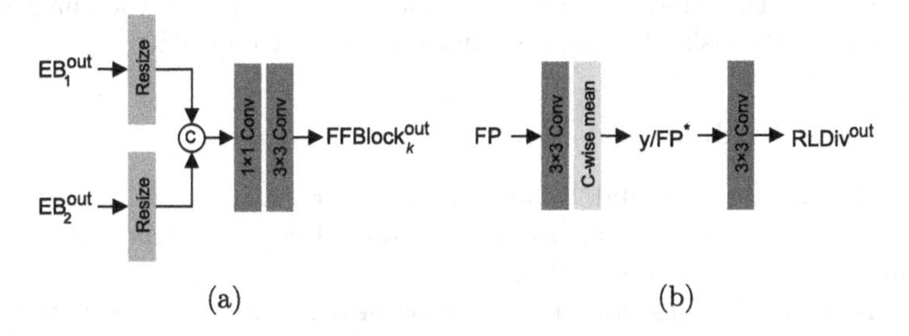

Fig. 2. The architecture of submodules: (a) Feature Fusion Block (FFBlock), (b) Richardson-Lucy Division Block (RLDiv).

2.1 Correction Module and Bottleneck

The proposed correction module and bottleneck architectures consist of encoder blocks (EBs), decoder blocks (DBs), and multi-scale feature fusion blocks to facilitate efficient information exchange across different scales within the model.

Feature Encoding. The features of the volumetric input image $y \in \mathbb{R}^{C \times D \times H \times W}$ are obtained through the first encoder block EB_1 in the correction module, and then encoded by a convolutional layer with a stride 2. Subsequently, the downsampled features are concatenated with the encoded features of the forward projection f from the update module and then fed to the bottleneck encoder EB_2 to integrate the information from both modules.

Feature Fusion Block. Similarly to Cho *et al.* [2], we enhance the connections between encoders and decoders and allow information flow from different scales within the network through Feature Fusion Blocks (FFBlocks). The features from EB_1 and EB_2 are refined as follows,

$$\mathrm{FFBlock}_1^{\mathrm{out}} = \mathrm{FFBlock}_1\left(\mathrm{EB}_1^{\mathrm{out}}, (\mathrm{EB}_2^{\mathrm{out}})^{\uparrow}\right), \tag{6}$$

$$\mathrm{FFBlock}_2^{\mathrm{out}} = \mathrm{FFBlock}_2\left((\mathrm{EB}_1^{\mathrm{out}})^{\downarrow}, \mathrm{EB}_2^{\mathrm{out}}\right), \tag{7}$$

where up-sampling ($\uparrow$) and down-sampling ($\downarrow$) is applied to allow for feature concatenation. The multi-scale features are then combined and processed by 1×1 and 3×3 convolutional layers, respectively, to allow the decoder blocks DB_1 and DB_2 to utilise information obtained on different scales. The structure of the blocks is shown in Fig. 2a.

Feature Decoding. Initially, the refined features are decoded in the bottleneck using a convolutional layer and residual block within the DB_2. Next, these features are expanded with a convolutional layer to match the dimensions in both the correction and update modules. The resulting features are then concatenated with the output of $\mathrm{FFBlock}_1$ and subsequently fed into decoder DB_2 within the correction module. The features are then mapped to the image dimensions resulting in mask M, which is summed with y to form $\tilde{z}$.

2.2 Update Module

Inspired by the forward and backward projector functions [8], we substitute the PSF convolution operations from Richardson-Lucy algorithm with learnable convolutional layers and residual blocks.

During forward projection (FP), shallow features are initially extracted by a single convolutional layer and then refined by a residual block. The output of f is then passed to Richardson-Lucy Division Block (RLDiv) which embeds the division of the raw image y by the channel-wise mean of the refined FP features. Next, we project the division result to a feature map to extract more information about the image. The visualisation of the process is in Fig. 2b. These features are then concatenated with the features extracted by the bottleneck

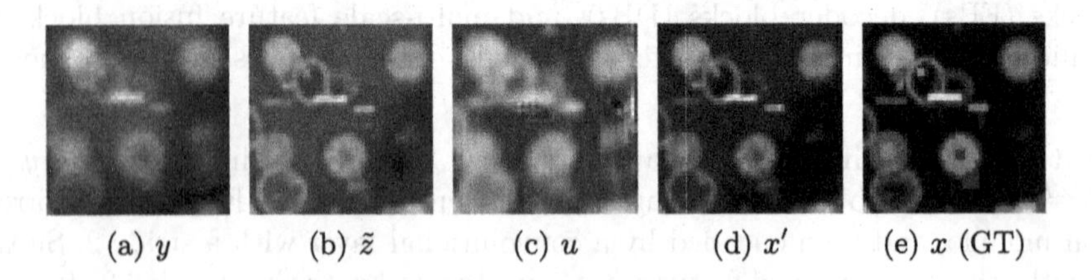

(a) y (b) $\tilde{z}$ (c) u (d) x' (e) x (GT)

Fig. 3. Overview of the deconvolution process given an input (a). The outputs of the correction module and the update module are shown in (b) and (c), respectively, and the final output obtained through their multiplication is in (d).

and combined by a convolutional layer which initiates the backward projection with b. The output is then summed with the output of RLDiv, forming a skip-connection, and passed through a residual block. The features are then refined by a convolutional layer and their channel-wise mean is taken to be the "update term" u, which is used to obtain the final model output x' through multiplication with $\tilde{z}$ (denoted as RLMul).

2.3 Loss Function

The entire model is trained end-to-end with a single loss function that combines the Mean Squared Error (MSE) and the Structural Similarity Index Measure (SSIM) as follows:

$$\mathcal{L}(x', x) = \mathrm{MSE}(x', x) - \ln\left(\frac{1 + \mathrm{SSIM}(x', x)}{2}\right), \quad (8)$$

where x is the ground truth sharp image and x' is the model estimation of x (Table 1).

3 Experiments

3.1 Setup

Datasets. We assess the performance of LUCYD on both simulated phantom objects and real microscopy images. To achieve this, we use five sets of 3D grayscale volumes generated by Li *et al.* [8], consisting of dots, solid spheres, and ellipsoidal surfaces, which are provided along with their sharp ground truth volumes of dimensions 128 × 128 × 128 (one exemplary image shown in Fig. 3e). To test the generalization capabilities of our method, we also include two blurry and noisy versions of the dataset, $\mathcal{D}_{\mathrm{nuc}}$ and $\mathcal{D}_{\mathrm{act}}$, which utilize different image degradation processes for embryonic nuclei and membrane data. Additionally, we generate a mixed dataset by applying permutations of three Gaussian blur intensities ($\sigma_b = [1.0, 1.2, 1.5]$) and three levels of additive Gaussian noise ($\sigma_n = [0, 15, 30]$) to the ground truth volumes, and then test the ability of the model to generalize to volumes blurred with Gaussian kernels ($\sigma_b = [0.5, 2.0]$) and additive Gaussian noise ($\sigma_n = [20, 50, 70, 100]$) levels outside of the training dataset. The model is trained on patches of dimensions 32 × 64 × 64 that are randomly sampled from the training datasets. Moreover, we evaluate the model trained using synthetic phantom shapes on a real 3D light-sheet image of a starfish (private data) and widefield microscopy image of U2OS cell (from the dataset of [8]), to explore the generalisation capabilities.

Table 1. Number of learnable parameters, comparing CARE, RLN and LUCYD.

CARE [17]	RLN [8]	LUCYD (ours)
1 M	15,900	24,964

Baseline and Metrics. We employ one classic U-Net-based fluorescence image restoration model CARE [17] and one RL-based convolutional model RLN [8] as baselines. We quantitatively evaluate the deconvolution performance on simulated data using two metrics: Structural Similarity Index Measure (SSIM) and Peak Signal-to-Noise Ratio (PSNR).

Table 2. Performance on synthetic datasets (SSIM/PSNR (dB)) degraded with blur and noise levels not present in the training dataset. The models are trained on phantom objects blurred with $\sigma_b = [1.0, 1.2, 1.5]$ and corrupted with Gaussian noise intensities $\sigma_n = [0, 15, 30]$.

Blur intensity σ_b	Noise level σ_n	CARE [17]	RLN [8]	LUCYD (ours)
0.5	20	0.9166/21.62	0.9571/25.60	**0.9725/26.85**
0.5	50	0.7589/15.96	0.8519/21.67	**0.9463/24.35**
0.5	70	0.6828/14.32	0.7235/18.52	**0.9040/21.78**
0.5	100	0.5856/12.56	0.5644/15.91	**0.7233/17.47**
2.0	20	0.8582/20.36	0.9040/22.34	**0.9271/23.49**
2.0	50	0.7057/16.35	0.7443/18.85	**0.8575/21.00**
2.0	70	0.6259/15.06	0.6051/16.69	**0.7995/19.38**
2.0	100	0.5154/13.08	0.4495/14.86	**0.6311/16.42**

Table 3. Performance on synthetic datasets (SSIM/PSNR (dB)) given varying training data.

	Train dataset	Test dataset	CARE [17]	RLN [8]	LUCYD (ours)
In-domain	$\mathcal{D}_{\text{nuc}}$	$\mathcal{D}_{\text{nuc}}$	0.7895/18.00	0.9247/26.43	**0.9525/28.57**
	$\mathcal{D}_{\text{act}}$	$\mathcal{D}_{\text{act}}$	0.7666/17.44	0.8966/26.10	**0.9450/27.83**
Cross-domain	$\mathcal{D}_{\text{nuc}}$	$\mathcal{D}_{\text{act}}$	0.7623/17.68	0.8841/24.33	**0.9024/24.82**
	$\mathcal{D}_{\text{act}}$	$\mathcal{D}_{\text{nuc}}$	0.7584/17.00	0.9081/27.23	**0.9336/27.63**

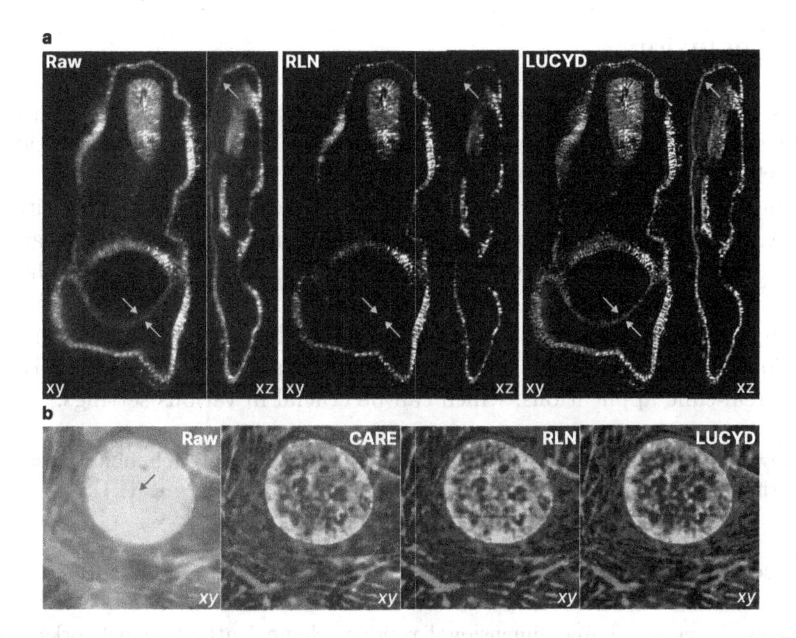

Fig. 4. Quantitative comparison of RLN and LUCYD on lateral and axial maximum-intensity projections of a starfish acquired by 3D light-sheet microscopy is shown in (a). Additional analysis of the deconvolution results of CARE, RLN and LUCYD trained on synthetic phantom objects in (b) shows patches of four-colour lateral maximum-intensity projections of a fixed U2OS cell acquired by widefield microscopy (from the dataset of [8]). LUCYD exhibits superior performance in recovering fine details and structures as compared to CARE and RLN, while simultaneously maintaining low levels of noise and haze surrounding the objects.

3.2 Results

In Table 2, we present the quantitative results of all three methods on simulated phantom objects degraded with blur and noise levels that were not present in the training dataset. LUCYD achieves the best performance even in cases where the amount of additive noise exceeds the maximum level included in the training dataset. This is in contrast to CARE and RLN, which did not demonstrate such exceptional generalisation capabilities and noise resistance. We further examine LUCYD's performance on datasets simulating widefield microscopy imaging of embryo nuclei and membrane data. As shown in Table 3, LUCYD outperforms CARE and RLN in both in-domain and cross-domain assessments, further supporting the model's capabilities in cross-domain applications.

We finally apply LUCYD on two real microscopy test samples, as illustrated in Fig. 4. On the 3D light-sheet image of starfish, LUCYD recovers more details and structures than RLN while maintaining low levels of noise and haze surrounding the object in both lateral and axial projections. On the other test sample of a fixed U2OS cell acquired by widefield microscopy, LUCYD also suppresses noise and haze to a higher degree compared to RLN and CARE and retrieves finer and sharper details.

4 Conclusion

In this paper, we introduce LUCYD, an innovative technique for deconvolving volumetric microscopy images that combines a classic image deconvolution formula with a U-shaped network. LUCYD takes advantages of both approaches, resulting in a highly efficient method capable of processing 3D data with high efficacy. We have demonstrated through experiments on both synthetic and real microscopy datasets that LUCYD exhibits strong generalization capabilities, as well as robustness to noise. These qualities make it an excellent tool for cross-domain applications in various domains, such as biology and medical imaging. Additionally, the lightweight nature of LUCYD makes it computationally feasible for real-time applications, which can be crucial in various settings.

Acknowledgements. Tomáš Chobola is supported by the Helmholtz Association under the joint research school "Munich School for Data Science - MUDS".

References

1. Chen, J., et al.: Three-dimensional residual channel attention networks denoise and sharpen fluorescence microscopy image volumes. Nat. Meth. **18**(6), 678–687 (2021). https://doi.org/10.1038/s41592-021-01155-x
2. Cho, S.J., Ji, S.W., Hong, J.P., Jung, S.W., Ko, S.J.: Rethinking coarse-to-fine approach in single image deblurring. In: Proceedings of the IEEE/CVF International Conference on Computer Vision, pp. 4641–4650 (2021)
3. Dell'Acqua, F., et al.: A modified damped Richardson-Lucy algorithm to reduce isotropic background effects in spherical deconvolution. Neuroimage **49**(2), 1446–1458 (2010). https://doi.org/10.1016/j.neuroimage.2009.09.033
4. Dey, N., et al.: Richardson–Lucy algorithm with total variation regularization for 3D confocal microscope deconvolution. Microsc. Res. Tech. **69**(4), 260–266 (2006). https://doi.org/10.1002/jemt.20294
5. Eichstädt, S., et al.: Comparison of the Richardson-Lucy method and a classical approach for spectrometer bandpass correction. Metrologia **50**(2), 107 (2013)
6. Guo, M., et al.: Rapid image deconvolution and multiview fusion for optical microscopy. Nat. Biotechnol. **38**(11), 1337–1346 (2020). https://doi.org/10.1038/s41587-020-0560-x
7. Kaderuppan, S.S., Wong, E.W.L., Sharma, A., Woo, W.L.: Smart Nanoscopy: a review of computational approaches to achieve super-resolved optical microscopy. IEEE Access **8**, 214801–214831 (2020). https://doi.org/10.1109/ACCESS.2020.3040319
8. Li, Y., et al.: Incorporating the image formation process into deep learning improves network performance. Nat. Meth. **19**(11), 1427–1437 (2022)
9. Lucy, L.B.: An iterative technique for the rectification of observed distributions. Astron. J. **79**, 745 (1974). https://doi.org/10.1086/111605
10. Qiao, C., et al.: Evaluation and development of deep neural networks for image super-resolution in optical microscopy. Nat. Meth. **18**(2), 194–202 (2021). https://doi.org/10.1038/s41592-020-01048-5
11. Richardson, W.H.: Bayesian-based iterative method of image restoration∗. J. Opt. Soc. Am. **62**(1), 55–59 (1972)

12. Sage, D., et al.: DeconvolutionLab2: an open-source software for deconvolution microscopy. Methods **115**, 28–41 (2017). https://doi.org/10.1016/j.ymeth.2016.12.015
13. Tan, K., Li, W., Zhang, Q., Huang, Y., Wu, J., Yang, J.: Penalized maximum likelihood angular super-resolution method for scanning radar forward-looking imaging. Sensors **18**(3), 912 (2018). https://doi.org/10.3390/s18030912
14. Vizcaíno, J.P., Saltarin, F., Belyaev, Y., Lyck, R., Lasser, T., Favaro, P.: Learning to reconstruct confocal microscopy stacks from single light field images. IEEE Trans. Comput. Imaging **7**, 775–788 (2021). https://doi.org/10.1109/TCI.2021.3097611
15. Wagner, N., et al.: Deep learning-enhanced light-field imaging with continuous validation. Nat. Meth. **18**(5), 557–563 (2021). https://doi.org/10.1038/s41592-021-01136-0
16. Wallace, W., Schaefer, L.H., Swedlow, J.R.: A workingperson's guide to deconvolution in light microscopy. Biotechniques **31**(5), 1076–1097 (2001). https://doi.org/10.2144/01315bi01
17. Weigert, M., et al.: Content-aware image restoration: pushing the limits of fluorescence microscopy. Nat. Meth. **15**(12), 1090–1097 (2018). https://doi.org/10.1038/s41592-018-0216-7
18. Wiener, N.: Extrapolation, Interpolation, and Smoothing of Stationary Time Series. The MIT Press, Cambridge (1949). https://doi.org/10.7551/mitpress/2946.001.0001

Deep Unsupervised Clustering for Conditional Identification of Subgroups Within a Digital Pathology Image Set

Mariia Sidulova[1,3], Xudong Sun[2], and Alexej Gossmann[1(✉)]

[1] U.S. Food and Drug Administration, Center for Devices and Radiological Health, Silver Spring, MD, USA
alexej.gossmann@fda.hhs.gov
[2] Institute of AI for Health, Helmholtz Munich, Munich, Germany
[3] Department of Biomedical Engineering, George Washington University, Washington, D.C., USA

Abstract. Consideration of subgroups or domains within medical image datasets is crucial for the development and evaluation of robust and generalizable machine learning systems. To tackle the domain identification problem, we examine deep unsupervised generative clustering approaches for representation learning and clustering. The Variational Deep Embedding (VaDE) model is trained to learn lower-dimensional representations of images based on a Mixture-of-Gaussians latent space prior distribution while optimizing cluster assignments. We propose the Conditionally Decoded Variational Deep Embedding (CDVaDE) model which incorporates additional variables of choice, such as the class labels, as conditioning factors to guide the clustering towards subgroup structures in the data which have not been known or recognized previously. We analyze the behavior of CDVaDE on multiple datasets and compare it to other deep clustering algorithms. Our experimental results demonstrate that the considered models are capable of separating digital pathology images into meaningful subgroups. We provide a general-purpose implementation of all considered deep clustering methods as part of the open source Python package DomId (https://github.com/DIDSR/DomId).

Keywords: Domain Identification · Deep Clustering · Subgroup Identification · Variational Autoencoder · Generative Model

1 Introduction

Machine learning (ML), specifically deep learning (DL), algorithms have shown exceptional performance on numerous medical image analysis tasks [2]. Never-

Supplementary Information The online version contains supplementary material available at https://doi.org/10.1007/978-3-031-43993-3_64.

H. Greenspan et al. (Eds.): MICCAI 2023, LNCS 14227, pp. 666–675, 2023.
https://doi.org/10.1007/978-3-031-43993-3_64

theless, comprehensive reviews highlight major issues of generalizability, robustness, and reproducibility in medical imaging AI/ML [9,15]. For a generalizability assessment, reporting only aggregate performance measures is not sufficient. Due to model complexity and limited training data, ML performance often varies across data subgroups or domains, such as different patient subpopulations or varied data acquisition scenarios. Aggregate performance measures (e.g., sensitivity, specificity, ROC AUC) can be dominated by the larger subgroups, masking the poor ML model performance on smaller but clinically important subgroups [11]. Thus, achieving (through training) and demonstrating (as part of testing) satisfactory ML model performance across relevant subgroups is crucial before the real-world clinical deployment of a medical ML system [13].

However, a challenging situation arises when relevant subgroups are unrecognized. One solution to this issue is to apply a clustering algorithm to the data, with the goal of identifying the unannotated subgroups. The main objective of unsupervised clustering is to group data points into distinct classes of similar traits. However, due to the complexity and high dimensionality of the medical imaging data and the resulting difficulty in establishing a concrete notion of similarity, extracting low-dimensional characteristics becomes the key to establishing the best criteria for grouping. Unsupervised generative clustering aims to simultaneously address both domain identification and dimensionality reduction. Deep unsupervised clustering algorithms could map the medical imaging data back to their causal factors or underlying domains, such as image acquisition equipment, patient subpopulations, or other meaningful data subgroups. However, there is a practical need to be able to guide the deep clustering model towards the identification of grouping structures in a given dataset that have not been already annotated. To that end, we propose a mechanism that is intended to constrain the model towards identifying clusters in the data that are not associated with given variables of choice (already known class labels or subgroup structures). The resulting algorithmic cluster assignments could then be used to improve ML algorithm training, or for generalizability and robustness evaluation.

2 Methods

We provide a PyTorch-based implementation of all deep clustering algorithms described below (VaDE, CDVaDE, and DEC) in the open source Python package DomId that is publicly available under https://github.com/DIDSR/DomId.

2.1 Variational Deep Embedding (VaDE)

Variational Deep Embedding (VaDE) [6] is an unsupervised generative clustering approach based on Variational Autoencoders [10]. In our study, VaDE is deployed as a deep clustering model using Convolutional Neural Network (CNN) architectures for the encoder $g(\mathbf{x}; \phi)$ and the decoder $f(\mathbf{z}; \boldsymbol{\theta})$. The encoder learns to compress the high-dimensional input images $\mathbf{x}$ into lower-dimensional latent representations $\mathbf{z}$. Using a Mixture-of-Gaussians (MOG) prior distribution for the latent representations $\mathbf{z}$, we examine subgroups or domains within the dataset, revealed by the individual Gaussians within the learned latent space, and how $\mathbf{z}$ affects the generation of $\mathbf{x}$. The model can be used to perform inference, where observed images $\mathbf{x}$ are mapped to corresponding latent variables $\mathbf{z}$ and their cluster/domain assignments c. We denote the latent space dimensionality by d (i.e., $\mathbf{z} \in \mathbb{R}^d$), and the number of clusters by D (i.e., $c \in \{1, 2, \ldots, D\}$). The trained decoder CNN can also be used to generate synthetic images from the algorithmically identified subgroups.

VaDE is optimized using Stochastic Gradient Variational Bayes [10] to maximize a statistical measure called the Evidence Lower Bound (ELBO). We denote the true data distribution by $p(\mathbf{z}, \mathbf{x}, c)$ and the variational posterior distribution by $q(\mathbf{z}, c|\mathbf{x})$. The ELBO of VaDE can be written as

$$\begin{aligned} \mathcal{L}_{ELBO}(\mathbf{x}) &= E_{q(\mathbf{z},c|\mathbf{x})}\left[\log \frac{p(\mathbf{z}, \mathbf{x}, c)}{q(\mathbf{z}, c|\mathbf{x})}\right] \\ &= E_{q(\mathbf{z},c|\mathbf{x})}[\log p(\mathbf{x}|\mathbf{z}) + \log p(\mathbf{z}|c) \\ &\quad + \log p(c) - \log q(\mathbf{z}|\mathbf{x}) - \log q(c|\mathbf{x})], \end{aligned} \tag{1}$$

where $p(\mathbf{x}|\mathbf{z})$ is modeled by the decoder CNN, and $q(\mathbf{z}|\mathbf{x})$ is modeled by the encoder CNN $g(\mathbf{x}; \phi)$ as

$$q(\mathbf{z}|\mathbf{x}) = \mathcal{N}\left(\mathbf{z}; \tilde{\boldsymbol{\mu}}, \operatorname{diag}\left(\tilde{\boldsymbol{\sigma}}^2\right)\right), \quad \left(\tilde{\boldsymbol{\mu}}, \log \tilde{\boldsymbol{\sigma}}^2\right) = g(\mathbf{x}; \phi).$$

Finally, the cluster assignments can be determined via

$$q(c|\mathbf{x}) \approx p(c|\mathbf{z}) = \frac{p(c)p(\mathbf{z}|c)}{\sum_{c'=1}^{D} p(c')p(\mathbf{z}|c')}, \tag{2}$$

$$p(c) = \operatorname{Cat}(\boldsymbol{\pi}), \quad p(\mathbf{z}|c) = \mathcal{N}\left(\mathbf{z}; \boldsymbol{\mu}_c, \operatorname{diag}\left(\boldsymbol{\sigma}_c^2\right)\right), \tag{3}$$

where the probability distributions $p(c)$ and $p(\mathbf{z}|c)$ come from the MOG prior of the latent space, with the respective distributional parameters $\boldsymbol{\pi}$, $\boldsymbol{\mu}_c$, $\boldsymbol{\sigma}_c^2$ (for $c \in \{1, 2, \ldots, D\}$) optimized by maximizing the ELBO of Eq. (1). Note that Eq. (2) follows from the observation that in order to maximize the ELBO in Eq. 1, the KL Divergence between $q(c|\mathbf{x})$ and $p(c|\mathbf{z})$ needs to be equal to 0. We refer to [6] for details.

In all our experiments, we apply VaDE with CNN architectures for the encoder and decoder. The CNN encoder consists of convolution layers with 32, 64, 128 filters, respectively, followed by a fully-connected layer. Respectively, the CNN decoder consists of a fully-connected layer followed by transposed convolution layers with the number of input/output channels decreasing as 128, 64, 32, 3. Batch normalization and the leaky ReLU activation functions are used.

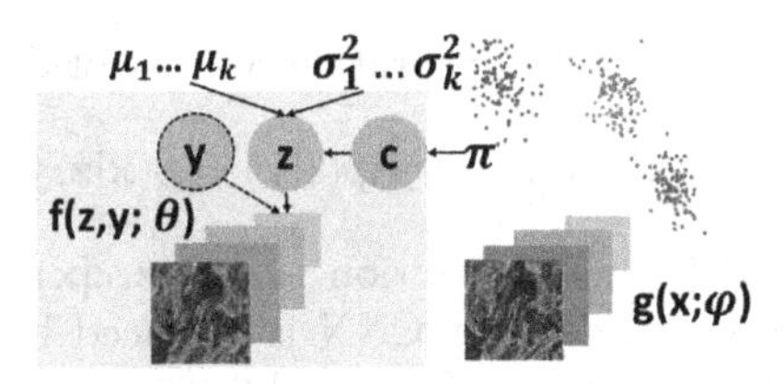

Fig. 1. Diagram of CDVaDE: $\mathbf{z}$ distribution is driven by Gaussian mean and covariance parameters $\boldsymbol{\mu}_k$ and $\boldsymbol{\sigma}_k^2$, prior cluster probabilities π_k, and conditioning variables $\mathbf{y}$.

2.2 Conditionally Decoded Variational Deep Embedding (CDVaDE)

We propose the Conditionally Decoded Variational Deep Embedding (CDVaDE) model as an extension to VaDE as shown in Fig. 1. The generative process of CDVaDE differs from VaDE in that it concatenates additional variables $\mathbf{y}$ to the latent representation $\mathbf{z}$. For example, $\mathbf{y}$ may contain the available class labels or already known subgroup structures, which do not need to be discovered. It is assumed that these additional variables $\mathbf{y}$ are available at training and test time. Specifically, the generative process of CDVaDE takes the form

$$p(c) = \mathrm{Cat}(\boldsymbol{\pi}) \tag{4}$$

$$p(\mathbf{z}|c) = \mathcal{N}\left(\mathbf{z}; \boldsymbol{\mu}_c, \mathrm{diag}\left(\boldsymbol{\sigma}_c^2\right)\right), \tag{5}$$

$$\left(\boldsymbol{\mu}_{xy}, \log \boldsymbol{\sigma}_{xy}^2\right) = f(\mathbf{z}, \mathbf{y}; \phi), \tag{6}$$

$$p(\mathbf{x}|\mathbf{z}, \mathbf{y}) = \mathcal{N}\left(\mathbf{x}; \boldsymbol{\mu}_{xy}, \mathrm{diag}\left(\boldsymbol{\sigma}_{xy}^2\right)\right) \tag{7}$$

Since our goal is to find clusters c that are unassociated with the available variables $\mathbf{y}$ of choice and to learn latent representations $\mathbf{z}$ that do not contain information about $\mathbf{y}$, the generative process of CDVaDE assumes that $\mathbf{z}, c$ are jointly independent of $\mathbf{y}$.

The changes compared to the generative process of VaDE can also be regarded as imposing a structure on the model, where the encoder learns hidden representations of the image $\mathbf{x}$ conditioned to the additional variables $\mathbf{y}$ (i.e., $q(\mathbf{z}|\mathbf{x}, \mathbf{y})$), but acts as an identity function with respect to $\mathbf{y}$ (i.e., $\mathbf{y}$ can be regarded as being simply concatenated to the latent space representations $\mathbf{z}$). The decoder then translates this data representation in the form of $(\mathbf{z}, \mathbf{y})$ to the input space (i.e., $p(\mathbf{x}|\mathbf{z}, \mathbf{y})$). Given that the underlying VAE architecture seeks to efficiently compress the input data $\mathbf{x}$ into a learned representation, this incentivizes the model to exclude information about $\mathbf{y}$ from the learned variables $\mathbf{z}$ and c.

The ELBO of CDVaDE can be derived as follows,

$$\begin{aligned}\mathcal{L}_{ELBO}(\mathbf{x}|\mathbf{y}) &= E_{q(\mathbf{z},c|\mathbf{x})}\left[\log \frac{p(\mathbf{z}, \mathbf{x}, c|\mathbf{y})}{q(\mathbf{z}, c|\mathbf{x})}\right] \\ &= E_{q(\mathbf{z},c|\mathbf{x})}[\log p(\mathbf{x}|\mathbf{z}, \mathbf{y}) + \log p(\mathbf{z}|c) \\ &\quad + \log p(c) - \log q(\mathbf{z}|\mathbf{x}) - \log q(c|\mathbf{x})],\end{aligned} \tag{8}$$

where we use the fact that by the generative process of CDVaDE it holds that

$$p(\mathbf{x}, \mathbf{z}, c|\mathbf{y}) = p(\mathbf{x}|\mathbf{z}, \mathbf{y})p(\mathbf{z}|c, \mathbf{y})p(c|\mathbf{y}) = p(\mathbf{x}|\mathbf{z}, \mathbf{y})p(\mathbf{z}|c)p(c), \tag{9}$$

and we adopt from VaDE the assumption that $q(\mathbf{z}, c|\mathbf{x}) = q(\mathbf{z}|\mathbf{x})q(c|\mathbf{x})$ holds. Hence, once the base VaDE decoder CNN is replaced by its modified version $f(\mathbf{z}, \mathbf{y}; \boldsymbol{\theta})$ in CDVaDE, there are no further differences between the ELBO loss function of Eq. (8) compared to Eq. (1).

While in this work we present our conditioning mechanism as an extension to VaDE, it can be combined with any deep clustering algorithm that follows an encoder-decoder architecture. In all our experiments, we use the same CNN architectures for the encoder and decoder as in VaDE (see Sect. 2.1).

2.3 Deep Embedding Clustering (DEC)

Deep Embedding Clustering (DEC) [14] is a popular state-of-the-art clustering approach that combines a deep embedding model with k-means clustering. In this study, we include comparisons of VaDE and the proposed CDVaDE to DEC, because it is a model that belongs to a different family of deep clustering algorithms which are not based on variational inference. In our DEC experiments, we use the same autoencoder architecture and the same initialization as for the VaDE.

2.4 Related Works in Medical Imaging

A number of studies have been conducted with several approaches of deep clustering for medical imaging data. Typically, clustering is performed on top of features extracted with the use of an encoder neural network, and the cluster assignments are determined by using conventional clustering algorithms, such as k-means, on top of the learned latent representations [1,5,7,12]. In contrast, this work investigates models which enforce a clustering structure in the latent space through the use of a MOG prior distribution, as well as guidance of the clustering model via the proposed conditioning mechanism.

3 Experiments

3.1 Colored MNIST

The Colored MNIST is an extension to the classic MNIST dataset [3], which contains binary images of handwritten digits. The Colored MNIST includes colored images of the same digits, where each number and background have a color assignment. We present results of the experiments with five distinct colors and five digits of MNIST (0–4). To enhance computational efficiency and expedite experiments, we utilized only 1% of the MNIST images, which were sampled at random. This simple dataset can be used to investigate whether a given clustering algorithm will categorize the images by color or by the digit label and whether the proposed conditioning mechanism of CDVaDE can successfully guide the clustering away from the categorization we want to avoid (e.g., condition the model to avoid clustering by color, in order to distinguish the digits in an unsupervised fashion). We compare CDVaDE to the deep clustering models VaDE and DEC that do not incorporate such conditioning. We use latent space dimensionality $d = 20$ for all models.

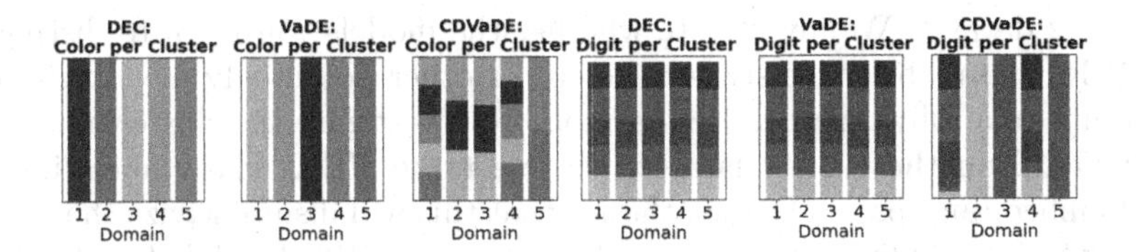

Fig. 2. Both VaDE and DEC cluster the Colored MNIST digits by the color, while CDVaDE clusters are associated with the digit label. Bar graphs labeled "Color" – each color represents a specific color of Colored MNIST digits. In the "Digit" plots, colors correspond to digits labels.

In Fig. 2 a summary of the results for the experiments on the colored MNIST dataset is presented. The results demonstrate that by allowing for the incorporation of additional information, particularly color labels, the proposed CDVaDE model is more sensitized to learning other underlying features, which allows for distinguishing between the different digits in this particular example. Notably, both VaDE and DEC end up clustering the data by color, as it is the most striking distinguishing characteristic of these images. On the other hand, the predicted domains of CDVaDE have no association with color, and the data are separated by the shapes in the images, distinguishing some of the digit labels (albeit imperfectly). This example serves as a proof of concept for the proposed conditioning mechanism of CDVaDE.

3.2 Application to a Digital Pathology Dataset

HER2 Dataset. Human epidermal growth factor receptor 2 (HER2 or HER2/neu) is a protein involved in normal cell growth, which plays an important role in the diagnosis and treatment of breast cancer [8]. The dataset consists of 241 patches extracted from 64 digitized slides of breast cancer tissue which were stained with HER2 antibody. Each tissue slide has been digitized at three different sites using three different whole slide imaging systems, evaluated by 7 pathologists on a 0–100 scale, and following clinical practice labeled as HER2 Class 1, 2, or 3 (based on mean pathologists' scores with cut-points at 33 and 66). We use a subset of this dataset consisting of 672 images (the remainder is held out for future research). Because the intended purpose is finding subgroups in the given dataset only, a separate test set is not used. The dimensions of the images vary from 600 to 826 pixels, and we scale all data to a uniform size of 128×128 pixels before further processing. We refer to [4,8] for more details about this dataset.

This retrospective human subject dataset has been made available to us by the authors of the prior studies [4,8], who are not associated with this paper. Appropriate ethical approval for the use of this material in research has been obtained.

Deep Clustering Models Applied to the HER2 Dataset. We evaluate the performance and behavior of the DEC, VaDE, and CDVaDE models on the HER2 dataset. We investigate whether the models will learn to distinguish the HER2 class labels, the scanner labels, or other potentially meaningful data subgroups in a fully unsupervised fashion. To investigate the clustering abilities of CDVaDE on the HER2 dataset, we inject the HER2 class labels into the latent embedding space. We hypothesize that this will disincentivize the encoder network from including information related to the HER2 class labels in the latent representations z. Thus, with CDVaDE we aim to guide the clustering towards identifying subgroup structures that are not associated with the HER2 classes, and potentially were not previously recognized. The dimensionality of the latent embedding space was set to $d = 500$ for all three models.

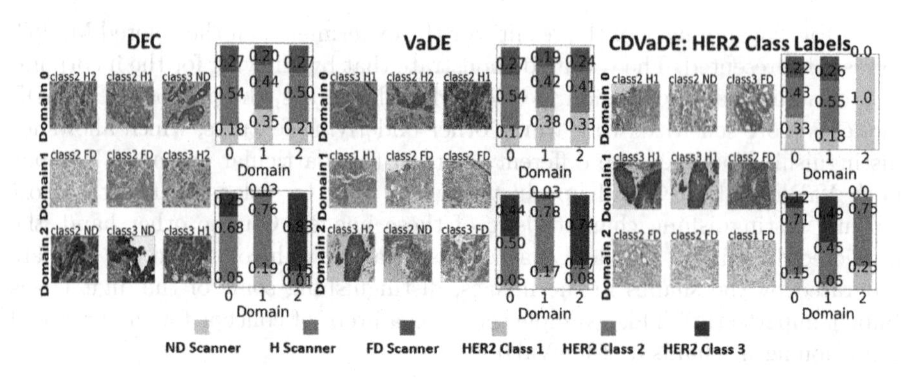

Fig. 3. Results summary for VaDE, CDVaDE, and DEC, all with $D = 3$. Example images from the identified clusters are visualized for each method. Distributions of HER2 class and scanner labels are shown per cluster (i.e., predicted domain).

Figure 3 demonstrates that even without scrutinizing, one can observe a strong visual separation between the algorithmically identified image domains for both VaDE and DEC experiments. For example, in the first predicted domain by VaDE in Fig. 3 images tend to have slightly visible boundaries but a comparatively uniform light appearance overall. In the second predicted domain, images have less visible boundaries and more pail staining. In the third predicted domain, images have more visible staining and sharper edges compared to the other domains.

As illustrated by the bar graphs in Fig. 3, there is an association between HER2 class 2 and predicted domain 2, as well as between HER2 class 3 and predicted domain 3. Similarly to the VaDE model, the DEC model has also shown the ability to separate between HER2 class 2 and HER2 class 3. To investigate these observations further, we look at the distribution of the ground truth HER2/neu scores within each of the predicted domains. The boxplots in Fig. 4 show that both the VaDE and DEC models tend to separate high HER2/neu scores from the lower ones. The Pearson's correlation coefficient between the clustering assignments c of VaDE and the HER2/neu scores is 0.46. The correlation coefficient between the DEC clusters and the HER2/neu scores is 0.71. However, neither VaDE nor DEC clusters are associated to the scanner labels.

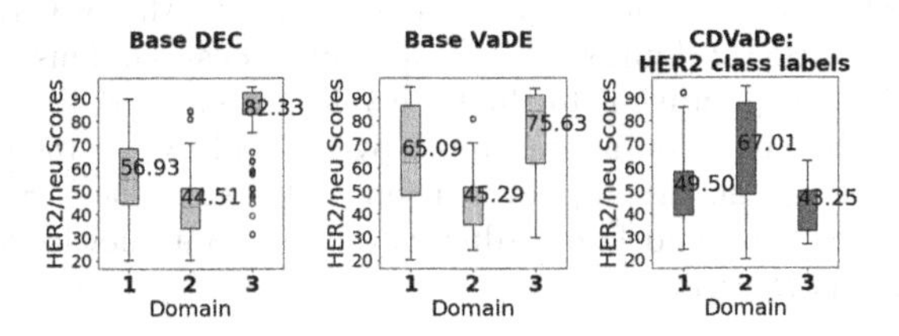

Fig. 4. Boxplots of HER2/neu scores per predicted domain for all experiments.

We investigate the proposed CDVaDE model with the goal of identifying meaningful data subgroups which are not associated with the already known HER2 class labels. As visualized in Fig. 3, the predicted domains are again clearly visually disparate. However, as intended, there is a weaker association with the HER2 class labels and a stronger association with the scanner labels, compared to the results of VaDE and DEC. In Fig. 4, HER2/neu median scores of the three clusters move closer together, illustrating the decrease of association with HER2 class labels, as intended by the formulation of the CDVaDE model. The correlation coefficient between the CDVaDE cluster assignments and the HER2/neu scores is 0.39. While the CDVaDE model does not achieve full independence between the identified clusters and the HER2 labels, it decreases this association compared to VaDE and DEC. Moreover, the clusters identified by CDVaDE are distinctly different from those of VaDE, with a 0.43 proportion of agreement between the two algorithms (after matching the two sets of cluster assignments using the Hungarian algorithm).

4 Conclusion

We investigated deep clustering models for the identification of meaningful subgroups within medical image datasets. The proposed CDVaDE model incorporates a conditioning mechanism that is capable of guiding the clustering model away from subgroup structures that have already been annotated and towards the identification of yet unrecognized image subgroups/domains. Our experimental findings on the HER2 digital pathology dataset surmise that VaDE and DEC are capable of finding, in an unsupervised fashion, image subgroups related to the

HER2 class labels, while CDVaDE (conditioned on the HER2 labels) identifies visually distinct subgroups that have a weaker association to the HER2 labels. Because the CDVaDE clusters do not clearly correspond to the scanner labels either, future work involves a review by a pathologist to see whether these subgroups capture meaningful but unannotated characteristics in the images. While CDVaDE can be used as an exploratory tool to unveil unknown subgroups in a given dataset, developing specialized quantitative evaluation metrics for this unsupervised task is inherently difficult and will also be a focus in our future work.

Acknowledgments. The authors would like to thank Dr. Marios Gavrielides for providing access to the HER2 dataset and for helpful discussion. This project was supported in part by an appointment to the Research Participation Program at the U.S. Food and Drug Administration administered by the Oak Ridge Institute for Science and Education through an interagency agreement between the U.S. Department of Energy and the U.S. Food and Drug Administration. XS acknowledges support from the Hightech Agenda Bayern.

References

1. Ahn, E., Kumar, A., Feng, D., Fulham, M., Kim, J.: Unsupervised feature learning with k-means and an ensemble of deep convolutional neural networks for medical image classification. arXiv preprint arXiv:1906.03359 (2019)
2. Barragán-Montero, A., et al.: Artificial intelligence and machine learning for medical imaging: a technology review. Physica Med. **83**, 242–256 (2021)
3. Deng, L.: The MNIST database of handwritten digit images for machine learning research. IEEE Sig. Process. Mag. **29**(6), 141–142 (2012)
4. Gavrielides, M.A., Gallas, B.D., Lenz, P., Badano, A., Hewitt, S.M.: Observer variability in the interpretation of HER2/*neu* immunohistochemical expression with unaided and computer-aided digital microscopy. Arch. Pathol. Lab. Med. **135**(2), 233–242 (2011). https://doi.org/10.5858/135.2.233
5. Gossmann, A., Cha, K.H., Sun, X.: Performance deterioration of deep neural networks for lesion classification in mammography due to distribution shift: an analysis based on artificially created distribution shift. In: Medical Imaging 2020: Computer-Aided Diagnosis, vol. 11314, p. 1131404. SPIE (2020). https://doi.org/10.1117/12.2551346
6. Jiang, Z., Zheng, Y., Tan, H., Tang, B., Zhou, H.: Variational deep embedding: an unsupervised and generative approach to clustering. In: IJCAI (2017)
7. Kart, T., Bai, W., Glocker, B., Rueckert, D.: DeepMCAT: large-scale deep clustering for medical image categorization. In: Engelhardt, S., et al. (eds.) DGM4MICCAI/DALI -2021. LNCS, vol. 13003, pp. 259–267. Springer, Cham (2021). https://doi.org/10.1007/978-3-030-88210-5_26
8. Keay, T., Conway, C.M., O'Flaherty, N., Hewitt, S.M., Shea, K., Gavrielides, M.A.: Reproducibility in the automated quantitative assessment of HER2/neu for breast cancer. J. Pathol. Inform. **4**(1), 19 (2013)
9. Kim, D.W., Jang, H.Y., Kim, K.W., Shin, Y., Park, S.H.: Design characteristics of studies reporting the performance of artificial intelligence algorithms for diagnostic analysis of medical images: results from recently published papers. Korean J. Radiol. **20**(3), 405–410 (2019). https://doi.org/10.3348/kjr.2019.0025

10. Kingma, D.P., Welling, M.: Auto-encoding variational bayes. In: ICLR (2013). arxiv.org/abs/1312.6114v10
11. Oakden-Rayner, L., Dunnmon, J., Carneiro, G., Re, C.: Hidden stratification causes clinically meaningful failures in machine learning for medical imaging. In: CHIL 2020, pp. 151–159. ACM (2020). https://doi.org/10.1145/3368555.3384468
12. Perkonigg, M., Sobotka, D., Ba-Ssalamah, A., Langs, G.: Unsupervised deep clustering for predictive texture pattern discovery in medical images. arXiv preprint arXiv:2002.03721 (2020)
13. Vokinger, K.N., Feuerriegel, S., Kesselheim, A.S.: Mitigating bias in machine learning for medicine. Commun. Med. **1**(1), 25 (2021). https://doi.org/10.1038/s43856-021-00028-w
14. Xie, J., Girshick, R., Farhadi, A.: Unsupervised deep embedding for clustering analysis. In: Balcan, M.F., Weinberger, K.Q. (eds.) Proceedings of the 33rd International Conference on Machine Learning. Proceedings of Machine Learning Research, New York, USA, vol. 48, pp. 478–487. PMLR (2016). https://proceedings.mlr.press/v48/xieb16.html
15. Yu, A.C., Mohajer, B., Eng, J.: External validation of deep learning algorithms for radiologic diagnosis: a systematic review. Radiol. Artif. Intell. **4**(3), e210064 (2022). https://doi.org/10.1148/ryai.210064

Weakly-Supervised Drug Efficiency Estimation with Confidence Score: Application to COVID-19 Drug Discovery

Nahal Mirzaie, Mohammad V. Sanian, and Mohammad H. Rohban(✉)

Sharif University of Technology, Tehran, Iran
rohban@sharif.edu

Abstract. The COVID-19 pandemic has prompted a surge in drug repurposing studies. However, many promising hits identified by modern neural networks failed in the preclinical research, which has raised concerns about the reliability of current drug discovery methods. Among studies that explore the therapeutic potential of drugs for COVID-19 treatment is RxRx19a. Its dataset was derived from High Throughput Screening (HTS) experiments conducted by the Recursion biotechnology company. Prior research on hit discovery using this dataset involved learning healthy and infected cells' morphological features and utilizing this knowledge to estimate contaminated drugged cells' scores. Nevertheless, models have never seen drugged cells during training, so these cells' phenotypic features are out of their trained distribution. That being said, model estimations for treatment samples are not trusted in these methods and can lead to false positives. This work offers a first-in-field weakly-supervised drug efficiency estimation pipeline that utilizes the mixup methodology with a confidence score for its predictions. We applied our method to the RxRx19a dataset and showed that consensus between top hits predicted on different representation spaces increases using our confidence method. Further, we demonstrate that our pipeline is robust, stable, and sensitive to drug toxicity.

Keywords: Drug Discovery · COVID-19 · High Throughput Screening · Out-of-Distribution Detection · Weakly-supervised

1 Introduction

COVID-19 is an infectious disease caused by Severe Acute Respiratory Syndrome Coronavirus 2 (SARS-CoV-2); that is a novel virus from the genus of Betacoronovirus of coronavirus genera. In extreme cases, host cells activate intensive

N. Mirzaie and M. V. Sanian—Contributed equally to this work.

Supplementary Information The online version contains supplementary material available at https://doi.org/10.1007/978-3-031-43993-3_65.

H. Greenspan et al. (Eds.): MICCAI 2023, LNCS 14227, pp. 676–685, 2023.
https://doi.org/10.1007/978-3-031-43993-3_65

immune responses for defense, leading to Acute respiratory distress syndrome (ARDS) and multiple organ failure [2,18]. Though modern vaccines against this disease effectively prevent severity, hospitalization, and death, the SARS-CoV-2 virus has continued to mutate, and there is still a chance of showing severe symptoms, especially for those with chronic diseases [2]. Yet, we have a few medications that affect this tough disease. Hence search for treatment has never stopped.

Designing a new drug is time-consuming, while drug repurposing can be a real game-changer when it comes to a pandemic. Every newly designed drug candidate should go through a filter of preclinical research, clinical trials, and FDA reviews to gain safety permission to be on the market. Many drugs filter out before reaching the market, while this journey takes about 7–12 years for those few that last [19]. Drug repurposing accelerates this process by reusing approved chemical compounds for a new target. Because drug safety and efficiency have initially been tested in another clinical study, getting approval for a new purpose is faster for approved medicines [12,14,23].

High Throughput Screening (HTS) has proven its worth in facilitating drug repurposing [8]. HTS methodology relies on the fact that chemical compounds change the morphology of cells. In particular, chemical compounds with similar molecular structures are expected to induce similar cellular morphological changes [3,15]. These changes can be tracked and analyzed to identify the compound hits capable of reversing the cell morphologies impacted by the disease.

RxRx19a is an extensive HTS experiment conducted by the Recursion biotechnology company to investigate potential therapeutics of approved drugs for COVID-19 treatment [7]. To our knowledge, RxRx19a is the biggest public fluorescent microscopy dataset exploring approved drugs' effects on SARS-CoV-2 virus. Based on the substances that are added to the cultured cells, samples in this study can be categorized into negative controls (mock and UV-irradiated SARS-CoV-2), positive controls (active SARS-CoV-2), and treatments (contaminated drugged).

Previous studies for hit discovery on this dataset used positive and negative control samples to train a model to discriminate healthy from infected cell image representations [7,11,16]. The trained model was then used to estimate treatment scores. Although adequate evidence ensures that positive and negative controls are separate in embedding space, the top hit-score drugs published by these works are rarely overlapped. Besides, previous algorithms falsely predicted many treatments as effective, including toxic doses. We argue that this phenomenon is due to the fact that although cellular morphological changes caused by drugs are not visible to human eyes, they are so drastic that can confuse the model. In other words, treatment samples that have never been seen in the training of models are out-of-distribution samples to the trained model distribution, and we need a measure of confidence to rely on model judgment on them [5,10,22].

Here, we present a novel weakly supervised confident hit prediction pipeline that estimates disease scores regardless of drug side effects and provides a confidence score for its prediction. We applied our model on both CellProfiler features [17], and deep neural network embeddings and showed that consensus between discovered hits increases. Our contributions are:

- We are the first to address the hit-discovery methodology's out-of-distribution problem and solve it with a novel weakly-supervised confident hit prediction pipeline.
- We exhaustively evaluate our pipeline and show that our results are robust, stable, and sensitive to the cell death and drug toxicity.
- We provide the CellProfiler features, deep learning embeddings, and their related pipelines for nowadays the biggest fluorescent microscopy dataset to explore drug efficiency.

2 Dataset

Our work uses RxRx19a dataset, which consists of 305,520 site-level 1024×1024 pixels images captured by fluorescent microscopes in multiple high throughput screening experiments. These experiments were conducted on two tissue types, HRCE and VERO, to explore the efficiency of 1672 FDA or EMA-approved drugs with 6 to 8 different doses against SARS-CoV-2. A drug with a specific dosage was added post-seeding for preparing treatment samples, and then samples were contaminated with the active SARS-CoV-2 virus. Then, samples were fixed and stained with 5 fluorescent colors to indicate DNA, RNA, actin cytoskeleton, Golgi apparatus, and endoplasmic reticulum organelles in cells. RxRx19a is publicly available and licensed under the Creative Commons Attribution 4.0 International License (http://creativecommons.org/licenses/by/4.0/).

3 Methods

Assume $\mathcal{X}$ is a set of all single-cell images in our dataset. We partitioned $\mathcal{X}$ based on information of wells into three classes $\mathcal{X}_{ctrl_+}, \mathcal{X}_{ctrl_-}$, and $\mathcal{X}_t$. These are sets of all single-cell images from positive control wells (active SARS-CoV-2), negative control wells (mock and UV-irradiated SARS-CoV-2), and contaminated drugged wells (treatment), respectively. We also define $\mathcal{X}_{ctrl} = \mathcal{X}_{ctrl_+} \cup \mathcal{X}_{ctrl_-}$ as a set of all control samples. The function $f(x_i)$, $f : \mathcal{X} \rightarrow \mathcal{H}$, maps each image x_i to its embedding h_i. Same as partitions in $\mathcal{X}$ space, we consider $\mathcal{H}_{ctrl_+}, \mathcal{H}_{ctrl_-}$, and $\mathcal{H}_t$ as:

$$\mathcal{H}_{ctrl_+} = \{f(x^+) \mid \forall\, x^+ \in \mathcal{X}_{ctrl_+}\} \tag{1}$$

$$\mathcal{H}_{ctrl_-} = \{f(x^-) \mid \forall\, x^- \in \mathcal{X}_{ctrl_-}\} \tag{2}$$

$$\mathcal{H}_t = \{f(x^t) \mid \forall\, x^t \in \mathcal{X}_t\} \tag{3}$$

A function $d : \mathcal{H} \rightarrow [0, 1]$ returns a disease score for a given embedding, which is a score number indicating adversity of the disease. The lower and closer to 0, the healthier the cell is, and vice versa. For the sake of simplicity, we ignore experimental errors and consider all positive controls cells infected

$\forall h^+ \in \mathcal{H}_{ctrl_+}, d(h^+) = 1$, and with the same logic, all negative cells as healthy $\forall h^- \in \mathcal{H}_{ctrl_-}, d(h^-) = 0$. Yet, the disease score for $\forall h_t \in \mathcal{H}_t$ is unknown. If a drug was effective, somehow, it could hinder the virus's activities in the cell after being contaminated. It could be through a direct effect on the virus, entry channels, or cells that make them more resistant to the virus. As a result, with a successful drug, the host cell's morphological features were close to the morphological characteristics of a healthy cell. That is, $d(h_t)$ is close to 0 for a successful drug. In this study, we aim to estimate d (Fig. 1).

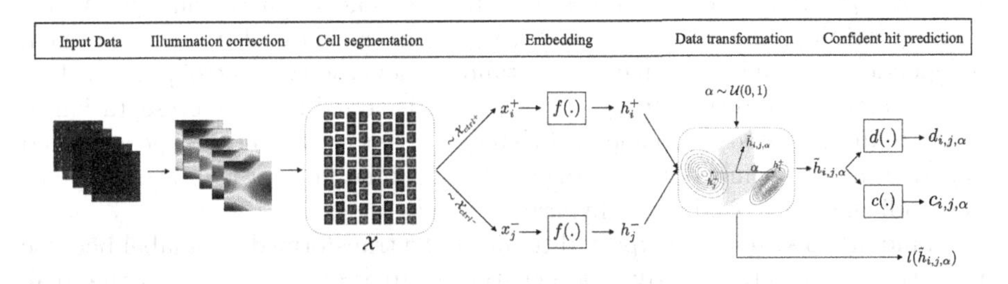

Fig. 1. Our weakly supervised confident hit predictor pipeline consists of image preprocessing, feature embedding, data transformation, and confident hit prediction sections. In image preprocessing, we correct field-of-view illumination in images, followed by the cell segmentation. The embedding function $f(.)$ calculates features for two random single-cell images from positive and negative control wells. A random convex combination of these two embeddings generates a transformed embedding $\tilde{h}$. We then calculate a disease score and a confidence score for $\tilde{h}$, which are evaluated with the weak label $l(\tilde{h})$ during training.

3.1 Image Preprocessing and Embedding

We used a retrospective multi-image method to correct field-of-view illumination and calculate illumination functions for each channel across plates [3]. For the cell segmentation, we used CellProfiler to calculate cell locations and boxes [17]. After this process, artifacts and highly saturated pixels were removed. We deployed two methods of rule-based feature extraction and representation learning to extract embeddings for the single-cell images. For the rule-based feature extraction pipeline, we used CellProfiler. Plate-level normalization and Typical Variation Normalization (TVN) were applied to reduce the batch effects [1]. For the representation learning, we adopted a pre-trained ResNet18 [6]. The model was pre-trained on the ImageNet dataset. Nevertheless, both methods embed single-cell images into feature vectors; throughout the paper, $f(.)$ could refer to either.

3.2 Data Augmentation with Weak Labels

We propose a novel data augmentation based on [4] at our embedding level that generates simulated drugged samples with weak labels. Not only is our augmentation method a game-changer in network evaluation, but it also helps the model learn meaningful disease-related characteristics. A transformed sample $\tilde{h}$ is calculated by convex combinations of positive and negative controls and then perturbed by random noise to mimic both a drug's efficiency and side effects. Our data transformation can be formulated as follows:

$$\tilde{h}_{i,j,\alpha} = \alpha h_i^+ + (1-\alpha)h_j^- + \gamma n_{\perp v} \tag{4}$$

where $\alpha \sim \mathcal{U}(0,1)$ is a random number between $[0,1]$, and $h_i^+ \sim \mathcal{H}_{ctrl_+}$ and $h_j^- \sim \mathcal{H}_{ctrl_-}$ are arbitrary samples from positive and negative controls. A normally distributed noise $n \sim \mathcal{N}(\mu_{\mathcal{H}_{ctrl}}, \sigma_{\mathcal{H}_{ctrl}})$ is calculated for each dimension independently based on the mean and standard deviation of control profiles. This noise is orthogonal to $v = \bar{\mathcal{H}}_{ctrl_+} - \bar{\mathcal{H}}_{ctrl_-}$ and is added to the representation to simulate drug-related morphological changes. v is a vector between positive and negative samples' embedding averages, and the final noise n is in a hyperplane perpendicular to v. And γ is a hyperparameter.

Random noise n is not expected to affect the transformed data label because it is orthogonal to the positive and negative samples axis. We consider the weak label for $\tilde{h}_{i,j,\alpha}$ only depended on α, $l(h_i^+)$ and $l(h_j^-)$:

$$l(\tilde{h}_{i,j,\alpha}) := \alpha l(h_i^+) + (1-\alpha)l(h_j^-) = \alpha \tag{5}$$

It is remarkable that with this method, we can augment our dataset indefinitely, and overcome the poor generalization issues that are caused by the limited control data. Further, this augmented dataset $\tilde{\mathcal{H}}$ which includes control samples $\mathcal{H}_{ctrl}$ is used to train and evaluate our confident hit predictor.

3.3 Confident Hit Predictor

Machine learning models' predictions about samples from out of their distribution cannot be trusted [5,10]. This issue is crucial when it comes to safety-critical applications like drug discovery. False negatives cause missing effective treatment, while false positives result in wasting money and time in follow-up experiments. That being said, the worst drawback is that we cannot trust our model judgments on unknown perturbations anymore, as they could act as out-of-distribution samples. In our study, unfortunately, there are not any ground truth that can help us to quantify treatment efficiency against SARS-CoV-2 in-vitro. Although the model can learn deep phenotypic characteristics for healthy and infected cells during training, drug-related morphological features of drugged cells are unknown. To lean on model disease score estimation for a treatment, we need to ask about the model's confidence in the prediction.

We offer a first-in-field confident hit predictor that utilizes the idea of hint and confidence [5] to calculate a measure of certainty about its hit-score estimations. During training, our model receives hints to predict disease scores. Larger hints lead to lower confidence scores. Using this method, we adjust the prediction probabilities using interpolation between the predicted disease score and the weak label probability distribution. The adjusted prediction is then subjected to MSE loss calculation as follows:

$$\mathcal{L} = MSE\left(c(\tilde{h})d(\tilde{h}) + (1 - c(\tilde{h}))l(\tilde{h}),\ l(\tilde{h})\right) - \lambda \log(c(\tilde{h})), \tag{6}$$

where $\tilde{h}$ is an augmented transformed hidden features and $l(\tilde{h}_i)$ is the weak label, and $d(\tilde{h}_i)$ is predicted diseas score. A function $c(.) \in [0, 1]$ calculates the confidence score. The adjusted prediction $c(\tilde{h})d(\tilde{h}) + (1 - c(\tilde{h}))l(\tilde{h})$ is closer to the hint score $(l(\tilde{h}_i))$ when the confidence score is low, and vice versa. To prevent the model from always asking for hints and getting stuck in a trivial solution where c always returns 0, a confidence loss, $-\log(c(\tilde{h}))$ is added to the loss.

4 Experiments and Results

4.1 Experimental Settings

We developed our CellProfiler pipeline with Luigi workflow manager, which was, in our experience, faster and more maintainable in locally hosted systems. For TVN, we use PCA whitening that embeds our representation into 1024 dimensions. Multilayer perceptrons (MLP) are used to estimate the confidence and disease scores. To train our model, we used the zero-shot learning method to test our model's reliability. For this purpose, we left out inactive UV-irradiated SARS-CoV-2 samples in the training process. The model should estimate inactive virus samples as healthy, while inactive viruses can still slightly change the cellular morphology.

4.2 Representation Quality Assurance

Replicate Reproducibility Test: It is expected to see replicates of the same biological perturbation are significantly similar to each other than a random set of profiles [3]. We used the Pearson correlation to calculate the similarity of two representations. The distribution of the mean Pearson correlation between all replicates of a biological perturbation is compared against a null distribution. We observed that the replicate correlation of 98.99% of drugs are significant using this criterion (see Fig. 2a, S1).

4.3 Disease Scores Quality Assurance

Stability of Disease Scores: To test the model's stability, we randomly omitted a portion of the control samples before creating a new augmented dataset and training a new model. Then, we compared the correlation between prime and new scores and repeated the test five times. Our model showed enhanced stability due to weak label data augmentation, as opposed to the unstable on-disease score algorithm used in RxRx19a (see Fig. 2b).

Robustness of Disease Scores: We show the robustness of our method for computing disease scores by accurately predicting scores for new samples in a zero-shot learning setup [20]. The model was trained on an augmented dataset and tested on unseen ultraviolet-irradiated SARS-CoV-2 samples. The model predicted scores of this group near zero, and these scores' distribution is similar to the other negative controls (see Fig. 2c, S2).

4.4 Confidence Scores Quality Assurance

Calibration of Confidence Scores: We have noticed that the entropy of disease score predictions rises as the samples are further away from being entirely

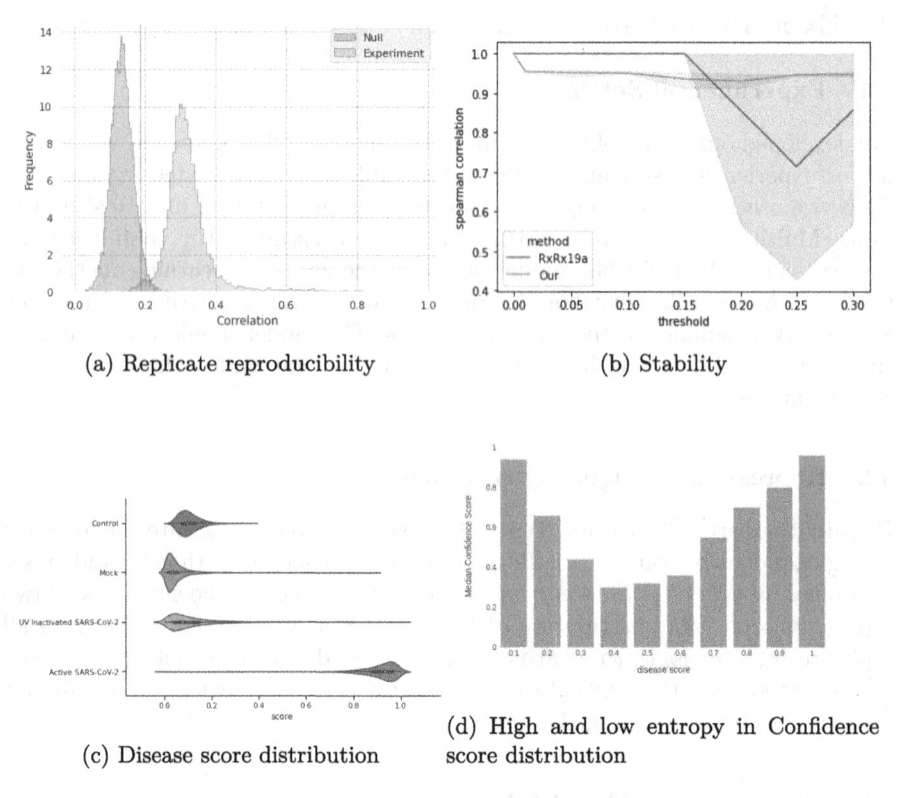

(a) Replicate reproducibility

(b) Stability

(c) Disease score distribution

(d) High and low entropy in Confidence score distribution

Fig. 2. a) Replicate reproducibility test. The blue distributions are the null distribution. The red vertical line indicates the upper 95% of the null distribution. **b)** Stability test of disease scores with respect to the percentage of missing data. We can see our proposed method (orange) is more stable against missing data, while the RxRx19a origin algorithm fluctuates adversely after missing more than 15% of data. **c)** Disease score distributions of our weakly supervised confident hit predictions for negative and positive controls. We observed that the model properly predicts unseen inactive ultraviolet-irradiated SARS-CoV-2 scores in a zero-shot learning process. **d)** Calibration of Confidence scores. We bucketize disease scores into 10 bins, and calculate median of confidence score in each bin. This plot shows model uncertainty and high entropy for disease score between [0.3, 0.7]. (Color figure online)

healthy or diseased. This reveals that the model demonstrates measurable uncertainty on OOD samples (see Fig. 2d).

4.5 Evaluation

We hypothesize that using confidence will reduce false positives in the hit discovery. Because ground truth labels for drugs in our dataset are not accessible, we cannot directly show false positive rate reduction. Instead, we can show that when we applied the confidence method consensus among top scores, the overlap between discovered hits based on different representations increased. We run our pipeline in both with and without confidence setups for three representations: our CellProfiler features, our single-cell image embedding, and the original proprietary RxRx19a embedding. When we used the confidence score, the mean Jaccard similarity between the top 10 drugs increased from 0.13 to 0.2 (see Table 1).

Table 1. The Jaccard similarity between the top 10 score drugs increased when we applied the confidence method to our model.

	Without Confidence			With Confidence		
	Cellprofiler	ResNet	RxRx19a	Cellprofiler	ResNet	RxRx19a
Cellprofiler	1	0.11	0.11	1	0.17	0.11
ResNet	0.11	1	0.17	0.17	1	0.33
RxRx19a	0.11	0.17	1	0.11	0.33	1
Mean		0.13			0.20	

Top score drugs predicted by our pipeline are **Remdesivir** [9,13] ($d = 0.74$, $c = 0.66$), **Aloxistatin** [23] ($d = 0.72$, $c = 0.71$), **GS-441524** [21] ($d = 0.67$, $c = 0.63$), **Albendazole** ($d = 0.61$, $c = 0.69$), and **Cinnarizine** ($d = 0.60$, $c = 0.66$) (see Tables S2, S3). As the disease score drops, the model confidence increases. That model confidence for the known therapeutic, Remdesivir, is 0.66, while the model is pretty sure ($c > 0.98$) that *Fluvoxamine* and *Hydroxychloroquine* are ineffective. We suspect that effective drugs can cause a sequence of cellular morphological changes unknown to the model. However, when we have ineffective drugs, the well characterized morphological changes that are caused by the SARS-CoV-2 infection dominates. That is why the model is more sure about ineffective drugs' disease scores than effective ones.

5 Conclusion

To safely conclude about hit discovery scores based on cellular morphological features, we need to be concerned about inevitable out-of-distribution phenotypes.

We explored one possible solution and proposed a confidence-based weakly-supervised drug efficiency estimation pipeline that was trained to be unsure about out-of-distribution samples. Further, because truth values for drug efficiency scores are unknown, we indirectly defined a metric to calculate false positive rate reduction and showed that this metric improves in the confidence-based setup. We also enhanced our drug efficiency estimation pipeline with a unique weakly-supervised data transformation, simulating contaminated drugged cell phenotypes. Finally, we assessed our pipeline's robustness and stability.

Acknowledgments. We express our gratitude to Dr. Forbes J. Burkowski, Dr. Vahid Salimi, and Dr. Ali Sharifi Zarchi for their priceless guidance and help in validating the output of the models.

Code and Data Availability. The code repository is available at https://github.com/rohban-lab/Drug-Efficiency-Estimation-with-Confidence-Score.

Raw Cellprofiler features are available at http://hpc.sharif.edu:8080/HRCE/, and normalized well level features at https://doi.org/10.6084/m9.figshare.23723946.v1.

References

1. Ando, D.M., McLean, C.Y., Berndl, M.: Improving phenotypic measurements in high-content imaging screens. bioRxiv (2017). https://doi.org/10.1101/161422, https://www.biorxiv.org/content/early/2017/07/10/161422
2. Aslan, A., Aslan, C., Zolbanin, N.M., Jafari, R.: Acute respiratory distress syndrome in COVID-19: possible mechanisms and therapeutic management. Pneumonia **13**(1), 14 (2021)
3. Caicedo, J.C., et al.: Data-analysis strategies for image-based cell profiling. Nat. Methods **14**(9), 849–863 (2017)
4. Caicedo, J.C., McQuin, C., Goodman, A., Singh, S., Carpenter, A.E.: Weakly supervised learning of Single-Cell feature embeddings. Proc. IEEE Comput. Soc. Conf. Comput. Vis. Pattern Recognit. **2018**, 9309–9318 (2018)
5. DeVries, T., Taylor, G.W.: Learning confidence for out-of-distribution detection in neural networks. arXiv preprint arXiv:1802.04865 (2018)
6. He, K., Zhang, X., Ren, S., Sun, J.: Deep residual learning for image recognition. CoRR abs/1512.03385 (2015). https://arxiv.org/abs/1512.03385
7. Heiser, K., et al.: Identification of potential treatments for COVID-19 through artificial intelligence-enabled phenomic analysis of human cells infected with SARS-CoV-2. bioRxiv (2020). https://doi.org/10.1101/2020.04.21.054387, https://www.biorxiv.org/content/early/2020/04/23/2020.04.21.054387'
8. Li, Y., et al.: High-throughput screening and evaluation of repurposed drugs targeting the SARS-CoV-2 main protease. Signal Transduction Targeted Therapy **6**(1), 356 (2021)
9. Li, Y., et al.: Remdesivir metabolite GS-441524 effectively inhibits SARS-CoV-2 infection in mouse models. J. Med. Chem. **65**(4), 2785–2793 (2021)
10. Liu, W., Wang, X., Owens, J., Li, Y.: Energy-based out-of-distribution detection. In: Advances in Neural Information Processing Systems (2020)

11. Mascolini, A., Cardamone, D., Ponzio, F., Di Cataldo, S., Ficarra, E.: Exploiting generative self-supervised learning for the assessment of biological images with lack of annotations. BMC Bioinform. **23**(1), 295 (2022)
12. Mirabelli, C., et al.: Morphological cell profiling of SARS-CoV-2 infection identifies drug repurposing candidates for COVID-19. Proc. Natl. Acad. Sci. **118**(36), e2105815118 (2021). https://doi.org/10.1073/pnas.2105815118, https://www.pnas.org/doi/abs/10.1073/pnas.2105815118
13. Pruijssers, A.J., et al.: Remdesivir inhibits SARS-CoV-2 in human lung cells and chimeric SARS-CoV expressing the SARS-CoV-2 RNA polymerase in mice. Cell Rep. **32**(3), 107940 (2020)
14. Pushpakom, S., et al.: Drug repurposing: progress, challenges and recommendations. Nat. Rev. Drug Discov. **18**(1), 41–58 (2018)
15. Rohban, M.H., et al.: Virtual screening for small-molecule pathway regulators by image-profile matching. Cell Syst. **13**(9), 724–736.e9 (2022)
16. Saberian, M.S., et al.: DEEMD: drug efficacy estimation against SARS-CoV-2 based on cell morphology with deep multiple instance learning. IEEE Trans. Med. Imaging **41**(11), 3128–3145 (2022). https://doi.org/10.1109/TMI.2022.3178523
17. Stirling, D.R., Swain-Bowden, M.J., Lucas, A.M., Carpenter, A.E., Cimini, B.A., Goodman, A.: Cell Profiler 4: improvements in speed, utility and usability. BMC Bioinform. **22**(1), 433 (2021)
18. Torres Acosta, M.A., Singer, B.D.: Pathogenesis of COVID-19-induced ARDS: implications for an ageing population. Eur. Respir. J. **56**(3), 2002049 (2020). https://doi.org/10.1183/13993003.02049-2020, https://erj.ersjournals.com/content/56/3/2002049
19. Van Norman, G.A.: Drugs, devices, and the FDA: Part 1: an overview of approval processes for drugs. JACC: Basic Transl. Sci. **1**(3), 170–179 (2016). https://doi.org/10.1016/j.jacbts.2016.03.002, https://www.sciencedirect.com/science/article/pii/S2452302X1600036X
20. Xian, Y., Lampert, C.H., Schiele, B., Akata, Z.: Zero-shot learning - a comprehensive evaluation of the good, the bad and the ugly. CoRR abs/1707.00600 (2017). https://arxiv.org/abs/1707.00600'
21. Yan, V.C., Muller, F.L.: Advantages of the parent nucleoside GS-441524 over Remdesivir for COVID-19 treatment. ACS Med. Chem. Lett. **11**(7), 1361–1366 (2020)
22. Yang, J., Zhou, K., Li, Y., Liu, Z.: Generalized out-of-distribution detection: a survey. CoRR abs/2110.11334 (2021). https://arxiv.org/abs/2110.11334
23. Yousefi, H., Mashouri, L., Okpechi, S.C., Alahari, N., Alahari, S.K.: Repurposing existing drugs for the treatment of COVID-19/SARS-CoV-2 infection: a review describing drug mechanisms of action. Biochem. Pharmacol. **183**, 114296 (2020)

Author Index

H. Greenspan et al. (Eds.): MICCAI 2023, LNCS 14227, pp. 687–690, 2023.
https://doi.org/10.1007/978-3-031-43993-3

Printed in the United States
by Baker & Taylor Publisher Services